FOSSIL FUEL COMBUSTION

FOSSIL FUEL COMBUSTION
A SOURCE BOOK

Edited by

William Bartok
Energy and Environmental Research Corporation
Whitehouse, New Jersey

and

Adel F. Sarofim
Department of Chemical Engineering
Massachusetts Institute of Technology
Cambridge, Massachusetts

A WILEY-INTERSCIENCE PUBLICATION
John Wiley & Sons, Inc.
NEW YORK / CHICHESTER / BRISBANE / TORONTO / SINGAPORE

Library of Congress Cataloging in Publication Data:
Fossil fuel combustion: a source book/edited by William Bartok and
Adel F. Sarofim.

   p. cm.
"A Wiley-Interscience publication."
Includes bibliographical references.
ISBN 0-471-84779-8
   1. Combustion. 2. Fossil fuels—Combustion.   I. Bartok, William,
1930–   II. Sarofim, Adel F.

QD516.F62 1989
621.402'3—dc20   89-22571
                                  CIP

Printed in the United States of America

10 9 8 7 6 5 4 3 2 1

PREFACE

It is anticipated that major changes will be needed in the design and operation of combustion equipment to accommodate eventual changes in the availability and combustion of fuels. The trend in fuel composition is toward higher concentrations of aromatics, nitrogen, and sulfur heteroatoms and toward the increased use of heavier liquids, low calorific value gases, and solid fuels. Other changes in combustion equipment design will be influenced by consideration of cycle efficiencies or by environmental regulations. The transitions in fuel composition and combustor operating conditions could well take place under crisis conditions over time scales too short to permit the traditional slow evolution in design in response to changing conditions. There will be a need for models that account for the complex chemical and physical phenomena to guide future developments.

During the past decades there have been significant advances in the understanding of the chemistry of combustion and its interaction with diffusional processes. Models are now available that can predict flame speeds for simple fuels, pollutant formation in a variety of practical combustor configurations, and an increasing number of combustion properties of interest. The understanding of the combustion chemistry of complex fuels and of the aerodynamics and mixing in practical combustion equipment is still incomplete, and the combustion engineer in the short term will need to continue to rely on empirical information such as flame speeds, flame stretch, and flammability limits in approaching problems. Theory is, however, already useful for helping to extrapolate the existing data base to new situations, and, with time, theoretical models will be used increasingly, first to complement experimentation, and ultimately to eliminate empiricism in the solution of many combustion problems.

This source book is intended to provide the principles that govern fundamental flame properties. It provides a detailed discussion of combustion chemistry because the basis for many changes in combustion equipment design and operation will be responsive to changes in the fuel composition. Flame properties such as flame velocity are determined by an interplay between flame chemistry and diffusional processes. The discussion of these properties shows this interrelationship qualitatively, relying still on empirical measurements for specific solutions to problems. To facilitate computations based on combustion principles, an appendix containing data on fuels and combustion related properties is provided.

The subject matter is divided into three major parts: I. Combustion Chemistry; II. Flame Phenomena, Diffusional Processes, and Turbulent Reactive Flow; and III. Heterogeneous Combustion. These parts are supplemented by the Appendix.

Part I, *Combustion Chemistry*, consists of chapters on the following topics:

1. *Interface between Fuels and Combustion*, by J. P. Longwell. This chapter sets the scene for the particular perspective of relating fuel properties to combustion equipment performance and design.

2. *Evaluation of Chemical Thermodynamics and Rate Parameters for Use in Combustion Modeling*, by D. Golden. This chapter provides a general framework for predicting the rates of elementary reaction steps of importance in gas phase combustion.

3. *The Phenomenology of Modeling Combustion Chemistry*, by F. L. Dryer. This chapter treats the methods available for the kinetic modeling of hydrocarbon oxidation reactions ranging from the use of elementary reaction steps to global rates, thus setting the stage for the modeling of hydrocarbon flame properties.

4. *Chemistry of Gaseous Pollutant Formation and Destruction*, by C. T. Bowman. This chapter presents knowledge of chemical reactions that affect the formation and destruction of three types of major gaseous pollutants generated in combustion processes, *viz.*, carbon monoxide, oxides of nitrogen, and oxides of sulfur.

5. *Soot and Hydrocarbons in Combustion*, by B. S. Haynes. This chapter deals with the chemistry of other classes of air pollutants, polycyclic aromatic hydrocarbons (PCAH), which in addition to being potential carcinogens may also be the precursors of soot, and soot itself, including the nucleation, surface growth, and agglomeration of soot particles.

Part II, *Flame Phenomena, Diffusional Processes, and Turbulent Reactive Flow* treats the following topics related to gaseous combustion:

6. *Premixed Flames and Detonations*, by R. A. Strehlow. This chapter covers the phenomena associated with premixed flames, *viz.*, flame velocities, flammability limits, quench diameters, ignition energies, detonations, and explosives.

7. *Diffusion Flames*, by M. Gerstein. This chapter reviews the properties of diffusion flames, starting with the Burke-Schumann laminar diffusion flame theory and includes the treatment of evaporating drops, flame height, flame chemistry, smoke, and turbulent diffusion flames.

8. *Turbulent Reacting Flows*, by F. A. Williams. This chapter reviews the state of the art in modeling turbulent combustion phenomena, including both "tractable problems" (such as profiles of mean temperature, concentration, overall rates of conversion, etc., in a reactor without

recirculation, under well-characterized inlet conditions; speed of propagation of a premixed turbulent flame) and "intractable problems" (such as turbulent flame speeds under the transient conditions that prevail in internal combustion engines) that require empiricism.

Part III, *Heterogeneous Combustion* deals with the combustion of liquid spray droplets and the combustion of coal and char. The individual chapters are as follows:

9. *Fuel Atomization, Droplet Evaporation, and Spray Combustion*, by A. H. Lefebvre. This chapter treats systematically the processes that govern liquid fuel jet atomization into droplets, droplet evaporation, and the combustion of ensembles of droplets in sprays, with descriptions of practical atomizers, vaporizers, and combustors.

10. *Coal and Char Combustion*, by L. D. Smoot. This chapter addresses solid fuel properties, coal particle ignition and devolatilization processes, the combustion of moist pulverized coal and coal water slurries, and the modeling of coal combustion processes.

The Appendix, *Data on Fuel and Combustion Properties*, by P. C. Wu and H. C. Hottel, contains a compendium of data in tabular and graphical forms and formulas for estimating fuel and combustion properties of use to practicing combustion researchers and engineers.

Acknowledgement. The Editors wish to express their appreciation to the members of the Exxon course committee (William S. Blazowski, Anthony M. Dean, Irwin L. Goldblatt, J. Carl Pirkle, Jr., Lawrence A. Ruth, and Michael E. Tomsho) whose enthusiasm and unflagging attention helped significantly in organizing the Exxon Combustion Fundamentals course on which this book is based.

WILLIAM BARTOK

Energy and Environmental Research Corporation
Whitehouse, New Jersey

ADEL F. SAROFIM

Massachusetts Institute of Technology
Cambridge, Massachusetts

CONTRIBUTORS

C. T. BOWMAN, Department of Mechanical Engineering, Stanford University, Stanford, CA 94305

FREDERICK L. DRYER, Department of Mechanical and Aerospace Engineering, Princeton University, Princeton, NJ 08554

MELVIN GERSTEIN, Department of Mechanical Engineering, University of Southern California, Los Angeles, CA 90089-1453

DAVID M. GOLDEN, Department of Chemical Kinetics, SRI International, Menlo Park, CA 94025

B. S. HAYNES, Department of Chemical Engineering, University of Sydney, New South Wales, NSW 2006 Australia

HOYT C. HOTTEL, Department of Chemical Engineering, Massachusetts Institute of Technology, Cambridge, MA 02139

ARTHUR H. LEFEBVRE, School of Mechanical Engineering, Purdue University, West Lafayette, IN 47907

JOHN P. LONGWELL, Department of Chemical Engineering, Massachusetts Institute of Technology, Cambridge, MA 02139

L. DOUGLAS SMOOT, Department of Chemical Engineering, Brigham Young University, Provo, UT 84602

ROGER A. STREHLOW, Department of Aeronautical and Astronautical Engineering, University of Illinois, Urbana, IL 61801

F. A. WILLIAMS, Department of Applied Mechanics and Engineering Sciences, University of California, San Diego, La Jolla, CA 92093

PAO-CHEN WU,* Department of Chemical Engineering, Massachusetts Institute of Technology, Cambridge, MA 02139

*Present address: Aspen Technology Inc., 251 Vassar Street, Cambridge, MA 02139.

CONTENTS

PART I   COMBUSTION CHEMISTRY   1

1 Interface between Fuels and Combustion   3
    John P. Longwell

2 Evaluation of Chemical Thermodynamics and Rate Parameters
    for Use in Combustion Modeling   49
    David M. Golden

3 The Phenomenology of Modeling Combustion Chemistry   121
    Frederick L. Dryer

4 Chemistry of Gaseous Pollutant Formation and Destruction   215
    C. T. Bowman

5 Soot and Hydrocarbons in Combustion   261
    B. S. Haynes

PART II   FLAME PHENOMENA, DIFFUSIONAL PROCESSES,
          AND TURBULENT REACTIVE FLOW   327

6 Premixed Flames and Detonations: Combustion Safety
    and Explosions   329
    Roger A. Strehlow

7 Diffusion Flames   423
    Melvin Gerstein

8 Turbulent Reacting Flows   459
    F. A. Williams

PART III   HETEROGENEOUS COMBUSTION   527

9 Fuel Atomization, Droplet Evaporation, and Spray Combustion   529
    Arthur H. Lefebvre

10 Coal and Char Combustion   **653**
   L. Douglas Smoot

Appendix Data on Fuel and Combustion Properties   **783**
   Pao-chen Wu and Hoyt C. Hottel

INDEX   **845**

FOSSIL FUEL COMBUSTION

PART I
Combustion Chemistry

1 Interface between Fuels and Combustion

JOHN P. LONGWELL
Department of Chemical Engineering
Massachusetts Institute of Technology
Cambridge, Massachusetts

1. History and future of fuel use
2. Composition and properties of major fuels
    2.1. Gas
    2.2. Gasoline
    2.3. Distillate fuels
    2.4. Fuel oils
    References
    Glossary

This introductory discussion presents some background on the history and on projections of future fuel use, and a discussion of the characteristics of the major fuels.

1. HISTORY AND FUTURE OF FUEL USE

The past 30–40 years have been a time of stability in the types of fuel resources used and in fuel properties. The period following World War II was one of introduction of modern refining conversion technologies and development of the modern automotive and aircraft propulsion systems. This was a time of great activity in establishment of fuel specifications and, with few exceptions, these specifications have remained almost unchanged until the last ten years when the requirements for lower emissions from combustion systems and more efficient utilization began forcing modifications in both fuel composition and combustion equipment.

   If we look further into the past and into the future, it is apparent that such stability is only temporary. Changes in both available sources of fuel and

3

TABLE 1. The Use of Fuels in the United States Annual Consumption in Quads (10^{15} Btu)

Year	Total Energy Consumption			Fraction of Total Energy (Includes Hydro, Nuclear, etc.)	Supplied by Solid Fuels
	Biomass	Natural Oil	Coal and Gas		
1850	2.1	0.2	0.0	2.3	1.0
1875	2.9	1.4	0.0	4.3	1.0
1900	2.0	7.0	0.5	9.9	0.91
1925	2.0	15.0	5.7	23.0	0.72
1950	1.2	13.0	19.5	35.0	0.57
1980 (4)	1.3	16.5	57.5	81.7	0.22
2000 (7)	5.9	27.2	47.3	93.4	0.35

Reprinted with permission from H. C. Hottel and J. B. Howard, *New Energy Technology*. MIT Press, Cambridge, MA, 1971. © MIT Press, 1971.

end-use technology have major impact. Some historical perspective can be gained from Table 1.

In 1850 wood accounted for over 90% of the nation's fuel supply, and coal made up most of the rest. Except for small amounts of whale oil and other animal and vegetable oils used for illumination, all fuel was in the solid form, with its attendant problems of inconvenience, high air pollution, and thermal inefficiency in the combustion equipment of the day. By 1900 petroleum, natural gas, and hydroelectric power were making small contributions, but solid fuels still contributed about 90% of the energy used. After 1900 the use of solid fuels declined rapidly; in 1980, with coal the dominant solid, these fuels contributed only 22% of the total, mainly for electricity and industrial steam. Use of solid fuels is, however, increasing, and their fractional contribution will grow.

Oil and gas, by 1980, supplied 73% of the total. This dominant role was due to their abundance and low cost and, equally important, their ease of handling and the low cost and high performance of their end-use equipment, such as home heating units, airplanes, and automobiles. These end-use advantages are such that the use of liquid and gaseous fuels is expected to continue and eventually to exceed the supply of petroleum and natural gas. Conversion of solid fuels to the liquid and gaseous form is therefore expected to ultimately become a major industry.

Some further perspective on the future of fuel supplies in the United States is afforded by considering estimates of fossil fuels believed to exist in recoverable form (Table 2).

"Recoverable reserves" are known resources that can be extracted economically. "Recoverable resources" refer to all deposits known or believed to exist in potentially economically extractable form.

U.S. oil and gas consumption in 1980 totaled 57.5 quads per year. Since the estimated U.S. resources amount to only about 17 years' supply, the need for a

TABLE 2. Fossil Fuel Reserves and Resources in the United States[a]

	Heat of Combustion in Quads	
Mineral	Recoverable Resources	Ultimately Recoverable Resources
Natural gas	230	500
Oil	200[b]	470
Oil shale		13,000
Peat[c]	700	1400
Coal	6000	79,000

[a] From Ref. 2.
[b] No recoverable reserve estimates are given for oil shale because commercial exploitation has not yet taken place.
[c] From Ref. 3.

combination of importation, conservation, and substitution of other resources is quite apparent.

The ultimate resource estimates can, in some cases, be viewed as underestimates. Very large quantities of natural gas, for example, are believed to be present in the geopressured brines of the Gulf Coast and in certain low-permeability formations. It is quite possible that technological advances might make some part of these reserves economically accessible.

The oil shale resource estimates, similarly, omit the very large deposits of Devonian oil shale found in the eastern and midwestern regions of the United States, because extracting oil from these shales will be more costly than from the western oil shales. However, in the long term these shales may be an important resource.

Biomass, or plant matter, might also be an important contributor. Estimates vary widely, depending to a considerable extent on judgments about the amount of agriculture and forest land that can be dedicated to crops for fuel use. One estimate,[4] a "national commitment" scenario for 2010, is shown in Table 3.

Since total energy consumption estimates for that year range from 70 to 100 quads, such a contribution would be perhaps 10% of the total. Here the

TABLE 3

Source	Quads/Year
Municipal waste	1.9
Agricultural waste	3.5
Energy crops	3.4
Total	8.8

starting material is, again, a solid that suffers from some of the same problems as coal. Conversion to gas or liquid fuels will be desirable for many uses.

Imports are an important part of our supply of liquid and gaseous fuel, and world resources therefore must be considered. The world's ultimately recoverable oil resources are estimated at about 10,000 quads, or 20 times the U.S. resources.[5] The expected slow demand growth for the rest of this century indicates that if no major disruptions occur, productive capacity can keep pace, or exceed, demand during this period. Around the year 2000 it is expected that demand will approach production capacity and that prices will rise, making production of liquid fuels from solid fuel sources more attractive.

Importation of gas in the United States and in the rest of the industrialized world is expected to grow. Although the prospects for future gas discoveries are better than for oil and additions to reserves currently exceed consumption, much of the new gas is expected to be found in areas remote from major industrial markets, so transportation facilities will continue to present a major barrier. Here again, it is expected that resource limitations will ultimately control output in the future.

The worldwide use of coal is expected to increase.[6] The United States is estimated to have about 25% of the world's coal resources and reserves. Approximately 90% of the world's coal is held by the United States, the Soviet Union, China, and Australia. Resource limitations will not present a serious problem for many decades, and coal presents a major opportunity for world trade.

The world's largest oil shale deposits are believed to be in the United States and Brazil, with major deposits in Australia and China. Because production in excess of national needs seems unlikely, production and use can be expected to serve domestic consumption.

Conversion of solid fuels to liquids or gaseous fuels and, for remote gas, conversion of gas to liquid fuel is expected to grow from its currently relatively small production to a major industry. Timing depends on the ability of world and U.S. petroleum production capacity to meet demand and on the price determined by the supply-demand-price relationships. It appears probable that after the turn of the century significant supplies of these "synthetic fuels" will be produced, greatly increasing the diversity of fuel sources and compositions.

Liquid fuels are of special interest in combustion since their composition, manufacture, and use involve more options than gas and solid fuels. A projection of demand for the major liquid fuels is shown in Table 4.

Although a major decrease in consumption of motor gasoline is projected, the pressure for increased efficiency, performance, and reduced emissions is expected to maintain interest in improving the combustion of this fuel at a high level. The growth shown for distillate fuel reflects an expectation that the use of the diesel engine for automotive transportation will grow to major proportions. This growth is, however, inhibited by inability to control emissions or lack of customer acceptance. The ratio of gasoline to jet fuel and distillate changes drastically in this projection. Estimates of the ratio for the

TABLE 4. U.S. Liquid Fuel Demand by Product[a]

	1960	1980	2000[b]
Gasoline	4.8	6.6	4.8
Jet fuel	0.4	1.1	1.3
Mid-distillates			
(diesel and #2 heating oil)	2.1	3.0	3.7
Residual fuel oil	1.5	2.5	2.5
Other	1.8	4.0	4.4
Totals	9.8	17.1	16.7
Ratio gasoline/jet fuel			
and distillate	1.6	1.6	1.0

[a]From Dukek.[7]
[b]Average of industry opinion, 1984.

year 2000 vary between 0.6 and 1.2, reflecting differing projections on the extent to which diesel fuel will replace gasoline. The higher number currently appears more probable.

Heavy fuel oil consumption goes through a maximum. A rapid growth between 1970 and 1980 occurred due to substitution of fuel oil for natural gas in industrial and power generation use. Substitution of coal or coal derived gas for this fuel is expected to reduce fuel oil demand. The combustion of heavy fuels, in the near term, remains a very important area for combustion research. Coal pyrolysis liquids, bottoms from coal liquefaction, and low hydrogen content petroleum residue will probably be burned to supply heat from these fuels since they are expensive to upgrade into higher-quality products.

Fuel use in Europe follows quite a different pattern as shown in Table 5.

The relatively low gasoline and large fuel oil production is striking. As petroleum becomes increasingly scarce, it seems probable that the European and U.S. product distributions will approach each other more closely since the use of fuel oil for industrial heating and power generation will probably be replaced by the alternatives, coal and nuclear. The gasoline to distillate ratios can also be expected to approach each other for the two regions.

TABLE 5. Liquid Fuel Use in Western Europe, 1979

Fuel	% of Total
Mo-gas and naphtha	19
Distillates	34
Heavy fuel oil	38
Other	9
Total	100

Gasoline and naptha/distillate ratio   0.56

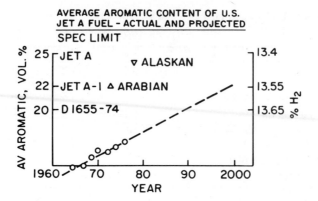

FIGURE 1. Average aromatic content of U.S. jet-A fuel—actual and projected.

Production of distillate fuels by residuum and heavy gas oil conversion produces relatively low hydrogen content cracked stocks unless relatively expensive hydroprocessing is used. As the proportion of distillate products grows and manufacture of residual fuel oil decreases, the fraction of cracked stocks, which run about 35% aromatics, grows. Increasing aromatics content increases the formation of soot in present distillate combustion equipment and causes both operational and environmental problems. Since soot can be totally consumed in ideally designed combustion equipment, accommodation of lower hydrogen content fuels is an important combustion R & D goal.

Figure 1 shows the actual and projected growth in jet fuel aromatics content.[7] The steady increase is primarily the result of increased use of high aromatics content crudes. As it becomes necessary to exploit the world's heavy crude deposits and as low hydrogen content materials, such as coal-derived liquids, enter the refineries, this effect of raw material composition will become more severe; however, this increase in aromatics will probably be limited by hydrogenation.

The above projections are based primarily on projections of resource availability, cost, and economic and population growth. The effect of increasing fuel cost is shown dramatically by the projected decrease in gasoline consumption. Although mileage standards have been set by regulation, the small car demand, in response to price increases since 1973, has exceeded the regulatory goals. Energy conservation in home heating will reduce oil and gas consumption, at the cost of investment in home modification. Similarly, further heat conservation in industry will require investment driven by increased fuel cost and government pressure. In general, combustion technology is available for major gains in heat conservation. There are exceptions such as the need for improved control of combustion and heat transfer (with low-grade fuels) in high-performance furnaces in steam cracking. Overall, economic and supply driving forces can give major reductions in fuel use while maintaining normal industrial and personal activities.

Environmental constraints are important because combustion systems are the major source of air pollutants in many areas and because, through research, major improvements seem possible. Combustion-generated air pollution has been a matter of public concern, starting with the first major uses of coal for heating. The burning of soft coal in England resulted in national legislation to reduce air pollution from this source in 1273. The problem of control by legislation was not seriously addressed in the United States until the 1940s when air quality problems in the Los Angeles basin and in industrial areas such as Pittsburgh led to local action. The history of Federal control began with the air pollution control act of 1955 which set up research, training, and technical assistance programs. Public concern with this problem and a corresponding complex legislative and regulatory structure has since grown to the point where constraints on manufacturer and use of fuels are important and promise to become more severe in the future as we turn to our coal, oil shale, and biomass resources to replace natural gas and petroleum.

Improved fuels and improved performance of combustion equipment can contribute to alleviating air pollution problems and the accompanying constraints on manufacturers and use of fuels.

Combustion generated pollutants can be categorized as follows:

Combustible

Unburned fuel
Hydrocarbons from fuel pyrolysis
Polycyclic aromatic hydrocarbons (PCAH) and other mutagens
Soot
HCN
CO

Combustion Products

SO_2/SO_3
$NO/NO_2/N_2O$
Inorganic particulates
CO_2

The combustible pollutants can be reduced to extremely low concentrations under well-controlled combustion conditions with adequate oxygen temperature and residence time. This can be easily achieved in continuous flow combustors, such as furnaces. These unburned materials are found in greatest quantity in cyclic systems where wall quenching and quenching by rapid expansion, as in water-cooled reciprocating internal combustion engines; in domestic oil burners that are allowed to cool between heating cycles; in gas turbine combustors with too rapid quenching by secondary air; in furnaces with air leaks or unbalanced burners; fireplaces and stoves with poor control or air supply and temperature; etc. Study of details of these quenching phenomena and of means of avoiding them, though still retaining the desired operating characteristics, remains an important opportunity for advances in combustion technology.

Combustion products refers to oxidized species that are formed in the combustion process: Among these, nitrogen oxides are, in principle, avoidable because of the low concentrations that are thermodynamically stable at low temperatures in the presence of oxygen or under rich mixture conditions. The problem is most intractable in reciprocating engines where, with hydrocarbon fuels, high temperature oxidizing conditions are unavoidable. For continuous systems, more control of the temperature-time oxidation history is feasible, and substantial progress in control technology is being made.

Although sulfur contained in fuel cannot be destroyed, combustion science and technology is very much involved since production of the more harmful SO_3 can be minimized and since sulfur can be retained and removed as solid inorganic compounds in coal ash or in added limestones or dolomite.

Where inorganic compounds are present in fuel, emission of inorganic particulates can be a problem. Of special interest are the submicron particulates found by vaporization and condensation processes. Vanadium oxide, from petroleum fuel oils, will vaporize at flame temperature and will condense on cooling as will SiO found in high-temperature, rich mixture zones. In the latter case the SiO subsequently oxidizes to form SiO_2 which then condenses. Many metals can also participate in this vaporization-condensation process.

Carbon dioxide is now suspect as a pollutant in that its continued increase in the atmosphere will cause changes in the earth's heat balance with accompanying climate changes. Combustion of fossil fuels inescapably increases the CO_2 inventory. Here there is little that combustion technology can contribute except to use fuels that produce maximum useful heat for the amount of CO_2 released. An example would be direct coal combustion for power generation as opposed to combustion of gas manufactured from coal with a loss of approximately 30% of the original coal heating value.

An approach to minimizing or eliminating combustion-generated emissions is to select fuels which allow "clean" burning. Hydrogen, for example, will obviously produce none of the above pollutants except for nitrogen oxides. These can be easily controlled to very low levels by carrying out combustion at sufficiently low temperatures to avoid fixation of atmospheric nitrogen. Advocates of the "hydrogen economy" are correct in identifying this as the ideal fuel (once it is purchased and delivered to the combustion chamber). The energy and economic cost of hydrogen manufacture and the problems of storage and transportation systems have made its near and mid-term use of marginal interest in this country. In the very long term, hydrogen manufactured from water using nuclear or fission energy may well find its place, especially if control of CO_2 emissions proves necessary.

Of more immediate interest is the use of CO/H_2 mixtures produced from coal, for industrial heat and for power generation. Removal of sulfur, nitrogen,

TABLE 6. Ambient Air Quality Standard-Microgram/m^3

Pollutant	Federal		Federal[a]		
(Annual Mean)	Primary	Secondary	Class I	Class II	Class III
SO$_2$	80	None	2	20	40
NO$_x$	100	100	None	None	50
Particulates	75	60	5	19	37
Non CH$_4$ hydrocarbons	160	160	None	None	80

[a]Allowable incremental change in ambient concentration over an established baseline.

and inorganic compounds from the fuel can be essentially complete and combustion of the manufactured gas can be controlled to almost any desired level of emissions. Compared to direct coal combustion without emission control, a high economic, efficiency, and complexity penalty is paid.

Use of methanol in spark ignition engines offers the potential for control of nitrogen oxides, by lean operation or water dilution, and elimination of PCAH, soot, sulfur oxides, and inorganic particulates. Methanol has a high octane number and, except for its low heating value per unit volume and more restrictive requirements for fuel-air mixture preparation, appears to be an ideal fuel. On a cost and thermal efficiency basis, it can compete with gasoline manufactured by other coal liquefaction processes; however hydrocarbons from shale oil are expected to be cheaper. Although these ideal fuels will, in some cases, offer both high performance and low emissions, use of hydrocarbon fuels and coal will be dominant for many years and advances in combustion technology are needed to allow use of materials, which can be produced at lower cost and lower energy consumption, with acceptable emissions.

Technological strategies for dealing with combustion-generated emissions are strongly influenced by the nature of legislation and regulations aimed at their control. Overlying the regulatory system are a set of National Ambient Air Quality Standards (NAAQS) that specify the maximum atmospheric concentration for chosen pollutants. These are summarized in Table 6.

Two federal standards are given. The primary standards are aimed at protecting human health, and the secondary standards are aimed at protecting public welfare such as atmospheric clarity, the health of nearby forests, etc. The health-related data, on which these standards are based, is very limited and subject to criticism. In addition the pollutants chosen for control are not, in some cases, those considered responsible for the health or public welfare effects of most concern. An example is the particulates standard which currently does not discriminate between large dust particles and submicron particulates that can enter the lung and are transported long distances. Similarly, control of SO$_2$ deals indirectly with the acid rain sulfate particulate problems since SO$_2$, as such, is much less of a problem. One can therefore expect future changes in these standards as the problems caused by pollutants become better understood.

Superimposed on these air quality standards is a requirement for prevention of significant deterioration (PSD) and the areas with cleaner air than required by NAAQS are placed into these categories:

I. Areas, such as national parks and wilderness areas, where almost no change from current air quality will be allowed.

II. Areas where moderate change will be allowed, but where stringent air quality constraints are desirable.

III. Areas where major industrial development is foreseen, contamination up to half the level of the secondary NAAQS will be permitted.

States can, with EPA approval, redesignate class II areas to III areas. These PSD definitions and requirements are relatively new and are not well defined. The western states currently have very clean air and are mostly in class I or II. The relatively small increments allowed for these areas can seriously constrain development of major industrial installations.

The states are required to develop plans to meet federal requirements in these various air quality control regions. The states can also set more severe standards but cannot allow concentrations of pollutants that exceed federal standards.

The oil shale area in Colorado, for example, is a class II area with currently quite low air pollution levels. In addition the state of Colorado has set category II and III SO_2 increases at 10 and 15 $\mu g/m^3$ (1980), approximately half the federal standard.

Federal standards are also set for emissions from new equipment such as automobiles, aircraft, and power plants. In general, these new source standards take into account the capabilities of the best available technology. However, in the case of automobile emission standards, the standards were such that significant cost or operating penalties were incurred, and the regulatory process has been one of continuing controversy and evolution.

An example for new large industrial steam generators and the electric utility steam generating units is shown in Table 7.

For industrial units, the requirements reflect fuel composition since coal tends to contain more sulfur and organic nitrogen than oils. Gas normally

TABLE 7. Allowed Emissions lb/10^6 Btu (1983)

	SO_2		NO_2	
	Industrial	Utility	Industrial	Utility
Coal	1.2	1.2[a]	0.7	0.6[c]
Oil	0.8	0.8[b]	0.3	0.3
Gas	—	—	0.2	0.2

[a]90% reduction also required or 70% when < 0.6 lb/10^6 Btu.
[b]90 reduction also required or 0.2 lb/10^6 Btu.
[c]0.5 lb/10^6 Btu for subbituminous coal or shale- and coal-derived liquids.

contains only trace amounts of sulfur, and there is therefore no requirement. NO_x in gas-fired systems arises from fixation of atmospheric nitrogen and has a fairly strict requirement, which tends to preclude use of air preheat and which requires careful design to meet.

Although the NO_x requirements for utilities are comparable to those for industrial units, the SO_2 requirements are complicated by a set of requirements aimed at discouraging use of low-sulfur coal as an optional control strategy since stack gas scrubbing or its equivalent in performance is required in all cases.

For nonattainment areas it is possible to arrange to shut down an existing source of pollution and to substitute a new source; however, the new installation is still required to operate with the "lowest achievable" emission rate. For attainment areas, best available control technology, taking economics and efficiency into account, must be used. The "bubble concept" allows averaging of emission sources within an industrial operation in order to optimize cost and efficiency. These approaches are not applicable to industrially developing areas such as the western shale area. It has been estimated[8] that, by using best available technology, production will be limited to 0.4 MB/D and that the first few installations, unless very carefully planned, will tend to preempt much of the allowable emission of pollutants.

Most of the air pollutants are combustion based, and though compromises in air quality may be in the national interest, it is clear that major reductions in combustion-generated emissions will increase our ability to augment our liquid fuel supplies with these resources. Similarly, in highly industrialized areas, new equipment offering significant reduction in emissions can allow additional economic growth and modernization.

The current policy of delegating air quality responsibility to individual states does not deal adequately with transport of pollutants between states. Sulfur oxides, for example, can be transported over very great distances and particulates from volcanic eruptions can blanket a hemisphere of the earth. The current practice of dispersing pollutants by use of very tall stacks to avoid locally high concentrations, while meeting current regulations, increase regional control problems. Of special current interest is the "acid rain" problem. The high pH encountered in northeastern United States is attributed, in part, to transport of sulfur and nitrogen oxides from coal burning midwestern areas. Emission control at the source has been recommended; however, application of stack-gas scrubbing or use of low-sulfur western coal is economically disadvantageous for the emitting states who are now meeting federal standards. The new source standards previously discussed, if met by all sources, have the potential of alleviating the problem; however, few new coal-burning furnaces are being built and the long life of existing systems offers little hope for major improvement during this century. A clear opportunity exists for sulfur control technology which can be used with high-sulfur coal in existing furnaces with reasonable investment and operating cost. Sulfur capture by calcium included with the fuel is an approach of current special interest. Since an important fraction of rain acidity is attributed to nitric acid, having its origin in combustion generated NO_x, acid rain may provide an additional incentive for further advances in NO_x control.

2. COMPOSITION AND PROPERTIES OF MAJOR FUELS

The major fuels consist of complex and widely varying mixtures of compounds. Since the performance in the design of combustion equipment depends strongly on both the physical and chemical properties of fuels, knowledge of these properties is essential to application of knowledge of combustion phenomena to combustion problems. Modern analytical techniques are capable of greatly extending our knowledge of fuel composition, and detailed analysis of fuel composition is a current and continuing activity; however, with the exception of gaseous fuels, basic knowledge of the chemistry and physics of combustion and computational capabilities is not yet able to take full advantage of detailed knowledge of fuel composition. The information most commonly used for complex fuels consists of lumped properties that have been found to be useful for characterizing the fuel-combustor interaction and that are not time-consuming or expensive to determine. An example is octane number, which rates a fuel's tendency to knock in an Otto-cycle engine under carefully standardized conditions. Standardization is an important feature of these measurements and nationally (or internationally) agreed on techniques have been developed for analyses related to fuel specifications.

In this discussion, the basis for combustion-related fuel properties and specifications will be presented along with a few limited examples of more detailed composition studies. For this purpose, fuels are grouped into the following categories:

1. Gas
2. Gasoline
3. Distillates
4. Fuel oils
5. Coal

Composition and properties of coal are discussed in Chapter 10. The other major fuels are discussed below.

2.1. Gas

This category refers to fuels that are normally delivered to the combustion system in the gaseous state. This includes liquefied petroleum gas (LPG) which is transported and stored in the liquid state but which is vaporized before mixing with air for combustion.

TABLE 8. Properties of Gaseous Fuel Components

Fuel Component	Molecular Weight	Heat of Combustion, at 25°C and 1 atm					
		Gross			Net		
		kcal/g-mol	Btu/lb-mol	Btu/ft³[a]	kcal/g-mol	Btu/lb-mol	Btu/ft³[a]
Hydrogen	2.016	68.318	122972	313.79	57.798	104036	265.47
Carbon monoxide	28.01	67.636	121745	310.66	67.636	121745	310.66
Methane	16.04	212.80	383040	977.41	191.76	345168	880.77
Ethane	30.07	372.82	671076	1712.40	341.26	614268	1567.44
Ethylene	28.05	337.15	606870	1548.57	316.11	568998	1451.93
Acetylene	26.04	310.62	559116	1426.71	300.10	540180	1378.39
Propane	44.10	530.60	955080	2437.09	488.53	879354	2243.87
Propylene	42.08	491.99	885582	2259.76	460.43	828774	2114.81

Fuel Component	Maximum Adiabatic Flame Temperature[b] (No Dissociation) K	Maximum Laminar Burning Velocity cm/sec	Flammability Limit in Air			
			Lean		Rich	
			Vol %	Equivalent Ratio	Vol %	Equivalent Ratio
Hydrogen	2525	264.8	4.0	0.14	75.0	2.54
Carbon monoxide	2660	39.0	12.5	0.42	74.0	2.51
Methane	2325	33.8	5.0	0.53	15.0	1.58
Ethane	2380	40.1	2.9	0.51	13.0	2.30
Ethylene	2565	68.3	2.7	0.41	36.0	5.52
Acetylene	2910	141.0	2.5	0.32	80.0	10.36
Propane	2390	39.0	2.0	0.50	9.5	2.36
Propylene	2505	43.8	2.0	0.45	11.7	2.64

[a]Calculated at 25°C and 1 atm for ideal gas.
[b]Base temperature: 298 K, pressure: 1 atm.

15

Table 8 lists the major components of gaseous fuels along with some of their combustion-related properties. A more complete presentation of these properties can be found in the Appendix.

Along with the above combustible compounds, practical gas mixtures contain varying amounts of nitrogen, carbon dioxide, oxygen, and water vapor. Water vapor pressure is usually adjusted to be quite low to avoid condensation and freezing, and oxygen is kept below the inflammability limit.

Volumetric heating value is important because pipeline capacity and design of equipment for mixing air and fuel depend on the volume of gas handled. Two heating values are reported. The "higher heating value" assumes that the heat of condensation of product water is recovered, and "net heating value" assumes that it is not. Since the heat of condensation is seldom recovered, the net value is the most informative for the user; however, for commercial purposes, the higher value is generally used. Comparison of hydrogen and carbon monoxide illustrates the problem. Hydrogen rates slightly higher on a HHV basis (313.8 Btu/ft^3 vs. 310.7) but will supply significantly less useful heat as judged by net heating value (265.5 vs. 310.7). Methane, the other major gas component, has a volumetric higher heating value of 1060 Btu/ft^3 and is classified as "high Btu gas" (> 400 Btu/ft^3). The presence of higher molecular weight hydrocarbons increases volumetric heating value and the diluents—nitrogen, oxygen, and carbon dioxide—of course reduce it. Gases in the 0–200 Btu/ft^3 range are classified as "low Btu gas." Gas in the 200–400 Btu/ft^3 range is classified as "medium Btu gas."

Flame temperature is important because of its effect on combustion rate and on the amount of heat that can be recovered above a given sink temperature. Fuels with high flame temperatures are clearly more valuable for use in applications where heat is transferred to a high-temperature sink and their relative value can be rated by taking this into account.

Of the common fuel gases, carbon monoxide has the highest flame temperature. Both carbon monoxide and hydrogen have significantly higher flame temperature than methane, so a medium Btu gas consisting of a mixture of carbon monoxide and hydrogen provides a higher-temperature heat source than the common high Btu gas methane. Acetylene has an exceptionally high flame temperature but, because of its expense and difficult storage and handling, is only used for special purposes such as welding. It is, however, found as a minor component in some fuel gas streams coming from high-temperature pyrolysis processes and is an important intermediate in rich mixture hydrocarbon combustion.

Laminar burning velocity goes through a maximum with equivalence ratio and these maximum values can be found in Table 8. The high reactivity and diffusivity of hydrogen results in a very high burning velocity.

Flammability limits are determined for upward propagation in a static system. The lean limit for saturated hydrocarbons is approximately 0.5 equivalence ratio. The more reactive unsaturated hydrocarbons can be burned leaner; however, the adiabatic flame temperature at the lean limit is close to 1120°C (2050°F) for all saturated hydrocarbons. CO and olefins will burn at somewhat lower temperature, and hydrogen is unique in its extremely lean inflammability limit which is due to its ability to diffuse rapidly into a nearby flame front, where the hydrogen has burned out and to its high reactivity. Since the lean blowout in practical equipment approximates the static measurement of inflammability, it serves as a useful guide of ease of combustion as well as of safety of air-fuel mixture. The lean limit flame temperature for a given fuel is approximately independent of the means of achieving the flame temperature so that the effect on lean inflammability of diluent gases and preheat can be easily taken into account.

The rich inflammability limit for saturated hydrocarbons is 2.2–2.4, with methane being lower at 1.5. Here hydrogen is about the same as carbon monoxide (2.5). The higher reactivity of hydrogen is balanced by its ability to diffuse into the reaction zone to make it richer than the average mixture. Acetylene is very high (10.3), and it can in fact be decomposed as a monopropellant at higher pressures.

A medium Btu gas consisting of CO and H_2 will clearly have substantial combustion advantages over methane, the major disadvantage being toxicity, a potential for higher NO formation because of its higher flame temperature, and the cost of storing, transporting, and handling a higher volume of gas. A medium Btu gas containing diluents, such as nitrogen, will have a lower flame temperature for a given equivalence ratio and also a narrower inflammability range.

Low Btu gases (< 200 Btu/ft^3) contain major amounts of noncombustible gas and become increasingly difficult to burn as flame temperature approaches that characteristic of the fuel blend inflammability limit. Mixtures with a heating value less than 100 Btu/ft^3 are considered quite difficult to burn. Mixtures of substantially lower heating value can, however, be burned by preheating the combustible mixture through either internal or external heat recycle. Catalytic combustion, with preheat, can give complete combustion in very low heating value gaseous fuels, with the automobile catalytic converter representing an extreme.

Production of low Btu gases from catalyst regeneration, in-situ combustion processes, coal mine methane recovery, refining, and synthetic fuels processes, etc., is expected to increase in the future, and improved techniques for the combustion of very low Btu gas are needed.

2.2. Gasoline

Gasoline is a highly specialized fuel designed specifically for use in the Otto-cycle engine. The performance of this engine-fuel system is strongly dependent on the composition of the gasoline. The following characteristics depend on fuel composition:

1. Quick starting at low temperature.
2. Smooth operation in a minimum time after a cold start.

3. Fast acceleration.
4. Freedom from stalling:
   a. due to vapor lock,
   b. due to carburetor icing.
5. Freedom from noise or damage due to knock or surface ignition.
6. Good mileage.
7. Protection against harmful deposits and rusting.

In order to offer these characteristics, gasoline has evolved into a complex mixture of hydrocarbons, ranging from butane to C_{10} with a carefully chosen mix of paraffins, olefins, aromatics, and, in some cases, alcohols. Nonhydrocarbon components such as tetraethyl lead and methyl-t-isobutyl ether, to control knock, and a variety of additives to control rusting, deposits, icing, etc., are also used.

A glossary of terms relating to gasoline performance can be found at the end of this chapter.

The first four gasoline performance characteristics to be discussed are related to its volatility. Characterization of a gasoline's volatility requires a distillation under standardized conditions (ASTM D86). The vapor temperatures as a function of amount evaporated are recorded and plotted graphically. Figure 2 shows such a plot with the ranges normally encountered. A second measure of fuel volatility is its Reid Vapor Pressure (RVP) (ASTM D323). RVP is defined as the equilibrium vapor pressure of a gasoline measured in a closed vessel at 100°F and at a vapor to liquid volume of 4 : 1. ASTM3710 is a replacement test that makes use of a gas chromatograph.

The crucial volatility characteristics of gasoline can be divided into three categories: hot weather performance, cold weather performance, and nonseasonal performance. Generally, hot weather performance becomes important after an engine is warmed up and its underhood temperatures have reached equilibrium. Cold weather performance is significant during the initial 15 minutes of operation after a prolonged engine shutdown. Nonseasonal factors are generally independent of temperature.

These performance features of gasoline as a function of volatility are illustrated in Figure 2. The four performance parameters are cold starting, warm-up, vapor lock, and engine cleanliness (as affected by back-end volatility). The entire distillation range is involved in these four performance characteristics. Front-end volatility affects both cold starting and vapor lock. The mid-fill volatility governs warm-up performance. The high boiling, aromatic portion of the fuel is a major contributor to combustion chamber deposits that are responsible for increasing the octane number requirement of the engine.

Cold Starting. At ambient temperatures of 20°F or lower, cold starting is a most important performance feature. At low ambient temperatures, the me-

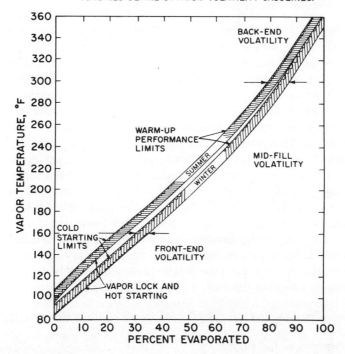

1. COLD STARTING AND WARM-UP PERFORMANCE LIMIT USE OF LOW VOLATILITY GASOLINE.
2. STARTING, WARM-UP, AND VAPOR LOCK CONSIDERATIONS COMBINED DEFINE OPTIMUM VOLATILITY GASOLINES.

FIGURE 2. Cold starting and warm-up performance limit use of low-volatility gasoline. Starting, warm-up, and vapor lock considerations combined define optimum volatility gasolines.

chanical and electrical condition of the automobile becomes critical. The critical features of the automotive system for starting are battery strength, starter motor power, engine oil viscosity, and ignition system condition. The ambient temperature greatly affects battery strength since the available electrical energy from the battery at 0°F is less than half of that available at 80°F.

In order for an engine to start, a flammable mixture of vaporized gasoline and air must first reach the cylinders. All engines are equipped with a choke that increases the manifold vacuum. The increased manifold vacuum draws up to 20 times the amount of liquid gasoline normally necessary for firing. The maximum air-fuel ratio (A/F) that will ignite in the engine is about 18/1. If an air-fuel ratio of 3/1 is supplied as a result of choking action, about 15% of the liquid fuel must evaporate at existing manifold conditions to permit starting. Since the time during which the fuel can vaporize is very short, the resulting A/F mixture is considerably leaner (i.e., higher) than that which occurs under equilibrium conditions. However, since so-called equilibrium tests are the easiest, most accurate, and the fastest tests available, cold starting fuel parameters have been correlated with fuel characteristics, such as RVP and ASTM distillation.

Warm-up Performance. The most critical time during warm-up occurs when the automatic choke begins to open. The first accelerations after starting are relatively satisfactory because the increased manifold vacuum due to choking draws extra fuel, and hence sufficient vapor to the engine. Performance drops off as the automatic choke begins to open and the manifold vacuum decreases. If the choke opens early, the A/F ratio becomes excessively lean, and power is lost. Some car manufacturers design chokes to open rapidly to attain maximum economy and minimum air pollution even though car performance may fall off badly after the first two or three minutes of warm-up. As is the case with all other fuel volatility performance features, warm-up is a function of the amount of fuel vaporized under the ambient temperature and pressure conditions encountered. The A/F ratio based on the total fuel supplied with the choke open is the primary variable. As a result warm-up performance is affected by the temperature needed to vaporize 50–90% of the fuel present. Thus warm-up is related to the volatility of the middle portion of the fuel.

Warm-up performance has been found to depend mainly on the 50% and 90% evaporated points of the fuel.

Vapor Lock. The fuel flow rate is affected by all of the following parameters:

- Fuel demand of the engine.
- Ability of fuel system to handle vapor.
- Temperature and pressures of the gasoline in the system.
- Volatility characteristics of the gasoline.

Fuel demand and vapor handling capacity are functions of car design (i.e., carburetor, fuel pump, and vapor return line) and operating conditions. Gasoline is subjected to temperatures and pressures controlled by car design (i.e., placement of fuel lines), operating and ambient conditions. The amount of vapor formed under any particular set of conditions depends on fuel volatility.

The critical months of the year are the spring months when ambient temperatures are rapidly increasing. Gasoline volatility must not only be adjusted seasonally but also geographically to ensure supply of the correct blend to specific areas. Ideal gasoline specifications also depend on altitude because gasoline, like any other liquid, vaporizes at lower temperatures when atmospheric pressure is reduced.

Avoidance of vapor lock and carburetor icing calls for lower volatility, whereas starting and warm-up calls for higher volatility. The balance will depend on temperature, humidity, and altitude. Gasoline properties are therefore adjusted according to season and geographical location.

Carburetor Icing. Gasoline marketed in cold weather must have high volatility for ease of starting; however, carburetor icing is aggravated by high-volatility fuels.

The problem of carburetor icing is observable only for the first 10–15 minutes of driving time following a prolonged engine shutdown. After 15 minutes of operation, the problem disappears. The mechanism of carburetor icing involves, as the name implies, the formation of ice crystals in the carburetor. Fuel is metered through the carburetor jets and vaporized by incoming air in the throttle plate area. Since the latent heat of vaporization for the fuel is supplied by the incoming air and the metal carburetor surfaces, the carburetor surfaces are cooled well below the dew point of the incoming air stream and can be cooled well below the freezing point of water (e.g., an incoming air stream of 40°F will cool the throttle plate to 12°F). Water in the air stream condenses and freezes on the throttle plate and surrounding carburetor throat area. When the throttle plate is closed, as it is idle, ice formation on the throttle plate and carburetor walls act as an air seal. Air supply to the engine is either severely or totally restricted, and a stall occurs. After 10–15 minutes of constant operation, the engine itself supplies sufficient heat to overcome the formation of ice in the carburetor.

There are four parameters that define the problem of carburetor icing. In order of decreasing importance, these are ambient temperature and humidity, fuel volatility, mode of engine operation, and underhood engine configuration.

The temperature and humidity of the ambient air can completely limit the amount of ice formed in the carburetor. Dry air provides insufficient moisture to form an ice seal to restrict air flow. Hot air provides sufficient heat to prevent or severely limit ice formation. Extremely cold air, by its nature, does not carry enough water to form an effective ice seal. These extremes define a large gray area in which stalling due to carburetor icing can occur. Field tests show that a temperature range between 30°F and 50°F, combined with a relative humidity above 62%, constitute icing weather.

Underhood configuration affects icing to the extent that heat can be transferred from the hot engine (e.g., exhaust manifold) to the cold carburetor surfaces. An engine that provides sufficient air passage between the hot engine and cold carburetor will experience icing for a shorter time period than an engine that provides no air passage between the hot engine and cold carburetor.

Anti-icing additives are used to overcome this problem. Compounds that exhibit the physical properties of either cryscopes or surfactants are potential anti-icing additives. The two classes of compounds retard ice formation in different ways. A cryscope inhibits ice formation by lowering the freezing point of water. A surfactant inhibits ice buildup because of its ability to concentrate at surfaces, modify the properties of surfaces, and the properties of phase interfaces.

An acceptable cryscope must have the ability to substantially lower the freezing point of water when the cryscope is in low concentration and must also have a high partition coefficient in favor of water to ensure that the additive will be transported from the gasoline phase to the water phase. The requirement that the cryscope effectively suppress the freezing point of water in low concentration arises from two considerations. First, the cryscope is usually added to gasoline in tenths of a volume percent. Thus, to start with, there is a low concentration of the additive in gasoline. Second, because of the extremely short water-gasoline contact times in the carburetor, only a small

TABLE 9. Sample Gas Chromatograph Analysis of Gasoline

Compound	%	
iC_4	0.36	
nC_4	3.92	
Butenes	0.44	4.72
iC_5	5.14	
nC_5	1.24	
C_5 olefins	4.68	
Cyclopentene	0.30	
Cyclopentane	0.21	11.57
nC_6	1.08	
2 MC_6	3.78	
3 MC_6	3.95	
23 DMB	0.79	
22 DMB	0.05	
C_6 olefins	2.32	
C_6 di-olefins	3.23	
M cyclo C_5	2.16	
Cyclohexane	0.11	
Benzene	2.65	20.12
nC_7	1.38	
MC_6	4.11	
dMC_5	2.74	
M cyclohexane	0.75	
C_7 olefins	1.93	11.41
Toluene	15.46	15.46
nC_8	0.61	
nC_7	2.58	
Other C_8 saturates	3.67	6.86
Ethyl benzene	2.27	
p-Xylene	2.28	
m-Xylene	6.07	
o-Xylene	3.26	
Styrene	0.62	14.50
C_9 saturates	3.20	
C_9 aromatics	7.83	
C_{10}	6.06	
	101.73	

amount of a cryscope would enter the water phase. Isopropyl alcohol is one compound that possesses the necessary properties, and it has been proved to be a highly effective anti-icing compound.

In general, surfactants are compounds whose molecules possess both a polar and a nonpolar end. This property allows them to selectively gather at a surface; here they significantly alter the properties of the surface. High-speed photography has shown that a surfactant alters the ice crystal's growth, resulting in smaller crystals. These smaller crystals are more readily carried away in the air stream. It is also likely that adhesion forces are modified,

TABLE 10. Sample Gasoline Inspection, Premium Grade

Inspection	
Octane RON (Research Octane Number)	98.2
MON (Motor Octane Number)	90.0
RVP (Reid Vapor Pressure)	10.39
D-86 distillation initial boiling point °F	87
5%, °F	112
10%, °F	123
20%, °F	144
30%, °F	164
40%, °F	187
50%, °F	214
60%, °F	241
70%, °F	270
80%, °F	307
90%, °F	347
95%, °F	382
FBP, °F	407
Recovery, %	97.0
Loss, %	1.5
Residue, %	1.5
Manganese, g/gal	< 0.01
Lead, g/gal	3.12
Cu beaker gum, mg/100 mL	16.0
ASTM gum, washed, mg/100 mL	1.6
unwashed, mg/100 mL	5.4
Sulfur, ppm	217
API gravity @ 60 F*	59.9
Aromatics, %	26.7
FIA olefins, %	11.7
saturates, %	61.6
Breakdown time, min	960+
Potential gum, mg/100 mL	14.4
Cu strip corrosion @ 122	1A
Benzene, %	1.50
Solids, mg/gal	2.4

TABLE 11. Comparison of Various Tests for Determining Antiknock Quality of Gasolines

ASTM Designation, Common Name, Type of Test, Engine Used	D-2699 Research Method, Lab, Single Cylinder	D-2700 Motor Method, Lab, Single Cylinder	Uniontown Road, Multicylinder Car	Modified Uniontown Road, Multicylinder Car	Borderline Road, Multicylinder Car
Engine speed, RPM	Constant—600	Constant—900	Full throttle acceleration 500–2500/3000	Full throttle acceleration 500–2500/3000	Full throttle acceleration 500–2500/3000
Spark setting, degrees advance	Constant—13	Automatically varied with Compression ratio	Automatically varied with speed from approximately 0/10—20/30—idle setting varied to give desired knock intensity range then held constant for test fuel and bracketing reference fuels	Automatically varied with speed from approximately 0/10—20/30—Idle setting varied to give constant knock intensity on all fuels	Constant during acceleration—varied to give several knock die-out speeds for each fuel
Compression ratio	Varied to obtain desired knock intensity	Varied to obtain desired knock intensity	Constant, depending on car	Constant depending on car	Constant, depending on car
Air-fuel ratio	Varied for maximum knock—essentially constant at approximately 15:1	Varied for maximum knock—essentially constant at approximately 15:1	As supplied by car approximately — 11–13 being somewhat richer at low speeds than at high speeds	As supplied by car approximately — 11–13 being somewhat richer at low speeds than at high speeds	As supplied by car approximately — 11–13 being somewhat richer at low speeds than at high speeds

Coolant temperature, °F	209–215	209–215	160[a]	160°	160°
Intake air temperature, °F	125	75–125	Ambient + 35[oa]	Ambient + 35[oa]	Ambient + 35[a]
Intake air humidity, Grains H_2O/dry air	25–50	25–50	Ambient	Ambient	Ambient
Mixture temperature, °F	100°	300	Ambient + 50[oa]	Ambient + 50[oa]	Ambient + 50[oa]
Main variables for rating various antiknock quality fuels	Compression ratio	Compression ratio	Knock intensity	Spark advance	Spark advance—speed
Method of measuring knock intensity	Knockmeter	Knockmeter	Human ear	Human ear	Human ear
Method of interpolating to obtain antiknock rating	Compare ratio set for test fuel, interpolate on basis of knock intensity	Compare ratio set for test fuel; interpolate on basis of knock intensity	Idle spark constant. Interpolate on basis of maximum knock intensity regardless of whether maximums occur at same speed	Idle spark varied for constant maximum knock intensity regardless of whether maximums occur at same speed; interpolate idle spark settings	Spark setting for knock die-out plotted vs. speed. Interpolate on basis of spark setting usually reporting ratings at several speeds

[a] Temperature not controlled (approximate).

making it harder to coat the throttle plate and carburetor throat with ice. Of the two classes of compounds used as anti-icing additives, the cryscopes are generally felt to be the more effective.

Final Boiling Point. The final boiling point of gasoline is controlled to less than about 430°F. This was set historically by the occurrence of crankcase oil dilution if higher boiling components were used. Higher boiling fractions bring other problems such as increased engine deposits, increased production of particulates and polynuclear aromatics, and relatively low octane number. Since these higher boiling fractions can be used in distillate fuels, there is relatively little incentive to include them in gasoline and there has been little pressure to increase the final boiling point.

Gasoline Composition. Gasolines are blended to meet local requirements from a wide variety of components, and therefore vary in their compositions. Table 9 gives the composition of one gasoline and is included to provide some background on the distribution of individual compounds in gasoline.

Aromatics are necessarily a major fraction of the fuel (40.4%), and the relatively small number of possible compounds results in their having high individual concentrations. Olefins enter from cracked streams and are of higher octane number than paraffins of the same carbon skeleton. The high ratio of branched pentane and hexane indicates that isomerization of the C_5 and C_6 fraction was practiced. Relatively small amounts of C_{10} are found, corresponding to the usual gasoline final boiling point.

The inspections for a summer 1979 leaded grade gasoline are shown in Table 10 to illustrate the properties of a "typical" gasoline. Unleaded premium gasolines would have higher aromatic content to retain high octane number in the absence of tetraethyl lead.

Knock. The explosions and the sound caused by the resulting vibration of the motor parts, described as knock, are very much dependent on the structure of hydrocarbon fuels. Iso octane (224 tri methyl pentane) has an "octane number" of 100, but *n*-heptane has an "octane number" of 0. Most gasoline boiling range hydrocarbons, found in petroleum, have a strong tendency to knock and are suitable only for very low compression ratio engines. The major part of gasoline manufacturing cost is incurred by processes that alter molecular structure to increase octane number, making possible use of higher compression ratio engines which increases engine efficiency. Thus efficiency increase can counterbalance increased manufacturing cost.

Obviously, there is an optimum where manufacturing cost increases exceed the value of efficiency gains. This optimum changes with time as petroleum costs and manufacturing costs change. This optimum compression ratio level drops when lead tetraethyl is not permitted in gasoline.

Octane number is determined by matching the knocking tendency of a gasoline with that of a mixture of *n*-heptane and *i*-octane. The percentage of *i*-octane in the test mixture is the octane number.

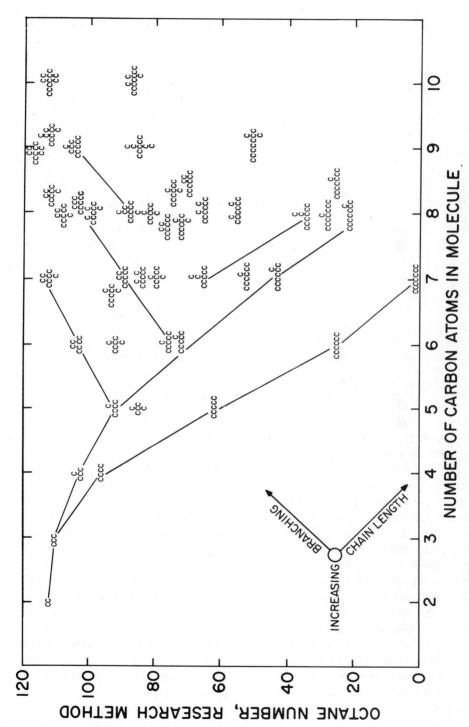

FIGURE 3. Antiknock level of paraffins. From American Petroleum Institute,[10] © 1958, ASTM. Reprinted with permission.

27

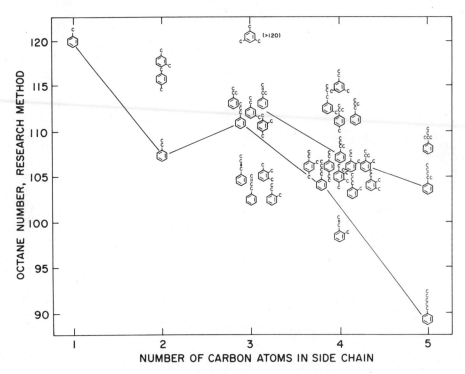

FIGURE 4. Antiknock level of aromatics. From American Petroleum Institute,[10] © 1958, ASTM. Reprinted with permission.

Table 11 gives the major features of ASTM tests for determining antiknock quality of gasolines. The "research" and "motor" methods are carried out in carefully standardized single-cylinder laboratory engines. The compression ratio in these engines can be continuously varied and is increased during a test to the point where knock is detected. The "research" method is carried out at lower speed and at lower mixture temperature than the more severe "motor" method. Road vehicles generally rate gasolines intermediate between these two methods, depending on the design of the vehicle.

Figures 3, 4, and 5 illustrate the variations of octane number with structure.

For paraffins of a given level of branching, the octane number decreases sharply with the number of carbon atoms in the molecule. Branching, however, greatly increases the octane number. Aromatics have a very high octane number compared to naphthenes or paraffins, and the refining process, Powerforming, is designed to convert these molecules to aromatics. Addition of paraffinic side chains to an aromatic reduces the octane number.

Octane numbers do not blend linearly, and important interactions occur.

Characteristics of typical refinery blending components are shown in Table 12.

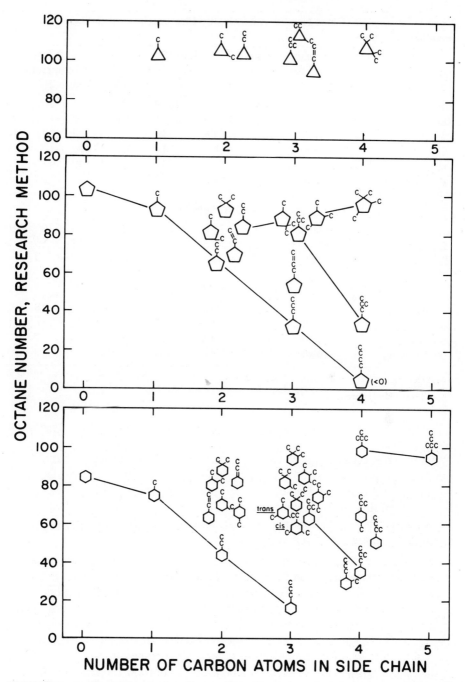

FIGURE 5. Antiknock level of cyclic paraffins. From American Petroleum Institute,[10] © 1958, ASTM. Reprinted with permission.

TABLE 12. Typical Octane Number and Volatility Characteristics of Refinery Blending Components

Component	Research Octane Number +3.0 cc Clear	Research Octane Number +3.0 cc TEL[a]/gal	Motor Octane Number +3.0 cc Clear	Motor Octane Number +3.0 cc TEL[a]/gal	RVP	Vol % Distilled at Temperature °F 158	Vol % Distilled at Temperature °F 212	Vol % Distilled at Temperature °F 302
n-Butane	96.0	104.4	98.0	104.4	65[b]	100	100	100
i-Butane	103.2	111.9	104.0	112.9	71[b]	100	100	100
n-Pentane	61.7	86.0	62.6	84.2	15.6	100	100	100
i-Pentane	93.3	104.1	88.3	104.6	20	100	100	100
Light virgin naphtha	65.1	82.8	65.1	82.8	7	40	85	100
C_4-light alkylate	96.0	108.0	94.0	98.0	6	10	42	96
Catalytic naphtha								
C_5	94.5	99.5	82.0	87.0	15	90	100	100
Light	91.5	97.5	79.5	85.0	7	12	55	91
Heavy	88.5	93.0	77.5	82.5	0.5	0	0	2
Reformate (mediumfeed)								
85 Reformate	85.0	95.5	78.6	87.8	5.2	2	15	89
90 Reformate	90.0	98.3	81.3	90.0	5.3	3	16	90
95 Reformate	95.0	100.6	84.5	91.6	5.5	5	17	90
100 Reformate	100.0	103.3	89.0	93.6	5.8	7	18	91

[a]Tetraethyl lead.
[b]Blending value (approximate).

Table 13 gives the octane number and boiling point of selected oxygenated compounds.

Methyl-t-butyl ether is of current interest as a blending agent capable of partly replacing the loss of lead tetra ethyl from gasoline. Methyl alcohol is very high and could well be of further interest as synthetic fuel; however, its limited solubility in wet gasoline greatly reduces its value as a blending agent. Ethanol also has a high octane number and is currently used as a blending agent. The others are of potential interest.

TABLE 13. Octane Number of Oxygenated Hydrocarbons

Compound	Boiling Point, °F	ROBN[a]	MOBN[b]
Methanol	149	124	94
Ethanol	173	125	86
t-Butanol	180	107	98
2 Ethylhexanol	364	73	66
Methyl-t-butyl ether	131	117	99
Furan	154	—	190

[a]Average in four pools, 92.5–98.6 RON and 84.0 MON at 10 vol % concentration.
[b]From American Petroleum Institute.[10]

Future Trends. The long life and large investment in automotive refining and distribution systems inhibits radical change in gasoline composition; however, over a long time, changes such as a special (lower octane number) fuel for stratified charge engines or a special high octane fuel for high compression ratio engines or fuels with major quantities of oxygenates could well be of importance. Another option might be development of systems specially designed to take advantage of the attractive properties of methanol.

2.3. Distillate Fuels

Distillate fuels are characterized by a flash point above the normal storage temperature and a final boiling point limited by freezing point or by tendency to produce soot.

The major commercial categories are

Kerosene (No. 1)
Turbo-jet fuel
Diesel fuel (No. 2)
Heating oil (No. 2)

Kerosene. Kerosene was the first petroleum fuel of importance. In the United States it was used to replace animal and vegetable oils for illumination. Safety required elimination of the volatile petroleum fractions that now go into gasoline, and these fractions were discarded until the automobile generated a major demand. As pointed out in the earlier section on supply and demand, because of the large gasoline and fuel oil markets in the past, adequate virgin petroleum fractions were available for use in distillates. Fuel specifications and combustion equipment have been optimized for this situation. The coming radical increase in the fraction of liquid fuels consumed as "distillates" and the introduction of tars, oil shale, and coal liquids for refinery feed indicate that evolution of both fuel specifications and combustion equipment will be in order.

Kerosene for domestic illumination and heating has become a minor product and is sufficiently similar to Jet A turbo-fuel that it will not be discussed separately.

Turbo-jet Fuel. The aviation gas turbine was developed during World War II and has since become the dominant commercial and military aviation power plant. This development took place at a time when gasoline consumption was growing rapidly and the difficulties of supplying fuels during the war period were fresh in the minds of both the military and commercial participants in setting the specifications for these new fuels. Ability to manufacture a maximum amount of jet fuel from a given amount of petroleum without serious interference with gasoline manufacture was a prime consideration. These considerations resulted in a mixture of gasoline boiling range naphtha and distillate fuel. Because of the naphtha dilution the boiling range of the distillate could include higher molecular weight fractions before the freezing point limitations were reached. The high volatility resulted in greater probability of ignition during ground handling or in accidents and a commercial kerosene type fuel of 100–120°F flash point was also developed because of this safety problem. Statistical studies of military and commercial experience with both of these fuels has shown a small but significant difference in safety, and more recently the military proposed phasing out the use of the volatile fuel (JP-4) in favor of a fuel quite similar to the commercial jet-A but with a lower freezing point.[6,11] This change occurred at a time of increasing concern

TABLE 14. Jet Fuel Specifications

	Jet-A Kerosene	U.S. Military JP-4 Wide-cut	ERBS Experimental fuel
Composition (max vol %)			
Aromatics	20[a]	25	~ 35 (12.8 ± 0.2%H)
Olefins	—	5	
Sulfur (wt %)	0.3	0.40	0.3
Naphthalenes (max)	3		report
Volatility			
Distillation IBP °F (°C)	—		report
10%	400 (204, max)		400 (max)
20	—	290 (145, max)	—
50	report	370 (190, max)	report
90	report	470 (245, max)	500 min
Final	572 (300)	520 (270)	report
Flash point °F (°C, min)	100 (37.8)	—	110 ± 10
Vapor pressure (lb Reid)	—	2–3	—
Fluidity			
Freezing Point, °F (max)	(−40)[b]	−72 (−58)	−20
Viscosity @ −20°C (cst max)	8	—	12 (max, −10°F)
Combustion			
Smoke Point (mm)	20 min	20 min	—
Thermal Stability			
Temperature, °C	260–245	260	240

[a]Aromatics to 25% max and/or smoke point to 18 min can be supplied if reported.
[b]ASTM Jet-A-1 is used by international airlines. Specifications are identical to Jet-A except that the freezing point is −47°C max.

about future supplies of jet fuel of jet-A quality and it is probable that JP-4 type fuels will continue to be used. Somewhat before the 1974 embargo, at the urging of industry advisors, NASA reactivated their interest in the question of appropriate jet fuel qualities for the future. To expedite handling and combustion studies, an experimental fuel (ERBS) was defined by an industry government workshop coordinated by NASA.[12] Abbreviated versions of these three fuel specifications are shown in Table 14. Complete commercial and military specifications can be found in Longwell and Grobman.[13]

From the viewpoint of combustion, the boiling point distribution and the composition are of greatest interest. Composition varies with boiling point, and the volatility is limited by freezing point and viscosity for the higher boiling materials and by flash point for the lower boiling materials. In the experimental (ERBS) fuel, raising the freezing point allows inclusion of higher boiling materials with their normally higher concentration of naphthalenes and other aromatics. The thermal stability requirement in this fuel will also tend to limit both olefin and aromatic content as well as final boiling point.

The conventional turbojet combustor injects liquid fuel through a pressure atomizer into a strongly back-mixed zone. A major spatial variation in fuel-air ratio is needed to achieve stable operation over a wide range of fuel flow rates. The resulting rich regions, especially at high-fuel throughput, produce soot which, at the high temperatures involved, radiates heat to the surrounding combustion chamber liner. Current engines suffer from linear damage due to heating, and linear replacement can result in an important maintenance cost. As the hydrogen content of a fuel decreases, more soot is formed and liner temperature increases as shown in Figure 6.[14]

The ERBS fuel would be troublesome in this particular system.

As shown in Figure 7, smoke also increases as fuel hydrogen content is reduced. A major retrofit for smoke reduction on takeoff was recently required. The combustors developed for this retrofit introduced more air into the recirculation zone that reduced smoke but also impaired flame stability at high altitude. NASA has sponsored the development of special combustors that

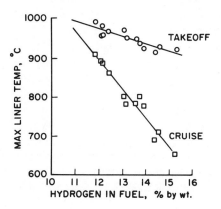

FIGURE 6. Effect of hydrogen content of fuel on maximum combustor liner temperature.

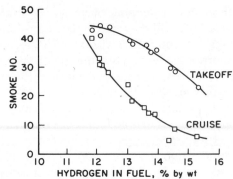

FIGURE 7. Effect of hydrogen content of fuel on smoke number.

incorporate some fuel-air premixing before entering the recirculation zone.[15] These experimental combustors are much less sensitive to fuel composition and serve to indicate that combustion research can do much to remove fuel composition limitations.

The composition referred to in the specifications is determined by elution from a silica gel column (ASTM D 1319). Fluorescent tracers are used to mark the boundary between compound types. The aromatic fraction includes all molecules containing one or more aromatic rings plus aromatic olefins, di-olefins, and compounds containing nitrogen, sulfur, or oxygen.

Table 15 offers some perspective on the fraction of aromatic carbon in alkylated aromatics as a function of molecular weight and boiling point.

Aromatics have been found to form more soot for a given equivalence ratio than paraffins, and one would expect that the fraction of carbon in aromatic rings would be a useful measure. The H/C ratio, or wt % hydrogen, is a measure of the amount of aromatic carbon, whereas the FIA aromatics are not. There are, however, differences between aromatics. Carbon in a naptha-lene structure has more tendency to produce soot than in benzene, for example. Olefins and naphthenes differ from paraffins, and the complexity and variability of jet fuel composition has resulted in use of tests where smoke number or luminometer number give a measure of soot production in a controlled laminar diffusion flame. There is still suspicion that naphthalenes may be particularly troublesome, and they are in fact limited to 3 vol % in some commercial specifications.

Figures 6 and 7 show a reasonable correlation between soot-related prob-lems and the wt % hydrogen in the fuel. Although it can be shown that individual compounds can deviate from such a correlation, the naturally occurring distribution of a wide variety of compounds is such that perfor-mance of commercial fuels is well correlated by % H. From the viewpoint of fuels manufacture, this is a useful specification since it relates directly to refinery requirements, whereas a laboratory burner test does not. In the case of the ERBS fuel, % H was chosen to characterize the fuel. Figure 8 gives an indication of the variation of hydrogen content with aromatics content.

TABLE 15. Properties of Alkylated Aromatics

Number of Side-Chain Carbons		Molecular Weights	Approximate Boiling Point, °F	H/C Atom Ratio	Wt % H	Aromatic C
Benzene	Naphthalene					
0		78	176	1.0	8.3	100
2		106	280	1.25	10.4	75
4		134	356	1.40	11.7	60
5		162	430	1.50	12.5	50
8		190	510	1.57	13.1	43
10		218	565	1.63	13.6	38
	0	128	412	0.80	6.7	100
	1	182	465	0.91	7.6	91
	2	156	510	1.00	8.3	83
	4	184	590	1.14	9.5	71
	6	212	650	1.25	10.4	63

Table 15 also offers some insight into the aromatic compounds included in the boiling range of jet fuels. The naphtha-jet (JP-4) can obviously include benzene although the concentration of the lower boiling aromatics will generally be small. Since no more than 10% of Jet-A can boil below 400°F, most of the single-ring aromatics would have five or more paraffinic carbons. Alkylated naphthalenes with 2–3 paraffinic carbons could be found in the higher boiling fractions (550°F). Increasing the final boiling point, as in the experimental ERBS fuel, greatly increases the number of possible polynuclear aromatic compounds.

The freezing point specification can limit final boiling point as shown in Figure 9.[14]

Although a broad variation between fuels exists, it is clear that raising the freezing point can result in a major increase in final boiling point. For commercial flights, the freezing problem arises on a very long, high-altitude flights. This is illustrated in Figure 10 for a long and unusually high 747 flight.[14]

For the current unheated fuel tanks, starting at −15°C, the fuel temperature can go below the Jet-A specification of −40°F. Heating the wing tanks can obviously solve the problem. The question arises as to whether all Jet-A should continue to be restricted in boiling range for the 1–2% of the flights involved.

Fuel thermal stability and viscosity are also adversely affected by increasing final boiling point; however, again there are design and fuels manufacture choices that will alleviate these problems if there is sufficient incentive to extend the range of fuel characteristics that can be accepted by modern aircraft.

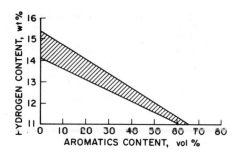

FIGURE 8. Variation of hydrogen content with aromatic content.

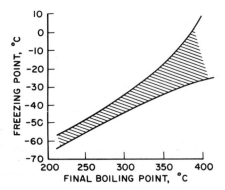

FIGURE 9. Typical fuel blend freezing points.

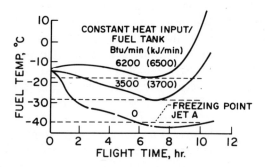

FIGURE 10. Fuel tank temperatures for 5000 nm flight with heating.

Diesel Fuel and Domestic Heating Oil. Although the diesel engine and the domestic oil burner carry out combustion under radically different conditions, the fuels used are quite similar and, to some extent, interchangeable.

The diesel engine was invented in 1892 and by 1925–1935 began to see substantial use in trucks, locomotives, and tractors. Displacement of steam-powered locomotives was rapid since the diesel engine offers an efficiency more than three times that of the steam engines used. From the viewpoint of efficiency, ease of fuel manufacture and reliable operation, the diesel engine-fuel system has been quite satisfactory; however, the diesel engine, in its present form, cannot meet the standards for NO_x being imposed on the gasoline-powered passenger cars. In addition approximately 0.5% of the fuel burned emerges as carbonaceous particulates on which are adsorbed compounds known to be carcinogenic. There is therefore a very active effort to greatly improve the understanding and technology of diesel fuel combustion. Modifications aimed at improving NO_x emissions generally increase particulates emissions, and vice versa. Particulates emissions decrease as the higher boiling fractions of conventional diesel fuel are removed. A fuel comparable to kerosene or Jet-A would, therefore, offer some improvement and, in addition, would improve cold starting. Such a change at a time when total distillate is a rapidly increasing fraction of liquid fuels would increase the difficulty of maintaining adequate hydrogen content in these fuels.

Domestic oil burners are not the subject of comparable scrutiny. Their NO_x emissions are relatively low, and their production of mutagenically active hydrocarbons is approximately a factor of 100 lower, per unit of fuel consumed, than that of the automotive diesel engine.[16] Emission of mutagenic material from wood burning is much greater. Even so, the emissions from domestic oil burners occur throughout the year in heavily populated areas, and improvement, preferably through improved burner design, may be needed.

In contrast to diesel and aviation fuel, relatively small growth for heating oil is projected. This results from application of heat conservation techniques and from competition from gas and electric heating.

Diesel Fuel Properties. Diesel fuels in many ways resemble home heating oils. They differ mainly in that a minimum cetane number, a controlled distillation range, and a maximum sulfur level are guaranteed to the consumer.

Diesel fuels are made from virgin, catalytically cracked, and occasionally thermally cracked stocks. Since blends of these components would tend to form gum and insoluble sediment in storage, chemical treatment is required. Virgin and light catalytic components were usually caustic and water washed; thermal and heavy catalytic fractions were acid treated in the past but now are hydrofined. Stocks with high mercaptan levels must be sweetened to improve odor.

Types of Diesel Fuel. Although all diesel engines require fuels with common characteristics, certain engines tend to be more critical of fuel quality than others. High-speed engines require a higher-quality fuel than do medium-speed engines, which in turn are more critical of fuel quality than low-speed units. Diesel engine classifications are discussed in Attachment I-1.

Table 16 shows the relationship between fuel quality and intended use. City buses require the highest-quality, most-volatile fuel to minimize air pollutants in metropolitan areas. Other high-speed engines such as automotive diesels and light marine engines use premium diesel fuel. The medium-speed railroad engine and mid-size marine and stationary engines are satisfied with a wider

TABLE 16

Fuel	Jet-A Spec.	Kerosene #1	Premium #2	Typical #2 Heating Oil	Railroad Diesel	Marine Diesel
API, gravity-min	—	40	37	34	34	28–23
Density, g./cc	—	—	—	—	—	—
Distillation, °F						
IBP	(325)R	325	350	325	350	350
50%	—	—	—	530	—	—
FBP	572	550	650	670	675	800
Pour point, °F (max)	−40	0	—	−5	—	—
Aromatics, vol %	20–25	—	(35)	(40)	—	—
Sulfur, wt %	0.3	0.12	0.30	0.40	0.50	1.2
Cetane, number-min	—	47	45	—	40	37
Flash point, °F	100	100+	—	150	—	—
Uses	Aircraft	City buses, heating, auto	Buses, trucks, auto	Home heating	RR, marine, stationary	Heavy marine large, stationary

boiling range fuel. Low-speed heavy marine and large stationary engines use the lowest-quality fuels. These large engines can also burn residual-type diesel fuels.

The specifications of these fuels are not closely controlled by international agreements between manufacturers, as in the case of turbo-jet fuels; however, ASTM has set minimum general standards for the major inspections for fuel oils, and in addition many states have set flash point requirements that range from 100°F upward. "Typical" inspections are shown in Table 16.

Desired Fuel Characteristics. Although diesel engines vary widely in size, speed, power output, and mechanical design, their fuel requirements are quite similar in many respects. Desirable characteristics for all diesel fuels, and the physical properties most directly related to them, are summarized below:

1. Flash point — For safe handling
2. Pour and cloud points — For pumpability at low temperature
3. Thermal stability — Minimizes plugging
4. Proper viscosity range — For easy atomization
5. Cetane number — For ease of ignition and start-up
6. Proper volatility — For complete combustion
7. Gravity — For good fuel economy
8. Low sulfur — Minimizes engine wear and deposits

The properties generally considered to be most indicative of diesel fuel quality are cetane number, volatility, viscosity, sulfur content, and API gravity.

Cetane Number. One of the most important properties of a diesel fuel is its ability to ignite spontaneously under the pressure and temperature conditions that exist in the engine cylinder. It is generally believed that ease of starting, detonation, and, to a certain extent, excessive smoke and engine carbonization are related to the cetane number.

The cetane number is an index of a fuel's ignition quality. It is measured in a single-cylinder, variable compression ratio diesel engine under fixed conditions of speed, load, jacket temperature, and inlet air temperature. The ignition quality of the test fuel is compared with mixtures of two reference hydrocarbon fuels: namely, cetane, a normal $C_{16}H_{34}$ paraffin, which is assigned a rating of 100 and alpha-methyl-naphthalene which is assigned a rating of 0.

The cetane number is defined as the percent of pure cetane in a mixture of cetane and alpha-methyl-naphthalene that matches the ignition quality of the fuel sample being tested. (When tests are conducted, however, hepta-methyl-nonane, cetane number 15, is used as the second reference fuel.)

In general, the more paraffinic a fuel, the better its ignition quality. Aromatic fuels (low-hydrogen content) are relatively poor in this respect. Cetane represents a paraffinic fuel of excellent ignition quality, whereas alpha-methyl-naphthalene typifies an aromatic fuel of very poor ignition quality.

If a cetane rating engine is not available, a cetane index may be derived from a relationship between API gravity and the volume average boiling point (VABP) of the test fuel. Cetane numbers (in the 30–60 range) can generally be estimated with an accuracy of 2 units, using the cetane index. The index correlation is acceptable for petroleum stocks but is highly questionable for nonpetroleum stocks, such as those derived from tars, shale oil, or coal liquids.

Volatility. Important distillation points for determining diesel fuel volatility are the 10%, 50%, 90%, and final boiling point. The higher the mid and/or final boiling points, the more difficult it is to vaporize the fuel completely.

When high boiling components cannot be vaporized in the short time available for combustion, sooty, incomplete burning results. It is very important in high speed engines to have volatile, readily vaporized fuels; the importance of fuel volatility decreases as speed decreases or, stated another way, as the time available for combustion increases.

Viscosity. Viscosity is directly related to the ease with which the fuel is atomized in the combustion chamber. Too high a viscosity will result in the fuel spray penetrating too far into the combustion chamber, wetting the piston and cylinder walls, and carbonizing on the hot combustion chamber surfaces. If fuel viscosity is too low, injector plungers and barrels that depend on the fuel for lubrication will tend to wear excessively.

Gravity. In general, the higher the density (or lower the API gravity), the higher will be the volumetric heating value of a fuel. High-density fuels, however, have lower cetane numbers because of their higher aromatic content.

Sulfur Content. During combustion sulfur compounds in the fuel are converted to sulfur oxides. In the presence of water, which is also formed during combustion, these oxides are potentially corrosive. Sulfurous and sulfuric acids condense on relatively cool engine parts and lead to corrosive wear. In addition, any sulfur trioxide (SO_3) formed in combustion attacks the lubricating film and forms undesirable varnishes and gums.

High-speed engines favor the formation of SO_3, although SO_2 is still the dominant form; slow-speed engines favor the formation of SO_2. SO_3 is the combustion product mainly responsible for lube oil deterioration and corrosive wear. High-speed engines often require as low as 0.4% sulfur for satisfactory operation. By way of contrast, low-speed engines have operated successfully on fuels containing over 4.0% sulfur.

Storage Stability. The increased use of cracked stocks in diesel fuels initially resulted in storage stability decline. However, proper selection and finishing methods were developed. Additives were also developed that controlled stability and sediment and gum formation in all types of diesel fuels. Because of these developments, fuel filter clogging and injector sticking are rarely encountered with present high-quality diesel fuels.

Pour and Cloud Points. Pour point, or the minimum temperature at which a fuel remains liquid, is of importance in mobile installations where the fuel must be handled at or near ambient temperatures. If the fuel is to be pumped through filters, the cloud points, or that temperature at which paraffin crystals begin to form, is the minimum temperature at which engine operation is permissible. Operation below the cloud point will result in clogging of the fuel filters by wax. As a result the final boiling point is frequently limited by pour and cloud point.

Other Fuel Properties. Other properties of diesel fuels related to performance are (1) carbon residue, (2) ash content, (3) acidity or neutralization number, and (4) sediment and water content.

Carbon residue content is an index of the carbonizing tendencies of a fuel. The higher the carbon residue value, the greater will be the tendency for deposits to form on the fuel injector tip or in other parts of the combustion chamber.

Ash-bearing materials in diesel fuels lead to increased ring and liner wear. Organic acids can attack metals in the fuel system to form metal soaps (surfactants) that clog filters and foul injectors. Sediment in fuel leads to filter clogging and injector fouling, whereas water is harmful because it promotes rusting of critical steel injection equipment.

Additives in Diesel Fuels. Additives are used today in diesel fuels as

- Stabilizers against gum and sediment formation in storage
- Cetane improvers
- Emulsion breakers
- Rust inhibitors
- Pour depressants

Although chemical finishing in the refinery is effective in improving the storage stability of unstable diesel fuel base stocks, finishing is relatively expensive and chemical finishing alone cannot provide the stability needed in present-day fuels. The most economical method of providing optimum stability in distillate fuel products involves the use of some chemical finishing plus a stabilizing additive.

Cetane improvers are effective in improving ignition quality of diesel fuels. Amyl nitrate, one of the better-known cetane improvers, is capable of raising the cetane number of a 40 cetane fuel to 45. Certain improvers have not been used widely in diesel fuels primarily because their cost has not been competitive with crude selection and refinery finishing.

Wax crystal modifiers have been developed that change the crystal structure, allowing the crystals to pass through filtration equipment. This can extend the operability limit as much as 10–15°F below the cloud point. Responsiveness of the fuel to the additive is dependent on the nature of the fuel, particularly the amount and composition of the wax-forming fraction.

Emulsion breaking additives are used in marine diesel fuels to prevent the formation of stable oil/water emulsion sludges. Although marine diesel fuels treated with emulsion breaker will emulsify with water, the additive causes a rapid separation of the oil and water phases.

Additives are used to provide rust inhibition of diesel fuels.

No. 2 Heating Oil. Domestic heating oil is tailored to give dependable performance in the typical home furnace or boiler.

Manufacturing specifications are defined that ensure good burning characteristics, freedom from sediment or treating residues, customer acceptance, and ease and safety in handling.

Domestic heating oils, boiling in the range of 325–675°F, include components from three sources: virgin stocks cut directly from crude, oils manufactured by catalytic cracking of heavier stocks, and thermally cracked streams. The volumes of these components blended into the heating oil pool vary considerably among refineries; however, the volume of thermally cracked stocks is normally small, usually less than 5%. Catalytically cracked components typically run between 30% and 70%.

Virgin and light cracked components usually require some chemical treatment because mercaptans are present. Caustic washing is used to remove mercaptans, organic acids, and other undesirable compounds; followed by water washing to remove traces of treating residues. The final finishing step is drying by filtration or coalescing.

Heavier catalytic and thermal stocks may require more severe treatment to remove olefins and other unstable compounds. Mild hydrogenation is commonly used and the industry average sulfur level has dropped below 0.4 wt % over the last two decades.

The various components are blended to meet manufacturing specifications and a stabilizing inhibitor is often added. Recently flow improving additives have been introduced that permit use of heating oils having slightly higher wax contents.

Desired Fuel Characteristics. Good product quality is dependent upon the following fuel properties:

1. Proper viscosity	Easy atomization
2. Proper volatility	Easy ignition
3. Low pour point	Acceptable pumpability
4. Low carbon residue	Clean combustion
5. Low sulfur level	Minimum odor and flue gas corrosivity
6. Good stability and low sediment	Minimum filter and nozzle plugging

Viscosity is important in atomization and a maximum limit is often specified.

Low carbon residue is particularly important for satisfactory performance in wall-flame vaporizing burners. This is measured by evaporating the oil, without burning, from a hot metal surface to simulate vaporization from a metal hearth ring.

A maximum sulfur content is usually specified to minimize acidic corrosion by flue gases. More restrictive sulfur specifications are sometimes imposed by environmental regulations designed to control SO_x emissions.

The presence of sediment in domestic heating oil will lead to filter plugging and fouling of the fine passages in high-pressure nozzles and other burner parts. A maximum limit is usually specified. In addition to the rust and dirt that might be present, organic sediment can form in storage from the oil itself. To measure this instability, the oil is artificially aged, and any sediment formed is measured. Artificial aging can be accomplished with heat, oxygen, or sunlight or a combination of these accelerators.

Because heating oil is used in the home, acceptable odor is important. Although the ultimate customer rarely sees the oil, water haze and color are considered important by the intermediate jobbers and dealers who handle the

TABLE 17. Hydrocarbon Analysis of No. 2 Fuels

Fuel	T-1	T-2	T-3
Composition, wt %			
Sulfur	0.24	0.25	0.06
Br #	0.61	0.61	0.64
Paraffins	31.5	31.3	35.3
Cycloparaffins	40.3	42.7	32.0
Alkylbenzenes	8.0	8.9	7.4
Indans and tetralins	5.0	4.6	6.5
Indene	1.3	1.2	1.0
Naphthalene	0.2	0.3	0.6
C_{11} + naphthalenes	4.2	3.0	7.3
Acenaphthylene	4.0	3.5	3.7
Tricyclicaromatics	2.2	1.9	2.1
Total	100	100	100
Naphthalene Distribution			
C10	0.20	0.29	0.6
C11	0.65	0.57	1.5
C12	1.19	0.70	2.1
C13	0.80	0.60	1.8
C14 +	1.49	1.10	1.9
Total	4.4	3.3	7.9

product, and are controlled by specifications. For safety in handling, a minimum flash point is specified. A minimum pour point specification ensures pumpability in cold weather.

The use of additives in heating oils to inhibit sediment formation and tank rust is common. Cracked components increase storage instability, forming sediment that tends to plug filters and nozzles. This difficulty is overcome by proper finishing, stability controls and the use of oil-soluble stabilizing inhibitors. These compounds usually have both antioxidant properties to slow undesirable reactions and dispersing characteristics to keep sediment particles small enough to pass through filters and nozzles. Most heating oils also contain a water-soluble rust inhibitor to prevent corrosion of the customer's tank.

In Table 17, a compound type analysis of three No. 2 fuel oils is shown.

Aromatic content varies between 26 and 33%, and approximately half are 2- or 3-ring compounds that are especially prone to form soot. The 4-, 5-, and 6-ring aromatics are excluded by the maximum boiling point which is set by the freezing point and viscosity limits. This provides some protection from inclusion of the mutagenic and carcinogenic polycyclic aromatics that are found in some cracked stocks and in coal tars.

2.4. Fuel Oils

Petroleum "fuel oils" contain varying proportions of residual oil, which is the undistillable portion of the original petroleum. These oils are used in applications, ranging from commercial heating (factories, apartment houses, office

TABLE 18. Representative Residual Fuels Analysis[a]

	Conventional High Sulfur	Intermediate Sulfur	Low Sulfur
Sulfur, %	2.2	0.96	0.50
Carbon, %	86.25	87.11	87.94
Hydrogen, %	11.03	11.23	11.85
H/C atom ratio	1.53	1.55	1.62
Nitrogen, %	—	—	—
API gravity	17.3	21.5	24.7
Viscosity SSU, @ 100°F	3138	586	225
Conradson carbon, %	12.51	5.64	2.43
Hexane, insoluble	10.33	4.22	2.25
Ash, %	0.08	0.04	0.02
Trace metals, ppm			
Vanadium	350	155	70
Nickle	41	20	10
Sodium	25	10	< 5
Iron	13	9	< 5

[a]From Goldstein and Siegmund.[15]

buildings, etc.) to major power generation installations where 25,000 gal/hr may be consumed in a dozen burners. As discussed earlier, these uses for oil will diminish as petroleum becomes increasingly scarce. Substitution by coal, manufactured gas, coal liquids, and nuclear will allow this reduction. Since petroleum residuum has a higher hydrogen content than unrefined coal liquid, it seems logical to refine it to higher-quality products and, where liquid fuel oils are required, to eventually substitute coal liquids.

The ASTM requirements for the various grades of fuel oil are given in the Appendix.

In this table the distillate fuels Nos. 1 and 2 are shown for comparison. There is no number three fuel oil, and Nos. 4, 5, 6, are distinguished primarily by their viscosity. The highest grade of fuel oil, No. 4, can normally be atomized and burned without preheating and is required to have a pour point

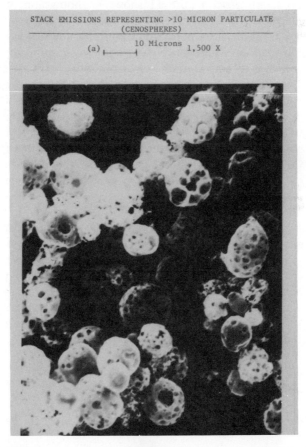

FIGURE 11. Stack emissions representing > 10 micron particulate (cenospheres).

of < 20°F. Sulfur content is set by local requirements. Sulfur and viscosity are controlled by blending distilled oils that may be desulfurized for sulfur control. Residuum desulfurization is also practiced in a few refineries to allow less use of distilled components. Number 6 oils are the lowest grade and may require preheating 230°F or higher to allow adequate atomization. Table 18 gives examples of No. 6 fuel oils.

Of special interest are the Conradson carbon and metal contents. Conradson carbon is the residue left after pyrolysis of the oil in a crucible under carefully controlled conditions. Evaporation of an oil drop during combustion results in a solid residue that corresponds approximately to the residue in this test. The remaining solid particle then burns in a manner similar to the combustion of the char remaining in coal combustion after devolatization has taken place. These solids have relatively low surface area and therefore burn slowly and are sometimes emitted as "stack solids." These solids are frequently in the form of hollow "cenospheres" that can be larger in diameter than the original oil drop. Figure 11 shows a photograph of typical stack solids.

Large drops resulting from poor atomization are the major source of these spheres. The low-sulfur oils are much less troublesome in this regard since they are of lower viscosity and contain less Conradson carbon. Distillate coal liquids have low Conradson carbon and, even though their hydrogen content is low, burn cleanly in equipment suitable for petroleum fuel oils. Their relatively high nitrogen content can, of course, present a NO_x emissions problem.

Petroleum fuel oils vary widely in organic nitrogen with a maximum of approximately 0.5 wt %. This nitrogen content is responsible for the greater NO_x emissions compared with natural gas-burning systems and has resulted in the somewhat greater amount of NO_x permitted by federal standards (0.3 lb/10^6 for oils vs. 0.2 for gas).

The metals content, though small, is of considerable importance. The presence of vanadium catalyzes the combustion of residual carbon and thus helps to control the emission of large carbonaceous spheres.[17] Vanadium, nickel, and sodium vaporize under flame conditions and upon condensation form submicron particulates that could well be of more environmental concern than the large cenospheres that fall to the ground near their source. This inorganic material is also highly corrosive to metals at high temperature and limits the usefulness of these fuels in stationary gas turbines and is troublesome in boilers.

REFERENCES

1. H. C. Hottel and J. B. Howard, *New Energy Technology*. MIT Press, Cambridge, MA, 1971.

2. *U.S. Energy Supply Prospects to 2010*. National Academy of Sciences, Washington, DC, 1979.

3. "Peat Draws Attention as Hydrocarbon Source." *Chem. Eng. News*, Nov. 7 (1977).

4. *U.S. Energy Demand Futures to 2010*. National Academy of Sciences, Washington, DC, 1979.

5. *Energy—Global Prospects 1985–2000*, Workshop on Alternative Energy Strategies. McGraw-Hill, New York, 1977.

6. *Coal Bridge to the Future*, World Coal Study. Ballinger, Cambridge, MA, 1980.

7. W. G. Dukek, "Future Outlook for Jet Fuel." Presentation at SAE Aerospace Meeting, 1977.

8. *An Assessment of Oil Shale Technologies*. Office of Technology Assessment, Washington, DC, 1980.

9. *Aviation Turbine Fuels—An Assessment of Alternatives*, p. 60. National Academy Press, Washington, DC, 1982.

10. American Petroleum Institute, "Knocking Characteristics of Pure Hydrocarbons," *API Proj. No. 45 ASTM Spec. Tech. Publ.* **STP-225** (1958).

11. "Jet Aircraft Hydrocarbon Fuels Technology." *NASA Conf. Publ.* **2033** (1977).

12. *Jet Fuel Specifications—Exxon*. 1978.

13. J. P. Longwell and J. Grobman, "Alternative Aircraft Fuels." *ASME Pap. No. 78-GT-59* (1978).

14. "Aircraft Research and Technology for Future Fuels." *NASA Conf. Publ.* **2146** (1980).

15. H. L. Goldstein and C. W. Siegmund, "Influence of Heavy Oil Composition and Boiler Combustion Conditions on Particulate Emission." *Environ. Sci. Technol.* **10**(12), 1109–1114 (1976).

16. *Soot in Combustion Systems and Its Toxic Properties*, p. 37. Plenum, New York, 1983.

17. K. C. Bachman and C. W. Siegmund, "The Effect of Fuel Oil Composition on Particulate Emission." *Int. Flame Res. Found. Symp., 1979* (1979).

GLOSSARY OF GASOLINE QUALITY RELATED TERMS

ASTM Distillation (ASTM D-86)   A distillation under standardized conditions. Vapor temperatures are read at stated volume percentages of the distillate collected.

Distillation Curve   A plot of vapor temperature as a function of percent distillate recovered.

Reid Vapor Pressure (ASTM D-323)   A measure of gasoline volatility. It is the vapor pressure (psia) at 100°F of a sample sealed in an enclosed bomb where the liquid to vapor is in the ratio of 1 : 4.

Front-end   The lower boiling portion of a gasoline.

Mid-fill   The middle boiling range of a fuel.

Back-end   The high boiling portion of a gasoline.

Choke   A butterfly valve used to restrict air flow, decreasing air pressure in the carburetor and manifold, thus richening the air-fuel mixture.

Air / Fuel Ratio (A / F) The ratio of air to fuel (by weight) charged to the engine. A lean A/F is a high ratio. There is a large amount of air present relative to fuel.

Manifold Vacuum Difference between ambient air pressure and gas pressure in the intake manifold.

Percent Evaporated Off (Distillation + Loss = D + L) The percentage of fuel distilled at a given temperature (e.g., 158°F, in an ASTM distillation). The sum of distillate recovered at that temperature plus the percentage unrecovered (i.e., uncondensed during the overall distillation).

Warm-Up Period of time until an initially cold engine gives a smooth and powerful acceleration.

Hesitation Loss of power when attempting to increase speed.

Flat Spot Prolonged hesitation during an acceleration.

Throttle Value that restricts air flow between carburetor and manifold. Opening this valve increases manifold pressure and thus engine speed.

Latin Square Pattern A statistical technique used to minimize differences between data sets.

Vapor Lock The blockage of fuel flow to the engine by bubbles of gasoline vapor lodged at critical points in the automotive fuel system.

Vapor / Liquid (V / L) Ratio Volume ratio of gasoline vapor to liquid measured in a closed vessel under standard conditions.

TEL Tetraethyl lead, the principle antiknock additive used in motor gasolines.

APRAC of CRC The Air Pollution Research Advisory Council of the Coordinating Research Council.

Positive Crankcase Ventilation (PCV) A system designed to feed vapors from the crankcase back through the induction system to be burned in the engine.

Induction System The system that produces and delivers the air-fuel mixture to the cylinders. It consists normally of the carburetor and manifold.

Theory (of Phosphorus) The units used to express the concentration of phosphorus, as an element, in gasoline. One theory is the amount of phosphorus to convert all the lead antiknock in a gasoline to lead orthophosphate.

Compression Ratio (CR) Total cylinder volume divided by cylinder volume not swept by the piston.

Octane Requirement Increase (ORI) Octane requirement of an engine (car) with mileage and deposits less than the initial octane requirement of the clean engine.

2 Evaluation of Chemical Thermodynamics and Rate Parameters for Use in Combustion Modeling

DAVID M. GOLDEN
Department of Chemical Kinetics
SRI International
Menlo Park, California

Prologue
1. Introduction
2. Estimation of thermochemical properties
    2.1. Additivity rules
    2.2. Structural estimates of entropy and heat capacity
    2.3. Enthalpy estimates
    2.4. Free radical group values
3. Problems using methods of Section 2
    3.1. Thermochemistry and equilibrium constants from additivity rules
    3.2. Structural estimates of entropy and heat capacity
4. Kinetics
    4.1. Transition state theory
    4.2. Reaction types
    4.3. Application of transition state theory
    4.4. Mechanistic problems—bound intermediates
5. Problems using methods of Sections 2 and 4
6. Conclusions
    References
    Bibliography
    Appendix: Answer Sheets

PROLOGUE

This chapter is best complemented by the monograph *Thermochemical Kinetics* by S. W. Benson. The mutual goal is the realization that quantitative chemical kinetics can be broadly understood within a general framework. Such a framework exists to provide a rational basis for guiding experiments. But it is important to realize that the framework is incomplete and requires further development. This chapter should enable readers to find the appropriate level of usefulness in attacking practical problems, such as combustion, and perhaps inspire readers to examine further the development of the theoretical and experimental basis of modern chemical kinetics.

NOTE (added January 1987): This chapter was first written in 1981, and then modified somewhat in 1983. In the interim there have, of course, been advances and changes in some of this material. If I were preparing it now, I would expand the sections on pressure dependence of unimolecular and complex reactions. For additional information, the reader is referred to Golden and Larson[1] and Cobos and Troe.[2]

1. INTRODUCTION

If a complex chemical and physical process, such as combustion, is to be understood at a fundamental level, a definition of understanding needs to be agreed upon.

Among the goals one might consider are the ability to (1) describe the entire mechanism, both chemical and physical transformations, reducing the problem to the solution of a set of coupled differential equations describing temporal and spatial concentration profiles of each species; (2) specify the physical parameters appropriate for the chemical transformations (rate constants as a function of temperature and pressure) and for the physical transformations (transport coefficients as a function of temperature and pressure); (3) specify the confidence limits for the above parameters; (4) compute the temporal and spatial behavior of all species, and thus any property of the system at a given time and position (with confidence limits); (5) perform a sensitivity analysis on the above responses of the system of differential equations, the goal being the identification of that input information that is most critical for the computation of an appropriate response.

In concert with these five goals, which define understanding of a complex system within specified confidence limits, two parallel activities are desirable. The first is an experimental effort aimed at reducing uncertainties in the key input parameters that have been revealed as most important by the sensitivity analysis. The second is a combined experimental and theoretical attack on the state of current knowledge, aimed at maximizing the amount of input information attainable within acceptable uncertainties without specific measurement.

Specific examples of combustion modeling abound. Review articles have begun to appear (see Chapter 3 and the list of references at the end of the chapter). A reasonable flame code enables the presentation of data in the format shown in Figure 1.[3]

Implied in such a presentation is the ability to calculate concentrations in a reacting chemical system as a function of temperature and pressure.

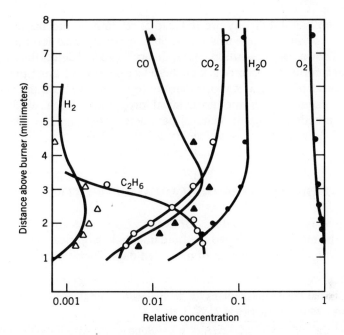

FIGURE 1. Computed Flame Profiles (*curves*) can be compared with laboratory measurements (*data points*) for low-pressure laminar flames. The profiles shown here were computed by Jürgen Warnatz of the Technical University in Darmstadt for a reaction mechanism with 58 elementary reactions, taking into account the diffusion of all 20 substances involved in the overall reaction. The experimental data, for an ethane-oxygen flame burning at one-tenth atmospheric pressure, were obtained by Robert M. Fristrom, William H. Avery, and C. Grunfelder of Johns Hopkins University. Laser analyses of similar flames by James H. Bechtel and his coworkers at the General Motors Research Laboratory have confirmed the mass-spectrometer profiles. From "The Chemistry of Flames," by W. C. Gardiner, Jr. Copyright © (1982) by *Scientific American*, Inc. All rights reserved.

This chapter discusses a consistent framework within which to evaluate such models as well as the tools for estimation of individual rate constant values.

The amount of detail required for the specification of a chemical mechanism is difficult to generalize. Most often the fundamental chemical reactions required are at the molecular level between species reacting from a thermally equilibrated bath. Sometimes more detail is required, and some specifics of a nonthermal energy distribution may have to be specified. Other times a global mechanism, involving the "lumping" of many molecular processes, will suffice.

In principle, the parameters desired could be obtained from *ab initio* quantum-mechanical calculations. A good potential surface and the ability to perform dynamical calculations will yield rate constants as a function of temperature and pressure. Although great strides are being made in this direction, the general difficulty precludes total reliance on this method. In fact various levels of approximation are available. A widely used level of approximation is known as "transition state theory" (TST). Since this statistical theory, which forms the framework of these lectures, places great premium on thermochemical properties, it is appropriate to commence with the consideration of the estimation of thermochemical properties. When we have fully discussed methods for thermochemical parameter estimation, we shall remind ourselves of how transition state theory arises, and then we will extend the methods for thermochemical estimation to include transition states.

2. ESTIMATION OF THERMOCHEMICAL PROPERTIES

The objective is to obtain values for ΔH_T°, ΔS_T°, and ΔC_{pT}° for chemical reactions. The methods employed are fundamentally empirical. They are based on extrapolation and codification of measured data. Thermodynamic data may

TABLE 1. Partial Bond Contributions for the Estimation of C_p°, S°, and ΔH_f° of Gas-Phase Species at 25°C, 1 atm[a]

Bond	C_p°	S°	ΔH_f°	Bond	C_p°	S°	ΔH_f°
C—H	1.74	12.90	−3.83	S—S	5.4	11.6	−6
C—D	2.06	13.60	−4.73	C_d—C^b	2.6	−14.3	6.7
C—C	1.98	−16.40	2.73	C_d—H	2.6	13.8	3.2
C—F	3.34	16.90	−52.5	C_d—F	4.6	18.6	−39
C—Cl	4.64	19.70	−7.4	C_d—Cl	5.7	21.2	−5.0
C—Br	5.14	22.65	2.2	C_d—Br	6.3	24.1	9.7
C—I	5.54	24.65	14.1	C_d—I	6.7	26.1	21.7
C—O	2.7	−4.0	−12.0	> CO—H[c]	4.2	26.8	−13.9
O—H	2.7	24.0	−27.0	> CO—C	3.7	−0.6	−14.4
O—D	3.1	24.8	−27.9	> CO—O	2.2	9.8	−50.5
O—O	4.9	9.1	21.5	> CO—F	5.7	31.6	−77
O—Cl	5.5	32.5	9.1	> CO—Cl	7.2	35.2	−27.0
C—N	2.1	−12.8	9.3	ϕ-H[d]	3.0	11.7	3.25
N—H	2.3	17.7	−2.6	ϕ-C[d]	4.5	−17.4	7.25
C—S	3.4	−1.5	6.7	(NO_2)—O[d]	—	43.1	−3.0
S—H	3.2	27.0	−0.8	(NO)—O[d]	—	35.5	9.0
C_ϕ—C_ϕ (biphenyl)	—	10.0		C_d—C_d	—	—	7.5

[a] See the text for corrections to entropy for symmetry and electronic contributions. C_p°, and S_0° estimated from the rule of additivity of bond contributions, are good to about ±1 cal/mol K, but they may be poorer for heavily branched compounds. The values of ΔH_f° are usually within ±2 kcal/mol but may be poorer for heavily branched species. Peroxide values are not certain by much larger amounts. All substances are in ideal gas state.

[b] C_d represents the vinyl group carbon atom. The vinyl group is here considered a tetravalent unit.

[c] > CO— represents the bond to carbonyl carbon, the latter being considered a bivalent unit. This is somewhat of a "fudge" on simple bond additivity.

[d] NO and NO₂ are here considered as univalent, terminal groups, but the phenyl group $\phi(C_6H_5)$ is considered as a hexavalent unit.

TABLE 2a. Group Values for ΔH_f°, S_{int}°, and $C_{p,T}^\circ$, Hydrocarbons[a]

Group	$\Delta H_{f\,298}^\circ$	$S_{int\,298}^\circ$	C_p°					
			300	400	500	600	800	1000
$C-(H)_3(C)$	−10.20	30.41	6.19	7.84	9.40	10.79	13.02	14.77
$C-(H)_2(C)_2$	−4.93	9.42	5.50	6.95	8.25	9.35	11.07	12.34
$C-(H)(C)_3$	−1.90	−12.07	4.54	6.00	7.17	8.05	9.31	10.05
$C-(C)_4$	0.50	−35.10	4.37	6.13	7.36	8.12	8.77	8.76
$C_d-(H)_2$	6.26	27.61	5.10	6.36	7.51	8.50	10.07	11.27
$C_d-(H)(C)$	8.59	7.97	4.16	5.03	5.81	6.50	7.65	8.45
$C_d-(C)_2$	10.34	−12.70	4.10	4.61	4.99	5.26	5.80	6.08
$C_d-(C_d)(H)$	6.78	6.38	4.46	5.79	6.75	7.42	8.35	8.99
$C_d-(C_d)(C)$	8.88	−14.6	(4.40)	(5.37)	(5.93)	(6.18)	(6.50)	(6.62)
$[C_d-(C_B)(H)]$	6.78	6.38	4.46	5.79	6.75	7.42	8.35	8.99
$C_d-(C_B)(C)$	8.64	(−14.6)	(4.40)	(5.37)	(5.93)	(6.18)	(6.50)	(6.62)
$[C_d-(C_t)(H)]$	6.78	6.38	4.46	5.79	6.75	7.42	8.35	8.99
$C_d-(C_B)_2$	8.0							
$C_d-(C_d)_2$	4.6							
$C-(C_d)(C)(H)_2$	−4.76	9.80	5.12	6.86	8.32	9.49	11.22	12.48
$C-(C_d)_2(H)_2$	−4.29	(10.2)	(4.7)	(6.8)	(8.4)	(9.6)	(11.3)	(12.6)
$C-(C_d)(C_B)(H)_2$	−4.29	(10.2)	(4.7)	(6.8)	(8.4)	(9.6)	(11.3)	(12.6)
$C-(C_t)(C)(H)_2$	−4.73	10.30	4.95	6.56	7.93	9.08	10.86	12.19
$C-(C_B)(C)(H)_2$	−4.86	9.34	5.84	7.61	8.98	10.01	11.49	12.54
$C-(C_d)(C)_2(H)$	−1.48	(−11.69)	(4.16)	(5.91)	(7.34)	(8.19)	(9.46)	(10.19)
$C-(C_t)(C)_2(H)$	−1.72	(−11.19)	(3.99)	(5.61)	(6.85)	(7.78)	(9.10)	(9.90)
$C-(C_B)(C)_2(H)$	−0.98	(−12.15)	(4.88)	(6.66)	(7.90)	(8.75)	(9.73)	(10.25)
$C-(C_d)(C)_3$	1.68	(−34.72)	(3.99)	(6.04)	(7.43)	(8.26)	(8.92)	(8.96)
$C-(C_B)(C)_3$	2.81	(−35.18)	(4.37)	(6.79)	(8.09)	(8.78)	(9.19)	(8.96)
$C_t-(H)$	26.93	24.7	5.27	5.99	6.49	6.87	7.47	7.96
$C_t-(C)$	27.55	6.35	3.13	3.48	3.81	4.09	4.60	4.92
$C_t-(C_d)$	29.20	(6.43)	(2.57)	(3.54)	(3.50)	(4.92)	(5.34)	(5.50)
$C_t-(C_B)$	(29.20)	6.43	2.57	3.54	3.50	4.92	5.34	5.50
$C_B-(H)$	3.30	11.53	3.24	4.44	5.46	6.30	7.54	8.41
$C_B-(C)$	5.51	−7.69	2.67	3.14	3.68	4.15	4.96	5.44
$C_B-(C_d)$	5.68	−7.80	3.59	3.97	4.38	4.72	5.28	5.61
$[C_B-(C_t)]$	5.68	−7.80	3.59	3.97	4.38	4.72	5.28	5.61
$C_B-(C_B)$	4.96	−8.64	3.33	4.22	4.89	5.27	5.76	5.95
C_a	34.20	6.0	3.9	4.4	4.7	5.0	5.3	5.5
$C_{BF}-(C_B)_2(C_{BF})$	4.8	−5.0	3.0	3.7	4.2	4.6	5.2	5.5
$C_{BF}-(C_B)(C_{BF})_2$	3.7	−5.0	3.0	3.7	4.2	4.6	5.2	5.5
$C_{BF}-(C_{BF})_3$	1.5	1.4	2.0	2.9	3.5	4.0	4.7	5.1

[a] C_d represents double-bonded C atom, C_t the triple-bonded C-atom, C_B the C atom in a benzene ring as allenic C atom. By convention, group values for $C-(X)(H)_3$ will always be taken as those for $C-(C)(H)_3$ is any other polyvalent atom such as C_d, C_t, C_B, O, and S. C_{BF} represents a carbon atom in a fused ring such as naphthalene, anthracene, etc. $C_{BF}-(C_{BF})_3$ represents the group in graphite.

TABLE 2b. Non-Next-Nearest Neighbor Corrections

Group	$\Delta H_{f\,298}^{\circ}$	$S_{int\,298}^{\circ}$	C_P°						
			300	400	500	600	800	1000	1500
Alkane *gauche* correction	0.80								
Alkene *gauche* correction	0.50								
cis-Correction	1.00^a	b	−1.34	−1.09	−0.81	−0.61	−0.39	−0.26	0
ortho Correction	0.57	−1.61	1.12	1.35	1.30	1.17	0.88	0.66	−0.05
1,5 H repulsion^c	1.5								

^{a}When one of the groups is *t*-butyl *cis cis* correction = 4.00, when both are *t*-butyl, *cis* correction = ~ 10.00, and when there are two *cis* corrections around one double bond, the total correction is 3.00.

^{b} + 1.2 for but-2-ene, 0 for all other 2-enes, and −0.6 for 3-enes.

^{c}These refer to repulsions between the H atoms attached to the 1,5 C atoms in such compounds as 2,2,4,4-tetramethyl pentane, and then only to the methyls close to each other.

be obtained by several methods and are found in several tabulations, some of which appear in the References at the end of the chapter.

2.1. Additivity Rules

Given the chemical reaction

$$aA + bB + \cdots \leftrightarrows pP + qQ + \cdots$$

a change in some property Φ is given by

$$\Delta\Phi = p\Phi_p + q\Phi_Q + \cdots - a\Phi_A - b\Phi_B - \cdots$$

Atom Additivity. A simple disproportionation reaction of diatomic molecules may be written

$$A_2 + B_2 \leftrightarrows 2AB$$

Any property for which $\Delta\Phi = 0$ obeys the law of atomic additivity. The property Φ is in reality an atomic property. It is very unlikely that thermochemical quantities would obey such a simple law, but the uncertainty limits might be adequate. To rephrase, $\Delta\Phi = 0 \pm \delta$ signifies a property for which the additivity relationship is obeyed within limits of $\pm\delta$. If atom additivity were adequate, a periodical table with one entry for each atom would allow the prediction of the property Φ for any molecule. This would be marvelously simple and would require an extremely small data base. Properties of any two hydrocarbons would be sufficient to predict all of the other hydrocarbons.

In reality, atom additivity is totally unsatisfactory for the estimation of enthalpies and only of marginal value in estimation of entropy and heat capacity.

TABLE 2c. Corrections to Be Applied to Ring-Compound Estimates[a]

Ring (σ)	$\Delta H^{\circ}_{f\,298}$	$S^{\circ}_{\text{int }298}$	C°_p						
			300	400	500	600	800	1000	1500
Cyclopropane (6)	27.6	32.1	-3.05	-2.53	-2.10	-1.90	-1.77	-1.62	(-1.52)
Cyclopropene (2)	53.7	33.6							
Cyclobutane (8)	26.2	29.8	-4.61	-3.89	-3.14	-2.64	-1.88	-1.38	-0.67
Cyclobutene (2)	29.8	29.0	-2.53	-2.19	-1.89	-1.68	-1.48	-1.33	-1.22
Cyclopentane (10)	6.3	27.3	-6.50	-5.5	-4.5	-3.8	-2.8	-1.93	-0.37
Cyclopentene (2)	5.9	25.8	-5.98	-5.35	-4.89	-4.14	-2.93	-2.26	-1.08
Cyclopentadiene (2)	6.0	28.0	-4.3						
Cyclohexane (6)	0	18.8	-5.8	-4.1	-2.9	-1.3	1.1	2.2	3.3
Cyclohexene (2)	1.4	21.5	-4.28	-3.04	-1.98	-1.43	-0.29	0.08	0.81
Cyclohexadiene 1,3	4.8								
Cyclohexadiene 1,4	0.5								
Cycloheptane (1)	6.4	15.9							
Cycloheptene	5.4								
Cycloheptadiene, 1,3	6.6								
Cycloheptatriene 1,3,5 (1)	4.7	23.7							
Cycloöctane (8)	9.9	16.5							
cis-Cycloöctene	6.0								
trans-Cycloöctene	15.3								
Cycloöctatriene 1,3,5	8.9								
Cycloöctatetraene	17.1								
Cyclononane	12.8								
cis-Cyclononene	9.9								
trans-Cyclononene	12.8								
Cyclodecane	12.6								
Cyclododecane	4.4								
Spiropentane (4)	63.5	67.6							
Bicycloheptadiene	31.6								
Biphenylene	58.8								
Bicycloheptane (2,2,1)	16.2								
Bicyclo-(1,1,0)-butane (2)	67.0	69.2							
Bicyclo-(2,1,0)-pentane	55.3								
Bicyclo-(3,1,0)-hexane	32.7								
Bicyclo-(4,1,0)-heptane	28.9								
Bicyclo-(5,1,0)-octane	29.6								
Bicyclo-(6,1,0)-nonane	31.1								
Methylene cyclopropane	41								

[a] Note that in most cases the ΔH°_f correction equals ring-strain energy.

TABLE 3a. Oxygen-Containing Compounds

Group	$\Delta H^\circ_{f\,298}$	$S^\circ_{int\,298}$	C_p°						
			300	400	500	600	800	1000	1500
$O(H_2)$	−57.8	45.1	8.0	8.4	9.2	9.9	11.2		
$O(H)(C)$	−37.9	29.07	4.3	4.4	4.8	5.2	6.0	6.6	
$O(H)(C_B)$	−37.9	29.1	4.3	4.5	4.8	5.2	6.0	6.6	
$O(H)(O)$	−16.3	27.85	5.2	5.8	6.3	6.7	7.2	7.5	8.2
$O(H)(CO)$	−58.1	24.5	3.8	5.0	5.8	6.3	7.2	7.8	
$O(C)_2$	−23.2	8.68	3.4	3.7	3.7	3.8	4.4	4.6	
$O(C)(C_d)$	−30.5	9.7							
$O(C)(C_B)^a$	−23.0								
$O(C)(O)$	−4.5	[9.4]	3.7	3.7	3.7	3.7	4.2	4.2	4.8
$O(C)(CO)$	−43.1	8.4							
$O(C_d)_2$	−33.0	10.1							
$O(C_B)_2$	−21.1								
$O(C_d)(CO)$	−45.2								
$O(C_B)(CO)$	−36.7								
$O(O)(CO)$	−19.0								
$O(CO)_2$	−46.5								
$O(O)_2$	[19.0]	[9.4]	[3.7]	[3.7]	[3.7]	[3.7]	[4.2]	[4.2]	[4.8]
$CO(H)_2$	−26.0	52.3	8.5	10.5	13.4	14.8	17.0		
$CO(H)(C)$	−29.1	34.9	7.0	7.8	8.8	9.7	11.2	12.2	
$CO(H)(C_B)^b$	−29.1								
$CO(H)(C_d)$	−29.1								
$CO(H)(C_t)$	−29.1								
$CO(H)(CO)$	−25.3								
$CO(H)(O)$	−32.1	34.9	7.0	7.9	8.8	9.7	11.2	12.2	
$CO(C)_2$	−31.4	15.0	5.6	6.3	7.1	7.8	8.9	9.6	
$CO(C)(C_B)$	−30.9								
$CO(C_B)_2$	−25.8								
$CO(C)(O)$	−35.1	14.8	6.0	6.7	7.3	8.0	8.9	9.4	
$CO(C)(CO)$	−29.2								
$CO(C_d)(O)^c$	−32.0								
$CO(C_B)(O)$	−36.6								
$CO(C_B)(CO)$	−26.8								
$CO(O_2)$	−23.9								
$CO(O)(CO)$	−29.3								
$C(H)_3(O)^d$	−10.08	30.41	6.19	7.84	9.40	10.79	13.03	14.77	17.58
$C(H)_2(O)(C)$	−8.1	9.8	4.99	6.85	8.30	9.43	11.11	12.33	
$C(H)_2(O)(C_d)$	−6.5								
$C(H)_2(O)(C_B)$	−8.1	9.7							
$C(H)_2(O)(C_t)$	−6.5								
$C(H)_2(O)(CO)$									
$C(H)_2(O)_2$	−16.1								
$C(H)(O)(C)_2$	−7.2	−11.0	4.80	6.64	8.10	8.73	9.81	10.40	
$C(H)(O)_2(C)$	−16.3								

TABLE 3a. *(Continued)*

Group	$\Delta H_{f\,298}^{\circ}$	$S_{int\,298}^{\circ}$	C_p°						
			300	400	500	600	800	1000	1500
$C(O)(C)_3$	−6.6	−33.56	4.33	6.19	7.25	7.70	8.20	8.24	
$C(O)_2(C)_2$	−18.6								
$C(H)_3(CO)^e$	−10.08	30.41	6.19	7.84	9.40	10.79	13.02	14.77	17.58
$C(H)_2(CO)(C)$	−5.2	9.6	6.2	7.7	8.7	9.5	11.1	12.2	
$C(H)_2(CO)(C_d)$	−3.8								
$C(H)_2(CO)(C_B)$	−5.4								
$C(H)_2(CO)(C_t)$	−5.4								
$C(H)_2(CO)_2$	−7.6								
$C(H)(CO)(C)_2^f$	−1.7	−12.0							
$C(CO)(C)_3$	1.4								
$C_d(O)(H)^g$	8.6	8.0	4.2	5.0	5.8	6.5	7.6	8.4	9.6
$C_d(O)(C)^h$	10.3								
$C_d(O)(C_d)^i$	8.9								
$C_d(O)(CO)$	11.6								
$C_d(H)(CO)$	5.0								
$C_d(CO)(C)$	7.5								
$C_B(O)$	−0.9	−10.2	3.9	5.3	6.2	6.6	6.9	6.9	
$C_B(CO)$	3.7								

[a] $O(H)(C_d)\equiv O(H)(C_B)\equiv O(H)(C_t)\equiv O(H)(C)$, assigned.
[b] $CO(H)(O)=CO(H)(C)$, assigned.
[c] $CO(C)(C_d)=CO(C_B)(O)$, assigned.
[d] $C(H)_3(O)\equiv C(H_3)(C)$, assigned.
[e] $C(H)_e(CO)=C(H)_3(C)$, assigned.
[f] $C(H)(CO)(C_2)$, estimated.
[g] $C_d(H)(O)=C_d(H)(C)$, assigned.
[h] $C_d(O)(C)=C_d(C)(C_d)$.
[i] $C_d(O)(C_d)=C_d(C)(C_d)$.

Bond Additivity. A next level in the hierarchy that can be used for estimation of thermochemical properties is envisaged in terms of the reaction:

$$A-X-A + B-X-B \leftrightarrows 2A-X-B$$

If for this reaction $\Delta\Phi = 0 \pm \delta$, the law of bond additivity is obeyed. Properties are associated with the chemical bonds, which are the same on both sides of the equation. For many relatively simple molecules containing C, H, O, N, S, F, Cl, Br, and I, bond additivity is acceptably accurate for entropy and heat capacity ($\delta \approx \pm 1$-2 cal mol^{-1} K^{-1}). Heats of formation can be estimated with less accuracy ($\delta \approx \pm 3$ kcal mol^{-1}) for many species. Table 1 gives bond contributions to thermochemical properties. These entries required a much greater data base than the atom additivity contributions. Thus from

TABLE 3b. Non-Nearest Neighbor and Ring Corrections

Strain	$\Delta H_{f\,298}^{\circ}$	$S_{int\,298}^{\circ}$	C_p°						
			300	400	500	600	800	1000	1500
Ether-oxygen *gauche*	0.5								
Di-tertiary ethers	7.8								
Oxygen *gauche*	0								
Oxygen *ortho*	0								
[oxirane ring structure]	26.9	30.5	−2.0	−2.8	−3.0	−2.6	−2.3	−2.3	
[oxetane ring structure]	25.7	27.7	−4.6	−5.0	−4.2	−3.5	−2.6	+0.2	
[tetrahydrofuran ring structure]	5.9								
[tetrahydropyran ring structure]	0.5								
[ring structure]	0.2								
[dioxane ring structure]	3.3								
[trioxane ring structure]	6.6								
[methyl-tetrahydrofuran ring structure]	4.7								
[dioxolane ring structure]	6.0								
[dihydrofuran ring structure]	−5.8								
[dihydropyran ring structure]	1.2								
[anhydride ring structure]	4.5								
[anhydride ring structure]	0.8								
[anhydride ring structure]	3.6								

TABLE 3b. *(Continued)*

Ring Correction	$\Delta H_{f\,298}^{\circ}$	Ring Correction	$\Delta H_{f\,298}^{\circ}$
[benzene fused 5-membered anhydride structure]	16.6	[cyclobutane–O structure]	22.6
		Cyclopenanone	5.2
		Cyclohexanone	2.2
		Cycloheptanone	2.3
[benzodioxane structure]	2.0		
		Cyclooctanone	1.5
[xanthene-type structure]	2.3	Cyclononanone	4.7
		Cyclodecanone	3.6
		Cycloundecanone	4.4
[octahydrobenzofuran structure]	11.4	Cyclododecanone	3.0
[decalone structure]	*cis* 15.3 / *trans* 20.9	Cyclo(C_{15})anone	2.1
		Cyclo(C_{17})anone	1.1
[β-lactone/oxetanone structure]	23.9		
[structure]	22.0		

methane and ethane properties, we obtain the C—H and C—C bond contributions and may predict properties of any alkane.

Example of Bond Additivity. Using Table 1,

$$\Delta H_{f,\,300}^{\circ}(CH_3CH_2OH)$$

$$= 5\,C\!-\!H + C\!-\!C + C\!-\!O + O\!-\!H$$

$$= 5(-3.83) + 2.73 + (-12.0) + (-27) = -55.42 \text{ kcal mol}^{-1}$$

$$\text{Observed} = -56.2 \text{ kcal mol}^{-1}$$

$$|\Delta| = 0.8 \text{ kcal mol}^{-1}$$

Group Additivity. The next step in the additivity hierarchy is defined through the reaction

$$A\!-\!XY\!-\!A + B\!-\!XY\!-\!B \rightleftharpoons A\!-\!XY\!-\!B + B\!-\!XY\!-\!A$$

TABLE 4a. Group Contributions to $C_{p,T}^\circ$, S°, and ΔH_f° at 25°C and 1 atm for Nitrogen-Containing Compounds

Group	$\Delta H_{f\,298}^\circ$	$S_{\text{int }298}^\circ$	C_p°						
			300	400	500	600	800	1000	1500
C—(N)(H)$_3$	−10.08	30.41	6.19	7.84	9.40	10.79	13.02	14.77	17.58
C—(N)(C)(H)$_2$	−6.6	9.8[a]	5.25[a]	6.90[a]	8.28[a]	9.39[a]	11.09[a]	12.34[a]	
C—(N)(C)$_2$(H)	−5.2	−11.7[a]	4.67[a]	6.32[a]	7.64[a]	8.39[a]	9.56[a]	10.23[a]	
C—(N)(C)$_1$	3.2	34.1[a]	4.35[a]	6.16[a]	7.31[a]	7.91[a]	8.49[a]	8.50[a]	
N—(C)(H)$_2$	4.8	29.71	5.72	6.51	7.32	8.07	9.41	10.47	12.28
N—(C)$_2$(H)	15.4	8.94	4.20	5.21	6.13	6.83	7.90	8.65	9.55
N—(C)$_3$	24.4	−13.46	3.48	4.56	5.43	5.97	6.56	6.67	6.50
N—(N)(H)$_2$	11.4	29.13	6.10	7.38	8.43	9.27	10.54	11.52	13.19
N—(N)(C)(H)	20.9	9.61	4.82	5.8	6.5	7.0	7.8	8.3	9.0
N—(N)(C)$_2$	29.2	−13.80							
N—(N)(C$_B$)(H)	22.1								
N$_1$—(H)	16.3								
N$_1$—(C)	21.3								
N$_1$—(C$_B$)[b]	16.7								
N$_A$—(H)	25.1	26.8	4.38	4.89	5.44	5.94	6.77	7.42	8.44
N$_A$—(C)	27								
N—(C$_d$)(C)(H)	15.4								
N—(C$_d$)(C)(N)	30								
N—(N$_1$)(C)(H)	21								
N—(C$_d$)(H)$_2$	4.8								
N—(C$_d$)(C)$_2$	24.4								
N—(C$_d$)(H)(N)	21.5								
N—(C$_B$)(H)$_2$	4.8	29.71	5.72	6.51	7.32	8.07	9.41	10.47	12.28
N—(C$_B$)(C)(H)	14.9								
N—(C$_B$)(C)$_2$	26.2								
N—(C$_B$)$_2$(H)	16.3								
N$_A$—(N)	23.0								
C$_B$—(N)	−0.5	−9.69	3.95	5.21	5.94	6.32	6.53	6.56	
CO—(N)(H)	−29.6	34.93	7.03	7.87	8.82	9.68	11.16	12.20	
CO—(N)(C)	−32.8	16.2	5.37	6.17	7.07	7.66	9.62	11.19	
N—(CO)(H)$_2$	−14.9	24.69	4.07	5.74	7.13	8.29	9.96	11.22	
N—(CO)(C)(H)	−4.4	3.9[a]							
N—(CO)(C)$_2$									
N—(CO)(C$_B$)(H)	+0.4								
N—(CO)$_2$(H)	−18.5								
N—(CO)$_2$(C)	−5.9								
N—(CO)$_2$(C$_B$)	−0.5								
C—(N$_A$)(C)(H)$_2$	−6.0								
C—(N$_A$)(C)$_2$(H)	−3.4								
C—(N$_A$)(C)$_3$	−1.5								
C—(CN)(C)(H)$_2$	22.5	40.20	11.10	13.40	15.50	17.20	19.7	21.30	
C—(CN)(C)$_2$(H)	25.8	19.80	11.00	12.70	14.10	15.40	17.30	18.60	

TABLE 4a. *(Continued)*

Group	$\Delta H_{f\,298}^{\circ}$	$S_{\text{int }298}^{\circ}$	C_p°						
			300	400	500	600	800	1000	1500
C—(CN)(C)$_3$	29.0	−2.80							
C—(CN)$_2$(C)$_2$		28.40							
C$_d$—(CN)(H)	37.4	36.58	9.80	11.70	13.30	14.50	16.30	17.30	
C$_d$—(CN)$_2$	34.1								
C$_d$—(NO$_2$)(H)		44.4	12.3	15.1	17.4	19.2	21.6	23.2	25.3
C$_B$—(CN)	35.8	20.50	9.8	11.2	12.3	13.1	14.2	14.9	
C$_t$—(CN)	63.8	35.40	10.30	11.30	12.10	12.70	13.60	14.30	15.30
C—(NCO)	−10.2	48.9	15.4						
C—(NO$_2$)(C)(H)$_2$	−15.1	48.4^a							
C—(NO$_2$)(C)$_2$(H)	−15.8	26.9^a							
C—(NO$_2$)(C)$_3$		3.9^a							
C—(NO$_2$)$_2$(C)(H)	−14.9								
O—(NO)(C)	−5.9	41.9	9.10	10.30	11.2	12.0	13.3	13.9	14.5
O—(NO$_2$)(C)	−19.4	48.50							
O—(C)(CN)	2.0	39.5	10.0						
C—(C$_d$)(CN)	7.5	43.1	13.0						
O—(C$_B$)(CN)	7.0	29.2	8.3						

TABLE 4b. Corrections to Be Applied to Ring-Compound Estimates

Ethyleneimine	27.7	31.6^a							

NH

Azetidine	26.2^a	29.3^a							

NH

Pyrrolidine	6.8	26.7	−6.17	−5.58	−4.80	−4.00	−2.87	−2.17	

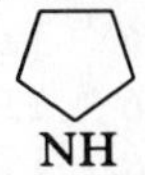

NH

Piperidine	1.0

NH

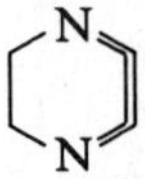	3.4

NH

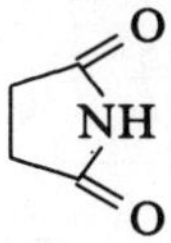	8.5

^{a}Estimates by authors.
^{b}For *ortho* or *para* substitution in pyridine add −1.5 kcal/mole per group. N$_1$ stands for imino N atom; N$_A$ represents azo N atom. C—(N$_1$)(C)(H)$_2$≡C—(N)(C)(H)$_2$; assigned. C—(N$_1$)(C)$_2$(H)=C—(N)(C)$_2$(H): C$_d$—(N$_1$)(H)≡C$_d$—(C$_d$)(H).

TABLE 5a. Halogen-Containing Compounds. Group Contribution $\Delta H_{f\,298}^\circ$, S_{298}°, and $C_{p,T}^\circ$, Ideal Gas at 1 atm

Group	$\Delta H_{f\,298}^\circ$	$S_{int\,298}^\circ$	C_p° 300	400	500	600	800	1000
C—(F)$_3$(C)	−161	42.5	12.7	15.0	16.4	17.9	19.3	20.0
C—(F)$_2$(H)(C)	(−102.3)	39.1	9.9	12.0		15.1		
C—(F)(H)$_2$(C)	−51.5	35.4	8.1	10.0	12.0	13.0	15.2	16.6
C—(F)$_2$(C)$_2$	−99	17.8	9.9	11.8	13.5			
C—(F)(H)(C)$_2$	−49.0	(14.0)						
C—(F)(C)$_3$	−48.5							
C—(F)$_2$(Cl)(C)	−106.3	40.5	13.7	16.1	17.5			
C—(Cl)$_3$(C)	−20.7	50.4	16.3	18.0	19.1	19.8	20.6	21.0
C—(Cl)$_2$(H)(C)	(−18.9)	43.7	12.1	14.0	15.4	16.5	17.9	18.7
C—(Cl)(H)$_2$(C)	−16.5	37.8	8.9	10.7	12.3	13.4	15.3	16.7
C—(Cl)$_2$(C)$_2$	−22.0	22.4	12.2					
C—(Cl)(H)(C)$_2$	−14.8	17.6	9.0	9.9	10.5	11.2		
C—(Cl)(C)$_3$	−12.8	5.4	9.3	10.5	11.0	11.3		
C—(Br)$_3$(C)		55.7	16.7	18.0	18.8	19.4	19.9	20.3
C—(Br)(H)$_2$(C)	−5.4	40.8	9.1	11.0	12.6	13.7	15.5	16.8
C—(Br)(H)(C)$_2$	−3.4							
C—(Br)(C)$_3$	−0.4	−2.0	9.3	11.0				
C—(I)(H)$_2$(C)	8.0	43.0	9.2	11.0	12.9	13.9	15.8	17.2
C—(I)(H)(C)$_2$	10.5	21.3	9.2	10.9	12.2	13.0	14.2	14.8
C—(I)(C)$_3$	13.0	0.0	9.7					
C—(I)$_2$(C)(H)	(26.0)	(54.6)	(12.2)	—	(16.4)	(17.0)		
C—(Cl)(Br)(H)(C)		45.7	12.4	14.0	15.6	16.3	17.9	19.0
N—(F)$_2$(C)	−7.8							
C—(Cl)(C)(O)(H)	−21.6	15.9	(9.0)	(9.9)	(10.5)	(11.2)		
C—(I)(O)(H)$_2$	3.8	40.7						
C$_d$—(C)(Cl)	−2.1	15.0						
C$_d$—(F)$_2$	−77.5	37.3	9.7	11.0	12.0	12.7	13.8	14.5
C$_d$—(Cl)$_2$	−1.8	42.1	11.4	12.5	13.3	13.9	14.6	15.0
C$_d$—(Br)$_2$		47.6	12.3	13.2	13.9	14.3	14.9	15.2
C$_d$—(F)(Cl)		39.8	10.3	11.7	12.6	13.3	14.2	14.7
C$_d$—(F)(Br)		42.5	10.8	12.0	12.8	13.5	14.3	14.7
C$_d$—(Cl)(Br)		45.1	12.1	12.7	13.5	14.1	14.7	14.7
C$_d$—(F)(H)	−37.6	32.8	6.8	8.4	9.5	10.5	11.8	12.7
C$_d$—(Cl)(H)	−1.2	35.4	7.9	9.2	10.3	11.2	12.3	13.1
C$_d$—(Br)(H)	11.0	38.3	8.1	9.5	10.6	11.4	12.4	13.2
C$_d$—(I)(H)	24.5	40.5	8.8	10.0	10.9	11.6	12.6	13.3
C$_t$—(Cl)		33.4	7.9	8.4	8.7	9.0	9.4	9.6
C$_t$—(Br)		36.1	8.3	8.7	9.0	9.2	9.5	9.7
C$_t$—(I)		37.9	8.4	8.8	9.1	9.3	9.6	9.8

TABLE 5b. Arenes

Group	$\Delta H^\circ_{f\,298}$	$S^\circ_{int\,298}$	C_p°					
			300	400	500	600	800	1000
C_B—(F)	−42.8	16.1	6.3	7.6	8.5	9.1	9.8	10.2
C_B—(Cl)	−3.8	18.9	7.4	8.4	9.2	9.7	10.2	10.4
C_B—(Br)	8.5	21.6	7.8	8.7	9.4	9.9	10.3	10.5
C_B—(I)	22.5	23.7	8.0	8.9	9.6	9.9	10.3	10.5
C—$(C_B)(F)_3$	−162.7	42.8	12.5	15.3	17.2	18.5	20.1	21.0
C—$(C_B)(Br)(H)_2$	−5.1							
C—$(C_B)(I)(H)_2$	8.4							

If for this reaction $\Delta\Phi = 0 \pm \delta$, the law of group additivity is obeyed. Properties are associated with an atom and its ligands. This is usually a very useful level of approximation. The data base required is relatively large but still accessible. Sometimes combination of bond and group additivity are necessarily employed.

The notation for groups is generally transparent. Usually a central atom is surrounded by its nearest neighbors. Occasionally, certain ligands are treated as pseudoatoms, and some peculiarities of notation have become embedded (i.e., instead of $C_d(C_d)(H)_2$, just $C_d(H)_2$). The value for any group $C(H)_3(X)$ is the same as $C(H)_3(C)$. Some examples are

$$CH_3CH_3; \quad 2C(C)(H)_3$$
$$CH_3CH_2CH_3; \quad 2C(C)(H_3) + C(C)_2(H)_2$$
$$CH_3OH; \quad C(O)(H)_3 + O(C)(H)$$
$$H_2C{=}CH_2; \quad 2C_d(H)_2$$
$$CH_3COCH_3; \quad 2C(H)_3(CO) + CO(C)_2$$
$$CH_3CH_2NO_2; \quad C(H)_3C + C(C)(H)_2(NO_2)$$

In order to describe all alkanes using group additivity, the four groups, $C(C)(H)_3$, $C(C)_2(H)_2$, $C(C)_3(H)$, and $C(C)_4$, would be required.

We find it profitable to tabulate group contributions to $\Delta H^\circ_{f,298}$, S°_{298}, and $C^\circ_{p,T}$ for various temperatures. We can calculate ΔH° and ΔS° for chemical

TABLE 5c. Corrections for Non-Next-Nearest Neighbors

ortho (F)(F)	5.0	0	0	0	0	0	0	0
ortho (Cl)(Cl)	2.2							
ortho (alk)(halogen)[a]	0.6							
cis (halogen)(halogen)	−0.3							
cis (halogen)(alk)	−0.8							

[a] Halogen = Cl, Br, and I only. The *gauche* correction = 1.0 kcal for Cl, Br, I; none for X—Me, and none for F—halogen.

TABLE 6a. Organometallic Compounds[a]

Metal	Groups	$\Delta H^\circ_{f\,298}$	Remarks
Tin	C—(Sn)(H)$_3$	−10.08	C—(Sn)(H)$_3$≡C—(C)(H)$_3$, assigned
	C—(Sn)(C)(H)$_2$	−2.18	
	C—(Sn)(C)$_2$(H)	3.38	
	C—(Sn)(C)$_3$	8.16	
	C—(Sn)(C$_B$)(H)$_2$	−7.77	
	C$_B$—(Sn)	5.51	C$_B$—(Sn)≡C$_B$—(C), assigned
	C$_d$—(Sn)(H)	8.77	C$_d$—(Sn)(H)≡C$_d$—(C)(H), assigned
	Sn—(C)$_4$	36.2	
	Sn—(C)$_3$(Cl)	−9.8	
	Sn—(C)$_2$(Cl)$_2$	−49.2	
	Sn—(C)(Cl)$_3$	−89.5	
	Sn—(C)$_3$(Br)	−1.8	
	Sn—(C)$_3$(I)	9.9	
	Sn—(C)$_3$(H)	34.8	
	Sn—(C$_d$)$_4$	36.2	Sn—(C$_d$)$_4$≡Sn—(C)$_4$, assigned
	Sn—(C$_d$)$_3$(Cl)	−8.2	
	Sn—(C$_d$)$_2$(Cl)$_2$	−50.7	
	Sn—(C$_d$)(Cl)$_3$	−82.2	
	Sn—(C)$_3$(C$_d$)	37.6	
	Sn—(C$_B$)$_4$	26.2	
	Sn—(C)$_3$(C$_B$)	34.9	
	Sn—(C)$_3$(Sn)	26.4	
Lead	C—(Pb)(H)$_3$	−10.08	C—(Pb)(H)$_3$≡C—(C)(H)$_3$, assigned
	C—(Pb)(C)(H$_2$)	−1.7	
	Pb—(C)$_4$	72.9	
Chromium	O—(Cr)(C)	−23.5	O—(Cr)(C)≡O—(Ti)(C), assigned
	Cr—(O)$_4$	−64.0	
Zinc	C—(Zn)(H)$_3$	−10.08	C—(Zn)(H)$_3$≡C—(C)(H)$_3$, assigned
	C—(Zn)(C)(H)$_2$	−1.8	
	Zn—(C)$_2$	33.3	
Titanium	O—(Ti)(C)	−23.5	O—(Ti)(C)≡O—(P)(C), assigned
	Ti—(O)$_4$	−157	
	N—(Ti)(C)$_2$	39.1	N—(Ti)(C)$_2$≡N—(P)(C)$_2$, assigned
	Ti—(N)$_4$	−123	
Vanadium	O—(V)(C)	−23.5	O—(V)(C)≡O—(Ti)(C), assigned
	V—(O)$_4$	−87.0	
Cadmium	C—(Cd)(H)$_3$	−10.08	C—(Cd)(H)$_3$≡C—(C)(H)$_3$, assigned
	C—(Cd)(C)(H)$_2$	−0.3	
	Cd—(C)$_2$	46.4	
Aluminum	C—(Al)(H)$_3$	−10.08	C—(Al)(H)$_3$≡C—(C)(H)$_3$, assigned
	C—(Al)(C)(H)$_2$	0.7	
	Al—(C)$_3$	9.2	

TABLE 6a. *(Continued)*

Metal	Groups	$\Delta H_{f\,298}^{\circ}$	Remarks
Germanium	C—(Ge)(C)(H)$_2$	−7.7	
	Ge—(C)$_4$	36.2	Ge—(C)$_4$≡Sn—(C)$_4$, assigned
	Ge—(Ge)(C)$_3$	15.6	
Mercury	C—(Hg)(H)$_3$	−10.08	C—(Hg)(H)$_3$≡C—(C)(H)$_3$, assigned
	C—(Hg)(C)(H)$_2$	−2.7	
	C—(Hg)(C)$_2$(H)	3.6	
	C$_B$—(Hg)	−1.8	C$_B$—(Hg)≡C$_B$—(O), assigned
	Hg—(C)$_2$	42.5	
	Hg—(C)(Cl)	−2.8	
	Hg—(C)(Br)	4.9	
	Hg—(C)(I)	15.8	
	Hg—(C$_B$)$_2$	64.4	
	Hg—(C$_B$)(Cl)	9.9	
	Hg—(C$_B$)(Br)	18.1	
	Hg—(C$_B$)(I)	27.9	

[a] No *gauche* corrections across C—M bond.

reactions at temperatures different than 298 K, via

$$\Delta H_T^{\circ} = \Delta H_{298}^{\circ} + \int_{298}^{T} \Delta C_p^{\circ}(T)\, dT$$

We assume that

$$\Delta C_p(T) = \frac{\Delta C_p(T) + \Delta C_p(298)}{2} = \overline{\Delta C_p}$$

$$\Delta H_T^{\circ} = \Delta H_{298}^{\circ} + \overline{\Delta C_p}(T - 298)$$

$$\Delta S_T^{\circ} = \Delta S_{298}^{\circ} + \int_{298}^{T} \Delta C_p(T)\, d\ln T = \Delta S_{298}^{\circ} + \overline{\Delta C_p}\ln\frac{T}{298}$$

It is also necessary to point out that all the additivity contributions for entropy are for *intrinsic* entropy rather than for real symmetry-corrected entropy. This is understood in terms of $S^{\circ} = S_{int}^{\circ} - R\ln\sigma$, where σ is the rotational symmetry number for the molecule. The external rotational symmetry correction takes into account that if all rotational permutations were weighted equally, then indistinguishable configurations would be counted more than once. Thus, while heteronuclear diatomic molecules have $\sigma = 1$, homonuclear diatomics have $\sigma = 2$, since a simple 180° rotation, which permutes the atoms, yields an indistinguishable configuration. For polyatomic species, the same general dictum holds. The symmetry number, σ, is the total number of ways that the molecule can be permuted to give an identical configuration by a simple rotation. Examples are CH_4 (12), benzene (12), and SF_6 (24).

The group contributions also need to be corrected for internal rotation symmetry. Thus CH_3 groups have a threefold rotational symmetry that has the effect of multiplying the external symmetry by 3 for every methyl group. Linear alkanes, which have an external $\sigma = 2$, also have two-end CH_3's, and therefore an overall symmetry number of $3 \times 3 \times 2 = 18$. Neopentane (C_5H_{12}) has an external symmetry of 12 (same as CH_4), and the four methyl groups

TABLE 6h Organophosphorus Groups[a]

Group	$\Delta H^\circ_{f\,298}$	$S^\circ_{int\,298}$	Remarks
$C-\overline{(P)}(H)_3$	-10.08	30.4	$C-(P)(H)_3 \equiv C-(C)(H)_3$, assigned
$C-(P)(C)(H)_2$	-2.47		
$C-(PO)(H)_3$	-10.08	30.4	$C-(PO)(H)_3-C-(C)(H)_3$, assigned
$C-(PO)(C)(H)_2$	-3.4		
$C-(P:N)(H)_3$	-10.08	30.4	$C-(P:N)(H)_3 \equiv C-(C)(H)_3$, assigned
$C-(N:P)(C)(H)_2$	19.4		
$C_B-(P)$	-1.8		$C_B-(P) \equiv C_B-(O)$, assigned
$C_B-(PO)$	2.3		$C_B-(PO) \equiv C_B-(CO)$, assigned
$C_B-(P:N)$	2.3		$C_B-(P:N) \equiv C_B-(CO)$, assigned
$P-(C)_3$	7.0		
$P-(C)(Cl)_2$	-50.1		
$P-(C_B)_3$	28.3		
$P-(O)_3$	-66.8		
$P-(N)_3$	-66.8		$P-(N)_3 \equiv P-(O)_3$, assigned
$PO-(C)_3$	-72.8		
$PO-(C)(F)_2$		46.7	
$PO-(C)(Cl)(F)$		50.8	
$PO-(C)(Cl)_2$	-123.0	53.0	
$PO-(C)(O)(Cl)$	-112.6		
$PO-(C)(O)_2$	-99.5		
$PO-(O)_3$	-104.6		
$PO-(O)_2(F)$	-167.7		
$PO-(C_B)_3$	-52.9		
$PO-(N)_3$	-104.6		$PO-(N)_3 \equiv PO-(O)_3$, assigned
$O-(C)(P)$	-23.5		$O \equiv (C)(P)-O-(C)_2$, assigned
$O-(H)(P)$	-58.7		
$O-(C)(PO)$	-40.7		$O-(C)(PO) \equiv O-(C)(CO)$, assigned
$O-(H)(PO)$	-65.0		
$O-(PO)_2$	-54.5		
$O-(P:N)(C)$	-40.7		$O-(P:N)(C) \equiv O-(C)(CO)$, assigned
$N-(P)(C)_2$	32.2		
$N-(PO)(C)_2$	17.8		
$P:N-(C)_3(C)$	0.50		$P:N-(C)_3(C) \equiv C-(C)_4$, assigned
$P:N-(C_B)_3(C)$	-25.7		
$P:N-(N:P)(C)_2(P:N)$	-15.5		
$P:N-(N:P)(C_B)_2(P:N)$	-22.9		
$P:N-(N:P)(Cl)_2(P:N)$	-58.2		
$P:N-(N:P)(O)_2(P:N)$	-43.4		

[a] No *gauche* corrections across the $X-P$, $X-PO$, and $X-P:N$ bonds (X represents C, O, and N).

contribute 3^4 for $\sigma = 12.3^4 = 972$. To reiterate,

$$S_T^\circ = S_{T,\text{intrinsic}}^\circ - R \ln \sigma$$

$$\sigma \to \text{symmetry number} = \sigma_{\text{external}} \times \sigma_{\text{internal}}$$

$$\sigma_{\text{external}} = \text{overall rotational symmetry}$$

Number of Identical Rotamers

Linear alkanes $\qquad \sigma_{\text{ext}} = 2$

$$\overset{\displaystyle N}{\underset{O \qquad O}{\diagup \quad \diagdown}} \qquad \sigma_{\text{ext}} = 2$$

CH_2 $\qquad\qquad \sigma_{\text{ext}} = 12$

Benzene $\qquad\quad \sigma_{\text{ext}} = 12$

$$\sigma_{\text{internal}} = \text{symmetry associated with internal rotors}$$

CH_3 $\qquad\qquad \sigma_{\text{int}} = 3$

Linear alkanes $\qquad \sigma = \sigma_{\text{ext}}\sigma_{\text{int}} = 2 \times 3^2 = 18$

TABLE 6c. Organoboron Groups[a]

Group	$\Delta H_{f\,298}^\circ$	Remarks
$C-(B)(H)_3$	-10.1	$C-(B)(H)_3 \equiv C-H(H)_3$, assigned
$C-(B)(C)(H)_2$	-2.22	
$C-(B)(C)_2(H)$	1.1	
$C-(BO_3)(H)_3$	-10.1	$C-(BO_3)(H)_3 \equiv C-(C)(H)_3$, assigned
$C-(BO_3)(C)(H)_2$	-2.2	
$C_d-(B)(H)$	15.6	
$B-(C)_3$	0.9	
$B-(C)(F)_2$	-187.9	
$B-(C)_2(Cl)$	-42.7	
$B-(C)_2(Br)$	-26.9	
$B-(C)_2(I)$	-8.9	
$B-(C)_2(O)$	29.3	$B-(C)_2(O) \equiv N-(C)_2(N)$, assigned
$B-(C_d)(F)_2$	-192.9	$B-(C_d)(F)_2 \equiv B-(C)(F)_2$, assigned
$B-(O)_3$	24.4	$B-(O)_3 \equiv B-(N)_3$, assigned
$B-(O)_2(Cl)$	-19.7	
$B-(O)(Cl)_2$	-61.2	
$B-(O)_2(H)$	19.9	
$B-(N)_3$	24.4	$B-(N)_3 \equiv N-(C)_3$, assigned
$B-(N)_2(Cl)$	-23.8	
$B-(N)(Cl)_2$	-67.9	
$BO_3-(C)_3$	-208.7	
$O-(B)(H)$	-115.5	
$O-(B)(C)$	-69.4	
$N-(B)(C)_2$	-9.9	
$B-(S)_3$	24.4	
$S-(B)(C)$	-14.5	
$S-(B)(C_B)$	-7.8	

[a] The *gauche* corrections across the $C-B$ bond are $+0.8$ kcal/mol.

TABLE 7a. Sulfur-Containing Compounds: Group Contributions to $\Delta H_{f\,298}^{\circ}$, S_{298}°, and $C_{p,T}^{\circ}$

Group	$\Delta H_{f\,298}^{\circ}$	$S_{int\,298}^{\circ}$	C_p° 300	400	500	600	800	1000
C—(H)$_3$(S)^a	−10.08	30.41	6.19	7.84	9.40	10.79	13.02	14.77
C—(C)(H)$_2$(S)	−5.65	9.88	5.38	7.08	8.60	9.97	12.26	14.15
C—(C)$_2$(H)(S)	−2.64	−11.32	4.85	6.51	7.78	8.69	9.90	10.57
C—(C)$_3$(S)	−0.55	−34.41	4.57	6.27	7.45	8.15	8.72	8.10
C—(C$_B$)(H)$_2$(S)	−4.73							
C—(C$_d$)(H)$_2$(S)	−6.45							
C—(H)$_2$(S)$_2$	−6.0 ± 3							
C$_B$—(S)^b	−1.8	10.20	3.90	5.30	6.20	6.60	6.90	6.90
C$_d$—(H)(S)^c	8.56	8.0	4.16	5.03	5.81	6.50	7.65	8.45
C$_d$—(C)(S)	10.93	−12.41	3.50	3.57	3.83	4.09	4.41	5.00
S—(C)(H)	4.62	32.73	5.86	6.20	6.51	6.78	7.30	7.71
S—(C$_B$)(H)	11.96	12.66	5.12	5.26	5.57	6.03	6.99	7.84
S—(C)$_2$	11.51	13.15	4.99	4.96	5.02	5.07	5.41	5.73
S—(C)(C$_d$)	13.0							
S—(H)(C$_d$)	6.1							
	9 ± 4							
S—(C$_d$)$_2$	13.5	16.48	4.79	5.58	5.53	6.29	7.94	9.73
S—(C$_B$)(C)	19.16							
S—(C$_B$)$_2$	25.90							
S—(S)(C)	7.05	12.37	5.23	5.42	5.51	5.51	5.38	5.12
S—(S)(C$_B$)	14.5							
S—(S)$_2$	3.01	13.4	4.7	5.0	5.1	5.2	5.3	5.4
S—(S)(H)	1.9							
C—(SO)(H)$_3$^{d}	−10.08	30.41	6.19	7.84	9.40	10.79	13.02	14.77
C—(C)(SO)(H)$_2$	−7.72							
C—(C)$_3$(SO)	−3.05							
C—(C$_d$)(SO)(H)$_2$	−7.35							
C$_B$—(SO)^e	2.3							
SO—(C)$_2$	−14.41	18.10	8.88	10.03	10.50	10.79	10.98	11.17
SO—(C$_B$)$_2$	−12.0							
SO—(O)$_2$	−51 ± 3							
C—(SO$_2$)(H)$_3$^{f}	−10.08	30.41	6.19	7.84	9.40	10.79	13.02	14.77
C—(C)(SO$_2$)(H)$_2$	−7.68							
C—(C)$_2$(SO$_2$)(H)	−2.62							
C—(C)$_3$(SO$_2$)	−0.61							
C—(C$_d$)(SO$_2$)(H)$_2$	−7.14							
C—(C$_B$)(SO$_2$)(H)$_2$	−5.54							
C$_B$—(SO$_2$)^g	2.3							
C$_d$—(H)(SO$_2$)	12.5							
C$_d$—(C)(SO$_2$)	14.5							
SO$_2$—(C$_d$)(C$_B$)	−68.6							
SO$_2$—(C$_d$)$_2$	−73.6							

TABLE 7a. *(Continued)*

Group	$\Delta H_{f\ 298}^{\circ}$	$S_{int\ 298}^{\circ}$	C_p° 300	400	500	600	800	1000
$SO_2-(C)_2$	-69.74	20.90	11.52					
$SO_2-(C)(C_B)$	-72.29							
$SO_2-(C_B)_2$	-68.58							
$SO_2-(SO_2)(C_B)$	-76.25							
$SO_2-(O)_2$	-101							
$CO-(S)(C)^h$	-31.56	15.43	5.59	6.32	7.09	7.76	8.89	9.61
$S-(H)(CO)$	-1.41	31.20	7.63	8.09	8.12	8.17	8.50	8.24
$C-(S)(F)_3$		38.9						
$CS-(N)_2^i$	-31.56	15.43	5.59	6.32	7.09	7.76	8.89	9.61
$N-(CS)(H)_2$	12.78	29.19	6.07	7.28	8.18	8.91	10.09	10.98
$S-(S)(N)^j$	-4.90							
$N-(S)(C)_2$	29.9							
$SO-(N)_2^k$	-31.56							
$N-(SO)(C)_2$	16.0							
$SO_2-(N)_2^l$	-31.56							
$N-(SO_2)(C)_2$	-20.4							
$O-(H)(SO_2)$	-38.0							

[a] $C-(S)(H)_3 \equiv C-(C)(H)_3$, assigned.
[b] $C_B-(S) \equiv C_B-(O)$, assigned.
[c] $C_d-(S)(H) \equiv C_d-(O)(H)$, assigned.
[d] $C-(SO)(H)_3 \equiv C-(CO)(H)_3$, assigned.
[e] $C_B-(SO) \equiv C_B-(CO)$, assigned.
[f] $C-(SO_2)(H)_3 \equiv C-(SO)(H)_3$.
[g] $C_B-(SO_2) \equiv C_B-(CO)$, assigned.
[h] $CO-(S)(C) \equiv CO-(C)_2$, assigned.
[i] $CS-(N)_2 \equiv CO-(C)_2$, assigned.
[j] $S-(S)(N) \equiv O-(O)(C)$, assigned.
[k] $SO-(N)_2 \equiv CO-(C)_2$, assigned.
[l] $SO_2-(N)_2 \equiv SO-(N)_2$, assigned.

Group additivity in its purest form serves us well but still falls short in several ways. We could attempt to correct this by going up the hierarchy one more step to an additivity that includes next-nearest-neighbor interactions, but the data base required would be so large as to be impractical. Thus we adapt and compromise by giving up the purity of the group additivity definition and using additive correction factors. These corrections are sometimes only to the heat of formation, but sometimes entropy and heat capacity need corrections as well. Corrections are tabulated for *cis* interactions, *gauche* interactions, and for cyclic compounds.

These corrections are simply additive, and the only difficult one to visualize might be the *gauche* correction. These are identified by drawing Newman

TABLE 7b. Corrections to be Applied to Organosulfur Ring Compounds

Ring (σ)	$\Delta H^\circ_{f\,298}$	$S^\circ_{int\,298}$	C_p° 300	400	500	600	800	1000
thiirane (S-triangle) (2)	17.7	29.5	-2.9	-2.6	-2.7	-3.0	-4.3	-5.8
Cyclo-S₃ (6)	22.9							
S (4-ring) (2)	19.4	27.2	-4.6	-4.2	-3.9	-3.9	-4.6	-5.7
Cyclo-S₄ (8)	18.4							
S (5-ring) (2)	1.7	23.6	-4.9	-4.7	-3.7	-3.7	-4.4	-5.6
Cyclo-S₆ (6)	5.3							
S (6-ring) (1)	0	16.1	-6.2	-4.3	-2.8	-0.7	0.9	-1.3
Cyclo-S₈	-1.1							
S (6-ring, one double bond) (1)	3.9							
S (5-ring, one double bond) (2)	5.0							
S (5-ring, two double bonds) (1)	2.0							
SO₂ (5-ring) (2)	5.7							
S (5-ring, two double bonds) (2)	-16.3	23.6	-4.9	-4.7	-3.7	-3.7	-4.4	-5.6

projections and counting the number of nonhydrogen interactions. For example, for 2-methyl butane, we have

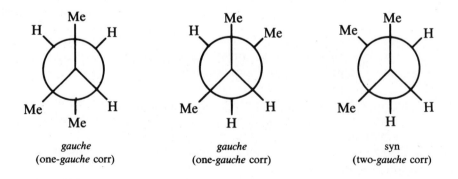

gauche	gauche	syn
(one-*gauche* corr)	(one-*gauche* corr)	(two-*gauche* corr)

Tables of group values are given in Tables 2 through 7. Table 8 demonstrates the application of group additivity.

TABLE 8. Examples of Group Additivity

	ΔH_f°	S_{int}°	$C_{p,300}^\circ$	$C_{p,800}^\circ$
Ethane: $CH_3CH_3 \rightarrow 2\, C(C)(H)_3$				
$C(C)(H)_3$	-10.2	30.4	6.2	13.0
$\therefore C_2H_6$	-20.4	60.8	12.4	26.0
Experimental	-20.2	60.6^a	12.7	25.8
Butene-1: $CH_3CH_2CH{=}CH_2 \rightarrow C(C)(H)_3 + C(C)(C_d)(H)_2 + C_d(H)(C) + C_d(H)_2$				
$C(C)(H)_3$	-10.2	30.4	6.2	13.0
$C(C)(C_d)(H)_2$	-4.8	9.8	5.1	11.2
$C_d(H)(C)$	8.6	8.0	4.2	7.7
$C_d(H)_2$	6.3	27.6	5.1	10.1
	$\overline{-0.1}$	$\overline{75.8}$	$\overline{20.6}$	$\overline{42.0}$
Experimental	0.0	76.0^a	20.6	41.8

^{a} Corrected for symmetry.

2.2. Structural Estimates of Entropy and Heat Capacity

Statistical Thermodynamics. If molecular properties, such as molecular weight, bond lengths and bond angles, and molecular frequencies are known, the entropy and heat capacity may be calculated. (In most cases the ideal gas approximation is valid.) Appropriate formulas are presented in this subsection.

Since we always know the molecular weight, we may compute translational contributions to S and C_p accurately. With some reasonable guesses as to bond lengths and angles, we can compute I_m^3, the product of inertia. (Simple codes are available.) The S and C_p errors associated with small changes in I_m^3 are small. We will check this with some calculations.

The biggest error that we will make in entropy (and C_p) calculations is in the vibrational contribution. Each molecule has 3N-6 vibrations, and larger molecules with low-lying frequencies can be a problem. It is important to understand that even in these latter cases, the size of the entropy error will not be overwhelming, and the advantage of being able to compute limits is enormous. Of course we could measure entropy calorimetrically if an accurate value is of importance, but we will also wish to extend these methods to species such as free radicals and to transition states, where experimental values are not accessible.

Quantitative examples include the fact that changing the C—C distance in *n*-butane from 1.54 to 2.54 Å, increases the value of I_m^3 by a factor of 3. This increases the rotational entropy by 1.1 cal mol^{-1} K^{-1}.

Q = molar partition function

q = molecular partition function

$$Q = \frac{1}{N_0!} q^{N_0}; \quad N_0 = \text{Avagadro's number}$$

Molar Thermodynamic Quantities

Helmholtz free energy $A = -kT \ln Q$

Gibbs free energy $G = A + PV$

Internal energy $E = kT\left(\dfrac{\partial \ln Q}{\partial \ln T}\right)_v = -k\left(\dfrac{\partial \ln Q}{\partial (1/T)}\right)_v$

Enthalpy $H \equiv E + PV$

Entropy $S = \dfrac{E - A}{T}$

$= k \ln Q + k\left(\dfrac{\partial \ln Q}{\partial \ln T}\right)_v$

Heat capacity
   at constant volume $C_v = \left(\dfrac{\partial E}{\partial T}\right)_v = k\left(\dfrac{\partial \ln Q}{\partial \ln T}\right)_v + k\left(\dfrac{\partial^2 \ln Q}{\partial (\ln T)^2}\right)_v$

$= \dfrac{k}{T^2}\left(\dfrac{\partial^2 \ln Q}{\partial (1/T)^2}\right)_v$

Ideal gases $C \equiv C_v + PV = C_v + R$

Evaluation of Molar Partition Functions and Molar Thermodynamic Quantities in Terms of Molecular Parameters. For the translational contribution to thermochemical properties for one mole of ideal gas at a standard state of 1 atm:

$$Q^\circ_{\text{trans}} = \frac{(nRT)^{N_0}}{N_0!}\left(\frac{2mkT}{h^2}\right)^{3N_0/2}$$

$$E^\circ_{\text{trans}} = \tfrac{3}{2}N_0 kT = \tfrac{3}{2}RT$$

$$C^\circ_{V(\text{trans})} = \tfrac{3}{2}R; \quad C^\circ_{p(\text{tran})} = \tfrac{5}{2}R$$

$$S^\circ_{\text{trans}} = 37.0 + \frac{3}{2}R\ln\left(\frac{M}{40}\right) + \frac{5}{2}R\ln\left(\frac{T}{298}\right)$$

$$+ R\ln(n)$$

where N_0 = Avogadro's number 6.02×10^{23} molecules.

In the last equation for S, M is the molecular weight (amu) and n is the number of optical isomers.

For rotational degrees of freedom, we have the following:

1 Linear molecule (2 degrees of freedom)

$$Q_{\text{rot-2}D} = \left(\frac{8\pi^2 lkT}{\sigma_e h^2} \right)^{N_0}$$

$$E_{\text{rot-2}D} = RT$$

$$C_{V(\text{rot-2}D)} = R$$

$$S_{\text{rot-2}D} = 6.9 + R \ln\left(\frac{I}{\sigma_e} \right) + R \ln\left(\frac{T}{298} \right)$$

2 Nonlinear molecule (3 degrees of freedom)

$$Q_{\text{rot-3}D} = \left[\frac{\pi^{1/2}}{\sigma_e} \left(\frac{8\pi^2 I_m kT}{h^2} \right)^{(3/2)} \right]^{N_0}$$

$$E_{\text{rot-3}D} = \frac{3}{2} RT$$

$$C_{V(\text{rot-3}D)} = \frac{3}{2} R$$

$$S_{\text{rot-3}D} = 11.5 + \frac{R}{2} \ln\left(\frac{I_m^3}{\sigma_e^2} \right) + \frac{3}{2} R \ln\left(\frac{T}{298} \right)$$

where σ_e is the external symmetry number of the molecule.

For each vibrational degree of freedom, we have

$$Q_{\text{vib}} = \left(1 - e^{-h\nu/kT} \right)^{-N_0}$$

with the appropriate thermodynamic contributions to be evaluated from the foregoing equations. ν is the frequency of the vibration, and in general, the contributions of a vibrational degree of freedom to S, C_p, and E tend to be small. For large molecules, these contributions may become sizable. The only exception to this occurs when $h\nu \ll kT$, so that we find

$$Q_{\text{vib}} \xrightarrow[\substack{\text{classical} \\ \text{limit}}]{} \left(\frac{kT}{h\nu} \right)^{N_0}$$

$$E_{\text{vib}} \longrightarrow RT$$

$$C_{\text{vib}} \longrightarrow R$$

$$S_{\text{vib}} \longrightarrow R \ln\left(\frac{kT}{h\nu} \right) + R$$

In the opposite extreme, when $kT \ll h\nu$, then $Q_{vib} \to 1$ and all of the thermodynamic functions for vibration approach zero. In the case of a one-dimensional rotor,

$$Q_{rot\text{-}1D} = \left[\frac{\pi^{1/2}}{\sigma_i} \left(\frac{8\pi^2 I_r kT}{h^2} \right)^{1/2} \right]^{N_0}$$

$$E_{rot\text{-}1D} = \frac{1}{2} RT$$

$$C_{V(rot\text{-}1D)} = \frac{1}{2} R$$

$$S_{rot\text{-}1D} = 4.6 + R \ln \left(\frac{I_r^{1/2}}{\sigma_i} \right) + \frac{R}{2} \ln \left(\frac{T}{298} \right)$$

TABLE 9. Absolute Entropy of a Harmonic Oscillator as a Function of Frequency and Temperature[a]

Frequency (cm^{-1})	Temperature (K)							
	300	400	500	600	800	1000	1200	1500
50	4.8	5.4	5.9	6.3	6.8	7.3	7.7	8.1
75	4.1	4.7	5.0	5.4	6.0	6.5	6.9	7.3
100	3.4	4.1	4.5	4.8	5.6	5.9	6.3	6.7
125	3.1	3.6	4.1	4.4	5.2	5.5	5.9	6.3
150	2.7	3.3	3.7	4.1	4.7	5.0	5.4	5.9
200	2.2	2.7	3.1	3.5	4.1	4.4	4.8	5.3
250	1.8	2.3	2.7	3.0	3.6	4.1	4.5	4.9
300	1.4	1.9	2.3	2.7	3.3	3.6	4.1	4.5
350	1.2	1.7	2.1	2.4	3.0	3.4	3.8	4.2
400	1.0	1.4	1.8	2.1	2.7	3.2	3.5	3.8
500	0.7	1.1	1.4	1.8	2.3	2.7	3.0	3.5
600	0.5	0.8	1.0	1.4	1.9	2.3	2.7	3.1
700	0.3	0.6	0.8	1.2	1.7	2.1	2.4	2.8
800	0.2	0.5	0.8	1.0	1.4	1.8	2.1	2.6
900	0.2	0.4	0.6	0.8	1.2	1.6	1.9	2.3
1000	0.1	0.3	0.5	0.7	1.1	1.4	1.8	2.1
1200	0.0	0.1	0.3	0.5	0.8	1.0	1.4	1.8
1500	0.0	0.1	0.2	0.3	0.5	0.8	1.1	1.4
2000	0.0	0.0	0.0	0.1	0.3	0.5	0.7	1.0
2500	0.0	0.0	0.0	0.0	0.1	0.3	0.5	0.7
3000	0.0	0.0	0.0	0.0	0.1	0.1	0.3	0.5
3500	0.0	0.0	0.0	0.0	0.0	0.1	0.2	0.3

[a] $x = 1.44\bar{\nu}/T$ determines $S°$, so that value of $S°$ must be the same for similar ratios of $\bar{\nu}/T$. Thus $S°$ of 500 cm^{-1} at 1000 K is the same as $S°$ of 250 cm^{-1} at 500 K or 1000 cm^{-1} at 2000 K. For $x \ll 1$, $S° = R + R \ln x = 1.99 + 4.575 \log x$.

As can be seen from Table 9, only the lower frequencies contribute much to the entropy at 300 K, and small errors in assignment will still cause only small errors. The difference between a 200 cm^{-1} and 100 cm^{-1} frequency is only 1.2 cal mol^{-1} K^{-1} at 300 K.

The assignment of vibrational frequencies is an exercise in describing internal coordinates. The vibrational assignment "locates" each atom in space.

Consider the *trans* form of *n*-butane, C_4H_{10} (3N $-$ 6 = 36):

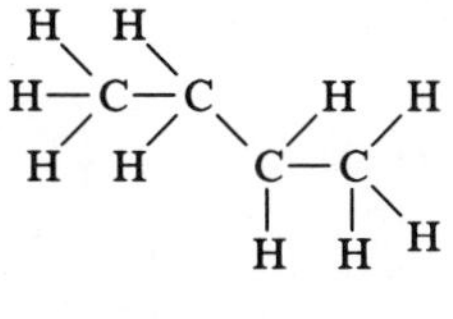

We can break the assignment up into parts as follows:

13-STRETCHES

10 C—H	3100 cm^{-1}
3 C—C	1000 cm^{-1}

BENDS

Three (a) Heavy Atoms	
2 C—C—C	420 cm^{-1}
1 *Et–Et* torsion	
Eight (b) CH$_2$ groups	
2 H—C—H scissors	1450 cm^{-1}
2 CH$_2$ wag	1150 cm^{-1}
2 CH$_2$ twist	1150 cm^{-1}
2 CH$_2$ rock	700 cm^{-1}
Twelve (c) CH$_3$ groups	
6 H—C—H scissors	1450 cm^{+1}
4 CH$_3$ rocks, wag, twist	1150 cm^{-1}
2 CH$_3$ torsion	—

The assignment of frequencies is from Table 10. The torsions are somewhat more complex. The easiest concept is to consider the groups (in the above case, CH$_3$ and C$_2$H$_5$) to be rotating against a very heavy mass. Thus we may compute a moment of inertia for this rotor and thus a partition function and appropriate thermochemical quantities corresponding to a free rotor. If the rotation is hindered, it behaves like a torsion at low temperatures $V \gg RT$ and like a rotation at higher temperatures.

Table 11 gives values for the free rotational entropy at several temperatures for many common groups. Corrections for the barrier to rotation can be made from Table 12,[4] and barrier values found in Table 13.[5,6] Heat capacity is treated in Table 14.

In the example of *n*-butane, we see that the CH$_3$ rotors would contribute 3.6 eu each if there were no barrier, while with a barrier of 3 kcal mol^{-1}, this is

TABLE 10. Frequencies Assigned to Normal and Partial Bond Bending and Stretching Motions

Bond Stretches	Frequency (ω cm^{-1})[a]	Bond Bends	Frequency (ω cm^{-1})
C=O	1700	H–C–H	1450
C–C	1400	H·C–H	1000
C–O ethers	1100	(H–C–C) t, w[b]	1150
C–O acids, esters C·O	1200 710	H·C–C	800
C=C	1650	H–O–C	1200
C–C	1300	H·O–C	840
C–C	1000	H–C≡C	1150
C·C[c]	675	H·C·C	1150
C–H; O–H; N–H S–H	3100 2600	C–C≡C	420
C·H	2200	C·C≡C	290
C–F	1100	C·C·C	420
C·F	820	O·C≡O	420
C–Cl	650	·O·C·O·	420

TABLE 10. *(Continued)*

Bond Stretches	Frequency (ω cm^{-1})^a	Bond Bends	Frequency (ω cm^{-1})
C · Cl	490	C=C=C	850
C—I	500	C=C, C (C=C angle bend)	635
C · I	375	(H–C–C) r^b	700
C—Br	560	(H·C·C) r^b	700
C · Br	420	(H–C=C) o.p	700
		C–C–Cl	400
		C–C·Cl	280
		C–C–Br	360
		C–C·Br	250
		C–C–I	320
		C–C·I	220
		C–C–C	420
		C–C·C	300
		C–O–C	400

TABLE 10. *(Continued)*

Bond Stretches	Frequency (ω cm^{-1})[a]	Bond Bends	Frequency (ω cm^{-1})
		C—O···C (bend)	280
		C—C—O (bend)	400
		C—C···O (bend)	280
		O—C—O (bend)	400

[a] Note that bending frequencies are surprisingly consistent with the relation, $\omega_1/\omega_2 = (\mu_2/\mu_1)^{1/2}$. Deviations from this relation seldom exceed 50 cm^{-1}. Here the reduced mass $\mu = [M_A M_B/(M_A + M_B)]$ for the bend $(A - R - B)$.

[b] Methyl and methylene wags and twists, whose frequencies range within 1000–1300 cm^{-1}, have been equated with $\left(\begin{array}{c} \mathrm{C} \\ \mathrm{H} \diagdown \mathrm{C} \end{array} \right) t, w$ bends and assigned a mean value of 1150 cm^{-1}. Methylene rocks have lower frequencies (i.e., ~ 700 cm^{-1}), which correspond closely to the out-of-plane $\left(\begin{array}{c} \mathrm{C} \\ \mathrm{H} \diagdown \mathrm{C} \end{array} \right)$ bends in olefins.

[c] Single dots, as in C · C, are meant to represent one-electron bonds.

TABLE 11. Approximate Moments of Inertia and Entropies for Some Free Rotors

	I^a	$S_{\text{free rot}}$[b,c] 300 K	600 K	1000 K
—CH$_3$	3.0	3.6	4.3	4.8
—CH$_2$	1.8	3.1	3.8	4.3
—C$_2$H$_5$	34.0	6.0	6.7	7.2
—i-Propyl	56.0	6.4	7.1	7.6
—t-Butyl	100.0	7.0	7.7	8.2
—Phenyl	88.0	6.7	7.4	7.9
—Benzyl	170.0	7.5	8.2	8.7
—OH	1.0	2.4	3.1	3.6

[a] Units amu-Å^2. Computed from approximate structures assuming rotor is connected to an infinite mass.

[b] $S_{\text{free rot}} = 4.6 + \dfrac{R}{2} \ln\left[\dfrac{I}{9} \cdot \dfrac{T}{300} \right]$.

[c] All rotors have foldedness ("symmetry") of 3. Individual rotamers must be computed separately.

TABLE 12. Decrease in Entropy of Free Rotor as Functions of Barrier Height (V), Temperature (T, K) and Partition Function, $Q_f{}^a$

	$1/Q_f$		
V/RT	0.0	0.2	0.4
0.0	0.0	0.0	0.0
1.0	0.1	0.1	0.1
2.0	0.4	0.4	0.4
3.0	0.8	0.8	0.7
4.0	1.1	1.1	1.0
5.0	1.4	1.4	1.2
6.0	1.7	1.6	1.4
8.0	2.0	2.0	1.7
10.0	2.3	2.2	1.9
15.0	2.7	2.6	2.2
20.0	3.1	2.9	2.4

[a] Values listed are $\Delta S = S_f^\circ - S_h^\circ$. From Lewis and Randall.[4]

TABLE 13. Some Characteristics Torsion Barriers, V (kcal / mol) to Free Rotation about Single Bonds^a

Bond	V	Bond	V
CH_3—CH_3	2.9	CH_3—OH	1.1
CH_3—C_2H_5	2.8	CH_3—OCH_3	2.7
CH_3—isopropyl	3.6	CH_3—NH_2	1.9
CH_3—t-butyl	4.7	CH_3—$NHCH_3$	3.3
CH_3—CH=CHCH$_3$ *cis*	0.75[b]	CH_3—$N(CH_3)_2$	4.4
trans	1.95	CH_3—SiH_3	1.7
CH_3—vinyl	2.0	CH_3—SiH_2CH_3	1.7
CH_3—CH_2F	3.3	CH_3—PH_2	2.0
CH_3—CF_3	3.5	CH_3—SH	1.3
CF_3—CF_3	4.4	CH_3—SCH_3	2.1
CH_3—CH_2Cl	3.7	CH_3—CHO	1.2
CH_2Cl—CF_3	5.8	CH_3—$COCH_3$	0.8
CCl_3—CCl_3	15	CH_3—allene	1.6
CH_3—CH_2Br	3.6	CH_3—(isobutene)	2.2
CH_3—CH_2I	3.2	CH_3—CO(OH)	0.5
CH_3—phenyl	0	CH_3—(epoxide ring)	2.6
CH_3—$CCCH_3$	0	$(CH_3)_2N$—$COCH_3$	20.0
CH_3—NO_2	0	CH_3—OCHO	1.2
CH_3O—NO	9.0	CH_3—ONO_2	2.3
CH_3O—NO_2	9.0	CH_3—O-vinyl	3.4
CH_3O—$CO(CH_3)$	13		

[a] For a more complete compilation, see Dale[5] and Orville-Thomas.[6]
[b] There seems to be a quite general "*cis* effect," such that F, Cl, CH_3, and CN, *cis* to a CH_3, lower the CH_3 barrier by ~ 1.4 kcal.

TABLE 14. Molar Heat Capacity C_p° for Internal Rotor as a Function of Barrier (V), Temperature, Partition Function $Q_f{}^a$

V/RT	(1/Q_f)				
	0.0	0.2	0.4	0.6	0.8
0.0	1.0	1.0	1.0	1.0	1.0
0.5	1.1	1.1	1.1	1.0	1.0
1.0	1.2	1.2	1.2	1.1	1.1
1.5	1.5	1.4	1.3	1.2	1.1
2.0	1.7	1.7	1.5	1.4	1.2
2.5	1.9	1.9	1.7	1.5	1.3
3.0	2.1	2.0	1.8	1.6	1.3
4.0	2.3	2.2	2.0	1.7	1.4
6.0	2.3	2.2	1.9	1.5	1.2
8.0	2.2	2.1	1.7	1.3	0.9
10.0	2.1	2.0	1.5	1.0	0.7
15.0	2.1	1.8	1.2	0.7	0.4
20.0	2.0	1.7	1.0	0.5	0.2

From Lewis and Randall.[4]
$^a Q_f = 3.6/\sigma\{I_r, T/100\}^{1/2}$ with I_r in amu-Å and T in K. σ = symmetry of barrier.

reduced by 1.4 eu to a total of 2.2 eu each. This is the same as a 200 cm^{-1} torsion. The ethyl rotor must be treated differently, since it is not rotating against a very heavy group but only another ethyl. Thus I_r is $I_{Et}/2$. Thus, a free rotor would contribute 5.3 eu, and a 3 kcal mol^{-1} barrier reduces this to 3.9 eu, the same as an 85–90 cm^{-1} torsion.

Molecular Model Compounds. The structural considerations make it apparent that model compounds offer another estimation method.

The *n*-butane molecule is similar in many ways to $CH_3OCH_2CH_3$, $CH_3NH—CH_2CH_3$, or $CH_3CH_2CH_2OH$. The changes are simple to make and due almost entirely to symmetry and H-motions. The most difficult estimate would be changing rotational barriers, and we have seen that this is a small effect. See Tables 15–17.

Many of the preceding comments are prologue to the estimation of free radical and transition state properties. Many radicals can be thought of as being derived from molecules by removal of an atom. That is, if we wish to have the properties of the radical $R\,\dot{}$, we may ask how we expect them to differ from the molecule $R—H$! In Table 18 it can be seen that C_2H_5 radical is "derived" from C_2H_6 with the only important changes being due to symmetry $18 \rightarrow 6$ for a gain of $R \ln 3 = 2.2$ eu, and the additional entropy due to the

TABLE 15. Some Examples of Structural Similarities in C_p° and S°

Molecule (σ total)		C_p° $(C_{vib})^c$			$S^{\circ\,d}$		
		300 K	600 K	1000 K	300 K	600 K	1000 K
A. N≡N	(2)	7.0(0.1)	7.2 (0.3)	7.8 (0.9)	45.8	50.7	54.5
C≡O	(1)	7.0(0.1)	7.3 (0.4)	7.9 (1.0)	47.3	52.2	56.0
H—C≡N	(1)	8.6(1.7)	10.6 (3.7)	12.3 (5.3)	48.3	54.9	60.7
H—C≡C—H	(2)	10.6(3.7)	13.9 (7.0)	16.3 (9.4)	48.0	56.6	64.3
B. O=C=O	(2)	8.9(2.0)	11.3 (4.4)	13.0 (6.1)	51.1	58.1	64.3
O=N≡N	(1)	9.3(2.4)	11.6 (4.7)	13.1 (6.2)	52.6	59.8	66.1
C. H_2C=O	(2)	8.5(0.6)	11.5 (3.6)	14.8 (6.9)	52.3	59.1	65.8
H_2C=CH_2	(4)	10.3(2.4)	16.9 (9.0)	22.4(14.5)	52.4	61.7	71.8
H_3C—C≡CH	(3)	14.6(6.7)	21.8(13.9)	27.7(19.8)	59.4	71.9	84.5
HO—OH	$(2)^a$	10.3(2.4)	13.3 (5.4)	15.0 (7.1)	55.7	63.9	71.2
D. O—N=O	$(2)^b$	8.9(1.0)	11.0 (3.1)	12.5 (4.6)	57.4	64.2	70.2
O_3	(2)	9.4(1.5)	11.9 (4.0)	13.2 (5.3)	57.1	64.5	71.0
F—O—F	(2)	10.4(2.5)	12.5 (4.6)	13.3 (5.4)	59.2	67.2	73.8
F—CH_2—F	(2)	10.3(2.4)	15.7 (7.8)	20.0(12.1)	59.0	67.9	77.0
E. O—Cl—O	$(2)^b$	10.9(3.0)	12.7 (4.8)	13.4 (5.5)	61.5	71.9	78.7
O=S—O	(2)	9.5(1.6)	11.7 (3.8)	13.0 (5.1)	59.4	66.7	73.0
CH_3—S—CH_3	(18)	17.8(9.9)	27.0(19.1)	35.2(27.3)	68.3	78.7	94.6

^{a} H_2O_2 exists in a skew form with optically active isomers. Hence S° includes a term $R \ln 2$ due to entropy of mixing.
^{b} S° includes a term $R \ln 2$ due to electron spin.
^{c} Values in parentheses are $C_p^\circ - C_{p(trans)}^\circ - C_{p(rot)}^\circ - R$.
^{d} These are absolute entropies. To obtain intrinsic entropies, add $R \ln(\sigma/n_i g_e)$, where g_e is the electronic degeneracy.

electronic state degeneracy for free radicals, accounting for a gain of $R \ln 2 = 1.4$ eu. (Of course, free radicals, such as allyl or benzyl which have a special delocalized character, will have to be treated a little differently.)

It is also apparent from the foregoing that the largest effect to be expected in forming a cyclic species from an open chain molecule, will be the change of torsional motions into ring motions. For our purposes, it is sufficient to notice that the net entropy loss per rotor will usually be of the order 3–4 eu.

TABLE 16

Species	CH_3CH_3	CH_3NH_2	CH_3OH	CH_2=CH_2	$CH_3\dot{C}H_2$	CH_3F
σ, spin	18,0	3,0	3,0	4,0	6,$\frac{1}{2}$	3,0
S_{298}°	54.9	58.1	57.3	52.5	58.1	53.3
S_{298}° (intrinsic)	60.6	60.3	59.5	55.3	60.3	55.5

TABLE 17

Species	$CH_3CH_2CH_3$	CH_3NHCH_3	CH_3OCH_3	CH_3CH_2F
σ, spin	18, 0	9, 0	18, 0	3, 0
S°_{298}	64.6	65.3	63.7	63.3
S°_{298} (intrinsic)	70.3	69.7	69.4	65.5
Species	$CH_3-CH=CH_2$	$\dot{C}H_2-CH=CH_2$	$CH_2=C=CH_2$	CH_3CH_2F
σ, spin	3, 0	$2, \frac{1}{2}$	4, 0	3, 0
S°_{298}	63.9	63.0 (63.2)^a	58.3	63.3
S°_{298} (intrinsic)	66.1	63.0 (63.2)^a	61.1	65.5

^{a}Calculated from Table 19 (radical additivities).

TABLE 18. Estimates of $S^\circ_{300}(\dot{C}_2H_5)$ and $C_{p\,300}$ by Differences from C_2H_6

	ΔS°	ΔC_p°
Translation	Negligible (-0.1)	None
Rotation^a	Negligible (-0.27)	None
Vibration		
One C—H stretch (3100 cm^{-1})	0	0
Two H—C—H bends (1450 cm^{-1})	0	-0.2
Hindered rotation: reduced moment	-0.25	
Barrier change from 3 kcal → 2 kcal	$+0.5$	0
Symmetry and spin	$+3.6$	None
Totals	$+3.5$	-0.2

^{a}If the C—C· bond is 1.45 Å rather than 1.54 Å, there is an additional correlation of about -0.3 due to lower moment.

2.3. Enthalpy Estimates

In general, we are restricted in estimating enthalpies of stable species to those for which we may apply additivity rules. (Some quantum-mechanical calculational schemes do give good results as well.) However, certain enthalpy characteristics are transferable and useful in estimations.

Bond Dissociation Energies (BDE). The enthalpy of the reaction:

$$A - B \rightarrow A^{\cdot} + B^{\cdot}$$

has come to be called the "bond dissociation energy," or BDE. $DH_T^\circ(A - B)$ means the BDE at temperature T.

Part of the chemist's ability to codify reactions comes from the fact that BDEs are largely transferable. Thus, in alkanes, all primary C—H bonds are 98 kcal mol^{-1}, all secondary are 95, and all tertiary, 94 kcal mol^{-1}. Note that we may codify these by listing radical heats of formation and/or BDEs themselves.

When the bond homolysis in question involves electronic rearrangement in the radical, as in the reaction

$$H_2C{=}CH{-}CH_3 \rightarrow H_2C \overset{\cdots}{-} CH \overset{\cdots}{-} CH_2 + H$$

we find that the delocalization causes a stabilization that is a transferable quantity. The above BDE in propylene is 86 kcal mol^{-1}, and the stabilization due to "allylic delocalization" (3 electrons in an orbital mode from 3 p-orbitals) is about 12 kcal mol^{-1}. Many experimental values of BDEs have been determined over the past 20 years and many prototypical values exist.[7,8]

Pi-Bond Energies. A specific type of bond dissociation energy can be defined by hypothetical reactions of the type

$$H_2C{=}H \rightarrow H_2\overset{\bullet}{C}{-}\overset{\bullet}{C}H_2$$

The enthalpy change here is called D_π, the pi-bond energy. These transferable quantities make thinking about addition reactions particularly simple. The enthalpy change for the reaction:

$$H + C_2H_4 \rightarrow C_2H_5^{\bullet}$$

is the difference between the primary C—H bond being formed (98 kcal mol^{-1}) and the pi-bond energy being destroyed (59 kcal mol^{-1}). The reaction is 39 kcal mol^{-1} exothermic.

We find that C$=$O pi-bonds are much stronger than $\ce{\backslash C=C/}$, and that C$=$N pi-bonds are about the same as $\ce{\backslash C=C/}$. These facts all are useful ways to classify knowledge that we wish to bring to bear when trying to understand chemical mechanisms.

Strain Energy. We see quite often that the incremental energy associated with ring compounds in which the bond angles are bigger or smaller than in the normal open chain compounds is a rough constant. This is often called "strain energy." The strain energy in cyclopropane is the difference between 3 C—(C)(H$_1$)$_2$ and ΔH_f(cyclo—C$_3$H$_6$). This number seems to be modestly constant for many three-membered rings. This is most useful when considering transition states.

2.4. Free Radical Group Values

Table 19 is a group table for free radicals.

TABLE 19. Free-Radical Group Additivities[a]

Radical	ΔH_f°	S°	C_p° 300	400	500	600	800	1000	1500
$[\cdot C{-}(C)(H)_2]$	35.82	30.7	5.99	7.24	8.29	9.13	10.44	11.47	13.14
$[\cdot C{-}(C)_2(H)]$	37.45	10.74	5.16	6.11	6.82	7.37	8.26	8.84	9.71
$[\cdot C{-}(C)_3]$	38.00	−10.77	4.06	4.92	5.42	5.75	6.27	6.35	6.53
$[C{-}(C\cdot)(H)_3]$	−10.08	30.41	6.19	7.84	9.40	10.79	13.02	14.77	17.58
$[C{-}(C\cdot)(C)(H)_2]$	−4.95	9.42	5.50	6.95	8.25	9.35	11.07	12.34	14.25
$[C{-}(C\cdot)(C)_2(H)]$	−1.90	−12.07	4.54	6.00	7.17	8.05	9.31	10.05	11.17
$[C{-}(C\cdot)(C)_3]$	1.50	−35.10	4.37	6.13	7.36	8.12	8.77	8.76	8.12
$[C{-}(O\cdot)(C)(H)_2]$	6.1	36.4	7.9	9.8	10.8	12.8	15.0	16.4	—
$[C{-}(O\cdot)(C)_2(H)]$	7.8	14.7	7.7	9.5	10.6	12.1	13.7	14.5	—
$[C{-}(O\cdot)(C)_3]$	8.6	−7.5	7.2	9.1	9.8	11.1	12.1	12.3	—
$[C{-}(S\cdot)(C)(H)_2]$	32.4	39.0	9.0	10.6	12.4	13.6	15.8	17.4	—
$[C{-}(S\cdot)(C)_2(H)]$	35.5	17.8	8.5	10.0	11.6	12.3	13.8	14.6	—
$[C{-\cdot}(S\cdot)(C)_3]$	37.5	−5.3	8.2	9.8	11.3	11.8	12.2	12.3	—
$[\cdot C{-}(H)_2 C_d]$	23.2	27.65	5.39	7.14	8.49	9.43	11.04	12.17	14.04
$[\cdot C{-}(H)(C)(C_d)]$	25.5	7.02	4.58	6.12	7.19	8.00	9.11	9.78	10.72
$[\cdot C{-}(C)_2(C_d)]$	24.8	−15.00	4.00	4.73	5.64	6.09	6.82	7.04	7.54
$[C_d{-}(C\cdot)(H)]$	8.59	7.97	4.16	5.03	5.81	6.50	7.65	8.45	9.62
$[C_d{-}(C\cdot)(C)]$	10.34	−12.30	4.10	4.71	5.09	5.36	5.90	6.18	6.40
$[\cdot C{-}(C_B)(H)_2]$	23.0	26.85	6.49	7.84	9.10	9.98	11.34	12.42	14.14
$[\cdot C{-}(C_B)(C)(H)]$	24.7	6.36	5.30	6.87	7.85	8.52	9.38	9.84	10.12
$[\cdot C{-}(C_B)(C)_2]$	25.5	−15.46	4.72	5.48	6.20	6.65	7.09	7.10	6.94
$[C_B{-}C\cdot]$	5.51	−7.69	2.67	3.14	3.68	4.15	4.96	5.44	5.98
$[C{-}(\cdot CO)(H)_3]$	−5.4	66.6	12.74	14.63	16.47	18.17	21.14	23.27	—
$[C{-}(\cdot CO)(C)(H)_2]$	−0.3	45.8	12.7	14.5	15.8	16.8	19.2	20.7	—
$[C{-}(\cdot CO)(C)_2(H)]$	2.6	(23.7)	(11.5)	(12.8)	(14.3)	(15.5)	(17.4)	(18.5)	—
$[\cdot N{-}(H)(C)]$	(55.3)	30.23	5.38	5.67	5.89	6.09	6.60	6.97	7.74
$[\cdot N{-}(C)_2]$	(58.4)	10.24	3.72	4.13	4.38	4.53	4.86	4.95	4.91
$[C{-}(\cdot N)(C)(H)_2]$	−6.6	9.8	5.25	6.90	8.28	9.39	11.09	12.34	—
$[C{-}(\cdot N)(C)_2(H)]$	−5.2	−11.7	4.67	6.32	7.64	8.39	9.56	10.23	—
$[C{-}(\cdot N)(C)_3]$	(−3.2)	34.1	4.35	6.16	7.31	7.91	8.49	8.50	—
$[\cdot C{-}(H)_2(CN)]$	(58.2)	58.5	10.66	12.82	14.48	15.89	18.08	19.80	—
$[\cdot C{-}(H)(C)(CN)]$	(56.8)	40.0	9.1	11.4	13.1	14.4	16.3	17.4	—
$[\cdot C{-}(C)_2(CN)]$	(56.1)	19.6	8.8	10.4	11.3	12.3	13.7	14.5	—
$[\cdot N{-}(H)(C_B)]$	38.0	27.3	4.6	5.4	6.0	6.4	7.2	7.7	8.6
$[\cdot N{-}(C)(C_B)]$	42.7	(6.5)	(3.9)	(4.2)	(4.7)	(5.0)	(5.6)	(5.8)	(5.9)
$[C_B{-}N\cdot]$	−0.5	−9.69	3.95	5.21	5.94	6.32	6.53	6.56	—
$[C{-}(CO_2\cdot)(H)_3]$	−47.5	71.4	14.4	17.8	20.4	23.1	27.1	29.6	—
$[C{-}(CO_2\cdot)(H)_2(C)]$	−41.9	49.8	15.5	18.5	20.3	22.3	27.5	27.2	—
$[C{-}(CO_2\cdot)(H)(C)_2]$	−39.0	−12.1	(4.5)	(6.0)	(7.2)	(8.0)	(9.3)	(10.1)	(11.2)
$[C{-}(N_A)(H)_3]$	−10.08	30.41	6.19	7.84	9.40	10.79	13.02	14.77	17.58
$[C{-}(N_A)(C)(H)_2]$	−5.5	9.42	5.50	6.95	8.25	9.35	11.07	12.34	14.25
$[C{-}(N_A)(C)_2(H)]$	−3.3	−12.07	4.54	6.00	7.17	8.05	9.31	10.05	11.17
$[C{-}(N_A)(C)_3]$	−1.9	−35.10	4.37	6.13	7.36	8.12	8.77	8.76	8.12
$[N_A{-}C]$	32.5	8.0	4.0	4.4	4.7	4.8	5.1	5.3	5.2
$[N_A{-}(N_A\cdot)(C)]$	74.2	36.1	7.8	8.2	8.4	8.6	8.9	9.0	9.0

[a]Values in parentheses are best guesses.

3. PROBLEMS USING METHODS OF SECTION 2

3.1. Thermochemistry and Equilibrium Constants from Additivity Rules

1. Using bond additivity (Table 1) compute C_p°, S°, and ΔH_f° for methyl ethyl ketone (MEK) at 25°C ($CH_3 \overset{\overset{O}{\|}}{C} CH_2CH_3$).

2. Using group additivity (Tables 2–7), compute C_p°, S°, and ΔH_f° for MEK at 25°C ($CH_3 \overset{\overset{O}{\|}}{C} CH_2CH_3$).

3. Calculate the equilibrium constant for the reaction:

$$C(CH_3)_3CH_2OCH_2\varnothing + HC(CH_3)_2(c\text{-}C_3H_5) \leftrightarows$$
$$(ortho) - CH_3 - C_6H_4 - CH_2OH$$
$$+ HC(CH_3)_2CH_2C(CH_3)_2C(CH_3)CH_2$$

at 300 K and at 800 K.

3.2. Structural Estimates of Entropy and Heat Capacity

1. Compute the entropy of *trans*-butane at 25°C from statistical thermodynamics. The product of inertia of *trans*-butane is 4.3×10^5 amu^3-Å^6.

2. Estimate the entropy of the following compounds from model compounds: C_2H_5NO, $CH_3OOOOCH_3$.

3. Using various estimation techniques, calculate the equilibrium constant at 25°C for the reaction: [BDE HO_2-H = 87.2 kcal mol]

$$CH_3CH = CH_2 + O_2 \leftrightarows HO_2 + CH_2 \overset{\cdots}{-} CH \overset{}{-} CH_2$$

4. Calculate the enthalpy change at 300 K for the reactions:

(a) $\phi \cdot + O_2 \leftrightarows \phi O \cdot + O$

(b) $\phi O \cdot \rightarrow$ $+ CO$

(c) $+ O_2 \leftrightarows$ $+ O$

(d) $\leftrightarows$

4. KINETICS

4.1. Transition State Theory (TST)

We will rely on transition state theory as the major building block in our attempts to understand reaction kinetics at the molecular level. In the 50 years since this theory was first introduced, a good deal of theoretical work has gone into testing its validity and limitations. We will allude to some of these, but the usefulness for our purposes will remain beyond questioning.

We may think of the reactants and products of a chemical reaction as the same physical system that occupies different regions of phase space (the $6N$-dimensional space of position and momentum). In general, there is only a small region of this space containing all the particles that we cannot clearly label as reactants or products. This helps us to picture a reacting system in terms of dynamics along potential energy surface. A simple triatomic reaction of the type

$$A + BC \rightarrow AB + C$$

occurs on a surface such as depicted in Figure 2.[9] Furthermore we may make special note of the coordinate representing a minimum energy pathway from reactants to products. This pathway is called the "reaction coordinate." A general profile of energy along this coordinate is shown in Figure 3. We will assume that this coordinate is truly separable in the distance along the top of the barrier. We will describe all systems whose position on the surface is described by a value of the reaction coordinate in this length δ as being in the transition state (or activated complex), which we will name $X^\ddagger$. At chemical equilibrium,

$$A + BC \rightleftarrows X^\ddagger \rightleftarrows AB + C$$

the rates of the forward and reverse processes are equal. The flux of systems through the transition state from "left to right" is the same as from "right to left." Under these (equilibrium) conditions, the rate of left to right crossing is

$$R(1 \rightarrow r) = \text{frequency}(1 \rightarrow r) \times \text{concentration}(1 \rightarrow r)$$

$$\text{Frequency}(1 \rightarrow r) = \frac{\text{velocity}(1 \rightarrow r)}{\delta}$$

$$\text{Velocity}(1 \rightarrow r) = \langle V + \rangle_{1-D} = \left(\frac{2kT}{\pi m^\ddagger} \right)^{1/2}$$

$$\text{Concentration}(1 \rightarrow r) = \frac{[X^\ddagger]}{2} \qquad \text{half in each direction}$$

$$R(1 \rightarrow r) = [X^\ddagger]\left(\frac{kT}{2\pi m^\ddagger} \right)^{1/2} \delta^{-1}$$

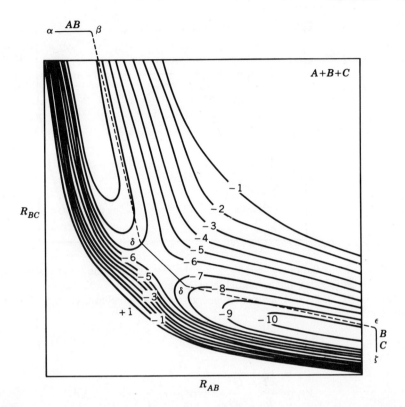

FIGURE 2. Potential-energy function for bimolecular atom-transfer reaction with five straight-line segments defined on this figure and used in the next figure. Taken from Johnston.[9]

At equilibrium $[X^{\ddagger}] = K^{\ddagger\prime}[A][BC]$, where $K^{\ddagger\prime}$ is the (arbitrary) name of the equilibrium constant:

$$\frac{R(1 \to r)}{[A][BC]} \equiv k_{bi} = K^{\ddagger\prime}\left(\frac{kT}{2\pi m^{\ddagger}}\right)^{1/2}\delta^{-1}$$

$$K^{\ddagger\prime} = \frac{Q^{\ddagger\prime}}{Q_A Q_{BC}}\exp\left\{-\frac{\Delta E_0^{\ddagger}}{RT}\right\}$$

If the reaction coordinate is truly separable,

$$Q^{\ddagger\prime} = Q^{\ddagger} \cdot Q_x$$

$$Q_x = Q_{1-D}(\text{translation}) = \left(\frac{2\pi m^{\ddagger}kT}{h^2}\right)^{1/2}\delta$$

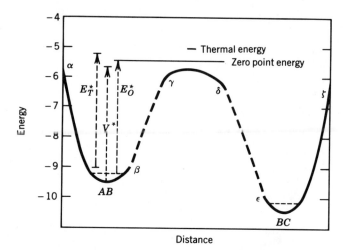

FIGURE 3. Potential-energy profile along the five straight-line segments of Figure 2. This figure illustrates the potential energy at activation, V^*; activation energy is absolute zero, E_0^*; and activation energy at any temperature, E_T^*. Taken from Johnston.[9]

Therefore

$$k_{bi} = \frac{kT}{h} \frac{Q^{\ddagger}}{Q_A Q_{BC}} \exp\left\{-\frac{\Delta E_0^{\ddagger}}{RT}\right\} \equiv \frac{kT}{h} K^{\ddagger}$$

$K^{\ddagger}$ is what is usually thought of as the equilibrium constant between reactants and transition state. It should be remembered that this is for the case where the reaction coordinate is removed.

If we now assume that the flux in the forward direction is unaffected by the flux in the reverse direction, we may accept the above expression for k_{bi} even far from equilibrium. The thermodynamic notation and language may be extended to the expression

$$k_{bi} = \frac{kT}{h} e^{-\Delta G_T^{\ddagger}/RT} = \frac{kT}{h} e^{\Delta S_T^{\ddagger}/R - \Delta H_T^{\ddagger}/RT}$$

This allows discussion of rate constants in terms of the "entropy and enthalpy of activation." Note that both of these quantities are temperature dependent, and thus we expect pure Arrhenius behavior only over limited temperature ranges.

$$k = A E^{-E_a/RT}$$

$$\log A = \log \frac{ek\langle T \rangle}{h} + \frac{\Delta S_{\langle T \rangle}^{\ddagger}}{2.3R}$$

$$E_a = \Delta H_{\langle T \rangle}^{\ddagger} + R\langle T \rangle$$

More generally, since the transition state theory expression for a thermal rate constant is

$$k = \left(\frac{kT}{h}\right)\exp\left[-\frac{\Delta G_T^{0\ddagger}}{RT}\right]$$

(the units are $\sec^{-1}$ for first-order and $\text{atm}^{-1}\ \sec^{-1}$ for second-order rate constants) and

$$\Delta G_T^{0\ddagger} = \Delta H_{300}^{\ddagger} - T\Delta S_{300}^{0\ddagger} + \langle \Delta C_p^{\ddagger}\rangle\left[(T - 300) - T\ln\left(\frac{T}{300}\right)\right]$$

(In the ideal gas approximation, we can drop the standard state notation on $\Delta H^{\ddagger}$ and $\Delta C_p^{\ddagger}$.) If the empirical temperature dependence is represented by

$$k = AT^B\exp\left[-\frac{C}{T}\right]$$

$$A = \frac{k}{\left[h(300)^{\langle\Delta C_p^{\ddagger}/R\rangle}\right]}\exp\left[\frac{\left(\Delta S_{300}^{0\ddagger} - \langle\Delta C_p^{\ddagger}\rangle\right)}{R}\right]$$

$$B = \frac{\langle\Delta C_p^{\ddagger}\rangle + R}{R}$$

$$C = \frac{\Delta H_{300}^{\ddagger} - \langle\Delta C_p^{\ddagger}\rangle(300)}{R}$$

where

$k = $ Boltzmann's constant
$h = $ Planck's constant
$\Delta S_{300}^{0\ddagger} = $ entropy of activation at 300 K standard state of 1 atm
$\Delta H_{300}^{\ddagger} = $ enthalpy of activation at 300 K
$\langle\Delta C_p^{\ddagger}\rangle = $ average value of the heat capacity at constant pressure of activation over the temperature range $(300 - T)/K$

If we wish to express second-order rate constants in concentration units instead of pressure units, we must multiply by RT in the appropriate units. This has the effect of writing

$$k' = A'T^{B'}\exp\left[-\frac{C}{T}\right]$$

where $A' = AR$ and $B' = B + 1 = (\langle\Delta C_p\rangle^{\ddagger} + 2R)/R$.

The best general way to represent rate constants is the use of this type of three-parameter expression. Even here, it is important to remember that we have made the assumption of constant heat capacity over the represented temperature range. This may easily be a bad assumption, as can be seen from the thermochemical considerations in Section 1, implying the need for more than three parameters over a wide temperature range.

Often, one finds rate constant data expressed in two-parameter Arrhenius form that has been obtained from measurements in the range 300–700 K. Cohen[10] has recently addressed the question of extrapolating such data to combustion temperatures using considerations such as discussed in this chapter. If lower temperature data can be fit to a model transition state that is reasonably accurate, then the full temperature dependence of the heat capacity would follow as a consequence. However, there are many uncertainties in these methods, including the restriction to harmonic oscillator rigid-rotor thermodynamic functions. Also in some cases an actual rate constant might be faster than the most accurate transition state theory calculation as a result of tunneling. The extent of this correction is not yet clear. Cohen's predictions, which do not explicitly correct for tunneling, have been found to be reasonably accurate where tested.

As always, the methods discussed herein can be used in a predictive and evaluative way, but if sensitivity to specific processes requires highly accurate values, they must be obtained by experiment!

We have derived the transition state theory rate constant expression by concentrating on a bimolecular reaction, but it is applicable to unimolecular processes as well (high-pressure limit only *vida infra*). The derivation that we have made has been for a population at thermal equilibrium; we might call it *canonical* transition state theory.

This level of approximation will often be sufficient, but there will be some limitations. We will encounter situations of nonthermal energy distribution arising from reactions between species not in thermal equilibrium and in unimolecular reactions at low pressures. Both of these can be handled within the limits of *microcanonical* transition state theory. Also we will encounter reactions that appear to be simple fundamental steps but are actually complex processes involving the formation of a stable intermediate. Treatment of these systems will be addressed.

4.2. Reaction Types

There are really only two different types of elementary reactions in the gas phase. These are unimolecular and bimolecular. In fact many bimolecular processes are really the reverse of unimolecular events and are best considered as such.

Bimolecular

- *Metathesis reactions.* Involve the transfer of a single atom. These are the $A + BC$ type that we used as a model earlier. In general, A and C may be large groups ($CH_3 + C_2H_6 \rightarrow CH_4 + C_2H_5$).

- *Addition reactions*. Involve the making and breaking of bonds (HCl + $C_2H_4 \rightarrow C_2H_5Cl$; $H + C_2H_4 \rightarrow C_2H_5$).
- *Association reactions*. The combination of species (usually free radicals) to form a new single bond ($2CH_3 \rightarrow C_2H_6$).

Of these, only metathesis reactions are bimolecular in both directions and thus not pressure dependent.

Unimolecular

- *Isomerization reactions*. Spatial interchange of atoms. Involves making and breaking of bonds.
- *Simple fission*. Reverse of association reactions.
- *Complex fission*. Reverse of addition reactions.

4.3. Application of Transition State Theory

In principle, TST can be applied to predict rate constants when the potential energy surface is known. (Of course, if the potential energy surface is known, we can perform dynamical calculations that will allow us to test TST.) We locate the transition state, calculate its frequencies, etc., and compute the TST rate constant. In practice, we cannot hope for much of this knowledge, so we use TST as a guide for codification and extrapolation of rate data.

We use our knowledge as chemists to make intercomparisons. An example might be the metathesis reaction:

$$CH_3 + CH_3CH_3 \rightarrow CH_4 + C_2H_5$$

Since we know the structures of reactants and products, we infer that the transition state structure can be represented by $[CH_3 \cdots H \cdots C_2H_5]$. This only means that we expect a structure in which the H-atom is partially transferred. This inference alone allows us to narrow down the possible range of A-factors. We know that the upper limit is collision frequency ($\sim 10^{11.3}\ M^{-1}\ sec^{-1} \approx 10^{-9.5}\ cm^3\ mol^{-1}\ sec^{-1}$), and the lower limit is given by the tightest model we can make for the transition state. The tightest model of the transition state that we have chosen would be propane. We may then predict

$$S^{\ddagger} > S_{C_3H_8} + \underset{\text{(spin)}}{R \ln 2} + \underset{\text{(symmetry)}}{R \ln 2}$$

$$\Delta S^{\ddagger} > S_{C_3H_8} + R \ln 2 + R \ln 2 - S_{CH_3} - S_{C_2H_6}$$

We may use our thermochemical estimation method of other sources to find that $\Delta S^{\ddagger} > -34$ cal mol^{-1} K^{-1}. This value (standard state 1 atm) corresponds to (298 K):

$$A \geq \frac{ekT}{h} 10^{(\Delta S^{\ddagger} + 8.35)/4.58} = 10^{13.2 - 5.6} = 10^{7.6} \ M^{-1} \ \text{sec}^{-1}$$

The quantity 8.35 cal mol^{-1} K^{-1} is the conversion factor to units of M^{-1} sec^{-1}. In general, this quantity may be called S' and is given by

$$S' = R(\ln R'T + 1)$$
$$R = 1.99 \ \text{cal mol}^{-1} \ \text{K}^{-1}$$
$$R' = 0.082(1 \ \text{atm}) \ \text{mol}^{-1} \ \text{K}^{-1}$$

TABLE 20. $Cl + C_2H_6 \rightarrow \left[\begin{array}{c} H \quad CH_3 \\ \cdot \quad \cdot \quad / \\ Cl \quad \quad CH_2 \end{array} \right]^{\ddagger} \rightarrow HCl + C_2H_5^{\cdot}$

	$\Delta S_{300}^{\circ \ddagger a}$	$\Delta C_{p,300}^{\ddagger a}$	$\Delta C_{p,500}^{\ddagger a}$
Reference reaction [transition state = C_2H_5Cl]	−28.3	−2.8	−2.4
Spin	1.4	—	—
Symmetry	0	—	—
External rotation $(2 \times 2 \times 1.5)^b$	1.7	—	—
Reaction coordinate			
$[C-Cl(650 \ \text{cm}^{-1}) \rightarrow \nu^{\ddagger}]$	−0.4	−0.9	−1.5
$Cl \cdot H \ (2100 \ \text{cm}^{-1})$	0	0	0.2
$Cl \cdot H \cdot C \ (600 \ \text{cm}^{-1})$	1.0	1.0	1.6
$Cl-C-C(400 \ \text{cm}^{-1}) \rightarrow Cl \cdot H \cdot C(600 \ \text{cm}^{-1})$	−0.5	−0.2	−0.1
$Cl-C-(H)_2(1000 \ \text{cm}^{-1}) \rightarrow (Cl \cdot H)-C-(H)_2(700 \ \text{cm}^{-1})$	0.2	0.5	0.4
$Cl \overset{\cdot}{\underset{\longleftarrow}{}} \begin{array}{c} H \\ \diagdown \\ CH_2 \end{array} \diagup CH_3$ (torsion)^c	2.2	1.0	1.0
	> −22.7	−1.4	−0.8

^{a}cal mol^{-1} deg^{-1} (standard state, 1 atm).
^{b}Estimated increase in product of inertia.
$^c I_r$ calculated for 10° off linear with $r_{H \cdot Cl} = 1.5$ Å.

$$\log[A_{300}/M^{-1} \ \text{sec}^{-1}] \geq 13.2 + \frac{-22.7 + 8.35}{4.58} \geq 10.1$$

$$\log[A_{400}/M^{-1} \ \text{sec}^{-1}] \geq 13.3 + \frac{-22.7 - 1.1 \ln 4/3 + 8.92}{4.58} \geq 10.2$$

$$E_{300} = \Delta H_{300}^{\ddagger} + 2R(300) = \Delta H_{300}^{\ddagger} + 1.2 \ \text{kcal mol}^{-1}$$

$$E_{400} = \Delta H_{300}^{\ddagger} - 1.1(0.1) + 2R(0.400) = \Delta H_{300}^{\ddagger} + 1.5 \ \text{kcal mol}^{-1}$$

$$E_{400} - E_{300} = 0.3 \ \text{kcal mol}^{-1}$$

We now expect that the A-factor will lie between $10^{7.6}$ and $10^{11.3}$ M^{-1} $\sec^{-1}$. (The actual value turns out to be $10^{8.5}$ M^{-1} $\sec^{-1}$.) Furthermore we realize that the collision frequency is a high upper limit and that, more likely, we have narrowed the range to about a factor of 10^2. This alone may help in the evaluation of experimental results, but we can also rely on chemical knowledge. For instance, we have good reason to expect that transition states for reactions, $CH_3 + H - R \rightarrow CH_4 + R$, would be very similar. Thus we can take measured results in one case to predict others.

Another device for estimating transition state thermochemistry is illustrated by the use of surrogate models. The reaction

$$O + CH_4 \rightarrow OH + CH_3$$

has a transition $[O\text{----}H\text{----}CH_3]$ for which CH_3F is a good surrogate, since F is about the same mass as OH. An analysis of this reaction requires corrections to the CH_3F thermochemistry.

Using a good deal of experimental data and some common sense to estimate the frequencies associated with partial bonds, Table 10 can be produced. They are, of course, just estimates and can be varied somewhat. We shall see that in general there are one or two key frequencies to know. We can get these from experimental values and then transfer to analogous systems.

A complete example of an estimation of the A-factor for the reaction $Cl + C_2H_6 \rightarrow HCl + C_2H_5$ is shown in Table 20. Recent measurements yield $\log k/M^{-1} \sec^{-1} = 10.7 - 0.26/\theta;\ 220 \le T/K \le T/K\ 600$. This A-factor is greater than the lower limit by a factor of 3. Indicating a transition state with 2–3 cal mol^{-1} K^{-1} more entropy. This is a "loose" transition state! (Notice the small activation energy.)

The application of TST to unimolecular processes requires the same type of thinking as for bimolecular reactions. In general, we expect values of $\Delta S^{\ddagger}$ that do not reflect the large negative value associated with external degrees of freedom being lost. Values of $\Delta S^{\ddagger}$ can be negative when open chain species react via a cyclic transition state, restricting internal rotations. Thus eithyl-vinyl ether decomposition

has an A-factor of $10^{11.6}$ $\sec^{-1}$. We find unimolecular A-factors as high as $\sim 10^{17}$ $\sec^{-1}$, or even 10^{18} $\sec^{-1}$, for reactions involving a simple bond cleavage with little other electronic rearrangement. The value of the A-factor

TABLE 21a. Activation Entropy for C_2H_5Cl Pyrolysis at 600 K^a

Degrees of Freedom	$\Delta S^{\ddagger}$
Symmetry	$+R \ln 3 = +2.2$
CH$_3$ internal rotation → CH$_2$—C$\begin{smallmatrix}\text{CH}_3\\\text{H}\end{smallmatrix}$ torsion (400 cm^{-1})	-3.7
C—Cl stretch 650 cm^{-1} → (reaction coordinate)	-1.4
C—C—Cl bend → C—C · Cl bend (400 cm^{-1}) → (280 cm^{-1})	$+0.7$
H—C—H bend → H—C—C bend (1450 cm^{-1}) → (950 cm^{-1})	$+0.6$
Total	-1.6

^{a}Note that we have neglected the small contributions due to the ring H stretches.

TABLE 21b. H$\begin{smallmatrix}\text{H}\\\text{>}\\\text{H}\end{smallmatrix}$C—C$\begin{smallmatrix}\text{H}\\\text{<}\\\text{Cl}\end{smallmatrix}$H

Stretches (cm^{-1})

5 C—H (3100)
1 C—C (1000)
1 C—Cl (650)

Bends

Heavy atoms
1 C—C—Cl (400)

CH$_2$ group

1 H—C—H scissors	(1450)
1 CH$_2$ wag → (H—C—C)$_w$	(1150)
1 CH$_2$ twist →(H—C—C)$_t$	(1150)
1 CH$_2$ rock → (H—C—C)$_r$	(700)

CH$_3$ group

2 CH$_3$ rocks (H—C—C)$_{t,w}$	(1150)
3 H—C—H scissors	(1450)

Internal rotors

| CH$_3$—C | [4.3] |

TABLE 21c.

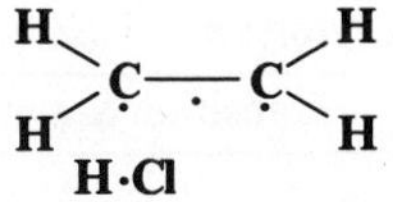

$$\begin{array}{ccc} \text{H} & & \text{H} \\ & \diagdown & \diagup \\ \text{H} \diagup & \overset{\bullet}{\text{C}} \,\cdots\, \overset{\bullet}{\text{C}} & \diagdown \text{H} \\ & \text{H·Cl} & \end{array}$$

Stretches

Simple

4C—H

Ring stretches

C——C	(1300)
C · Cl	(RC)
C · H	(2200)
H · Cl = H · C	(2200)

Bends

CH_2 groups

2 H—C—H scissors	(1450)
2 CH_2 wags → (H—C—C)$_{t,\,w}$	(1150)
2 CH_2 twists → (H—C—C)$_{t,\,w}$	(1150)
2 CH_2 rocks → (H—C—C)$_r$	(700)

Ring deformations

$$\text{Cl} \cdot \overset{\bullet}{\text{C}} - \text{C} \underset{\diagdown}{\overset{\diagup}{}} \begin{array}{l} \text{Cl} \cdot \text{C} {=} \text{C (280)} \\ \text{Cl} \cdot \text{C} {-} \text{C (280)} \end{array} > 280$$

(H · C$\overset{\bullet}{-}$C)$_{op}$ ≡ (H—C$\overset{\bullet}{-}$C)$_r$	(700)

for the reaction

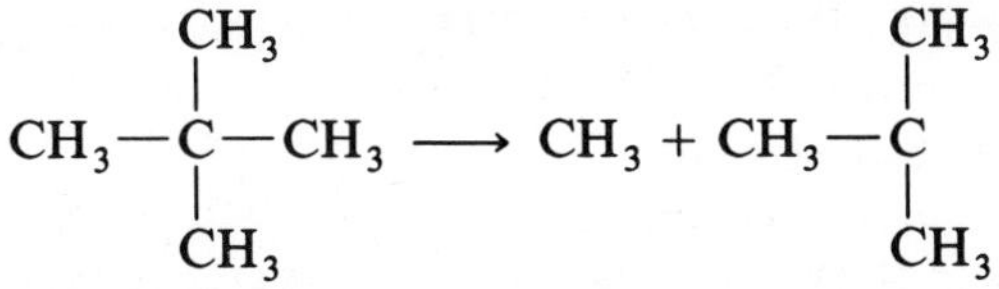

$$\begin{array}{ccc} \quad\ \ \text{CH}_3 & & \text{CH}_3 \\ \quad\ \ | & & | \\ \text{CH}_3{-}\text{C}{-}\text{CH}_3 \longrightarrow \text{CH}_3 + \text{CH}_3{-}\text{C} \\ \quad\ \ | & & | \\ \quad\ \ \text{CH}_3 & & \text{CH}_3 \end{array}$$

is $\sim 10^{17}$ sec^{-1}. (Temperature dependence must be kept in mind!)

An example of a calculation of the A-factor for the reaction $C_2H_5Cl \rightarrow C_2H_4 + HCl$ is shown in Table 21. The most important features here are the loss of internal rotation and the gain of low-frequency ring modes. Tables 21a–d show the details of the calculation.

An important observation that becomes apparent from the kind of quantitative evaluation that we have been discussing is that rate parameters for individual chemical reactions can be compared with each other for tests of

TABLE 21d. Correlations for EtCl Pyrolysis

Ground State		Transition State		$\Delta S^{\ddagger}$ (600 K)
Stretches				
5 C—H	(3100)	4 C—H	(3100)	0
		1 C·H	(2200)	0
1 C—H	(1000)	1 C$\overset{\cdot}{-}$C	(1300)	−0.2
1 C—Cl	(650)	1 C·Cl	(RC)	−1.3
Bends				
Heavy atoms				
1 C—C—Cl	(400)	Cl·C$\overset{\cdot}{-}$C	(280)	+0.7
CH$_2$ groups				
1 H—C—H		1 H—C—H		
2 (H—C—C)$_{t,w}$		2 (H—C—C)$_{t,w}$		0
1 (H—C—C)$_r$		1 (H—C—C)$_r$		
CH$_3$ group				
		1 H—C—H	(1450)	
2 (H—C—C)$_{t,w}$	(1150)	2 (H—C—C)$_{t,w}$	(1150)	+0.6
3 H—C—H	(1450)	1 (H—C—C)$_r$	(700)	+0.6
		1 H·Cl	(2200)	
Internal rotors				
CH$_3$——C	[4·3]	(H—C$\overset{\cdot}{-}$C)$_r$	(700) [1.2]	−2.1
				−2.3

consistency. Thus in a series of exothermic abstraction reactions such as

$$O + C_2H_6 \rightarrow OH + C_2H_5 \qquad \Delta H = -3 \text{ kcal mol}^{-1}$$

$$\log k/M^{-1} \text{ sec}^{-1} = 10.4 - \frac{6.4}{\theta}$$

$$OH + C_2H_6 \rightarrow H_2O + C_2H_5 \qquad \Delta H = -20 \text{ kcal mol}^{-1}$$

$$\log k/M^{-1} \text{ sec}^{-1} = 9.7 - \frac{2.1}{\theta}$$

$$CH_3 + C_2H_6 \rightarrow CH_4 + C_2H_5 \qquad \Delta H = -7 \text{ kcal mol}^{-1}$$

$$\log k/M^{-1} \text{ sec}^{-1} = 8.5 - \frac{10.8}{\theta}$$

We might expect (and we do find) that the *A*-factor differences primarily reflect the differences in rotational entropy lost in forming the transition state.

Unimolecular Reactions—Pressure Dependence. Thermal unimolecular processes are pressure dependent because only those molecules with sufficient energy to overcome the barrier to reaction may react. Thus the rate of energization is potentially the rate-controlling step. This is illustrated in a simple way by the "Lindemann mechanism":

$$A + M \underset{k_{-1}}{\overset{k_1}{\rightleftharpoons}} A^* + M$$

$$A^* \xrightarrow{k_2} \text{products}$$

The steps 1 and -1 are collisional activation and deactivation, which maintain the thermal energy distribution. If step 2 depletes energized species faster than collisions produce them, step 1 becomes rate controlling, introducing the factor of pressure through the concentration of M. Consider

$$(A^*)_{ss} = \frac{k_1(A)(M)}{k_{-1}(M) + k_2}$$

$$k_{\text{uni}} \cong \frac{1}{(A)} \frac{d \, \text{Prod}}{dt} = \frac{1}{(A)} k_2(A^*) = \frac{k_2 k_1(M)}{k_{-1}(M) + k_2}$$

If $k_{-1}(M) \gg k_2$ (i.e., high pressure), then $k_{\text{uni}} = k_2 k_1/k_{-1} \cong k_\infty$. If $k_{-1}(M) \ll k_2$ (i.e., low pressures), then $k_{\text{uni}} = k_1(M) \cong k_0$. Values between the low and high pressure-limiting values are said to be in the "fall-off" regime (see Figure 4).

Since k_2 is much more temperature dependent than k_{-1}, any reaction is further into the fall-off regime as the temperature is raised. Thus, for combustion problems, the considerations outlined here are imperative.

A very large effort has gone into understanding this problem. Rather than discuss the microscopic details, we will confine ourselves to the problem of estimating the rate constant for modeling purposes.

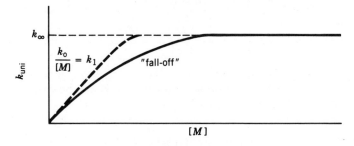

FIGURE 4

As can be seen from the Lindemann mechanism and the accompanying figure, if we know k_∞ and $k_0/[M]$, we would have the limiting values, and the need would be to estimate the fall-off curve. A convenient formalization of the Lindemann mechanism is

$$\frac{k_{uni}}{k_\infty} = \frac{k_0/k_\infty}{1 + k_0/k_\infty} = \frac{k_{-1}[M]}{k_{-1}[M] + k_2}$$

and to a first approximation, the curvature is known if k_0 and k_∞ are known.

For this chapter we will rest at this limit, under the assumption that our goal is the estimation of rate constants for a complex chemical system. If the sensitivity of our model to pressure-dependent rate constants is high, we will have to delve into the problem more seriously. At the moment this generally means applying what is known as RRKM theory, or an empirical formulation due to Troe and coworkers which has the form:

$$\frac{k_{uni}}{k_\infty} = \frac{k_0/k_\infty}{1 + k_0/k_\infty} F\left(\frac{k_0}{k_\infty}\right)$$

The correction factor F is dependent on the temperature and the nature of the molecule. Current efforts are toward finding simple ways to estimate F as a function of T as a next step in estimating k/k_∞.

We have already touched on estimating the A-factor for unimolecular processes in the high-pressure limit. As usual, the estimation of activation energies is more difficult. For simple bond scission reactions, the BDE is a good guess (see examples in Section 5). Other classes of reactions may allow a guess based on similar reactions having been measured.

Estimating the value of k_1, and thus k_0, is a bit more difficult. We begin by realizing the k_{-1} will be no higher than bimolecular collision frequency. Thus

$$k_1[M] = \beta\omega = \beta Z[M]$$

where β, the collision efficiency, lies between 0 and 1. Also, $K_1 \cong k_1/k_{-1}$ and $k_1 = k_{-1}K_1$. K_1 is the equilibrium constant:

$$K_1 = \frac{[A^*]_{eq}}{[A]_{eq}} = \frac{Q_A^*}{Q_A}$$

Remembering that a partition function for species X can be written in terms of the state density as

$$Q_X = \int_0^\infty \rho_X(E) e^{-E/kT} dE$$

we may now write

$$k_1 = k_{-1}K_1 = \frac{\beta Z}{Q_A} \int_{E_c}^{\infty} \rho_A(E) e^{-E/kT} dE$$

The partition function Q_A^* may be computed; Z and Q_A are also accessible, so the major problem is the estimation of β. (β is also temperature dependent.) This problem is also the subject of current research, but indications are that a good approximation is

$$\frac{\beta}{1 - \beta^{1/2}} \approx \frac{\langle \Delta E \rangle}{kT}$$

$\langle \Delta E \rangle$ is the average energy transferred in a collision. This quantity is roughly constant over the temperature range involved in a combustion model (caveat emptor!) ranging from values of tenths of kcal mol^{-1} for rare gases to tens of kcal mol^{-1} for larger colliders.

The partition function $Q_A^* = \int_{E_c}^{\infty} \rho_A(E) e^{-E/kT} dE$ may be calculated by numerical methods. For our purposes, a simple zero-order approximation may be used.

We will assume that[11]

$$k_1 = \frac{\beta Z \rho_{vib}(E_0) kT}{Q_A(vib)} e^{-E_0/kT} F_{rot} \cdot F_{int\ rot}$$

where

$$\rho_{vib}(E_0) = \frac{[E_0 + 0.9E_Z]^{s-1}}{(s-1)!} \prod_{r=1}^{s} (h\nu_i)$$

Linear Molecule

$$F_{rot} = \frac{[E_0 + 0.9E_Z]}{skT} \times \left[\frac{2.15(E_0/kT)^{1/3}}{2.15(E_0/kT)^{1/3} - 1 + [E_0 + 0.9E_Z]/skT} \right]$$

Nonlinear Molecule

$$F_{rot} = \frac{(s-1)!}{(s+\frac{1}{2})!} \left[\frac{E_0 + 0.9E_Z}{kT} \right]^{3/2}$$

$$\times \left[\frac{2.15(E_0/kT)^{1/3}}{2.5(E_0/kT)^{1/3} - 1 + [E_0 + 0.9E_Z]/(s+\frac{1}{2})kT} \right]$$

$$F_{int\ rot} = \frac{(s-1)!}{(s+r/2-1)!} \left(\frac{(E_0 + 0.9E_Z)}{kT} \right)^{r/2}$$

where r is the number of internal rotations.

This formulation will give us a chance to complete our estimations of k_{uni}. If the reaction is particularly sensitive, we may have to use more sophisticated theory.

A computer code exists for the calculation of $k_1(T)$ given the required molecular parameters, although hand calculation of a few values is not terribly difficult.

4.4. Mechanistic Problems—Bound Intermediates

A particular problem arises when the bimolecular reaction of interest proceeds along a reaction coordinate involving the formation of an intermediate bound complex. This type of chemical activation reaction, termed "complex," may be described by the mechanistic and rate equations that follow (steady-state assumption applied to Y^*):

$$A + B \underset{-1}{\overset{1}{\rightleftharpoons}} Y^* \overset{2}{\longrightarrow} C + D$$
$$3 \downarrow [M]$$
$$Y$$

$$k_{exp't} \equiv \frac{-1}{[A][B]} \frac{d[A]}{dT} = \frac{k_1(k_2 + k_3[M])}{k_{-1} + k_2 + k_3[M]} \cong k_1 * \varepsilon(T, M)$$

Here we see that several limiting cases are possible, amounting to different values of the "efficiency factor" $\varepsilon(T, M)$.

Case 1: $k_3[M] \ll k_2$

$$k_{exp't} = \frac{k_1 k_2}{k_{-1} + k_2} = k_1 * \varepsilon(T)$$

a) $\overset{k_{-1} \gg k_2}{\longrightarrow} K_1 k_2; \quad \varepsilon(T) = \dfrac{k_2}{k_{-1}} \ll 1$

b) $\overset{k_{-1} \ll k_2}{\longrightarrow} k_1; \quad \varepsilon(T) = 1$
(direct reaction)

Case 2: $k_3[M] \gg k_2$

$$k_{exp't} = \frac{k_1 k_3[M]}{k_{-1} + k_3[M]}$$

a) $\overset{k_3[M] \gg k_{-1}}{\longrightarrow} k_1; \quad \varepsilon(T, M) = 1$
(direct reaction)

b) $\overset{k_3[M] \ll k_{-1}}{\longrightarrow} K_1 k_3(M); \quad \varepsilon(T, M) = \dfrac{k_3[M]}{k_{-1}} \ll 1$

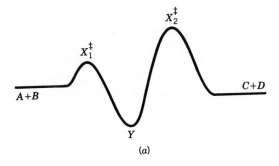

(a)

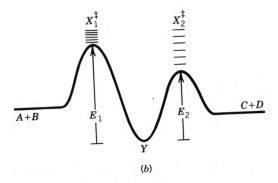

(b)

FIGURE 5

In the limit described by either case 1b or case 2a, the rate-controlling step is process (1), an association reaction, and the experimental rate constant may be described by TST for this reaction. (In some cases this may require some knowledge of the problem of describing simple bond fission processes by nonfixed transition states.)

In the limit described by case 1a, the experimental rate constant is not pressure dependent, but the efficiency factor $\varepsilon(T)$ is temperature dependent. Furthermore, since the magnitude of k_{-1} and k_2 are determined by both the respective entropy of activation (density of states) and the energy barriers, the inequality $k_{-1} \gg k_2$ can be maintained such that $\varepsilon(T)$ may either increase or decrease with temperature. In the former case a simple TST treatment would be adequate, since the potential surface is really being described as shown in Figure 5a. In the latter case, however, we have the situation which may be described by the surface shown in Figure 5b. In the figure the schematic representation of level densities is meant to illustrate the possibility that $k_2 < k_{-1}$ even though $E_2 < E_{-1}$. In this case $\varepsilon(T)$ will decrease with temperature, and the sign of the temperature dependence of $k_{\text{exp't}}$ will depend on the magnitude of the temperature dependence of k_1 itself.

The limit described by case 2b can be recognized as the simple case of an association reaction which, of course, is just the microscopic reverse of a unimolecular fission process.

The actual values of rate constants and Arrhenius parameters in these complex cases will not be described by simple canonical TST. In general, these rate constants will need to be calculated through use of some form of unimolecular rate theory if the appropriate potential parameters are obtainable, as well as theories of energy transfer to predict $k_3[M]$.

A simple approximation for the rate constants k_{-1} or k_2 is given by the QRRK (or "quantum version of Rice-Ramsperger-Kassel") theory. The formula is

$$k(E) = A_\infty \frac{n!(n - m + s - 1)!}{(n - m)!(n + s - 1)!}$$

where

n = total energy (in quanta)
m = critical energy (in quanta)
s = number of oscillators

Most hydrocarbons have an average frequency of ~ 1000 cm^{-1} (2.85 kcal mol^{-1}), which we can use as the measure of one quantum.

The Reaction OH + CO → Products. The reaction of OH with CO has been subjected to extensive experimentation and analysis. The system has been discussed by many workers. A picture of the potential surface might be that shown in Figure 6a, but the data may also be described by a system depicted as shown in Figure 6b. In these figures the fact that the barrier to H + CO$_2$ formation is lower is made up for by a tighter transition state (closer to linear),

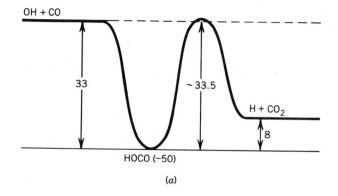

(a)

FIGURE 6

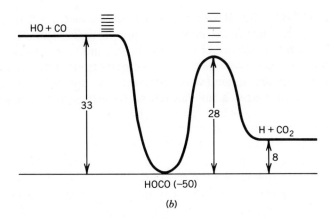

FIGURE 6. *(Continued)*

which means that the branching ratio $k_2/(k_{-1} + k_2) \sim 10^{-1.5}$. Also the absolute

$$HO + CO \underset{k_{-1}}{\overset{k_1}{\rightleftharpoons}} HOCO^* \overset{k_2}{\longrightarrow} H + CO_2$$
$$\downarrow{\scriptstyle k_3}\,[M]$$
$$HOCO$$

value of $k_2 \sim 10^{9.3}$ sec^{-1} (using quantum RRK theory) which means that at

TABLE 22. *A*-Factors for HOCO Decomposition^a

3456 ⟵ (O—H st)	3456 ⟶ reaction coordinate	
1900 ⟵ (C=O st)	1833 ⟶ 1833	
900 ⟵ (HOĊ bd)	1260 ⟶ 840	
Reaction coordinate ⟵ (Ċ—O st)	1071 ⟶ 1300	
425 ⟵ (OCO bd)	615 ⟶ 635(2)	
Free rotation ⟵ (HO-torsion)	615 ⟶ 840	
$\Delta S^{\ddagger}_{vib}$ (300 K) 2.4	0.3	
$\Delta S^{\ddagger}_{rot}$ (300 K) ~ 2	− 7.3	
$\log[A(300\ K)/s^{-1}]$ 14.2	11.7^b	
E_{act} $\|\Delta E_1\|$		

^{a}Data sources: For HOCO, ref. 12, with an additional $R \ln 2$ added to the entropy to account for mixing of *cis* and *trans* forms. Transition state frequencies are estimated, and moments of inertia are computed from guessed structures. For HO···C=O, the bond length was increased by 1 Å; for H···OCO the moments of inertia for CO_2 (from JANAF) were used.
^{b}A bent transition state raises the A-factor to $\sim 10^{12.9}$ s^{-1} and raises E_2 to ca. E_{-1}.

about 1 atm, $k_3[M]$ could be competitive for larger molecules. The question of pressure dependence may need more experimental work. Table 22[12] presents values used to compute transition state properties, and thus activation parameters. The values of k_1 computed from these parameters, when multiplied by the branching ratio, are in good agreement with the low-pressure rate data at all temperatures. Of course, a full calculation would be more precise.

5. PROBLEMS USING METHODS OF SECTIONS 2 AND 4

1. Estimate the A-factor for the reaction, $CH_3 + NH_3 \rightarrow CH_4 + NH_2$.

2. From knowledge of the fact that the rate constant for $O + CH_4 \rightarrow OH + CH_3$ (between 350 and 1000 K) is given by $\log k/M^{-1} \sec^{-1} = 10.32 - 9.04/\theta$, estimate k at 1500 and 2000 K.

3. Calculate the high-pressure A-factor for the reaction $\phi CH_3 \rightarrow \phi CH_2 + H$ from the fact that the A-factor for $\phi CH_2 CH_3 \rightarrow \phi CH_2 + CH_3$ is $10^{15.3} \sec^{-1}$ at 1000 K.

4. (a) Compute k_0 for the reaction $CH_4 + Ar \rightarrow CH_3 + H + Ar$ at 1000 K if β for $Ar \approx 0.1$, $Z = 3 \times 10^{-14}$, $Q_{vib} = 2$, $\nu = [3000(4), 1400(2)]$.

   (b) Make a reasonable estimate of k_∞ for the same reaction.

   (c) Compute the pressure at which simple L–H theory predicts $k = k_\infty/2$.

5. At 298 K, t-butyl radicals combine with Br atoms to give a vibrationally hot t-butyl bromide. Estimate the branching ratio this hot compound going back to t-butyl and Br-atom compared to going to HBr + isobutene (i.e., k_{-1} vs. k_2):

$$Br + \overset{\cdot}{\underset{\quad}{+}}\, \underset{k_{-1}}{\overset{k_1}{\rightleftharpoons}} \left(\overset{\quad}{\underset{\quad}{+}} Br\right)^* \overset{k_2}{\longrightarrow} \,\overset{}{=}\!\!< + HBr$$

6. The following rate constants have been used as part of a model for methanol oxidation. (Units are $cm^3\ mol^{-1}\ sec^{-1}$ for A and $kcal\ mol^{-1}$ for E).

	Log A	E
$CH_3OH + O_2 \rightarrow CH_2OH + HO_2$	13.6	50.9
$CH_3OH + OH \rightarrow CH_2OH + H_2O$	12.6	2.0
$CH_3OH + O \rightarrow CH_2OH + HO$	12.2	2.3
$CH_3OH + H \rightarrow CH_2OH + H_2$	13.5	7.0
$CH_3OH + H \rightarrow CH_3 + H_2O$	12.7	5.3
$CH_3OH + CH_3 \rightarrow CH_2OH + CH_4$	11.3	9.8
$CH_3OH + HO_2 \rightarrow CH_2OH + H_2O_2$	12.8	19.4

As a final exercise, comment on the absolute values and the relative values of these parameters.

6. CONCLUSIONS

This chapter has been intended as a presentation of the general framework for quantitative understanding of elementary rate data. This overview allows the prediction, extrapolation, and evaluation of rate data for use in combustion models.

The literature contains many models for combustion systems. These are generally presented as a series of elementary chemical reactions and their Arrhenius parameters. Often the values are taken individually from experimental data. This seemingly enlightened approach has dangerous pitfalls leading to inconsistent rate parameters. TST considerations may not be able to distinguish factors of 2 or 3, but general features should be evaluated with the rudimentary level of theory in mind.

REFERENCES

1. D. M. Golden and C. W. Larson, *Symp.* (*Int.*) *Combust.* [*Proc.*] **20**, 594 (1985).
2. C. J. Cobos and J. Troe, *J. Chem. Phys.* **83**, 1010 (1985).
3. W. C. Gardiner, Jr., *Sci. Am.* **246**(2), 110 (1982).
4. G. N. Lewis and M. Randall, *Thermodynamics*, 2nd ed. McGraw-Hill, New York, 1961.
5. J. Dale, *Tetrahedron* **22**, 3373 (1966).
6. W. J. Orville-Thomas, ed., *Internal Rotation in Molecules*. Wiley, New York, 1974.
7. *Handbook of Chemistry and Physics*. Chemical Rubber Co., Cleveland, Ohio, 1980.
8. D. F. McMillen and D. M. Golden, *Annu. Rev. Phys. Chem.* **33**, 493 (1982).
9. H. S. Johnston, *Gas-Phase Reaction Rate Theory*. Ronald Press, New York, 1966.
10. N. Cohen, *Int. J. Chem. Kinet.* **18**, 99 (1986), and references therein.
11. J. Troe, *J. Chem. Phys.* **66**, 4758 (1977).
12. W. C. Gardiner, Jr. et al., *Chem. Phys. Lett.* **53**, 134 (1978).

BIBLIOGRAPHY

Thermochemistry (see also references in *Thermochemical Kinetics*)

American Petroleum Institute, "Selected Values of Properties of Hydrocarbons," API Proj. No. 44. Carnegie Institute of Technology, Pittsburgh, Pennsylvania, 1953 (currently issued in loose-leaf form by Thermodynamics Research Center, Texas A & M University, College Station).

J. D. Cox and G. Pilcher, *Thermochemistry of Organic and Organometallic Compounds*. Academic Press, London, 1970.

Handbook of Chemistry and Physics. Chemical Rubber Co., Cleveland, Ohio, 1980.

JANAF Thermochemical Tables, NSRDS-NBS 37. U.S. Govt. Printing Office, Washington, D.C., 1970.

J. B. Pedley and J. Rylane, *Sussex—N.P.L.: Computer Analyzed Thermochemical Data: Organic and Organomettalic Compounds*, University of Sussex, 1977.

R. Shaw and S. W. Benson, in *The Chemistry of the Carboxylic Acids and Esters* (S. Patai, ed.). Wiley, New York, 1970.

R. Shaw, D. M. Golden, and S. W. Benson, "Thermochemistry of Some Six-Membered Cyclic and Polycyclic Compounds Related to Coal." *J. Phys. Chem.* **81**, 1716 (1977).

S. E. Stein and D. M. Golden, "Aromatic Hydrocarbon Radicals." *J. Org. Chem.* **42**, 839 (1977).

S. E. Stein, D. M. Golden, and S. W. Benson, "Predictive Scheme for Thermochemical Properties of Polycyclic Aromatic Hydrocarbons." *J. Phys. Chem.* **81**, 314 (1977).

D. R. Stull, E. F. Westrum, Jr., and G. C. Sinke, *The Chemical Thermodynamics of Organic Compounds*. Wiley, New York, 1969.

Kinetics (*Tabulations*)

D. L. Baulch, D. D. Drysdale, D. G. Horne, and A. C. Lloyd, *Evaluated Kinetic Data for High Temperature Reactions*. CRC Press, Cleveland, Ohio, 1972.

D. L. Baulch, D. D. Drysdale, J. Duxbury, and S. J. Grant, *Evaluated Kinetic Data for High Temperature Reactions*. Butterworth, London, 1973.

D. L. Baulch, J. Duxbury, S. J. Grant, and D. C. Montague, *Evaluated Kinetic Data for High Temperature Reactions*. Butterworth, London, 1976.

D. L. Baulch et al., "Evaluated Kinetic and Photochemical Data for Atmospheric Chemistry." *J. Phys. Chem. Ref. Data* **9**(2), 295 (1980).

D. L. Baulch et al., "Evaluated Kinetic and Photochemical Data for Atmospheric Chemistry: Supplement I." *J. Phys. Chem. Ref. Data* **11**(2), 327 (1982).

D. L. Baulch et al., "Evaluated Kinetic and Photochemical Data for Atmospheric Chemistry: Supplement II." *J. Phys. Chem. Ref. Data* **13**(4), 1259 (1984).

S. W. Benson and H. E. O'Neal, *Kinetic Data on Gas-Phase Unimolecular Reactions*, NSRDS-NBS 21. U.S. Govt. Printing Office, Washington, D.C., 1970.

N. Cohen, "The Use of Transition State Theory to Extrapolate Rate Coefficients for Reactions of OH with Alkanes." *Int. J. Chem. Kinet.* **14**, 1339 (1982).

W. B. DeMore et al., *Chemical Kinetic and Photochemical Data for Use in Stratospheric Modeling*, JPL 82-57. Jet Propulsion Laboratory, Pasadena, California, 1982.

Journal of Physical Chemistry Entire issue, Jan. 11, Vol. 83, No. 1 (1979), particularly pp. 108 and 114.

J. A. Kerr and M. J. Parsonage, *Evaluated Kinetic Data on Gas-Phase Addition Reactions*. CRC Press, Cleveland, Ohio, 1972.

J. A. Kerr and M. J. Parsonage, *Evaluated Kinetic Data on Gas-Phase Hydrogen Transfer Reactions of Methyl Radicals*. Butterworth, London, 1976.

J. A. Kerr and E. Ratajczak, *Second Supplementary Tables of Bimolecular Gas Reactions*. Department of Chemistry, The University, Birmingham, England, 1973.

J. A. Kerr and E. Ratajczak, *Third Supplementary Tables of Bimolecular Gas Reactions*. Department of Chemistry, The University, Birmingham, England, 1977.

E. Ratajczak and A. F. Trotman-Dickenson, *Supplementary Tables of Bimolecular Gas Reactions*. Institute of Science and Technology, University of Wales, Cardiff, 1970.

A. F. Trotman-Dickenson and G. S. Milne, *Tables of Bimolecular Gas Reactions*, NSRDS-NBS 9. U.S. Govt. Printing Office, Washington, D.C., 1967.

J. Warnatz, "Chemistry of Stationary and Non-Stationary Combustion." In *Modelling of Chemical Reaction Systems* (K. H. Ebert, P. Deuflhard, and W. Jäger, eds.). Springer-Verlag, Berlin, 1981.

J. Warnatz, "Survey of Rate Coefficients in the C/H/O System." In *Chemistry of Combustion Reactions* (W. C. Gardiner, Jr., ed.). Springer-Verlag, New York, 1983, p. 197.

C. K. Westbrook and F. L. Dryer, "Chemical Kinetics Modeling of Hydrocarbon Combustion." *Prog. Energy Combust. Sci.*, **10**, 1 (1984).

F. Westley, *Tables of Recommended Rate Constants for Chemical Reactions Occurring in Combustion*, NSRDS-NBS 67. U.S. Govt. Printing Office, Washington, D.C., 1980.

The bibliography cannot possibly be complete and is meant as a good entry into the appropriate literature. Evaluations and compilations of data can be very useful.

APPENDIX: ANSWER SHEETS

A.1. PROBLEMS USING METHODS OF SECTION 2

1.1. Thermochemistry and Equilibrium Constants from Additivity Rules

1. Using bond additivity (Table 1), compute C_p^0, S^0, and ΔH_f° for methyl ethyl ketone (MEK) at 25°C ($CH_3\overset{\overset{\displaystyle O}{\|}}{C}CH_2CH_3$).

	C_p^0/eu	S^0/eu	ΔH_f°/kcal mol^{-1}
8 C—H	8 (1.74)	8 (12.90)	8 (−3.83)
1 C—C	1.98	−16.40	2.73
2 $\diagdown$CO—C $\diagup$	2 (3.7)	2 (−0.6)	2 (−14.4)
	23.3	85.6	− 56.7
−R ln 9		−4.4	
		81.2	
Observed	24.7	80.8	− 57.0

2. Using group additivity (Tables 2–7), compute C_p^0, S^0, and ΔH_f° for

MEK at 25°C ($CH_3\overset{\overset{\displaystyle O}{\|}}{C}CH_2CH_3$).

	C_p^0/eu	S^0/eu	ΔH_f°/kcal mol^{-1}
$2(C-(CO)(H)_3) = 2C-(C)(H)_3$	2 (6.19)	2 (30.41)	2 (−10.20)
$CO-(C)_2$	5.6	15.0	−31.4
$C-(CO)(C)(H)_2$	6.2	9.6	−5.2
	24.2	85.4	−57.0
$-R \ln 9$		−4.4	
		81.0	
Observed	24.7	80.8	−57.0

3. Calculate the equilibrium constant for the reaction:

$$C(CH_3)_3CH_2OCH_2\phi + HC(CH_3)_2(C-C_3H_5)$$
$$\rightleftarrows (ortho)-CH_3-C_6H_4-CH_2OH + HC(CH_3)_2CH_2C(CH_3)_2C(CH_3)CH_2$$

at 300 K and at 800 K.

First, some bookkeeping:

LEFT-HAND SIDE	RIGHT-HAND SIDE

(a) $CH_3-\overset{\overset{\displaystyle CH_3}{|}}{\underset{\underset{\displaystyle CH_3}{|}}{C}}-CH_2-O-CH_2-O$ $(\sigma = 3^3 \times 2)$

(c) benzene ring with $-CH_3$ and $-CH_2-OH$ $(\sigma = 3)$

(b) $H-\overset{\overset{\displaystyle CH_3}{|}}{\underset{\underset{\displaystyle CH_3}{|}}{C}}-\triangleleft$ $(\sigma = 3^2)$

(d) $H-\overset{\overset{\displaystyle CH_3}{|}}{\underset{\underset{\displaystyle CH_3}{|}}{C}}-CH_2-\overset{\overset{\displaystyle CH_3}{|}}{\underset{\underset{\displaystyle CH_3}{|}}{C}}-C\overset{\diagup CH_3}{\diagdown CH_2}$ $(\sigma = 3^5)$

(a)	(b)	(c)	(d)
$C-(C)(H)_3$	$2C-(C)(H)_3$	$C-(C_B)(H)_3$	$5C-(C)(H)_3$
$C-(C)_4$	$2C-(C)_3(H)$	$2C_B-C$	$C-(C)_3(H)$
$C-(C)(O)(H)_2$		$4C_B-H$	$C-(C)_2(H)_2$
$O-(C)_2$	$2C-(C)_2(H)_2$	$C-(C_B)(H)_2(O)$	$C-(C)_3(C_d)$
$C-(C_B)(O)(H)_2$	$\triangle$	$O-(C)(H)$	$C_d-(C)_2$
C_B-C	(no *gauche*?)	*ortho*	$C_d-(H)_2$
$4C_B-H$			2-*gauche*
2-*gauche*			1-alkene *gauche*

LEFT-HAND SIDE	RIGHT-HAND SIDE
$5C-(C)(H)_3$	$6C-(C)(H)_3$
$2C-(C)_2(H)_2$	$C-(C)_2(H)_2$
$2C-(C)_3(H)$	$C-(C)_3(H)$
$C-(C)_4$	$C-(C)_3(C_d)$
$C-(C)(O)(H)_2$	$C_d-(C)_2$
$C-(C_B)(O)(H)_2$	$C_d-(H)_2$
$O-(C)_2$	$2C_B-C$
C_B-C	$4C_B-H$
$5C_B-H$	$C-(C_B)(H)_2(O)$
$\triangle$	$O-(C)(H)$
2-*gauche*	2-*gauche*
	1-alkene *gauche*
	ortho

NET

$C-(C)_2(H)_2$	
$C-(C)_3(H)$	$C-(C)(H)_3$
$C-(C)_4$	$C-(C)_3(C_d)$
$C-(C)(O)(H)_2$	$C_d-(C)_2$
$O-(C)_2$	$C_d-(H)_2$
C_B-H	C_B-C
$\triangle$	$O-(C)(H)$
	alkene *gauche*
	ortho

LEFT-HAND SIDE	ΔH_f°	S_{int}	$C_{p,3}^0$	$C_{p,8}^0$
$C-(C)_2(H)_2$	-4.93	9.42	5.50	11.07
$C-(C)_3(H)$	-1.90	-12.07	4.54	9.31
$C-(C)_4$	0.50	-35.10	4.37	8.77
$C-(C)(O)(H)_2$	-8.1	9.8	4.99	11.11
$O-(C)_2$	-23.2	8.68	3.4	4.4
C_B-H	3.30	11.53	3.24	7.54
$\triangle$	27.6	32.1	-3.05	-1.77

- -

RIGHT-HAND SIDE				
$C-(C)(H)_3$	-10.20	30.41	6.19	13.02
$C-(C)_3(C_d)$	1.68	-34.72	3.99	8.02
$C_d-(C)_2$	10.34	-12.70	4.10	5.80
$C_d-(H)_2$	6.26	27.61	5.10	10.07
$C_B-(C)$	5.51	-7.69	2.67	4.96
$O-(C)(H)$	-37.9	29.07	4.3	6.0
alkene *gauche*	0.50	—	—	—
ortho	0.57	-1.61	1.12	0.88
RHS $-$ LHS	-16.47	6.0	4.5	-1.7

$$+R \ln \frac{\sigma_{\text{LHS}}}{\sigma_{\text{RHS}}} = R \ln \frac{3^5 \times 2}{3^5 \times 3} \qquad -0.8$$

$$= -0.8 \text{ eu} \qquad\qquad \overline{5.2}$$

$$\Delta H_{800} = \Delta H_{300} + \frac{(4.5 - 1.7)}{2}(0.3) = -16.47 + 0.42 = -16.05 \text{ kcal mol}^{-1}$$

$$\Delta S_{800} = \Delta S_{300} + \frac{(4.5 - 1.7)}{2} \ln \frac{800}{300} = 5.2 + 1.4 = 6.6 \text{ eu}$$

$$K_{300} = \exp\left\{ \frac{5.2}{1.99} + \frac{16.47}{1.99(0.3)} \right\} = 1.3 \times 10^{13}$$

$$K_{800} = \exp\left\{ \frac{6.6}{1.99} + \frac{16.05}{1.99(0.8)} \right\} = 6.6 \times 10^{5}$$

1.2. Structural Estimates of Entropy and Heat Capacity

1. Compute the entropy of *trans*-butane at 25°C from statistical thermodynamics. The product of inertia of *trans*-butane is 4.3×10^5 amu^3-Å^6.

$$S^0_{\text{trans}} = 37.0 + \frac{3}{2} R \ln \frac{58}{40} = 38.1$$

$$S_{\text{rot}} = 11.5 + \frac{R}{2} \ln \frac{4.3 \times 10^5}{(2)^2} = 23.1$$

See notes for frequencies.

$$S_{\text{vib}} \text{ for all but internal rotors } = 2.7$$

$$2\text{CH}_3 \text{ rotors at } v = 3 \text{ kcal mol}^{-1} = 4.4$$

$$1Et \text{ rotors at } v = 3 \text{ kcal mol}^{-1} = 3.9$$

$$\overline{72.2}$$

Real butane (a mixture of *trans* and *gauche*) is greater by ~ 1.4.

2. Estimate the entropy of the following compounds from model compounds: C_2H_5NO, $CH_3OOOOCH_3$.

$$C_2H_5\text{—N}=\text{O} \rightarrow C_2H_5\text{—CH}=\text{CH}_2 \qquad S^0 = 63.8$$

Subtract a little for CH_2 torsion; no symmetry change.

$$CH_3OOOOCH_3 \rightarrow CH_3(CH_2)_4CH_3 \qquad S^0 = 93.0 \text{ [rotation barriers?]}$$
$$+4.0?$$

$$\rightarrow CH_3O(CH_2)_2OCH_3$$

$$2[\text{C—(O)(H)}_3 + \text{O(C)}_2 + \text{C(C)(O)(H)}_2]$$

$$2[30.41 + 8.68 + 9.8] = 97.8 - R \ln 18 = 92$$

3. Using various estimation techniques, calculate the equilibrium constant at 25°C for the reaction: [BDE HO_2—H = 87.2 kcal mol and internuclear distance in O_2 is 1.2 Å].

$$CH_3CH=CH_2 + O_2 \rightleftharpoons HO_2 + CH_2\text{—CH—CH}_2$$

Compare with thermodynamic values in *Thermochemical Kinetics*.

$$CH_3CH=CH_2$$

	ΔH_f°	S_{int}^0
$C-(C)(H)_3$	-10.2	30.41
$C_d-(C)(H)$	8.59	7.97
$C_d-(H)_2$	6.26	27.61
Σ	4.65	65.99
$-R\ln 3$		63.8

$$O_2 \qquad \Delta H_f^\circ \equiv 0$$

$$S_{trans}^0 = 37.0 + \frac{3}{2}R\ln\frac{32}{40} = 36.3$$

$$S_{rot} = 6.9 + R\ln\left[\frac{11.52}{2}\right] = 10.4$$

$$[I = \mu r^2 = \frac{16\times16}{16+16}(1.2)^2 = 11.52]$$

$$S_{vib} \approx 0$$

$$S_{elec} = R\ln 3 = 2.2 \qquad S_{O_2} = 48.9$$

$$HO_2$$

$$\Delta H_f^\circ: HOOH \to HO_2 + H$$

$$\Delta H = 87.2 = X + 52.1 - (-32.6)$$

$$X = 2.5$$

$$S^0 \sim S^0(HOOH) + R\ln(\uparrow) + R\ln\sigma - R\ln n - \tau_{OH}$$
$$56 \quad + \quad 1.4 \quad - \quad 1.4 \quad + \quad 1.4 \quad - \quad 1.7$$
$$\sim 55.3$$

$$C_3H_5$$

$$\Delta H_f^\circ \text{ from } C_3H_5-H = \text{pri } C-H-\text{allyl stabilization}$$

$$C_3H_6 \to C_3H_5 + H$$

$$\Delta H = (98-12) = 86 = X + 52.1 - (4.9)$$

$$X \cong 39$$

S^0 from $\wedge \to \diagup\diagdown$.

$$S^0(C_3H_6) = 63.8$$

$CH_3 \longrightarrow CH=CH_2$	$3.6 - .9 = 2.7$ eu
$v=2$	
$\dot{C}H_2 \longrightarrow CH=CH_2$	$3.1 - 3.1 = 0$
$v=14$	
	-2.7

Spin increases entropy by 1.4

$$S^0(C_3H_5) \approx S^0(C_3H_6) - 2.7 + 1.4 = 62.5$$

[If we assume that rotor goes to $\tau \approx 800$ cm^{-1} as a maximum stiffness (i.e., same as propylene), there is a difference of 0.2 eu. If torsions both go to 400 cm^{-1} (i.e., $\overset{.}{C}$—C—CH$_3$), then allyl entropy could be as high as 64.5 eu.]

	CH$_3$CH$=$CH$_2$	+ O$_2$	$\rightleftarrows$	HO$_2$	+ Allyl		Δ
$\Delta H_f^°$	4.65	0		2.5	39		36.8
S^0	63.8	48.9		55.3	62.5		5.1

$$K = \exp\left[\frac{5.1}{1.99} - \frac{36.8}{1.99(0.3)}\right] = 2.2 \times 10^{-26}$$

This is low, but consider oxidation initiation at higher T's.

4. Calculate the enthalpy change at 300 K for the reactions:

   (a) $\phi{}^. + O_2 \rightleftarrows \phi O{}^. + O$

   (b) $\phi O{}^. \longrightarrow$ $+$ CO

   (c)  $+ O_2 \rightleftarrows$ $+ O$

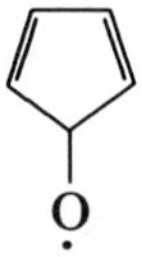

   (d) 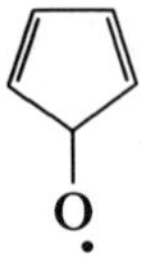 $\rightleftarrows$

   (a) $\phi{}^. + O_2 \rightarrow \phi O{}^. + O$
         $78.5 + 0 \rightarrow 10 + 59$

$$\Delta H_{300} = -10$$

   (b) $\phi O{}^. \rightarrow$ $+$ CO
         $10 \rightarrow 61 - 26$
         $\Delta H = 25$

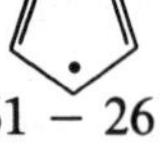

   (c)  $+ O_2 \rightarrow$ $+$ O
         $61 + 0 \rightarrow$ $(47) + 59$
         $\Delta H = 45$

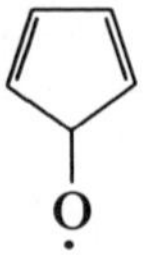

$$2C_d\!-\!(C_d)(H) \qquad 2 \times 6.8 = 13.6$$
$$2C_d\!-\!(C)(H) \qquad 2 \times 8.6 = 17.2$$
$$[C\!-\!(C_d)_2O \qquad\qquad -4.0]$$
$$O\!-\!(C)(H) \qquad\qquad -36.0$$

$$6.0$$
$$\overline{-5.2}$$

$$-5.2$$

$$\qquad 52$$

$$\text{—OH} \rightarrow \qquad \text{—O} \cdot + \text{H} \qquad \Delta H = 104$$

$$104 = 52 + X - (-5.2); \quad X = 47$$

$$47 \qquad\qquad\qquad 6$$

(d) $\rightleftarrows$ $\qquad\qquad \Delta H = -41$

$$C_d\!-\!(C)(H) \qquad\qquad 8.6$$
$$C\!-\!(C_d)(C)(H)_2 \qquad -4.8$$
$$C_d\!-\!(H)(CO) \qquad\qquad 5.0$$
$$C\!-\!(C)(CO)(H)_2 \qquad -5.2$$
$$(CO)\!-\!(C_d)(C) \qquad (-31.0)$$

$$5.9$$
$$\overline{-21.5}$$

$$\text{=O} \rightarrow \text{H} + \text{=O}$$

$$-21.5 \qquad 52 \qquad X \qquad \cong 95 - 15 = 80$$

$$52 + X + 22 = 80$$
$$X = 6$$

A.2. PROBLEMS USING METHODS OF SECTIONS 2 AND 4

1. Estimate the A-factor for the reaction, $CH_3 + NH_3 \rightarrow CH_4 + NH_2$ (at 300 K).

$$CH_3 + NH_3 \rightarrow [CH_3 \cdot H \cdot NH_2] \rightarrow CH_4 + NH_2$$

MODEL TRANSITION STATE	LINEAR TS	BENT TS
$H\!\!-\!\!C\!-\!N\text{----}H$ (with H, H, H)	$CH_3 \cdot H \cdot NH_2$	$CH_3 \quad NH_2$ (with H)
$3N - 6 = 15$	$3N - 6 = 18$	$3N - 6 = 18$

6 *Stretches*	7 *Stretches*	
3 CH	3 CH	3 CH
2 NH	2 NH	2 NH
CN	C · H (r.c)	C · H (r.c)
	N · H	N · H

9 *Bends*	11 *Bends*	11 *Bends*
CH$_3$: 3 HCH	CH$_3$: 3 HCH	CH$_3$: 3 HCH
2 rocks	2 rocks	2 rocks
NH$_2$: HNH	NH$_2$: HNH	NH$_2$: HNH
wag	wag	wag
rock	rock	rock

1 Hindered rotor	*1* Hindered rotor	*2* Hindered rotors
	2 C · H · N	*1* C · H · N

$$\frac{\Delta S^{\ddagger}_{300}}{-34.8}$$

Model $R_{\chi}[\text{CH}_3 \cdot \text{NH}_3 \rightarrow \text{CH}_3\text{NH}_2]$

Spin	($R \ln 2$)	1.4
External rotation	(2 × 2 × 1.5)	1.7
C—N (1100 cm^{-1}) → C · H(r.c.) + N · H(2200)		0.0
CH$_3$—NH$_2$ rotor → NH$_2$-free rotor		1.7
(2)C · H · N (600 cm^{-1})		1.0 linear
CH$_3$—NH$_2$ rotor → NH$_2$-free rotor + CH$_3$-free rotor		5.3
C · H · N (600 cm^{-1})		0.5 bent

$$-29.0 \text{ linear}; \ -25.9 \text{ bent}$$

$$\log A = 13.2\frac{-29.0 + 8.35}{4.58} = 8.7 \text{ linear}$$

$$= 13.2\frac{-25.9 + 8.35}{4.58} = 9.4 \text{ bent}$$

2. From knowledge of the fact that the rate constant for O + CH$_4$ → OH + CH$_3$ (between 350 and 1000 K) is given by $\log k/M^{-1} \text{ s}^{-1} = 10.32 - 9.04/\theta$, estimate k at 1500 and 2000 K.

Model TS CH$_3$F surrogate for CH$_3$O, which could have been obtained from CH$_3$OH by pulling H.

Activation Parameters for O + CH$_4$ (in cal mol^{-1} K^{-1})

	$\Delta S^{\ddagger}_{300}$	$\Delta C^{\ddagger}_{p,400}$	$\Delta C^{\ddagger}_{p,800}$	$\Delta C^{\ddagger}_{p,1250}$	$\Delta C^{\ddagger}_{p,1750}$
Model reaction	−29.7	−4.30	−3.60	−3.95	−4.25
(TS = CH$_3$F)					
Spin ($R \ln 2$)	1.37				
External rotation	1.93				
Translation	−0.10				
(C · F) 1100 → $r \times n$ coord.	−0.10	−0.63	−1.45	−1.72	−1.90
(O · H) 2000	0	+0.08	+0.74	+1.17	+1.43
2(O · H · C)600 (linear)	1.0	2 × 1.57	2 × 1.80	2 × 1.90	2 × 1.98
1(O · H · C)600 + internal rotation]	0.15 + 3.60	1.37 + 1.0	1.90 + 1.0	1.90 + 1.0	1.98 + 1.0
(bent)					
3(FCH)1200 → [O · H]CH, 700	+0.9	2.10	1.11	0.60	0.18
Linear TS	−23.05	0.00	+0.40	−0.10	−0.58
Bent TS	−19.95	−0.38	−0.40	−1.00	−1.56

Arrhenius Parameters for O + $\overset{3}{C}H_4 \rightarrow CH_3 + OH$

	300 (T/K)	500 (T/K)	1000 (T/K)	1500 (T/K)	2000 (T/K)
$\Delta S^{\ddagger}$ (cal mol^{-1} K^{-1})					
Linear	−23.05	−23.05	−22.77	−22.81	−22.98
Bent	−19.95	−20.14	−20.42	−20.82	−21.27
Log(A_T) (M^{-1} s^{-1})					
Linear	10.01	10.46	11.12	11.46	11.68
Bent	10.69	11.09	11.63	11.90	12.05
$\Delta H_T^{\ddagger}$ (kcal mol^{-1})					
Linear	7.52	7.52	7.72	7.67	7.38
Bent	8.46	8.38	8.18	7.68	6.90
E_T (kcal mol^{-1})					
Linear	8.72	9.52	11.72	13.67	15.38
Bent	9.66	10.38	12.18	13.68	14.90
k_2 (cm^{-3} s^{-1})					
Linear	7.4×10^{-18}	3.3×10^{-15}	6.1×10^{-13}	4.9×10^{-12}	1.7×10^{-11}
Bent		6.0×10^{-15}	1.6×10^{-12}	1.4×10^{-11}	4.4×10^{-11}
According to Roth and Just[6]	7.4×10^{-18}	3.6×10^{-15}	7.0×10^{-13}	5.8×10^{-12}	2.0×10^{-11}

3. Calculate the high-pressure A-factor for the reaction

$\phi CH_3 \rightarrow \phi CH_2 + H$ from the fact that the A-factor for $\phi CH_2CH_3 \rightarrow$ $\phi CH_2 + CH_3$ is $10^{15.8}$ s^{-1} at 1000 K.

$$\phi CH_2CH_3 \longrightarrow \phi CH_2 \cdots CH_3$$

$$\log A_{1000} \sim 15.8 \rightarrow \Delta S^{\ddagger} \sim 9.2 \text{ eu}$$

What changes can we expect?

External rotation $(2 \times 2 \times 1.5)$	$+1.7$
$\phi CH_2 - CH_3$ (1000 cm^{-1}) $\rightarrow$ r.c.	-1.4
$\phi CH_2 - CH_3$ (int rot) $\rightarrow$ (free rot)	$+0.2$
$\phi - CH_2CH_3$ $\left\{ \begin{array}{l} \text{(int rot } v = 3) \rightarrow (v = 12) \\ \text{increase } I \text{ by } (2.8)^2 \end{array} \right\}$	$+0.4$
$\left. \begin{array}{l} 2CH_3 \text{ rocks (1150 cm}^{-1}) \rightarrow \text{looser rocks} \\ C-C-C \text{ bend (380 cm}^{-1}) \rightarrow \text{looser bend} \\ C-C-C \text{ bend (150 cm}^{-1}) \rightarrow \text{looser bend} \end{array} \right\}$	$[9.1]$
	$\overline{9.2}$

$$2(1150) = 2.0 \rightarrow 2(280) \rightarrow 7.8$$
$$1 \ (380) = 3.3 \rightarrow 1 \ (90) \rightarrow 6.4$$
$$1 \ (150) = 1.9 \rightarrow 1 \ (90) \rightarrow 1.4$$
$$\overline{7.2} \qquad \qquad \sim \overline{15.6}$$

For $\phi CH_3 \rightarrow \phi CH_2 \cdots H$, what changes do we expect?

External rotation	(1.0)
$\phi CH_2 - H$ (3000 cm^{-1}) $\rightarrow$ r.c.	-0.1
$\phi \text{---} CH_3$ (int rot v = 0) $\rightarrow$ (v = 12)	-1.7
2HCH bends (1450 cm^{-1}) $\rightarrow$ 2 at $[\frac{1}{4}\nu = 350]$	$+5.2$
	$\overline{+4.4}$

It's hard to imagine that $\log A_{1100} \widetilde{>} 15.0$.

4. (a) Compute k_1 for the reaction $CH_4 + Ar \rightarrow CH_3 + H + Ar$ at 1000 K, if $E_0 = 3.6 \times 10^4 \ cm^{-1}$, β for $Ar \approx 0.1$, $Z = 3 \times 10^{+14} \ cm^3 \ mol^{-1} \ s^{-1}$, $Q_{vib} = 2$, $\nu = [3000(4), 1400(5) \ cm^{-1}]$.

$$k_1 = \frac{\beta Z \rho_{vib}(E_0) kT}{Q_{vib}} e^{-E_0/kT} \cdot F_{rot} \cdot F_{int \ rot}$$

$$\rho_{vib}(E_0) = \frac{(E_0 + 0.9E_z)^{s-1}}{(s-1)!} \prod_{i=1}^{s} (h\nu_i)$$

$$E_0 = 103 \ \text{kcal mol}^{-1} = 3.6 \times 10^4 \ cm^{-1}$$

$$E_z = \frac{1}{4}[4(3000) + 5(1400)] = 9.5 \times 10^3; \ 0.9E_z = 8.6 \times 10^3$$

$$\rho = \frac{(4.5 \times 10^4)^8}{8!}(3 \times 10^3)^4(1.4 \times 10^3)^5 = 9.7 \times 10^2 \ (cm^{-1})^{-1}$$

$$F_{rot} = \frac{8!}{9.5!}\left(\frac{4.5 \times 10^4}{702}\right)^{3/2}\left[\frac{2.15(51.4)^{1/3}}{2.15(51.4)^{1/3} - 1 + (4.5 \times 10^4)/(9.5)(702)}\right]$$

$$F_{rot} = \frac{4.13 \times 10^4}{1.13 \times 10^6}(64.1)^{3/2}[0.58] = 10.6$$

$$k_1 = \frac{(0.1)(3 \times 10^{14})(9.7 \times 10^2)(702)}{2}(e^{-51.4})(10.6)(1)$$

$$= 5.2 \times 10^{-3} \ \frac{cm^3}{mol - sec}$$

(b) Make a reasonable estimate of k_∞ for the same reaction.

$$k_\infty \sim 10^{15 - 104/\theta} = 2 \times 10^{-8} \ \text{at 1000 K}$$

(c) Compute the pressure at which simple L–H theory predicts $k = k_\infty/2$.

$$[M] = \frac{k_\infty}{k_1} + \frac{2 \times 10^{-8}}{5.2 \times 10^{-3}} = 4 \times 10^{-4} \ \frac{mol}{cm^3} \approx 10 \ \text{atm}$$

5. At 298 K, t-butyl radicals combine with Br atoms to give a vibrationally hot t-butyl bromide. Estimate the branching ratio this hot compound going back to t-butyl and Br-atom compared to going to HBr + isobutene (i.e., k_{-1} vs. k_2):

$$Br + \left. \right| \cdot \underset{k_{-1}}{\overset{k_1}{\rightleftharpoons}} \left(\left. \right| - Br\right)_* \overset{k_2}{\longrightarrow} \Big\rangle\!\!= + HBr$$

$$\Delta H_1 = \Delta H_f(t\text{-bu-Br}) - [\Delta H_f(t\text{-bu}) + \Delta H_f(Br)] \approx 68 \ \text{kcal mol}^{-1}$$

$$\Delta S_1 = 33.5 \ \text{eu}$$

Let's guess $A_1 \sim 10^{10.5}$, $\log \dfrac{A_1}{A_{-1}} = \dfrac{33.5 - 8.4}{4.6} = 5.5$; therefore $A_{-1} \cong 10^{16.0}$,

$A_2 \sim 10^{13.5}$, $E_2 = 42$ kcal

$$k = A \frac{n!(n - m + s - 1)!}{(n - m)!(n + s - 1)!}$$

$\langle \nu/\text{cm}^{-1} \rangle \approx 1000$

$n = \dfrac{68}{2.85} \sim 24$ quanta

$m_{-1} = 24$; $m_2 = \dfrac{42}{2.85} \approx 15$; $s = 36$

$$\frac{k_{-1}}{k_2} = \frac{10^{16.0}}{10^{13.5}} \frac{35!}{0!} \cdot \frac{9!}{42!} = 10^{2.5} \cdot 10^{-5.8} = 10^{-3.1}$$

6. The following rate constants have been used as part of a model for methanol oxidation. (Units are $\text{cm}^3 \text{ mol}^{-1} \text{ s}^{-1}$ for A and kcal mol^{-1} for E.)

	Log A	E
1. $CH_3OH + O_2 \rightarrow CH_2OH + HO_2$	13.6	50.9
2. $CH_3OH + OH \rightarrow CH_2OH + H_2O$	12.6	2.0
3. $CH_3OH + O \rightarrow CH_2OH + HO$	12.2	2.3
4. $CH_3OH + H \rightarrow CH_2OH + H_2$	13.5	7.0
5. $CH_3OH + H \rightarrow CH_3 + H_2O$	12.7	5.3
6. $CH_3OH + CH_3 \rightarrow CH_2OH + CH_4$	11.3	9.8
7. $CH_3OH + HO_2 \rightarrow CH_2OH + H_2O_2$	12.8	19.4

As a final exercise, comment on the absolute values and the relative values of these parameters.

1. A-factor is high—compare with 2.
   E is also high—$\Delta H \sim 45$ kcal.
2. A-factor looks good.
   E looks a touch high.
3. A-factor looks low; compare with 2.
   E could be low; O-atom usually has a few kcal barrier.
4. Looks good; A-factor about expected for an atom.
5. Unlikely; molecular gymnastics.
6. Looks normal.
7. A-factor looks high; compare 6.
   E looks high; $\Delta H \sim +7$

3 The Phenomenology of Modeling Combustion Chemistry

FREDERICK L. DRYER

Department of Mechanical and Aerospace Engineering
Princeton University
Princeton, New Jersey

1. Introduction
2. Types of chemical mechanisms
   2.1. Unbranched (straight) chain reactions
   2.2. Branched chain reactions
   2.3. Degenerate branching
3. Organic molecules, chemical kinetics, and combustion phenomena
   3.1. Low-temperature phenomena
   3.2. "Low-temperature" hydrocarbon oxidation mechanism
   3.3. Intermediate temperature reaction phenomena
   3.4. High-temperature reaction phenomena
4. Some thoughts about combustion applications
   4.1. Detailed chemical kinetic modeling
   4.2. Methodology for deriving and validating detailed mechanisms
5. Empirical modeling concepts
   5.1. One- and two-step global mechanisms
   5.2. Multistep global and quasi-global mechanisms
   5.3. Detailed kinetic models as sources of reduced mechanisms
6. Closing remarks
   References

This chapter was originally prepared as part of an Exxon Engineering Education lecture series in 1981. The article was revised in the spring of 1988 in preparation for publication as part of this book. This chapter should not be considered a "review" of the literature but a tutorial that draws from the literature data to provide a general perspective on the subject. The field of chemical kinetics is progressing so rapidly that after first gaining these general insights, it is important to refer to the most recent literature in forming the basis for further work. The reader is therefore cautioned in using detailed mechanisms and/or numerical data presented herein as resources in future efforts.

1. INTRODUCTION

Fuels, through their physical and chemical properties, have profound influences on defining the performance and emissions that can be achieved with various energy conversion technologies. Although the combustion properties of fuels have been studied empirically for many years, early numerical combustion modeling virtually ignored chemistry. In the last decade numerical modeling has rapidly become an essential part of many combustion research and development programs, and there has been an accelerating evolution from the use of single-step empirical representations, to the use of lumped (overall) multistep models, and finally to the inclusion of full detailed chemical kinetic mechanisms to better simulate chemistry interactions.[1-3]

Early numerical models were used primarily to assist in the interpretation of specific laboratory scale experiments. Indeed, computer analysis of idealized problems has often provided useful insights to combustion problems that have been too complex to approach on an experimental basis. In contrast, the goal of much of the more recent modeling work has been the construction of numerical representations that can assist directly in the design of practical combustor systems, namely, numerical data that would be predictive as well as interpretive. Given the wide range of operating parameters experienced in most conventional combustion situations, it is clear that a suitable physical and numerical model for each of the many physical and chemical processes (atomization, vaporization, turbulence, heat transfer, chemical kinetics, etc.) must be embodied in this representation. Such "submodels" should be able to predict both independent phenomena as well as coupled physical and chemical processes (i.e., turbulence and chemistry), if there is ever to be much credibility placed on their collective use in determining the combustion behavior of complete systems. However, this premise is not as simple to put into practice as it would seem, nor has it been a common underlying philosophy in developing submodels. For example, in some chemical models used for predicting flame propagation, first-order dependence of the reaction rate on fuel and oxidizer concentrations have been assumed a priori. More recently, it has been recognized that such an assumption can yield neither the proper pressure dependence nor the flammability limits for laminar flame propagation through hydrocarbon/air mixtures. In what follows, several techniques for describing the chemistry that occurs in combustion systems will be discussed. At the most complex level are detailed kinetic reaction mechanisms, namely, those descriptions which involve large sets of elementary reactions. Availability of large amounts of elementary kinetic data, improved techniques for estimating specific rate information, development of "stiff equation" solution techniques, and the introduction of more efficient parameter sensitivity approaches have contributed to the increasingly rapid development and application of detailed kinetic modeling. The assembly, validation, and use of detailed kinetic mechanisms continues to grow.[1,3] Detailed kinetic mechanisms provide useful tools for determining what elementary processes are most important and require improved definition,[4-9] in acting as benchmarks from which simplified empirical chemical mechanisms can be developed[10] and against which they may be tested,[11] and in studying fundamental issues such as oscillatory ignition phenomena[12-15] and the interaction of chemistry with fluid mechanics, diffusive transport, etc., in spatially well-defined systems such as laminar flames,[16-24] doubly premixed and pure diffusion flames,[25-28] and well-stirred reactors.[29, 30]

Kinetic mechanisms having substantial detail or simpler descriptions derived from them will probably be required for clearly understanding a number of practical combustion problems of current interest, including

1. transition to fast chemical reaction
   (a) spark ignition
   (b) compression ignition
   (c) engine knock
   (d) transition from deflagration to detonation
   (e) explosion
2. flame quenching and hydrocarbon emissions
   (a) by walls
   (b) by charge stratification
   (c) by rapid expansion
3. flame inhibition and extinction
4. flame nitrogen conversion
5. thermal NO_x production within-flame structures
6. SO_2/SO_3 conversion in postflame gases
7. soot formation particularly for aromatics
8. soot oxidation
9. catalytic processes

Yet we are faced with some serious difficulties in developing and applying the required mechanisms. First, detailed mechanisms have been developed and validated for only a relatively small number of simple fuel molecules. Principal efforts have concentrated on normal and iso-paraffins, ethylene, and acetylene. The associated, larger, olefin mechanisms which are required continue to create significant difficulties in constructing these models. Little effort has been devoted to hydrocarbon oxygenates, and only phenomenological mechanisms have appeared on other fuel structures such as aromatics. Thus there is a need for some type of alternative description for many complex fuel molecules.

Second, numerical models of combustors that consider two- and three-dimensional geometry cannot currently include detailed kinetic mechanisms in a direct way. This problem is primarily a result of the fact that even state of the art scientific computers have size and speed limitations. In the most direct method of formulation, the concentration of each chemical species in a reaction mechanism is represented by a single differential equation; common techniques of solution replace this differential equation with a finite difference equation. If the problem is one in which only chemical kinetics need be considered and spatial variations of physical quantities could be neglected, then the number of equations to be solved simultaneously at each time step is equal to the number of species, plus an equation for energy or enthalpy. Such sets of equations, which are coupled through nonlinear reaction rate terms of widely differing time constants, are already often difficult to solve in an efficient and accurate manner. When spatial variations of all quantities must be solved at each point in the physical domain, usually represented by a finite difference spatial grid, the total number of equations to be solved, the amount of computational time, and the total memory requirements increase very rapidly. All of the limits will be severely tested even when spatial variations are restricted to one dimension. When two or three spatial dimensions must be considered, the number of species considered is constrained to *less than ten* and more likely less than five. This constraint is particularly important when any type of parametric numerical study is envisioned. These simplified representations have frequently been derived directly from experimental data, but analytical models may also be developed from detailed mechanisms as well.

Finally, there are occasions where only a limited amount of description of the combustion chemistry is required; namely, only rates of energy release, flame propagation speeds, or flammability conditions are to be predicted. Even under these simple conditions, certain physical and thermodynamic constraints must be considered carefully.

It should be emphasized here that regardless of the apparent levels of sophistication of the various chemical models alluded to above, all levels of chemical kinetic descriptions are in a real sense only approximations. Even the most detailed elementary kinetic mechanisms are constructed on the basis of reproducing observable phenomena that are not all-encompassing. In some cases important species and even the conceptual description of the mechanism remain unidentified; furthermore the independent evaluation of specific rate constants (or overall reactions) over ranges of temperature and pressure adequate for combustion modeling is an ever-present source of difficulty. Thus it should not be assumed a priori that any validated mechanism can be dissected and its parts used in describing wholly independent phenomena, unless it was conversely constructed in such a manner. Indeed, the extension of mechanisms to parameter ranges outside those for which they were developed should be approached with much skepticism.

This chapter, then, will attempt to provide some general background and overview of combustion chemistry, and of the tools and methods available for developing, including, and understanding the role of chemistry in combustion analysis. The approach will be to first discuss the phenomenology of combustion chemistry, namely, the characteristics that are to be modeled. This will be followed by brief discussion of both detailed and semiempirical, simplified kinetic approaches, with emphasis on the methods used to develop such mechanisms and the state of the art in available kinetic models with examples of applications such as have appeared in recent and complementary publications.[1,3]

2. TYPES OF CHEMICAL MECHANISMS

Before beginning a discussion of combustion chemistry, we will quickly review some of the various characteristics typical of hydrocarbon oxidation systems. First, various types of chain reaction mechanisms are often experienced. The early concepts (pre-1930) of nonchain type schemes for hydrocarbon oxidation gave way under further scientific scrutiny to this concept. A wide and exhaustive study of chain theory was initiated in the late 1920s by Semenov and his coworkers in response to the fact that nonchain theories could not explain the rapidity of low-temperature hydrocarbon oxidation, the ignition character variation of hydrocarbon oxidation with temperature and pressure, "cool" flame behavior, nor the negative temperature coefficient of the low-temperature oxidation rate.[31]

The basic premise of chain theory is that active centers (free radicals and atoms) play a leading role in the destruction of reactant molecules. The free electron of the active center acts on the bonds in the molecule with which it is attempting to react, thus giving it a high chemical reactivity with valency-satisfied molecules. The indestructability of valency in such a reaction leads to the production of a new active center to replace the one that initially reacted and to continue the reaction process. The reactions are termed "chain propagation reactions." Thus a chain transformation arises in which only the generation of the first radical (chain initiation reactions) involves significant endothermicity, whereas ensuing destruction of the reactant molecules involves only small or no endothermicity. Such chains will be terminated by the reaction of two active centers again producing a valency-satisfied species (termination reactions). A third body is often needed in such reactions to remove from the reaction site energy that might otherwise lead to dissociation of the valency-satisfied species. This type of chain reaction termination is often referred to as "quadratic" breaking. Active centers can also be removed by absorption on surfaces or by reaction with impurities to form weakly reactive centers that later recombine ("linear" chain termination). Chain reaction mechanisms can be considered to be one of three general types: unbranched (straight), branched, or degenerate.

2.1. Unbranched (Straight) Chain Reactions

The unbranched chain mechanism is composed only of initiation, propagation, and termination reactions. A good example is the mechanism for the nonilluminated overall reaction

$$H_2 + Cl_2 = 2HCl:$$

$$Cl_2 + M = C\dot{l} + C\dot{l} + M \quad \text{initiation}$$

$$C\dot{l} + H_2 = HCl + \dot{H} \quad \text{propagation}$$

$$\dot{H} + Cl_2 = HCl + C\dot{l} \quad \text{propagation}$$

$$\text{etc.}$$

$$C\dot{l} + C\dot{l} = Cl_2 + M \quad \text{termination}$$

There are few, if any, relevant systems important in the field of combustion that can be considered straight chain processes.

2.2. Branched Chain Reactions

In 1926–1929 Semenov[32] discovered the very important fact that some (in fact most) chain reactions can undergo branching. Chain reactions result when an active center reacts with a molecule to form two active centers. Valency must continue to be conserved, and thus, instead of a single monovalent center being formed, the process can result in, for example, three new monovalent centers (or one mono- and one di-valent center). A good example of this process is supplied by some of the elementary reactions of the hydrogen/oxygen mechanism. In a certain range of temperatures and pressures, and in addition to some other elementary reactions, hydrogen oxidation proceeds through the sequence

$$H_2 + O_2 = 2\dot{O}H \quad \text{initiation}$$

$$H_2 + \dot{O}H = H_2O + \dot{H} \quad \text{propagation}$$

$$\dot{H} + O_2 = \dot{O}H + \ddot{O} \quad \text{branching}$$

$$H_2 = \ddot{O} = \dot{H} + \dot{O}H \quad \text{branching}$$

$$H_2O + \ddot{O} = \dot{O}H + \dot{O}H \quad \text{branching}$$

$$\dot{H} + \dot{O}H + M = H_2O + M \quad \text{termination}$$

$$\text{etc.}$$

Such mechanisms may yield either "stationary" (quasi-steady) or "nonstationary," self-accelerative behavior, depending on external parameters (pressure, temperature, heat evolution/transfer coupling) and their effects on the ratio of chain branching to chain termination. The shift of a chemical chain reacting system from stationary to self-accelerative behavior by change in the external parameters plays a significant role in such phenomena as reaction explosion and flammability limits.[33]

In most combustion reaction systems some of the reactions are chain branching, some are chain propagating, and others are chain terminating. However, the more important issue is the overall or net ability of the collection of reactions (i.e., the reaction mechanism) to result in depletion or growth of the total pool of reactive centers. It is convenient to define the branching factor as the ratio of all active centers formed to those reacted in the system, and it is the balance of chain branching with termination that controls the value of the branching factor. Glassman[33] presents a very simple conceptual analysis that demonstrates this concept and its relationship to pressure/temperature effects on explosion limits. The effectiveness of the chain branching can be emphasized by considering a system in which there is no chain branching, only chain propagation. If, in a bimolecular system, there are 10^8 collisions cm^{-3}-s^{-1}, each collision resulting in reaction of the active center and a molecule to yield product and another reactive center (a chain propagation reaction), initially only one chain active center cm^{-3}, and 10^{19} molecules-cm^{-3}, all molecules would be consumed in 10^{11} s, or approximately 30 years. On the other hand, if a chain branching factor of 2 were to exist (i.e., one radical creates two), for the same example the number of generations necessary to consume all molecules is $2^N = 10^{19}$, or $N = 62$. Thus all molecules are consumed in 62×10^{-8} s, or approximately a microsecond. For a branching factor of 1.01, the time is about 100 microseconds; thus as long as the branching factor is slightly greater than one, a reaction will proceed very rapidly. Such reactions will be termed in the present chapter to be "explosively fast." In fact the exponential growth character resulting from a branched reaction system can only be departed from by controlling the supply and/or depleting the amount of reactants. In this regard a branched chemical reaction system is very much like the nuclear fission process. The overall chemical reaction rate can be moderated similarly, namely, by affecting the total pool of reactive centers. Indeed, the role of the so-called flame inhibitors for fire suppression is to moderate chemical rate by controlling one of the principal branching active centers, hydrogen atoms.[34]

2.3. Degenerate Branching

Although straight and branched chain mechanisms can explain much of the early kinetic observations, such mechanisms cannot explain the fact that in a defined range of temperature and pressure, some hydrocarbon oxidations have been observed to proceed through a self-accelerative behavior similar to that found for explosive chain branching but at a self-acceleration rate that is three to four orders of magnitude slower. This effect manifests itself as a very long "chemical induction period," followed by stages of reaction that are very complex and dependent on initial pressure/temperature (discussed in detail below). A further development of the theory of branched chain reactions, termed degenerate branching, was put forward by Semenov[32] in the early 1930s to explain the chemical induction phenomena. The basic principle of this process is schematically described in overall terms as

$$
\text{Active center + reactants} \rightarrow \text{intermediates}
\begin{array}{l}
\overset{\text{(path I)}}{\nearrow} \quad
\begin{array}{l}
\text{non chain branching} \\
\text{(stable molecules)}
\end{array} \\
\underset{\text{(path II)}}{\searrow} \quad
\begin{array}{l}
\text{chain branching} \\
\text{(active centers)}
\end{array}
\end{array}
$$

The hypothesis put forth was that in certain circumstances, intermediate products formed in the reaction during the development of essentially an unbranched chain mechanism can undergo further transformations to give active centers with greater ease than the original reactants themselves (path II). It was further assumed that the intermediate substances were converted to an overwhelming degree to nonactive end products (path I), without the formation of additional new active centers, and only a small portion of the intermediates proceed along path II to form additional active centers. The existence of this process, along with the relative overall temperature dependences and exothermicity of the reaction process, leads to the very complex kinetic behavior associated with the chemical induction period, cool flames, and the negative temperature coefficient effects discussed below.

3. ORGANIC MOLECULES, CHEMICAL KINETICS, AND COMBUSTION PHENOMENA

For the same initial reactants, contained in a vessel, the temperature and pressure dependence of various elementary reactions can cause the behavior of the chemical oxidation mechanism of organic molecules to change from one of the above classes to another. Depending on the initial conditions, the same reactants can also yield different products and result in different overall stoichiometry and energetics. The result of this complex interaction is that although there is a consensus on many of the phenomenological details of mechanistic behavior, the translation of this information to specific detailed kinetic descriptions of the phenomena remains incomplete.

3.1. Low-Temperature Phenomena

Below about 423 K the oxidation of organic compounds is immeasurably slow unless either chemical or photochemical initiators are used to promote reaction. Benson[35] terms this temperature regime as one of chemical synthesis reactions; the most important intermediate species produced in terms of reaction mechanism in this regime are hydroperoxides of the type RCH_2OOH or $R'R''CHOOH$, where R, R', etc., denote the remainder of the organic molecule. (A compact review of organic structure of molecules typically occurring as initial fuels and reaction products appears in Glassman.[33]) The reaction is exothermic and can lead to spontaneous ignition if heat is not removed from the environment. Above 373 K, the hydroperoxides formed undergo homogeneous, autocatalytic, chain decomposition to produce predominantly alcohols ($RC-OH$), ketones ($RC=O$) (or aldehydes, ($R-CHO$)). Further reaction of the aldehydes, which are formed only from primary hydroperoxides (RCH_2OOH), rapidly produces acids ($RC=O-OH$).

Above about 573 K the gas phase oxidation of hydrocarbons is a slow process that predominantly yields olefins ($RC=CH_2$) and hydrogen peroxide (H_2O_2). The olefins formed are more easily oxidized than the initial hydrocarbon and tend to form epoxides ($RCH-CH_2O$) and unsaturated aldehydes ($RCH=CH-CHO$) along with CO and CO_2. Although the net reaction is nearly thermoneutral (and thus self-heating becomes unimportant), an interesting feature of reaction at these temperatures is the appearance of autocatalysis. The reaction begins at an immeasurably slow rate, eventually accelerates through degenerate branching processes to a maximum rate after which the depletion of reactants causes the rate to decline.

Another interesting characteristic of this reaction regime is the appearance of diffuse luminosity and, further, the production of slowly moving wave(s) of

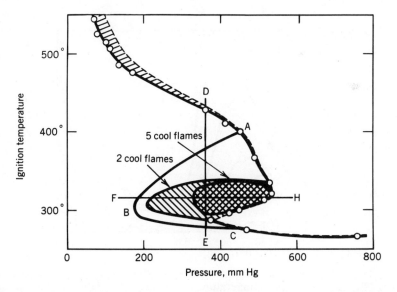

FIGURE 1. Kinetic features of the propane–oxygen system in the cool flame regions. To the left of the cool flame region are regions of slow reaction. Regions to the right of the cool flame region are where "hot" ignition occurs. Data are for a 500 cm^3 quartz vessel and equimolar mixtures of propane and oxygen. Reprinted with permission from S. W. Benson, "The Kinetics and Thermochemistry of Chemical Oxidation with Application to Combustion and Flames," © 1981, Pergamon Press PLC.[35]

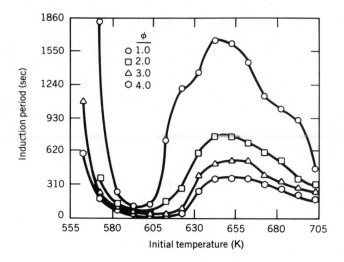

FIGURE 2A. Effect of initial temperature and equivalence ratio on the chemical induction period. Data are for a 1395 cm³ quartz vessel and mixtures of propane and air at an initial pressure of 600 Torr. Figure taken from Ref. 8. Reprinted with the permission of Gordon and Breach Science Publishers S. A. from R. D. Wilk, N. P. Cernansky, and R. S. Cohen, "The Oxidation of Propane at Low and Transition Temperatures," *Combustion Science and Technology* **49**, 41 (1986).

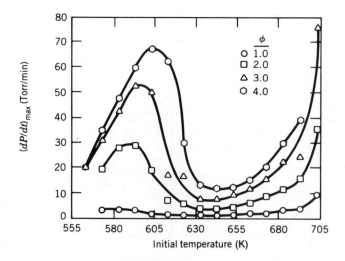

FIGURE 2B. Effect of initial temperature and equivalence ratio on the maximum rate of pressure rise. Data are for a 1395 cm³ quartz vessel and mixtures of propane and air at an initial pressure of 600 Torr. Figure taken from Ref. 8. Reprinted with the permission of Gordon and Breach Science Publishers S. A. from R. D. Wilk, N. P. Cernansky, and R. S. Cohen, "The Oxidation of Propane at Low and Transition Temperatures," *Combustion Science and Technology* **49**, 41 (1986).

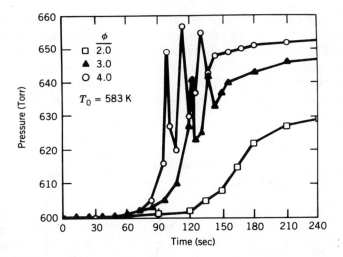

FIGURE 3A. Effect of equivalence ratio on the pressure histories and cool flames. Data are for a 1395 cm³ quartz vessel and mixtures of propane and air at an initial pressure of 600 Torr. Initial equivalence ratio = 3.0, and initial temperature = 583 K. Figure taken from Ref. 8. Reprinted with the permission of Gordon and Breach Science Publishers S. A. from R. D. Wilk, N. P. Cernansky, and R. S. Cohen, "The Oxidation of Propane at Low and Transition Temperatures," *Combustion Science and Technology* **49**, 41 (1986).

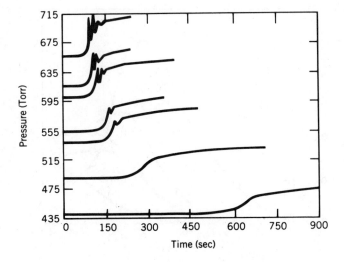

FIGURE 3B. Effect of equivalence ratio on the pressure histories and cool flames. Data are for a 1395 cm³ quartz vessel and mixtures of propane and air at an initial pressure of 600 Torr. Initial equivalence ratio = 3.0, and initial temperature = 583 K. Figure taken from Ref. 8. Reprinted with the permission of Gordon and Breach Science Publishers S. A. from R. D. Wilk, N. P. Cernansky, and R. S. Cohen, "The Oxidation of Propane at Low and Transition Temperatures," *Combustion Science and Technology* **49**, 41 (1986).

blue light traversing the reaction volume. Since early observation suggested that the traverse of these waves did not yield an increase in temperature of the overall mixture, they have been historically referred to as "cool" flames. These phenomena apparently do not occur with the simple alkanes, methane, or ethane, the simple olefin ethylene, the simple aromatics, benzene, toluene, ethyl benzene, or *p*-xylene (exhibits low-temperature chemical behavior but no

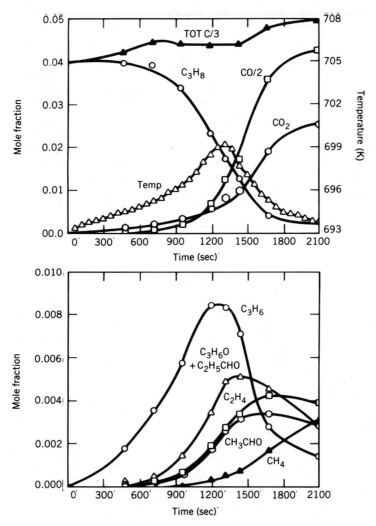

FIGURE 4A. Species concentration and temperature profiles as a function of reaction time. Data are for a 1395 cm^3 quartz vessel and a stoichiometric mixture of propane and air at an initial pressure of 600 Torr and at an initial temperature of 693 K. Figure taken from Ref. 8. Reprinted with the permission of Gordon and Breach Science Publishers S. A. from R. D. Wilk, N. P. Cernansky, and R. S. Cohen, "The Oxidation of Propane at Low and Transition Temperatures," *Combustion Science and Technology* **49**, 41 (1986).

cool flames) as the initial fuel,[35] but are observed for all higher carbon number paraffins, other higher carbon number olefins, other xylenes and higher carbon number alkylated aromatics, and all carbon number aldehydes. Careful temperature measurements now show that the local temperature rise associated with cool flames may be several hundred degrees Kelvin. Despite this local self-heating, the overall vessel reaction does not accelerate, and in fact the

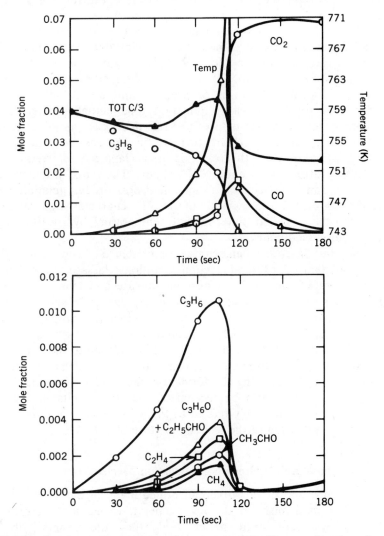

FIGURE 4B. Species concentration and temperature profiles as a function of reaction time. Data are for a 1395 cm^3 quartz vessel and a stoichiometric mixture of propane and air at an initial pressure of 600 Torr and at an initial temperature of 743 K. Figure taken from Ref. 8. Reprinted with the permission of Gordon and Breach Science Publishers S. A. from R. D. Wilk, N. P. Cernansky, and R. S. Cohen, "The Oxidation of Propane at Low and Transition Temperatures," *Combustion Science and Technology* **49**, 41 (1986).

appearance of one cool flame may be followed by another, even several more, at short and even long (1 minute) time intervals. The occurrence, number, and spacing of the cool flames is sensitive to the choice of initial fuel, vessel size, temperature, pressure, and reaction stoichiometry. Although quasi-steady cool flame phenomena have been produced in flow systems,[36] early experiments in closed vessels showed that passage of one or more cool flames was often followed by explosion (i.e., very rapid reaction) of the remaining reactive mixture.

Figure 1 displays a pressure-temperature explosion diagram and the typical cool flame features developed from the closed vessel oxidation of an equi-molar mixture of propane and oxygen.[37, 35] Figures 2a, 2b, 3a, 3b, 4a and 4b display some recent experimental data[8] obtained on the low-temperature reaction of various mixtures of propane and air at different initial temperatures. A region in which induction time actually increases (and reaction rate in terms of rate of pressure rise decreases) as initial reaction temperature is increased is clearly evident in Figures 2a and 2b. The temperature regime in which this result occurs is classically referred to as the "negative temperature coefficient" (NTC) regime. The range of temperatures over which this transition occurs is a function of pressure but generally ceases at about 650–700 K. It is the cessation of the NTC characteristics that defines a transition to the so-called "intermediate temperature hydrocarbon oxidation regime" discussed below.

Figures 3a and 3b clearly demonstrate the existence of and changes in the number of cool flames for several reaction conditions. Passage of cool flames is noted by very rapid time scale changes in pressure and temperature at the sensing positions within the closed reaction vessel. Figures 4a and 4b show the stable intermediate/product evolution for stoichiometric propane/air mixtures at 693 K and 743 K initial temperature and 600 Torr initial reaction pressure. Qualitatively, it should be noted that in addition to the intermediate species characteristics described above, there are very significant productions of CO and CO_2 accompanying the formation and oxidation of the reactant intermediates. It should also be recognized that the results in Figures 2 through 4 are not quantitatively related to Figure 1 because of differences in stoichiometry and dilution of the reacting mixture by nitrogen.

3.2. "Low-Temperature" Hydrocarbon Oxidation Mechanism

Although Semenov and other workers identified many of the kinetic features of the reaction mechanism in the low-temperature regime,[32] according to Benson,[38] it was not until about 1965 that the general observed features—"cool" flames, NTC effect, and induction period character—were fitted into any general oxidation mechanism. The general mechanism is as follows, with other reactions having to be added as temperature increases (e.g., reaction 12) and as the initial fuel hydrocarbon molecule becomes larger than

3 to 5 carbon atoms (e.g., reactions 13–15).

$$RH + O_2 + M = \dot{R} + H\dot{O}_2 + M \tag{1}$$

$$\dot{R} + O_2 = R\dot{O}_2 \tag{2}$$

$$\dot{R} + O_2(+M) = \text{olefin} + H\dot{O}_2(+M) \tag{3}$$

$$RH + R\dot{O}_2 = ROOH + \dot{R} \tag{4}$$

$$R\dot{O}_2 = R'CHO + R''O \tag{5}$$

$$RH + H\dot{O}_2 = H_2O_2 + \dot{R} \tag{6}$$

$$ROOH = R\dot{O} + \dot{O}H \tag{7}$$

$$RH + \dot{O}H = \dot{R} + H_2O \tag{8}$$

$$R'CHO + O_2 = R'\dot{C}O + H\dot{O}_2 \tag{9}$$

$$R\dot{O}_2 \xrightarrow{\text{wall}} \text{destruction} \tag{10}$$

$$H_2O_2 \xrightarrow{\text{wall}} \text{destruction} \tag{11}$$

$$H_2O_2 + M = \dot{O}H + \dot{O}H + M \tag{12}$$

$$R\dot{O}_2 = \dot{R}OOH \tag{13}$$

$$\text{Olefin} + H\dot{O}_2 = \text{epoxide} + \dot{O}H \tag{14}$$

$$RO\dot{O}H + O_2 = \text{aldehyde} + \text{ketone} + 2\dot{O}H \tag{15}$$

It will be seen later that some of these reactions are not elementary processes (e.g., reaction 15) but represent the result of several elementary processes.

In the case of alkanes, the initiation reaction for the mechanism is

$$RH + O_2 = \dot{R} + H\dot{O}_2 \tag{1}$$

The reaction is typically slow, reaction-site selective, and quite endothermic (activation energies are the order of 45 to 55 Kcal-mol^{-1}). This process can result in several alkyl radical isomers. For example, in the case of propane, the formation of both iso-propyl and n-propyl radicals will occur. The formation of iso-propyl radicals is favored over that of normal-propyl radicals, by the fact that the secondary C—H bond energy is approximately 3 Kcal less than the primary C—H bond energy. However, there are three times as many primary C—H bonds to undergo reaction. Thus both the relative bond energies of the available reaction sites and the relative number of sites must be taken into account to obtain the overall rate of relative formation of the different alkyl radicals. Similar arguments would hold if the molecule were larger and the even more reactive tertiary C—H bond were to be present (such as in branched alkanes). Despite the production of two radicals, the endothermicity of the reaction does not permit it to play a significant role in the overall destruction of the initial reactant; that is, it is primarily an initiation reaction.

The propagation reactions at low temperature are of two essential types;

$$\dot{R}O_2 = R\dot{O}_2 \qquad (2)$$

$$\dot{R} + O_2(+M) = \text{olefin} + H\dot{O}_2(M) \qquad (3)$$

Thermal decompositions of radicals, R, are typically highly endothermic processes, becoming important only at higher reaction temperatures ($T > 750$ K) and therefore play no role in the low-temperature regime. Although the formation of peroxy radicals by molecular oxygen attack is always faster than that of olefin formation, the production of peroxy radicals becomes more favored still at lower temperatures. In general, Benson[38] has shown that the formation of peroxy radicals is rapidly reversible above about 473 K; that is, the reverse reaction to the reformation of the initial radical and molecular oxygen is rapid above these temperatures. Because of these facts, other $R + O_2$ processes prevail above 500 K such as those resulting in the formation of conjugate olefins. Recently, studies by Slagle et al.[39] support the suggestions of Fish[40] that these olefin forming processes are not separate parallel processes but rather the result of a second decomposition channel involving the RO_2^* formed in the initial addition step:

$$\dot{R} + O_2 = R\dot{O}_2^* \qquad (3a)$$

$$R\dot{O}_2^* + M = R\dot{O}_2 + M \qquad (3b)$$

$$R\dot{O}_2^* = \text{olefin} + H\dot{O}_2 \qquad (3c)$$

Since reaction (3a) is highly reversible above 500 K, RO_2^* is constantly formed and decomposed. Rather than decomposing into the initial reactants, some of that formed yields olefins and HO_2, until there is complete conversion of R to olefin. In the case of propane the reactions would be

$$\dot{C}_3H_7 + O_2 = C_3H_7O\dot{O}^*$$

$$C_3H_7O\dot{O}^* + M = C_3H_7O\dot{O} + M$$

$$C_3H_7O\dot{O}^* = C_3H_6 + H\dot{O}_2$$

It is known that when O_2 attaches to the radical, the bonds form about a 90° angle with the carbon atoms.[33] The realization of this steric condition will facilitate understanding of certain reactions to be depicted later.

The peroxy radical (RO_2) abstracts an H from any fuel molecule or other hydrogen donor to form the hydroperoxide (ROOH):

$$RH + R\dot{O}_2 = ROOH + \dot{R} \qquad (4)$$

The amount of hydroperoxide that will form depends upon the competition of reaction (4) with reaction (3) (which forms stable olefin and HO_2). The decomposition of RO_2 formed in reaction (2) can yield molecular products such as aldehydes and ketones (reaction 5), whereas the HO_2 formed in reaction (3) leads to the formation of hydrogen peroxide, H_2O_2, through,

$$RH + H\dot{O}_2 = \dot{R} + H_2O_2 \qquad (6)$$

Destruction of the initial reactant is primarily a result of the attack of RO_2, HO_2, and OH formed by the degenerate branching step, reaction (7).

Well into the NTC regime, the dissociation of H_2O_2 formed in reaction (6) becomes a significant source of OH radicals through

$$H_2O_2 + M = \dot{O}H + \dot{O}H + M \qquad (12)$$

However, in the low-temperature regime this dissociation is very slow, and the fate of the H_2O_2 is termination (usually heterogeneous) to form water and oxygen:

$$H_2O_2 \xrightarrow{\text{wall}} \text{destruction } (H_2O, O_2) \qquad (11)$$

The abstraction of H will occur preferentially at various C—H bonds depending on their relative bond strengths. Again, the weakest C—H bonding is on a tertiary carbon, and if such C atoms exist, oxygen will preferentially attack these positions. If no tertiary carbon atoms exist, then the next weakest C—H bonds are those on the second carbon atoms from the ends of the chain. Although these statements are true for most radical attack on the initial fuel, the energetics of the hydroxyl radical reaction,

$$RH + \dot{O}H = \dot{R} + H_2O \qquad (8)$$

cause this particular abstraction to be, for all intents and purposes, nonselective. The buildup of ROOH and R′CHO—the degenerate branching species—is required before chain branching occurs to a sufficient degree to dominate the system, and the required time for this buildup results in the chemical induction times noted in low-temperature hydrocarbon combustion experiments.

Controversy exists as to the relative importance of ROOH versus aldehydes (reactions 5 and 8) as the important intermediates; however, recent work[41] would indicate that the hydroperoxide step strongly dominates. Aldehydes are quite important as fuels in the cool flame region, but they do not appear to lead to the important chain-branching step as readily as hydroperoxide.

A recent analysis by Benson[35] sheds further light on the phenomenological result as the relative importance of reactions (2) and (3) changes,

$$\dot{R} + \dot{O}_2 = RO_2 + 31 \text{ Kcal} \tag{2}$$

$$\dot{R} + O_2(+M) = \text{olefin} + H\dot{O}_2(+M) + 14 \text{ Kcal} \tag{3}$$

$$RH + R\dot{O}_2 = ROOH + \dot{R} \tag{4}$$

Reaction (2) is rapidly reversible above 473 K, whereas reaction (3) (the composite of 3a through 3c) is effectively irreversible. In fact the reaction between HO_2 and olefin is primarily an addition reaction to form $RCH-CHO_2H$. The latter is appreciably faster since it has a lower activation energy and leads to epoxidation:

$$
\underset{\begin{array}{c} | \\ H \end{array}}{\overset{\begin{array}{c} H \\ | \end{array}}{R-C}}=\underset{\begin{array}{c} | \\ H \end{array}}{\overset{\begin{array}{c} H \\ | \end{array}}{C}}-H + H\dot{O}_2 = R-\underset{\diagdown}{\overset{\begin{array}{c} H \\ | \end{array}}{C}}\underset{\diagup}{\overset{\begin{array}{c} H \\ | \end{array}}{C}}-H + \dot{O}H \tag{14}
$$
$$O$$

Although reaction (2) is much faster than reaction (3) at all temperatures, reaction (2) becomes important as the temperature increases, and reaction (3) becomes the dominant route, actually depleting the hydrocarbon radical. Treating these two reactions as independent processes, Benson performed a steady-state analysis of the rates of production of ROOH and the olefin to obtain the following ratio:

$$\frac{d[\text{olefin}]/dt}{d[ROOH]/dt} = \frac{d[\text{olefin}]}{d[ROOH]} = \frac{k_3[R][O_2]}{k_4[RO_2][RH]} \tag{16}$$

Under cool flame conditions it has been shown that $[RO_2]$ and $[R]$ are essentially in equilibrium via reaction (2) so that it is possible to write

$$\frac{[RO_2]}{[R]} = K_2[O_2] \tag{17}$$

where K_2 is the equilibrium constant for reaction (2). Combining these equations,

$$\frac{d[\text{olefin}]}{d[ROOH]} = \frac{k_3}{k_4 K_2[RH]} \tag{18}$$

which is a somewhat paradoxical result, since O_2 and all radical concentrations have disappeared from the ratio. By assuming that the H is abstracted from a secondary carbon atom and by assigning values to the constants in this equation, Benson found that

$$\frac{d[\text{olefin}]}{d[\text{ROOH}]} = \frac{10^{\{5.7-(8.47/rT)\}}}{[\text{RH}]} \tag{19}$$

where r is the universal gas constant (kcal-mol^{-1}-K^{-1}), T the temperature (K), 8.47 in units of kcal-mol^{-1}, and [RH] in mol-L^{-1}.

This last equation predicts that the olefin production becomes more important than the ROOH production as the temperature rises. Thus, when a cool flame begins and the mixture temperature rises as it proceeds, the olefin production becomes prevalent and dominates the step leading to degenerate branching. This results in cessation of chemical reaction on a time scale comparable to that available in the cool flame thermal wave. Indeed, recent detailed kinetics calculations[13,14] show that heat transfer losses and subsequent cooling (causing the local mixture temperature to again decrease) result in the observed multiple cool flames in low-pressure reactors. The appropriate analysis will be considerably more complex if reactions (2) and (3) are treated as truly sequential (i.e., as reactions 3a–3c), as was again recently advanced by Slagle et al.,[39] but the qualitative result, that the competition of the two possible product channels results in the overall reaction rate decreasing with temperature, will be similar.

Equation (19) is both temperature and pressure dependent, and thus it appears that the beginning of the NTC region that separates low-temperature and intermediate temperature regimes will be at different temperatures as ambient pressure changes. As many conventional combustion systems operate at pressures well in excess of ambient, how such transitions vary with pressure can be very important.

Figure 5[42] shows the behavior of the "turnover" temperature (the temperature at which the NTC regime begins) for a stoichiometric n-butane-air mixture as a function of pressure. The low-temperature, static reactor data of Wilk[8] were used to define the magnitude of equation (19) at the turnover condition in his experiments, and equation (19) was then used to extrapolate this condition to other temperatures and pressures. There are currently no fundamental studies of the kinetics of the negative temperature coefficient regime at high pressure. However, recent modeling of the autoignition time delay of adiabatically compressed combustible gas mixtures[43,44] qualitatively confirm the pressure dependent character of the turnover temperature represented in Figure 5. It will be seen later that as pressure is increased, the turnover temperature preedicted in Figure 5 will approach conditions at which other chemistry that is less pressure dependent may be active.

With larger carbon number hydrocarbon molecules, an important isomerization reaction to form the hydroperoxy radical

$$\dot{R}O_2 = \dot{R}OOH \tag{13}$$

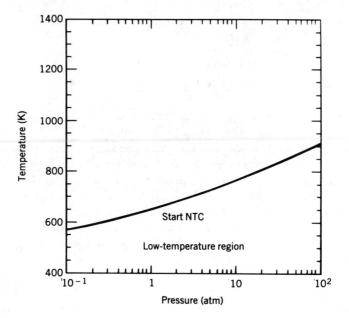

FIGURE 5. "Turnover" temperature as a function of reaction pressure for a stoichiometric mixture of butane and air. Data are obtained from Eq. (19) fitted to replicate the measured value reported in Ref. 8 at 600 Torr. Figure courtesy of Ref. 42.

will also occur, and Benson[35,38] notes that with six or more carbon atoms, this reaction may compete substantially with reaction (4), yielding the following complex path structure:

$$\text{RH} \xrightarrow[+\text{HO}_2]{+\text{OH}} \dot{\text{R}} \xrightarrow[2]{+\text{O}_2} \dot{\text{RO}}_2 \xrightarrow[4]{+\text{RH}} \text{ROOH} \xrightarrow[7]{} \dot{\text{RO}} + \dot{\text{OH}} \text{ (chain branching)}$$

with branches:

$$\dot{\text{ROOH}} \xrightarrow[15]{+\text{O}_2} \text{products (chain branching)} \quad (13)$$

$$\text{Olefin} + \dot{\text{HO}}_2 \text{ (steady reaction)} \quad (3)$$

More specifically, the overall reaction

$$\dot{\text{ROOH}} \xrightarrow{+\text{O}_2} \text{products} \qquad (15)$$

typically proceeds through the following general sequence of elementary reaction processes:

$$\dot{R}OOH \xrightarrow[15]{+O_2} \dot{O}OR'-R''OOH \xrightarrow[16]{} HOOR'-R''=O + \dot{O}H$$

$$13a\downarrow \qquad\qquad\qquad\qquad\qquad\qquad\qquad 17\downarrow$$

$$RO + \dot{O}H \qquad\qquad\qquad\qquad \dot{O}R'''CH\dot{O}OH$$

(Epoxide)

$$18\downarrow$$

$$R'''{}'CHO + R'''{}''C=O$$

(Aldehydes) (Ketones)

Depending on the structure of the radical $\dot{O}OR'-R''OOH$, internal isomerization may occur (16) or the radical may abstract an H atom from another hydrocarbon (19) to yield ketones, aldehydes, and hydroxyl radicals:

$$\dot{O}OR'-R''OOH \xrightarrow{+RH} aldehyde + ketone + \dot{O}H + \dot{O}H \qquad (19)$$

The specific ketones and aldehydes which are produced will depend on which reaction route is followed and the structure produced by the addition of oxygen to the isomerized ROOH radical.

Although reaction (13) is endothermic and the reverse of reaction (13) is fast, reaction (15) can be faster still, so that the isomerization (reaction 13) becomes rate determining. At 600 K Benson[35] suggests that reaction (13) can be some 30 times faster than reaction (4), and therefore, this additional sequence of reactions resulting from oxygen addition to the ROOH radical will contribute substantially to branching. Thus the competition between olefin production (reaction 3) and chain-branching processes resulting from the production of RO_2 becomes more severe at higher temperatures when isomerization of ROOH is significant. It has been suggested that the greater tendency for long-chain hydrocarbons to knock as compared to smaller-chain and branched-chain molecules is a result of this internal, isomerization-branching mechanism.[35,38] Recent modeling of engine processes suggests that this isomerization sequence may be important in autoignition chemistry of paraffin hydrocarbons as small as n-butane.[44] These effects should be partially accounted for in the results presented in Figure 5 since experimental data at low pressure were used to generate the figure. However, the fact that oxygen concentration will be substantially increased at higher pressure may cause the turnover temperature to extend to even higher values as pressure increases.

Finally, one will note in the earlier figures that there is production of some CO_2 along with CO in this temperature range. This production is strongly associated at the lower temperature with the reaction:

$$CH_3\dot{C}O + H\dot{O}_2 = CO_2 + \dot{C}H_3 + \dot{O}H \qquad (20)$$

with some contribution also coming from

$$CO + H\dot{O}_2 = CO_2 + \dot{O}H \tag{21}$$

Reaction (20) is rather unusual in that it represents a route to carbon dioxide formation that does not require previous formation of CO such as in reaction (21) or (22):

$$CO + \dot{O}H = CO_2 + \dot{H} \tag{22}$$

At higher temperatures where the formation of OH becomes prevalent, reaction (22) completely dominates the formation of carbon dioxide.

As temperatures are increased in the intermediate temperature regime, the thermal decomposition,

$$CH_3\dot{C}O + M = \dot{C}H_3 + CO \tag{23}$$

becomes much faster than reaction (20), preventing direct carbon dioxide formation. Since reaction (21), and reactions (24) and (25),

$$CO + \dot{O} + M = CO_2 + M \tag{24}$$

$$CO + O_2 = CO_2 + \dot{O} \tag{25}$$

have small rate constants in comparison to reaction (22), and the rate constant for reaction (8),

$$RH + \dot{O}H = \dot{R} + H_2O \tag{8}$$

is typically much greater than that for reaction (22), the presence of hydrocarbon species generally inhibits the formation of CO_2 at higher temperatures.

3.3. Intermediate Temperature Reaction Phenomena

As temperatures are raised, H_2O_2 production becomes more significant (reactions 3 and 6), and hydrogen peroxide begins to decompose (reaction 12) before it can undergo transport to a heterogeneous destruction site (reaction 11). With the continued increase in the reverse of reaction (2), and the increasing importance of reaction (12), HO_2 and OH become the dominant chain-carrying radicals. Functionally, hydroperoxy radicals replace RO_2, while H_2O_2 replaces ROOH, and the dominant oxidation mechanism at low pressures and high temperatures is

$$H_2O_2 + M = \dot{O}H + \dot{O}H + M \tag{12}$$

$$RH + \dot{O}H = \dot{R} + H_2O \tag{8}$$

$$\dot{R} + O_2 = \text{olefin} + H\dot{O}_2 \tag{3}$$

$$RH + H\dot{O}_2 = \dot{R} + H_2O_2 \tag{6}$$

Termination of the reaction chain is provided by

$$H\dot{O}_2 + H\dot{O}_2 = H_2O_2 + O_2 \tag{26}$$

It is the increasing dominance of this mechanism over that described earlier which apparently is responsible for the termination of the NTC regime observed in Figure 2, and the beginning of the intermediate temperature oxidation regime. Benson[35,38] has referred to the inception of this regime as the occurrence of "hot ignition." Reaction (12) is bimolecular,[45] whereas the decomposition of ROOH is likely to be in its high-pressure limit at these temperatures and pressures. Therefore the hot ignition condition should occur at lower temperatures as pressure is increased. Thus, based on earlier discussion, it appears that not only does the turnover temperature increase, but the range of temperature over which the NTC regime exists may become smaller as the environmental pressure is increased. Indeed, it is possible that at some higher pressure, the negative temperature coefficient regime may disappear completely.

At temperatures above about 800 K (and well within the intermediate temperature regime), the oxidation chemistry of most hydrocarbons begins to change very significantly. Alkyl radicals formed through abstraction of an H atom by an active center will begin to decompose to yield smaller hydrocarbon radicals and small olefins (primarily with carbon numbers less than 3). Special consideration of methyl, ethyl, and propyl radicals formed in this process is warranted as these are the only radicals likely to yield H atoms through further reaction.

In the case of methyl radicals, further decomposition will not occur until temperatures are very high (> 1700 K). At low temperatures the reaction

$$\dot{C}H_3 + O_2 \longrightarrow \text{Products}$$

proceeds as described earlier, that is,

$$\dot{C}H_3 + O_2 = CH_3\dot{O}_2 \tag{27}$$

$$CH_3\dot{O}_2 = CH_2O + \dot{O}H \tag{28}$$

$$\dot{C}H_3 + O_2 = CH_3\dot{O} + \ddot{O} \tag{29}$$

with the equilibrium for the first reaction largely shifted toward reactants at temperatures above 850 K (where beta scission, discussed below, becomes prevalent). Methyl radicals therefore tend to react with themselves to form C_2 species (C_2H_6 and C_2H_4), to abstract hydrogen from other molecules to form CH_4, and to react with other radicals. At intermediate temperatures and higher pressures (see later discussion), the prominent radical–radical reaction route is

$$\dot{C}H_3 + H\dot{O}_2 = CH_3\dot{O} + \dot{O}H \tag{30}$$

followed by

$$CH_3\dot{O} + O_2 = CH_2O + H\dot{O}_2 \tag{31}$$

and at higher temperatures,

$$CH_3\dot{O} + M = CH_2O + \dot{H} + M \tag{32}$$

Between 850 and 1000 K, there appears to be no significant reaction of methyl radicals and oxygen, whereas at higher temperatures the highly endothermic reaction,

$$\dot{C}H_3 + O_2 = CH_3\dot{O} + \ddot{O} \tag{29}$$

occurs.[1] Thus methyl radicals can yield H atoms only through reactions (30) and (32), that is, through relative circuitous routes at higher temperatures.

For ethyl radicals, C_2H_5, which are produced by beta scission of larger hydrocarbon radicals or by H abstraction from ethane formed from methyl radical recombination. Thermal decomposition, that is,

$$\dot{C}_2H_5 + M = C_2H_4 + \dot{H} + M \tag{33}$$

will begin to compete with

$$\dot{C}_2H_5 + O_2 = C_2H_4 + H\dot{O}_2 \tag{34}$$

at higher temperatures. Depending on the oxygen mole fraction, and the total pressure, reaction (34) will be more rapid than (33) up to about 1200 K (for $O_2 = 21\%$). Similar behavior is true for isopropyl radicals and t-butyl radicals. However, it will be seen below that the formation of even small quantities of H atoms from reactions paths such as (30)–(32) and (33) may be important.

For alkyl radicals with carbon number greater than 3, unimolecular loss of an H atom to form an olefin is highly disfavored in comparison to cleavage of a C—C bond because of the relative bond strengths. The higher carbon number conjugate olefins that are formed result primarily from the abstraction of an H atom by molecular oxygen.

For larger carbon number alkyl radicals, cleavage of the C—C bond is favored at the location one C—C bond away from the radical carbon site.[46] When more than one C—C bond cleavage site is possible, the scission that leads to the largest alkyl radical will be preferred.[47] The only radicals that will tend to support the loss or abstraction by oxygen of another H atom will be those for which no such scission site exists, such as ethyl, iso-propyl, and t-butyl radicals, as mentioned above. This type of C—C bond scission process has been referred to in the literature as "beta" scission or "alpha" cleavage.

Thus, for *n*-butyl radicals, one would expect

$$\cdot C-C-C-C-C-H = \dot{C}_2H_5 + C_2H_4 \qquad (35)$$

and

$$H-C-\dot{C}-C-C-H = \dot{C}H_3 + C_3H_6 \qquad (36)$$

to be the important reactions that compete with

$$\dot{n}\text{-}C_4H_9 + O_2 = 1\text{-butene} + \dot{H}O_2 \qquad (37)$$

and

$$2\text{-}\dot{C}_4H_9 + O_2 = 2\text{-butene} + \dot{H}O_2 \qquad (38)$$

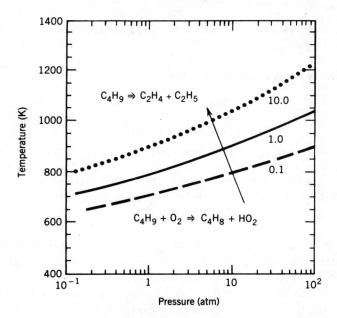

FIGURE 6. Temperature for a constant ratio of the rate of H-atom abstraction by molecular oxygen and the rate of radical beta scission for *n*-butyl radicals as a function of reaction pressure. Results are for a stoichiometric mixture of butane and air. Figure courtesy of Ref. 42.

A. Primary n-Octyl Radical (6 Possible #1 Positions)

$$C-C-C-C-C-C-C-C \cdot \rightarrow C_2H_4 + \dot{C}_6H_{13}$$
$$C_6H_{13} \rightarrow C_2H_4 + n\text{-}C_4H_9$$
$$n\text{-}C_4H_9 \rightarrow C_2H_4 + C_2H_5$$

Net products: $3C_2H_4, \dot{C}_2H_5$

B. Secondary n-Octyl Radical (4 Possible #2 Positions)

$$C-C-C-C-C-C-\underset{\cdot}{C}-C \rightarrow C_3H_6 + n\text{-}\dot{C}_5H_{11}$$
$$n\text{-}C_5H_{11} \rightarrow C_2H_4 + n\text{-}C_3H_7$$
$$n\text{-}C_3H_7 \rightarrow C_2H_4 + CH_3$$

Net products: $C_3H_6, 2C_2H_4, \dot{C}H_3$

C. Secondary n-Octyl Radical (4 Possible #3 Positions)

$$C-C-C-C-C-\underset{\cdot}{C}-C-C \rightarrow I-C_7H_{14} + CH_3, I-C_4H_8 + n\text{-}\dot{C}_4H_9$$
$$n\text{-}C_4H_9 \rightarrow C_2H_4 + C_2H_5$$

Net major products: $I-C_4H_8, C_2H_4, \dot{C}_2H_5$

D. Secondary n-Octyl Radical (4 Possible #4 Positions)

$$C-C-C-C-\underset{\cdot}{C}-C-C-C \rightarrow I-C_6H_{12} + \dot{C}_2H_5, I-C_5H_{10} + n\text{-}\dot{C}_3H_7$$
$$n\text{-}C_3H_7 \rightarrow C_2H_4 + \dot{C}H_3$$

Net major products: $I-C_5H_{10}, C_2H_4, \dot{C}H_3$

FIGURE 7. Beta scission fragmentation patterns of *n*-octyl radicals. Figure taken from Ref. 48. Reprinted with permission from Gordon and Breach Science Publishers S. A.

One must also take into account the possibility of internal isomerization reactions after the initial addition process in reactions (37) and (38).

Figure 6[42] displays the ratio of the rate of reaction (35) to the rate of (37) as a function of pressure (M = air). It can be seen that conjugate olefin formation can be prevented by beta scission processes in the intermediate temperature range, at temperatures only about 100 K higher than the turnover temperature. Beta scission processes may therefore compete with internal isomerization of RO_2 to the ROOH radical in the negative temperature regime.

A. Tertiary Iso-Octyl Radical (Only 1 Position Possible)

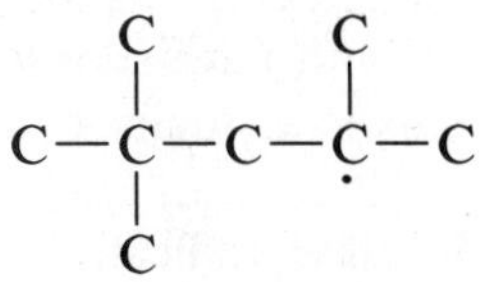

$$C-\overset{\overset{\displaystyle C}{|}}{\underset{\underset{\displaystyle C}{|}}{C}}-C-\overset{\overset{\displaystyle C}{|}}{\underset{\displaystyle \cdot}{C}}-C \rightarrow iso-C_4H_8 + tert-\dot{C}_4H_9$$

Net products: $Iso-C_4H_8$, $tert-\dot{C}_4H_9$

B. Secondary Iso-Octyl Radical (Only 2 Possible #3 Positions)

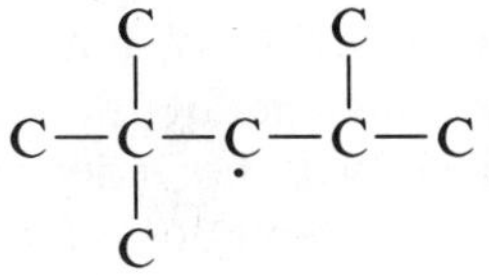

$$C-\overset{\overset{\displaystyle C}{|}}{\underset{\underset{\displaystyle C}{|}}{C}}-\overset{}{\underset{\underset{\displaystyle \cdot}{}}{C}}-\overset{\overset{\displaystyle C}{|}}{C}-C \xrightarrow{Major} \begin{array}{l} C_7H_{14} + \dot{C}H_3 \\ 4,4\text{-dimethyl-2-pentene} \end{array}$$

$$\xrightarrow{Major} \begin{array}{l} C_7H_{14} + \dot{C}H_3 \\ 2,4\text{-dimethyl-2-pentene} \end{array}$$

Net products: $2C_7H_{14}, 2\dot{C}H_3$

C. Primary Iso-Octyl Radical (9 Possible #1 Positions)

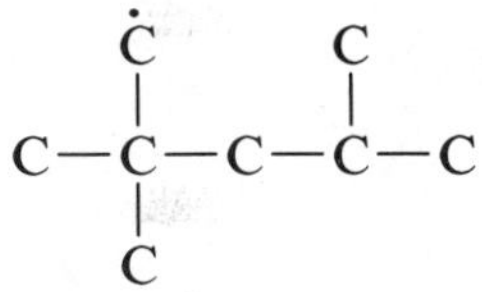

$$C-\overset{\overset{\displaystyle \dot{C}}{|}}{\underset{\underset{\displaystyle C}{|}}{C}}-C-\overset{\overset{\displaystyle C}{|}}{C}-C \rightarrow iso-C_4H_8 + iso-\dot{C}_4H_9$$

$$iso-C_4H_9 \rightarrow C_3H_6 + \dot{C}H_3$$

Net major products: $iso-C_4H_8, C_3H_6, \dot{C}H_3$

D. Primary Iso-Octyl Radical (6 Possible #5 Positions)

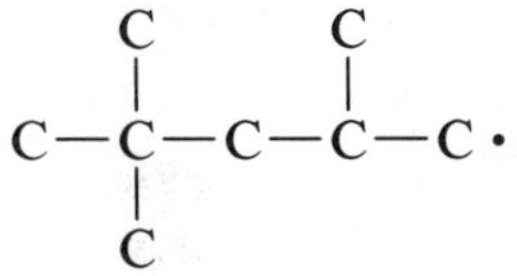

$$C-\overset{\overset{\displaystyle C}{|}}{\underset{\underset{\displaystyle C}{|}}{C}}-C-\overset{\overset{\displaystyle C}{|}}{C}-C\cdot \rightarrow C_3H_6 + neo-\dot{C}_5H_{11}$$

$$neo-C_5H_{11} \rightarrow iso-C_4H_8 + \dot{C}H_3$$

Net major products: $iso-C_4H_8, C_3H_6, \dot{C}H_3$

FIGURE 8. Beta scission fragmentation patterns of iso-octyl radicals. Figure taken from Ref. 48. Reprinted with permission from Gordon and Breach Science Publishers S. A.

Figures 7 and 8 show the expected fragmentation patterns based upon beta scission rules for *n*-octyl and iso-octyl radicals.[48] Brezinsky and Dryer[48,49] have studied these oxidations at 1100 K and have shown that at atmospheric pressure, cleavage of these alkyl radicals is virtually complete before significant oxidation has taken place.

However, above about 500 K, any alkyl radical, C_nH_{2n+1}, with carbon number greater than 4 may undergo rapid reversible isomerization (H atom migration) across 5 carbon atoms and more slowly across 4 and 6 carbon atoms.[50,51] Benson[52] suggests that this is the most rapid of any alkyl radical reactions at this temperature. Under this assumption all positional isomers for the radical (e.g., in Figures 7 and 8) will be essentially in equilibrium with one another. This fact implies that even if product distributions generated by the fission of each alkyl radical differ, the total product distribution formed from the decomposition of all alkyl radicals will be independent of the initial abstraction site. Furthermore the rate of formation of products will depend only on the total rate of alkyl radical production, independent of which species was originally responsible for the abstraction of an H atom and the site at which the abstraction occurred. Whether isomerization is indeed as rapid relative to scission reactions at higher temperatures and pressures remains in question.[48]

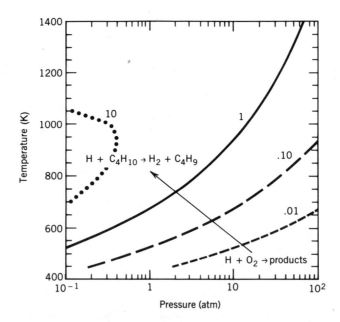

FIGURE 9. Temperature for a constant ratio of the rate of H-atom abstraction from *n*-butane and the total rate of reaction of H-atoms with molecular oxygen as a function of reaction pressure. Data are for a stoichiometric mixture of butane and air. Figure courtesy of Ref. 42.

At intermediate temperatures, the yield of hydrogen atoms from reactions such as (30)–(32) and (33) yield HO_2 through

$$\dot{H} + O_2 + M = H\dot{O}_2 + M \qquad (39)$$

Not until conditions are such that the reaction rate of

$$\dot{H} + O_2 = \dot{O}H + \ddot{O} \qquad (40)$$

becomes comparable with that of the major competing H atom loss processes, reactions (39) and (41),

$$RH + \dot{H} = \dot{R} + H_2 \qquad (41)$$

are O atoms available in significant quantities to react with the hydrocarbon species or their fragments. Figure 9[42] displays the ratio of the rate for reaction (41) divided by the sum of the rates for reactions (39) and (40) for a stoichiometric butane-air mixture at different pressures and temperatures. It is interesting to note the large range of temperatures over which reaction (41) dominates, particularly for the pressures (< 1 atmosphere) that have been more commonly studied experimentally.

Figures 10–14 show the typical oxidative chemical characteristics for several alkanes at these temperatures and at one atmospheric pressure that graphically demonstrate the consequences of the interactions of the reactions discussed above.[48,49,53,54] These data were derived using a flow reactor technique[53,54] and correspond to the "postinduction" phase of the oxidation, that is, chemical reactions leading to initiation of fuel disappearance had already occurred in the upstream regions of the reactor during mixing of the vaporized fuel with large amounts of nitrogen and oxygen. Chemical induction chemistry is typically very fast (with the exception of small carbon number species and aromatics) and can be modified or entirely precluded in conventional combustion by back-mixing or diffusion of partially reacted species and radicals with unreacted mixture.

Under even relatively oxygen-rich conditions, there appears to be an initial isoenergetic region in which the initial fuel is converted to unsaturated hydrocarbons, carbon monoxide, and water. The conversion of alkanes to alkenes is clearly an endothermic process. However, the hydrogen made available by this process is oxidized to water releasing just enough energy to nearly balance this endothermicity. Subsequently, and with some energy release, the (primarily) olefinic intermediates formed are oxidized to carbon monoxide and water, but further oxidation of the carbon monoxide to carbon dioxide is inhibited by the more favored reaction of hydroxyl radicals with these hydrocarbon and their fragments. Finally, the large amount of carbon monoxide formed is oxidized to carbon dioxide, releasing most of the heat of

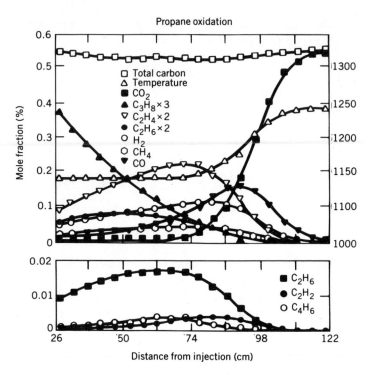

FIGURE 10. Species evolution for the reaction of propane and oxygen as a function of reaction distance. Data are from the Princeton Flow Reactor, for an initial equivalence ratio of 0.82, an initial reaction temperature of 1138 K, and atmospheric pressure. A distance of 1.0 cm corresponds to approximately 0.90 msec of reaction time. Figure from Ref. 53.

reaction of the overall process,

$$C_nH_{2n+2} + \left(\frac{3n + 1}{2} \right)O_2 \rightarrow nCO_2 + (n + 1)H_2O$$

Figures 11–14 show the effects of isomeric structure on the oxidation characteristics, and Figures 13 and 14 also demonstrate the resultant intermediate product distribution differences produced by the beta scission routes shown in Figures 7 and 8. Note, however, that the general phenomenology of the oxidations are very similar, although the characteristic time scales for conversion of the initial fuel to intermediates and conversion of the intermediates to carbon monoxide are quite different.

Figures 15 and 16 show the effects of increasing equivalence ratio on the oxidation characteristics. One should note the appearance of acetylene, allene, and even aromatic compounds as intermediate species. These species result

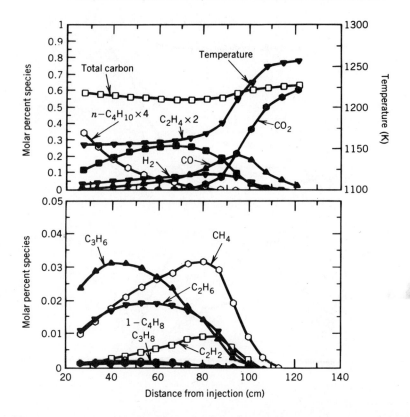

FIGURE 11. Species evolution for the reaction of n-butane and oxygen as a function of reaction distance. Data are from the Princeton Flow Reactor, for an initial equivalence ratio of 0.80, an initial reaction temperature of 1155 K, and atmospheric pressure. A distance of 1.0 cm corresponds to approximately 0.67 msec of reaction time. Figure from Ref. 54.

from the oxidation chemistry of the simple olefins and the interaction of the species formed with themselves and other hydrocarbon fragments, particularly the more stable hydrocarbon radicals such as C_2H_3 (vinyl), C_3H_5 (allyl), and C_4H_7 (t-butyl), as well as acetylene and oxidative reaction intermediates. This chemistry is discussed in Westbrook and Dryer, Gardiner, Frenklach et al., and Westmoreland.[1,3,4,5,7] One should be aware that many facets of the elementary kinetics of these reactions are still not fully understood and are continuing to be studied. Clearly this chemistry is important to controlling the fuel-rich combustion of alkanes and the formation of larger molecular species (soot precursors) and gas phase soot itself.[4,5]

Finally, it is worth noting some of the more recent progress that has occurred in defining the intermediate temperature chemistry of aromatics. Studies of aromatic hydrocarbons have occurred sporadically over the last 50

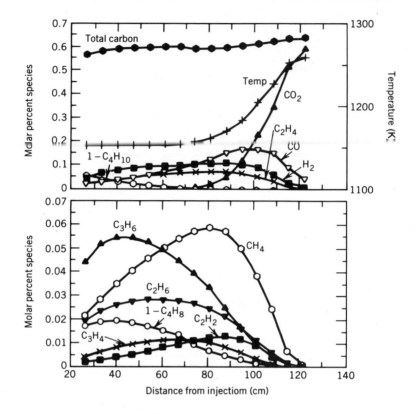

FIGURE 12. Species evolution for the reaction of iso-butane and oxygen as a function of reaction distance. Data are from the Princeton Flow Reactor, for an initial equivalence ratio of 0.70, an initial reaction temperature of 1182 K, and atmospheric pressure. A distance of 1.0 cm corresponds to approximately 1.07 msec of reaction time. Figure from Ref. 54.

years, but efforts have been few compared to consideration given to aliphatic hydrocarbon kinetics. Recent interest in synthetic liquid fuels derived from coal and other fossil forms as well as degrading qualities of petroleum have led to increasing quantities of aromatics in all fuels. The removal of lead as a means of improving fuel octane rating further emphasized the use of aromatics in gasoline, and both autoignition and soot production provide practical motivation to understand this chemistry.

It has generally been shown that several competing processes are required to account for the product distributions observed for the oxidation of different aromatic species.[33,55] In particular, for alkyl substituted aromatics, oxygen should preferentially abstract hydrogen from the alkyl group because the CH bond energies in these positions are much less than those in the ring. Therefore one would expect that the initiation of reaction for alkyl-substituted aromatics should be much faster than that of benzene or other nonsubstituted aromatics.

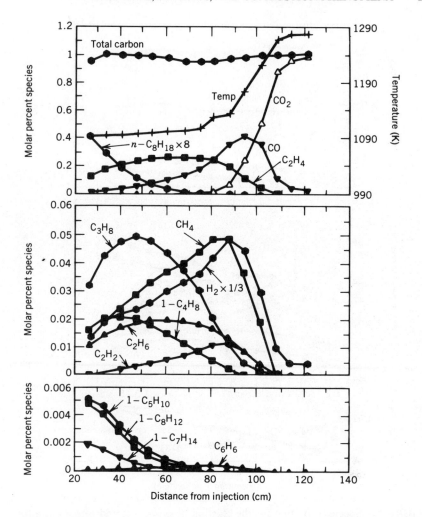

FIGURE 13. Species evolution for the reaction of *n*-octane and oxygen as a function of reaction distance. Data are from the Princeton Flow Reactor, for an initial equivalence ratio of 0.90, an initial reaction temperature of 1094 K, and atmospheric pressure. A distance of 1.0 cm corresponds to approximately 1.04 msec of reaction time. Figure of Ref. 48. Reprinted with permission from Gordon and Breach Science Publishers S. A.

The subsequent chemistry of the aromatic radicals thus formed are complex and not fully understood.

Early work primarily on benzene, toluene, and the xylenes at temperatures below 925 K led to proposed mechanism for phenyl radical oxidation that proceeded through the formation of hydro-peroxide,

$$\dot{C}_6H_5 + O_2 = C_6H_5O\dot{O}$$

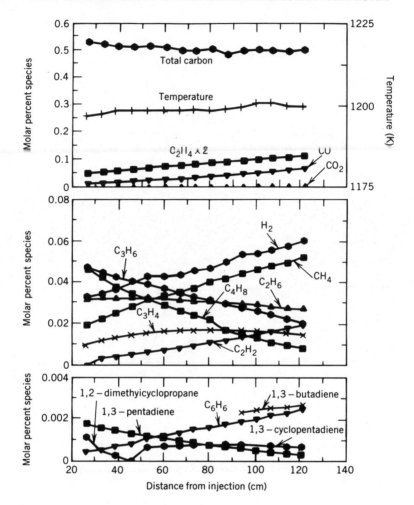

FIGURE 14. Species evolution for the reaction of iso-octane and oxygen as a function of reaction distance. Data are from the Princeton Flow Reactor, for an initial equivalence ratio of 1.00, an initial reaction temperature of 1197 K, and atmospheric pressure. A distance of 1.0 cm corresponds to approximately 0.79 msec of reaction time. (Iso-octane consumed previous to 20 cm position.) Figure from Ref. 48.

The peroxide was suggested to undergo ring fragmentation by hydroxyl-radical addition. More recent work suggests that hydro-peroxides are unstable, namely, the turnover temperature is about 575 K, thus excluding this mechanism at higher temperatures. A review of flow reactor and flame work at higher temperatures leads to the phenomenological mechanisms described below.[55]

Figure 17 details some typical flow reactor results obtained for lean oxidation conditions of toluene.[56] As toluene is consumed, benzaldehyde

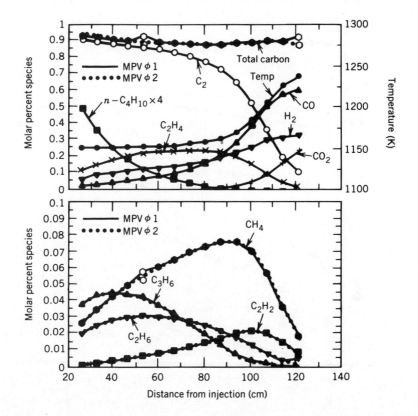

FIGURE 15. Species evolution for the reaction of n-butane and oxygen as a function of reaction distance. Data are from the Princeton Flow Reactor, for an initial equivalence ratio of 1.70, an initial reaction temperature of 1150 K, and atmospheric pressure. A distance of 1.0 cm corresponds to approximately 1.08 msec of reaction time. MPV#1 and MPV#2 represent two separate chemical analyses of the samples taken from the reactor. Figure from Ref. 54.

formation is followed by the appearance of benzene, then by C_5 and finally by C_4 hydrocarbon compounds. Carbon monoxide appears early in the reaction and increases rapidly as the benzene and C_3 and C_4 hydrocarbons oxidize. From these results, as well as studies on ethyl benzene and propyl benzene,[57] it is apparent that the side chains oxidize first in alkylated aromatics.

Furthermore, from these results it appears that both cleavage of H atoms from the methyl group and substitution of the methyl group by hydrogen atoms initiate the reaction sequence. From these data and their similitude to that for benzene oxidation (Figure 18), an overall mechanism for aromatic oxidation has been proposed and refined,[55-57] including the phenomenology of the ring rupture itself (Table 1).[55]

Pyrolytic cleavage and oxidation of the side chain are seen to lead to a phenyl radical that is proposed to react primarily with molecular oxygen and

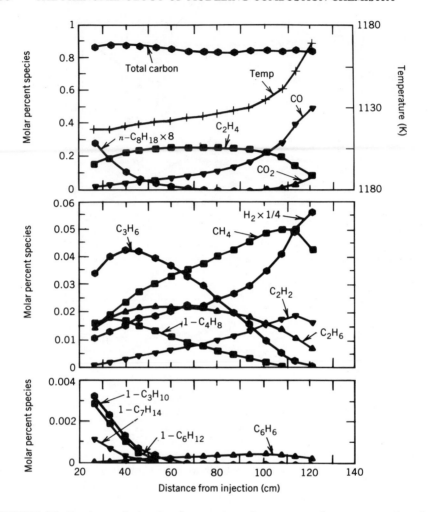

FIGURE 16. Species evolution for the reaction of *n*-octane and oxygen as a function of reaction distance. Data are from the Princeton Flow Reactor, for an initial equivalence ratio of 1.4, an initial reaction temperature of 1116 K, and atmospheric pressure. A distance of 1.0 cm corresponds to approximately 1.05 msec of reaction time. Figure from Ref. 54.

HO_2. Although similar reactions for alkyl radicals are endothermic and comparatively slow, the reaction of molecular oxygen with phenyl radicals is exothermic and quite rapid. Bittner[58] has suggested that in low pressure flames this reaction is superseded by the faster reaction of phenyl with oxygen atoms, a result that must be attributed to the high temperatures, low pressures, and diffusive characteristics of the flame studied.

As in the mechanism proposed by Bittner, the isomerization of the phenoxy radical leads to the expulsion of CO and formation of a cyclopentadienyl

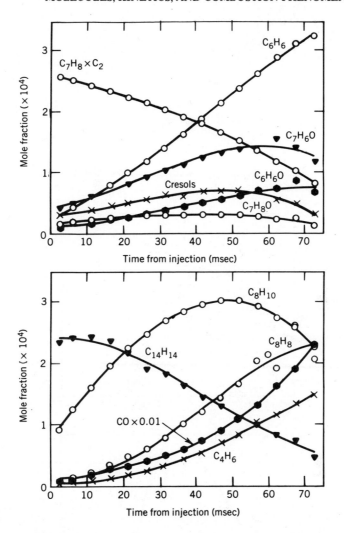

FIGURE 17. Species evolution for the reaction of toluene and oxygen as a function of reaction time. Data are from the Princeton Flow Reactor, for an initial equivalence ratio of 0.63, an initial reaction temperature of 1115 K, and atmospheric pressure. Reaction time determined from flow reactor cross-sectional geometry with plug-flow assumptions and a total flow rate of 1.10 mol-sec^{-1}. Reprinted with permission from K. Brezinsky, "The High Temperature Oxidation of Aromatic Hydrocarbons, © 1985, Pergamon Press PLC.[55]

radical. Subsequent reaction of the cyclopentadienyl radical with oxygen atoms, HO_2 and/or OH (reaction with molecular oxygen is considerably endothermic), leads to further expulsion of CO and ring rupture to form butadienyl radicals and butadiene. Studies of butadienyl radicals and butadiene suggest that the butadienyl radical generally forms vinyl acetylene and, through further reactions, allene and CO. Although butadiene itself can yield

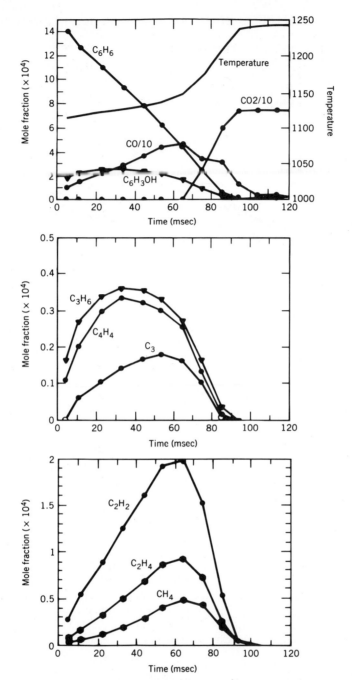

FIGURE 18. Species evolution for the reaction of benzene and oxygen as a function of reaction time. Data are from the Princeton Flow Reactor, for an initial equivalence ratio of 1.0, an initial reaction temperature of 1115 K, and atmospheric pressure. Reaction time determined from flow reactor cross-sectional geometry with plug-flow assumptions and a total flow rate of 0.56 mol-sec^{-1}. Reprinted with permission from K. Brezinsky, "The High Temperature Oxidation of Aromatic Hydrocarbons, © 1985, Pergamon Press PLC.[55]

TABLE 1. A Mechanism for Intermediate Temperature Oxidation of Benzene and / or Phenyl Radical[a]

Initiation processes $\longrightarrow$ Radical pool (H, OH, O, HO$_2$)

$$\phi H + OH \Big\langle \begin{array}{l} \longrightarrow \phi + H_2O \quad \text{Phenyl radical} \\[4pt] \xrightarrow{\ ?\ } \phi OH + H \quad \text{Phenoxy radical} \end{array}$$

$$\phi H + O \Big\langle \begin{array}{l} \longrightarrow \phi OH \\ \longrightarrow \phi O + H \end{array} \quad \text{Phenoxy radical}$$

$\phi H + H \rightarrow H\phi H \rightarrow \phi H + H \qquad$ "Nonreaction"
Cyclohexadienyl
radical

$\phi + O_2 \rightarrow \phi O + O$

$\phi + HO_2 \rightarrow \phi O + OH$

$\phi + O \xrightarrow{\text{minor}} \phi O$

$\phi + OH \xrightarrow{\text{minor}} \phi OH$

$$\phi OH + \left\{ \begin{array}{l} OH \\ HO_2 \\ O_2 \\ H \\ O \end{array} \right\} \rightarrow \phi O + \left\{ \begin{array}{l} HOH \\ H_2O_2 \\ HO_2 \\ H_2 \\ OH \end{array} \right\}$$

$\phi O \rightarrow C_5H_5 + CO$
Cyclopentadienyl
radical

$C_5H_5 + HO_2 \longrightarrow C_5H_5O + OH$

$C_5H_5 + O \rightarrow C_5H_5O$
Cyclopentadienonyl radical

$C_5H_5 + OH - C_5H_4OH$ $\qquad$ + H
Cyclopentadienolyl radical
$\qquad \xrightarrow{\ ?\ } C_4H_4 \qquad$ + HCO
Vinyl acetylene formyl radical

$C_5H_5 + H \rightarrow C_5H_6$
Source $\qquad$ Cyclopentadiene
$$\quad + O \Big\langle \begin{array}{l} \longrightarrow C_5H_5O + H \\ \longrightarrow C_5H_5O \end{array}$$
$\qquad \qquad \qquad$ Cyclopentenone

$$\begin{array}{lll} C_5H_5O & \rightarrow & C_4H_5 \quad +CO \\ & & \text{Butadienyl} \\ & & \text{radical} \\ C_3H_6O & \rightarrow & C_4H_5 \quad +CO \\ & & \text{Butadiene} \end{array} \Bigg\} \text{Ring is opened}$$

Reprinted with permission from K. Brezinsky, "The High Temperature Oxidation of Aromatic Hydrocarbons, © 1985, Pergamon Press PLC.[55]
[a] In this table only, the symbol ϕ represents the phenyl radical, C$_6$H$_5$.

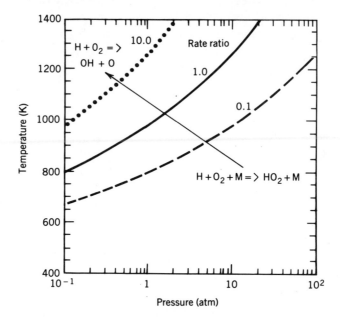

FIGURE 19. Temperature for a constant ratio of H-atom addition to oxygen and the H-atom-oxygen metathesis reaction as a function of pressure. Figure courtesy of Ref. 42.

butadienyl radicals, reaction with O atoms yields aldehydes and eventually the allyl radical. The addition reaction of butadiene and H atoms can lead to vinyl radicals and ethylene through fragmentation.

   Although aliphatic fragmentation and oxidation can be described by the phenomena discussed earlier, little is currently known about the oxidation chemistry of butadiene[59] and even less about that of vinylacetylene. These mechanisms will require further elaboration to be used in modeling combustion systems.

3.4. High-Temperature Reaction Phenomena

The transition between "intermediate" and "high-temperature" kinetic regimes is generally defined by the conditions under which reaction (39) dominates over reaction (40). This transition is not only a function of pressure but of the third body efficiencies of molecules present in the reaction system.[1] Figure 19 displays the ratio of the rates of reactions (39) and (40) as a function of pressure for nitrogen as the third body. It is seen that the transition to the high-temperature kinetic regime is in the temperature range of the atmospheric pressure flow reactor experiments described earlier but will occur at temperatures well above 1100 K at high pressures.

As peroxide decomposition to form OH radicals is fast above 800 K, OH and HO_2 remain as the principal abstracting radicals until this transition condition is reached. In the high-temperature regime, however, the reactions,

$$\dot{H} + O_2 = \dot{O}H + \ddot{O} \tag{42}$$

$$H_2 + \dot{O}H = H_2O + \dot{H} \tag{43}$$

$$H_2 + \ddot{O} = \dot{H} + \dot{O}H \tag{44}$$

$$H_2O + \ddot{O} = \dot{O}H + \dot{O}H \tag{45}$$

control the relative concentrations of O, OH, and H available to react with hydrocarbons through

$$RH + \dot{O}H = \dot{R} + H_2O \tag{46}$$

$$RH + \dot{H} = \dot{R} + H_2 \tag{47}$$

$$RH + \dot{O} = \dot{R} + \dot{O}H \tag{48}$$

Additional complexities may exist, depending on the species RH and the temperature because reactions with OH and O may be both abstractions, as noted above, and additions:

$$RH + \ddot{O} = \text{products} \tag{49}$$

$$RH + \dot{O}H = \text{products} \tag{50}$$

Such radical addition reactions occur with small olefins such as ethene and propene as well as with acetylene. Considerable experimental work and discussion continues on the possible transition of some radical-hydrocarbon reactions from addition to abstraction as temperature increases.[1, 7, 33, 60, 61]

As temperatures increase, chemical initiation results from

$$RH + M = \dot{R} + \dot{H} + M \tag{51}$$

$$RH + M = \dot{R}' + \dot{R}'' + M \tag{52}$$

as well as

$$RH + O_2 + M = \dot{R} + H\dot{O}_2 + M \tag{1}$$

Overall chemical rates occur on time scales comparable to residence times in flames. Finally, as very severe conditions are approached (1800 K at 1 atm), decomposition of reactants—including very stable ones such as aromatics,[62] and more stable radicals, such as phenyl, t-butyl, and methyl radicals—becomes a significant process. At lower temperatures decomposition primarily initiates chain reaction, whereas at high temperatures it may actually compete with the chain reaction sequence in determining the rate of disappearance of reactants.

The characteristic time for chemical reactions is considerably shorter than that needed for achieving intimate mixing of reactants; thus chemical rate is often limited by diffusive or mixing processes at these temperatures. Shock tube results at 1 atmosphere for *n*-heptane[63,64] and isooctane[64] as well as ignition delay studies,[65-67b] confirm that the general phenomenological characteristics of alkane oxidation are similar to those observed at lower temperatures in flow reactors. It should be noted that the "induction" period defined earlier and "ignition" delay measurements correspond closely only for methane. For the higher carbon number alkanes,[52,53] shock-tube ignition delay (especially when determined by pressure rise) corresponds closely to the characteristic time required before carbon monoxide oxidation becomes significant. Major reaction of the initial fuel (and even the olefins formed) logically should occur more rapidly than the oxidation of carbon monoxide, since the reaction rate constants of the reactions with hydroxyl radicals are one to two orders of magnitude faster than those for hydroxyl radicals with carbon monoxide.

4. SOME THOUGHTS ABOUT COMBUSTION APPLICATIONS

It is worthwhile considering briefly the relative importance of each of the regimes of chemistry discussed above in defining combustion phenomena. Many combustion phenomena involve transport effects and time transient behavior of temperature and often pressure (e.g., engine knock). For example, let us consider the homogeneous reaction of a premixed hydrocarbon-air mixture undergoing compression ignition. If the characteristic chemical times produced in a regime are large in comparison to characteristic residence times available at the necessary prevailing environmental conditions, then even though a system passes through conditions that would yield certain chemical behavior for homogeneous isothermal, isobaric reaction, there is not time for that chemistry to substantially contribute to the observed phenomena. On the other hand, if chemical and residence times are of the same order, substantial chemistry may occur and result in some conversion of initial reactants. Intermediates and products produced, together with the remaining reactants, compose a dynamically changing "initial mixture" for later reaction. As the system moves to a new set of environmental conditions that define a different reaction regime, the result is to modify initial conditions for that chemical regime in comparison to those that would be present if no reaction had occurred previously. Finally, transport of species and energy can cause drastic changes in the chemical initiation, rates of oxidation, and chemical behavior from that which might be expected under isothermal, homogeneous conditions. It is for these reasons that the construction of comprehensive chemical kinetic mechanisms is important.

4.1. Detailed Chemical Kinetic Modeling

We wish to briefly review here some of the principles and techniques that have evolved for developing elementary kinetic mechanisms. It is important to have some exposure to these subjects whether there is an intent to develop a "new" mechanism from the available elementary kinetic resources, to choose a specific mechanism from the literature for application to a similar or different problem than that for which it was developed, or to only contemplate the published results and their relationship to another problem of interest.

In theory, an elementary kinetic mechanism for a specific combustion event should consist of all the chemical species that are present, together with all representations of the possible elementary reactions among these species. In practice, the construction of an elementary mechanism requires identification of those species and specific elementary reactions that have rates of magnitude sufficient to influence the observations of interest. From the previous section, it is clear that the oxidation of larger fuel molecules represents a fragmentation and oxidation process to intermediate hydrocarbons, oxygenation of the smaller molecular weight species, and, ultimately, formation of final oxidation products. This last stage may be further complicated by addition reactions leading to larger unsaturated hydrocarbons, ring structures, and soot precursors.

For some of the smaller fuel molecules such as methane, this process is somewhat more complicated in that fragments may also yield higher molecular weight hydrocarbon intermediates before proceeding to oxidation products. Yet this general observation can serve to great advantage in developing kinetic mechanisms. For example, moist carbon monoxide oxidation involves hydrogen–oxygen kinetics. The resulting $CO-H_2-O_2$ reaction mechanism is part of that for formaldehyde oxidation, and the CH_2O mechanism is in turn part of the CH_4 oxidation mechanism. This hierarchical process[68] can be continued, building each mechanism on a foundation of validated reaction mechanisms for simpler molecules. In fact, following such a sequential path can simplify the task of the next mechanism validation, since, in principle, only those reactions and rates that have been added to account for the next level of complexity should require special attention. However, the definitions of some of the rates and reaction products are often so imprecise that some modifications may be necessary.

Absolute specific rate constants for some elementary reactions often vary strongly with changes in temperature and pressure. Indeed, multiple product channels for which the relative importance of each may vary with pressure and temperature can exist for the same elementary reactants. Only recently have a few elementary reactions received sufficient, direct experimental study to define behavior for the range of environmental conditions experienced in combustion systems.[69, 70] The use of transition state theory or other theoretical methods can also aid in establishing rate estimates for reactions, extrapolating experimental rate measurements and suggesting uncertainty limits where direct measurements are unavailable.[71,72] A number of important compilations and discussions of elementary rate data for reactions involved in hydrocarbon oxidation are available,[1,3,73,74] but it is currently a shortcoming of the elementary kinetics field that a state of the art source of rate data such as that maintained for thermochemistry[75,76] is not available. Neither are the compilations in references 3, 75 and 76 all-inclusive or necessarily consistent with one another.

One should resist indiscriminately aligning computed results with observations by assigning rate expressions that differ dramatically with published data. However, it is frequently difficult to identify whether a deficient reaction mechanism is incomplete with regard to the correct evaluation of the specific rate data, the specific reaction products and species involved, or the thermochemistry itself. Indeed, errors in thermochemistry can be as important as erroneous forward reaction rate data, and it should be mentioned that further improvements in thermochemical data, including that for larger species, are needed for the more complex detailed models.

Pressure is also a significant issue where unimolecular decomposition and recombination reactions are important. Troe[77–79] has dealt extensively with methods to describe pressure-dependent reactions. A major result of this work has been the development of a rationale for dealing with the transition region while retaining the proper limiting expressions at high and low pressures. This topic is particularly important in constructing comprehensive mechanisms for combustion modeling since temperature and pressure ranges often span the falloff region for numerous reactions. Falloff expressions for many of these reactions have not been embodied in an analytical manner in current reaction mechanisms and are needed to improve detailed models. Although the simple Lindeman mechanism continues to be the principal functional method used for describing pressure dependence in computational programs, Golden and Larson[80,81] have recently proposed a nine-parameter method as an alternative approach. Unfortunately, parameter evaluations are available for only a limited number of reactions.

Finally, elementary reaction rate modifications constantly occur that require the detailed mechanisms to be continually updated and revalidated as time proceeds. This fact is probably the one area that is most often neglected. Indeed, it has been suggested that changes in only the last few years require that much of the modeling for simple hydrocarbons be done all over again.[1,3,82] An example of why such reworking is needed is described later in this section.

The number of species in the mechanism grows rapidly with increasing fuel molecule size. The mechanism for $CO-H_2-O_2$ necessarily contains about 10 species, that for CH_4 contains about 25 species, that for CH_3OH, 27 species, that for propane, about 41. Extensively more species are required if the model is to include the negative temperature coefficient and intermediate temperature regimes.[83–85] Detailed mechanisms for aromatic fuel molecules have not yet been developed. Although some of the principal elements for constructing these models have been discussed,[55] understanding of aromatic oxidation mechanisms remains in a formulative stage. The important abstraction and ring-breaking reactions must be more clearly identified and their rates determined before serious detailed modeling studies can begin.

It would be redundant and beyond the scope of this chapter to consider specific reactions, to critically review elementary rate or thermochemical data, or to describe preferred models for the various hydrocarbons mentioned above. Other review articles deal with these issues in greater detail.[1,3,73-76]. Instead, the remainder of this section will give a brief overview of mechanism development.

4.2. Methodology for Deriving and Validating Detailed Mechanisms

In developing kinetic submodels for numerical prediction of practical combustion phenomena, relative ranges of the independent variables that far exceed those available in any particular experiment must be considered. The use of several intermediate and high-temperature sources of data and the hierarchical nature of $H-C-O$ oxidation chemistry in developing "comprehensive" detailed kinetic models have been applied to methanol,[86] ethylene,[87] propane,[88] butane,[89] and n-octane.[90] Mechanisms for other practical fuels can be constructed in a similar manner, yielding kinetic models that are useful in the analysis of many types of practical combustion systems.

Several important observations concerning mechanism development can be made by considering the methanol mechanism developed in this manner.[86] In earlier work[91,92] several subsets of elementary reactions that are important to methanol oxidation were developed and verified. In particular, the fuel lean oxidation of moist carbon monoxide was extensively investigated, and it was verified that the mechanism reduced to that commonly used in predicting atmospheric pressure high-temperature oxidation.[91] A detailed methane oxidation model adequately reproduced lean, intermediate temperature flow reactor data for carbon monoxide and methane oxidation.[91] However, under rich conditions, considerable limitations were observed and attributed to inadequacies in the proposed oxidation scheme for C_2H_4. In a later paper that expanded the ethylene reaction set through addition of a more recent ethylene oxidation mechanism,[93] the resulting chemistry was found to predict shock-tube ignition delay measurements for methane/ethane mixtures.[92] To the inclusive reaction set thus defined, a set of 11 additional reactions, based on the work of Bowman[94] and Aronowitz et al.,[95] involving methanol and CH_2OH were added to yield a mechanism of 84 reactions to describe methanol oxidation (Table 2).

Reverse reaction rates were calculated from the specific forward rates and the appropriate equilibrium constants. Thermochemical data for the chemical species were calculated from the JANAF Tables[75] for most species and from Bahn[96] for CH_2OH. However, the failure to include either the thermochemistry or the reverse rates in the publication leaves the mechanism ill-defined for those works in which it has been employed since its development. Inclusion of

TABLE 2. The Westbrook and Dryer Comprehensive Mechanism for Methanol Oxidation

	Reaction	$\log A$	n	E_a	Reference[b]
		Rate[a]			
2.1	$CH_3OH + M \rightarrow CH_3 + OH + M$	18.5	0	80.0	This study
2.2	$CH_3OH + O_2 \rightarrow CH_2OH + HO_2$	13.6	0	50.9	Aronowitz et al. (1978)
2.3	$CH_3OH + OH \rightarrow CH_2OH + H_2O$	12.6	0	2.0	This study
2.4	$CH_3OH + O \rightarrow CH_2OH + OH$	12.2	0	2.3	LeFevre et al. (1972)
2.5	$CH_3OH + H \rightarrow CH_2OH + H_2$	13.5	0	7.0	This study
2.6	$CH_3OH + H \rightarrow CH_3 + H_2O$	12.7	0	5.3	This study
2.7	$CH_3OH + CH_3 \rightarrow CH_2OH + CH_4$	11.3	0	9.8	Gray and Herod (1968)
2.8	$CH_3OH + HO_2 \rightarrow CH_2OH + H_2O_2$	12.8	0	19.4	Aronowitz et al. (1978)
2.9	$CH_2OH + M \rightarrow CH_2O + H + M$	13.4	0	29.0	This study
2.10	$CH_2OH + O_2 \rightarrow CH_2O + HO_2$	12.0	0	6.0	Aronowitz et al. (1978)
2.11	$CH_4 + M \rightarrow CH_3 + H + M$	17.1	0	88.4	Hartig et al. (1971)
2.12	$CH_4 + H \rightarrow CH_3 + H_2$	14.1	0	11.9	Baldwin et al. (1970a)
2.13	$CH_4 + OH \rightarrow CH_3 + H_2O$	3.5	3.08	2.0	Zellner and Steinert (1976)
2.14	$CH_4 + O \rightarrow CH_3 + OH$	13.2	0	9.2	Herron (1969)
2.15	$CH_4 + HO_2 \rightarrow CH_3 + H_2O_2$	13.3	0	18.0	Skinner et al. (1972)
2.16	$CH_3 + HO_2 \rightarrow CH_3O + OH$	13.2	0	0.0	Colket (1975)
2.17	$CH_3 + OH \rightarrow CH_2O + H_2$	12.6	0	0.0	Fenimore (1969)
2.18	$CH_3 + O \rightarrow CH_2O + H$	14.1	0	2.0	Peeters and Mahnen (1973)
2.19	$CH_3 + O_2 \rightarrow CH_2O + O$	13.4	0	29.0	Brabbs and Brokaw (1975)
2.20	$CH_2O + CH_3 \rightarrow CH_4 + HCO$	10.0	0.5	6.0	Tunder et al.
2.21	$CH_3 + HCO \rightarrow CH_4 + CO$	11.5	0.5	0.0	Tunder et al.
2.22	$CH_3 + HO_2 \rightarrow CH_4 + O_2$	12.0	0	0.4	Skinner et al. (1972)
2.23	$CH_3O + M \rightarrow CH_2O + H + M$	13.7	0	21.0	Brabbs and Brokaw (1975)
2.24	$CH_3O + O_2 \rightarrow CH_2O + HO_2$	12.0	0	6.0	Engleman (1976)
2.25	$CH_2O + M \rightarrow HCO + H + M$	16.7	0	72.0	Schecker and Jost (1969)
2.26	$CH_2O + OH \rightarrow HCO + H_2O$	14.7	0	6.3	Bowman (1975)
2.27	$CH_2O + H \rightarrow HCO + H_3$	12.6	0	3.8	Westenberg and deHaas (1972a)
2.28	$CH_2O + O \rightarrow HCO + OH$	13.7	0	4.6	Bowman (1975)
2.29	$CH_2O + HO_2 \rightarrow HCO + H_2O_2$	12.0	0	8.0	Lloyd (1974)
2.30	$HCO + OH \rightarrow CO + H_2O$	14.0	0	0.0	Bowman (1970)
2.31	$HCO + M \rightarrow H + CO + M$	14.2	0	19.0	Westbrook et al. (1977)
2.32	$HCO + H \rightarrow CO + H_2$	14.3	0	0.0	Niki et al. (1969)
2.33	$HCO + O \rightarrow CO + OH$	14.0	0	0.0	Westenberg and deHaas (1972b)
2.34	$HCO + HO_2 \rightarrow CH_2O + O_2$	14.0	0	3.0	Baldwin and Walker (1973)
2.35	$HCO + O_2 \rightarrow CO + HO_2$	12.5	0	7.0	Westbrook et al. (1977)
2.36	$CO + OH \rightarrow CO_2 + H$	7.1	1.3	-0.8	Baulch and Drysdale (1974)
2.37	$CO + HO_2 \rightarrow CO_2 + OH$	14.0	0	23.0	Baldwin et al. (1970b)
2.38	$CO + O + M \rightarrow CO_2 + M$	15.8	0	4.1	Simonaitis and Heicklen (1972)
2.39	$CO_2 + O \rightarrow CO + O_2$	12.4	0	43.8	Gardiner et al. (1971)
2.40	$H + O_2 \rightarrow O + OH$	14.3	0	16.8	Baulch et al. (1973a)
2.41	$H_2 + O \rightarrow H + OH$	10.3	1	8.9	Baulch et al. (1973b)
2.42	$H_2O + O \rightarrow OH + OH$	13.5	0	18.4	Baulch et al. (1973b)
2.43	$H_2O + H \rightarrow H_2 + OH$	14.0	0	20.3	Baulch et al. (1973b)

TABLE 2. *(Continued)*

Reaction	log A	n	E_a	Reference[b]
2.44 $H_2O_2 + OH \rightarrow H_2O + HO_2$	13.0	0	1.8	Baulch et al. (1973b)
2.45 $H_2O + M \rightarrow H + OH + M$	16.3	0	105.1	Baulch et al. (1973b)
2.46 $H + O_2 + M \rightarrow HO_2 + M$	15.2	0	-1.0	Baulch et al. (1973b)
2.47 $HO_2 + O \rightarrow OH + O_2$	13.7	0	1.0	Lloyd (1974)
2.48 $HO_2 + H \rightarrow OH + OH$	14.4	0	1.9	Baulch et al. (1973b)
2.49 $HO_2 + H \rightarrow H_2 + O_2$	13.4	0	0.7	Baulch et al. (1973b)
2.50 $HO_2 + OH \rightarrow H_2O + O_2$	13.7	0	1.0	Lloyd (1974)
2.51 $H_2O_2 + O_2 \rightarrow HO_2 + HO_2$	13.6	0	42.6	Lloyd (1974)
2.52 $H_2O_2 + M \rightarrow OH + OH + M$	17.1	0	45.5	Baulch et al. (1973b)
2.53 $H_2O_2 + H \rightarrow HO_2 + H_2$	12.2	0	3.8	Baulch et al. (1973b)
2.54 $O + H + M \rightarrow OH + M$	16.0	0	0.0	Moretti (1965)
2.55 $O_2 + M \rightarrow O + O + M$	15.7	0	115.0	Jenkins et al. (1967)
2.56 $H_2 + M \rightarrow H + H + M$	14.3	0	96.0	Baulch et al. (1973b)
2.57 $C_2H_6 \rightarrow CH_3 + CH_3$	19.4	-1	88.3	Pacey (1973)
2.58 $C_2H_6 + CH_3 \rightarrow C_2H_5 + CH_4$	-0.3	4	8.3	Clark and Dove (1973)
2.59 $C_2H_6 + H \rightarrow C_2H_5 + H_2$	2.7	3.5	5.2	Clark and Dove (1973)
2.60 $C_2H_6 + OH \rightarrow C_2H_5 + H_2O$	13.8	0	2.4	Greiner (1970)
2.61 $C_2H_6 + O \rightarrow C_2H_5 + OH$	13.4	0	6.4	Herron and Huie (1973)
2.62 $C_2H_5 \rightarrow C_2H_4 + H$	13.6	0	38.0	Lin and Back (1966)
2.63 $C_2H_5 + O_2 \rightarrow C_2H_4 + HO_2$	12.0	0	5.0	Cooke and Williams (1971)
2.64 $C_2H_5 + C_2H_3 \rightarrow C_2H_4 + C_2H_4$	17.5	0	35.6	Benson and Haugen (1967)
2.65 $C_2H_4 + O \rightarrow CH_3 + HCO$	13.0	0	1.1	David et al. (1972)
2.66 $C_2H_4 + M \rightarrow C_2H_3 + H + M$	17.6	0	98.2	Just et al. (1977)
2.67 $C_2H_4 + H \rightarrow C_2H_3 + H_2$	13.8	0	6.0	Benson and Haugen (1967)
2.68 $C_2H_4 + OH \rightarrow C_2H_3 + H_2O$	14.0	0	3.5	Baldwin et al. (1966)
2.69 $C_2H_4 + O \rightarrow CH_2O + CH_2$	13.4	0	5.0	Peeters and Mahnen (1973)
2.70 $C_2H_3 + M \rightarrow C_2H_2 + H + M$	16.5	0	40.5	Benson and Haugen (1967)
2.71 $C_2H_2 + M \rightarrow C_2H + H + M$	14.0	0	114.0	Jachimowski (1977)
2.72 $C_2H_2 + O_2 \rightarrow HCO + HCO$	12.6	0	28.0	Gardiner and Walker (1968)
2.73 $C_2H_2 + H \rightarrow C_2H + H_2$	14.3	0	19.0	Browne et al. (1969)
2.74 $C_2H_2 + OH \rightarrow C_2H + H_2O$	12.8	0	7.0	Vandooren and Van Tiggelen (1977)
2.75 $C_2H_2 + O \rightarrow C_2H + OH$	15.5	-0.6	17.0	Browne et al. (1969)
2.76 $C_2H_2 + O \rightarrow CH_2 + CO$	13.8	0	4.0	Vandooren and Van Tiggelen (1977)
2.77 $C_2H + O_2 \rightarrow HCO + CO$	13.0	0	7.0	Browne et al. (1969)
2.78 $C_2H + O \rightarrow CO + CH$	13.7	0	0.0	Browne et al. (1969)
2.79 $CH_2 + O_2 \rightarrow HCO + OH$	14.0	0	3.7	Benson and Haugen (1967)
2.80 $CH_2 + O \rightarrow CH + OH$	11.3	0.68	25.0	Mayer et al. (1967)
2.81 $CH_2 + H \rightarrow CH + H_2$	11.4	0.67	25.7	Mayer et al. (1967)
2.82 $CH_2 + OH \rightarrow CH + H_2O$	11.4	0.67	25.7	Peeters and Vinckier (1975)
2.83 $CH + O_2 \rightarrow CO + OH$	11.1	0.67	25.7	Peeters and Vinckier (1975)
2.84 $CH + O_2 \rightarrow HCO + O$	13.0	0	0.0	Jachimowski (1977)

From Ref. 86. Reprinted with permission from Gordon and Breach Science Publishers S. A.

[a]Units are cm-mol-s-kcal. $k = AT^n \exp(-Ea/rT)$

[b]References refer to sources noted in the original article.

such data in published reaction mechanisms is important and has recently become a common practice.

Throughout the study,[86] only modifications of the reactions dealing with methanol, namely, reactions (2.1)–(2.11) in Table 2 were made. Thus, if others were properly integrated into the entire mechanism, the complete set of reactions should continue to adequately model carbon monoxide and methane as well as methane/ethane ignition delay properties described by the previous reaction sets. This procedural approach graphically represents the simplifying character resulting from the hierarchical nature of the hydrocarbon oxidation.

The subsequent modeling study resulted in modifications to the rate expressions (initially taken from Bowman[94] and Aronowitz et al.[95]) for reactions (2.1), (2.3), (2.5), (2.6), and (2.9), yielding the values reported in the paper. The dissociation reactions (2.1) and (2.9) for methanol and the hydroxymethyl radical respectively were assumed to be bimolecular. However, these reactions, as well as other bimolecular dissociation reactions included in Table 2 may actually be in the falloff region for some of the temperatures and pressures considered. Thus some pressure dependence must be embodied in the model rate constants in an implicit manner since it reproduced experimental data over a pressure range of 1 to 5 atmospheres.

The experimental data used to develop and validate the reaction mechanism covered a wide range of conditions, with different elementary reactions tending to dominate the methanol and hydroxymethyl consumption in different regimes. For example, the methanol thermal decomposition, reaction (2.1), was found to be completely unimportant in intermediate temperature flow reactor and laminar flame environments but was important for shock-tube ignition delay conditions. Reaction (2.3),

$$CH_3OH + \dot{O}H = \dot{C}H_2OH + H_2O \qquad (2.3)$$

was found to be the primary reaction consuming methanol in fuel-lean conditions, whereas reactions (2.5) and (2.6),

$$CH_3OH + \dot{H} = \dot{C}H_2OH + H_2 \qquad (2.5)$$

$$CH_3OH + \dot{H} = \dot{C}H_3 + H_2O \qquad (2.6)$$

dominated under fuel-rich conditions. Similarly, reaction (2.10),

$$\dot{C}H_2OH + O_2 = CH_2O + H\dot{O}_2 \qquad (2.10)$$

was found to be the principal reaction for CH_2OH removal in fuel-lean experiments, whereas reaction (2.9),

$$\dot{C}H_2OH + M = CH_2O + \dot{H} + M \qquad (2.9)$$

was of equal importance with reaction (2.10) in fuel-rich conditions.

The computed results for lean flow reactor experiments were most sensitive to the rates of reactions (2.3) and (2.10) and relatively insensitive to the remaining CH_3OH and CH_2OH reactions, whereas in rich flow reactor conditions reactions (2.5), (2.6), and (2.9) predominated. Reaction (2.1) was found to be important only in the shock-tube calculations. Thus in each regime it was possible to take advantage of the weighted sensitivities of particular reactions to independently determine the rates necessary to describe the experimental observations. In none of the experiments were the weighted sensitivities of reactions (2.2), (2.4), (2.7), or (2.8) such that their rates could be determined with any accuracy.

The development of a suitable reaction rate set for the specific hierarchical subset in a comprehensive model (in this case reactions 2.1–2.11) is not a matter that can always be carefully regimented nor uniquely performed as described above. The methanol study was considerably simplified by the weighted sensitivity of various experiments to limited numbers of the reactions. In developments of comprehensive mechanisms for ethylene,[87] propane,[88] n-butane[89] and n-octane[90] oxidation, no similar weighting existed. In such cases parameters have to be determined parametrically while working with several sets of experimental data and conditions.

Even when care is taken, the resultant model remains nonunique, limited by the available validation information and, as stated before, subject to further updating as additional fundamental data and experimental observations become available. For example, the relative importance of the two product branches possible for

$$CH_3OH + \dot{H} = \dot{C}H_2OH + H_2 \qquad (2.5)$$

$$CH_3OH + \dot{H} = \dot{C}H_3 + H_2O \qquad (2.6)$$

in the above mechanism were in question at the time of the initial mechanism development. Products of reaction (2.5) are much more reactive than those of (2.6). Rapid decomposition of CH_2OH yields H atoms that can provide chain branching, whereas CH_3 is a relatively unreactive radical species. Computed profiles for methanol and hydrogen were found to be very sensitive to both the ratio and the sum of these rate constants, yet little data existed for the ratio of the product branch at high temperature. Aders and Wagner[97] and Aders[98] reported low-temperature measurements (250–650 K) and concluded that reaction (2.6) was the principal branch. However, the intermediate and high-temperature experiments could not be satisfactorily modeled with this condition. Calculations supported that although the sum of specific rate constants was approximately that predicted by extrapolation of the Aders data, the principal path at high temperature was apparently reaction (2.5).

After this work was published, it was shown through independent experiments that reaction (2.6) was in fact much slower than (2.5) under all conditions,[99] and that even Westbrook and Dryer had used a much too large rate constant for (2.6). However, the Westbrook and Dryer mechanism has continued to be used with little modification in other work since,[100-107] and in fact, the methanol oxidation submechanism (2.1)–(2.14) has continued to appear unchanged in other comprehensive hydrocarbon mechanisms developed since 1983.[88-90] A frequent error alternatively committed is the addition (rather than substitution) of more recent thermochemical or rate data without investigation of the ability to reproduce the data originally used in deriving and validating the mechanism.

In the case of the methanol mechanism discussed above, the new measurements of the rate for reaction (2.6) causes the Westbrook and Dryer mechanism to no longer replicate experimental observations,[108] and along with other rate data and information that has become available, a revised comprehensive mechanism needs to be developed. It should also be apparent from this discussion that reaction submechanisms should be cautiously added to other reaction mechanisms in building larger reaction systems.[7]

However, the advantages of developing comprehensive models are adequately demonstrated by considering the 19-step mechanism originally developed by Bowman[94] to predict methanol oxidation in shock tubes. It is clear that one should not a priori expect this mechanism to predict flow tube results at intermediate temperatures. However, it is common for authors to draw mechanisms from the literature without consideration for the ranges of parameters to which they apply. Thus it is interesting to consider whether even gross features of the experiment are adequately reproduced by this model. Results of these comparative calculations appear as continuous curves in Figures 20 and 21. The fuel dissappearance and heat release rates are much slower in the computational models than in the experiment. Furthermore, the experimental data show a production of substantial quantities of hydrogen, while the mechanism predicts little if any H_2 formation. These general observations reemphasize the dangers of extending a reaction mechanism to conditions outside the range of parameters for which it was validated[109] and further underscore the need for comprehensive model development for combustion purposes.

Evaluation of the relative importance of specific reactions using "brute force" fitting techniques requires extensive parametric studies of the assembled mechanism, namely, each rate parameter must be varied independently over the estimated range of its uncertainty. Choice of added reactions and important species also requires extensive parametric study to elucidate the effects on computed results. This process can be substantially aided by employing reaction path analysis, namely, determining mathematically the reaction(s) that control the formation and destruction of each species.

In addition, experimental development and validation of kinetic mechanisms can now be approached in a much more efficient manner through the use of any of several available mathematical methods for performing "sensitivity" analyses. The concept is to mathematically determine the dependence of predictions of a solution for a set of differential equations on each of the

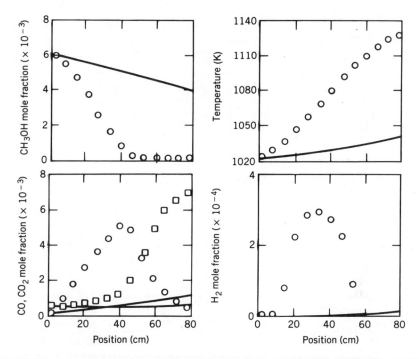

FIGURE 20. Comparison of predictions using methanol oxidation mechanism derived from shock-tube data with intermediate temperature flow reactor data. Dashed lines are predicted data. Solid lines are experimental results. From Ref. 86. Reprinted with permission from Gordon and Breach Science Publishers S. A.

system parameters, such as rate constants. Although this concept is not new to the thermochemical kinetic modeling field, the general application of the approach to kinetics is only now emerging as a result of improved, very efficient methods for obtaining sensitivity analysis information.

Sensitivity analysis procedures may be classified as either stochastic or deterministic in nature.[110] Consider the interpretation of system sensitivities in terms of first-order elementary sensitivity coefficients, $(\partial C_i/\partial a_j)$, where C_i is the concentration of the ith species at time t and a_j is the jth input parameter. The gradient, evaluated for a set of nominal parameter values, a, is called a local sensitivity coefficient and typifies the deterministic approach to sensitivity analysis. Theoretical treatments have included the "direct method,"[111] the Green's function method,[112] and Taylor series expansion methods.[113] Although the sensitivity coefficients, $\partial C_i/\partial a_j$, provide direct information on the effect of a small perturbation in each parameter about its nominal value on each concentration, they do not necessarily indicate the effect of simultaneous, large variations in all parameters on each specie concentration. An analysis that accounts for simultaneous parameter variations of arbitrary magnitude is termed a "global" sensitivity analysis. This

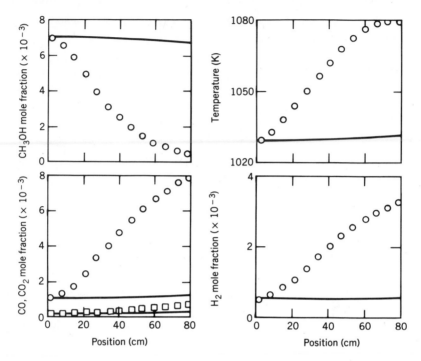

FIGURE 21. Comparison of predictions using methanol oxidation mechanism derived from shock-tube data with intermediate temperature flow reactor data. Dashed lines are predicted data. Solid lines are experimental results. From Ref. 86. Reprinted with permission from Gordon and Breach Science Publishers S. A.

analysis produces coefficients that have an average measure of sensitivity over the entire admissible range of parameter variation and thus provides an essentially different measure of sensitivity than that derived from local elementary sensitivity coefficients. Examples of this stochastic approach are the FAST method,[114] Monte Carlo methods,[115] and Pattern methods.[116] Both local and global analyses are useful in studying the behavior of a system since each have advantages and disadvantages. For an excellent review of these different approaches, see the papers by Tilden et al.[117] and Rabitz et al.[118]

Computational codes are currently in various stages of development to calculate first-order sensitivity coefficients for all of the parameters in zero-dimensional, homogeneous kinetic problems simultaneous to the problem solution itself.[119-121] Sensitivity analysis has a number of applications in chemical kinetic modeling[122] but perhaps the most evident are applications to error analysis. Due to the complex mixing of parameter dependence inherent in kinetics, the question of error propagation cannot unequivocally be resolved by the "brute force" method of varying a single rate parameter at a time. Beyond this concern is the more fundamental question of what features of a mechanism are responsible for the particular characteristics of the solution.[123]

The complexity of this question becomes more evident when it is recognized that it is possible mathematically to interchange the role of the independent and dependent variables of the experimental problem. Application of Legendre-type transformations similar to those performed in thermodynamics[124] offer a method to achieve this result.

Recently, Yetter[110] and Yetter et al.[125,126] have extensively demonstrated the use of Green's function sensitivity analysis, including first-order, second-order, and derived terms, in the study of the $CO/H_2/O_2$ and H_2-O_2 reaction systems.

The simplest application of sensitivity analysis consists of calculating the elementary sensitivity gradients. As an example,[110] a model consisting of 52 chemical reactions and 11 species of $CO/H_2/O_2$ oxidation, see Table 3, was integrated using LSODE, a stiff ODE solver developed by Hindmarsh,[127] and sensitivity coefficients were found using AIM.[119]

Figure 22 shows the concentration profiles resulting from a calculation, with initial conditions simulating those found in intermediate temperature flow tube experiments. The sensitivity coefficients for those rate constants that affect the calculated CO profile by more than one order of magnitude are shown in Figure 23. In this calculation the quantity of interest is $(\partial CO)/\partial k_j$, where k_j represents a specific forward, k_{jf}, or backward, k_{jb}, rate constant for the jth elementary reaction listed in Table 3. The forward and backward rate constants are related through the equilibrium constant of each specific reaction. Since only two of these three parameters are independent, the effect of an equilibrium constant K_j on specie concentrations C_i can be obtained through chain rule differentiation:

$$\frac{\partial \ln C_i}{\partial \ln K_j} = \left(\frac{1}{2}\right) n \frac{\partial \ln C_i}{\partial \ln k_{jf}} - m \frac{\partial \ln C_i}{\partial \ln k_{jb}}$$

For k_{jf} varied, k_{jb} fixed: $n = 2, m = 0$

For k_{jb} varied, k_{jf} fixed: $m = 2, n = 0$

K_j varied with k_{jf}, k_{jb} varied equally: $n = 1, m = 1.$

Finally, the net effect of perturbing either rate constant while fixing K_j is

$$\frac{\partial \ln C_i}{\partial \ln K_{j\,\text{net}}} = \frac{\partial \ln C_i}{\partial \ln k_{jf}} + \frac{\partial \ln C_i}{\partial \ln k_{jb}}$$

Note these equations and results of Figure 23 are developed as normalized sensitivity coefficients, thus removing any artifact caused by the relative numerical magnitudes of various C_i or k_j. Similar results are produced for each species, each and every rate constant, and each calculation as continuous functions of time. It is clear that for large mechanisms, with n species and m reactions ($n \ll m$), one is quickly overwhelmed with the amount of available

TABLE 3. CO—H_2—O_2 Mechanism Used for Demonstration of Gradient Sensitivity Analysis Application to Kinetics

	Reaction	k_f^a	k_b	I^b	UF^c
3.1, 3.2^d	$HCO + H = CO + H_2$	$3.32(-10)^e$	$2.85(-27)$	f	2
3.3, 3.4	$HCO + OH = CO + H_2O$	$1.66(-10)$	$6.07(-30)$	f	3.5
3.5, 3.6	$O + HCO = CO + OH$	$5.00(-11)$	$4.76(-28)$	f	2
3.7, 3.8	$HCO + O_2 = CO + HO_2$	$5.00(-12)$	$4.30(-18)$	f	1.5
3.9, 3.10	$CO + HO_2 = CO_2 + OH$	$5.12(-15)$	$2.95(-26)$	f	2
3.11, 3.12	$CO + OH = H + CO_2$	$3.19(-13)$	$1.31(-15)$	f	1.5
3.13, 3.14	$CO_2 + O = CO + O_2$	$7.36(-22)$	$1.41(-21)$	b	3
3.15, 3.16	$H + O_2 = O + OH$	$1.87(-13)$	$2.30(-11)$	f	2
3.17, 3.18	$H_2 + O = H + OH$	$5.62(-13)$	$6.36(-13)$	f	2
3.19, 3.20	$O + H_2O = OH + OH$	$2.74(-14)$	$6.82(-12)$	f	2.5
3.21, 3.22	$H + H_2O = OH + H_2$	$1.28(-14)$	$3.05(-12)$	b	2
3.23, 3.24	$H_2O_2 + OH = H_2O + HO_2$	$6.03(-12)$	$1.18(-17)$	f	1.5
3.25, 3.26	$HO_2 + O = O_2 + OH$	$3.60(-11)$	$4.11(-22)$	f	2
3.27, 3.28	$H + HO_2 = OH + OH$	$1.87(-10)$	$2.30(-19)$	f	1.5
3.29, 3.30	$H + HO_2 = H_2 + O_2$	$4.16(-11)$	$4.13(-22)$	f	2

3.31, 3.32	$OH + HO_2 = H_2O + O_2$	2.18(−11)	9.18(−25)	f	2.5
3.33, 3.34	$H_2O_2 + O_2 = HO_2 + HO_2$	2.24(−19)	1.05(−11)	b	3
3.35, 3.36	$HO_2 + H_2 = H_2O_2 + H$	2.33(−16)	5.07(−13)	b	2
3.37, 3.38	$O_2 + M = O + O + M$	3.37(−32)	5.13(−34)	b	3
3.39, 3.40	$H_2 + M = H + H + M$	3.07(−29)	8.27(−33)	f	2
3.41, 3.42	$OH + M = O + H + M$	3.00(−28)	2.76(−32)	b	30
3.43, 3.44	$H_2O_2 + M = OH + OH + M$	1.81(−16)	2.55(−32)	f	2
3.45, 3.46	$H_2O + M = H + OH + M$	5.00(−29)	3.19(−31)	f	2
3.47, 3.48	$HO_2 + M = H + O_2 + M$	2.91(−18)	7.18(−33)	b	3
3.49, 3.50	$CO_2 + M = CO + O + M$	5.37(−32)	2.50(−33)	b	4
3.51, 3.52	$HCO + M = H + CO + M$	4.61(−14)	8.85(−34)	f	1.5

From Ref. 110. Used by permission of the author.

[a]Units are cm molecule s kcal. Rate constants are evaluated at 1100 K; see ref. 110 for rate constant references.

[b]f indicates direction of reaction for which rate constant data were used.

[c]UF is the uncertainty factor, $k_{max} = k \times UF$, $k_{max} = k/UF$. The uncertainty factors are based on J. Warnatz, *Sandia Lab.* [*Tech. Rep.*] **SAND83-8606** (1983), and on W. Tsang, *Combustion Kinetic Data Survey.* National Bureau of Standards, Washington, D.C., 1983.

[d]Number associated with forward rate constant, followed by number associated with reverse rate constant.

[e]In this and Table 5, numbers in parentheses denote powers of ten.

175

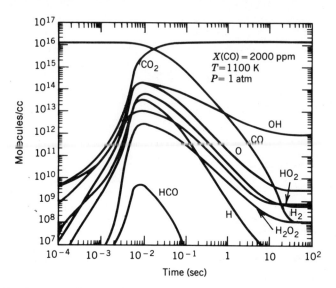

FIGURE 22. Concentration profiles as a function of time calculated using the mechanism in Table 3. Initial conditions: temperature $= 1100$ K, pressure $= 1$ atm, $[CO] = 1.337 \times 10^{16}$ mol-cm^{-3}, $[O_2] = 1.867 \times 10^{17}$ mol-cm^{-3}, $[H_2O] = 6.686 \times 10^{16}$ mol-cm^{-3}. Reprinted by permission of Elsevier Science Publishing Co., Inc. from "Some Interpretive Aspects of Elementary Sensitivity Gradients in Combustion Kinetics Modeling," by R. A. Yetter, F. L. Dryer, and H. Rabitz, *Combustion and Flame*, **59**, 107, © 1985 by The Combustion Institute.

information. For example, 10 species and 100 chemical reactions would yield 1000 first-order elementary sensitivity coefficients as functions of reaction time for just the sensitivities of species concentrations to each rate constant alone. This result clearly emphasizes the importance of efficient data handling and summary, as well as the importance of interactive connection through sensitivity analysis of numerical model development and laboratory experiment in guiding research direction.

Several interesting features are easily pointed out from the first-order coefficients displayed in Figure 23. Rank ordering alone delineates which of the many routes in the mechanism are important to the prediction of CO disappearance for these initial conditions. By rank ordering, it is apparent that only reactions (3.11), (3.48), (3.15), (3.16), (3.19), and (3.20) in Table 3 have significant influence on this calculated result. It is notable here that the importance of specific reactions changes dramatically with extent of the reaction. Second, by looking at forward and backward rate sensitivity, it is easily noted which elementary reactions achieve microscopically balanced conditions. For example, in Figure 23 it is evident that reactions (3.15) and (3.16) as well as (3.19) and (3.20) of Table 3 are microscopically balanced reactions. On the other hand, reaction set (3.11) and (3.12) of Table 3 is far away from the micro-balanced condition. Furthermore the sensitivity of reactions not explicitly contained in Table 3 can be assessed from the same results providing no new species need be introduced. This somewhat surprising fact

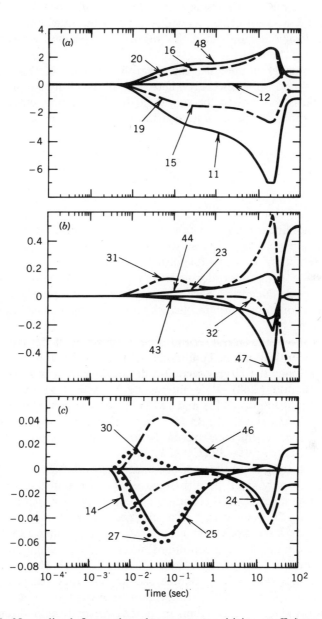

FIGURE 23. Normalized first-order elementary sensitivity coefficients of the CO concentration with respect to various reaction rate constants listed in the kinetic mechanism appearing in Table 3. Reprinted by permission of Elsevier Science Publishing Co., Inc. from "Some Interpretive Aspects of Elementary Sensitivity Gradients in Combustion Kinetics Modeling," by R. A. Yetter, F. L. Dryer, and H. Rabitz, *Combustion and Flame*, **59**, 107, © 1985 by The Combustion Institute.

follows because although missing reactions effectively have rate constants of zero value, the gradient of results with regard to the rate constants, need not be zero. Edelson[128] demonstrated this result for a system of oscillating reactions.

In local sensitivity analysis, the determination of second-order partial derivatives is important, especially where the parameter uncertainties are large. This situation is more often the rule than the exception in chemical kinetics since uncertainties in specific rate constants can easily approach an order of magnitude or more. Coefficients such as

$$\frac{\partial^2 C_i}{\partial k_j^2}$$

indicate the nonlinearity in variation of predicted specie concentration C_i with specific rate constants, k_j. If these second-order coefficients are not small, the range of prediction accuracy for variance from the normal values for rate constants made using first-order sensitivity analysis can be seriously compromised. Coffee and Heimrl[129] have compared first- and first-plus-second-order Taylor expansion sensitivity results with exact numerical solutions for the H_2/O_2 system.

Figure 24 shows the largest second-order coefficients of this type for the CO profile and identical initial conditions of Figures 22 and 23. In Table 3 reaction (3.11) has the largest second-order coefficient followed by reactions (3.15), (3.19), and (3.48). It is interesting to note that the second-order coefficients for reactions (3.15) and (3.19) are considerably larger than those for reactions (3.16) and (3.20). In fact there exists a period of time for which these second-order coefficients exceed the first-order magnitudes, for example, $(k_{11}^2/CO)(\partial^2 CO/\partial k_{11}^2) > \partial \ln CO/\partial k_{11}$ for 5 msec $< t <$ 10 msec and for $t > 30$ msec. During these time periods linear first-order sensitivity results will be accurate for only small perturbations in K_{11}. Examples of other features that can be investigated with second-order sensitivity coefficients are discussed for the $CO/H_2/O_2$ systems in Yetter and Yetter et al.[110,125,126]

The previous discussions have all addressed the problem input as specific rate constant data and initial conditions, whereas the problem output was the observable specific concentrations as a function of time. However, the information desired frequently is not these profiles but some physically definable features of the profiles such as chemical induction time (or other characteristic time), position shape or maximum in species concentration profile, and the periodicity of system oscillation. "Derived" sensitivity[110,126] analysis attempts to address such issues by defining new dependent variables or by interchanging dependent and independent variables in the analysis. A very interesting example to which such an analysis can be applied is related to the determination of the degree to which a given initial species affects shock-tube characteristic time-delay measurements.[110]

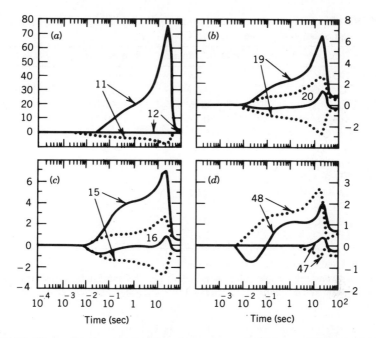

FIGURE 24. Normalized second-order elementary sensitivity coefficients of the CO concentration with respect to various reaction rate constants listed in the kinetic mechanism appearing in Table 3. Reprinted by permission of Elsevier Science Publishing Co., Inc. from "Some Interpretive Aspects of Elementary Sensitivity Gradients in Combustion Kinetics Modeling," by R. A. Yetter, F. L. Dryer, and H. Rabitz, *Combustion and Flame*, **59**, 107, © 1985 by The Combustion Institute.

Dean et al.[130] defined the induction time, t_i, for the reaction of the $CO-H_2-O_2$ mixtures studied as the time at which the tangent to the measured CO_2^* light emission versus time curve intercepted the time axis, t' as the time at which the measured CO_2 concentration equaled 8×10^{15} molecules-cm^{-3}, and t'' as the time at which the CO_2 concentration equaled 2.4×10^{16} molecules-cm^{-3}. Important questions are, how and which reactions in Table 3 affect these measured quantities, and how are these measured quantities affected by changes in the initial gas mixture and temperature?

Table 4 shows the results of a feature sensitivity analysis whose objectives are t_i, t', and t'', where the relative values of the sensitivity coefficients define which reactions most affect the quantity in question. Note that different reactions may be important to these quantities than are important to the precision of CO concentration as a function of time.

Table 5 shows the results of a similar feature sensitivity analysis for the induction time, t_i, as a function of initial conditions. Note that a 10% uncertainty in the induction time could result from only a 3%, a 1%, or a 0.05% uncertainty in the initial CO, O_2, or H_2 concentration. An uncertainty of approximately 50 K in initial temperature would cause a similar uncertainty in t_i. Although impurities have long been recognized to affect shock-tube

TABLE 4. Results of a Feature Sensitivity Analysis for the Relative Effects of Elementary Reactions on the Shock-Tube Characteristic Times t_i, t', t'' [a]

α_j	$\partial \ln t_i / \partial \ln \alpha_j$	$\partial \ln t' / \partial \ln \alpha_j$	$\partial \ln t'' / \partial \ln \alpha_j$
$CO + OH - CO_2 + H$	-0.28	-0.42	-0.57
$CO_2 + H - CO + OH$	—	—	0.03
$CO + O_2 - CO_2 + O$	-0.16	-0.10	-0.06
$H + O_2 - OH + O$	-0.17	-0.21	-0.26
$OH + O - H + O_2$	0.01	0.05	0.14
$O + H_2 - OH + H$	-0.36	-0.34	-0.22
$OH + H - O + H_2$	—	0.01	0.03
$O + H_2O - OH + OH$	—	-0.02	-0.08
$OH + OH - O + H_2O$	0.01	0.04	0.09
$H_2 + OH - H_2O + H$	-0.02	—	0.02
$H_2 + O_2 - H + HO_2$	-0.01	-0.01	—
$CO + O + M - CO_2 + M$	0.01	-0.02	-0.09

From Ref. 110. Used by permission of the author.
[a] Results are for constant volume calculations emulating the shock-tube experiments reported in Ref. 130. Shock-tube initial conditions: 0.49% H_2, 1.01% O_2, 3.28% CO, balance Ar, and 2050 K initial temperature.

TABLE 5. Results of a Feature Sensitivity Analysis for the Effects of Initial Condition Perturbations on the Shock-Tube Characteristic Time, t_i [a]

c_{j0}	$\partial t_i / c_{j0}$ (s mol^{-1} cm^{-3})	Δc_{j0} for 10% Change in t_i (ppm)	% of Initial Mixture
CO	$4.68(-22)$	6770	3
O_2	$1.21(-21)$	2620	1
H_2O	$4.76(-21)$	670	—
H_2	$2.96(-20)$	100	0.05
H	$7.03(-18)$	0.45	—
HO_2	$5.79(-18)$	0.55	—
H_2O_2	$1.02(-17)$	0.31	—
OH	$5.12(-18)$	0.62	—
	$\partial t_i / \partial T$	ΔT for 10% change in t_i	
	(s K^{-1})	(K)	
T	$1.67(-03)$	58	2.8

From Ref. 110. Used by permission of the author.
[a] Conditions are as described in Table 4.

induction (ignition) time measurements, these results clearly show such measurements in shock tubes are much less desirable (and perhaps imprecise) for validating kinetic mechanisms than are concentration-energy release variations with time.

What hopefully has been demonstrated here is that the application of local sensitivity analysis techniques can provide significant advances in kinetic model development and interpretation. However, one must be cautioned that the benefits are not simply achieved in that although these analyses are very powerful interpretative tools, they do not in themselves lead to absolute quantitative identification of answers; that is, there is currently no absolute guarantee that consideration of higher-order sensitivity analyses will not lead to different conclusions.

5. EMPIRICAL MODELING CONCEPTS

The concept of overall (global) reaction kinetics is a direct result of the complexity of most chemical reactions and the complicated fluid-mechanical situation in which some knowledge of heat release and chemical rates is necessary. Historically, the assumption invoked has been that the course of chemical kinetic events may be described in terms of a few of the principal reactants and products (C_i) in one or more global functional relations, each with much the same form as an elementary reaction process. Typically, the form of a global relation is

$$C_1 + C_2 \rightarrow C_3 + C_4 + \cdots \tag{53}$$

with the rate for each relation expressed defined by

$$-[\dot{C}_1] = k_{ov}[C_1]^{n1}[C_2]^{n2} \cdots \tag{54}$$

k_{ov}, the overall specific rate constant, is expressed in the Arrhenius form, where

$$k_{ov} = f(T) A \exp\left(-\frac{E}{rT}\right) \tag{55}$$

The n_i's appearing in each term of Eq. (54) are defined as the order of Eq. (54) with respect to C_i but are not necessarily equal to the stoichiometric coefficients appearing in Eq. (53). The summation of all of the n_i's in (54) is termed the *overall reaction order*. The product of $f(T)$ and A is termed the *overall frequency factor*, and E is referred to as the *overall activation energy* (in cal-mol^{-1} for $r = 1.987$ cal-mol^{-1}-K^{-1}) for the relation, Eq. (53). The brackets surrounding a particular C_i defines the use of the concentration of C_i in mol-cm^{-3}.

Such relations imply nothing about the actual kinetic mechanism (in terms of elementary reactions), although the parameters in the strictly empirical rate expressions sometimes are governed by a single elementary step (or a number of steps) that basically controls the rate of the chemical process. Under what circumstances such overall relations are usable is largely dependent on both the detailed kinetic behavior of the reaction and the physical environment in which the expression is derived. For example, Levy and Weinberg[131] concluded that use of a single relation such as Eq. (53) is not generally applicable to chemical measurements taken in flames. However, these authors did not attempt to separate chemical and diffusive effects from one another in developing their results, and it will be seen later that under some circumstances a single relation can be used to estimate flame propagation rate and mixture limits.

Where a particular rate-determining step or sequence of steps in the detailed chemical reaction mechanism exist and the physical circumstances of the application are similar to those from which the empirical model was derived, the approach can be a valid and vastly simplifying idea. However, extension of the model to experimental conditions outside the range of parameters for which validation has been performed should be done with great reservation. Unfortunately, there is currently little hope of avoiding this problem in the case of hydrocarbon combustion chemistry, particularly in two and three dimensional numerical simulations.

5.1. One- and Two-Step Global Mechanisms

One- and two-step global mechanisms have commonly been used in many combustion modeling exercises where a description of heat release or flame propagation velocity is required. For example, Butler et al.[132] represented the combustion of n-octane as

$$2C_8H_{18} + 25O_2 \rightarrow 16CO_2 + 18H_2O \qquad (56)$$

$$(\dot{y}_f) = -9.38 \times 10^{11}(y_f)(y_o)\exp\left(\frac{-15{,}780}{T}\right) \text{ g cm}^{-3} \text{ sec} \qquad (57)$$

where the subscripts f and o refer to fuel and oxygen and $\dot{y}$ indicates the rate of change in partial mass density (y_i) due to chemical reaction. This representation is typical in that the description of energy release requires both an overall stoichiometric relation and a rate expression.

Both carbon monoxide and methane oxidation have received considerable attention as fuels for which single-step global reaction parameters have been derived directly from experimental data. Numerous global modeling studies have been performed directly from experimental data. Numerous global modeling studies have been performed on carbon monoxide, and these have been reviewed by Dryer[133] and Howard et al.[134] Methane ignition and oxidation kinetics also have been expressed in this manner, often in conjunction with developing detailed mechanisms, and many of these studies are also discussed in Dryer and Matula et al.[133, 135]

The characteristics of the postinduction reaction of methane-oxygen mixtures as studied by Dryer and Glassman[136] encourage speculation that the rate of hydrocarbon reaction can be expressed by a simple global expression of the form of Eq. (53). The rate of reaction in the postinduction phase of the lean methane oxidation in a flow reactor was found to be described well by the overall expression:

$$-\left[\dot{CH_4}\right] = 10^{13.2} \exp\left(\frac{-48{,}400}{rT}\right)[CH_4]^{0.7}[O_2]^{0.8} \qquad (58)$$

However, it should be noted that the parameters of this equation are significantly different from those found by investigators who have studied the induction (ignition delay) phase of this reaction in shock tubes and flow reactors. A review of available data through 1973[135] predicts the rate of the induction reaction to be inhibited by the concentration of methane. Seery and Bowman[137] empirically correlated the ignition delay time as

$$(\text{Rate})^{-1} \simeq (\text{Reaction time})$$

$$= 7.65 \times 10^{-18} \exp\left(+\frac{51{,}400}{RT}\right)[CH_4]^{0.4}[O_2]^{-1.6} \qquad (59)$$

and developed a detailed elementary mechanism that reasonably predicted the relation. Figure 25 shows a comparison of the overall rate constant derived by Dryer[133] and the results calculated from parameters predicted by detailed analytical studies of Bowman.[138] The analytical overall rate constant was calculated from

$$k_{ov} = -\frac{\left[\dot{CH_4}\right]}{[CH_4]^{0.7}[O_2]^{0.8}} \qquad (60)$$

Clearly there are two phases of this reaction that are not modeled by the same global parameters. Indeed, the experimental flow reactor data of Dryer show similar behavior.[46, 133]

Dryer[133] completed the modeling of the postinduction methane oxidation using a two-step irreversible global representation:

$$CH_4 + \left(\tfrac{3}{2}\right)O_2 \rightarrow CO + H_2O \qquad (61)$$

$$CO + \left(\tfrac{1}{2}\right)O_2 \rightarrow CO_2 \qquad (62)$$

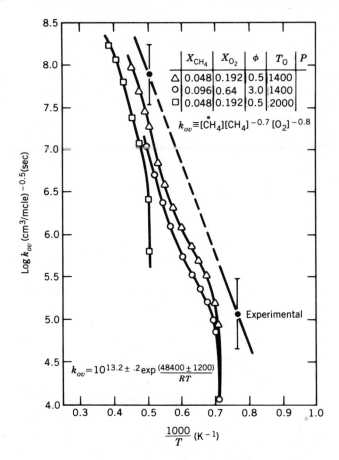

FIGURE 25. Comparison of overall rate constants for methane disappearance. Experimental values (solid curves) are from F. L. Dryer and I. Glassman, "Combustion Chemistry of Chain Hydrocarbons," *Prog. Astronaut. Aeronaut.* **62**, 55 (1979). © American Institute of Aeronautics and Astronautics; reprinted with permission. Calculated values (dashed curves) were derived using the mechanism of Bowman.[138]

with the reaction rates described by

$$\left[\dot{CH}_4\right] = 10^{13.2}\exp\left(\frac{-48,000}{rT}\right)[CH_4]^{0.7}[O_2]^{0.8} \tag{63}$$

$$\left[\dot{CO}_2\right] = 10^{14.6}\exp\left(\frac{-40,000}{rT}\right)[CO][H_2O]^{0.5}[O_2]^{0.25} \tag{64}$$

Application of this two-step global mechanism to the post induction flow reactor data produces the comparison plots shown in Figure 26. Considering the simplicity of the model, the agreement between experimental and computed results is excellent. While this encouraging result suggests that similar

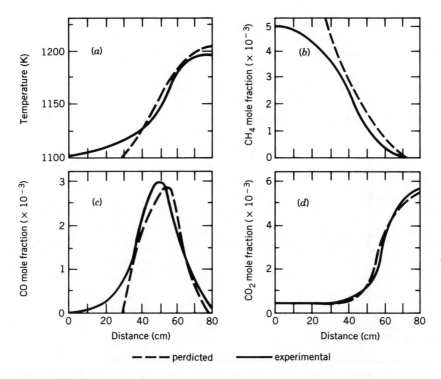

FIGURE 26. Comparison of predictions using global methane mechanism with experimental flow reactor data. Dashed lines are predicted data. Solid lines are experimental results. The original version of this material was first published by the Advisory Group for Aerospace Research and Development, North Atlantic Treaty Organisation (AGARD/NATO) in Conference Proceedings No. 275, *Combustion Modelling*, published in February 1980.

simplified mechanisms could be useful in modeling flame propagation, etc., it does not follow that this mechanism and its parameter values would directly apply to such problems or other temperature ranges.[10]

Westbrook and Dryer[139] have discussed some of the properties of simple one- and two-step reaction mechanisms for modeling laminar pre-mixed flame properties. It was shown that a simple single-step irreversible reaction

$$\text{Fuel} + n_1 O_2 \rightarrow n_2 CO_2 + n_3 H_2 O \tag{65}$$

(n_i's are determined by the choice of fuel) with a rate expression defined as

$$[\text{fuel}] = A \exp\left(\frac{-E_a}{rT}\right)[\text{fuel}]^a [O_2]^b \tag{66}$$

could be used to estimate flame propagation reasonably well when transport effects were considered independently. Transport coefficients were fixed at values determined in previous numerical modeling of flame propagation using detailed chemical kinetics,[101] and Eqs. (65) and (66) were used to replace the detailed chemical mechanism. The rate parameters in Eq. (66) were used to provide agreement between computed and experimental results. The effective activation energy of the rate expression was found to affect primarily the computed flame thickness, a parameter for which no experimental data were available. An activation energy of 30 kcal mol^{-1} was chosen arbitrarily based on available experimental determinations.[140, 141]

With E_a, a, and b held fixed, the pre-exponential A was varied until the model correctly predicted the flame speed for an atmospheric pressure, stoichiometric fuel-air mixture, such as 40 cm/sec for C_8H_{18}/air. The resulting rate expression was then used to predict flame speeds for fuel-air mixtures at other equivalence ratios and pressures.

The sensitivity of the computed flame speeds to the fuel concentration exponent required that a be approximately 0.25. In addition, if $a + b$ was about 1.75, then the global mechanism also reproduced the desired pressure dependence of the flame speed for pressures greater than or equal to atmospheric. In principal, the oxidizer concentration exponent, b, determined the lean flammability limit, leaving three conditions (ϕ_R, ϕ_L, and pressure exponent) to be satisfied by two constants a and b. Fortunately, computed results on the lean side of stoichiometric were relatively insensitive to variations in the oxygen concentration exponent, b. Thus it was possible to satisfy all of the observed behavior with one set of concentration exponents.

Table 6 presents the parameters that were found to give the best results for several types of fuels, including alkanes, alkenes, and aromatics. Also shown are the computed flammability limits and the experimental limits from Dugger et al.[142] or Lewis and von Elbe.[143] For each fuel, the use of a smaller value for the fuel concentration exponent leads to a lower predicted value for the rich flammability limit.

Except for methane, the values of a and b were selected to give a flame velocity pressure dependence of $p^{-0.125}$. For methane, the laminar flame pressure dependence ($p^{-0.05}$) indicates that $a + b = 1.0$, while the observed rich flammability limit of $\phi_R = 1.6$ requires $a = -0.3$. Methane oxidation in shock tubes[144] and in the turbulent flow reactor[133, 136] was characterized by a considerably higher overall activation energy of about 48.4 kcal mol^{-1}. The flame thickness in the model, with $E_a = 30$ kcal mol^{-1}, was about 30% greater than when $E_a = 48.4$ kcal mol^{-1}, but the best concentration exponents, the flame speeds and their dependence on pressure and equivalence ratio, were essentially the same for both models.

The fuel concentration exponent for CH_4 from Table 6 has a negative value, -0.3. Technically, the fuel acts as an inhibitor, similar to observations for methane ignition in shock tubes.[136] Note that this is in considerable disagreement with parameters derived from flow reactor data.[133, 136] From a numerical point of view this negative coefficient can create problems since the rate of methane consumption increases without limit as the methane concentration approaches zero. Other semiempirical models for hydrocarbon oxida-

TABLE 6. Single-Step Reaction Rate Parameters, Giving Best Agreement between Experimental Flammability Limits (ϕ'_L and ϕ'_R) and Computed Flammability Limits (ϕ_L and ϕ_R)[a]

Fuel	A	E_a	a	b	ϕ'_L	ϕ'_R	ϕ_L	ϕ_R
CH_4	1.3×10^8	48.4	-0.3	1.3	0.5	0.5	1.6	1.6
CH_4	8.3×10^5	30.0	-0.3	1.3	0.5	0.5	1.6	1.6
C_2H_6	1.1×10^{12}	30.0	0.1	1.65	0.5	0.5	2.7	3.1
C_3H_8	8.6×10^{11}	30.0	0.1	1.65	0.5	0.5	2.8	3.2
C_4H_{10}	7.4×10^{11}	30.0	0.15	1.6	0.5	0.5	3.3	3.4
C_5H_{12}	6.4×10^{11}	30.0	0.25	1.5	0.5	0.5	3.6	3.7
C_6H_{14}	5.7×10^{11}	30.0	0.25	1.5	0.5	0.5	4.0	4.1
C_7H_{16}	5.1×10^{11}	30.0	0.25	1.5	0.5	0.5	4.5	4.5
C_8H_{18}	4.6×10^{11}	30.0	0.25	1.5	0.5	0.5	4.3	4.5
C_8H_{18}	7.2×10^{12}	40.0	0.25	1.5	0.5	0.5	4.3	4.5
C_9H_{20}	4.2×10^{11}	30.0	0.25	1.5	0.5	0.5	4.3	4.5
$C_{10}H_{22}$	3.8×10^{11}	30.0	0.25	1.5	0.5	0.5	4.2	4.5
CH_3OH	3.2×10^{12}	30.0	0.25	1.5	0.5	0.5	4.1	4.0
C_2H_5OH	1.5×10^{12}	30.0	0.15	1.6	0.5	0.5	3.4	3.6
C_6H_6	2.0×10^{11}	30.0	-0.1	1.85	0.5	0.5	3.4	3.6
C_7H_8	1.6×10^{11}	30.0	-0.1	1.85	0.5	0.5	3.2	3.5

From Ref. 139. Reprinted with permission from Gordon and Breach Science Publishers S. A.
[a]Units are cm-sec-mol-kcal-K.

tion also have this difficulty,[145] and there are several possible solutions to the problem. A reverse reaction can be used that provides an equilibrium fuel concentration at some small level, preventing the rate expression from becoming too large. The rate expression can also be artificially truncated at some predetermined value. However, in some cases it may be preferable to sacrifice some of the generality provided by the rate parameters in order to keep the rate expression conveniently bounded. In the present work it was possible to reproduce the flame speed dependence on equivalence ratio for methane at a given pressure with concentration exponents that did not satisfy the constraint $a + b = 1.75$ and therefore will not reproduce the correct dependence of flame speed on pressure.

A most important point can be made by recalling the model of Butler et al.[132] for octane combustion discussed earlier. In propagating flames, first-order fuel and oxidizer dependence would require that

$$k_{ov} = 1.15 \cdot 10^{14} \exp\left(-\frac{30,000}{rT}\right)[C_8H_{18}]^{1.0} \cdot [O_2]^{1.0}$$

to achieve a flame speed of 40 cm sec^{-1} for a stoichiometric mixture, $\phi = 1.0$, and atmospheric pressure. However, with this expression computed flame speeds for fuel-rich mixture are much too fast, and an extrapolation of the curve gives a rich flammability limit, ϕ_R, of approximately 10. The maximum flame speed of nearly 55 cm sec^{-1} occurs near $\phi = 2$, in considerable disagreement with experiment results. The inadequacy of the assumption that $a = b = 1$ for rate expressions, one that has often been made for mathematical simplifications, is clear. The important point to be remembered is that it is important to define the constraints of the empirical modeling assumptions one chooses.

Although single-step mechanisms can predict flame speeds reasonably well over considerable ranges of conditions, the approach has several flaws that can be important in certain applications. The irreversible reaction to the specified products (CO_2 and H_2O) must overpredict the amount of chemical energy released (and the final gas composition) in an adiabatic environment. At adiabatic flame temperatures typical of hydrocarbon fuels ($T > 2000$ K), substantial amounts of dissociation products exist in equilibrium with the combustion products CO_2 and H_2O. This equilibrium substantially lowers the total heat release and the adiabatic flame temperature below the values predicted by Eq. (65).

The overestimate of adiabatic flame temperature by the single-step mechanism grows with increasing equivalence ratio, and the problem can be corrected in a number of ways, all of which appropriately modify the total heat of reaction.

For example, the adiabatic equilibrium condition of the postflame gases can be approximated by considering only the two additional species CO and H_2 rather than a full adiabatic equilibrium solution.[139] The effects of incomplete conversion to CO_2 and H_2O, as well as the sequential nature of the hydrocarbon oxidation, can be implemented into a simplified model by considering the reaction to occur in two overall steps.[99] For n-paraffin fuels this results in

$$C_nH_{2(n+1)} + \frac{2(n+1)}{2}O_2 \rightarrow (n+1)H_2O + nCO \qquad (67)$$

$$CO + \frac{1}{2}O_2 = CO_2 \qquad (68)$$

If the second reaction is considered to be reversible, then carbon dioxide dissociation can be embodied within the simplified model. Westbrook and Dryer[139] demonstrated this approach by developing a reverse reaction expression that together with the global forward rate expression of Dryer and Glassman,[136] Eq. (64), predicted proper heat release and pressure dependence of the CO/CO_2 equilibrium:

$$k_b = 5 \cdot 10^8 \exp\left(-\frac{40,000}{rT}\right)[CO_2]^{1.0} \qquad (69)$$

Note that this specification actually requires only two overall rate expressions be defined since the third is defined by the equilibrium constant for Eq. (68).

TABLE 7. Parameters for Two-Step Reaction Mechanism, Giving Best Agreement between Experimental and Computed Flammability Limits[a]

Fuel	A	E_a	a	b
CH_4	2.8×10^9	48.4	-0.3	1.3
CH_4	1.5×10^7	30.0	-0.3	1.3
C_2H_6	1.3×10^{12}	30.0	0.1	1.65
C_3H_8	1.0×10^{12}	30.0	0.1	1.65
C_4H_{10}	8.8×10^{11}	30.0	0.15	1.6
C_5H_{12}	7.8×10^{11}	30.0	0.25	1.5
C_6H_{14}	7.0×10^{11}	30.0	0.25	1.5
C_7H_{16}	6.3×10^{11}	30.0	0.25	1.5
C_8H_{18}	5.7×10^{11}	30.0	0.25	1.5
C_8H_{18}	9.6×10^{12}	40.0	0.25	1.5
C_9H_{20}	5.2×10^{11}	30.0	0.25	1.5
$C_{10}H_{22}$	4.7×10^{11}	30.0	0.25	1.5
CH_3OH	3.7×10^{12}	30.0	0.25	1.5
C_2H_5OH	1.8×10^{12}	30.0	0.15	1.6
C_6H_6	2.4×10^{11}	30.0	-0.1	1.85
C_7H_8	1.9×10^{11}	30.0	-0.1	1.85

From Ref. 139. Reprinted with permission from Gordon and Breach Science Publishers S. A.
[a]Same units as in Table 6.

The resulting mechanism with the constants of Table 7 predicted flame speeds in close agreement with those predicted by the single-step models in Table 6 over the same ranges of equivalence ratio and pressure. In addition the $CO-CO_2$ equilibrium improved the mechanism by providing a better estimate of the adiabatic flame temperature and postflame CO concentration. Further refinement in expressing equilibrium effects on burned gas temperature lead to additional improvements, particularly for the fuel-rich case where hydrogen becomes a significant product in the postflame gases. Following the procedure just discussed, two additional rate expressions and reversible global reactions could be added to include hydrogen as a reaction product.

Reitz and Bracco[146] have recently suggested another approach for achieving this result, that of developing a "local equilibrium model." This work reemphasizes the arguments of Westbrook and Dryer[139] that irreversible mechanisms cannot in themselves reproduce thermodynamic equilibrium steady-state composition of final products and thus the appropriate adiabatic energy release. On the claim that reversible global kinetics introduces too many independent parameters (in specifying the required overall rate constants), Reitz and Bracco suggested a formulation approach similar to that used in describing vibrational relaxation[147] in nonequilibrium flow.

In this method the rates of reaction are defined by a first-order expression about local equilibrium:

$$(\dot{p}\overline{Y}) = \overline{w} = \overline{w}^* - J^*(Y - \overline{Y}^*) \tag{70}$$

where $\overline{Y}$ and $\overline{w}$ are N component vectors composed of the mole fractions and mass rates of change with time of each specie. The Jacobian matrix $J*$ is

$$J* = \left. \frac{\partial w^{(k)}}{\partial Y^{(1)}} \right|_{p,T}^{*} \tag{71}$$

where p, T are the instantaneous gas density and temperature. The principal assumption made by the authors for the purposes of global modeling with one irreversible reaction is that $J*$ is a diagonal matrix with only one nonzero component T, where $T^{-1} = k_{ov}(W_{\text{fuel}}/Y_{\text{fuel}})$. Equation (70) then becomes

$$\left(p\dot{Y}^{k} \right) = -T^{-1}(Y^{(k)} - Y^{*(k)}) \tag{72}$$

with $Y^{*(k)}$ given by the thermodynamic equilibrium solution for the instantaneous local pressure and temperature.

Reitz and Bracco also developed a simple algebraic solution for thermodynamic equilibrium that considers the species CO, CO_2, H_2, H_2O, and O_2, thus vastly simplifying the required computations. The method appears to be a satisfactory alternative to the reversible reaction method demonstrated by Westbrook and Dryer,[139] and is valid over a wide range of conditions in that it considers both H_2 and CO as final products. Furthermore the authors argue that it is computationally more efficient.

The inference of the above results is that one- and two-step global chemistry can be used under limited circumstances for modeling flames if the global parameters are fitted to provide appropriate behavior for the problem of interest. Coffee et al.[148,149] have made the point that accurate burning velocity, heat release profile, and flame temperature over very wide ranges of equivalence ratio require fitting parameters to be a function of equivalence ratio as well as fuel type. Although this fact is true by definition at some desired level of accuracy, one should always consider the trade-off between desirable tractability and the consequent sacrifices in achieving an accuracy beyond that necessary for the particular problem at hand.

Yetter and Dryer[10] have recently amplified that single-step chemistry has general utility by showing that single-step chemistry can be derived to reproduce the overall rate of reaction characteristics for $CO-H_2-O_2$ mixtures under homogeneous, laminar flame, or stirred reactor conditions. However, an important observation was that the global rate parameters were vastly different for the same reactive mixture in each of the different environments. The global models were compared with detailed kinetic results for each condition, and it was shown that the primary reason for changes in the global parameters is that the relative importance of different elementary reactions was influenced in each environment by the diffusive mass and energy transport conditions.

5.2. Multistep Global and Quasi-global Mechanisms

Although one-/two-step global and one-step local equilibrium mechanisms provide adequate approaches to describing energy release and burned gas compositions, they are incapable of reproducing the intermediate chemical behavior of hydrocarbon oxidation, and consequently the interactions necessary to be universally applicable for different environments. To generally characterize at least some of the intermediate chemistry in the oxidation of alkanes, a minimum of four overall reactions (or characteristic times) is required:[145]

$$C_nH_{2n+2} + \frac{1-x}{2}O_2 \rightarrow \frac{n}{2}C_2H_4 + xH_2 + (1-x)H_2O \qquad (73)$$

$$C_2H_4 + \frac{4-2x}{2}O_2 \rightarrow 2CO + 2xH_2 + (2-2x)H_2O \qquad (74)$$

$$CO + \frac{1}{2}O_2 \rightarrow CO_2 \qquad (75)$$

$$H_2 + \frac{1}{2}O_2 \rightarrow H_2O \qquad (76)$$

where x is a function of equivalence ratio and the concentration of C_2H_4 represents the mass of all hydrocarbon intermediates that are formed during the oxidation process. These species are mainly the 1-olefins and are given the properties of the simple olefin, C_2H_4, as a result of the experimental observation described earlier that the dominant intermediate is ethene. The first reaction may be considered to be the result of the oxidative pyrolysis of the aliphatic fuel to the olefin and hereafter will be referred to as the fuel pyrolysis step. At very lean stoichiometries, x approaches zero, and as the stoichiometry becomes richer, x increases.

Another possible approach would be to set x equal to one in the first reaction, that is, assume no water was directly produced in the alkane destruction:

$$C_nH_{2n+2} \rightarrow \frac{n}{2}C_2H_4 + H_2 \qquad (77)$$

$$C_2H_4 + O_2 \rightarrow 2CO + H_2 \qquad (78)$$

$$CO + \frac{1}{2}O_2 \rightarrow CO_2 \qquad (79)$$

$$H_2 + \frac{1}{2}O_2 \rightarrow H_2O \qquad (80)$$

Still other methods of modeling are to use the local equilibrium method with multiple time constants for each of the irreversible global reaction steps,[150] or to use global reaction steps along with a limited mechanism of elementary reactions (quasi-global modeling).[151]

Hautman et al.[145] investigated a multistep global modeling concept for paraffin oxidation using Eqs. (77)–(80) with semiempirical overall rate expressions derived from intermediate temperature flow reactor oxidation of propane. The model adequately reproduced the major species profiles in the flow reactor and characteristic times for shock-tube oxidation, but the reaction rate expressions were formulated in a manner that caused computational difficulties (certain concentration dependences in the rate expressions have negative exponents). More important, the mechanism was an irreversible one that could not therefore reproduce equilibrium product distributions at high temperatures or rich conditions. Proscia[53,152] generalized this approach for n-alkanes by defining reversible reaction rate expressions for each of the Eqs. (77)–(80). Reversibility for the rate equations for Eqs. (77) and (78) is required to eliminate numerical floating point division errors during computations, whereas reversibility for the rate equations for Eqs. (79) and (80) approximate equilibrium product distributions. The actual rate expressions for homogeneous reaction of n-butane, n-octane, iso-butane, and iso-octane mixtures developed for modeling flow reactor data appear in Proscia.[53] Results are, however, of limited value because determined quantities apply only to the limited experimental parameter ranges available in the flow reactor.

One might also consider an early approach proposed by Edelman and Fortune,[151] namely, combining the use of both global expressions and elementary reactions (quasi-global modeling). In their original approach Edelman and Fortune chose to approximate the higher paraffin oxidation to carbon monoxide and hydrogen as a unidirectional global reaction

$$C_nH_{2n+2} + \frac{n}{2}O_2 \rightarrow nCO + (n+1)H_2O \tag{81}$$

with the rate given by

$$[C_n\dot{H}_{2n+2}] = k_{ov}[C_nH_{2n+2}]^{0.5}[O_2]^{1.0} \tag{82}$$

and combined these equations with a number of elementary reactions from the hydrogen/oxygen and carbon monoxide/oxygen reaction mechanisms. Edelman[151,153–156] has used this concept extensively for modeling both plug flow and well-stirred conditions, and Westbrook and Dryer[139] have evaluated the technique for laminar flame modeling. Because all of the important elementary reactions and species of the $CO-H_2-O_2$ system are included in the mechanism, this approach has the potential to provide an accurate values for the equilibrium postcombustion composition and temperature. Since thermal NO_x production in flames depends primarily on burned gas characteristics, addition of the extended Zeldovich mechanism to the quasi-global scheme should be capable of providing reasonable estimates of NO_x formation.

Similar attributes can be obtained from one-, two-, or multistep global mechanisms with coupled partial equilibrium reactions.

Modified mechanisms based on the quasi global concept have also have appeared in literature. In attempting to predict gas turbine NO_x emissions, Mellor[157] replaced Eq. (81) with

$$C_nH_n + \left(\frac{n}{2} + \frac{m}{4}\right)O_2 \rightarrow CO + \left(\frac{m}{2}\right)H_2O \qquad (83)$$

and defined the rate constant to be infinite. This modification was based upon the results of Marteney.[158] However, Bowman, in comments to Edelman et al.,[153] showed that "infinite" quasi-global kinetics do not offer any significant advantages over the partial equilibrium approach for prediction of NO_x emissions. It should be noted that these calculations also show that quasi-global finite and quasi-global infinite kinetics are equally capable of estimating NO_x emissions for residence times that are long in comparison to the time necessary to complete hydrocarbon combustion to its equilibrium product distribution. Thus it should be remembered that in many cases and particularly for lean oxidation, prediction of NO_x emissions may not be a sensitive enough test to judge the qualities of a proposed hydrocarbon oxidation model.

Roberts et al.[159] have also derived a combustion mechanism based upon the results of Edelman[151] by replacing Eq. (81) with

$$C_8H_{18} + O_2 \rightarrow 2C_4H_8O \qquad (84)$$

$$2C_4H_8O + 3O_2 \rightarrow 8CO + 8H_2O \qquad (85)$$

$$C_8H_{16} + OH \rightarrow H_2CO + CH_3 + 2C_2H_2 \qquad (86)$$

to describe the initial fuel disappearance. The model also significantly modified and extended the set of elementary reactions included by Edelman and Fortune, added a number of intermediate (nonelementary) reactions of species such as HCO, H_2CO, CH_3, and C_2H_2 and was the first to include chemical coupling of fuel disappearance with the elementary mechanism through the radical species. (The complete mechanism appears in Kollrack and Aceto.[160]) The reaction orders in the associated rate correlations were defined to be equal to the stoichiometric coefficients of Eqs. (84)–(86) and the rate constants themselves were obtained by matching the fuel disappearance predicted by Eq. (82). The mechanism was used in numerical calculations for gas turbine combustors to predict the formation of nitric oxide,[159–161] the effects of water addition on NO_x emissions,[162] carbon monoxide production in the primary zone[163] (including droplet effects[164]), and emissions produced by methanol and jet fuels.[165]

Dryer and Glassman[46] have shown that although original quasi-global modeling can apparently predict overall reaction times of normal and cyclo-paraffins, it does not predict intermediate product evolution and destruction well at flow reactor temperatures. This results from the fact that the model has no mechanism through which intermediate hydrocarbons can be evolved and destroyed or through which the carbon monoxide oxidation can be inhibited by the presence of hydrocarbon species. Because of these difficulties, energy release is also not properly modeled. Finally, one should expect the rate parameters for fuel consumption in the quasi-global model to depend on the effects of transport.[10]

Westbrook and Dryer[139] corroborated similar deficiencies for applications to laminar flames. Using the quasi-global modeling concept, they determined a global reaction rate for the fuels discussed earlier. The computed flame speeds as functions of equivalence ratio and pressure were essentially indistinguishable from those found from the single-step and two-step mechanisms discussed earlier. The principal advantage was a further improvement in defining the burned gas composition and temperature. However, the flame structure and species concentrations in the flame zone were not well predicted by the quasi-global mechanism. Thus, though the quasi-global modeling currently available for hydrocarbon oxidation appears to have several deficiencies, the concept upon which it is developed is quite appropriate for combustion modeling. Correct modeling of carbon monoxide/hydrogen oxidation in detail can estimate both the major energy release step of alkyl-hydrocarbon oxidation, and the radical intermediates (OH, O, H) necessary to predict the Zeldovich NO_x production and quenching of carbon monoxide. By adding elementary SO_2/SO_3 chemistry, the mechanism also could define the SO_2/SO_3 conversion that occurs during dilution or cooling of combustion gases. However, a global model for CO oxidation coupled with partial equilibrium might be sufficient for calculation of the Zeldovich NO_x production when one is concerned only with long residence times at high temperature. But the more difficult problem is how to predict empirically the ignition delay period for the carbon monoxide conversion. From the qualitative hydrocarbon oxidation description assembled earlier, it is apparent that the initiation of CO_2 formation and the formation of soot precursor species coincides with the oxidation of olefins formed from the initial hydrocarbon, and not with that of the initial hydrocarbon itself. These two conversion steps may indeed have different temperature and concentration dependences, and thus it is likely that accurate modeling of hydrocarbon oxidation over the required temperature range will necessitate empirical prediction of each of these hydrocarbon conversion processes: the conversion of primary fuel to olefins, and the conversion of olefins to CO. In this regard quasi-global modeling may offer little benefit over multistep global reaction schemes.

5.3. Detailed Kinetic Models as Sources of Reduced Mechanisms

When detailed kinetic mechanisms are available, a more suitable approach is to derive the semiempirical overall rate expressions for the global mechanism step by comparison with model calculations using the detailed kinetics and the global model.[10] Kiehne et al.[166] recently advanced Proscia's[53] empirical efforts by applying this approach to propane flames. A modified, reversible, four-step global model was developed through comparison with the laminar flame calculations using the detailed kinetic mechanism for propane oxidation of Westbrook and Pitz.[88] It was found that such a model could accurately reproduce the primary features, including major species profiles, temperature profiles, and energy release rate profiles, quench distance, and ignition delay for both freely propagating and wall quenched laminar flame configurations. Table 8 presents the modified reversible four-step mechanism that Kiehne et al. developed, while Figures 27 compares laminar flame calculations using the Westbrook and Pitz[88] detailed mechanism, the four-step reversible mechanism developed for flow reactor conditions by Proscia,[53] and the mechanism appearing in Table 8. As expected from earlier discussions here and in Yetter

TABLE 8. Summary of the Modified Four-Step, Reversible Mechanism for Laminar Flames of Propane and Air

$$2C_3H_8 \rightarrow 3C_2H_4 + 2H_2$$
$$C_2H_4 + O_2 \rightarrow 2H_2 + 2CO$$
$$2CO + O_2 \rightarrow 2CO_2$$
$$2H_2 + O_2 \rightarrow 2H_2O$$
$$3C_2H_4 + 2H_2 \rightarrow 2C_3H_8$$
$$2H_2 + 2CO \rightarrow C_2H_4 + O_2$$
$$2CO_2 \rightarrow 2CO + O_2$$
$$2H_2O \rightarrow 2H_2 + O_2$$

With rate expressions:

$$R_1 = f_1(P)2.089 \times 10^{17} \exp(-49,600/RT)[C_3H_8]^{0.50}[O_2]^{1.07}[C_2H_4]^{0.40}$$
$$R_2 = f_2(P)2 \times 10^{13} \exp(-50,000/RT)[C_2H_4]^{0.90}[O_2]^{1.18}[C_3H_8]^{-0.37}$$
$$R_3 = S(\phi)1.5 \times 10^{13} \exp(-40,000/RT)[CO]^{1.0}[O_2]^{0.25}[H_2O]^{0.50}$$
$$R_4 = 3.311 \times 10^{13} \exp(-38,100/RT)[H_2]^{0.85}[O_2]^{1.42}[C_2H_4]^{-0.56}$$
$$R_5 = 4.920 \times 10^8 \exp(-49,600/RT)[C_3H_8]^{0.127}[O_2]^{1.07}[C_2H_4]^{0.40}$$
$$R_6 = 2.25 \times 10^9 \exp(-50,000/RT)[C_2H_4]^{0.528}[O_2]^{1.18}[C_3H_8]^{-0.37}$$
$$R_7 = 4.16 \times 10^{16} T^{-1/2} \exp(-106,950/RT)[CO_2]^{1.0}[O_2]^{-0.25}[H_2O]^{0.50}$$
$$R_8 = 6.12 \times 10^{15} T^{-1/2} \exp(-100,586/RT)[H_2]^{-0.153}[O_2]^{0.916}[C_2H_4]^{-0.563}[H_2O]^{1.0},$$

where

$$f_1(P) = 6.434P^{-0.8116}$$
$$f_2(P) = 1.115 - 1.125e^{-0.251}$$
$$S(\phi) = \min[1.0, 16.0 \exp(-2.48\phi)]$$

From Ref. 166. Reprinted with permission from Gordon and Breach Science Publishers S. A.

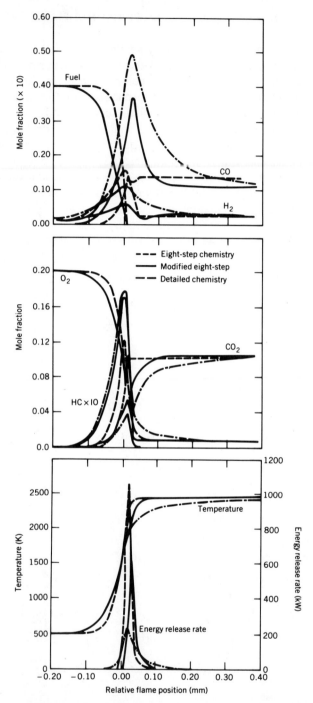

FIGURE 27. Comparison of predictions for the profiles of temperature, energy release rate, and species profiles for a freely propagating laminar stoichiometric propane-air flame. Initial pressure = 10 atm, initial temperature = 500 K. From Ref. 166. Reprinted with permission from Gordon and Breach Science Publishers S. A.

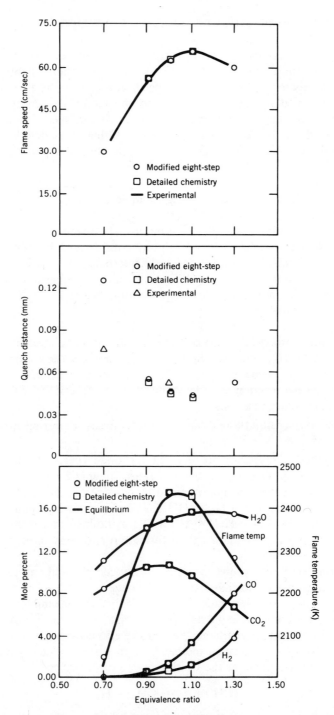

FIGURE 28. Comparison of predicted and experimental flame speeds, quench distances, and the equilibrium postflame mole percentages and temperatures as a function of equivalence ratio. Initial pressure = 10 atm, initial temperature = 500 K. From Ref. 166. Reprinted with permission from Gordon and Breach Science Publishers S. A.

and Dryer,[10] it is not surprising that rate expressions for flow reactor, homogeneous conditions do not work as well for laminar flame modeling. Figure 28 compares the laminar flame speeds and quench distances calculated, using the detailed kinetic mechanism and the modified four-step reversible mechanism with experimental data from Metgelchi and Keck[167] and Daniel.[168] The use of the multistep global mechanism rather than detailed chemistry clearly leads to substantial reduction of computational requirements without loss in accuracy.

The work of Proscia,[152] and Kiehne et al.[166] indicate that this same technique should be extendable to higher molecular weight hydrocarbons for which detailed kinetic mechanisms exist, but they lend no guidance as to what functional relationships should be used for the reaction equations or the reaction rate expressions. There has been considerable interest recently in developing systematic methods for reducing detailed kinetic mechanisms. Paczko et al.[11] and Peters and Williams[169] have developed a scheme that results in simple three- and four-step reaction mechanisms. The strategy[11] consists of several steps:

1. Together with steady-state relations for all intermediates with the exception of H, H_2, and CO, the fastest reactions are determined. With the elimination of the steady-state species, these reactions represent the quasi-global reaction scheme, while the rate expressions remain those of the elementary reactions.

2. Based on sensitivity analysis or path analysis, a starting mechanism is derived.

3. Simplified expressions are derived for these species by analyzing the relative magnitude of the reaction rates. Algebraic relations for the OH and O concentrations are obtained by considering the reactions $O + H_2 \rightarrow OH + O$ and $O + H_2 \rightarrow OH + H$ and the partial equilibrium of the reaction $OH + O = H_2O + H$ for OH. All other species are assumed to be determined by equating their formation and consumption rates.

The resulting rate expressions are algebraically complicated and their reaction orders differ from the stoichiometric coefficients.

Paczko et al. and Peters and Williams[11,169] provide examples of the technique for the derivation multistep mechanisms for methane and methanol oxidation. In the case of methane[169] a starting elementary reaction mechanism for flame propagation through methane air mixtures (Table 9) was defined, and the fastest reactions for the C_1 chain were identified to be

$$CH_4 + H \rightarrow CH_3 + H_2$$
$$CH_3 + O \rightarrow CH_2O + H$$
$$CH_2O + H \rightarrow CHO + H_2$$
$$CHO + M \rightarrow CO + H + M$$

TABLE 9. The Elementary Reaction Mechanism and Associated Rate Coefficients Used for Reduction to the Four-Step Model for Methane Oxidation in a Stoichiometric Laminar Flame

	Reaction	B^b	α^b	$E^b(= RT_a)$
9.1	$O_2 + H \rightarrow OH + O$	2.00×10^{14}	0.00	70.30
9.1[b]	$OH + O \rightarrow O_2 + H$	1.40×10^{13}	0.00	3.20
9.2	$O + H_2 \rightarrow H + OH$	1.50×10^7	2.00	31.60
9.2[b]	$H + OH \rightarrow O + H_2$	6.73×10^6	2.00	22.35
9.3	$OH + H_2 \rightarrow H + H_2O$	1.00×10^8	1.60	13.80
9.3[b]	$H + H_2O \rightarrow OH + H_2$	4.62×10^8	1.60	77.50
9.4	$OH + OH \rightarrow H_2O + O$	1.50×10^9	1.14	0.42
9.4[b]	$H_2O + O \rightarrow OH + OH$	1.49×10^{10}	1.14	71.14
9.5[c]	$H + O_2 + M \rightarrow OH_2 + M$	2.30×10^{18}	-0.80	0.00
9.6	$HO_2 + H \rightarrow OH + OH$	1.50×10^{14}	0.00	4.20
9.7	$HO_2 + H \rightarrow H_2 + O_2$	2.50×10^{13}	0.00	2.90
9.8	$HO_2 + H \rightarrow H_2O + O$	3.00×10^{13}	0.00	7.20
9.9	$HO_2 + OH \rightarrow H_2O + O_2$	6.00×10^{13}	0.00	0.00
9.10	$CO + OH \rightarrow CO_2 + H$	4.40×10^6	1.50	-3.10
9.10[b]	$CO_2 + H \rightarrow CO + OH$	4.96×10^8	1.50	89.71
9.11	$CH_4 + H \rightarrow H_2 + CH_3$	2.20×10^4	3.00	36.60
9.11[b]	$H_2 + CH_3 \rightarrow CH_4 + H$	8.83×10^2	3.00	33.53
9.12	$CH_4 + OH \rightarrow H_2O + CH_3$	1.60×10^6	2.10	10.30
9.13	$CH_3 + O \rightarrow CH_2O + H$	7.00×10^{13}	0.00	0.00
9.14	$CH_3 + OH \rightarrow CH_2O + H + H$	9.00×10^{14}	0.00	64.80
9.15	$CH_3 + OH \rightarrow CH_2O + H_2$	8.00×10^{12}	0.00	0.00
9.16[d]	$CH_3 + H \rightarrow CH_4$	6.00×10^{16}	-1.00	0.00
9.17	$CH_2O + H \rightarrow CHO + H_2$	2.50×10^{13}	0.00	16.70
9.18	$CH_2O + OH \rightarrow CHO + H_2O$	3.00×10^{13}	0.00	5.00
9.19	$CHO + H \rightarrow CO + H_2$	2.00×10^{14}	0.00	0.00
9.20	$CHO + OH \rightarrow CO + H_2O$	1.00×10^{14}	0.00	0.00
9.21	$CHO + O_2 \rightarrow CO + HO_2$	3.00×10^{12}	0.00	0.00
9.22[c]	$CHO + M \rightarrow CO + H + M$	7.10×10^{14}	0.00	70.30
9.23	$CH_3 + H \rightarrow CH_2 + H_2$	1.80×10^{14}	0.00	63.00
9.24	$CH_2 + O_2 \rightarrow CO_2 + H + H$	6.50×10^{12}	0.00	6.30
9.25	$CH_2 + O_2 \rightarrow CO + OH + H$	6.50×10^{12}	0.00	6.30
9.26	$CH_2 + H \rightarrow CH + H_2$	4.00×10^{13}	0.00	0.00
9.26[b]	$CH + H_2 \rightarrow CH_2 + H$	2.79×10^{13}	0.00	12.61
9.27	$CH + O_2 \rightarrow CHO + O$	3.00×10^{13}	0.00	0.00
9.28	$CH_3 + OH \rightarrow CH_2 + H_2O$	1.50×10^{13}	0.00	20.93
9.29	$CH_2 + OH \rightarrow CH_2O + H$	2.50×10^{13}	0.00	0.00
9.30	$CH_2 + OH \rightarrow CH + H_2O$	4.50×10^{13}	0.00	12.56
9.31	$CH + OH \rightarrow CHO + H$	3.00×10^{13}	0.00	0.00

[a] Reactions 9.1–9.22 were used in the development of the model presented in the text.
[b] Here cm, mol, K, and kJ are the units.
[c] Catalytic efficiencies were taken from Ref. 3.
[d] The high-pressure value k_∞ is given here; tail-off curves, $k/k_\infty = (1 + 21.5 \times 10^{10} T^3/p^{0.6})^{-1}$, where p is in atm and T in K, were approximated from Fig. 30 in Ref. 3.

Together with the partial equilibrium reactions

$$H + OH \rightarrow O + H_2$$
$$H + H_2O \rightarrow OH + H_2$$

the summing of the above reactions yields:

$$CH_4 + 2H + H_2O \rightarrow CO + 4H_2 \tag{87}$$

Similarly, the combination of the reactions

$$CO + OH = CO_2 + H$$
$$H + H_2O = OH + H_2$$

yields the overall water-gas shift reaction

$$CO + H_2O = CO_2 + H_2 \tag{88}$$

and the combination

$$O_2 + H + M \rightarrow HO_2 + M$$
$$OH + HO_2 \rightarrow H_2O$$
$$H + H_2O \rightarrow OH + H_2$$

results in the overall recombination

$$2H + M \rightarrow H_2 + M \tag{89}$$

Finally, a sum of the steps

$$O_2 + H = OH + O$$
$$O + H_2 = OH + H$$
$$OH + H_2 = H_2O + H$$

(with the third reaction taken twice) produces the overall reaction

$$O_2 + 3H_2 = 2H_2O + 2H \tag{90}$$

In terms of the elementary rate constants k_i and equilibrium constants K_i for the above elementary reactions as numbered in Table 9, the reaction rates for

Eqs. (81)–(84) are given by

$$\text{Rate}_{(81)} = k_{11}[CH_4][H] \tag{91}$$

$$\text{Rate}_{(82)} = \left(\frac{k_{10}}{K_3}\right)\left(\frac{[H]}{[H]_2}\right)\left([CO][H_2O] - \frac{[CO_2][H_2]}{K_a}\right) \tag{92}$$

$$\text{Rate}_{(83)} = k_5[O_2][H][M] = K_{II}[O_2][H] \tag{93}$$

$$\text{Rate}_{(84)} = k_1[H]\left([O_2] - \frac{([H]^2[H_2O]^2)}{([H]^3 K_{IV})}\right) \tag{94}$$

where the partial equilibrium results

$$[OH] = \frac{[H][H_2O]}{[H_2]K_3}$$

and

$$[O] = \frac{[H][OH]}{[H_2]K_2}$$

$$K_2 = 2.23 \exp\left(-\frac{1112}{T}\right)$$

$$K_3 = 0.216 \exp\left(\frac{7658}{T}\right)$$

have been introduced, and K_{II} and K_{IV} are defined as

$$K_{II} = 0.035 \exp\left(\frac{3562}{T}\right)$$

$$K_{IV} = 1.48 \exp\left(\frac{6133}{T}\right)$$

K_{III} involves the three-body reaction efficiencies and is defined as

$$K_{III} = \left(\frac{k_5 W_{ave}}{rT}\right)\left(\frac{s_i Y_i}{W_i}\right)$$

where

W_i = molecular weight of component i
s_i = third body efficiency of component i
Y_i = mass fraction of component i
P = pressure in atmospheres

For postflame composition of stoichiometric methane air flames,

$$K_{\mathrm{III}} = \frac{1.6k_5}{rT},$$

and the coefficient 1.6 results from consideration of the third-body efficiencies of N_2, CO_2, and H_2O (0.4, 1.5, 6.5) for reaction (9.11) of Table 9.

Paczko et al.[11] demonstrate the process for methane as the fuel by using a slightly different starting detailed reaction than that used by Peters and Williams,[169] and, as noted above, all other species are defined by algebraic relations developed from the steady-state equations. This process leads to rather complex algebraic relations that are solved along with the differential equations.

As can be seen, the methodology remains somewhat less rigorous than would be desirable for general use. However, the approach apparently yields comparable results to the flame propagation velocity and major species profiles derived using the detailed kinetic mechanism.[11]

Lam and Goussis[170-172] have recently suggested the use of "computational singular perturbation" techniques to simplify detailed chemical kinetic mechanisms. The basic idea of computational singular perturbation is that the large number of meaningful elementary reactions in a complex reaction system can be grouped into separate reaction groups, each identified with a single characteristic time scale. The theory is beyond the scope of the present discussions, but the approach apparently provides an exact algorithm for the determination of this grouping. Furthermore the reaction rates for each of the groups are directly related through complex functionals to specific rates for the elementary reactions. No intuition is needed to develop these groupings, but as might be expected, they may be a function of extent of reaction and environment (to date only homogeneous reaction environments have been considered). Clearly the technique holds promise as a guiding tool to reduced mechanism development from detailed mechanisms as do other systematic techniques that are now only emerging.[173-175]

6. CLOSING REMARKS

Physical and chemical regimes of interest to combustion science are so broad that extended validation of chemistry models is necessary to provide reaction mechanisms with sufficient generality to be reliable interpretive and predictive tools. With such a wide range of operating parameters, many more factors become more important than those apparent in developing more specialized mechanisms. For example non-Arrhenius temperature dependence of reaction rates, pressure falloff considerations, and third body (chaperon) efficiencies need additional consideration. In spite of the greater difficulty involved in validating such mechanisms, the wide applicability makes them an essential and growing part of combustion research. Recent innovations in experimental diagnostics, in numerical methods, and mathematical techniques for comparing analytical and experimental results (sensitivity analysis) are rapidly accelerating our knowledge to develop and apply these improved mechanisms. Formal techniques are also under development to permit embodiment of the necessary elements of chemical kinetics in complex numerical models of combustion processes. For the engineer, this should result in improved guidance of engineering development, but it is unlikely that numerical modeling will never entirely supersede the need for hardware development and optimization. This chapter has conveyed some appreciation for some of the general problems and background one must be aware of to follow and apply the rapidly expanding research and literature on combustion chemistry.

ACKNOWLEDGMENTS

The author gratefully acknowledges the support of the Exxon Corporation in funding the preparation of the original material that forms the basis of the chapter. He also gratefully acknowledges past support of research efforts by the Air Force Office of Scientific Research, Mobil Research Corporation, the National Science Foundation, and the United States Department of Energy. The author wishes to express his thanks to Prof. David Gutman, Drs. Kenneth Brezinsky, Richard Yetter, Tom Norton, and Michael Vermeersch for their helpful discussions. Finally, a special thanks is offered to his colleague, Prof. Irvin Glassman, for his long-term collaboration and his many contributions to the thoughts and opinions expressed here.

REFERENCES

1. C. K. Westbrook and F. L. Dryer, "Chemical Kinetic Modeling of Hydrocarbon Combustion." *Prog. Energy Combust. Sci.* **10**, 1 (1984).
2. C. T. Bowman, "Chemical Kinetic Models for Complex Reacting Flows." *Ber. Bunsenges. Phys. Chem.* **90**, 934 (1986).
3. W. C. Gardiner, Jr., ed., *Combustion Chemistry*. Springer-Verlag, New York, 1984.
4. M. Frenklach, D. W. Clary, W. C. Gardiner, Jr., and S. E. Stein, "Shock-Tube Pyrolysis of Acetylene: Sensitivity Analysis of the Reaction Mechanism for Soot Formation." *Proc. Int. Symp. Shock Waves Shock Tubes, 15*th (D. Bershader and R. Hanson, eds.), pp. 295–309, Stanford Univ. Press, 1985.
5. M. Frenklach, D. W. Clary, W. C. Gardiner, Jr., and S. E. Stein, "Effect of Fuel Structure on Pathways to Soot." *Symp. (Int.) Combust. [Proc.]* **21**, 1067 (1988).
6. T. M. Sloane, "The Effect of Selective Energy Deposition on the Homogeneous Ignition of Methane and Its Implication for Flame Initiation and Combustion Enhancement." *Combust. Sci. Technol.* **42**, 131 (1985).

7. P. R. Westmoreland, "Experimental and Theoretical Analysis of Oxidation and Growth Chemistry in a Fuel Rich Acetylene Flame." Ph.D. Thesis, Department of Chemical Engineering, Massachusetts Institute of Technology, Cambridge, May, 1986.

8. R. D. Wilk, "Preignition Oxidation Characteristics of Hydrocarbon Fuels." Ph.D. Thesis, Department of Mechanical Engineering and Mechanics, Drexel University, Philadelphia, Pennsylvania, June, 1986.

9. R. A. Yetter, H. Rabitz, F. L. Dryer, R. C. Brown, and C. E. Kolb, "Kinetics of High Temperature B/O/H/C Chemistry." *Combust. Flame* (1989), in press.

10. R. A. Yetter and F. L. Dryer, "Complications of One-Step Kinetics for Moist CO Oxidation." *Symp. (Int.) Combust. [Proc.]* **21**, 749 (1988).

11. G. Paczko, P. M. Lefdal, and N. Peters, "Reduced Reaction Schemes for Methane, Methanol, and Propane Flames." *Symp. (Int.) Combust. [Proc.]* **21**, 739 (1988).

12. P. G. Lignola, E. Reverchon, R. Autuori, A. Insola, and A. M. Silvestre, "Propene Combustion in a CSTR." *Combust. Sci. Technol.* **44**, 1 (1985).

13. P. G. Lignola and E. Reverchon, "Dynamics of *n*-Heptane and Iso-octane Combustion Processes in a Jet Stirred Flow Reactor Operated under Pressure." *Combust. Flame* **62**, 1772 (1986).

14. K. Chinnick, C. Gibson, and J. F. Griffiths, "Isothermal Interpretations of Oscillatory Ignition during Hydrogen Oxidation in an Open System. I. Analytical Predictions and Experimental Measurements of Periodicity." *Proc. R. Soc. London, Ser. A* **405**, 117 (1986).

15. K. Chinnick, C. Gibson, and J. F. Griffiths, "Isothermal Interpretations of Oscillatory Ignition during Hydrogen Oxidation in an Open System. II. Numerical Analysis." *Proc. R. Soc. London, Ser. A* **405**, 129 (1986).

16. R. J. Blint, "The Relationship of Laminar Flame Width to Flame Speed." *Combust. Sci. Technol.* **49**, 79 (1986).

17. T. P. Coffee, "Kinetic Mechanisms for Pre-mixed Laminar Steady State Hydrogen Nitrous Oxide Flames." *Combust. Flame* **65**, 53 (1986).

18. J. Warnatz, "Chemistry of High Temperature Combustion of Alkanes Up to Octane." *Symp. (Int.) Combust. [Proc.]* **20**, (1985).

19. G. Dixon-Lewis, "Towards a Quantitatively Consistent Scheme for the Oxidation of Hydrogen, Carbon Monoxide, Formaldehyde and Methane Flames." *Workshop Model. Chem. React. Syst., 2nd* (J. Warnatz and W. Jäger, eds.), pp. 265–291, Springer-Verlag, Berlin, 1987.

20. L. R. Thorne, M. C. Branch, D. W. Chandler, R. J. Kee, and J. A. Miller, "Hydrocarbon/Nitric-Oxide Interactions in Low Pressure Flames." *Symp. (Int.) Combust. [Proc.]* **21**, 965 (1988).

21. J. Vandooren, L. Oldenhove de Guerstechin, and P. J. Van Tigglen, "Kinetics of a Lean Formaldehyde Flame." *Combust. Flame* **64**, 127 (1986).

22. C. K. Westbrook and J. M. Miller, eds., Modeling of Laminar Flame Propagation in Premixed Flames." *Combust. Sci. Technol., Spec. Issue*, p. 34 (1984).

23. J. O. Olsson, I. B. M. Olsson, and L. L. Andersson, "Lean Premixed Laminar Methanol Flames: A Computational Study." *J. Phys. Chem.* **91**, 4160 (1987).

24. M. D. Smooke, H. Rabitz, Y. Reuven, and F. L. Dryer, "Application of Sensitivity Analysis to Premixed Hydrogen Flames." *Combust. Sci. Technol.* **59**, 295 (1988).

25. V. Giovangigli and M. D. Smooke, *Extinction of Strained Premixed Laminar Flames with Complex Chemistry*, Rep. ME-103-86. Department of Mechanical Engineering, Yale University, New Haven, Connecticut, October, 1986.

26. M. D. Smooke, J. A. Miller, and R. J. Kee, "Solution of Premixed and Counterflow Diffusion Flame Problems by Adaptive Boundary Value Methods." In *Numerical Methods for Two-Point Boundary Value Problems* (A. Ascher and R. Russell, eds.), Vol. 5, p. 303. Birkhaeuser, Cambridge, Massachusetts, 1985.

27. J. H. Miller, W. G. Mallard, and K. C. Smyth, "Chemical Production Rates of Intermediates in a Methane/Air Diffusion Flame." *Symp. (Int.) Combust. [Proc.]* **21**, 1057 (1988).

28. M. D. Smooke, I. K. Puri, and K. Seshadri, "A Comparison between Numerical Calculations and Experimental Measurements of the Structure of a Counterflow Diffusion Flame Burning Diluted Methane in Diluted Air." *Symp. (Int.) Combust. [Proc.]* **21**, 1783 (1988).

29. P. Glarborg, J. A. Miller, and R. J. Kee, "Kinetic Modeling and Sensitivity Analysis of Nitrogen Oxide Formation in Well Stirred Reactors." *Combust. Flame* **65**, 177 (1986).

30. C. K. Westbrook, W. J. Pitz, M. M. Thornton, P. C. Malte, and A. L. Crittenden, "A Kinetic Modeling Study of n-Pentane Oxidation in a Jet Stirred Reactor." *Combust. Flame* **72**, 45 (1988).

31. V. Ya. Shtern, *The Gas Phase Oxidation of Hydrocarbons*, pp. 53–118. Macmillan, New York, 1964.

32. N. N. Semenov, *Some Problems in Chemical Kinetics and Reactivity*, Chapter VII. Princeton University Press, Princeton, New Jersey, 1958.

33. I. Glassman, *Combustion*, 2nd ed., pp. 51–105. Academic Press, San Diego, California, 1987.

34. C. K. Westbrook, "Inhibition of Hydrocarbon Oxidation in Laminar Flames and Detonations by Halogenated Hydrocarbons." *Symp. Int. Combust. [Proc.]* **19**, 127 (1983).

35. S. W. Benson, "The Kinetics and Thermochemistry of Chemical Oxidation with Application to Combustion and Flames." *Prog. Energy Combust. Sci.* **7**, 125 (1981).

36. F. W. Williams and R. S. Sheinson, "Manipulation of Cool and Blue Flames in the Winged Vertical Flow Tube." *Combust. Sci. Technol.* **7**, 85 (1973).

37. D. M. Newitt and L. S. Thornes, *J. Chem. Soc.*, p. 1656 (1937); see also Ref. 35.

38. S. W. Benson, "Cool Flames and Oxidation: Mechanism, Thermochemistry, and Kinetics." *Oxid. Commun.* **2**, 169 (1982).

39. I. R. Slagle, J.-Y. Park, and D. Gutman, "Experimental Investigation of the Kinetics and Mechanism of the Reaction of n-Propyl Radicals with Molecular Oxygen from 297 K to 535 K." *Symp. (Int.) Combust. [Proc.]* **20**, 733 (1985).

40. A. Fish, "Rearrangement & Cyclization Reactions of Organic Peroxy Radicals." In *Organic Peroxides* (D. Swerm, ed.), Vol. 1, pp. 141–198. Wiley (Interscience), New York, 1970.

41. K. C. Salooja, "The Degenerate Chain Branching Intermediates in Hydrocarbon Combustion: Some Evidence from Studies on the Isomeric Hexanes." *Combust. Flame* **9**, 219 (1965).

42. M. Vermeersch and F. L. Dryer, Work in Progress under U.S. Dept. of Energy Grant No. DEFG04-87AL3371. Fuels Research Laboratory, Mechanical and Aerospace Engineering Department, Princeton University, Princeton, New Jersey, July, 1988.

43. H. Hu and J. Keck, "Autoignition of Adiabatically Compressed Combustible Gas Mixture." *Soc. Automot. Eng. Int. Fuels Lubricants Meet. Expos.*, Toronto, Ontario, Nov. 2–5, *1987*, SAE Pap. No. 872110.

44. R. M. Green, N. P. Cernansky, W. J. Pitz, and C. K. Westbrook, "The Role of Low Temperature Chemistry in the Autoignition of *n*-Butane." *Soc. Automot. Eng. Int. Fuels Lubricants Meet. Expos.*, Toronto, Ontario, Nov. 2–5, 1987, SAE Pap. No. 872108.

45. E. Meyer, H. A. Olschewiski, J. Troe, and H. G. Wagner, "Investigation of N_2H_4 and H_2O_2 Decomposition in Low and High Pressure Shock Waves." *Symp. (Int.) Combust. [Proc.]* **12**, 345 (1969).

46. F. L. Dryer and I. Glassman, "Combustion Chemistry of Chain Hydrocarbons." *Prog. Astronaut. Aeronaut.* **62**, 55 (1979).

47. F. W. McLafferty, *Interpretation of Mass Spectra.* University of Science Books, Mill Valley, California, 1980.

48. F. L. Dryer and K. Brezinsky, "A Flow Reactor Study of the Oxidation of *n*-Octane and Iso-Octane." *Combust. Sci. Technol.* **45**, 199 (1986).

49. K. Brezinsky and F. L. Dryer, "A Flow Reactor Study of Isobutylene and an Isobutylene/*n*-Octane Mixture." *Combust. Sci. Technol.* **47**, 225 (1986).

50. E. A. Hardwidge, C. W. Larson, and B. S. Rabinovitch, "Isomerization of Vibrationally Excited Alkyl Radicals by Hydrogen Atom Migration." *J. Am. Chem. Soc.* **92**, 378 (1970).

51. C. W. Larson, P. T. Chua, and B. S. Rabinovitch, "Identity Scrambling and Isomerization Networks in Systems of Excited Alkyl Radicals." *J. Phys. Chem.* **76**, 2507 (1972).

52. S. W. Benson, "Combustion, A Chemical and Kinetic View." *Symp. (Int.) Combust. [Proc.]* **21**, 703 (1988).

53. D. J. Hautman, "Pyrolysis and Oxidation Mechanisms of Propane," MAE Rep. No. 1471-T. Ph.D. Thesis, Department of Mechanical and Aerospace Engineering, Princeton University, Princeton, New Jersey, 1979.

54. W. M. Proscia, "High Temperature Flow Experiments and Multi-Step Overall Kinetics for the Oxidation of High Carbon Number Paraffin Hydrocarbons," MAE Rep. No. 1625-T. M.S.E. Thesis, Department of Mechanical and Aerospace Engineering, Princeton University, Princeton, New Jersey, 1983.

55. K. Brezinsky, "The High Temperature Oxidation of Aromatic Hydrocarbons." *Prog. Energy Combust. Sci.* **12**, 1 (1985).

56. K. Brezinsky, T. A. Litzinger, and I. Glassman, "The High Temperature Oxidation of the Methyl Side Chain of Toluene." *Int. J. Chem. Kinet.* **16**, 1053 (1984).

57. T. A. Litzinger, K. Brezinsky, and I. Glassman, "The Oxidation of Ethylbenzene near 1060 K." *Combust. Flame* **63**, 251 (1986); also see "Reactions of *n*-Propyl Benzene during Gas Phase Oxidation." *Combust. Sci. Technol.* **50**, 117 (1986).

58. J. D. Bittner and J. B. Howard, "Composition and Reaction Mechanisms in Near Sooting Premixed Benzene/Oxygen Argon Flame." *Symp.* (*Int.*) *Combust.* [*Proc.*] **18**, 1105 (1981).

59. K. Brezinsky, E. Burke, and I. Glassman, "The High Temperature Oxidation of Butadiene." *Symp.* (*Int.*) *Combust.* [*Proc.*] **20**, 613 (1985).

60. R. R. Baldwin and R. W. Walker, "Elementary Reactions in the Oxidation of Alkenes." *Symp.* (*Int.*) *Combust.* [*Proc.*] **18**, 819 (1981).

61. F. P. Tully and J. E. M. Goldsmith, "Kinetic Study of the Hydroxyl Radical-Propene Reaction." *Chem. Phys. Lett.* **116**, 345 (1985).

62. R. D. Kern, H. J. Singh, M. A. Easlinger, and P. W. Winkler, "Product Profiles Observed During the Pyrolyses of Toluene, Benzene, Butadiene and Acetylene." *Symp.* (*Int.*) *Combust.* [*Proc.*] **19**, 1351 (1983).

63. G. S. Levinson, "High Temperature Preflame Reactions of *n*-Heptane." *Combust. Flame* **9**, 63 (1959).

64. C. R. Orr, "Combustion of Hydrocarbons behind a Shock Wave." *Symp.* (*Int.*) *Combust.* [*Proc.*] **9**, 1034 (1963).

65. A. Burcat, K. Scheller, and R. W. Crossley, "Shock Tube Ignition of Ethane-Oxygen-Argon Mixtures." *Combust. Flame* **18**, 115 (1972).

66. A. Burcat, A. Lifshitz, K. Scheller, and G. B. Skinner, "Shock Tube Investigation of Propane-Oxygen-Argon Mixtures." *Symp.* (*Int.*) *Combust.* [*Proc.*] **13**, 745 (1971).

67a. D. F. Cooke and A. Williams, "Shock Tube Studies of Methane and Ethane Oxidation." *Combust. Flame* **24**, 254 (1975).

67b. R. D. Hawthorne and A. C. Nixon, "Shock Tube Ignition Delay Studies of Endothermic Fuels." *AIAA J.* **4**, 513 (1956).

68. C. K. Westbrook and F. L. Dryer, "Chemical Kinetics and Modeling of Combustion Processes." *Symp.* (*Int.*) *Combust.* [*Proc.*] **8**, 749 (1981).

69. K. Mahmud, P. Marshall, and A. Fontijn, "An HTP Kinetics Study of the Reaction of $O^{(3p)}$ Atoms with Ethylene from 290 K to 1510 K." *J. Phys. Chem.* **91**, 1568 (1987); also K. Mahmud and A. Fontijn, "An HTP Kinetics Study of the Reaction of $O^{(3p)}$ Atoms with Acetylene from 290 K to 1510 K." *ibid.*, p. 1918.

70. J. W. Sutherland, J. V. Michael, A. N. Pirraglia, F. L. Nesbitt, and R. B. Klemm, "Rate Constant for the Reaction of $O^{(3p)}$ with H_2 by the Flash Photolysis-Shock Tube and Flash Photolysis-Resonance Flourescence Techniques: 504 K $<$ T $<$ 2495 K." *Symp.* (*Int.*) *Combust.* [*Proc.*] **21**, 929 (1988).

71. S. W. Benson, *Thermochemical Kinetics*, 2nd ed. Wiley, New York, 1976.

72. D. M. Golden, this volume, chapter 2.

73. W. Tsang and R. F. Hampson, "Chemical Kinetic Data Base for Combustion Chemistry. Part 1. Methane and Related Compounds." *J. Phys. Chem. Ref. Data* **15**, 1087 91986); "Part 2. Methanol." *ibid.* **16**, 471 (1987).

74. F. Westley, R. J. Cvetanovic, and J. T. Herron, "Compilation of Chemical Kinetic Data for Combustion Chemistry. Part 1. Non-aromatic C, H, O, N, and S Containing Compounds (1971–1982)." *Natl. Stand. Ref. Data Ser.* (*U.S.*), *Natl. Bur. Stand.* **NSRDS-NBS 73**, Part 1 (1987); "Part 2 (1983)." *ibid.*, Part 2 (1987).

75. D. K. Stull and H. Prophet, eds., *JANAF Thermochemical Tables*, NSRDS-NBS 3. U.S. Govt. Printing Office, Washington, DC, 1971; also Dow Chemical Co.

Midland Michigan, distributed by Clearing House for Federal Scientific and Technical Information (PB168370), 1965; also see M. W. Chase, Jr., C. A. Davies, J. R. Downey, Jr., D. J. Fulrip, R. A. McDonald, and A. N. Syverud, "JANAF Thermochemical Tables, Third Edition." *J. Phys. Chem. Ref. Data* **14**, Suppl. 1 (1985).

76. R. J. Kee, F. M. Rupley, and J. A. Miller, *The Chemkin Thermodynamic Data Base*, Sandia Rep. No. SAND87-8215. Sandia National Laboratories, Livermore, California, 1987.

77. J. Troe, "Thermal Dissociation and Recombination of Polyatomic Molecules." *Symp. (Int.) Combust. [Proc.]* **15**, 667 (1975).

78. J. Troe, "Theory of Thermal Unimolecular Reactions at Low Pressures. I. Solutions of Master Equation. II. Strong Collision Rate Constants, Applications." *J. Chem. Phys.* **66**, 4745, 4758 (1977).

79. J. Troe, "Contributions to Symposium on Current Status of Kinetics of Elementary Gas Reactions: Predictive Power of Theory and Accuracy of Measurement." *J. Phys. Chem.* **83**, 114 (1979).

80. D. M. Golden and C. W. Larson, "Rate Constants for Use in Modeling." *Symp. (Int.) Combust. [Proc.]* **21**, 595 (1988).

81. C. W. Larson, R. Patrick, and D. M. Golden, "Pressure and Temperature Dependence of Unimolecular Bond Fission Reactions: An Approach for Combustion Modelers." *Combust. Flame* **58**, 229 (1984).

82. F. L. Dryer, *Some Recent Advances in Combustion Chemistry*. Eastern States Section/Combustion Institute Meeting, San Juan, Puerto Rico, Dec. 15–17, 1986.

83. E. W. Kaiser, C. K. Westbrook, and W. J. Pitz, "Acetaldehyde Oxidation in the Negative Temperature Coefficient Regime: Experimental and Modeling Results." *Int. J. Chem. Kinet.* **18**, 655 (1986).

84. R. D. Wilk, N. P. Cernansky, W. J. Pitz, and C. K. Westbrook, "Propene Oxidation at Low and Intermediate Temperatures: A Detailed Chemical Kinetic Study." *Combust. Flame* **77**, 145 (1989).

85. W. J. Pitz, R. D. Wilk, C. K. Westbrook, and N. P. Cernanasky, *The Oxidation of n-Butane at Low and Intermediate Temperatures: An Experimental and Modeling Study*. Western States Section/Combustion Institute Meeting, Salt Lake City, Utah, Mar. 21–22, 1988; also available as *Lawrence Livermore Lab. [Rep.] UCRL* **UCRL-98402** (1988).

86. C. K. Westbrook and F. L. Dryer, "A Comprehensive Mechanism for Methanol Oxidation," *Combust. Sci. Technol.* **20**, 125 (1979).

87. C. K. Westbrook, F. L. Dryer and K. P. Shug, "A Comprehensive Mechanism for the Pyrolysis and Oxidation of Ethylene." *Symp. (Int.) Combust. [Proc.]* **19**, 153 (1983).

88. C. K. Westbrook and W. J. Pitz, "A Comprehensive Chemical Kinetic Reaction Mechanism for Oxidation and Pyrolysis of Propane and Propene." *Combust. Sci. Technol.* **37**, 117 (1984).

89. W. J. Pitz, C. K. Westbrook, W. M. Proscia, and F. L. Dryer, "A Comprehensive Chemical Kinetic Reaction Mechanism for the Oxidation of *n*-Butane." *Symp. (Int.) Combust. [Proc.]* **20**, 831 (1985).

90. E. I. Axelsson, K. Brezinsky, F. L. Dryer, W. J. Pitz, and C. K. Westbrook, "Chemical Kinetic Modeling of the Oxidation of Large Alkane Fuels: n-Octane and Iso-Octane." *Symp. (Int.) Combust. [Proc.]* **21**, 783 (1988); also see "A Detailed Kinetic Mechanism for Oxidation of n-Octane and Iso-Octane." *Lawrence Livermore Lab. [Rep.] UCRL* **UCRL-94449** (1986).

91. C. K. Westbrook, J. Creighton, J. Lund, and F. L. Dryer, "A Numerical Model of Chemical Kinetics of Combustion in a Turbulent Flow Reactor." *J. Phys. Chem.* **81**, 2542 (1977).

92. C. K. Westbrook, "An Analytical Study of the Shock Tube Ignition of Mixtures of Methane and Ethane." *Combust. Sci. Technol.* **20**, 5 (1979).

93. C. J. Jachimowski, "An Experimental and Analytical Study of Acetylene and Ethylene Oxidation behind Shock Waves." *Combust. Flame* **29**, 55 (1977).

94. C. T. Bowman, "A Shock Tube Investigation of the High Temperature Oxidation of Methanol." *Combust. Flame* **25**, 343 (1975).

95. D. Aronowitz, R. J. Santoro, F. L. Dryer, and I. Glassman, "Kinetics of the Oxidation of Methanol: Experimental Results, Mechanistic Concepts, and Global Modeling." *Symp. (Int.) Combust. [Proc.]* **17**, 633 (1979).

96. G. S. Bahn, "Approximate Thermochemical Tables for some C-H and C-H-O Species." *NASA [Contract. Rep.] CR* **NASA-CR-2178** (1972).

97. W. K. Aders and H. G. Wagner, "Untersuchungen zur Reaction von Wasserstoffatomen mit Athanol und tert. Butanol." *Z. Phys. Chem.* **74**, 224 (1971).

98. W. K. Aders, "Reactions of H-Atoms with Alcohols." In *Combustion Institute European Symposium*, (F. J. Weinberg, ed.), p. 19. Academic Press, New York, 1973.

99. K. Hoyermann, R. Sievert, and H. G. Wagner, "Mechanism of the Reaction of H Atoms with Methanol." *Ber. Bunsenges. Phys. Chem.* **85**, 149 (1981).

100. C. K. Westbrook and F. L. Dryer, "Prediction of Laminar Flame Properties of Methanol/Air Mixtures." *Combust. Flame* **37**, 171 (1980).

101. C. K. Westbrook, A. A. Adamczyk, and G. A. Lavoie, "A Numerical Study of Laminar Flame Wall Quenching." *Combust. Flame* **40**, 81 (1981).

102. S. M. Schoenung and R. K. Hanson, "CO and Temperature Measurements in a Flat Flame by Laser Absorption Spectroscopy and Probe Techniques." *Combust. Sci. Technol.* **24**, 227 (1981).

103. M. Cathonnet, J. C. Boettner, and H. James, "Etude de l'oxidation et de l'auto-inflammation du methanol dans le domaine de températures 500-600 C." *J. Chim. Phys.* **79**, 475 (1982).

104. L. L. Andersson, B. Christenson, A. Hoglund, J. O. Olsson, and L. G. Rosengren, "The Structure of Premixed Laminar Methanol-Air Flames: Experimental and Computational Results." *Prog. Astronaut. Aeronaut.* **95**, 164 (1985).

105. W. R. Leppard, "A Detailed Kinetics Simulation of Engine Knock." *Combust. Sci. Technol.* **43**, 1 (1985).

106. S. Koda and M. Tanaka, "Ignition of Premixed Methanol/Air in a Heated Flow tube and the Effect of NO_2 Addition." *Combust. Sci. Technol.* **47**, 165 (1986).

107. J. O. Olsson, L. S. Karlson, and L. L. Anderson, "Addition of Water to Pre-mixed Laminar Methanol-Air Flames: Experimental and Computational Results." *J. Phys. Chem.* **90**, 1458 (1986).

108. T. S. Norton and F. L. Dryer, "Some New Observations on Methanol Oxidation Chemistry." *Combust. Sci. Technol.* **63**, 107 (1989).

109. P. R. Westmoreland, J. B. Howard, and J. P. Longwell, "Tests of Published Mechanisms by Comparison with Measured Laminar Flame Structure in Fuel-Rich Acetylene Combustion." *Symp. (Int.) Combust. [Proc.]* **21**, 773 (1988).

110. R. A. Yetter, "An Experimental/Numerical Study of Carbon Monoxide-Hydrogen-Oxygen Kinetics with Applications of Gradient Sensitivity Analysis," Rep. No. MAE 1720-T. Ph.D. Thesis, Mechanical and Aerospace Engineering Department, Princeton University, Princeton, New Jersey, 1985.

111. R. P. Dickenson and R. J. Gelinas, "Sensitivity Analysis of Ordinary Differential Equation System—A Direct Method." *J. Comput. Phys.* **21**, 123 (1976); also see A. M. Dunker, "The Direct Decoupled Method for Calculating Sensitivity Coefficients in Chemical Kinetics." *J. Chem. Phys.* **81**, 2385 (1984).

112. J.-T. Hwang, E. P. Dougherty, S. Rabitz, and H. J. Rabitz, "The Green's Function Method of Sensitivity Analysis in Chemical Kinetics." *J. Chem. Phys.* **69**, 5180 (1978).

113. R. W. Atherton, R. B. Schainker, and E. R. Ducot, "On the Statistical Sensitivity Analysis of Models for Chemical Kinetics." *AIChE J.* **21**, 441 (1973).

114. R. I. Cukier, C. M. Fortium, K. E. Schuler, A. G. Perschek, and J. H. Schaibly, "Study of the Sensitivity of Coupled Reaction Systems to Uncertainties in Rate Coefficients. I. Theory." *J. Chem. Phys.* **59**, 3873 (1973).

115. R. S. Stolarksi, D. M. Butler, and R. D. Rindel, "Uncertainty Propagation in a Stratospheric Model. II. Monte Carlo Analysis of Imprecision due to Reaction Rates." *JGR, J. Geophys. Res.* **83**, 3074 (1978).

116. M. C. Dodge and T. A. Hect, "Rate Constant Measurements Needed to Improve a General Kinetic Mechanism for Photochemical Smog." *Int. J. Chem. Kinet.* **1**, 155 (1975).

117. J. Tilden, V. Constanza, G. McRae, and J. Seinfeld, "Sensitivity Analysis of Chemically Reacting Systems." *Springer Ser. Chem. Phys.* **18**, 69–91 (1981).

118. H. Rabitz, M. A. Kramer, and D. Dacol, "Sensitivity Analysis in Chemical Kinetics." *Annu. Rev. Phys. Chem.* **34**, 419 (1983).

119. M. Kramer, J. Calo, H. Rabitz, and R. Kee, *AIM: The Analytically Integrated Magnus Method for Linear and Second Order Sensitivity Coefficients*, Sandia Rep. SAND82-8231. Sandia National Laboratories, Livermore, California, 1982.

120. M. Kramer, R. Kee, and H. Rabitz, *CHEMSEN: A Computer Code for Sensitivity Analysis of Elementary Chemical Reactions*, Sandia Rep. SAND82-8230. Sandia National Laboratories, Livermore, California, 1982.

121. A. E. Lutz, R. J. Kee, and J. A. Miller, *SENSKIN: A Fortran Program for Predicting Homogeneous Gas Phase Chemical Kinetics with Sensitivity Analysis*, Sandia Rep. SAND87-UC-4. Sandia National Laboratories, Livermore, California, 1988.

122. M. Demiralp and H. Rabitz, "Chemical Kinetic Functional Sensitivity Analysis: Derived Sensitivities and General Applications." *J. Chem. Phys.* **75**, 1810 (1980).

123. M. Skumanich and H. Rabitz, "Feature Sensitivity Analysis in Chemical Kinetics." *Comments J. Mol. Sci.* **2**, 79 (1982).

124. R. Larter, H. Rabitz, and M. Kobayashi, "Derived Sensitivity Densities in Chemical Kinetics: A Computational Approach with Applications." *J. Chem. Phys.* **79**, 692 (1983).

125. R. A. Yetter, F. L. Dryer, and H. Rabitz, "Some Interpretive Aspects of Elementary Sensitivity Gradients in Combustion Kinetics Modeling." *Combust. Flame* **59**, 107 (1985).

126. R. A. Yetter, L. A. Eslava, F. L. Dryer, and H. Rabitz, "Elementary and Derived Sensitivity Information in Chemical Kinetics." *J. Phys. Chem.* **88**, 1497 (1984).

127. A. C. Hindmarsh, "LSODE and LSODEI, Two Initial Value Ordinary Differential Equation Solvers." *ACM SIGNUM Newsl.* **15**, 10 (1980).

128. D. Edelson, "Mechanistic Determination of the Belonsov-Zhabotinsky Oscillations. IV. Sensitivity Analysis." *Int. J. Chem. Kinet.* **13**, 1175 (1981).

129. T. P. Coffee and J. M. Heimrl, *Sensitivity Analysis for Premixed Laminar, Steady State Flames*, Techn. Rep. ARBRL-TR02457. Army Ballistic Research Laboratory, Aberdeen Proving Ground, Maryland, 1983.

130. A. M. Dean, D. C. Steiner, and E. E. Wang, "A Shock Tube Study of the $H_2/O_2/-CO/Ar$ and $H_2/N_2O/CO/Ar$ Systems. Measurements of the Rate Constant for $H + N_2O = N_2 + OH$." *Combust. Flame* **32**, 73 (1978).

131. A. Levy and F. J. Weinberg, "Optical Flame Structure Studies: Examination of Reaction Rate Laws in Lean Ethylene-Air Flames." *Combust. Flame* **3**, 229 (1959).

132. T. D. Butler, L. D. Cloutman, J. K. Dukowicz, J. D. Ramshaw, and R. B. Krieger, "Toward a Comprehensive Model for Combustion in a Direct-Injection Stratified-Charge Engine." In *Combustion Modeling in Reciprocating Engines* (J. N. Mattavi and C. A. Amann, eds.), pp. 231–264. Plenum, New York, 1980.

133. F. L. Dryer, "High Temperature Oxidation of CO and CH_4 in a Turbulent Flow Reactor," Rep. AMS 1034-T. Ph.D. Thesis, Aerospace and Mechanical Sciences Department, Princeton University, Princeton, New Jersey, 1972.

134. J. B. Howard, G. C. Williams, and D. H. Fine, "Kinetics of Carbon Monoxide Oxidation in Postflame Gases." *Symp. (Int.) Combust. [Proc.]* **14**, 975 (1973).

135. R. A. Matula, H. L. Gangloff, and K. L. Maloney, "Ignition Delay in Hydrocarbon Systems." *Symp. Hydrocarbon Combust. Chem., Am. Chem. Soc., Div. Pet. Chem.*, Dallas, Texas, *1973*.

136. F. L. Dryer and I. Glassman, "High Temperature Oxidation of CO and CH_4." *Symp. (Int.) Combust. [Proc.]* **14**, 987 (1973).

137. D. J. Seery and C. T. Bowman, "An Experimental and Analytical Study of Methane Oxidation behind Shock Waves." *Combust. Flame* **14**, 37 (1970).

138. C. T. Bowman, "Non-equilibrium Radical Concentrations in Shock-Initiated Methane Oxidation." *Symp. (Int.) Combust. [Proc.]* **15**, 869 (1975).

139. C. K. Westbrook and F. L. Dryer, "Simplified Reaction Mechanisms for the Oxidation of Hydrocarbon Fuels in Flames." *Combust. Sci. Technol.* **27**, 31 (1981); also see T. P. Coffee, "Comments on . . . ." *ibid.* **43**, 333 (1985).

140. J. B. Fenn and H. F. Calcote, "Activation Energies in High Temperature Combustion." *Symp. (Int.) Combust. [Proc.]* **4**, 231 (1953).

141. P. L. Walker and C. C. Wright, "Hydrocarbon Burning Velocities Predicted by Thermal versus Differential Mechanisms." *J. Am. Chem. Soc.* **74**, 3769 (1952).

142. G. L. Dugger, D. M. Simon, and M. Gerstein, "Laminar Flame Propagation. Chapter IV." *Natl. Advis. Comm. Aeronaut., Rep.* **1300** (1952).

143. B. Lewis and G. von Elbe, *Combustion, Flames, and Explosions.* Academic Press, New York, 1952.

144. W. M. Heffington, G. W. Parks, K. G. P. Salzman, and S. S. Penner, "Studies of Methane Oxidation Kinetics." *Symp. (Int.) Combust. [Proc.]* **16**, 997 (1977).

145. D. J. Hautman, F. L. Dryer, K. P. Shug, and I. Glassman, "A Multiple Step Overall Kinetics Mechanism for the Oxidation of Hydrocarbons." *Combust. Sci. Technol.* **25**, 219 (1981).

146. R. D. Reitz and F. V. Bracco, "Toward a Formulation of a Global Local Equilibrium Kinetic Model for Laminar Hydrocarbon Flames." *Numer. Fluid Mech.* **5**, 130 (1982).

147. W. G. Vincenti and C. H. Kruger, *Introduction to Physical Gas Dynamics.* Wiley, New York, 1967.

148. T. P. Coffee, A. J. Koltar, and M. S. Miller, "The Overall Reaction Concept in Premixed, Laminar, Steady State flames. I. Stoichiometries." *Combust. Flame* **54**, 155 (1983).

149. T. P. Coffee, A. J. Koltar, and M. S. Miller, "The Overall Reaction Concept in Premixed, Laminar, Steady State Flames. II. Initial Temperatures and Pressures." *Combust. Flame* **58**, 59 (1984).

150. R. D. Reitz and F. V. Bracco, "Global Kinetics and Lack of Thermodynamic Equilibrium." *Combust. Flame* **53**, 141 (1983).

151. R. B. Edelman and O. F. Fortune, "A Quasi Global Chemical Model for the Finite Rate Combustion of Hydrocarbon Fuels with Application to Turbulent Burning and Mixing in Hypersonic Engines and Nozzles." *AIAA Pap.* **69–86** (1986).

152. W. Proscia and F. L. Dryer, *Evaluation of a Multi-Step Overall Kinetic Mechanism for n-Butane Oxidation*, Pap. No. WSSCI 83-23. Western States Section/Combustion Institute Meeting, Pasadena, California, Apr. 11–12, 1983.

153. R. B. Edelman, O. Fortune, and G. Weilerstein, "Some Observations on Flows Described by Coupled Mixing and Kinetics." In *Emissions form Continuous Combustion Systems* (W. Cornelius and W. G. Agnew, eds.), pp. 55–106. Plenum, New York, 1972.

154. V. S. Engleman, W. Bartok, J. P. Longwell, and R. B. Edelman, "Experimental and Theoretical Studies of NO_x Formation in a Jet Stirred Combustor." *Symp. (Int.) Combust. [Proc.]* **14**, 755 (1973).

155. R. B. Edelman, "The Relevance of Quasi-Global Kinetics Modeling as an Aid in Understanding the I.C. Engine Process." In *Workshop on the Numerical Simulation of Combustion for Application to Spark Ignition and Compression Ignition Engines* (A. A. Boni, ed.), Appendix 4.1B. Science Applications Inc., La Jolla, California, 1975.

156. R. B. Edelman and P. T. Harsha, *Current Status of Laminar and Turbulent Gas Dynamics in Combustors*, Invit. Surv. Pap. Central States Section/Combustion Institute Meeting, NASA Lewis Research Center, Cleveland, Ohio, Mar. 29–30, 1977.

157. A. M. Mellor, "Current Kinetic Modeling Techniques for Continuous Flow Combustors." In *Emissions form Continuous Combustion Systems* (W. Cornelius and W. G. Agnew, eds.), pp. 23–53. Plenum, New York, 1972.

158. P. J. Marteney, "Analytical Study of the Kinetics of Formation of Nitrogen Oxide in Hydrocarbon-Air Combustion." *Combust. Sci. Technol.* **1**, 461 (1970).

159. R. Roberts, L. D. Aceto, R. Kollrack, D. P. Teixeira, and J. M. Bonnell, "An Analytical Model for Nitric Oxide Formation in a Gas Turbine Combustor." *AIAA J.* **10**, 820 (1972).

160. R. Kollrack and L. D. Aceto, "Nitric Oxide Formation in Gas Turbine Combustors." *AIAA J.* **11**, 664 (1973).

161. R. J. Mador and R. Roberts, "A Pollutant Emission Prediction Model for Gas Turbine Combustors." *AIAA Pap.* **74–1113** (1974).

162. R. Kollrack and L. D. Aceto, "The Effects of Liquid Water Addition in Gas Turbine Combustors." *J. Air Pollut. Control Assoc.* **23**, 116 (1973).

163. L. D. Aceto and R. Kollrack, "Primary Zone Carbon Monoxide Levels for Gas Turbines. I. Pre-mixed Combustion." *AIAA J.* **12**, 463 (1974).

164. L. D. Aceto and R. Kollrack, "Primary Zone Carbon Monoxide Levels for Gas Turbines. II. Liquid Fuel Combustion." *AIAA J.* **12**, 465 (1974).

165. H. G. Adelman, L. H. Browning, and R. K. Pefley, "Pre-tested Emissions from Methanol and Jet Fueled Gas Turbine Combustors." *AIAA J.* **14**, 793 (1976).

166. T. M. Kiehne, R. D. Mathews, and D. E. Wilson, "An Eight-Step Kinetics Mechanism for High Temperature Propane flames." *Combust. Sci. Technol.* **54**, 1 (1987).

167. M. Metgelchi and J. C. Keck, "Laminar Burning Velocity of Propane-Air Mixtures at High Temperature and Pressure." *Combust. Flame* **38**, 143 (1980).

168. W. A. Daniel, "Flame Quenching at the Walls of an Internal Combustion Engine." *Symp. (Int.) Combust. [Proc.]* **6**, 886 (1957).

169. N. Peters and F. A. Williams, "The Asymptotic Structure of Stoichiometric Methane-Air Flames." *Combust. Flame* **68**, 185 (1987).

170. S. H. Lam, "Singular Perturbation for Stiff Equations Using Numerical Methods." In *Recent Advances in Aerospace Sciences* (C. Casci, ed.), p. 3. Academic Press, Orlando, Florida, 1985.

171. S. H. Lam and D. A. Goussis, *Understanding Complex Chemical Kinetics with Computational Singular Perturbation*, MAE Rep. No. 1799. Department of Mechanical and Aerospace Engineering, Princeton University, Princeton, New Jersey, Jan. 1988 [to be presented at *Symp. (Int.) Combust. [Proc.]* **22** (1988).]

172. S. H. Lam and D. A. Gousis, "Basic Theory and Demonstration of Computational Singular Perturbation for Stiff Equations." *Proc. IMACS World Congr. Sci. Comput.*, Paris, France, July 18–22, *1988* (to be published).

173. J.-Y. Chen, "A General Procedure for Constructing Reduced Reaction Mechanisms with Given Independent Relations." *Combust. Sci. Technol.* **57**, 88 (1988).

174. G. Li and H. Rabitz, "A General Analysis of Exact Lumping in Chemical Kinetics." *Chem. Eng. Sci.* **44**, 1413 (1988).

175. G. Li and H. Rabitz, "A General Analysis of Approximate Lumping in Chemical Kinetics" (to be published).

4 Chemistry of Gaseous Pollutant Formation and Destruction

C. T. BOWMAN

Department of Mechanical Engineering
Stanford University
Stanford, California

1. Introduction
2. Formation and destruction of CO in flames
3. Formation and destruction of nitrogen oxides in flames
   3.1. Thermal NO formation
   3.2. Prompt NO formation
   3.3. Fuel-NO formation
   3.4. NO_2 formation
   3.5. N_2O formation
4. Sulfur oxides in flames
   References

1. INTRODUCTION

The five principal classes of pollutants emitted from combustion sources are carbon monoxide, nitrogen oxides, sulfur oxides, organic compounds (unburned and partially burned hydrocarbons), and particulates (soot, flyash, and aerosols). Table 1 lists typical estimates of emissions of these five classes of pollutants attributable to combustion sources.

The emission levels from a specific combustion device depend on the interaction between physical and chemical processes occurring within the device. Local stoichiometry, determined by physical processes governing fuel-air contacting, is of particular importance in determining pollutant formation and destruction in flames. In general, the concentrations of the various pollutant species in the effluent gases from combustion devices differ from calculated equilibrium levels for exhaust conditions, indicating the importance of reaction kinetics in determining pollutant emissions.

TABLE 1. Estimates of Pollutant Emissions from Controllable Combustion Sources in the United States in 1977[a] (10^9 kg / yr)

Source Category	Particulates	SO_x	NO_x	HC^b	CO
Transportation	1.1	0.8	8.8	10.7	83.2
Highway	0.8	0.4	6.6	9.2	74.3
Nonhighway	0.3	0.4	2.2	1.5	8.9
Stationary fuel combustion	4.8	22.0	13.0	0.3	1.2
Electric utilities	3.4	17.6	7.2	0.1	0.3
Industrial/commercial/					
residential	1.4	4.4	5.8	0.2	0.9
Solid waste burning	0.5	< 0.1	0.1	0.9	3.0
Miscellaneous	0.2	< 0.1	< 0.1	0.2	1.5
Forest managed burning	0.1		< 0.1	0.1	1.0
Agricultural burning	0.1		< 0.1	0.1	0.5
Total	6.6	22.9	22.0	12.1	88.9

[a]From USEPA.[1]
[b]Volatile organic compounds.

For some pollutants—notably carbon monoxide, organic compounds, and soot—the formation and destruction reactions are an intimate part of the combustion process. Understanding the chemistry of these pollutant species requires some knowledge of the hydrocarbon fuel combustion mechanism. For other pollutant species—notably nitrogen oxides and sulfur oxides—the formation and destruction reactions are not part of the combustion process. However, the reactions involving these pollutants occur in an environment established by the combustion reactions, and hence the pollutant chemistry is intimately connected to the combustion process.

In this chapter the current understanding of the principal chemical mechanisms involved in the formation and destruction of three of the classes of pollutants (CO, NO_x, and SO_x) will be reviewed, and kinetic models for prediction of pollutant emissions will be discussed. Our present level of understanding of the fundamental processes governing pollutant formation and destruction in flames is, in some instances, incomplete. But enough is known to allow formulation of models that are useful in the design of combustion equipment.

2. FORMATION AND DESTRUCTION OF CO IN FLAMES

Carbon monoxide is a major pollutant emitted from engines used in the transportation sector, Table 1. In Figure 1 the equilibrium CO mole fractions in the products of combustion are shown for methane-air as a function of temperature and stoichiometry. For stoichiometric and lean mixtures and for typical exhaust gas temperatures (less than 1000 K), equilibrium CO mole

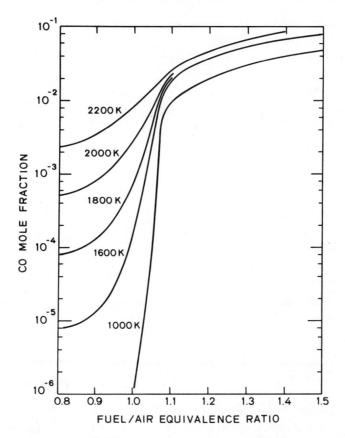

FIGURE 1. Equilibrium CO mole fractions for premixed methane-air combustion at a pressure of 1 atm.

fractions are less than 1 ppmv, a level that would not represent a significant pollution problem. However, typical concentrations in the exhaust products of spark ignition, compression ignition, and gas turbine engines significantly exceed the equilibrium values for the exhaust conditions but are lower than values corresponding to equilibrium at the maximum combustion temperature.

The explanation for the high levels of CO in the exhaust products of these engines is found in the chemical mechanisms for formation and oxidation of CO in flames. In lean and stoichiometric premixed hydrocarbon flames, the CO concentration increases rapidly in the flame zone to a maximum value that generally exceeds the equilibrium value for adiabatic combustion, Figure 2.[2] Following the maximum, the CO concentration decreases slowly toward the equilibrium value. A similar behavior is observed in the reaction zone of laminar hydrocarbon diffusion flames, Figure 3.[3] These experimental observations suggest that both the formation and destruction of CO in flames are

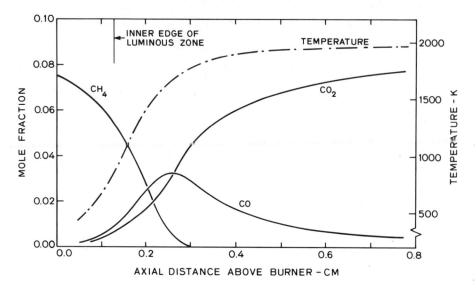

FIGURE 2. Measured species mole fraction and temperature profiles in a 0.1 atm premixed fuel-lean methane-oxygen flame, fuel/air equivalence ratio = 0.17.[2]

kinetically controlled. Hence kinetic models for both CO formation and destruction are required in models for predicting CO emissions from combustors.

Carbon monoxide formation is one of the principal reaction paths in the hydrocarbon combustion mechanism. The primary CO formation mechanism may be represented schematically by

$$RH \rightarrow R^{\cdot} \rightarrow RO_2 \rightarrow RCHO \rightarrow R\dot{C}O \rightarrow CO \qquad (1)$$

where RH represents the parent hydrocarbon fuel and R· is a hydrocarbon radical produced by removing one or more H-atoms from the fuel molecule. Subsequent oxidation of the hydrocarbon radical leads eventually to the formation of aldehydes, RCHO, which in turn react to form $R\dot{C}O$ radicals. The reaction of the acyl radical to produce CO may occur by thermal decomposition

$$R\dot{C}O \rightarrow CO + R\cdot \qquad (2)$$

or by

$$R\dot{C}O + \begin{bmatrix} O_2 \\ OH \\ O \\ H \end{bmatrix} \rightarrow CO + \cdots \qquad (3)$$

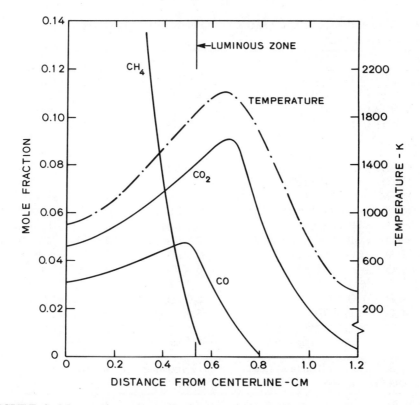

FIGURE 3. Measured species mole fraction and temperature profiles in an atmospheric pressure methane-air diffusion flame.[3]

The relative importance of reactions (2) and (3) in the CO formation process depends on the local flame conditions.

Several different approaches can be used to model CO formation in the combustion of hydrocarbon fuels. If a detailed kinetic mechanism for the combustion reaction is known, then the CO concentration profile may be calculated by integration of the set of coupled differential rate equations. At the present time detailed combustion mechanisms are known in sufficient detail to permit calculation of CO concentrations for a few hydrocarbon fuels, notably C_1–C_4 hydrocarbons[4,5] and methanol.[6] The detailed mechanisms for several of these fuels are discussed in Chapter 3. Preliminary mechanisms have been proposed for alkanes up to C_8;[7] however, these mechanisms require further refinement before they can be used to calculate CO formation rates over the range of conditions typically encountered in practical combustion devices.

Tables 2 and 3 summarize important CO-formation reactions in methane and methanol combustion, respectively. For both fuels, CO is formed via a series of rapid reactions similar to the sequence (1).

TABLE 2. Important Reactions Leading to CO Formation in Premixed Methane Combustion[a]

Reaction	A	B	$C(K)$
$CH_4 + OH \rightarrow CH_3 + H_2O$	1.6(3)	2.1	1.24(3)
$CH_4 + O \rightarrow CH_3 + OH$	1.2(4)	2.1	3.84(3)
$CH_4 + H \rightarrow CH_3 + H_2$	2.2(1)	3.0	4.40(3)
$CH_3 + O \rightarrow CH_2O + H$	7.0(10)	0	0
$CH_3 + O_2 \rightarrow CH_3O + O$	1.5(10)	0	14.4(3)
$CH_3O + M \rightarrow CH_2O + H + M$	1.0(11)	0	12.6(3)
$CH_3O + H \rightarrow CH_2O + H_2$	2.0(10)	0	0
$CH_3O + O_2 \rightarrow CH_2O + HO_2$	1.0(10)	0	3.61(3)
$CH_2O + M \rightarrow HCO + H + M$	5.0(13)	0	38.5(3)
$CH_2O + OH \rightarrow HCO + H_2O$	3.0(10)	0	0.60(3)
$CH_2O + O \rightarrow HCO + OH$	3.5(10)	0	1.77(3)
$CH_2O + H \rightarrow HCO + H_2$	2.5(10)	0	2.01(3)
$HCO + M \rightarrow CO + H + M$	2.5(11)	0	8.46(3)
$HCO + OH \rightarrow CO + H_2O$	5.0(10)	0	0
$HCO + O \rightarrow CO + OH$	3.0(10)	0	0
$HCO + H \rightarrow CO + H_2$	2.0(11)	0	0
$HCO + O_2 \rightarrow CO + HO_2$	3.0(9)	0	0

[a]From Warnatz.[4] Units = kgmol, m³, K, sec; $k = AT^B \exp(-C/T)$.

TABLE 3. Important Reactions Leading to CO Formation in Premixed Methanol Combustion[a]

Reaction	A	B	$C(K)$
$CH_3OH + O_2 \rightarrow CH_2OH + HO_2$	4.0(10)	0	25.6(3)
$CH_3OH + OH \rightarrow CH_2OH + H_2O$	4.0(9)	0	1.0(3)
$CH_3OH + O \rightarrow CH_2OH + OH$	1.6(9)	0	1.16(3)
$CH_3OH + H \rightarrow CH_2OH + H_2$	3.2(10)	0	3.5(3)
$CH_3OH + H \rightarrow CH_3 + H_2O$	5.0(9)	0	2.7(3)
$CH_2OH + M \rightarrow CH_2O + H + M$	2.5(10)	0	14.6(3)
$CH_2OH + O_2 \rightarrow CH_2O + HO_2$	1.0(9)	0	3.0(3)

[a]From Westbrook and Dryer.[6] $k = AT^B \exp(-C/T)$; units = kgmol, m³, K, sec. Subsequent reaction of CH_3 and CH_2O by the mechanism of Table 2.

The series of reactions represented by Eq. (1) occurs very rapidly at typical combustion temperatures. Hence it may be possible to model CO formation by a one-step reaction in which the hydrocarbon fuel reacts with molecular oxygen to form CO and H_2 or H_2O at some empirically determined rate,

$$C_nH_m + \frac{n}{2}O_2 \rightarrow nCO + \frac{m}{2}H_2 \tag{4}$$

or

$$C_nH_m + \left(\frac{m}{4} + \frac{n}{2}\right)O_2 \rightarrow nCO + \frac{m}{2}H_2O \tag{5}$$

This approach, termed a "global model," is described in Chapter 3.

The CO formed in the combustion process is oxidized to CO_2 at a rate that is relatively slow compared to the CO formation rate. The principal CO oxidation reaction at conditions typical of hydrocarbon flames is

$$CO + OH \rightleftarrows CO_2 + H \tag{6}$$

The rate coefficient for this reaction has been measured in numerous investigations. At temperatures in excess of 1500 K, the rate coefficient for reaction (6) exhibits a significant temperature dependence. However, at temperatures below 1000 K, the rate coefficient is nearly independent of temperature. Hence quenching of CO at lower temperatures occurs principally by depletion of radical species, such as OH, by recombination reactions. A recommended rate coefficient expression for reaction (6) is listed in Table 4.[8]

The reaction

$$CO + HO_2 \rightleftarrows CO_2 + OH \tag{7}$$

has been suggested as an alternative reaction path for CO oxidation. Computer simulations of CO oxidation[9] indicate that reaction (7) plays a significant role in CO oxidation for low temperatures (1000–1500 K) and at elevated pressures (> 10 atm). For these conditions, the HO_2 concentrations produced by the termolecular recombination reaction,

$$H + O_2 + M \rightarrow HO_2 + M \tag{8}$$

are comparable to the OH concentrations produced by the rapid bimolecular reactions,

$$H + O_2 \rightleftarrows OH + O \tag{9}$$

$$O + H_2 \rightleftarrows OH + H \tag{10}$$

$$O + H_2O \rightleftarrows OH + OH \tag{11}$$

$$H + H_2O \rightleftarrows OH + H_2 \tag{12}$$

Hence a complete kinetic model for CO oxidation should include reaction (7). A recommended rate coefficient for this reaction is listed in Table 4.[4] Oxidation of CO by the direct reaction with O_2,

$$CO + O_2 \rightarrow CO_2 + O \tag{13}$$

is very slow, even at flame temperatures; hence reaction (13) should not be an important reaction in hydrocarbon flames, which generally have large radical concentrations.

To model CO burnout in the postcombustion gas of hydrocarbon flames, the principal CO oxidation reactions must be coupled to a kinetic model

TABLE 4. Rate Coefficients for the CO + OH → CO$_2$ + H and CO + HO$_2$ → CO$_2$ + OH Reactions[a]

Number	Reaction	A	B	$C(K)$	Reference
(6)	CO + OH → CO$_2$ + H	1.5(4)	1.3	−0.385(3)	8
(7)	CO + HO$_2$ → CO$_2$ + OH	1.5(11)	0	11.8(3)	4

[a]Units: kgmol, m^3, K, sec; $k = AT^B \exp(-C/T)$.

describing the relaxation of various radical species (OH, H, O, HO$_2$) toward their respective equilibrium values. One approach to this problem is to integrate the complete set of differential rate equations for the reacting gas. A detailed kinetic model for CO oxidation in flames is given in Table 5.

Calculation of CO emissions from practical combustion devices using detailed chemistry requires coupling of the CO formation and oxidation mechanisms with a combustor flow model. A major disadvantage of this approach is the time-consuming nature of the calculations.

A simple way to couple chemistry with the flow field in practical devices has been suggested by Mellor.[10,11] This semiempirical approach, called the "characteristic time model," has successfully correlated data on pollutant emissions from gas turbine engines in terms of ratios of characteristic times for the physical and chemical processes occurring in those regions of the burner that govern combustor performance and emissions. Application of the model requires identification of the relevant physical and chemical processes and determination of appropriate time scales for these processes. In combustors burning gaseous fuels, the most important physical time is the mixing time. In the turbulent flow environment of most practical devices, this mixing time is determined from the ratio of a turbulent length scale, l, and a turbulent velocity fluctuation, u',

$$\tau_{\text{mix}} = \frac{l}{u'}$$

The choice of a specific length scale and velocity depends on a model of the flame structure and on the particular combustor parameter (emissions or flame stability) of interest. Typically, the appropriate length scale is taken as a significant macrodimension of the burner and the velocity is an appropriate mean velocity. In liquid-fueled burners a characteristic time for vaporization of the fuel spray may be needed to correlate the data. Chemical times are obtained from relevant global expressions for the reaction rate. Table 6 contains various physical and chemical time scales suggested by Mellor for correlation of emissions data for gas turbines. The characteristic times for CO emissions are based on the assumption that CO oxidation rates in the outer region of the flame determine CO levels in the combustion products. Although the characteristic time model has been applied only to gas turbines, in principle the concept should be applicable to other combustion devices.

TABLE 5. Detailed Kinetic Model for Premixed CO Oxidation[a]

Number	Reaction	A	B	$C(K)$
	$H + H + M \rightarrow H_2 + M$	6.4(11)	-1.0	0
	$H + OH + M \rightarrow H_2O + M$	2.2(16)	-2.0	0
(14)	$O + O + M \rightarrow O_2 + M$	1.0(11)	-1.0	0
(8)	$H + O_2 + M \rightarrow HO_2 + M$	1.0(9)	0	$-0.5(3)$
	$H_2O_2 + M \rightarrow OH + OH + M$	1.2(14)	0	22.9(3)
(9)	$H + O_2 \rightarrow OH + O$	1.2(14)	-0.91	8.3(3)
(10)	$O + H_2 \rightarrow OH + H$	1.5(4)	2.0	3.8(3)
(11)	$O + H_2O \rightarrow OH + OH$	1.5(7)	1.4	8.68(3)
(−12)	$OH + H_2 \rightarrow H_2O + H$	1.0(5)	1.6	1.66(3)
	$H + HO_2 \rightarrow OH + OH$	1.5(11)	0	0.5(3)
	$H + HO_2 \rightarrow H_2 + O_2$	2.5(10)	0	0.35(3)
	$OH + HO_2 \rightarrow H_2O + O_2$	2.0(10)	0	0
	$O + HO_2 \rightarrow OH + O_2$	2.0(10)	0	0
	$HO_2 + HO_2 \rightarrow H_2O_2 + O_2$	2.0(9)	0	0
	$H + H_2O_2 \rightarrow H_2O + OH$	1.0(10)	0	1.8(3)
	$H_2O_2 + OH \rightarrow H_2O + HO_2$	7.0(9)	0	0.7(3)
	$H_2O_2 + H \rightarrow HO_2 + H_2$	1.7(9)	0	1.89(3)
	$CO + O + M \rightarrow CO_2 + M$	5.3(10)	0	$-2.28(3)$
(6)	$CO + OH \rightarrow CO_2 + H$[b]			
(7)	$CO + HO_2 \rightarrow CO_2 + OH$[b]			
(13)	$CO + O_2 \rightarrow CO_2 + O$	2.5(9)	0	24.0(3)
	Including reactions involving HCO in Table 2.			

[a] From Warnatz.[4] Units: kgmol, m³, K, sec; $k = AT^B \exp(-C/T)$.
[b] See Table 4.

TABLE 6. Characteristic Times for CO and NO Emissions from Gas Turbine Combustors[a]

Parameter	$\tau_{mix}(sec)$[b]	$\tau_{chem}(sec)$[c]
CO emission	l_{CO}/V_a	$10^{-4} \exp(3880/T_{exh})$
NO emission	$l_{NO}/V_{\phi=1}$	$10^{-15} \exp(68000/T_{\phi=1})$

[a] From Mellor et al.[10,11]
[b] $l_{CO}^{-1} = d_{comb}^{-1} + l_{quench, CO}^{-1}$; $l_{NO}^{-1} = d_{comb}^{-1} + l_{quench, NO}^{-1}$, where

$$d_{comb} = \text{combustor diameter}$$

$$l_{quench} = \text{characteristic length for quenching}$$

$$(\text{geometry dependent})$$

V_a = mean air injection velocity; $V_{\phi=1}$ = mean flow velocity in the stoichiometric region of the combustor.
[c] T_{exh} = adiabatic burned gas temperature; $T_{\phi=1}$ = adiabatic burned gas temperature for stoichiometric equivalence ratio.

TABLE 7. Empirical Overall Rate Expressions for CO Oxidation[a]

$-d(CO)/dt =$	Temperature Range	Reference
$1.3 \times 10^{11}(CO)(H_2O)^{1/2}(O_2)^{1/2} \exp(-15105/T)$	840–2360	13
$4.0 \times 10^{11}(CO)(H_2O)^{1/2}(O_2)^{1/4} \exp(-20140/T)$	1030–1230	14

[a]Units: kgmol, m³, K, sec.

A limitation of the characteristic time model is the necessity of having some experimental data on the combustion flow field of interest to establish the semiempirical characteristic time relationships. Hence the model is a correlative tool rather than a predictive tool. The simplest approach for incorporating CO formation and oxidation in a predictive flow model is use of a two-step global model which, involves a one-step reaction forming CO and H_2O (or H_2) from the parent hydrocarbon fuel at an empirically determined rate by reactions (4) or (5). The CO formed then oxidizes at a rate calculated using an overall rate equation. Several empirical overall rate expressions for CO oxidation under premixed conditions are given in Table 7. Yetter et al.[12] have discussed the limitations of these overall rate expressions, noting particularly the variation in the activation energy with experimental conditions. The dependence of measured CO oxidation rates on the concentrations of O_2 and H_2O has been explained[2,15] in terms of equilibration of the reactions

$$O + H_2O \rightleftarrows OH + OH \qquad (11)$$

$$O + O + M \rightleftarrows O_2 + M \qquad (14)$$

When these two reactions are equilibrated,

$$(OH)_e = K(T) \cdot (H_2O)^{1/2} \cdot (O_2)^{1/4}$$

where $K(T) = K_{11}^{1/2}(T)/K_{14}^{1/4}(T)$ is a ratio of the equilibrium constants for reactions (11) and (14) and the subscript e designates the equilibrium concentration at the local temperature, T. If CO oxidation occurs principally by reaction (6), then, neglecting the back reaction,

$$-\frac{d(CO)}{dt} = k_6(CO)(OH)_e = \underbrace{k_6 K(T)(CO)(H_2O)^{1/2}(O_2)^{1/4}}_{k_{emp}} \qquad (15)$$

The simple kinetic model outlined above neglects the reverse of reaction (6); hence the model strictly applies only to the initial stages of the CO oxidation process. Westenberg[16] proposed a CO oxidation model which incorporates both the forward and reverse directions of reaction (6), while still assuming equilibration of reactions (11) and (14). In this model the CO concentration at any time t is given by

$$\frac{(CO) - (CO)_e}{(CO_2)_e} = \exp\left[-k_6(OH)_e\left[1 + \frac{(CO)_e}{(CO_2)_e}\right]t\right] \tag{16}$$

At $t = 0$, Eq. (16) reduces to

$$-\frac{d(CO)}{dt} = k_6(CO)(OH)_e,$$

which is identical to Eq. (15). An important assumption in these CO oxidation rate equations is the equilibration of reactions (11) and (14) at the local gas temperature. If the equilibrium assumption breaks down, then departures from calculated oxidation rates are to be expected. Combining the rate coefficient expression for k_6, Table 4, with equilibrium constant data, gives values of $k_{emp} = k_6K(T)$, which are approximately two orders of magnitude smaller than the global rate coefficients in Table 7. This observation suggests that the OH concentrations in the experiments significantly exceed local equilibrium values. Hence simple equilibrium models for radical concentrations cannot be used to interpret the results, and the reported global rate parameters are simply empirical fits to the experimental data. Use of global models for CO chemistry presumes that the local conditions in the practical device (temperature and stoichiometry) encompass those for which the empirical rate expression was obtained.

CO quenching in combustion gases occurs both by reduction in temperature and pressure and by depletion of radicals and since radical concentrations generally differ from equilibrium values, a desirable feature of a CO oxidation mechanism is a model which predicts the radical pool. Such a model would overcome some of the shortcomings of the kinetic models, discussed above. The detailed kinetic model, Table 5, provides this information. However, a more simplified kinetic mechanism is desirable for use in complex flow models. One such mechanism, based on the partial equilibrium concept, is outlined below.[17] In the model, the rapid bimolecular reactions (9)–(12), involving the H—O species, are assumed to be equilibrated at the local gas temperature. The pool of radical species (H, OH, O) changes at a rate governed by gas-phase or wall recombination reactions,

$$H + H \rightarrow H_2$$
$$H + OH \rightarrow H_2O$$
$$H + O_2 \rightarrow HO_2$$

All of these recombination reactions reduce the total number of moles of combustion gas. Hence a rate equation for the recombination reactions can be written as a single equation involving the total molar concentration,

$$-\frac{d(\text{M})}{dt} = \sum_i (R_{if} - R_{ib}) \tag{17}$$

where R_{if} and R_{ib} are the forward and reverse rates for the recombination reaction i.

The CO oxidation reactions may be incorporated in the model in two ways depending on local gas temperature and cooling rate, dT/dt. At sufficiently high temperatures (greater than 1300 K at one atmosphere) and sufficiently low cooling rates of the combustion gas $(-dT/dt < 3.4 \times 10^5 T^2 \exp(-19{,}800/T)$ K/sec$)$, the CO oxidation reaction (6) also may be assumed to be partially equilibrated. Hence the CO concentration is computed from equilibrium considerations and the only kinetic constraint on the system is Eq. (17); that is, the CO oxidation rate is governed solely by the recombination reactions. At sufficiently low temperatures and high cooling rates, the rates of reactions (6) and (7) become too slow to remain in partial equilibrium, and the CO kinetics must be incorporated in the overall model. In this instance the CO oxidation rate is determined both by the rates of the recombination reactions and by the rates of reactions (6) and/or (7).

Quantitative predictions of CO concentrations in the exhaust gas require that the initial concentration of CO in the combustion products be specified. One simple approach is to assume that all of the carbon in the fuel is converted to CO prior to the onset of CO oxidation. An alternative approach is to assume that the initial CO concentration is the equilibrium value for adiabatic combustion of the fuel-air mixture. Although this approach is reasonable for combustion systems with large residence times, in general, initial or maximum CO concentrations differ from the adiabatic equilibrium values and are determined by the kinetics of the CO formation reactions. In these instances a kinetic model for CO formation will be required in the overall CO kinetic model.

Figure 4 shows a comparison of measured CO mole fractions in quenched combustion gases with predicted values obtained from the quasi-global and the partial equilibrium oxidation models outlined above. In the experiment the products of lean combustion of a hydrocarbon fuel in air are rapidly quenched by a compact heat exchanger which provides a cooling rate on the order of 10^6 K/sec. The dashed line is the equilibrium CO mole fraction based on local temperature. Prior to the heat exchanger, the CO mole fraction approaches the local equilibrium value. The solid lines show predicted CO mole fractions. The quasi-global model, Eq. (16), significantly underpredicts the CO oxidation rate, due principally to the low equilibrium values of OH concentration after the quench. The partial equilibrium model, assuming that reaction (6) proceeds at a finite rate, gives predicted CO concentrations that are in reasonably good agreement with experimental results. Hardy and Lyon[18] have shown

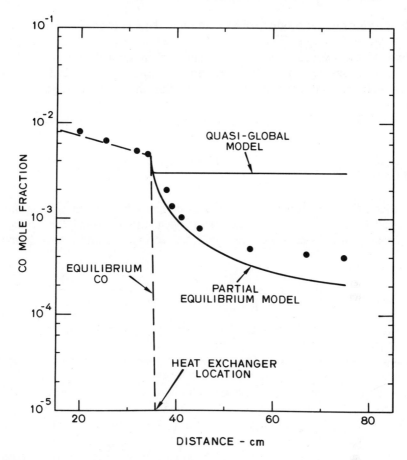

FIGURE 4. Measured and calculated CO mole fractions in lean quenched combustion products, fuel-air equivalence ratio = 0.91.[17]

similar discrepancies between measured CO oxidation rates and rates predicted by equilibrium quasi-global models under isothermal (nonquenched) conditions.

The validity of the rate-constrained partial equilibrium model, outlined above, depends on the partial equilibration of reactions (9)–(12). At sufficiently low temperatures and high cooling rates, a breakdown in the partial equilibrium assumption can be expected. In this instance accurate prediction of CO emissions can be obtained only from a detailed kinetic model or from an empirical model appropriate to the conditions in the combustion system of interest.

In summary, the CO emissions from practical combustion devices are governed by both the CO formation and oxidation rates. The CO formation reactions tend to be rapid, and in devices operating with excess air, CO oxidation rates in the product gases determine the CO levels in the exhaust. In this instance CO quenching occurs by reduction in oxidation rate due to cooling of the exhaust gas and by reduction in radical concentrations by gas-phase or wall recombination reactions. In general, the radical concentrations cannot be calculated by assuming equilibrium at the local gas temperature.

3. FORMATION AND DESTRUCTION OF NITROGEN OXIDES IN FLAMES

The three principal sources of nitrogen oxide emissions in combustion are (1) oxidation of atmospheric (molecular) nitrogen, often termed the *"thermal"* *NO formation mechanism*, (2) "prompt" NO formation, and (3) oxidation of

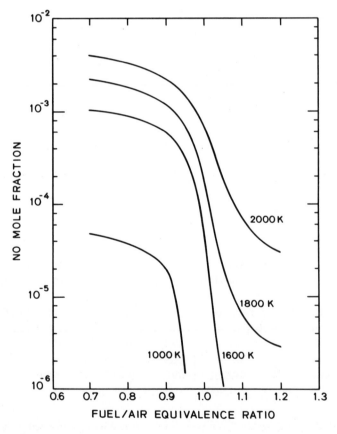

FIGURE 5. Equilibrium NO mole fractions for premixed methane-air combustion at a pressure of 1 atm.

nitrogen-containing compounds in the fuel, termed the *fuel-nitrogen mechanism*. The relative importance of these three sources to total nitrogen oxide emissions from a particular combustion device depends on operating conditions and on fuel composition.

Equilibrium NO mole fractions in the combustion products are shown as a function of temperature and stoichiometry in Figure 5 for methane-air. For typical exhaust gas temperatures (less than 1000 K), equilibrium NO mole fractions are less than 30 ppmv. Typical measured exhaust NO concentrations exceed the equilibrium values for exhaust conditions but are lower than values corresponding to equilibrium at the maximum combustion temperature, evidencing the importance of kinetic processes in establishing exhaust NO levels.

In the combustion of "clean" fuels (fuels not containing nitrogen compounds), under lean or stoichiometric conditions the thermal mechanism is the principal source of nitrogen oxide emissions. However, as the nitrogen content

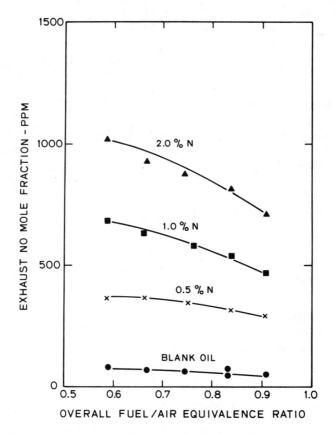

FIGURE 6. Measured exhaust NO mole fractions (dry basis, corrected to stoichiometric) for an oil-fired furnace with various amounts of pyridine (C_5H_5N) added to the fuel.[19]

TABLE 8. Typical Fuel Nitrogen and Fuel Sulfur Contents of Fossil Fuels[a]

Fuel	Nitrogen (wt %)	Total Sulfur (wt %)
Petroleum Crude Oil		
Residual oil	0.3–2.2	0.3–3.0
Heavy distillates (#3–5)	0.3–1.4	0.2–2.5
Light distillates (#1, 2)	0–0.4	0.01–0.4
Coal Liquids		
Crude (COED)	1.1	0.02
Utah heavy distillate	0.2	0.05
Utah light distillate	0.1	< 0.01
Sea coal (COED)	0.1	0.02
Synthoil (PERC)	0.8	0.2
Shale Liquids		
Residual oil	1.5–2.5	0.2–0.5
Heavy distillates	1.4–2.0	0.1–0.5
Light distillates	0.25–1.4	0.01–0.2
Anthracite Coal		
Hazelton, Pennsylvania	0.79[b]	0.47[b]
Bituminous Coals		
Black Creek, Alabama (MV)	1.47[b]	1.3[b]
Western Kentucky (HVA)	1.55	3.2
Pittsburgh Seam #8 (HVA)	1.65	0.84
Price, Utah (HVB)	1.54	0.66
Cadiz, Ohio (HVB)	1.07	7.40
Illinois #6 (HVC)	1.18	3.77
Four Corners, New Mexico (HVC)	1.23	1.03
Subbituminous Coals		
Hardin, Montana (B)	0.99[b]	1.07[b]
Shell, Texas (C)	1.13	1.02
Rosebud, Montana (B)	0.84	1.00
Colstrip, Montana (B)	1.38	0.63
Lignite		
Beulah, North Dakota (A)	0.96[b]	0.37[b]
Savage, Montana (A)	1.00	0.42
Scranton, North Dakota (A)	0.83	1.52
COED Char	1.67	2.67

[a] Various sources.
[b] Dry basis.

TABLE 9. The Importance of Prompt NO in NO_x Emissions from Practical Combustion Devices[a]

Device/Fuel	Estimated Percent NO_x Emission due to Prompt NO Mechanism
Utility boiler/natural gas	17
Gas turbine/natural gas	30
SI engine/gasoline	10
CI engine/diesel fuel	5
Utility boiler/coal (N = 1%)	< 5

[a] From Hayhurst and Vince.[20]

of the fuel increases, significant contributions from the fuel nitrogen mechanism to total nitrogen oxide emissions occur (Figure 6).[19] For nitrogen contents typical of fuel oils, shale oils, and pulverized coal (Table 8), fuel nitrogen generally will be the principal source of nitrogen oxide emissions.

Nitric oxide formation rates near the flame zone in hydrocarbon-air premixed and diffusion flames exceed those attributable to direct oxidation of molecular nitrogen by the "thermal" mechanism, especially under fuel-rich conditions. This rapidly formed NO has been termed *prompt NO*. Typical levels of prompt NO produced in stoichiometric premixed hydrocarbon-air flames range from 50–90 ppmv, with the highest levels occurring for fuels with low H/C ratios (e.g., C_2H_2 and C_6H_6). Table 9 lists estimates of the relative importance of prompt NO in nitrogen oxide emissions from practical combustion systems.[20]

For nitrogen-containing fuels, the contribution of prompt NO to total nitrogen oxide emissions from conventional combustion systems is negligible. However, with improved techniques for reducing emissions due to the thermal and fuel nitrogen mechanisms, the relative importance of prompt NO to nitrogen oxide emissions may increase.

Experimental studies have shown that a substantial fraction of the nitrogen oxide emissions from some combustion sources, particularly gas turbines, can be NO_2 (Figure 7).[21] Existing regulations on emissions from stationary and mobile sources do not distinguish between NO and NO_2. However, from the viewpoint of environmental effects, the form of the nitrogen oxide emitted by the combustion source is important since NO and NO_2 interact in the smog-producing atmospheric photolytic cycle in different ways. Large concentrations of NO_2 as a primary pollutant can reduce the induction times for production of photochemical smog since the relatively slow NO $\rightarrow$ NO_2 oxidation step is bypassed. Furthermore the form of nitrogen oxide in the exhaust gas will impact specification of control techniques, since different techniques have different effectiveness in removing NO or NO_2.

Recent measurements[22,23] have revealed moderate levels of N_2O in the stack gases from fossil-fuel-fired power plants. Nitrous oxide (N_2O) plays an

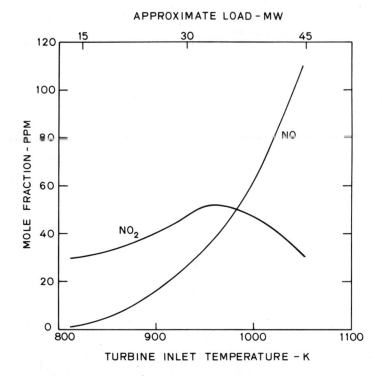

FIGURE 7. NO and NO_2 exhaust mole fractions for a gas turbine operating on natural gas for various loads. From Ref. 21. Reprinted with permission from Gordon and Breach Science Publishers S. A.

important role in the chemistry of the upper atmosphere, and hence it affects solar radiative heat transfer to the surface of the earth. Studies in laboratory-scale furnaces[24,25] show that N_2O levels in the combustion products are highest for unstaged combustion of nitrogen-containing fuels, with approximately one mole of N_2O being produced for every three moles of NO. Staged combustion, which utilizes an initial fuel-rich combustion zone followed by secondary injection of combustion air, reduces both NO and N_2O emissions. In unstaged coal and oil flames, up to 12% of the fuel nitrogen was converted to N_2O, whereas in staged flames, approximately 3% was converted to N_2O.

In this section the three NO formation mechanisms, outlined above, will be discussed, together with the mechanisms for conversion of NO to NO_2 and for formation and removal of N_2O. Although only those reactions that specifically involve nitrogen species will be considered, it is important to realize that these reactions take place in an environment established by the combustion of the fuel and that often there is an important coupling between the fuel combustion reactions and the nitrogen chemistry. A more recent survey of nitrogen chemistry in flames is provided in Ref. 60.

TABLE 10. Rate Coefficients for Thermal NO Formation Reactions[a]

Reaction Number	Reaction	A	B	$C(K)$
18	$O + N_2 \rightarrow NO + N$	1.8(11)	0	38.4(3)
-19	$O + NO \rightarrow N + O_2$	3.8(6)	1.0	20.8(3)
-20	$H + NO \rightarrow OH + N$	2.6(11)	0	25.4(3)

[a]From Hanson and Salimian.[26] Units: kgmol, m³, K, sec; $k = AT^B \exp(-C/T)$.

3.1. Thermal NO Formation

The three principal reactions that comprise the thermal NO formation mechanism are

$$O + N_2 \rightleftarrows NO + N \tag{18}$$

$$N + O_2 \rightleftarrows NO + O \tag{19}$$

$$N + OH \rightleftarrows NO + H \tag{20}$$

The rate coefficients for both the forward and reverse reactions have been measured over a wide temperature range (Table 10).[26]

Invoking a steady-state approximation for the N-atom concentration* and assuming that reaction (9) is partially equilibrated, the NO formation rate due to the thermal mechanism may be expressed,

$$\frac{d(NO)}{dt} = 2k_{18}(O)(N_2) \frac{1 - (NO)^2/K(O_2)(N_2)}{1 + k_{-18}(NO)/[k_{19}(O_2) + k_{20}(OH)]} \tag{21}$$

where $K = (k_{18}/k_{-18})(k_{19}/k_{-19}) =$ equilibrium constant for the reaction $N_2 + O_2 \rightleftarrows 2NO$. Calculation of the NO formation rate requires values of the local temperature and the local concentrations of O_2, N_2, O, and OH.

The characteristic times for NO formation by reactions (18)–(20) generally exceed the characteristic combustion times so that it often is feasible to decouple the thermal NO formation process from the combustion process. Using this approximation, NO formation rates are calculated from Eq. (21) using local equilibrium values of temperature and concentrations of O_2, N_2, O, and OH.

From Eq. (21) it can be seen that the maximum NO formation rate is

$$\frac{d(NO)}{dt} = 2k_{18}(O)(N_2)$$

*The steady-state approximation assumes that the rate of production of N-atoms by reaction (18) is equal to the rate of removal by reactions (19) and (20) so that $d(N)/dt \cong 0$.

corresponding to $(NO) \ll (NO)_{equilibrium}$. Assuming equilibration of the combustion reactions, the O-atom concentration may be calculated from equilibrium considerations,

$$(O)_e = \frac{K_0}{(RT_e)^{1/2}}(O_2)_e^{1/2}$$

where K_0 is the equilibrium constant for

$$\frac{1}{2}O_2 \rightleftarrows O$$

Using the expression for k_{18} from Table 10 and $K_0 = 3.97 \times 10^8 \exp(-31090$ $K/T_e)$ atm$^{1/2}$, the maximum NO formation rate may be written

$$\frac{d(NO)}{dt} = 1.45 \times 10^{20} T_e^{-1/2} \exp[-69460 \, K/T](O_2)_e^{1/2}(N_2)_e \text{ kgmol/m}^3 \cdot \text{sec}$$

$$(22)$$

The strong dependence of the NO formation rate on the combustion gas temperature and the somewhat weaker dependence on oxygen concentration is evident from Eq. (22). Conventional methods for control of NO_x emissions produced by the thermal mechanism generally involve modification of the combustion process to reduce either the combustion gas temperature or the availability of oxygen. Both approaches are directed toward reduction of the NO formation rate rather than enhancing the NO removal rate, which is very slow at typical exhaust gas temperatures.

Temperature reductions can be obtained in various ways, including exhaust gas recirculation, introduction of dilution air downstream of the primary combustion zone, water injection, and reduction in fuel-air ratio. In piston engines variations in spark or fuel injection timing will alter the temperature-time history of the combustion products and subsequent variations in NO_x emissions will be observed. While reduced temperatures can produce significant reductions in NO_x emissions, the reduced temperatures also may result in quenching of the CO oxidation reactions, thereby causing an increase in CO emissions. Hence the effectiveness of temperature reduction for thermal NO_x control often is limited by increases in CO emissions as well as operational problems introduced by temperature reduction, such as flame stability.

Reduction of available oxygen may be achieved by operating with a fuel-rich primary combustion zone. Although NO_x emissions generally will be reduced, the products of rich combustion will contain increased levels of CO. To burn out this CO, additional air must be injected downstream of the primary combustion zone so that the overall fuel-air equivalence ratio is less than one (overall excess air). In addition some provision should be made for extraction of energy between the primary combustion zone and the air injection point, either by heat transfer or by work extraction, so that the final combustion gas temperature will be lower than the adiabatic combustion temperature for the overall fuel-air ratio.

Calculation of thermal NO_x emissions from combustion devices burning clean fuels (natural gas, producer gas, and low-nitrogen oils) requires coupling of reactions (18)–(20) with a combustor flow model. A simple approach to this coupling for continuous flow devices is the characteristic time model, discussed in Section 2. Appropriate chemical and physical time scales for thermal NO_x emissions for gas turbines are given in Table 6. These times are based on the view that NO_x emissions are determined by the maximum NO formation rates in the flame, that is, at rates appropriate to the maximum (stoichiometric) burned gas temperature. An alternative approach to calculate NO formation rates in the burned gas is to assume complete decoupling of the thermal NO reactions from the combustion process. A combustor flow model is used to calculate the mean temperature and equilibrium species concentration distributions in the combustor. Equation (21) is used with these distributions to calculate local NO formation rates.

3.2. Prompt NO Formation

Investigations of NO formation in and near the flame zone of hydrocarbon-air flames indicate that NO formation rates can exceed those predicted by Eq. (22). In lean and slightly rich flames, this discrepancy can be traced in part to a breakdown in the assumption of equilibration of the $O \rightleftarrows \frac{1}{2}O_2$ reaction. Near the flame zone radical concentrations significantly exceed equilibrium values, and the overshoot of radical concentrations, particularly O and OH, result in higher NO formation rates. Hence prediction of NO formation rates near the combustion zone requires coupling of the NO formation process to the radical-producing reactions. There are several approaches for this coupling. The most direct approach is the simultaneous integration of the rate equations for reactions (18)–(20) and the rate equations describing the oxidation of the fuel. Application of this approach requires knowledge of the detailed fuel oxidation mechanism, and as noted previously, detailed mechanisms are known for only a few fuels. Various approximate models for fuel combustion, termed global or quasi-global models, have been developed. However, application of these models to predict NO formation rates near the combustion zone is questionable since these models frequently do not accurately predict radical concentrations.[27] A simplified approach for estimating the radical concentrations near the combustion zone is based on a partial equilibration of the rapid bimolecular H—O reactions (9)–(12). The concentrations of O and OH then can be related to the concentrations of stable species, such as H_2, O_2, or H_2O, which are readily measured.

In fuel-rich flames, the rapid NO formation rates near the flame cannot be explained by superequilibrium concentrations of O and OH since the radical concentrations required to produce observed NO formation rates are significantly larger than partial equilibrium values. Although O-atom concentrations can exceed partial equilibrium values for fuel-rich combustion, reactions in addition to those described above are required to model the rapid NO formation rates observed in these flames. Although there is some uncertainty in the detailed reaction mechanism for this prompt NO formation, there is a consensus that a principal product of the initial reactions is HCN (or CN radicals) and that the presence of hydrocarbon species is essential. Two plausible HCN formation reactions involving hydrocarbon fragments are

$$CH + N_2 \rightarrow HCN + N \tag{23}$$

$$CH_2 + N_2 \rightarrow HCN + NH \tag{24}$$

The rate coefficients for reactions (23) and (24) are uncertain; however, indirect rate coefficient determinations are available (Table 11).[28-31] Since the hydrocarbon fragments in these reactions are products of the initial stages of the fuel pyrolysis and oxidation process, the formation of prompt NO is closely coupled to the hydrocarbon oxidation reactions. Hence predictions of prompt NO depend on satisfactory predictions of the concentrations of hydrocarbon radicals, which may require a reasonably complete hydrocarbon reaction model.

Subsequent reaction of the N-atom produced in reaction (23) to form NO likely would occur by the thermal mechanism, whereas HCN and NH would react by reactions important in the fuel nitrogen conversion mechanism, discussed below. It may be possible to model prompt NO formation in a quasi-global sense in that the hydrocarbon radicals can be assumed to form in a single step from the parent fuel species, or from a hydrocarbon derived from the fuel species,

$$RH + H \rightarrow CH_i + \cdots$$

with the CH_i reacting with N_2 to form HCN and NH_{i-1},

$$CH_i + N_2 \rightarrow HCN + NH_{i-1}$$

This approach would circumvent the need for a complete kinetic model of the hydrocarbon chemistry. Appropriate global kinetic parameters could be obtained by comparison with experimental data on HCN and prompt NO formation in premixed hydrocarbon-air flames.

3.3. Fuel-NO Formation

Fuel-nitrogen is a principal source of NO_x emissions in combustion of fossil fuels. The extent of conversion of fuel nitrogen to NO_x is nearly independent of the identity of the parent fuel nitrogen compound but is strongly dependent

TABLE 11. Mechanisms and Rate Coefficients for Hydrocarbon-N_2 and Hydrocarbon-NO Reactions[a]

Reaction Number	Reaction	A	B	C(K)	Temperature	Reference
23	$CH + N_2 \rightarrow HCN + N$	1.9(8)	0	6.8(3)	1600–2100	28
24	$CH_2 + N_2 \rightarrow HCN + NH$	1.0(10)	0	37.3(3)	1600–2100	28
	$CH + NO \rightarrow HCN + O$	1.0(11)	0	0	1600–2100	28
	$CH_2 + NO \rightarrow HCN + OH$	1.4(9)	0	−0.55(3)	1600–2100	28
	$CH_3 + NO \rightarrow$ products	1.3(11)	0	0	295	29
	$\rightarrow HCN + H_2O$	2.0(8)	0	0	2000	30
	$\rightarrow HCN + H_2O$	<3.2(5)	—	—	1220	31

[a]Units: kgmol, m^3, K, sec; $k = AT^B \exp(-C/T)$.

237

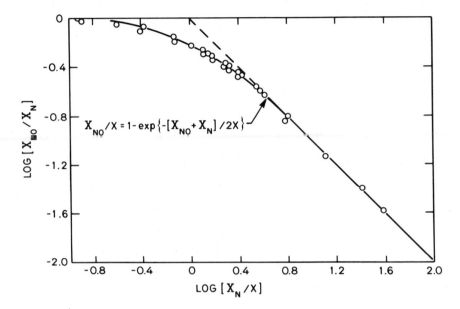

FIGURE 8. Correlation of the fraction of fuel-nitrogen converted to NO in atmospheric-pressure, premixed, ethylene flames: X_N = mole fraction of added fuel-nitrogen, X_{NO} = mole fraction of NO in the combustion products, X = a parameter characteristic of the flame = $f(T, \Phi)$ but independent of X_N.[32]

on the local combustion environment (temperature and stoichiometry) and on the initial fuel nitrogen concentration in the reactants. Figure 8[32] shows a correlation of the fraction of fuel nitrogen converted to NO with flame conditions and initial fuel nitrogen concentration in a laminar premixed flame. The data correlation was obtained for a range of stoichiometries, combustion gas temperatures, nitrogen compounds, and initial fuel nitrogen concentrations. The parameter X is characteristic of the flame and is independent of [N]. For small values of X_N/X, that is, low levels of fuel-N and lean flames, nearly complete conversion of fuel-N to NO is observed. As the levels of fuel-N increase, decreasing fractional conversion to NO is observed. The correlation equation in Figure 8 can be explained by a simple reaction mechanism in which NO is both formed and removed by reaction of a nitrogen-containing species, I, which itself is formed from the parent fuel nitrogen compound,

$$\text{NO} \qquad (25)$$

$$\text{Fuel-N} \rightarrow \text{I} \nearrow^{+OX} \searrow_{+NO}$$

$$\text{N}_2 \qquad (26)$$

In this simplified kinetic model, the parameter X in the correlation equation is

$$X = \frac{k_{25}(\text{OX})}{k_{26}}$$

where (OX) is the concentration of the oxygen-containing species that reacts with I to form NO. Although the identity of OX and I cannot be determined unambiguously, available data suggest that OX is the OH radical and that I is an NH_i species. Assuming that OX = OH, then $k_{25}/k_{26} = 2.5 \pm 0.5$ for a range of stoichiometries and postflame temperatures. The OH concentrations in postflame gases typically exceed equilibrium values. Hence application of the correlation equation to estimate fuel-N conversion in flames requires a method for calculating the OH concentration. A partial equilibrium model, similar to that employed in the CO oxidation model, might be satisfactory. Although the data in Figure 8 were obtained in a laminar premixed flame, similar trends of fractional conversion of fuel-N to NO with nitrogen content of the fuel are found for a range of fossil and synthetic fuels in practical combustors (Figure 9). Hence the correlation equation may be useful in estimating NO_x emissions from fuel nitrogen in NO_x emissions from fuel nitrogen in practical devices.

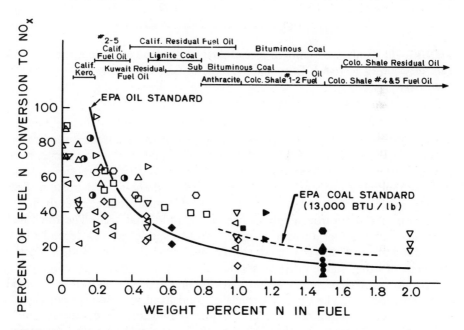

FIGURE 9. Fractional conversion of fuel-nitrogen in fossil fuel combustion in practical devices.[59] Open symbols—oil; solid symbols—coal; semisolid symbols—laminar premixed flames.[32]

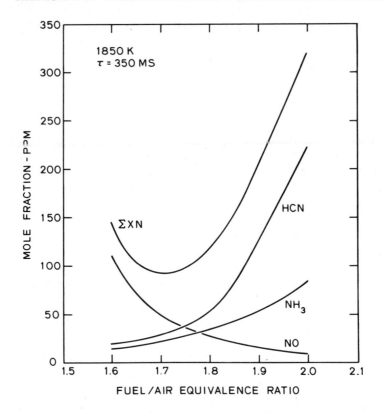

FIGURE 10. Stable fixed nitrogen species product distribution for atmospheric pressure methane-air flames: $\Sigma X_N \equiv X_{HCN} + X_{NH_3} + X_{NO}$; fuel nitrogen additive = 600 ppm NH_3.[33]

In addition to NO, fuel-nitrogen is a source of other gaseous nitrogenous species, such as HCN and NH_3, if the local stoichiometry is sufficiently rich. Figure 10[33] shows the dependence of the stable fixed nitrogen species product distribution on equivalence ratio in premixed methane-air combustion. Secondary injection of air, as employed in staged combustion, can result in the oxidation of these nitrogen species producing additional NO_x. The fractional conversion of the fuel nitrogen to fixed nitrogen species also is dependent on stoichiometry and, for rich mixtures, is much less than one, indicating the presence of other stable nitrogen species in the product gases, principally N_2. There also is a stoichiometry that gives a minimum concentration of fixed nitrogen products. The observed levels of total fixed nitrogen species significantly exceed the equilibrium levels for the combustion gas conditions, indicating the importance of chemical kinetics in the fuel-nitrogen conversion process.

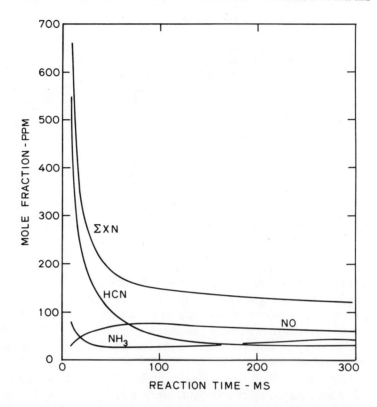

FIGURE 11. Stable fixed nitrogen species and total fixed nitrogen profiles in a rich, premixed, methane-air flame: $T = 1850$ K, $\Phi = 1.7$, fuel-nitrogen additive = 600 ppm NH_3.[33]

Although product distributions of N-containing species are useful in understanding the fuel nitrogen conversion process, detailed species concentration profiles in controlled combustion environments are needed to develop and validate a fuel-nitrogen reaction mechanism. Figure 11 shows one set of stable fixed nitrogen species profiles and the total fixed nitrogen profile for a rich methane-air flame. The sequence of kinetic processes in this flame are (1) a rapid and nearly quantitative conversion of the parent fuel-nitrogen species (NH_3 in this flame) to HCN, and (2) a subsequent decay of HCN and reformation of NH_3 and formation of NO and N_2 (as evidenced by the decay of the total fixed nitrogen pool).

Although some details of the kinetic mechanism for fuel-nitrogen conversion in flames are unresolved at the present time, there is a consensus that the reactions proceed as illustrated in Figure 12.

The development of a detailed kinetic model is hampered by the fact that the identity and distribution of the chemical types of nitrogen compounds in

FUEL-NITROGEN MECHANISM

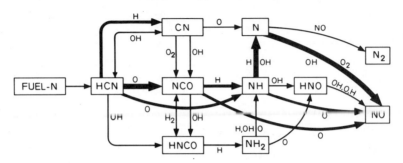

FIGURE 12. Schematic diagram of the principal reaction paths in the fuel nitrogen conversion process in flames.

fossil fuels are not well established. This uncertainty, coupled with the lack of data for reaction paths and rate coefficients for pyrolysis and oxidation of complex nitrogenous species, requires the introduction of simplifications in the kinetic model and restricts detailed reaction sets to those relevant to pyrolysis and oxidation of HCN and NH_3 and to the reduction of NO in hydrocarbon flame environments.

The justification for limiting the kinetic model to simple nitrogenous species is derived from data on inert and oxidative pyrolysis of model fuel-nitrogen compounds, from pyrolysis studies of coal and fuel oil and from fundamental flame studies of NO formation from fuel-nitrogen species added to simple hydrocarbon fuels.

Investigation of the inert pyrolysis of several model fuel-nitrogen compounds (pyridine, quinoline, pyrrole, and benzonitrile) at temperatures in the range 1100–1400 K shows that the nitrogen present in the model compound is converted primarily to HCN with small amounts forming NH_3.[34,35] At 1373 K, HCN accounted for 100% of the nitrogen in pyridine, 82% of the nitrogen in benzonitrile, and about 55% of the nitrogen in quinoline. The nitrogen product distributions identified in the oxidative pyrolysis of pyridine and benzonitrile are qualitatively similar to those obtained under inert conditions, but with increased yields of NH_3. Extrapolation of the data obtained in the temperature range 1100–1400 K to temperatures of interest in combustion (1800 K) yields half-lives for pyridine, pyrrole, and quinoline of 60 μsec, 200 μsec, and 840 μsec, respectively. The half-life for pyridine is in reasonable agreement with high-temperature shock-tube data, which indicate a half-life of 15–30 μsec at 1800 K.[36] Studies of the inert pyrolysis of coal and fuel oil also show that a significant amount of the nitrogen present was converted to HCN with lesser amounts appearing as NH_3.

Investigations of the fate of simple fuel-nitrogen compounds in fuel-rich hydrocarbon flames, such as discussed above, show rapid and nearly quantitative conversion to cyano compounds, which appear as HCN in the sampled combustion gases.

The results of the above studies indicate that some kinetic model is required for pyrolysis and oxidation of the parent fuel-nitrogen species in any detailed kinetic model for conversion of fuel-nitrogen in flames. In certain cases, such as when the fuel-nitrogen compound is a simple one such as NH_3 or pyridine, it may be reasonable to assume that the HCN yield is quantitative and instantaneous. However, data for quinoline and pyrrole decomposition indicate that the assumption of very short times for conversion of fuel-nitrogen to HCN may not be justified for more complex fuel-nitrogen compounds. Since the detailed reaction paths leading to HCN formation are not known, a global modeling approach may be useful. In this approach, it is assumed that HCN is formed from the parent fuel-nitrogen compound in a single-step reaction,

$$Fuel\text{-}N \rightarrow HCN$$

at an empirically determined rate. However, at the present time there is insufficient data to obtain these rate expressions for most fuel-nitrogen species.

Available data also show that it is necessary to consider reactions describing the pyrolysis and oxidation of HCN and NH_3 in a detailed kinetic model for fuel nitrogen conversion in flames.

In an oxygen-free environment, HCN reacts principally by reactions important in the thermal decomposition mechanism. The principal HCN decomposition steps are

$$HCN + M \rightarrow CN + H + M \tag{27}$$

$$HCN + H \rightarrow CN + H_2 \tag{28}$$

$$HCN + CN \rightarrow C_2N_2 + H \tag{29}$$

with subsequent reaction of the C_2N_2 produced in reaction (29) to form CN radicals.

In the presence of oxygen-containing radicals (O, OH), the following reactions contribute to HCN removal:

$$
\begin{array}{ll}
& CN + OH \tag{30} \\
HCN + O \longrightarrow NCO + H \tag{31} \\
& NH + CO \tag{32}
\end{array}
$$

$$
\begin{array}{ll}
& CN + H_2O \tag{33} \\
HCN + OH & \\
& HOCN + H \tag{34}
\end{array}
$$

TABLE 12. Mechanisms and Rate Coefficients for HCN Removal Reactions[a]

Reaction Number	Reaction	A	B	$C(K)$	Temperature	Reference
27	$HCN + M \rightarrow CN + H + M$	5.7(13)	0	58.9(3)	2200–2700	37
−28	$CN + H_2 \rightarrow HCN + H$	7.5(10)	0	0	2700–3500	38
29	$HCN + CN \rightarrow C_2N_2 + H$	1.0(10)	0	0	2700–3100	39
30	$HCN + O \rightarrow CN + OH$	< 4.0(8)	0	0	2000–2500	40
31	$HCN + O \rightarrow NCO + H$	5.9(9)	0	4.0(3)	540 900	41
		7.3(10)	0	7.5(3)	1800–2500	42
32	$HCN + O \rightarrow NH + CO$	2.2(10)	0	7.7(3)	1500–2600	40
33	$HCN + OH \rightarrow H_2O + CN$	1.4(10)	0	5.5(3)	850–2600	43
34	$HCN + OH \rightarrow HOCN + H$	2.0(8)	0	0	1950–2380	44

[a] Units: kgmol, m^3, K, sec; $k = AT^B \exp(-C/T)$.

Rate coefficient data for the reactions of HCN are tabulated in Table 12.[37–44] Available data suggest that reaction (33) is partially equilibrated at flame temperatures and that reactions (31) and (32) are the principal paths for HCN removal. The CN reactions paths in flames are

$$CN + OH \rightarrow NCO + H$$
$$CN + O \rightarrow CO + N$$
$$CN + O_2 \rightarrow NCO + O$$

and the reverse of reaction (28). However, these reactions do not appear to play an important role in the fuel-nitrogen conversion process. Reaction of the NCO radical occurs by

$$NCO + H \rightarrow NH + CO$$
$$NCO + H_2 \rightarrow HNCO + H$$
$$NCO + O \rightarrow NO + CO$$

with reaction of HNCO occurring by

$$HNCO + H \rightarrow NH_2 + CO$$

Subsequent reaction of the NH_i species formed in the above reactions occurs by means of the NH_3 mechanism described below.

The reaction mechanism for NH_3 oxidation is complex, and some uncertainties exist in the reaction paths and rate coefficients. A detailed kinetic model for NH_3 oxidation under premixed conditions is given in Table 13. This detailed mechanism contains several important reaction subsets. One of these subsets comprises the bimolecular reactions that provide for the interconver-

TABLE 13. Detailed Kinetic Model for Premixed NH_3 Oxidation[a]

Reaction Number	Reaction	A	B	$C(K)$
	$NH_3 + M \rightarrow NH_2 + H + M$	2.5(13)	0	47.2(3)
	$NH_3 + OH \rightarrow NH_2 + H_2O$	5.8(10)	0	4.06(3)
	$NH_3 + O \rightarrow NH_2 + OH$	2.0(10)	0	4.47(3)
	$NH_3 + H \rightarrow NH_2 + H_2$	1.3(11)	0	10.8(3)
	$NH_2 + OH \rightarrow NH + H_2O$	5.0(8)	0.5	1.0(3)
	$NH_2 + O \rightarrow NH + OH$	6.8(9)	0	0
	$NH_2 + O \rightarrow HNO + H$	6.3(11)	-0.5	0
	$NH_2 + H \rightarrow NH + H_2$	1.9(10)	0	0
	$NH_2 + O_2 \rightarrow HNO + OH$	1.8(9)	0	7.5(3)
	$NH + OH \rightarrow HNO + H$	1.0(9)	0.5	1.0(3)
	$NH + OH \rightarrow N + H_2O$	5.0(8)	0.5	1.0(3)
	$NH + O \rightarrow NO + H$	6.3(8)	0.5	0
	$NH + O \rightarrow N + OH$	6.3(8)	0.5	4.0(3)
	$NH + H \rightarrow N + H_2$	5.0(10)	0	1.0(3)
	$NH + O_2 \rightarrow HNO + O$	1.0(10)	0	6.0(3)
	$NH + NH_2 \rightarrow N_2H_2 + H$	3.2(10)	0	0.5(3)
	$HNO + OH \rightarrow NO + H_2O$	1.3(9)	0.5	1.0(3)
	$HNO + O \rightarrow NO + OH$	5.0(8)	0.5	1.0(3)
	$HNO + H \rightarrow NO + H_2$	1.3(10)	0	2.0(3)
	$HNO + M \rightarrow NO + H + M$	1.8(13)	0	24.5(3)
	$N_2H_2 + M \rightarrow N_2H + H + M$	1.0(13)	0	25.0(3)
	$N_2H_2 + H \rightarrow N_2H + H_2$	1.0(10)	0	0.5(3)
	$N_2H + M \rightarrow N_2 + H + M$	2.0(11)	0	10.0(3)
35	$NH_2 + NO \rightarrow N_2 + H_2O$	6.3(16)	-2.5	0.95(3)
36	$NH_2 + NO \rightarrow N_2 + H + OH$	6.3(16)	-2.5	0.95(3)
	Including $H_2 - O_2$ reactions, Table 5			
	Including thermal NO reactions, Table 10			
	Including N_2O reactions, Table 15			

[a]From Hanson and Salimian.[26] Units: kgmol, m³, K, sec; $k = AT^B \exp(-C/T)$.

sion of the amine species,

$$NH_i + \begin{bmatrix} O \\ H \\ OH \end{bmatrix} \rightleftarrows NH_{i-1} + \begin{bmatrix} OH \\ H_2 \\ H_2O \end{bmatrix}$$

where $i = 1, 2, 3$. High-temperature rate data for this reaction subset are fairly well established. Available experimental data and kinetic simulations indicate that at typical combustion temperatures, these bimolecular reactions are partially equilibrated so that the relative concentrations of the species in the NH_i pool often may be calculated from equilibrium considerations at the local gas temperature.

A second important reaction subset comprises reactions of NH_i forming, eventually, NO, involving the direct oxidation of NH_i species, for example,

$$NH + O \rightarrow NO + H$$

or

$$N + O_2 \rightarrow NO + O \tag{19}$$

and a second route involving HNO as an intermediate reaction product, for example,

$$NH + OH \rightarrow HNO + H$$

$$NH + O_2 \rightarrow HNO + O$$

$$NH_2 + O \rightarrow HNO + H$$

$$NH_2 + O_2 \rightarrow HNO + OH$$

Subsequent reaction of HNO would occur by

$$HNO + M \rightarrow NO + H + M$$

$$HNO + H \rightarrow NO + H_2$$

$$HNO + OH \rightarrow NO + H_2O$$

$$HNO + O \rightarrow NO + OH$$

A third important reaction subset comprises reactions that involve NH_i species and lead eventually to the formation of N_2,

$$N + NO \rightarrow N_2 + O \tag{18}$$

and

$$
\begin{array}{ll}
& \nearrow \; N_2 + H_2O \qquad (35)\\
NH_2 + NO & \\
& \searrow \; N_2 + H + OH \qquad (36)
\end{array}
$$

The precise paths followed by reactions (35) and (36) are not known with certainty; however, rate data for the overall reactions are given in Table 13. It is likely that, as written, they are not elementary reactions but probably involve unmeasured nitrogen-containing intermediate species, such as N_2H, which subsequently reacts to form N_2. The following three-reaction sequence, involving N_2H, is equivalent to the overall reaction (36):

$$NH_2 + NO \rightarrow N_2H + OH$$
$$N_2H + NO \rightarrow HNO + N_2$$
$$HNO + M \rightarrow H + NO + M$$
$$\overline{NH_2 + NO \rightarrow N_2 + H + OH}$$

There also is evidence for the importance of N_2H_i species in NH_3 pyrolysis. In pyrolysis, reactions involving N_2H_i species such as

$$NH_2 + NH_2 \rightarrow N_2H_3 + H$$
$$N_2H_3 + M \rightarrow N_2H_2 + H + M$$
$$N_2H_2 + M \rightarrow N_2H + H + M$$
$$N_2H + M \rightarrow N_2 + H + M$$
$$\overline{NH_2 + NH_2 \rightarrow N_2 + 4H}$$

provide a source of radicals that significantly enhance the overall NH_3 decomposition rate.

The reaction sequences outlined above have two important characteristics:

1. Two species in the fixed nitrogen pool react to produce N_2, resulting in a depletion of the total fixed nitrogen pool.
2. The sequences provide for net production of radical species, which play an important role in the propagation steps of the overall fuel-nitrogen conversion mechanism.

At elevated temperatures in combustion devices, reactions between hydrocarbons and NO are observed, with HCN as the principal stable reaction product,

$$CH_i + NO \rightarrow HCN + OH_{i-1}$$

Rate data for reactions of several hydrocarbon radicals with NO are listed in Table 11. The initial products of these reactions have not been determined, and it is likely that HCN formation involves some intermediate nitrogen-containing species. The important characteristic of this reaction sequence is that it provides a route to recycle NO in the fuel-nitrogen conversion process (Figure 12). These reactions should be especially important in situations where the hydrocarbon radical concentrations are large as, for example, in the flame zones of hydrocarbon flames and in the postflame zone of very fuel-rich flames (e.g., in the products of the rich combustion zone in staged combustion and reburning). If these reactions are incorporated into a detailed model for fuel-nitrogen chemistry, then some mechanism for the prediction of hydrocarbon radical concentrations is required. Since detailed hydrocarbon chemistry models are available only for a few fuels, it may be necessary to use a quasi-global kinetic model for hydrocarbon radicals similar to that proposed to model the precursors to prompt NO formation.

Three methods for reduction of exhaust NO_x emissions from combustion devices burning fuels with high fuel-nitrogen content are staged combustion, reburning, and selective reduction of NO in the product gas by reaction with NH_3 or other amine species. Kinetic models of these processes are useful in establishing optimum conditions for NO_x reduction.

In staged combustion the fuel is burned in a rich primary stage at an overall stoichiometry that gives a minimum in the concentration of total fixed nitrogen species (Figure 13).[45] Secondary air is injected downstream from the primary stage to provide for burnout of CO and hydrocarbons in the primary stage products. The second stage typically operates with excess air, and in the lean second-stage combustion environment, a significant fraction of the fixed nitrogen species is converted to NO (Figure 13). In reburning, fuel is burned in

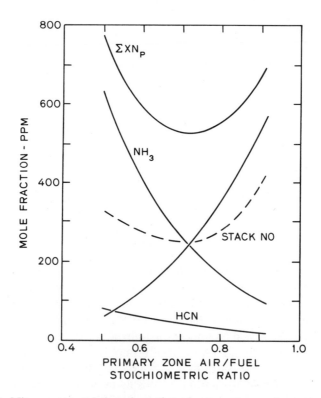

FIGURE 13. Nitrogenous emissions from the primary and secondary stages for staged combustion of pulverized coal.[45]

a lean primary stage, and additional fuel is injected downstream in a fuel-rich secondary zone to reduce the NO formed in the primary stage. As in staged combustion, additional air is injected downstream from the rich stage. To model NO emissions in staged combustion and reburning, the fuel-nitrogen reactions, outlined above, must be coupled to a kinetic model for hydrocarbon combustion. Detailed model calculations of rich premixed flames, simulating the primary zone of staged combustion, have been carried out with only limited success, due in part to uncertainties in the hydrocarbon-nitrogen chemistry. Further work is required in establishing and validating detailed models for fuel-nitrogen conversion in rich flame environments.

Selective reduction of NO in combustion gases by reaction with injected NH_3 has been demonstrated. Two different approaches are employed in practice: one in which the reduction occurs entirely by gas-phase reactions and a second where reduction occurs, at least in part, on the surface of a catalyst. Several detailed gas-phase reaction mechanisms for NO removal in $NO-NH_i$-combustion gas mixtures have been proposed which give relatively good agreement with experimental observations.[46, 47] These models have several features in common. In all of the models the nitrogen chemistry is a subset of the fuel-nitrogen mechanism discussed above. The principal reaction subsets in the nitrogen chemistry model include the bimolecular reactions for interconversion of NH_i species, reactions of NH_i to form NO, and reactions involving NH_i and NO that eventually form N_2. The nitrogen chemistry model is coupled with some kinetic model for the $O-H$ chemistry. There is a consensus that NO is removed principally by reaction with NH_2, for example, by reactions (35) and (36). There also is agreement that radicals, particularly OH, play a major role in the process, since it is the reaction between NH_3 and OH that replenishes the NH_2 which reacts with NO. At the low temperatures where the gas-phase reduction of NO occurs ($T \sim 1200$ K), the $O-H$ chemistry is not fast enough to maintain adequate OH concentrations, and some OH must be produced by reactions involving NH_i and NO, such as reaction (36). Available results suggest that at 1200 K, about 30–50% of the overall NH_2 + NO reaction occurs by reaction (36).

3.4. NO_2 Formation

As noted above, the NO formed during combustion can react subsequently with hydrocarbon fragments to form cyano species or with amine species to form N_2. In addition to these reactions, NO also can react with various oxygen-containing species to form NO_2. Chemical equilibrium considerations indicate that for typical flame temperatures ($T > 1500$ K), $(NO_2)/(NO)$ ratios are negligible. However, significant NO_2 concentrations have been measured in gas turbine exhausts, Figure 7, and in situ measurements of NO_x concentrations in turbulent diffusion flames[48, 49] indicate that there are relatively large $(NO_2)/(NO)$ ratios near the combustion zone. In probe sampling studies of one-dimensional premixed hydrocarbon-air flames, significant levels of NO_2

TABLE 14. Reactions Forming and Removing NO_2[a]

Reaction Number	Reaction	A	B	$C(K)$
37	$NO + HO_2 \rightarrow NO_2 + H$	2.1(9)	0	$-0.24(3)$
38	$NO_2 + O \rightarrow NO + O_2$	1.0(10)	0	0.30(3)
39	$NO_2 + H \rightarrow NO + H$	3.5(11)	0	0.74(3)

[a]From Hanson and Salimian.[26] Units: kgmol, m^3, K, sec; $k = AT^B \exp(-C/T)$.

have been found in the flame zone, with apparent conversion of the NO_2 back to NO in the postflame region.[50]

The experimental observations are consistent with the following gas-phase mechanism for NO_2 formation and destruction:

$$NO + HO_2 \rightarrow NO_2 + OH \tag{37}$$

$$NO_2 + O \rightarrow NO + O_2 \tag{38}$$

$$NO_2 + H \rightarrow NO + OH \tag{39}$$

Reaction (37) is fast at room temperature, having a rate coefficient, $k > 10^9$ $M^{-1}sec^{-1}$ (Table 14). In the initial low-temperature regions of hydrocarbon flames, significant HO_2 concentrations are found, suggesting that reaction (37) is a plausible route for NO_2 formation. Reactions (38) and (39) also are fast (Table 14) and in the presence of high radical concentrations, they would rapidly reconvert the NO_2 back to NO.

In sampling probes, reactions also may occur on interior probe surfaces, for example,

$$NO + O \xrightarrow{\text{wall}} NO_2$$

Hence, in regions of the flame with large radical concentrations, probe reactions resulting in NO_2 formation cannot be ruled out.

From the above discussion, it is clear that NO_2 can exist only as a transient species at flame temperatures. If NO_2 is to persist in the flow, then there must be some means of quenching the NO_2 formed in the flame. This quenching might occur in turbulent diffusion flames by turbulent mixing of hot and cold fluid elements, which might serve to quench the NO_2-removing reactions (by reduction in radical concentrations).

3.5. N_2O Formation

Measured NO in the products of fuel-lean, low-temperature combustion in stirred reactors exceeds that predicted by the thermal mechanism, discussed above, indicating the importance of alternative NO-formation reactions.[51] At

TABLE 15. Rate Coefficients for N_2O Reactions[a]

Reaction Number	Reaction	A	B	$C(K)$
-40	$N_2O + M \rightarrow N_2 + O + M$	6.9(20)	-2.5	32.7
41	$N_2O + O \rightarrow N_2 + O_2$	1.0(11)	0	14.1(3)
42	$N_2O + O \rightarrow NO + NO$	6.9(10)	0	13.4(3)
43	$N_2O + H \rightarrow N_2 + OH$	7.6(10)	0	7.6(3)
44	$NH + NO \rightarrow N_2O + H$	4.3(10)	-0.5	0
45	$NH_2 + NO \rightarrow N_2O + H_2$	5.0(10)	0	12.4(3)
46	$NCO + NO \rightarrow N_2O + CO$	1.0(10)	0	$-0.2(3)$

[a] From Hanson and Salimian[26] and Glarborg et al.[28] Units: kgmol, m^3, K, sec; $k = AT^B \exp(-C/T)$.

low temperatures, the rate of the reaction

$$O + N_2 + M \rightarrow N_2O + M \tag{40}$$

can exceed that of the rate-limiting step, reaction (18), of the thermal mechanism. The N_2O formed in reaction (40) subsequently reacts principally via,

$$
N_2O + O \nearrow^{N_2 + O_2 \quad (41)}_{\searrow NO + NO \quad (42)}
$$

$$N_2O + H \rightarrow N_2 + OH \tag{43}$$

The N_2O mechanism provides an alternative path for NO formation under conditions where the NO formation rate due to the thermal mechanism is becoming slower. Rate coefficients for reactions (40)–(43) are listed in Table 15.

Significant levels of N_2O have been measured in the products of combustion of nitrogen-containing fuels. The principal N_2O formation reactions in these flames are believed to involve NO and various nitrogen-containing radicals,

$$NH + NO \rightarrow N_2O + H \tag{44}$$

$$NH_2 + NO \rightarrow N_2O + H_2 \tag{45}$$

$$NCO + NO \rightarrow N_2O + CO \tag{46}$$

Rate coefficients for these reactions are listed in Table 15. Reactions (44) and (46) are fast and can result in rapid N_2O formation near the flame zone. The

N_2O formed in these reactions subsequently reacts to form N_2, principally via,

$$N_2O + H \rightarrow N_2 + OH \qquad (43)$$

Reaction (43) is fast at flame temperatures; hence N_2O should be rapidly removed in the postflame gas. For N_2O to persist in the combustion products, reaction (43) must be quenched either by cooling or by reduction in radical concentrations.

In summary, NO_x emissions from practical combustion devices comprise NO, NO_2, and N_2O. The emissions of these species are governed by both formation and removal rates. A fundamental understanding of the nitrogen chemistry in flames forms the basis for development of control techniques, such as staged combustion and reburning. Although chemistry plays a major role in NO_x emissions, fluid dynamic effects, which alter local flame conditions, must also be considered in devising pollution controls.

4. SULFUR OXIDES IN FLAMES

Various sulfur oxides, designated SO_x, are a major pollutant species emitted from stationary combustion sources using fossil fuels (Table 1). The origin of these sulfur oxides is the fuel, which contains sulfur in amounts ranging from a trace to more than 7% by weight (Table 8). Coals, particularly eastern bituminous coals, contain high levels of sulfur. The sulfur in coal exists principally in two forms: inorganic, mainly FeS_2, and organic, incorporated into the various organic structures present in coal. Trace amounts of various sulfate species also are present in coal. Many of the sulfur-containing compounds in fuel oil have been identified, including thiols, organic sulfides and disulfides, and thiophenes. The sulfur-containing compounds in gaseous fuels, including natural gas and gas produced by coal and char gasification, include H_2S and lesser amounts of CS_2 and COS. The sulfur content of all of these fuels can be reduced by physical and chemical processing, and fuel desulfurization is one of the principal techniques for control of SO_x emissions from combustion sources.

In combustion of sulfur-containing fuels, virtually all of the sulfur appears as various sulfur oxides in the combustion products. Some of these oxides are gaseous species, such as SO_2 and SO_3, while others are aerosols, comprising various primary sulfates such as H_2SO_4. In the combustion of fuels with significant inorganic content, such as coal and heavy oils, some of the sulfur oxides, mainly inorganic sulfates, will be incorporated in the flyash. For temperatures in excess of 1000 K, equilibrium considerations show that the principal sulfur oxide is SO_2, with lesser amounts of SO_3, primary sulfates and, under rich conditions, H_2S (Table 16). Experimental data from laboratory flames indicate that, at flame temperatures, SO_2 is the principal sulfur oxide, but that SO_3 levels are larger than local equilibrium values (Figure

TABLE 16. Typical Equilibrium SO_x Product Distributions in Combustion[a]

Temperature (K)	SO_2	SO_3	H_2SO_4
500	—	0.978	0.022
700	0.050	0.940	—
900	0.464	0.536	—
1100	0.757	0.243	—
1300	0.933	0.067	—
1500	0.988	0.012	—
1700	0.998	0.002	—

[a] No. 6 fuel oil, 2.8 wt % sulfur, 2% excess air.

14).[52] At typical exhaust temperatures ($T < 800$ K), equilibrium considerations indicate that the principal sulfur oxides are SO_3 and the primary sulfates. However, stack sampling data show that SO_x emissions from stationary powerplants are mainly SO_2, with only small amounts of SO_3 and only trace amounts of primary sulfates.

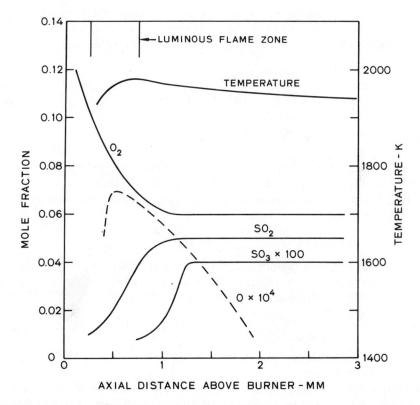

FIGURE 14. SO_x profiles in a premixed H_2S—O_2—N_2 flame. Reprinted with permission from A. Levy, E. L. Merryman, and W. T. Reid, "Mechanisms of Formation of Sulfur Oxides in Combustion." *Environ. Sci. Technol.* **4**, 653, © 1970, American Chemical Society.

Most detailed studies of sulfur chemistry in flames have been carried out using simple model fuel-sulfur compounds such as H_2S, CS_2, COS, and CH_3SH. The relevance of these studies to the oxidation of the complex organic sulfur compounds found in coal and heavy oils is due, in part, to the fact that fuel pryolysis studies show simple sulfur compounds as the principal stable sulfur-containing products. Although the mechanisms for the oxidation of simple sulfur compounds are not completely understood, there is a consensus that, at flame temperatures, the reaction proceeds by a sequence of steps,

$$\text{Fuel-S} \overset{\overset{\displaystyle \dot{R}S}{\longrightarrow}}{\underset{\longrightarrow}{}} \text{SO} \to SO_2 \to SO_3$$

where $\dot{R}S$ is a sulfur-containing radical such as HS, CS, CH_3S, or S. Available data indicate that SO_2 formation in the flame is rapid, occurring on a timescale comparable to that of the fuel oxidation reactions. The rapid progress of the SO_2 formation process is due principally to rapid bimolecular reactions between sulfur species and O—H radicals. Measurements in the postflame zone of premixed hydrogen-air flames doped with small amounts of H_2S[53] show that several of these bimolecular reactions, as well as the bimolecular reactions involving O—H species, are partially equilibrated, including

$$
\begin{aligned}
\text{SO} + \text{OH} &\leftrightarrows SO_2 + \text{H} \\
H_2O + \text{H} &\leftrightarrows \text{OH} + H_2 \\
\hline
\text{SO} + H_2O &\leftrightarrows SO_2 + H_2
\end{aligned}
\tag{47}
$$

Hence, the relative SO_2 and SO concentrations can be calculated from thermodynamic considerations at the local gas temperature.

The principal reaction responsible for SO_3 formation in flames is

$$SO_2 + O + M \to SO_3 + M \tag{48}$$

Reaction (48) proceeds rapidly near the flame zone, where the O-atom concentrations significantly exceed final equilibrium values. Some of the SO_3, formed in reaction (48) is subsequently reconverted to SO_2 by

$$SO_3 + O \to SO_2 + O_2 \tag{49}$$

$$SO_3 + H \to SO_2 + OH \tag{50}$$

The SO_3 levels observed in the flame are determined by the rates of the formation and removal steps. Some rate data are available for these reactions (Table 17).[54-56] Reactions (49) and (50) are relatively slow at flame tempera-

TABLE 17. Rate Coefficients for SO_3 Reactions[a]

Reaction Number	Reaction	A	B	$C(K)$	Reference
48	$SO_2 + O + M \rightarrow SO_3 + M$	7.4(8)	—	—	54
		1.0(9)	—	—	55
49	$SO_3 + O \rightarrow SO_2 + O_2$	1.2(9)	0	4.8(3)	55
50	$SO_3 + H \rightarrow SO_2 + OH$	1.0(9)	—	—	56

[a] Units: kgmol, m^3, K, sec; $k = AT^B \exp(-C/T)$.

tures. Hence SO_3 formed rapidly near the flame zone by reaction (48) is only slowly removed, and SO_3 concentrations can exceed equilibrium levels appropriate to the local gas temperature. The rate coefficient for reaction (48) is only weakly dependent on temperature so that SO_3 levels in the quenched combustion gases are determined by quenching of reaction (48), principally by reduction of the O-atom concentration.

The presence of sulfur species in flames, particularly SO_2, increases the rate of O—H radical recombination. Reaction (48), followed by reaction (49), increases the rate of O-atom recombination. SO_2 also plays a role in H-atom and OH recombination via

$$H + SO_2 + M \rightarrow HSO_2 + M \tag{51}$$

$$H + HSO_2 \rightarrow H_2 + SO_2 \tag{52}$$

$$OH + HSO_2 \rightarrow H_2O + SO_2 \tag{53}$$

Hence sulfur species tend to depress radical concentration overshoots in the flame zone and facilitate the rapid approach to total equilibrium in the postflame gases.

As noted above, the details of the reaction mechanism for SO_x formation from the parent fuel-sulfur species are not known. However, by virtue of the rapidity of the SO_x formation process, approximate kinetic models that bypass the need for a detailed fuel-sulfur oxidation model may be postulated to estimate the gaseous SO_x product distribution in the exhaust products. Three principal assumptions are involved in the proposed model: (1) the fuel-sulfur compounds are minor species so that the major stable species concentrations are those due to combustion of the hydrocarbon fuel, (2) the bimolecular O—H reactions and reaction (47) are partially equilibrated in the postflame gases, and (3) SO_3 concentrations may be calculated from reaction (48) and either or both of reactions (49) and (50). The SO_x pool ($SO + SO_2 + SO_3$) then will vary as the overall reaction approaches equilibrium by means of O—H radical recombination reactions, including reaction (17) and reactions (48) and (53).

TABLE 18. Reduction in NO_x Emissions Formed by the Thermal Mechanism due to Sulfur Addition to Premixed Hydrocarbon Flames[a]

	NO_x/NO_x (No Additive)		
cΦ	2.5% SO_2	4.9% SO_2	5.0% H_2S
1.25	0.874	0.840	0.886
1.10	—	0.827	0.854
1.00	0.860	0.643	0.768
0.97	—	0.740	0.811
0.91	—	0.755	0.806
0.86	0.883	0.861	—

[a] From Wendt and Ekmann.[57]

Reactions of fuel-sulfur and fuel-nitrogen are closely coupled to the fuel oxidation reactions and both sulfur-containing radicals and nitrogen-containing radicals compete for available O—H radicals with the hydrocarbon radicals. Most studies of sulfur or nitrogen chemistry in flames have been carried out in the absence of the other species. Because of the close coupling of the sulfur and nitrogen chemistry and the O—H radical pool in flames, interactions between fuel-sulfur chemistry and fuel-nitrogen chemistry may be expected. Furthermore combustion modifications designed for control of NO_x emissions formed from fuel-nitrogen, for example, staged combustion, may impact the SO_x product distribution due to variations in the O—H radical pool.

As noted above, the presence of sulfur species in flames tends to reduce the O—H radical pool in the flame zone. This reduction in O-containing radicals will reduce the rates of reactions converting molecular nitrogen to NO by the thermal mechanism. This effect has been observed in premixed methane-air flames where the addition of SO_2 and H_2S produced reductions in the NO concentrations in the postflame gases of up to 30% (Table 18).[57] Perturbation of the O—H radical pool by sulfur species also should affect the fuel-nitrogen reaction. Available data from rich premixed laminar hydrocarbon-air flames doped with model fuel-nitrogen compounds show both enhancement and inhibition of NO formation from fuel-nitrogen depending on equivalence ratio.[58] For moderately fuel-rich flames, $\Phi < 1.64$, addition of sulfur compounds tends to reduce NO_x emissions, whereas for very fuel-rich flames, $\Phi > 2.1$, sulfur addition tends to increase NO_x emissions. At the present time a reaction mechanism to explain the effects of sulfur of fuel-nitrogen conversion to NO is not available. Since the equivalence ratio range of the premixed flame studies encompasses that generally used in the primary stage of staged combustion for NO_x control from fuel-nitrogen, the presence of sulfur in the fuel may impact the selection of the optimum primary stage equivalence ratio for NO_x control.

REFERENCES

1. USEPA, *National Air Quality and Emissions Trends Report, 1977*. U.S. Environmental Protection Agency, Washington, D.C., 1978.

2. R. M. Fristrom and A. A. Westenberg, *Flame Structure*. McGraw-Hill, New York, 1965.

3. R. E. Mitchell, A. F. Sarofim, and L. A. Clomburg, "Experimental and Numerical Investigation of Confined Laminar Diffusion Flames." *Combust. Flame* **37**, 227 (1980).

4. J. Warnatz, "Rate Coefficients in the C/H/O System." In *Combustion Chemistry* (W. C. Gardiner, ed.), pp. 197–360. Springer-Verlag, New York, 1984.

5. C. K. Westbrook and F. L. Dryer, "Chemical Kinetic Modeling of Hydrocarbon Combustion." *Prog. Energy Combust. Sci.* **10**, 1 (1984).

6. C. K. Westbrook and F. L. Dryer, "A Comprehensive Mechanism for Methanol Oxidation." *Combust. Sci. Technol.* **20**, 125 (1979).

7. J. Warnatz, "Chemistry of High Temperature Combustion of Alkanes up to Octane." *Symp. (Int.) Combust. [Proc.]* **20**, 845 (1985).

8. D. L. Baulch and D. D. Drysdale, "An Evaluation of the Rate Data for CO + OH → CO$_2$ + H Reaction." *Combust. Flame* **23**, 215 (1974).

9. C. K. Westbrook and F. L. Dryer, "Chemical Kinetics and Modeling of Combustion Processes." *Symp. (Int.) Combust. [Proc.]* **18**, 749 (1981).

10. A. M. Mellor, "Characteristic Time Emissions Correlations and Sample Optimization: GT-309 Gas Turbine Combustion." *J. Energy* **1**, 244 (1977).

11. J. H. Tuttle, M. B. Colket, R. W. Bilger, and A. M. Mellor, "Characteristic Times for Combustion and Pollutant Formation in Spray Combustion." *Symp. (Int.) Combust. [Proc.]* **16**, 209 (1977).

12. R. A. Yetter, F. L. Dryer, and H. Rabitz, "Complications of One-Step Kinetics for Moist CO Oxidation." *Symp. (Int.) Combust. [Proc.]* **21**, 749 (1988).

13. J. B. Howard, G. C. Williams, and D. H. Fine, "Kinetics of Carbon Monoxide Oxidation in Postflame Gases." *Symp. (Int.) Combust. [Proc.]* **14**, 975 (1973).

14. F. L. Dryer and I. Glassman, "High Temperature Oxidation of CO and CH$_4$." *Symp. (Int.) Combust. [Proc.]* **14**, 987 (1973).

15. I. Glassman, *Combustion*, p. 226. Academic Press, New York, 1977.

16. A. A. Westenberg, "Kinetics of NO and CO in Lean, Premixed Hydrocarbon-Air Flames." *Combust. Sci. Technol.* **4**, 59 (1971).

17. A. M. Morr and J. B. Heywood, "Partial Equilibrium Model for Predicting Concentration of CO in Combustion." *Acta Astronaut.* **1**, 949 (1974).

18. J. E. Hardy and R. K. Lyon, "Isothermal Quenching of the Oxidation of Wet CO." *Combust. Flame* **39**, 317 (1980).

19. G. B. Martin and E. E. Berkau, "An Investigation of the Conversion of Various Fuel Nitrogen Compounds to Nitrogen Oxides in Oil Combustion." *70th Nat. AIChE Meet., 1971*.

20. A. N. Hayhurst and I. M. Vince, "Nitric Oxide Formation from N_2 in Flames: The Importance of Prompt NO." *Prog. Energy Combust. Sci.* **6**, 35 (1980).

21. G. M. Johnson and M. Y. Smith, "Emissions of Nitrogen Dioxide from a Large Gas Turbine Power Station." *Combust. Sci. Technol.* **19**, 67 (1978).

22. D. Pierotti and R. A. Rasmussen, "Combustion as a Source of Nitrous Oxide in the Atmosphere." *Geophys. Res. Lett.* **5**, 265 (1976).

23. R. F. Weiss and H. Craig, "Production of Atmospheric Nitrous Oxide by Combustion." *Geophys. Res. Lett.* **5**, 751 (1976).

24. W. M. Hao, S. C. Wofsy, M. B. McElroy, W. F. Farmayan, M. A. Togan, J. M. Beer, M. S. Zahniser, J. A. Silver, and C. E. Kolb, "Nitrous Oxide Emission from Coal, Oil, and Gas Furnace Flames." Poster from the 21st International Combustion Symposium, Munich, West Germany, 1986.

25. J. C. Kramlich, R. K. Nihart, S. L. Chen, D. W. Pershing, and M. P. Heap, "Behavior of N_2O in Staged Pulverized Coal Combustion." *Combust. Flame* **48**, 101 (1982).

26. R. K. Hanson and S. Salimian, "Survey of Rate Constants in the N—H—O System." In *Combustion Chemistry* (W. C. Gardiner, Jr., ed.), pp. 361–421. Springer-Verlag, New York, 1984.

27. C. T. Bowman, comment in *Emissions from Continuous Combustion Systems*, pp. 98–102. Plenum, New York, 1972.

28. P. Glarborg, J. A. Miller, and R. J. Kee, "Kinematic Modeling and Sensitivity Analysis of Nitrogen Oxide Formation in Well-Stirred Reactors." *Combust. Flame* **65**, 177 (1986).

29. H. E. van der Bergh and A. B. Callear, *Trans. Faraday Soc.* **67**, 2017 (1971).

30. B. S. Haynes, "Kinetics of Nitric Oxide Formation in Combustion." *Prog. Astronaut. Aeronaut.* **62**, 359 (1978).

31. A. C. Baldwin and D. M. Golden, "Reactions of Methyl Radicals of Importance in Combustion Systems." *Chem. Phys. Lett.* **55**, 350 (1978).

32. C. P. Fenimore, "Formation of Nitric Oxide from Fuel Nitrogen in Ethylene Flames." *Combust. Flame* **19**, 289 (1972).

33. Y. H. Song, W. Bartok, D. W. Blair, and V. J. Siminski, "Conversion of Fixed Nitrogen to N_2 in Rich Combustion." *Symp. (Int.) Combust. [Proc.]* **18**, 53 (1981).

34. A. E. Axworthy, G. R. Schneider, M. D. Shuman, and V. H. Dayan, *Chemistry of Fuel Nitrogen Conversion to Nitrogen Oxides in Combustion*, Rep. EPA 600/2-76-034. U.S. Environmental Protection Agency, Washington, D.C., 1976.

35. T. J. Houser, M. Hull, R. M. Alway, and T. Biftu, "Kinetics of Formation of HCN during Pyridine Pyrolysis." *Int. J. Chem. Kinet.* **12**, 569 (1980).

36. C. L. Proctor and N. M. Laurendeau, *A Shock Tube Study of the Inert Pyrolysis of Pyridine*, Pap. 78-55. Western States Section/Combustion Institute Meeting, Pasadena, California, 1978.

37. P. Roth and T. Just, "Measurement of the Thermal Decomposition of HCN behind Shock Waves." *Ber. Bunsenges. Phys. Chem.* **80**, 171 (1976).

38. A. Szekely, R. K. Hanson, and C. T. Bowman, "High-Temperature Determination of the Rate Coefficient for the Reaction $H_2 + CN \rightarrow H + HCN$." *Int. J. Chem. Kinet.* **15**, 915 (1983).

39. A. Szekely, R. K. Hanson, and C. T. Bowman, "Shock Tube Determination of the Rate Coefficient for the Reaction $CN + HCN \rightarrow C_2N_2 + H$." *Int. J. Chem. Kinet.* **15**, 1237 (1983).

40. A. Szekely, R. K. Hanson, and C. T. Bowman, "Shock Tube Study of the Reaction between Hydrogen Cyanide and Atomic Oxygen." *Symp. (Int.) Combust. [Proc.]* **20**, 647 (1985).

41. R. A. Perry and C. F. Melius, "The Rate and Mechanism of the Reaction of HCN with Oxygen Atoms over the Temperature Range 540–900 K." *Symp. (Int.) Combust. [Proc.]* **20**, 639 (1985).

42. P. Roth, R. Lohr, and H. D. Hermanns, "Shock Wave Study of the Kinetics of the Reaction HCN + O." *Ber. Bunsenges. Phys. Chem.* **84**, 835 (1980).

43. A. Szekely, R. K. Hanson, and C. T. Bowman, "High Temperature Determination of the Rate Coefficient for the Reaction $H_2O + CN \rightarrow HCN + OH$." *Int. J. Chem. Kinet.* **16**, 1609 (1984).

44. B. S. Haynes, "The Oxidation of Hydrogen Cyanide in Fuel-Rich Flames." *Combust. Flame* **28**, 113 (1977).

45. S. L. Chen, M. P. Heap, R. Nihart, D. W. Pershing, and D. P. Rees, *The Influence of Fuel Composition on the Formation and Control of NO_x in Pulverized Coal Flames*, Pap. 80-13. Western States Section/Combustion Institute Meeting, Pasadena, California, 1980.

46. J. A. Miller, M. C. Branch, and R. J. Kee, "A Chemical Kinetic Model for the Selective Reduction of NO by NH_3." *Combust. Flame* **43**, 81 (1981).

47. R. K. Lyon, "Kinetics and Mechanism of Thermal $DeNO_x$: A Review," 194th Annual ACS Meeting, *Div. of Fuel Chem.* **32**, 433 (1987).

48. R. W. Schefer and R. F. Sawyer, "Lean Premixed Recirculating Flow Combustion for Control of Oxides of Nitrogen." *Symp. (Int.) Combust. [Proc.]* **16**, 119 (1977).

49. N. P. Cernansky and R. F. Sawyer, "NO and NO_2 Formation in a Turbulent Hydrocarbon/Air Diffusion Flames." *Symp. (Int.) Combust. [Proc.]* **15**, 1039 (1975).

50. E. L. Merryman and A. Levy, "Nitrogen Oxide Formation in Flames: The Role of NO_2 and Fuel Nitrogen." *Symp. (Int.) Combust. [Proc.]* **15**, 1073 (1975).

51. P. C. Malte and D. T. Pratt, "The Role of Energy-Releasing Kinetics in NO_x Formation: Fuel-Lean, Jet-Stirred CO-Air Combustion." *Combust. Sci. Technol.* **9**, 721 (1974).

52. A. Levy, E. L. Merryman, and W. T. Reid, "Mechanisms of Formation of Sulfur Oxides in Combustion." *Environ. Sci. Technol.* **4**, 653 (1970).

53. C. H. Muller, K. Schofield, M. Steinberg, and H. P. Broida, "Sulfur Chemistry in Flames." *Symp. (Int.) Combust. [Proc.]* **17**, 867 (1979).

54. E. L. Merryman and A. Levy, "Enhanced SO_3 Emissions from Staged Combustion." *Symp. (Int.) Combust. [Proc.]* **17**, 727 (1979).

55. C. F. Cullis and M. F. R. Mulcahy, "The Kinetics of Combustion of Gaseous Sulfur Compounds." *Combust. Flame* **18**, 225 (1972).

56. C. P. Fenimore and G. W. Jones, *J. Phys. Chem.* **69**, 3593 (1961).

57. J. O. L. Wendt and J. M. Ekmann, "Effects of Fuel Sulfur Species on Nitrogen Oxide Emissions from Premixed Flames." *Combust. Flame* **25**, 355 (1975).

58. J. O. L. Wendt, J. T. Morcomb, and T. L. Corley, "Influence of Fuel Sulfur on Fuel Nitrogen Oxidation Mechanisms." *Symp. (Int.) Combust. [Proc.]* **17**, 671 (1979).

59. J. H. Pohl and A. F. Sarofim, "Fate of Coal Nitrogen during Pyrolysis and Oxidation," *Proc. Stationary Source Combust. Symp.* EPA Rep. 600 12-76-15a (1976).

60. J. A. Miller and C. T. Bowman, "Mechanism and Modelling of Nitrogen Chemistry in Combustion," *Prog. Energy Combust. Sci.*, in press (1989).

5 Soot and Hydrocarbons in Combustion

B. S. HAYNES

Department of Chemical Engineering
University of Sydney
New South Wales, Australia

1. Major processes in hydrocarbon combustion
   1.1. Combustion of aliphatics
   1.2. Combustion of benzene
   1.3. Mechanisms of hydrocarbon emissions from engines
2. Polycyclic aromatic hydrocarbons
   2.1. Introduction
   2.2. Studies of PAH kinetics in combustion
   2.3. Laminar diffusion flames
   2.4. Turbulent diffusion flames
   2.5. Fuel PAH
   2.6. Conclusions
3. Soot formation
   3.1. Structure of soot
   3.2. Fundamental studies of soot formation
   3.3. Premixed flames
   3.4. Sooting structure of laminar diffusion flames
   3.5. Turbulent diffusion flames
   3.6. Influence of additives
   3.7. Conclusions
   References

Organic and carbonaceous emissions from combustion systems represent not only a fuel loss but also, in some cases, a significant pollution problem. In this chapter the combustion chemistry of the hydrocarbons is considered in relation to such problems.

The thermal reactions of hydrocarbons lead primarily to smaller molecules than were present at the outset. In Section 1 some of these degradative pathways of pyrolysis and combustion are briefly considered. The minor routes via condensation reactions to polycyclic aromatic hydrocarbons (PAH) and soot are considered separately in Sections 2 and 3, respectively.

1. MAJOR PROCESSES IN HYDROCARBON COMBUSTION

1.1. Combustion of Aliphatics

The major reaction pathways in aliphatic hydrocarbon combustion have been studied extensively. These pathways are discussed elsewhere[1] in detail and are referred to only briefly here.

The mechanisms at low and high temperatures are somewhat different. Characteristic of the low-temperature reactions, such as occur in cool flames, is the formation of many oxygenated species—alcohols, aldehydes, and acids. The formation of olefinic hydrocarbons is another feature of these reactions.[2]

At higher temperatures (> 1000 K) the radicals H, O, and OH become responsible for chain branching and the mechanism of reaction is markedly different from that in cool flames. Ignition and induction period phenomena are controlled by the interaction of chain-branching and chain-terminating reactions. The unimolecular decomposition of the hydrocarbon fuel and its direct reaction with O_2 are important here. However, once ignition is established, flame propagation occurs by radical attack on the fuel and by radical reactions not involving the fuel, such as $O + H_2 \rightarrow OH + H$.

Warnatz,[3-6] among others, has examined the combustion mechanisms of simple aliphatics by computer kinetic modeling. A flow diagram for the oxidation of CH_4 in stoichiometric methane/air flames at atmospheric pressure is shown in Figure 1, where the thickness of the arrows represents the relative importance of the pathways.[5] The dominant route is via C_1-species, but even under these conditions the parallel C_2-pathway is important. As conditions are made more fuel rich, the methyl radical recombination

$$CH_3 + CH_3 \rightarrow C_2H_6 \tag{1}$$

becomes even more favorable, and as much as 80% of the fuel may pass through the C_2 route in slightly fuel-rich methane flames.[3]

Higher hydrocarbon analogues of reaction 1 occur to form C_3H_8 and C_4H_{10}, but these species quickly return to the C_2 pathway via hydrogen abstraction to form alkyl radicals that in turn decompose thermally by elimination of an alkene, for example,

$$C_2H_5 + CH_3 \rightarrow C_3H_8 \rightarrow C_3H_7 \rightarrow C_2H_4 + CH_3 \tag{2}$$

Clearly the mechanisms of combustion of all the aliphatics are closely related. Indeed, it is the rate-controlling decomposition of the CH_3 and C_2H_5 radicals that is responsible for the similarity of all alkane and alkene flame properties.[6]

Recently, the occurrence of a synthesis route to higher hydrocarbons via acetylene has been found to constitute an important pathway, especially in fuel-rich mixtures.[4,7]

$$C_2H_2 + C_2H \rightleftharpoons C_4H_2 + H \tag{3}$$

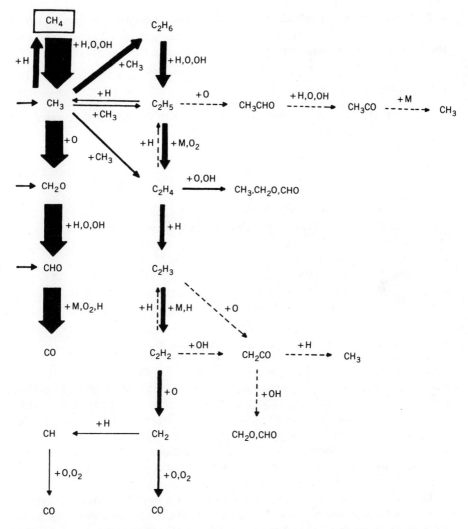

FIGURE 1. Reaction pathways in stoichiometric combustion of methane with air at 1 bar. The thickness of the arrows is proportional to the calculated reaction rates integrated over the whole flame front.[5] From Warnatz.[5]

Extension to the polyacetylenes has also been demonstrated experimentally.[8-13]

$$C_2H_2 \leftrightarrow C_4H_2 \leftrightarrow C_6H_2 \leftrightarrow C_8H_2 \tag{4}$$

The formation of higher, unsaturated hydrocarbons is therefore an integral part of the kinetics of conversion of even the simplest hydrocarbons to CO, CO_2, H_2, and H_2O in a flame. This is especially true for fuel-rich combustion, and if conditions are made rich enough, all the oxygen is consumed so that chain branching effectively ceases before the last hydrocarbons are consumed. These hydrocarbons therefore break through the primary reaction zone into the postflame region where they are consumed only slowly.[12,14,15]

1.2. Combustion of Benzene

The chemistry of the combustion of aromatics is not well known. The dominant route for benzene combustion appears to involve an early ring-opening:[10,11,16-18]

$$C_6H_6 + O \rightarrow C_6H_6O \rightarrow C_5H_6 + CO \qquad (5)$$

This is followed by the usual aliphatic chemistry, as is shown in Figure 2, for selected species concentration profiles obtained in very fuel-rich benzene and acetylene flames.[11] When the aliphatic side chains are attached to the benzene, they may be stripped prior to rupture of the ring.[19]

Judging from pyrolysis studies,[20-23] condensation reactions to form polyphenyls[20] or polynuclear aromatic hydrocarbons may occur, but these processes do not appear to constitute major pathways in combustion.

1.3. Mechanisms of Hydrocarbon Emissions from Engines

The internal combustion engine is responsible for the bulk of hydrocarbon emissions from combustion systems. Incomplete combustion of hydrocarbons can occur when the fuel-air charge is poorly mixed so that the mixture is either too lean or too rich to establish ignition. In the first case the low-temperature oxidation mechanisms may partly transform the fuel; in the second, more pyrolysis can be expected. Poor charge homogeneity is more of a problem in diesel engines, particularly in those employing direct injection of the fuel, than it is in carburetted engines.

The influence of various engine parameters on hydrocarbon emissions has been treated exhaustively by Patterson and Henein.[24] However, the actual mechanism of survival of hydrocarbons through the combustion cycle in spark ignition engines has recently received further attention.

For many years it was believed that the chief cause of hydrocarbon emissions from spark ignition engines was "wall quench," whereby the heat and radical sink provided by the cylinder walls was thought to be sufficient to quench combustion. The hydrocarbons between the quenched flame front and the wall were believed to be emitted from the cylinder by being "scraped" from the walls during the exhaust cycle.[25]

However, it has now become apparent that this process of wall quenching does not lead to significant emission of hydrocarbons. Thus, in combustion bomb experiments,[26,27] the yield of hydrocarbons surviving the passage of the combustion wave does not depend on the wall surface area but on the volume of small niches and crevices. Sampling of the quench layer in an experimental

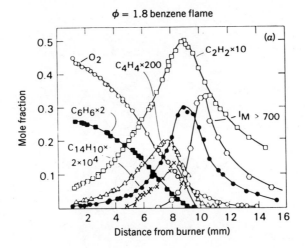

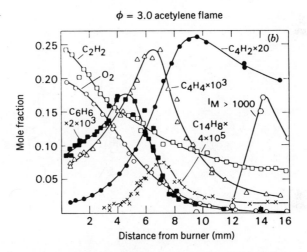

FIGURE 2. Various species mole fractions and high-mass signals ($I_{M > m'}$ scale arbitrary)[11] in (a) a near-sooting ($\phi = 1.8$) benzene/oxygen/argon flame and (b) a sooting ($\phi = 3.0$) acetylene/oxygen/argon flame. Pressure = 2.67 kPa; cold gas velocity = 0.5 m/sec. From Bittner and Howard.[11]

engine has confirmed that less than 10% of the hydrocarbons in the exhaust are attributable to wall quenching.[28]

More recently, computations of flame propagation up to a wall have confirmed that wall quenching of the flames does not produce a quench layer in which hydrocarbons survive[29]—surely, the flame is quenched a short distance (of the order of 0.1 mm) from the wall, but the diffusion of the unburned hydrocarbons from the wall region into the hot gases ensures that these will be consumed completely within about 1 msec. Since, in an engine,

TABLE 1. Polycyclic Aromatic Hydrocarbons Typically Found in Flames[a]

Name	Structure	Name	Structure
Indene		Benzacenaphthylene	
Naphthalene		Pyrene	
Methylnaphthalene		Benzofluorene	
Biphenyl		Methylfluoranthrene	
Biphenylene		Methylpyrene	

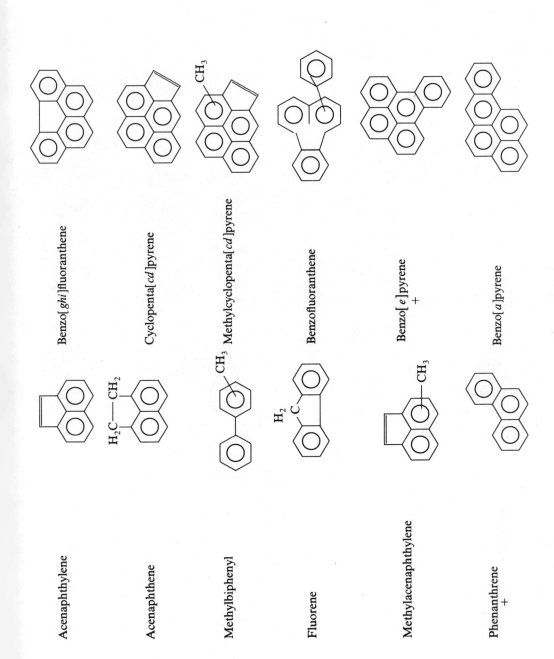

Acenaphthylene

Benzo[*ghi*]fluoranthene

Acenaphthene

Cyclopenta[*cd*]pyrene

Methylbiphenyl

Methylcyclopenta[*cd*]pyrene

Fluorene

Benzofluoranthene

Methylacenaphthylene

Benzo[*e*]pyrene
+

Phenanthrene
+

Benzo[*a*]pyrene

267

TABLE 1. (Continued)

Name	Structure	Name	Structure
Anthracene		Perylene	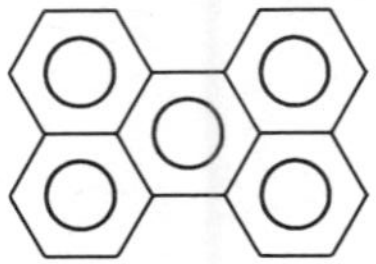
Methylphenanthrene	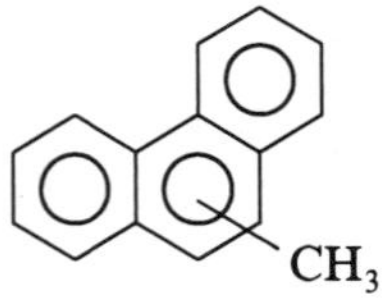	Indeno[1,2,3-*cd*]pyrene	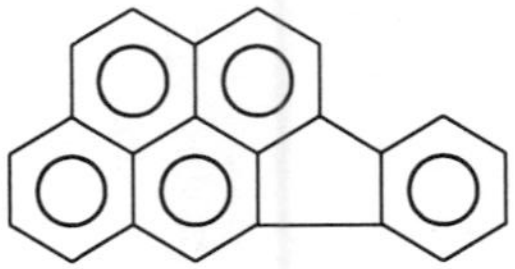
4H-cyclopenta[*def*]phenanthrene	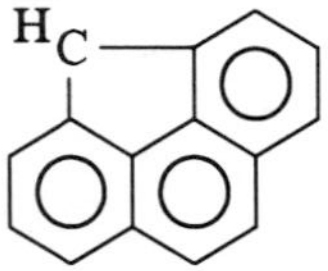	Benzo[*ghi*]perylene + Anthanthrene	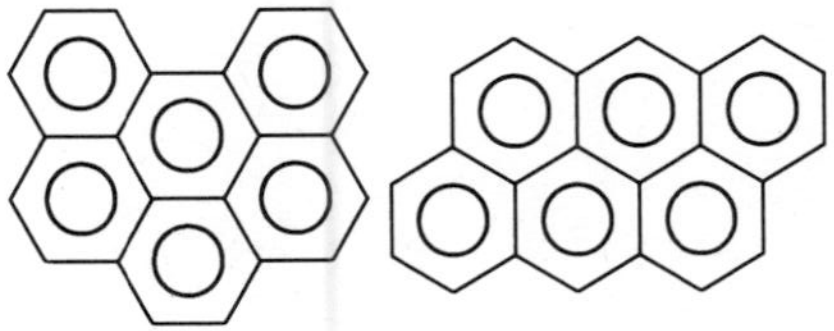
2-phenylnaphthalene			
Fluoranthene	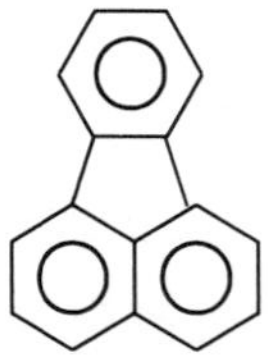	Coronene	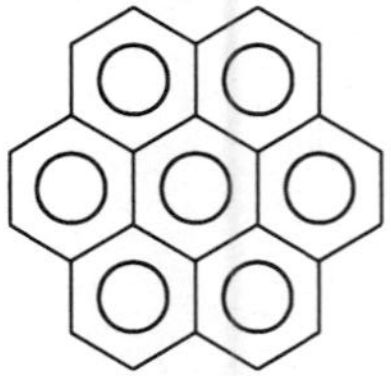

*a*From Prado et al.[78]

268

TABLE 2. Mutagenicity of Selected PAH and Soot Extracts[a]

Compound	Bacterial Test		Human Cell Test	
	PMS Activation	No Activation	PMS Activation	No Activation
Benzo[a]pyrene	1	> 20	0.25	> 20
Phenanthrene	> 53	> 53	9	> 9
1-Methylphenanthrene	15	> 50	0.58	> 19
9-Methylphenanthrene	7.7	> 190	0.77	> 19
Fluoranthene	1	> 10	0.4	> 10
Cyclopenta[cd]pyrene	1.4	> 9	1.6	> 20
Cyclopenta[cd]pyrene epoxide	1.5	0.17	4.8	0.097
Kerosene soot extract	17	> 100	15	> 24
Diesel soot extract	90	0.8	70	> 100
Residential oil burner soot extract	100	40	NT	NT

[a]From Longwell.[33] Concentration required for significant mutant fractions, μg/mL.

times for diffusion and reaction of this wall layer are of the order of 10 to 30 msec, it is not surprising that "wall quench" does not contribute much to hydrocarbon emissions in practice.

At this stage the question of the mechanism of hydrocarbon survival is open. Ring crevice storage phenomena and absorption/desorption of fuel from oil layers and surface deposits are possible mechanisms.[30, 31] Homogeneous fuel-rich quenching has also been suggested.[28]

2. POLYCYCLIC AROMATIC HYDROCARBONS

2.1. Introduction

Polycyclic aromatic hydrocarbons (PAH) are produced by most practical combustion systems. Structural representations for some PAH species are presented in Table 1.

The main source of concern with this class of compound is that some members are known mutagens, co-carcinogens, or carcinogens.[32] Recently, rapid screening tests based on bacterial mutation induced by a specific compound have become popular. For example, the concentration of a test material required to produce a statistically significant mutant fraction is shown in Table 2 for the bacterium salmonella typhimurium and for human white blood cells.[33] Most pure PAH compounds are only active in the presence of an oxidative enzyme such as contained in liver extract (PMS), but polar PAH (e.g., nitrated or oxygenated) may require no activation, as shown in Table 2 for cyclopenta(cd)pyrene epoxide. Tests such as these may provide an early warning signal for possible health effects in humans although, as seen from Table 2, there is no simple correlation between bacterial and human mutagenicity. Nor is the relationship of mutagenic activity to carcinogenesis clear.[32, 34]

The entries in Table 2 show a wide range of effects with varying species. This specificity means that we should ideally like to understand the formation and behavior of individual PAH as functions of combustion conditions. However, the information required to do this is simply not available, and for the most part, the PAH are treated as a single group. Fortunately, there is some justification for this approach in that, although some PAH species may be relatively favored under some conditions and not under others, the overall makeup of the PAH class is relatively constant as conditions change.[32, 35]

Benzo[a]pyrene is frequently chosen as the reference component in characterizing total PAH. However, it should be noted that benzo[a]pyrene usually constitutes only a minor fraction (probably less than 1%) of the PAH. In fact it now appears that concentrations of this component alone are generally too low to be of any significance to the biological activity of the whole soot sample.[32]

Emissions of PAH. Practically all combustion systems are capable of producing and releasing some PAH. Estimated annual U.S. emissions of benzo[a]pyrene are tabulated in Table 3.[36] Major sources include domestic space heating, especially coal- and wood-fired applicances. Overall, automotive sources do not appear significant, but urban emissions may be dominated by this category. The increasing use of diesel engines in place of gasoline engines, with a concomitant rise in PAH emissions of tenfold or more,[32] is of particular concern.

Fate of PAH Emissions. In considering possible health effects, it is important to know something of the fate of the PAH surviving the high-temperature combustion zone to be released in the atmosphere. The most likely fate of gas-phase PAH emitted into the atmosphere is oxidation, either directly by oxygen, by photooxidation, or, in polluted environments by O_3, NO_2, and sulphur oxides. In many cases the initial products of these oxidation processes may themselves be carcinogenic. Laboratory studies indicate that, in darkness, gaseous PAH are fairly inert to O_2, but in sunlight, oxidation is rapid with half-lives of a few hours being typical.

However, the bulk of the PAH in the atmosphere are believed to be adsorbed to the surface of particulates supplied either by the combustion process itself (e.g., soot and PAH are frequently emitted together; coal combustion also yields inorganic ash particulates) or by the ambient atmosphere into which the combustion products exhaust.[37] The effect of adsorption to surfaces is to alter radically the oxidation behavior of the PAH, apparently stabilizing many of them indefinitely against photolytic degradation.[38] In this case the fate of the PAH in the atmosphere is very much that of the particles with which they are associated. The condensation of the PAH onto particulates concentrates them in the small size fractions, in the submicron range.

TABLE 3. Estimated Annual Emissions of Benzo[a]pyrene in the United States[a]

Source	Metric Tons/Year	%
Coal		
Heat and power generation	440–500	40.0
Coke production	220	18.0
Coal refuse burning	> 55	4.4
Subtotal	775	62.0
Petroleum and natural gas		
Oil/residential heat	2.2–3.3	0.26
Gas/residential heat	4.8	3.8
Petroleum catalytic cracking	6.6	0.53
Automotive/gasoline-powered	22–28	2.2
Automotive/diesel-powered	22–28	2.2
Subtotal	114	9.0
Other sources		
Wood heating	44	3.5
Enclosed incinerators	33–39	3.1
Open burning of waste	220–275	22.0
Subtotal	358	29.0
Total	1110–1250	100.0

[a]Based on Hoffmann and Wynder.[36]

Such particles have very long atmospheric residence time and are also capable of penetrating deep into the human lung before being deposited.

Sampling and Analysis. The strong association of PAH with particulates complicates their sampling and analysis. Generally, particulate samples are collected on a filter, and PAH are subsequently extracted with a solvent such as methylene chloride prior to analysis. Analysis techniques include gas-chromatography, mass spectrometry, liquid chromatography, and fluorescence spectroscopy.[39]

For ambient sampling, vapor-phase constituents are not generally scrubbed because of the large gas-volumes involved. Depending on the effective volatility of the PAH in question (smaller PAH are generally more volatile; different surfaces will bind the PAH to different extents), this may represent a sampling error. This problem may become more serious when sampling combustion products. In this case the particulate filter is often held at higher temperatures, above the dew point of the gases, to prevent steam and H_2SO_4 condensation. An illustration of the significance of temperature in the association of PAH

TABLE 4. PAH Retention on a Glass-fiber Filter. Fraction of Compound Collected on Filter at Various Temperatures[a]

Compound	Filter Temperature		
	40°C	85°C	200°C
Naphthalene	0.6	0.04	0.001
Biphenylene	0.9	0.7	0.001
Biphenyl	0.90	0.5	0.005
Fluorene	0.98	0.85	0.02
Phenanthrene and anthracene	0.90	0.7	0.05
3H-Cyclopenta[def]phenanthrene	0.97	0.85	0.02
Fluoranthene	1	0.8	0.4
Pyrene and benzacenaphthylene	1	0.8	0.3

[a] From Longwell.[35]

with particulates is shown in Table 4 by the measured partitioning of various PAH between soot-adsorbed and vapor states in the sampled combustion products of a sooting flame.[35] The soot has been collected in a filter maintained at a certain temperature, while vapor-phase PAH passing through the filter are scrubbed at lower temperatures downstream of the filter. At a filter temperature of 40°C, all but the smallest PAH are strongly associated with the soot, but at 200°C even larger molecules such as pyrene remain predominantly in the gas-phase.

The importance of temperature in the association of PAH with particles is also illustrated by the PAH-contents of fly ashes sampled (1) from within a coal-fired power-plant stack and (2) outside the stack where temperatures have fallen to near ambient. In the former case no individual polycyclics are detectable in the sample, whereas in the latter the total PAH concentration in the ash is at least three orders of magnitude greater.[40]

Obviously, great care must be exercised in establishing sampling procotols for PAH. Equally much caution should be applied to the interpretation of results obtained under conditions where the effects discussed have not been accounted for.

2.2. Studies of PAH Kinetics in Combustion

There have been many studies of the formation of PAH in the combustion and pyrolysis of both aliphatic and aromatic hydrocarbons, but the routes of reaction remain largely unclear. One reason for the uncertainty as to the details of the mechanism lies in the complexity of the reaction mixtures, in that a large number of PAH (see Table 1 for just a few of these) are invariably present together. A further complicating factor lies in the concomitant formation of soot in many cases. Finally, in much of the pyrolysis work, it is not clear that heterogeneous (wall and soot) effects can be excluded.

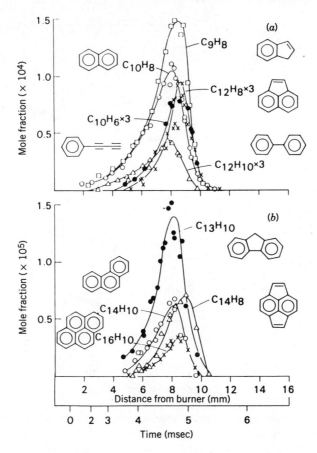

FIGURE 3. Mole fractions of polycyclic aromatic hydrocarbons in the low-pressure benzene/oxygen/argon flame of Figure 2a. From Bittner and Howard.[11]

These problems are largely overcome in the analysis of low-pressure flames by molecular beam sampling and mass spectrometry.[8-13,16,41] Bittner and colleagues have recently used this technique to determine the detailed chemical structure of fuel-rich benzene,[9,11] acetylene,[11,42] and butadiene[42] flame zones.

The behavior of the various smaller aromatic molecules in the primary reaction zone of a benzene flame just leaner than needed to produce soot (the same flame as depicted in Figure 2a) is shown in Figure 3.[11] The aromatics are clearly formed early in the primary zone which extends from $Z = 6$ to $Z = 12$ mm in Figure 3. Their concentrations reach a maximum before the species disappear rapidly toward the end of the flame zone. This behavior, albeit with higher species concentrations, has also been observed in sooting benzene flames.[16,43]

Qualitatively the same behavior is shown in Figure 4 (cf., Figure 2b) to occur also in acetylene flames, although now the concentrations of the individ-

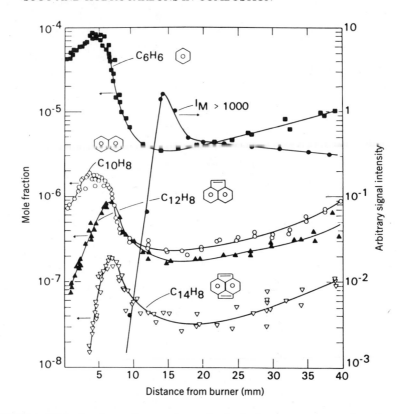

FIGURE 4. Mole fractions of aromatic species in the low-pressure acetylene/oxygen/argon flame of Figure 2*b*. From Bittner and Howard.[11]

ual aromatic species are some two orders of magnitude less than in the benzene flames. Just as the major degradative pathways in benzene flames become, upon rupture of the ring, the same as those in acetylene flames, so the behavior of aromatics in the acetylene flames, once ring closure occurs, appears to mirror that in benzene flames.

Results similar to these have been obtained for sooting, low-pressure propane-,[43] ethylene-,[44] and acetylene[43,44] oxygen flames. Probe studies in atmospheric-pressure methane flames do not resolve the primary reaction zone.[45,46]

Ring Formation by Aliphatics. The pyrolysis of acetylene at temperatures similar to those in the early flame region of Figure 4 (ca. 1000 K) has long been known to give rise to benzene.[47] The occurrence of Diels-Alder condensations has generally been assumed, but the rapidity of the reaction implied by Figure 4 is more consistent with a radical mechanism. Indeed, it has recently been proposed that the initial formation of ring compounds occurs through the addition of the 1,3-butadienyl radical to the various acetylene species occurring in this region: [42]

$$C_4H_5\cdot + R-C\equiv C-H \longrightarrow Ph-R + H \qquad (6)$$

for:

R = H	acetylene, C_2H_2 → benzene, C_6H_6
R = CH_3	methylacetylene, C_3H_4 → toluene, $C_6H_5CH_3$
R = C_2H	diacetylene, C_4H_2 → phenylacetylene, $C_6H_5C_2H$
R = C_2H_3	vinylacetylene, C_4H_4 → styrene, $C_6H_5C_2H_3$

Ionic mechanisms have also been proposed to explain the formation of aromatics in acetylene flames.[48-50] Here the propensity of simple aliphatic hydrocarbon ions, principally $C_3H_3^+$, to undergo condensation reactions is taken to produce larger aliphatic ions which then isomerize to more stable aromatic forms. However, this route was suggested not to be important in a recent study of ions in sooting acetylene flames.[51] The ionic structure of hydrocarbon flame zones is considered in more detail in Section 3.2.

Higher Aromatics from Benzene. In the profiles of PAH concentrations in benzene and acetylene flame zones (Figures 3 and 4), there is a clear trend toward larger molecules with increasing distance from the burner. The rapid growth of the PAH occurs in the region where simple aliphatics such as C_2H_2, C_4H_4, C_4H_3, and C_4H_2 are at their highest concentrations.

The occurrence of condensation reactions of the aromatic species with C_2, C_3, and C_4 hydrocarbons may therefore account for the formation of larger aromatics.

A similar correlation between higher molecular weight PAH formation and the simpler hydrocarbons is obtained from the low-pressure pyrolysis of toluene and benzene[21-23]—here the maximum yield of PAH occurs at temperatures of the order of 1500 K, where the mole fractions of the phenyl radical and simple hydrocarbons also maximize. Above 1700 K, C_2H_2 and C_4H_2 are favored over all these other species, including PAH.

Radical Mechanisms. Bittner and Howard[9-11] have considered these phenomena in some detail. The crucial feature of any condensation mechanism at these high temperatures is that additional complexes must be stabilized against unimolecular fragmentation back to simpler forms—in the case of PAH formation, what is further required is that stabilization by formation of six-membered rings should ultimately dominate over processes that may stabilize the adduct without forming six-membered rings.

In general, products of radical–radical reactions are likely to be too unstable to allow stabilization of the addition complex, whereas stable molecule–molecule reactions are probably too slow to be consistent with the observed rates.

As an example of the addition of a nonaromatic to benzene to yield a cyclohexadienyl-type adduct, consider the addition of the radical C_4H_5 (1,3-butadien-1-yl) to benzene:

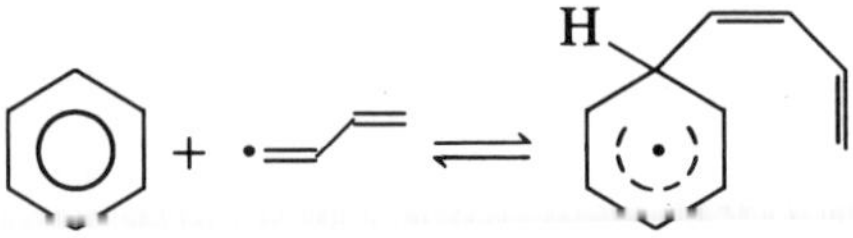

$$\Delta H = -110 \text{ kJ/mol}$$

(7)

Decomposition of this adduct to form a stable structure with an unsaturated side chain is favored over ring closure:

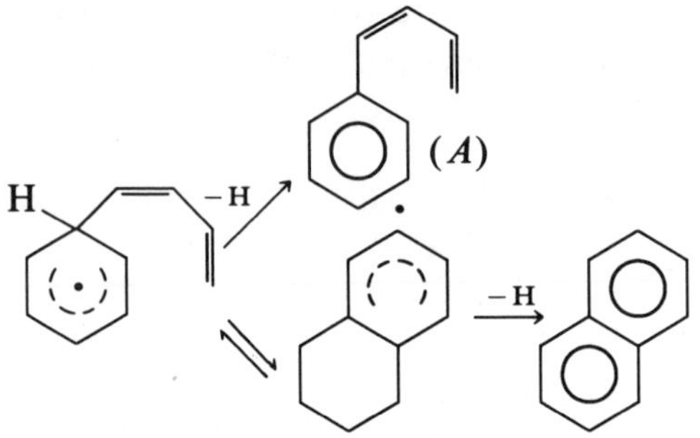

$$\frac{\Delta H}{105 \text{ kJ/mol}}$$

$$185 \text{ kJ/mol}$$

(8)

However, the stabilized species A may subsequently cyclize rapidly by an internal aromatic substitution to form naphthalene upon H-atom abstraction from the terminal carbon of the side chain. This is then the analogue of the mechanism proposed[42] for the formation of the first ring structures in aliphatic flames:

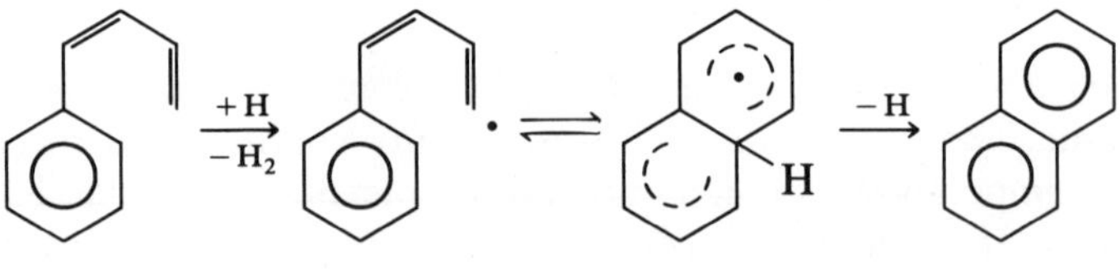

$$\Delta H = -110 \text{ kJ/mol}$$

(9)

The addition of aromatic radicals to nonaromatic molecular species can also lead to condensed-ring compounds. For example, addition of the benzyl radical to acetylene yields directly a side chain with a terminal radical site. This may cyclize to form indene in a manner similar to the naphthalene-forming steps discussed above:

$$\Delta H = -45 \text{ kJ/mol}$$

$$\Delta H = 110 \text{ kJ/mol}$$

$$(10)$$

In benzene flames the alkylated benzenes probably derive from reactions of the phenyl radical. In the case of phenyl radical additions to triple bonds, a vinyl-type radical is formed. For example,

$$\Delta H = -225 \text{ kJ/mol} \qquad (11)$$

The most likely unimolecular decomposition reaction of this adduct is the elimination of hydrogen to form an acetylenic side chain. However, if this side chain has polyacetylenic character, as in the above example, the interaction of the unpaired electron with the adjacent triple bond provides considerable stability to the vinyl-type adduct itself—so much so that it may persist long enough to add to acetylene in a bimolecular step:

$$+ C_2H_2 \longrightarrow \qquad \Delta H = -110 \text{ kJ/mol} \qquad (12)$$

Now the radical site is in the terminal position, and an internal aromatic substitution reaction to form the substituted condensed-ring compound can follow:

$$(13)$$

The stability of the vinyl-type radical B is such that it could also be formed from the addition of an H-atom to the stable aryl polyacetylene.

The types of pathway we have discussed here for the benzene derivatives obviously obtain also to the higher aromatics. The case of the larger benzyl-type species is worthy of further consideration, as summarized by the reaction sequence in Table 5. Methyl substitution to the α-position in naphthalene (A) and subsequent abstraction of a methyl hydrogen (B) yield a benzyl-type radical whose addition to the triple bond of acetylene yields an adduct easily stabilized by internal aromatic substitution (C, D). Further methyl substitution (E), hydrogen abstraction (F), C_2H_2 addition (G), and unimolecular expulsion of hydrogen (H) result in ortho- and peri-fused ring systems such as pyrene. Because the number of sites for, and the reactivity toward, methyl substitution and the degree of stability of the adduct—all of which increase with the size of the growing PAH—this process could be considered autocatalytic.

Discussed here are only a few of the possible pathways. The important feature is that starting from an initial aromatic structure provides a foundation on which larger aromatics can be built by the addition of nonaromatics such as C_2H_2. Crucial to these processes is the ability of the substituted ring to undergo internal aromatic substitution reactions and so stabilize the growing carbon framework.

Ionic Mechanisms. Gaseous hydrocarbon ions are known to be capable of rearranging readily to their most stable structures which, for large molecules, are those of the PAH.[48,52] Since hydrocarbon flame zones form ions (see the discussion in Section 3.2 for a more detailed description of this phenomenon), it is not surprising that a variety of PAH ions have been identified in rich flames of benzene and other fuels, as summarized in Table 6.

It has been suggested that the ability of the gaseous ions to rearrange provides a mechanism of PAH formation and growth not available to the neutrals.[48-50,52-54] For example, the sequence generated by continued C_2H_2 addition to $C_3H_3^+$ (the major small ion in these rich flames) has been proposed:

$$C_3H_3^+ \rightarrow C_5H_5^+ \rightarrow C_7H_7^+ \rightarrow C_9H_9^+ \quad \text{(etc.)} \tag{14}$$

TABLE 5. Methyl Substitution Pathways in PAH Formation[a]

TABLE 6. Ions Observed in Sooting Acetylene-Oxygen Flames[a]

Molecular Weight	Formula	Suggested Structure
39	$C_3H_3^+$	$HC \overset{\overset{\displaystyle H}{\displaystyle C}}{\triangle^+} CH$
51	$C_4H_3^+$	$>C=\overset{+}{C}-C\equiv C-$
53	$C_4H_5^+$	$>C=\overset{+}{C}-C=C<$
63	$C_5H_3^+$	$-C\equiv C-C\equiv C-\overset{\mid}{\underset{\mid}{C}}{}^+$
65	$C_5H_5^+$	$-C\equiv C-\underset{+}{C}=C-\overset{\mid}{\underset{\mid}{C}}-$
75	$C_6H_3^+$	$-C\equiv C-C\equiv C-\underset{+}{C}=C<$
77	$C_6H_5^+$	$-C\equiv C-\overset{+}{C}=\overset{\mid}{C}-\overset{\mid}{C}=C<$
79	$C_6H_7^+$	$>C=\overset{\mid}{C}-\overset{\mid}{\overset{+}{C}}=\overset{+}{C}-\overset{\mid}{C}=C<$ or (cyclohexadienyl cation ring)
89	$C_7H_5^+$	$-C\equiv C-C\equiv C-\overset{+}{C}=\overset{\mid}{C}-\overset{\mid}{C}-$
91	$C_7H_7^+$	(benzyl cation CH_2^+) or (open-chain cation)
103	$C_6H_7^+$	(phenyl–CH=CH$^+$)
115	$C_9H_7^+$	(phenyl–C≡C–CH$^+$)

TABLE 6. *(Continued)*

Molecular Weight	Formula	Suggested Structure
129	$C_{10}H_9^+$	
139	$C_{11}H_7^+$	
141	$C_{11}H_9^+$	
143	$C_{11}H_{11}^+$	
153	$C_{12}H_9^+$	
165	$C_{13}H_9^+$	
179	$C_{14}H_{11}^+$	
191	$C_{15}H_{11}^+$	

TABLE 6. (Continued)

Molecular Weight	Formula	Suggested Structure
203	$C_{16}H_{11}^+$	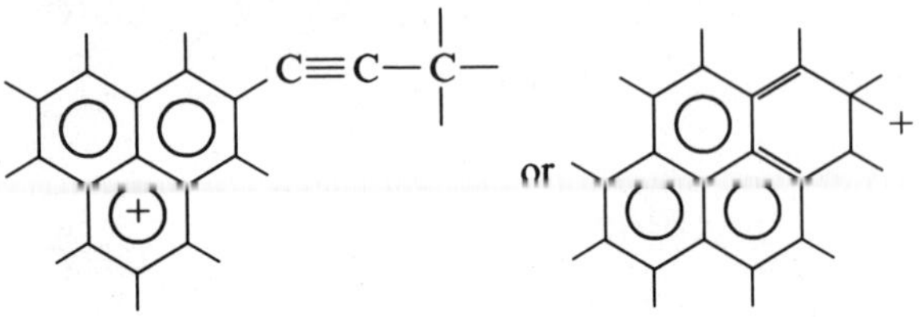
205	$C_{16}H_{13}^+$	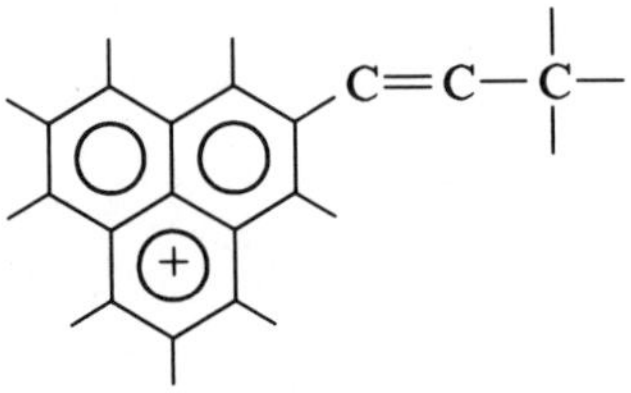
215	$C_{17}H_{11}^+$	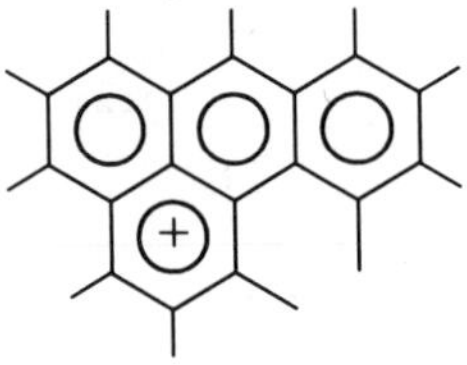
227	$C_{16}H_{11}^+$	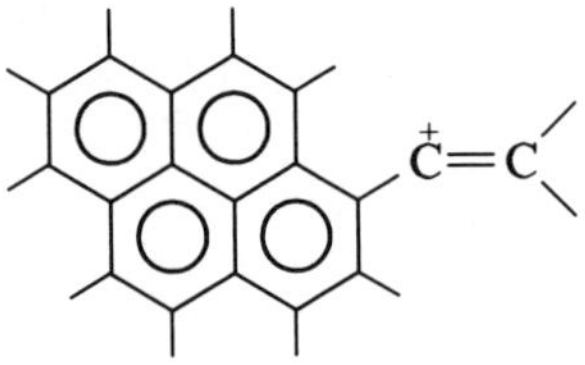
229	$C_{18}H_{13}^+$	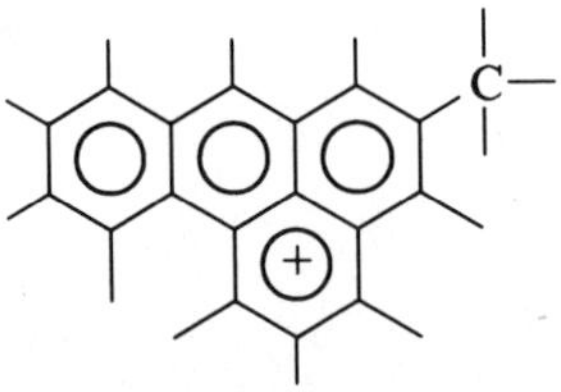
239	$C_{19}H_{11}^+$	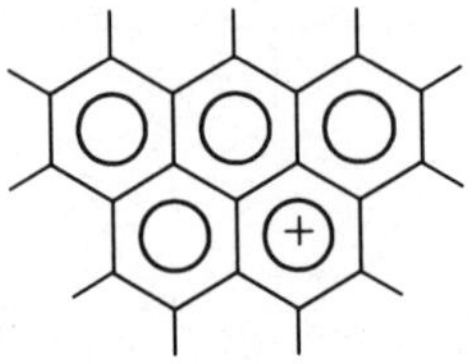

[a] From Calcote.[48]

Involvement of C_4H_2 also allows equations to be written for the formation of ions such as $C_{10}H_9^+$ (protonated naphthalene) and $C_{14}H_{11}^+$ (protonated anthracene).

Alternatively, reactions of existing PAH neutrals with $C_3H_3^+$ have been suggested to lead to large species, for example,

$$C_6H_6 + C_3H_3^+ \rightarrow C_7H_7^+ + C_2H_2$$

and (15)

$$C_9H_8 + C_3H_3^+ \rightarrow C_{10}H_9^+ + C_2H_2$$

Actual observations of the extent of ionization in flames[49-51,54-56] have indicated the concentrations of these PAH ions to be many orders of magnitude less than that of the corresponding neutrals, much as expected from equilibrium considerations.[54,56] Therefore, if these ionic growth mechanisms are to be substantially responsible for the PAH formation and growth, there should also be a mechanism operating to recycle the charge back from the larger species to the smaller. This seems unlikely, given that the larger PAH have, in general, lower ionization potentials than their smaller counterparts. In fact the presence of the PAH ions could be explained in terms of protonation of the corresponding neutrals and their growth, without requiring such drastic rearrangements, for example,

$$C_{10}H_8 + H_3O^+ \rightarrow C_{10}H_9^+ + H_2O \qquad \Delta H = -110 \, \text{kJ/mol} \qquad (16)$$

Destruction of Benzene and the PCAH. The main fate of the benzene fuel in Figure 2 is obviously ring rupture, fragmentation, and oxidation. Only a small fraction escapes this fate to undergo condensation to PAH within the primary reaction zone. These PAH themselves have all but disappeared by the end of the primary reaction zone—some of this loss may be due to growth to species of yet higher mass,[57] but decomposition by pyrolysis and oxidation must account for much of it.

Bittner and Howard[9-12,57] have considered a number of possible routes leading to the rupture of the ring in benzene flames. As is typical of all hydrocarbons flames, the dominant processes must be radical in character in order to account for the rapid rates of destruction observed. However, addition of an H-atom to and, more important, abstraction of an H-atom from the benzene ring are both considered unlikely to produce significant ring rupture. In fact, addition of OH (and perhaps of O) to the ring may be responsible for initiating the consumption of benzene: it is yet unclear how this mechanism operates, but the ring rupture process may involve expulsion of CO from the species C_6H_6O to form C_5H_6.[9,19]

There have been very few studies of the oxidation of other PAH at flame temperatures. Some flow-tube studies at moderate temperatures[58-60] do indi-

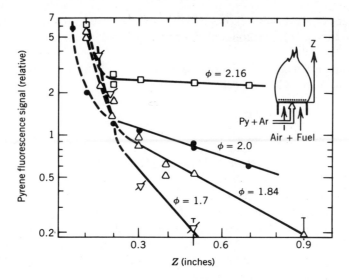

FIGURE 5. Pyrene laser fluorescence intensities in the postflame gases of ethylene/air flames at atmospheric pressure. Pyrene is added to these flames downstream of the primary reaction zone—that region up to 0.25″ is the mixing region for this injected pyrene—the data here are meaningless. Further downstream, the conditions are isothermal and well mixed.

cate appreciable reactivity of alkyl benzenes to O and OH, but no distinction can be made from these studies between addition and abstraction—by analogy with benzene, ring rupture may be less likely to proceed via H-atom abstraction than by O-atom addition.

Some preliminary results on the decomposition of pyrene injected into the postflame gases ($T \sim 1700$ K) of near-sooting and slightly sooting ethylene/air flames at atmospheric pressure are illustrated in Figure 5.[61] In the richest, sooting flame, no decomposition of the pyrene, as detected by laser-induced fluorescence, occurs over extended reaction times (~ 25 ms). As the flame is made leaner, significant removal of the pyrene occurs until, at $\Phi = 1.7$, the lifetime of the pyrene is of the order of 5–10 msec. An analysis of expected conditions in these flames indicates that the pyrene decomposition is brought about by OH radical attack with a bimolecular rate constant of the order of $10^{12 \pm 1}$ mol/cm³-sec.

Returning to the benzene flame described in Figures 2 and 3, it is apparent that the rapid decline in the PAH concentrations toward the end of the primary reaction zone is occurring in a region where high levels of OH, and of O-atoms, prevail. The same type of mechanism as is occurring for the pyrene decomposition shown in Figure 5 is probably occurring here.

Although the kinetics of the PAH oxidation are obscure, Figure 5 does indicate that pyrene is resistant to thermal degradation and radical (H-atom)

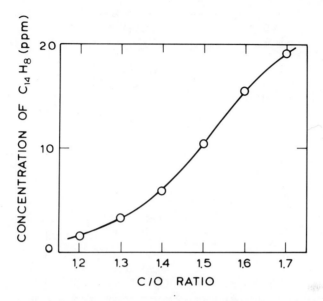

FIGURE 6. Downstream concentrations of a typical PAH ($C_{14}H_{8}$, cyclopentacenaphy-lene) in low-pressure, sooting acetylene/oxygen flames.[41] Redrawn from Homann and Wagner.[41]

attack in flames that are sufficiently rich. Some benzene[20] and PAH[62] pyroly-sis does occur at temperatures as low as 800 K in heated reactors. As discussed for benzene in Section 1.2, the PAH tend to form bi-aryl and polyaryl species (cf., biphenyl and polyphenyl), but the extremely long reaction times (many minutes), high concentrations of reactants, and the presence of surfaces in these studies do not permit direct comparison with the flame situation.

Beyond the Primary Reaction Zone. Inspection of the PAH profiles in the low-pressure acetylene flame of Figure 4 shows that, beyond the flame-zone peak, the concentrations of the PAH rise again. As shown in Figure 6 for a typical PAH ($C_{14}H_{8}$), the extent to which this occurs is very strongly a function of flame stoichiometry.[41] This phenomenon of postflame PAH forma-tion has been observed also in flames of other aliphatic fuels at low pressures and at atmospheric pressures.[43-46,63]

Recently, the application of visible laser light scattering techniques for soot particle size determinations has been found to produce a broad-band visible (max ~ 500 nm) fluorescence.[64-71] It is generally accepted that this fluores-cence indicates the presence of PAH,[70,71] and it now appears small (3- and 4-ring) PAH may be expected to absorb UV and visible light at high tempera-tures.[69,72,73] Although acenaphthylene, $C_{12}H_{8}$, has been suggested as a source of the fluorescence,[69] other species are likely to contribute. Even though the total fluorescence signal has unknown composition and temperature depen-

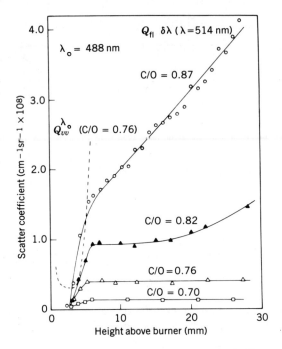

FIGURE 7. Fluorescence intensities in sooting, atmospheric-pressure ethylene/air flames. Excitation at 488 nm; emission at 514 nm. From Haynes et al.[65]

dence, its measurement does appear a useful qualitative probe for PAH in flames.[67,71] Thus the fluorescence intensities obtained upon laser irradiation (488 nm) of the sooting postflame gases of ethylene/air flames at atmospheric pressure (Figure 7) also show significant postflame PAH formation.

Curiously, significant postflame PAH formation has been observed only in sooting flames of aliphatic fuels. No significant concentrations of PAH survive the reaction zone in Figure 3, and this observation has also been made in probe studies of sooting benzene[16,43] and toluene[33] flames. Furthermore broad-band laser fluorescence also indicates that there is a very marked PAH peak in the primary reaction zone of sooting benzene flames at atmospheric pressure but that the concentration of fluorescing species in the postflame gases is very low and does not increase with distance from the flame.[65]

Higher temperatures are known to reduce PAH concentrations,[46] either by inhibiting their formation or by promoting their decomposition, and the benzene flames tend to be hotter than their aliphatic counterparts in the above comparisons.[43,65] Furthermore the concentrations of the aliphatics also are suppressed in benzene postflames even at high soot concentration[43]—as is well known (Section 3), aromatic fuels form soot at lower C/O ratios than do their counterparts.

Despite these uncertainties, it is clear that postflame PAH formation may lead to higher concentrations than achieved within the primary reaction zone[43] and so may be more important from the point of view of emissions. Unfortunately, however, the processes responsible are not known. Homann and Wagner[41] concluded that the PAH formed over extended periods in the postflame gases of their acetylene flames are by-products of the soot formation process. They suggested, as discussed further in Section 3.2, that at the end of the primary reaction zone, radical reactions between the various polyacetylenic species give rise to branched, strongly unsaturated species which may then cyclize to yield PAH possessing unsaturated radical side chains. These reactive PAH are then presumed to grow, to become the first soot nuclei, by further reaction with acetylene and the polyacetylenes, much as discussed in Section 2.2. However, in competition with these growth reactions are processes stripping away the side chains to leave the presumably unreactive PAH kernels such as pyrene as by-products.

2.3. Laminar Diffusion Flames

The chemical structure of an axi-symmetric, sooting, n-hexane diffusion flame in air is depicted schematically in Figure 8.[74,75] The true diffusion flame height, which is the height at which stoichiometric proportions of fuel-derived species and oxygen are achieved on the flame axis, is about 25 mm. Beyond this height the only combustibles surviving are CO and soot, the consumption of both of these being kinetically controlled.

The axial concentration profiles for a large number of species in this flame are shown in Figure 9—the general similarities in composition between this flame and the premixed flames just discussed are apparent. The dominant hydrocarbon in the downstream region is acetylene; benzene is present at a maximum concentration of 0.5%, the alkylated benzenes (toluene, phenylacetylene and styrene) at about 100 ppm, and the various polycyclics analyzed in this work at peak concentrations ranging from 0.1 ppm to 3 ppm for pyrene and 20 ppm for naphthalene. The concentrations of all these species peak at about 16 mm above the burner, some time before the end of the true diffusion flame and the peak in the soot loading.

Recently reported detailed measurements of PAH concentrations in a laminar diffusion flame of methane in oxygen[71] have indicated a good correlation of PAH concentrations with the local fuel mixture fraction. This may imply that the formation and decomposition of PAH is fast relative to diffusion and that the chemical processes are in a state of dynamic balance determined by the local mixture stoichiometry.

A comparison of these results with those from a variety of less well-defined flames indicates that peak local concentrations for acetylene of a few mole percent and for benzene of 0.1 to 0.5% are typical of diffusion flames.[76]

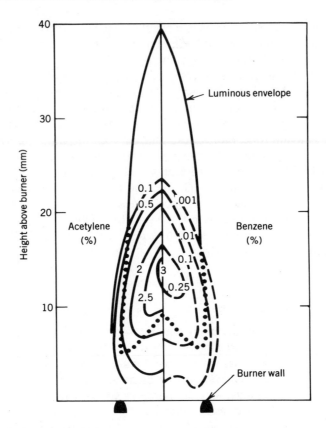

FIGURE 8. Profiles of benzene and acetylene concentrations in an *n*-hexane diffusion flame in air. Drawn from data in Spengler and Kern,[74] with permission from the publisher.

Unfortunately, there is no information available on the distribution of PAH in other flames.

Characterization of PAH emissions from smoking ethylene diffusion flames has been reported.[77] (As discussed in Section 3, a diffusion flame emits the familiar plume of smoke when the oxidation of the soot after the end of the true diffusion flame is quenched.) The effect of the oxygen index, OI, (oxygen mole fraction of the oxidant) is to bring about a monotonic reduction of PAH levels with increasing OI. The main problem with interpreting experiments of this sort lies in the fact that the emissions reported are the result of spatial integration of all the PAH-forming and PAH-removing processes occurring within the system. Broadly speaking, an increased OI increases the flame temperature and hence, on average, the temperatures everywhere in the system. As we have seen for the premixed flames, higher temperatures inhibit PAH formation and enhance oxidation, which is consistent with the observations of steadily declining PAH emissions with increasing OI. The somewhat

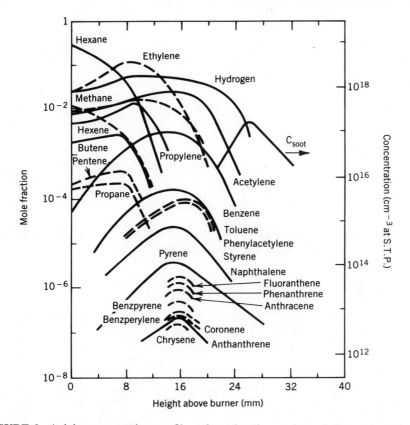

FIGURE 9. Axial concentration profiles of combustion and pyrolysis products in the *n*-hexane diffusion flame of Figure 8. Redrawn from Kern and Spengler,[75] with permission from the publisher.

different behavior for soot, whose emission first increases and then decreases with increasing OI, reflects competition between enhanced soot formation and enhanced oxidation with increasing temperatures (see Section 3).

2.4. Turbulent Diffusion Flames

There have been few studies reported of spatially resolved PAH concentration in turbulent diffusion flames.[68, 70, 78-80] In general, the observed profiles and concentrations are similar to those in the laminar flames discussed above.

Axial concentration profiles for total condensibles, PAH and soot obtained for swirled combustion of No. 2 light oil are shown in Figure 10.[79] Profiles for individual PAH species in this flame are shown in Figure 11, where it is seen that the smaller (two- and three-ring) aromatics arise very early, possibly from aromatics contained in the fuel. Further downstream, heavier polycyclics and soot are formed. Just as in the laminar diffusion flame of Figure 9, the

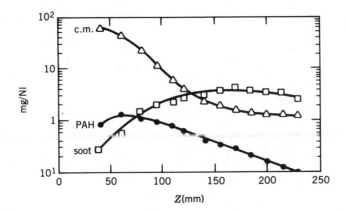

FIGURE 10. Axial concentration profiles of condensible matter (cm), total measured PAH, and soot in swirled light-oil combustion. From Ciajolo et al.[79]

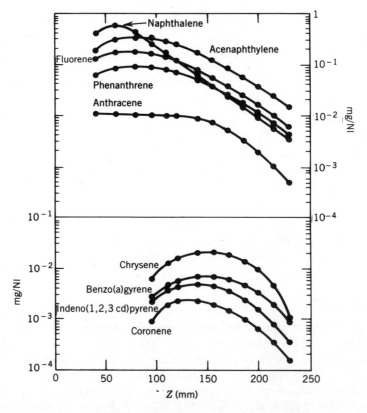

FIGURE 11. Individual PAH concentrations on the axis of the swirled light-oil combustor of Figure 10. From Ciajolo et al.[79]

aromatics burn out well before the soot, again indicating a greater degree of kinetic control in soot burnout.

2.5. Fuel PAH

Coal and many liquid fuels contain aromatic structures themselves—under some conditions these may become more or less directly a source of PAH emissions, as suggested above for the formation of the light PAH in Figure 11.

The distribution of PAH in the emissions from the combustion of a light distillate fuel oil in a domestic heating unit is similar to that of the fuel, except that the emissions are strongly dealkylated.[81] Because the fuel PAH and emissions in this work are similar to those PAH arising in any combustion system (mono- and di-cyclic aromatics predominating, with lower concentrations of polycyclics), it is not in fact clear what proportion of the emissions in this work derive from unburned fuel and what is formed within the flame.

In the case of PAH emissions from a turbulent laboratory-scale coal combustor, the PAH are much more strongly alkylated than they are for kerosene combustion in the same system.[82] Because stripping of alkyl side chains occurs fairly readily at high temperatures, the coal emissions probably derive from incompletely burned coal volatiles.

2.6. Conclusions

Fuels containing aromatic structures are capable of producing substantial concentrations of PAH in the primary reaction zone. For a well-characterized nonsooting benzene flame, as much as 0.1% of the benzene mass appears in polycyclic structures—under richer, sooting conditions, higher conversions are expected.

Much of the PAH material formed in the reaction zone is oxidized while still in the radical-rich region. The OH radical appears to be the chief oxidant downstream of this region.

Polycyclics are also formed in the flame zones of aliphatic fuels but at much reduced efficiency compared with benzene. The main PAH-forming mechanism for sooting aliphatic flames occurs downstream of the primary reaction zone—the mechanism of formation is not understood but appears to be related, indirectly at least, to soot formation. Many of the PAH are stable indefinitely in the rich postflame gases.

3. SOOT FORMATION

Soot is a carbonaceous solid produced in pyrolysis and combustion systems when conditions are such as to allow gas-phase condensation reactions of the fuel and its decomposition products to compete with further decomposition and oxidation. In this sense the formation and emission of soot is related to processes controlling the formation and emission of the PAH—however, as will become evident in the following discussion, there are some major differences as well as some close similarities in the kinetic phenomena leading to soot and PAH.

The formation of carbonaceous residues such as cenospheres by some fuels, notably coal and heavier fuel oils, is a problem separate from that of soot formation in that such residues are formed during combustion directly from the liquid phase rather than by gas-phase condensation reactions.

The formation and emission of soot are subjects of some concern for a variety of reasons. Soot particles are strongly absorbing and, within a combustor, can enhance significantly radiative heat transfer: witness the common candle. However, emission of soot represents not only a fuel loss but also a hazard to the environment. Soot agglomerates are typically in a size range (0.1 to 1 μm) which is collected with relatively low efficiency by electrostatic precipitators and which, coincidentally, is particularly effective in reducing atmospheric visibility.

A large variety of organic molecules is frequently associated with the surfaces of soot particles. As discussed in Section 2, the polycyclic aromatic hydrocarbons constitute an important fraction of these hydrocarbons. The small soot particles released into the atmosphere are easily inhaled deep into the respiratory tract, providing the ideal transport mechanism for ingestion of the adsorbed PAH.

The surface of emitted soot particles has also been shown to be active in the catalytic oxidation of SO_2 and NO in the atmosphere to SO_4^{2-} and NO_2 or NO_3^-, respectively.

3.1. Structure of Soot

Soot emitted from typical pyrolysis and combustion systems is found to consist of agglomerates of a number (ranging from unity to many hundreds) of "primary spheres" whose diameters are typically in the range 10 to 40 nm for a wide variety of conditions. These primary spheres typically exhibit a size distribution.

The internal structure of soot particles has been examined by high-resolution phase-contrast electron microscopy. Near the edge of the particle, bent carbon layers follow the shape of the particle surface. Inside the particles, lattice structures seem to be located more or less regularly around certain centers between which the structure is less ordered. Many dislocations and other lattice defects are present. The density of the particles may be less than 2 g/cm^3 due to large interplanar spacings.

Heat treatment improves the internal order of the particles, and the interplanar spacing approaches that of graphite. The rate of graphitization depends on temperature so that the age of a particle and its degree of order are closely linked. For soot particles at high temperatures, some ordering may occur as the particles continue to grow. A more detailed discussion of soot structure has been presented recently by Lahaye and Prado.[83]

Chemically, soot consists mainly of carbon but also contains 10 mol % and more of hydrogen. The hydrogen content of soot is highest (approaching a C/H atom ratio as low as 1) in nascent particles. At all stages most of the hydrogen can be extracted in organic solvents where it appears primarily in condensed aromatic ring compounds.

3.2. Fundamental Studies of Soot Formation

Developments in the detailed understanding of the physical and chemical mechanisms in soot formation are discussed in various published reviews.[17,84-89] In this section we discuss new and old work with emphasis where possible on recent developments.

Soot as an Aerosol. In dealing with soot formation during combustion we are considering the formation and behavior of a solid carbonaceous phase from gas-phase hydrocarbon species. The soot aerosol can be characterized by the total amount of material in the condensed phase, often expressed as the soot volume fraction, f_v (cm^3 soot/cm^3; a value of $f_v = 1 \times 10^{-7}$ corresponds to approximately 1% of the fuel appearing as soot); the number density of particles, N; and the characteristic dimension of the particles, d. The particles also possess a size distribution, but this is usually relatively narrow and for the most part we will consider here only an average size. For a given particle shape, the quantities f_v, N, and d are mutually dependent (for spherical particles, $f_v = \pi/6 \cdot Nd^3$), and any two are sufficient to characterize the system.

The time-resolved behavior of these various quantities in some premixed flat flames is illustrated in Figure 12.[65] The first particles appear, small ($d \sim 2$ nm) and at high number density ($\geq 10^{12}$ cm^{-3}), soon after the main reaction zone. The amount of soot, f_v, is initially only a small fraction of the final amount, growing steadily well downstream of the flame zone. As f_v grows, the particles also grow in size; at the same time their number density diminishes. Richer flames produce more soot and larger particles than lean flames. The particle number density curves are similar in all cases.

Obviously, soot formation begins with *particle inception*, whereby the first condensed phase material arises from the fuel-molecules via their oxidation and/or pyrolysis products. Once the particles are formed, the soot loading can increase by *surface growth*, which involves the attachment of gas-phase species to the surface of a particle and their incorporation into the particulate phase. Surface growth reactions lead to an increase in the amount of soot, but the number of particles (N) remains unchanged by this process. The opposite is true for growth by *coagulation*, where particles collide and coalesce, thereby decreasing N while f_v remains constant. Particle growth (increasing d) is the result of simultaneous surface growth reaction and coagulation. For spherical

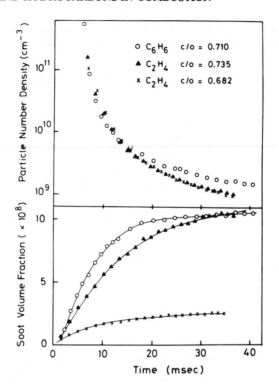

FIGURE 12. Soot volume fraction, average particle size, and number density in atmospheric pressure flat, premixed flames of ethylene and benzene with air. From Haynes et al.[65]

particles, this can be expressed as

$$d \ln d = \tfrac{1}{3} d \ln f_v - \tfrac{1}{3} d \ln N \tag{17}$$

The importance of this dependence of particle growth on the independent processes of surface growth and coagulation is not always appreciated in the older literature, nor unfortunately in some more recent publications. In the discussion of the effects of additives on soot (Section 3.5), it is shown that some additives bring about a reduction in diameter of the primary particles without changing their number density (f_v decreases), while others produce the same effect of smaller diameter without changing f_v (N increases). Clearly the soot aerosol is inadequately described by determination only of d, or of any *one* of the variables N, f_v, and d.

In the strictest sense the transition from particle inception to particle growth is probably not a sharp one. The growth of very large molecules by addition of smaller hydrocarbons is obviously not far removed from the addition of similar species to "particles." Similarly, large molecules can react

TABLE 7. Phenomenology of Soot Formation

	N	f_v
Source	Nucleation	Surface growth
Sink	Coagulation	Oxidation

among themselves in the same way as particles coagulate. However, in both premixed and diffusion flames, the zones of particle nucleation and of particle growth do appear to be spatially distinct. New particle formation is restricted to a narrow region very near the main reaction zone where temperatures and radical and primary ion concentrations are highest. Except under weakly sooting conditions (e.g., a premixed flame operating near the sooting limit), the generation of volume fraction by new particle formation is negligible compared with that contributed by surface growth. Surface growth occurs over extended periods as the gases move away from the primary reaction zone, even into relatively cool, otherwise unreactive regions. Therefore we will not attempt to define the interface between molecule and particle any more precisely than to accept, for example, the appearance of continuous yellow emission as signaling the occurrence of particles.

The stages of particle generation and growth constitute the soot-formation process. This is often followed by a phase of soot *oxidation* in which the soot is burnt in the presence of oxidizing species to form gaseous products such as CO and CO_2. The eventual emission of soot from any combustion device will depend on the balance between these processes of formation and burnout.

The phenomenology of soot formation can now be summarized as in Table 7. Clearly, in considering the influence of parameters such as temperature, stoichiometry, or fuel type on soot formation, it is important to distinguish between the various component phenomena if a coherent picture is to emerge.

Particle Inception. The smallest identifiable solid particles observed in luminous flames have diameters in the range 1.5 to 2 nm and masses of the order of 2000 amu at the start of the luminous zone.[90] Particles observed in flames just barely sooting have similar sizes.[91] It is the generation of such particles from the initial gas-phase reactants that constitutes the particle inception process. Typical particle production rates are of the order of 10^{15} cm^{-3} s^{-1}.

The chemical structure of (near-)sooting benzene and acetylene flames at low pressures is discussed in Section 2 (Figures 2–4). The major qualitative difference between the two types of flame with regard to soot formation lies in the observation that, in the benzene flame, the first soot particles arise already within, or just beyond, the primary reaction zone.[16,18] By contrast, in the acetylene flame, there is a definite induction time from the end of the primary zone to the appearance of particles—this time corresponds to the "dark space" between the blue-green primary zone and the continuum yellow emission signaling the presence of soot particles and is characteristic of aliphatic fuels. This behavior is shown qualitatively in Figures 2 and 4 by the location of the zones where rapid formation (and consumption) of large molecules (I_m − profiles) is occurring.[57]

Radical Mechanisms. As summarized in Figure 2, the chemical structures of the benzene and acetylene flames are very similar with regard to the major species' behavior. However, in the acetylene flame zone, the PAH concentrations are about two orders of magnitude less than they are in the benzene flame. Given that the benzene flame is nonsooting and the acetylene flame is sooting, this difference may be even greater under conditions more similar with regard to soot formation.[9-11]

In the case of the benzene flame, the high concentrations of PAH in the main reaction zone and the demonstrated ability of these species to grow by addition of unsaturated nonaromatics suggests immediately a mechanism of soot particle nucleation.[9-11,17,57] Thus the PAH growth processes apparent in Figure 3 for the nonsooting $\Phi = 1.8$ flame continue on to produce much larger species, as shown in Figure 13 for the curves of signal intensities correspond-

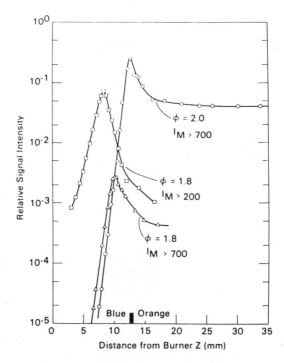

FIGURE 13. Relative signal intensities for high-mass species in near-sooting ($\phi = 1.8$) and sooting ($\phi = 2.0$) benzene/oxygen/argon flames at low pressures—the $\phi = 1.8$ flame is the same as that described in Figures 2 and 3. Visible soot emission occurs at $Z = 13$ mm in the $\phi = 2.0$ flame. From Bittner and Howard.[11]

ing to MW > 200 and then MW > 700. When the flame is made just slightly sooting (Φ = 2.0), the high mass (MW > 700) peak is some two orders of magnitude higher. The main distinction between the nonsooting and the sooting flames is apparently no more than the concentration of the high mass species and the size that they achieve,[57] which factors will determine their radiative properties.

The PAH growth reactions, in order to produce the spherical units as small as 1.5 nm, must eventually lead to a departure from planar PAH molecules. Qualitatively, growth into the third dimension should not be difficult because some of the PAH carry side chains and interactions of these with other PAH can be expected to produce three-dimensional structures. Such interactions are obviously analogous to particle coagulation, and once again the transition from molecule to particle is blurred.

For the acetylene flames, Homann and Wagner[17,41] did not consider the in-flame formation of PAH to be significant with regard to soot formation because of the low concentration of these PAH and the substantial delay between their disappearance and the appearance of higher mass species (Figure 4). Instead, they proposed that the initiation of soot formation in this flame occurs as a result of the high concentrations of polyacetylenes present at the end of the primary reaction zone where radical concentrations are at their maxima. Thus they envisaged the formation of branched-chain species by radical addition reactions to the polyacetylenes. Such reactions would preserve the radical character, and hence reactivity, of the growing molecules. They would lead to heavily branched, highly unsaturated radical species that could ultimately cyclize internally to yield aromatics with many radical-rich side chains.

On the other hand, Howard and Bittner[57] have argued that the delay from the disappearance of the simple PAH to the appearance of the high mass peak in acetylene flames simply reflects the lower concentrations and later appearances of PAH in those flames. The first nuclei are believed to be the result of the same aromatic and acetylenic addition processes occurring in the benzene flames. The major distinction between the two types of flame is therefore in the particle generation rates and the fraction of the final soot mass contributed by the nucleation process—this fraction may be as high as 20% in the low-pressure benzene flames and as low as 1% in their acetylene counterparts. In flames at atmospheric pressure also, the fraction of the ultimate soot yield formed during particle inception is substantially higher with an aromatic fuel (toluene) than with an aliphatic (ethylene).[92]

Ionic Mechanisms. A typical hydrocarbon flame zone produces chemi-ions,[93–95] and it has been a matter of conjecture as to whether such ions might influence the process of soot particle formation. Hydrocarbon ions appear a satisfying explanation for the rapidity of the particle inception process—ion-molecule reactions are very fast and rearrangements of these ions to their most stable forms occur readily. Calcote[93] has discussed recently the circumstantial evi-

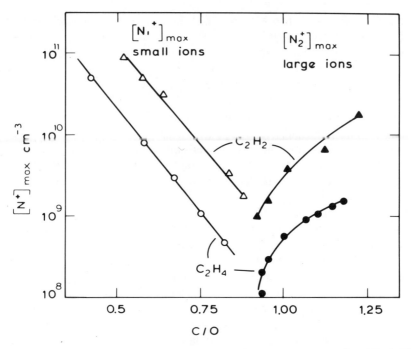

FIGURE 14. Peak ion concentrations in low-pressure (45 torr) acetylene and ethylene flames. The open symbols refer to N_1^+, the small ions whose concentrations peak in the primary reaction zone. The filled symbols are for N_2^+, the larger charged species (> 700 amu) arising downstream of the primary reaction zone. Redrawn from Delfau et al.[55]

dence regarding the importance of ions in soot formation. However, only in the past few years have direct sampling studies of large ions from sooting flames been able to provide new, important insights into the electrical aspects of soot formation.

It is now clear[49–51,55,56,96–99] that the onset of sooting brings with it a new mechanism of ion formation, independent of the classical chemi-ionization mechanism operating in all hydrocarbon flame zones. This is clearly seen in Figure 14 where $[N_1^+]_{max}$, the maximum concentrations of small (12–300 amu) positive ions arising in the flame zone (height $z = 8$ mm above the burner), and $[N_2^+]_{max}$, the peak concentrations of heavy (> 700 amu) ions occurring in the sooting region ($z \simeq 15$ mm), are compared for low-pressure acetylene- and ethylene-oxygen-nitrogen flames.[55] The value of $[N_1^+]_{max}$ is highest in lean, nonsooting flames and decreases steadily as the flame is made more fuel-rich. On the other hand, with the onset of sooting (which occurs at $\Phi \simeq 2.2$ for the acetylene series), $[N_2^+]_{max}$ rises rapidly to exceed $[N_1^+]_{max}$, increasing as the flame is made richer and the soot loading increases.

As a function of distance from the burner, the positive charge concentration in the sooting region first rises steeply to a maximum and then falls more

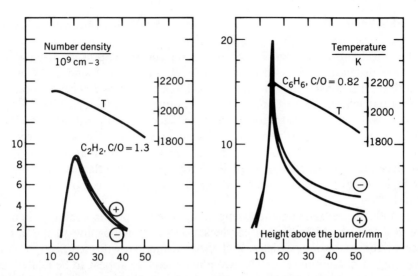

FIGURE 15. Ionization and temperature profiles in low-pressure (27 mbar) acetylene/ oxygen and benzene/oxygen flames having equal concentrations of soot in the burned gases. From Homann and Stroefer.[97]

gradually further downstream.[50,55,56,95] There is some discrepancy in the reported values of $[N_2^+]_{max}$—Homann[56] and Delfau et al.[55] find $[N_2^+] \simeq 5 \times 10^9$ typically, whereas Wersborg et al.[96] reported $[N_2^+] \simeq 10^{12}$ under similar conditions. The maximum concentration of these ions increases with increasing temperature at constant flame stoichiometry while soot yields decrease.[97] The nature of the fuel may also be important, as shown in Figure 15 where total ion concentrations in low-pressure acetylene- and benzene-oxygen flames are compared at conditions yielding the same final soot concentration.[97]

Mass spectra of the ions arising in the sooting region have been obtained by various electrical discrimination techniques.[49,50,55,56,96,98] The distribution of positive charge broadens and shifts to larger species—namely, from large hydrocarbon ions, to the smallest soot particles, to larger soot particles—as the height above the burner increases.[56] Although in the early regions there are many neutral molecules of the same masses as the charged species, the average size of the charged soot particles arising further downstream is slightly less than that of particles collected and observed under the electron microscope. The effect is particularly pronounced in benzene flames where all the charges reside on particles smaller than about 1.5 nm.[98,99]

Negatively charged species are also formed during the rapid ionization process early in the sooting region in about equal proportions with the positive ions.[56,98] However, the negative ions are bimodally distributed with respect to size, being mainly hydrocarbons of molecular weight about 180 and very small particles of molecular weights below 2000 (or nominal diameter less than 1.5 nm). The relative number of ions in the two modes varies in a complex fashion with the distance from the flame zone, but even well downstream no negative charges larger than about 1.5 nm occur.[98, 99]

The mechanism of formation of ions in the sooting region has been a point of some discussion. The very high levels of ionization found by Wersborg et al.[96] are not consistent with the thermal ionization of hydrocarbon molecules or soot particles in this region.[100] However, other investigations[55, 56, 97-99] show much lower ion concentrations, more in keeping with equilibrium levels. Nevertheless, ionization by thermal electron emission is a slow process unlikely to be important on the millisecond time scale of the ionization region.[101] The absence of charged soot particles from the benzene flames confirms that thermal ionization, even in these relatively hot flames, is not an important process.[98, 99]

In both the benzene and acetylene flames the second ion peak occurs in the region where the hydrocarbon condensation reactions leading to the appearance of the first soot particles, and their further growth (see Section 3) are occurring at maximum rates. The two phenomena do appear to be related, and this has led to speculation[48-50] that the ions provide the rapid mechanism of growth, largely by extension of the proposed ionic mechanism for the growth of PAH by $C_3H_3^+$ discussed in Section 2. This, however, leaves unanswered the question of mechanism of the second ion generation peak which, especially in the acetylene flames, occurs well past the primary reaction zone where the primary flame ions CHO^+ and $C_3H_3^+$ are formed. Also the number density of ions actually observed (ca. $5 \times 10^9 \, cm^{-3}$) appears to be substantially less than the particle number density in this early region ($> 10^{11} \, cm^{-3}$). Though it is true that concentration levels may not reflect rates of generation,[102] a purely ionic mechanism of particle generation would also have to account for this observation.

An alternative explanation for the second ionization peak is that it is itself a by-product of the surface additions of the hydrocarbons giving rise to the rapid particle growth. If, as seems possible, the growth species include acetylenic structures (Section 3.2), their addition reactions are likely to be strongly exothermic, leading perhaps to a kind of chemi-ionization.[56, 98]

The decrease in total ion concentration further downstream (Figure 15) is consistent with the recombination of the positive and negative charge carriers. The concentration of electrons in this region is unknown, but judging from Figure 15, it is not likely to be large. Even so, the rate of ion neutralization is sufficiently rapid to suggest some involvement by free electrons as charge carriers.[56]

In conclusion, although ions have excellent credentials as intermediates in the formation of soot nuclei, a number of important points remain unclear:

- There may, or may not, be insufficient ions present to account for the number of nuclei detected.
- The mechanism of formation of the second ion peak—crucial to any ionic mechanism—is not established.

• The mass distribution of the charged species, positive and negative, does not mirror that of the neutrals and depends on fuel type.

Coagulation and Agglomeration. In Figure 12 it is seen that from immediately beyond the particle formation zone the particle number density falls monotonically into the postflame zone. This phenomenon is a result of the collision, and sticking, of particles undergoing brownian motion and leads to an increase in the characteristic dimension of the particles without affecting the soot loading. Two types of collision can be identified: those in which two colliding (spherical) particles coalesce to form a single larger (spherical) particle, and those in which the particles stick together but do not fuse and so give rise to an aggregate in which the individual particles retain their identities. This occurs after about 4 ms in Figure 16, at which point the aggregate size measured by light scattering becomes larger than the spherical particle size measured by electron microscopy.[103]

The reason for this transition from coalescent to chain-forming collisions is not understood. The volume to be filled in to form spherical particles after noncoalescent collision is proportional to the total particle volume, so that a

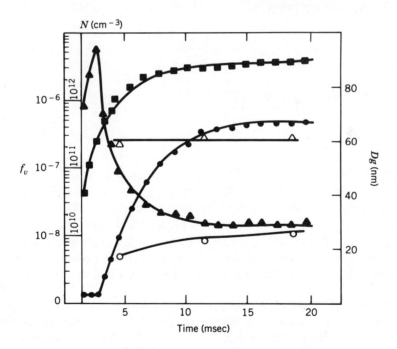

FIGURE 16. Profiles of soot volume fraction, f_v (■); number density, N, of agglomerates (▲) and of primary spheres (△); and diameter Dg of agglomerates (●) and of primary spheres (○) in a heavily sooting atmospheric-pressure/oxygen flame. From Prado et al.[103]

resymmetrization of the particles is much faster with smaller than with larger particles. In addition, surface growth rates are higher for small particles than for the chain-forming "old" particles, which are already tempered and exhibit generally reduced surface growth activity (see Section 3.2). The young soots are highly disordered and possess many radical sites, factors that may make the internal mobility of the younger particles significant.

Typically, it is found that particle collisions are effectively coalescent for particle diameters up to several tens of nanometers; beyond that size, they appear to be more chain forming.[41, 85, 103–106]

Theoretically, the decrease in particle number can be expected to occur according to the Smoluchowski equation:

$$\frac{dN}{dt} = -k(d)N^2 \tag{18}$$

where the rate constant k depends on the particle diameter d, or equivalently, on the mean volume $\bar{v}$. For particles small compared with gas mean free path (Knudsen number > 10), brownian motion, and k, are described in terms of free molecule theory; for soot particles in a typical flame environment, this requires that $d \gtrsim 60$ nm. For such a system in which uncharged spherical particles coalesce on every collision, this leads to a coagulation rate of[107–110]

$$\frac{dN}{dt} = -\frac{6}{5}k_{\text{theory}}f_v^{1/6}N^{11/6} \tag{19}$$

where

$$k_{\text{theory}} = \frac{5}{12}\left(\frac{3}{4\pi}\right)^{1/6} \cdot \left(\frac{6kT}{\rho}\right)^{1/2} \cdot G \cdot \alpha \tag{20}$$

and

> f_v is the particulate volume fraction,
>
> ρ is the density of the particles,
>
> G is a factor to take account of interparticle dispersion forces and can be expected to have a value of about 2 for spherical particles,[107, 108]
>
> α is a weak function of the particle size distribution, reflecting the variation in collision rates with different particle sizes. For monodisperse systems, $\alpha = 4\sqrt{2} = 5.66$; for the self-preserving size distribution, $\alpha = 6.55$.[108]

The process of coagulation leads to the establishment of a particle size distribution. For coalescent collisions in the absence of surface growth, a self-preserving size distribution is predicted.[107–110] This distribution is distorted and is no longer self-preserving when gas-to-particle conversion by surface growth also occurs. However, model computations have shown that this distortion is probably not great for typical soot aerosols,[63, 111, 112] and this has been confirmed experimentally.[63, 107, 111]

When the soot volume fraction f_v is constant, and coagulation alone determines particle growth, then for long reaction times ($N^{5/6}f_v^{1/6}k_{theory} \gg 1$, or $t \gtrsim 1$ msec for typical flame soots), the particle number becomes independent of the initial number N_0 and decreases as

$$N \sim \left(k_{theory}f_v^{1/6}t \right)^{-6/5} \qquad (21)$$

which shows a rather weak dependence on f_v (i.e., the number of particles is largely independent of the soot loading). Also, since k_{theory} is only a weak function of temperature, N at a given time should not be too strongly dependent on gas temperature. Here now is the explanation for the observation in Figure 12 that the particle number density curves in ethylene flames containing different amounts of soot, and in a benzene flame, are barely distinguishable. These measurements establish that coagulation is a physical process that does not depend on the nature of the parent molecules nor the manner in which they give rise to soot.

Ionizing metal additives such as caesium and potassium inhibit particle coagulation even where present at concentrations below 1 ppm.[113,114] Presumably, the presence of the metal promotes charging of the small soot particles which subsequently resist coagulation by coulombic repulsion. The actual yield of soot (f_v) is unchanged by the presence of the additives, but the particles are much smaller.

Surface Growth. In Figure 12, and similarly in Figure 16, the zone of new particle formation is narrow, giving rise to a rapid rise in the number of particles, up to a peak number density exceeding 10^{12} cm^{-3}. Although many particles are formed, they are very small, and the soot loading associated with them is relatively low. Only further downstream, where the number density is decreasing by coagulation, does the soot loading approach its final value. This is brought about by surface growth, whereby small gas-phase species attach to the surface of the particles and become incorporated into the soot structure.

It is evident, from the discussion of particle inception, that there is no clear chemical definition to the end of the nucleation stage and the appearance of the first soot particles. Similarly, because the particle inception processes are probably radical (or ion-molecule) additions of small, probably aliphatic, hydrocarbons to larger aromatic molecules, there can be no clear distinction between the end of mass addition by these processes and the onset of surface growth. However, in most systems the bulk of the soot mass is laid down in regions distinct from the zone of new particle formation.

The surface of hot soot particles readily accepts hydrocarbons from the gas-phase. This is clearly demonstrated by the appearance of macroscopic changes when soot particles are held (e.g., by electric fields[115] or by a stream of gas[41]) in the carbon formation zone. Tesner[89,116] pointed out that, for a given species, this can occur at rather lower temperatures than the generation of the particles themselves. Surface growth can also continue at hydrocarbon concentrations below the lower limit required for the inception of sooting.[85]

The high reactivity of the soot surface is such that the presence of soot can accelerate the decomposition of benzene[117] and acetylene[118] in pyrolysis. In methane pyrolysis the surface rate of deposition on soot particles is an order of magnitude higher than it is on alumina or graphite.[119]

There have been many studies, particularly in the Russian literature reviewed recently by Tesner,[89] of surface growth reactions. By and large, these studies pertain to pyrocarbon growth (well after the initial induction period in which the first few molecular layers are set down) rather than to soot particle growth. However, there are sufficient similarities that these results should be at least qualitatively applicable to the growth of soot particles as well. In general, acetylene and aromatics are more effective growth species than are the aliphatics; larger molecules are only slightly more effective than their smaller homologues. Thus in a flame environment simple hydrocarbons such as acetylene and methane should dominate surface growth reactions Tesner.[89] However, at least in the case of methane, it appears that higher, perhaps unsaturated hydrocarbons formed by pyrolysis are responsible.[119,120]

For their low-pressure acetylene/oxygen flame, Homann and Wagner[41] suggested that the main species being attached to the surface are acetylene and the polyacetylenes. More direct evidence for the role of acetylene as the major growth species in sooting atmospheric pressure flames has been reported recently by Harris and Weiner[92,121,122] who studied the kinetics of surface growth in a range of ethylene flames. As shown in Figure 17, the specific surface growth rate constant, k_s, defined in

$$\frac{df_v}{dt} = k_s[C_2H_2]S \tag{22}$$

where S is the aerosol surface, is found to be a function only of the reaction time in this series of flames for which the final soot volume fraction varies by more than an order of magnitude. The same behavior of k_s is obtained also for toluene flames.[92]

Figure 17 illustrates clearly a long-recognized[41] fact that surface growth rates evanesce at some point downstream of particle inception, even though the bulk gas environment is relatively unchanging. This phenomenon is usually attributed to the "tempering" or loss of reactivity by the particles as they age.[41]

It has recently been shown that the pseudo-first-order rate constant for the loss of surface growth reactivity, k_f, is constant as the particles age.[123-126] Here k_f is defined from

$$\frac{df_v}{dt} = -k_f(f_v^* - f_v) \tag{23}$$

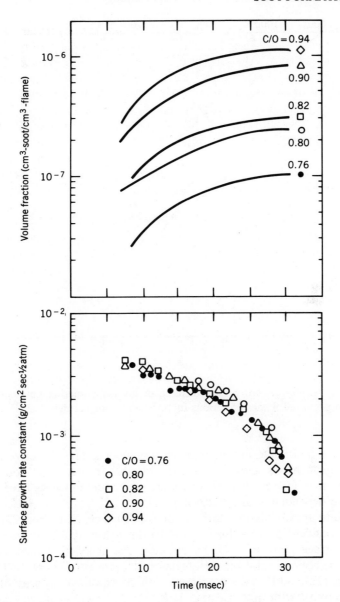

FIGURE 17. Surface growth rate constant, k_s (Eq. (22)) in ethylene/air flames producing a range of soot volume fractions. Redrawn from Harris and Weiner.[122]

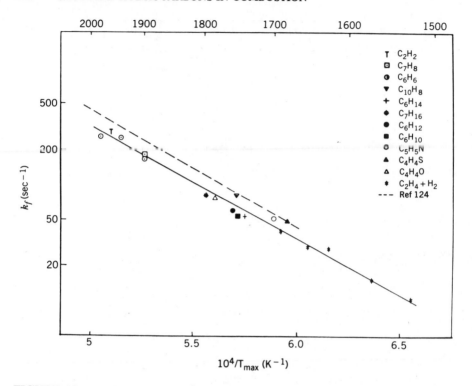

FIGURE 18. Pseudo-first-order rate constant, or inverse-tempering time constant, k_f (Eq. (23)) in premixed flames of various fuels. From Baumgaertner et al.[125]

where f_v^* is the ultimate soot volume fraction reached after effective cessation of surface growth. As shown in Figure 18, this inverse time constant for surface growth is independent of the nature of the fuel but does have a strong temperature dependence, about 200 kJ/mol.

To a crude approximation, the soot aerosol surface, S, in Eq. (22) is more or less independent of time (loss of surface through coagulation being largely offset by surface growth[112,121]). Furthermore the acetylene concentration in the postflame region is also relatively constant[121] (see also Figure 2). Therefore kinetic Eqs. (22) and (23) are consistent with an exponential decay of surface growth reactivity, with time constant $1/k_f$[127]

$$k_s = k_{s_0} \exp\left(-k_f t\right) \qquad (24)$$

Then the ultimate propensity to form soot, f_v^*, is

$$f_v^* = f_v \Big|_{\substack{\text{particle} \\ \text{inception}}} + \frac{k_{s_0} S [C_2H_2]}{k_f} \qquad (25)$$

The fraction of the total soot volume fraction contributed by particle inception is usually low, but the early flame-zone chemistry has a strong influence on $S = N_0 \pi d_0^2 / 4$, the total surface area of the first particles, and, to a lesser extent, on $[C_2H_2]$.[92,121,122] Therefore, although the particle inception process does not of itself produce much soot, it may have a substantial influence on the extent of surface formation through the generation of area. Earlier suggestions[124,128] based on results such as expressed by Eqs. (21) and (23), that particle inception and ultimate yield might be treated separately, are not confirmed by these results.

As discussed in Sections 2.2 and 3.2 the factors influencing the inception process are poorly understood. Obviously fuel type and presumably also pressure, temperature, and stoichiometry are important. Other aspects requiring further work in the area surface growth concern the mechanism of tempering and the confirmation of Eq. (22) over a wider range of conditions. The fact that ionizing additives lead to an increase in particle area apparently without affecting surface growth[113,124] is also unexplained.

Soot Oxidation. In all practical combustion systems the eventual emission of soot is much less than the peak soot loading achieved within the combustor. This difference is due to oxidation of the soot after it has been formed, left the fuel-rich environment, and encountered oxidizing conditions.

The oxidation of soot aerosol particles in a flame environment has proved rather difficult to follow experimentally. However, Radcliffe and Appleton[129] argued on the basis of structural similarities that the rates of oxidation of soot and of pyrographites should be the same. This is a considerable simplification in that, if care is taken to allow for diffusion resistance, studies of bulk samples of pyrographite may be used as a basis of our understanding of soot aerosol oxidation.

In the case of oxidation by O_2, the semiempirical Nagle and Strickland-Constable[130] formula has been shown to correlate graphite and soot oxidation over a wide range of temperatures and oxygen concentrations.[131]

$$\frac{\omega}{12} = \left(\frac{k_A P_{O_2}}{1 + k_Z P_{O_2}} \right) x + k_B P_{O_2} (1 - x) \quad \text{g-carbon/cm}^2 \text{ s} \qquad (26)$$

where x is given by

$$x = \left[1 + \frac{k_T}{k_B P_{O_2}} \right]^{-1} \qquad (27)$$

The various empirical rate coefficients for the model are listed in Table 8.

In the flame environment, about 10% of collisions of OH radicals with soot particles are effective in gasifying a carbon atom.[132-134] Such a high reaction probability means that the OH radical may dominate soot oxidation in flames

**TABLE 8. Empirical Rate Parameters for the Nagle and Strickland-Constable Model
($R = 1.987$ cal / mol K)[a]**

Rate Constant	Value	Units
k_A	$20 \exp(-30{,}000/RT)$	g-atom cm^{-2} sec^{-1} · atm^{-1}
k_B	$4.46 \times 10^{-3} \exp(-15{,}200/RT)$	g-atom cm^{-2} sec^{-1} · atm^{-1}
k_T	$1.51 \times 10^{-5} \exp(-97{,}000/RT)$	g-atom cm^{-2} sec^{-1}
k_Z	$21.3 \exp(4100/RT)$	atm^{-1}

[a]From Nagle and Strickland-Constable.[130]

even in the presence of significant concentrations of oxygen. This is confirmed in Figure 19 where experimental rates of soot oxidation (due to OH) are compared with those predicted for O_2 attack according to the formula of Nagle and Strickland-Constable.[130] Apparently, at all but the highest values of P_{O_2}, the contribution of O_2 to the oxidation of soot is minor. It appears, therefore, that the oxidation of soot in flames is brought about under fuel-rich and stoichiometric conditions by the OH radical, and under lean conditions by OH and O_2. In either case, soot oxidation rates can be expected to possess substantial apparent activation energies. The overall activation energy of the

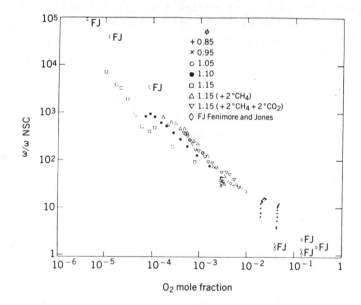

FIGURE 19. Measured soot oxidation rates relative to rate predicted by the Nagle and Strickland-Constable formula. The oxidation reaction is occurring in the burned gases of methane/oxygen/nitrogen flames of various stoichiometries, as indicated. Data of Fenimore and Jones[132] included for comparison. From Neoh et al.[134]

Nagle and Strickland-Constable formula is approximately that of k_A/k_Z or 140 kJ/mol, whereas the equilibrium OH concentration defined by $H_2O + 1/2O_2 \rightleftharpoons 2OH$ has a temperature dependence corresponding to 280 kJ/mol.

When substantial soot burnout occurs in the presence of O_2, there is a tendency for the soot aggregates to break up, leading to a sudden increase in the particle number density. Such behavior is not observed in fuel-rich oxidation (by OH). This phenomenon is attributed to the occurrence of internal burning by O_2—the more reactive OH is assumed at the surface of the aggregates.[135]

3.3. Premixed Flames

Given the complexity of the soot formation process, it is not surprising that much effort has been, and continues to be, devoted to the characterization of sooting propensity through simpler measures.

The Sooting Limit. The stoichiometry at which soot (luminosity) first appears in going from leaner flames to richer is called the *soot limit*. From a thermodynamic standpoint the first appearance of soot, as carbon, should occur when the mixture C/O atom ratio is one. However, the soot limit usually occurs at much lower values of C/O (frequently 0.5 to 0.6), and this critical C/O ratio is widely taken as a measure of the propensity of a fuel to form soot. The soot limit is also expressed in the literature in terms of the fuel/air equivalence ratio assuming complete combustion to CO_2 and H_2O (ϕ). The stoichiometric ratio relative to that required for partial combustion to CO and H_2 (ψ) is also used.

Some typical soot limits are listed in Table 9. However, it should be noted that the grading of the sooting tendency of different fuels depends to some extent on the measure of mixture stoichiometry at the soot limit. For example, the acetylene/air flame has generally been supposed to be anomalously resistant to sooting ($\phi_c = 2.0$ vs. 1.7 for ethane/air). However, compared in terms of Ψ_c, the sooting tendencies of acetylene and ethane are similar ($\Psi_c = 1.2$).[136]

Of the various means of expressing the sooting limit stoichiometry, the fuel-air equivalence ratio ϕ_c is of most direct practical significance. Calcote and colleagues[137-139] have recently proposed the use of a threshold soot index (TSI), defined as

$$TSI = a - b\phi_c \tag{28}$$

as a means of ranking sooting tendencies. The constants a and b are chosen for a given set of data to enable data obtained under different conditions to be compared. A plot of TSI versus carbon number is shown in Figure 20 for various classes of molecular structure.[139] Generally, TSI increases markedly with carbon number up to about six carbons and then more slowly for larger

TABLE 9. Soot Limits in Premixed Flames[a]

Species	Formula	ϕ_c	ψ_c	$(C/O)_c$
	Alkanes			
Ethane	C_2H_6	1.67	1.19	0.477
Propane	C_3H_8	1.91	1.34	0.573
Cyclopentane	C_5H_{10}	1.81	1.21	0.603
Cyclohexane	C_6H_{12}	1.81	1.21	0.603
Methylcyclohexane	C_7H_{14}	1.73	1.15	0.577
n-Heptane	C_7H_{16}	1.70	1.16	0.541
Ethylcyclohexane	C_8H_{16}	1.69	1.13	0.563
Cyclooctane	C_8H_{16}	1.84	1.23	0.613
Isooctane (2-2-4 trimethylpentane)	C_8H_{16}	1.80	1.22	0.576
	Alkenes			
Ethylene	C_2H_4	1.99	1.33	0.663
1-3 Butadiene	C_4H_6	1.69	1.08	0.615
Cyclopentene	C_5H_8	1.80	1.16	0.643
Cyclohexene	C_6H_{10}	1.69	1.09	0.596
1-Hexene	C_6H_{12}	1.84	1.22	0.613
4-Methylcyclohexene	C_7H_{12}	1.71	1.17	0.599
1-5 Cyclooctadiene	C_8H_{12}	1.89	1.20	0.687
	Alkynes			
Acetylene	C_2H_2	2.05	1.23	0.820
1-Pentyne	C_5H_8	1.89	1.22	0.675
1-Hexyne	C_6H_{10}	1.96	1.27	0.692
1-Heptyne	C_7H_{12}	1.96	1.27	0.686
2-Heptyne	C_7H_{12}	1.84	1.20	0.644
1-Octyne	C_8H_{14}	1.78	1.16	0.619
1-Decyne	$C_{10}H_{18}$	1.88	1.23	0.648
	Aromatics			
Benzene	C_6H_6	1.63	0.99	0.650
Toluene	C_7H_8	1.52	0.93	0.591
Styrene	C_8H_8	1.68	1.01	0.672
Ethylbenzene	C_8H_{10}	1.55	0.96	0.590
m-Xylene	C_8H_{10}	1.51	0.93	0.575
p-Xylene	C_8H_{10}	1.50	0.93	0.571
o-Xylene	C_8H_{10}	1.47	0.91	0.560
1-Phenyl-1-propyne	C_9H_8	1.53	1.04	0.626
Indene	C_9H_8	1.48	0.87	0.605
Tetralin	$C_{10}H_{12}$	1.48	0.91	0.569

[a]From Haynes and Wagner[87] and Olson and Pickens.[139]

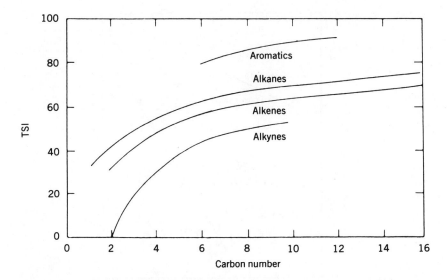

FIGURE 20. Representative TSI values for various classes of fuel structure in pre-mixed flames. From Olson and Pickens.[139]

molecules. In order of increasing TSI, the tendency of the various fuel types to soot is alkyne < alkene < alkane < aromatics. Alkylation of the aromatics has only a minor influence on this group.[139] A mixing rule for multicomponent TSI has been proposed.[138]

The effect of flame structure on the soot limit is generally slight, the limits of soot formation being similar in back-mixed combustion and in flat flames. For heavier fuels (with molecular weights about twice that of air, and above), the soot limit occurs at somewhat leaner conditions with bunsen-type flames than with flat flames—the explanation for this phenomenon lies in the unequal diffusion rates (magnitude and direction) of fuel and oxidant induced by curved concentration gradients in the bunsen-type flame.

The critical C/O ratio is only weakly dependent on pressure, as is consistent with the weak effects of back-mixing. Increasing temperatures generally allow slightly higher mixtures to be burned without the appearance of soot, but the effect is not marked, being less than 20% for a temperature change from 1800 to 2500 K.[136, 140] Some additives also appear to influence the critical C/O ratio—such effects are discussed in Section 3.6.

The Soot Yield. In premixed flames richer than the carbon (soot) limit, the yield of soot in premixed flames increases as the surface growth process becomes significant, and soon dominant, in determining the mass of soot. As shown in Figure 21, the final soot yield in a series of approximately isothermal

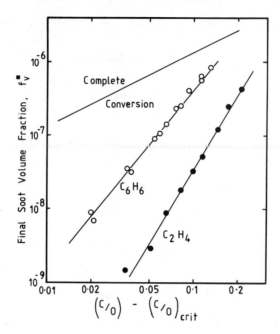

FIGURE 21. Soot volume fraction $f_v{}^*$ as a function of C/O ratio in excess of that required just to give soot in premixed flames at atmospheric pressure. From Haynes.[142]

(± 50 K) flames can be correlated as

$$f_v \sim \left\{ \left(\frac{C}{O} \right) - \left(\frac{C}{O} \right)_{\text{critical}} \right\}^n \tag{29}$$

where n is about 3.[65,125,141,142] If the increase in $f_v{}^*$ with C/O ratio is due largely to the strength of the particle formation process as a source of growth area (see Section 3.2), this indicates a high sensitivity of this process to the flame stoichiometry. The proportionality constant in the above expression varies markedly with fuel type, but even for benzene flames the fractional conversion to soot of incremental fuel carbon just beyond the critical C/O ratio is very low.

Increasing temperature in the range of 1700 to 2000 K at constant equivalence ratio causes a strong decrease in the final soot volume fraction.[87,124,143] In Eq. (25) k_f has an apparent activation energy of 200 kJ/mol—given that the temperature dependence of $f_v{}^*$ probably exceeds this amouint,[124,142] the particle formation process therefore appears weaker as a source of growth area as the temperature rises. The ultimate soot yield also increases with increasing pressure[87] possibly as condensation reactions such as surface growth are favored.

Conclusions. Despite the wealth of data available on soot limits, some caution must be exercised in translating this information to more general consideration of soot formation. The amount of soot formed is, in contrast to the sooting limit, a strong function of temperature and pressure, although different fuels produce substantially different amounts of soot for equal fuel increments beyond the critical ratio.

The influence of temperature and pressure on soot yield is not well characterized and is less well understood. Equation (25) may be a useful starting point for further work in this area.

3.4. Sooting Structure of Laminar Diffusion Flames

There have been a number of studies of the sooting structure of laminar diffusion flames[66, 74, 75, 144-148] (see also Haynes and Wagner[87]). It is generally observed that the sooting region begins just a few millimeters to the fuel side of the stoichiometric fuel-air interface (the flame zone, where maximum temperatures are achieved) and penetrates a further few millimeters into the fuel. For an axisymmetric burner, soot does not reach the axis until some distance downstream of the burner port, as shown in Figures 9 and 22.

Chemically, these flames are not well characterized with regard to soot formation. The results of Figures 9, 10, and 22 indicate that the hot pyrolyzing mixture on the fuel-side of the flame exhibits many of the features of the pre-sooting and sooting gases in laminar premixed flames. Thus, in the sooting region, high concentrations of acetylene are accompanied by small amounts of benzene and much lower concentrations of PAH. All these species begin to disappear as conditions approach stoichiometric. However, the soot persists longer on the centerline, indicating some kinetic control in burnout rates.

Soot Distribution. Laser light scattering and extinction measurements have been used recently to provide details on the soot volume and number distribution in laminar diffusion flames.[66, 145-148] Profiles for f_v, N (and d) across a flat Wolfhard-Parker diffusion flame are shown for the early soot-forming region in Figure 23. The particle number density N is highest nearest the reaction zone where temperatures and radical and ion concentrations are high, much as in the primary reaction and immediate postflame zones of premixed sooting flames. Peak rates of particle generation in this region appear to be of the order 10^{14} to 10^{15} cm^{-3} sec^{-1}.[146]

An important structural feature of all typical hydrocarbon/air flames now comes into play to limit new particle formation to a narrow zone fairly close to the main reaction zone. Because one volume of fuel requires more than one volume of oxidant to complete combustion, the stoichiometric fuel-air interface moves outward from the dividing streamline (that streamline emanating from the fuel-air partition of the burner) into the air side.[149] Streamlines from the air-side passing through the flame zone therefore see increasing fuel

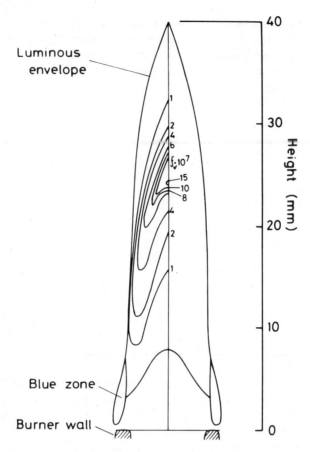

FIGURE 22. Distribution of soot volume fraction in the *n*-hexane diffusion flame of Figures 8 and 9. Redrawn from Spengler and Kern,[74] with permission from the publisher.

concentration and lower temperatures, and radical concentrations at greater distances from the burner.*

Thus, the new particles following the streamlines (their diffusivities are negligible) soon move into regions too unreactive to promote further significant new particle formation. This effect is reinforced by thermophoresis, which drives the particles down the temperature gradient deeper into the cooler fuel-rich interior of the flame. Because there is practically no dispersion of the particles, their depth of penetration into the interior at a given height above the burner is a measure of their age, those having been formed earliest penetrating farthest.

---

*If the situation were such that one volume of fuel should require less than one volume of oxidizer, streamlines would move from the fuel side to the air side—particles formed on the fuel side would immediately move into oxidizing regions, and very little soot would be formed. This may, in part, explain the suppression of soot in highly diluted fuels.

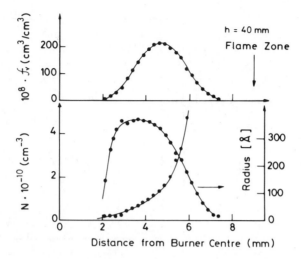

FIGURE 23. Soot loading, particle size, and number density profiles across the sooting region of a laminar ethylene/air diffusion flame established on a Wolfhard-Parker burner. The particle number density increases monotonically to exceed 10^{12} cm^{-3} at the edge of the sooting region nearer the air side. The fuel-air partition in the burner occurs at 3 mm from the center. The arrow marked "flame zone" indicates the position of the stoichiometric fuel-air interface at this height. From Haynes and Wagner.[66]

From the moment of their inception the particles are undergoing coagulation to decrease their number density. For a given height above the burner, particles farthest from their point of origin have penetrated deepest into the fuel and have had the greatest time for coagulation so that their number density is lowest. Conversely, nearest the particle generation zone, the particle number density is highest.

Although new particle formation ceases away from the flame zone, there is a distributed region of surface growth activity in the flame, extending from the particle generation zone into the fuel side to the point where temperatures are too low even for surface growth to occur or to where surface growth activity ceases as a result of particle tempering. In keeping with the diffusional structure of the flame, this growth region is narrower near the burner and broader farther downstream.

The soot loading occurring at a given point reflects the surface growth history of the particles—namely, qualitatively, the amount of time spent within the growth region. Thus, at a given height above the burner, the soot loading goes through a maximum. Particles near the generation zone are younger and have spent less time in the growth region. Particles further into the fuel region are those formed early in the flame where the transverse gradients are steeper—these particles have traversed quickly the narrow growth region and always seen cool unreactive regions thereafter. Thus they undergo

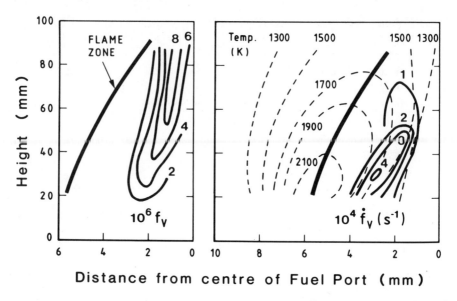

FIGURE 24. (*a*) Contours of soot volume fraction in an ethylene/air flat diffusion flame. (*b*) Temperature and soot volume fraction generation rate contours in the same flame. Redrawn from Kent and Wagner.[147]

very little surface growth and the soot loading associated with them remains small.

The location of the surface growth region in an overventilated flame (i.e., one burning in an excess of the stoichiometric air requirement) is shown in Figure 24.[146] The peak rates of soot mass addition are seen to occur near the base of the flame in fuel-rich regions, with temperatures in the range 1500 to 1700 K. Farther downstream the rates drop off, even in regions of temperature in the optimum range—presumably, surface growth rates drop as the hydrocarbons are consumed and diluted by-products and the particles begin moving into increasingly leaner regions as the stoichiometric interface moves back towards the centerline. Eventually, the particles approach and even pass through this interface—now, finally, rapid particle oxidation can begin.

However, as discussed in Section 3.2, soot oxidation rates are highly temperature dependent. Therefore, as shown recently by Kent and Wagner,[150,151] the ultimate fate of the soot depends entirely on the temperatures occurring in this region. If temperatures drop too low here (i.e., less than about 1300 K), soot burnout virtually ceases, and the flame emits smoke.

At higher pressures, the soot-forming regions are narrower but otherwise similar to those at atmospheric pressure. The soot loading increases with increasing pressure, corresponding approximately to a dependence of soot yield on pressure as $p^{0.7}$.[152]

Smoke Point. The "smoke point" is used as a global measure of the sooting tendency of a laminar diffusion flame, representing the net effects of formation and burnout integrated through the flame. It is determined experimentally by increasing the fuel flow and hence flame height until the flame begins to emit black smoke and is reported either as the minimum fuel flow or flame height at which this occurs.

For pure fuels with air, the sooting propensity in diffusion flames is found in terms of the smoke point to increase as

$$\text{paraffins} < \text{olefins} \ll \text{acetylene} < \text{benzene} < \text{naphthalene}^{153}$$

which, for the aliphatics, is quite a different ranking from that obtained with premixed flames. However, comparisons made at constant calculated adiabatic flame temperatures yield a somewhat different ranking, especially for acetylene which now behaves more like other aliphatics.[154] On the other hand, the apparently anomalous behavior of acetylene may result from the influence of stoichiometric requirements on the flame height.[137]

There have been numerous attempts to divine the influence of parameters such as temperature on the sooting behavior of laminar diffusion flames by observing changes in the smoke point as the fuel and/or oxidizer are diluted or enriched.[87,154-157] However, as discussed above, the regions where soot is formed and oxidized are distributed but distinct so that characterization of the entire flame in terms of one calculated adiabatic or other characteristic flame temperature must be incomplete.

This is shown explicitly in Figure 25 where the maximum centerline and soot temperatures obtained in a series of ethylene flames in air are plotted against the normalized flame height.[151] The normalizing procedure is not important in the present context, but it should be noted that the true diffusion flame height occurs in all cases at $Z \sim 0.6$[144,151]—in this region the peak soot yield is reached, and the three temperature measurements in each flame converge.

The larger flames (higher fuel flow rate) in Figure 25 lose relatively less heat to the burner and are therefore somewhat hotter at low heights. This in turn appears to be accompanied by higher soot formation rates here.[151,158] However, the combination of more soot and longer residence times in larger flames leads to disproportionately greater radiation losses so that, farther downstream, the largest flames cool faster and reach earlier the critical temperature for the quenching of soot oxidation (ca. 1300 K).[150]

In these terms the soot temperature in the burnout region is controlled by radiative heat losses from the soot formed. In accordance with this, the total heat loss from a flame of any fuel at the smoke point has been found to be always about 30% of the combustion heat release.[159] Furthermore Kent[160] has proposed a quantitative relationship between soot yield and smoke point

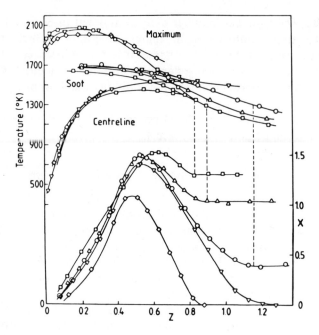

FIGURE 25. Temperature and normalized section-average soot volume fraction (X) profiles as a function of normalized flame height (Z) in ethylene/air diffusion flames. The maximum temperature occurs at the stoichiometric fuel-air interface. From Kent and Wagner.[151]

measurements, based on similarity modeling of soot formation profiles, flame shape, and radiant emission.

3.5. Turbulent Diffusion Flames

The mean distribution of soot within turbulent gaseous diffusion flames is very similar to that within their laminar counterparts.[161-165] This is true also for turbulent spray flames.[78-80,166] Thus soot is produced in fuel-rich regions where concentrations of hydrocarbons are high, whereas soot burnout occurs in downstream regions.[164]

Turbulent gaseous flames are distinguished by the fact that they emit smoke at low flows and stop emitting smoke as fuel flow increases. Nevertheless, the onset of smoking in turbulent flames appears to be controlled by the temperature history of the flame, and particularly in the burnout region, much as in laminar flames.[150,164,165] Because soot concentration and flame size are independent of nozzle velocity for moderately turbulent flames on a given nozzle,[164,165] radiation losses from these flames remain constant as the fuel flow is reduced. Therefore temperatures fall, particularly in the burnout region, and smoke is emitted.[165]

Although the emission of smoke is kinetically controlled by temperature, measurements of soot concentration fluctuations in the burnout region[161,163] have led to the suggestion that this region, prior to the freezing of the reaction as temperatures fall, is diffusion controlled. The largest fuel-rich eddies resist dissipation longest and so determine burnout rates.[161,165]

Preheating of the fuel in a gaseous turbulent flame leads to substantially higher soot formation rates.[164] The effect on burnout depends then on the resultant temperatures in that region—that is, on whether the increased soot loadings lead to heat losses exceeding the preheating effect. Fuel preheating may therefore inhibit smoke emission from faster flames and favor smoking in slower flames.[162,164]

The influence of fuel diluents in the formation zone appears to be largely thermal, leading to a reduction in local soot concentrations.[162,165]

3.6. Influence of Additives

There have been many studies of the influence of additives on soot formation.[87,167] Unfortunately, many of these were carried out on poorly defined systems, and not surprisingly, a host of theories has arisen.

Ionizing metal additives can cause soot particles to be charged and so inhibit the coagulation of these particles through their mutual coulombic repulsion.[113,114] The ultimate soot volume fraction remains largely unaffected by these additives. Barium can also ionize enough to inhibit coagulation, but this metal also influences the total soot yield—the mechanism for this latter effect is not clear but has been suggested to involve ionic nucleation or catalysis of soot oxidation and gasification. Similar results have been obtained with these metallic additives in diffusion flames.[168]

A number of gaseous additives such as SO_3, SO_2, H_2S, NH_3, H_2O, H_2, and N_2 have been studied for their effects on soot limits[84] and soot yield[141] in premixed flames, and on flame height in laminar diffusion flames.[169] The influence of all these gases, with the possible exception of SO_3, is to decrease the soot yield in premixed flames (roughly in descending order of effectiveness as $SO_3, H_2S, SO_2 > NH_3 > H_2O > H_2, N_2$), without affecting the surface growth tempering rate or the coagulation behavior. Results for the influence of SO_3 are conflicting—although Haynes et al.[141] found SO_3 to behave much like the other sulphur additives, earlier reports[84,87] indicate strong pro-soot effects with this additive, possibly due to fuel degradation by SO_3 in the unburned gases.[84]

In the diffusion flames inert additives (SO_2, CO_2, H_2, H_2O, N_2 He, etc.) are found to reduce the tendency to smoke. All of the additives are equally effective when considered in terms of their heat capacities, thus implying a thermal mechanism for their behavior, such as a reduction in the rate of pyrolysis of the fuel at lower temperatures.[169]

Other additives in the diffusion flame have pronounced pro-soot effects. These are all oxidants such as chlorine-containing compounds, oxygen, or

N_2O. In these cases enhanced pyrolysis of the fuel is assumed to occur in the presence of the additive.[169]

3.7. Conclusions

The formation and emission of soot in combustion systems involves many factors. Consideration of the problem in terms of the overall processes,

1. particle inception
2. coagulation
3. surface growth
4. oxidation

provides a framework for understanding the influence of these factors.

REFERENCES

1. F. L. Dryer, Chapter 3 of this volume.
2. G. McKay, *Prog. Energy Combust. Sci.* **3**, 105–126 (1977).
3. J. Warnatz, *Symp. (Int.) Combust. [Proc.]* **18**, 369–384 (1981).
4. J. Warnatz, H. Bockhorn, A. Moeser, and H. W. Wenz, *Symp. (Int.) Combust. [Proc.]* **19**, 197–209 (1982).
5. J. Warnatz, in *Combustion Chemistry* (W. C. Gardiner, Jr., ed.), pp. 197–360. Springer-Verlag, New York, 1984.
6. J. Warnatz, *Symp. (Int.) Combust. [Proc.]* **20**, 845–856 (1985).
7. J. A. Miller, R. E. Mitchell, M. D. Smooke, and R. J. Kee, *Symp. (Int.) Combust. [Proc.]* **19**, 181–196 (1982).
8. J. Bonne, K. H. Homann, and H. G. Wagner, *Symp. (Int.) Combust. [Proc.]* **10**, 503–512 (1965).
9. J. D. Bittner, "A Molecular Beam Mass Spectrometer Study of Fuel-Rich and Sooting Benzene—Oxygen Flames." Sc.D. Thesis, Massachusetts Institute of Technology, Cambridge, 1981.
10. J. D. Bittner and J. B. Howard, *Symp. (Int.) Combust. [Proc.]* **18**, 1105–1116 (1981).
11. J. D. Bittner and J. B. Howard, in *Particulate Carbon: Formation during Combustion* (D. C. Siegla and G. W. Smith, eds.), pp. 109–142. Plenum, New York, 1981.
12. J. D. Bittner and J. B. Howard, *Symp. (Int.) Combust. [Proc.]* **19**, 211–221 (1982).
13. J. D. Bittner, J. B. Howard, and H. B. Palmer, in *Soot in Combustion Systems and Its Toxic Properties* (J. Lahaye and G. Prado, eds.), pp. 95–125. Plenum, New York, 1983.
14. P. H. Kydd, *Combust. Flame* **3**, 133–142 (1959).
15. J. M. Levy, B. R. Taylor, J. P. Longwell, and A. F. Sarofim, *Symp. (Int.) Combust. [Proc.]* **19**, 167–1792 (1982).

16. K. H. Homann, M. Mochizuki, and H. G. Wagner, *Z. Phys. Chem.* [N.S.] 37, 299–313 (1963).

17. K. H. Homann, *Combust. Flame* 11, 265–287 (1967).

18. K. H. Homann, W. Morgeneyer, and H. G. Wagner, in *Combustion Institute European Symposium* (F. J. Weinberg, ed.), pp. 394–399. Academic Press, London, 1973.

19. C. Venkat, K. Brezinsky, and I. Glassman, *Symp. (Int.) Combust. [Proc.]* 19, 143–152 (1982).

20. C. T. Brooks, S. J. Peacock, and B. G. Reuben, *J. Chem. Soc., Faraday Trans. I* 75, 652–662 (1979).

21. R. D. Smith, *J. Phys. Chem.* 83, 1553–1563 (1979).

22. R. D. Smith, *Combust. Flame* 35, 179–190 (1979).

23. R. D. Smith and A. L. Johnson, *Combust. Flame* 51, 1–22 (1983).

24. D. J. Patterson and N. A. Henein, *Emissions from Combustion Engines and Their Control.* Ann Arbor Science Publ., Ann Arbor, Michigan, 1972.

25. J. B. Heywood, *Prog. Energy Combust. Sci.* 1, 135–164 (1976).

26. P. Bergner, H. Eberius, and H. Pokorny, *Proc. Alc. Fuels Technol. Int. Symp., 3rd*, Asilomar, California, May, *1979*.

27. A. A. Adamczyk, E. W. Kaiser, J. A. Cavalowsky, and G. A. Lavoie, *Symp. (Int.) Combust. [Proc.]* 18, 1695–1702 (1981).

28. J. A. Lorusso, E. W. Kaiser, and E. A. Lavoie, *Combust. Sci. Technol.* 25, 121–125 (1981).

29. C. K. Westbrook, A. A. Adamczyk, and G. A. Lavoie, *Combust. Flame* 40, 81–99 (1981).

30. E. W. Kaiser, A. A. Adamczyk, and G. A. Lavoie, *Symp. (Int.) Combust. [Proc.]* 18, 1881–1890 (1981).

31. E. W. Kaiser, J. A. Lorusso, G. A. Lavoie, and A. A. Adamczyk, *Combust. Sci. Technol.* 28, 69–73 (1982).

32. T. R. Barfknecht, *Prog. Energy Combust. Sci.* 9, 199–237 (1983).

33. J. P. Longwell, *Symp. (Int.) Combust. [Proc.]* 19, 1339–1350 (1982).

34. D. A. Kaden, R. A. Hites, and W. G. Thilly, *Cancer Res.* 39, 4152–4160 (1979).

35. J. P. Longwell, in *Soot in Combustion Systems and Its Toxic Properties* (J. Lahaye and G. Prado, eds.), pp. 37–56. Plenum, New York, 1983.

36. D. Hoffmann and E. L. Wynder, in *Air Pollution* (A. C. Stein, ed.), 3rd ed., Vol. II. Academic Press, New York, 1977.

37. D. F. S. Natusch, *Environ. Health Perspect.* 22, 79–90 (1980).

38. W. A. Korfmacher, D. F. S. Natusch, D. R. Taylor, G. Mamantov, and E. L. Wehry, *Science* 207, 763–765 (1980).

39. M. L. Lee and K. D. Bartle, in *Particulate Carbon: Formation during Combustion* (D. C. Siegla and G. W. Smith, eds.), pp. 91–106. Plenum, New York, 1981.

40. G. L. Fisher and D. F. S. Natusch, in *Analytical Methods for Coal and Coal Products*, (C. J. Karry, ed.), Vol. 3, pp. 489–542. Academic Press, New York, 1979.

41. K. H. Homann and H. G. Wagner, *Symp. (Int.) Combust.* [*Proc.*] **11**, 371–379 (1967).

42. J. A. Cole, J. D. Bittner, J. P. Longwell, and J. B. Howard, *Combust. Flame* **56**, 51–70 (1984).

43. H. Bockhorn, F. Fetting, and H. W. Wenz, *Ber. Bunsenges. Phys. Chem.* **87**, 1067–1073 (1983).

44. B. D. Crittenden and R. Long, *Combust. Flame* **20**, 359–368 (1973).

45. A. D'Alessio, A. F. Sarofim, F. Beretta, S. Masi, and C. Venitozzi, *Symp. (Int.) Combust.* [*Proc.*] **15**, 1427–1438 (1975).

46. A. DiLorenzo, A. D'Alessio, V. Cincotti, S. Masi, P. Menna, and C. Venitozzi, *Symp. (Int.) Combust.* [*Proc.*] **18**, 485–491 (1981).

47. G. M. Badger, G. E. Lewis, and I. M. Napier, *J. Chem. Soc.* pp. 2825–2827 (1960).

48. H. F. Calcote, *Combust. Flame* **42**, 215–242 (1981).

49. D. B. Olson and H. F. Calcote, *Symp. (Int.) Combust.* [*Proc.*] **18**, 453–464 (1981).

50. D. B. Olson and H. F. Calcote, in *Particulate Carbon: Formation during Combustion* (D. C. Siegla and G. W. Smith, eds.), pp. 177–206. Plenum, New York, 1981.

51. A. N. Hayhurst and H. R. N. Jones, *Symp. (Int.) Combust.* [*Proc.*] **20**, 1121–1218 (1985).

52. S. E. Stein, *J. Phys. Chem.* **82**, 566–571 (1978).

53. S. E. Stein, *Combust. Flame* **51**, 357–364 (1983).

54. P. Michaud, J. L. Delfau, and A. Barassin, *Symp. (Int.) Combust.* [*Proc.*] **18**, 443–451 (1981).

55. J. L. Delfau, P. Michaud, and A. Barassin, *Combust. Sci. Technol.* **20**, 165–177 (1979).

56. K. H. Homann, *Ber. Bunsenges. Phys. Chem.* **83**, 738–745 (1979).

57. J. B. Howard and J. D. Bittner, in *Soot in Combustion Systems and Its Toxic Properties* (J. Lahaye and G. Prado, eds.), pp. 57–93. Plenum, New York, 1983.

58. R. Atkinson and J. N. Pitts, *J. Phys. Chem.* **79**, 295–297 (1975).

59. R. A. Perry, R. Atkinson, and J. N. Pitts, *J. Phys. Chem.* **81**, 296–304 (1977).

60. K. Lorenz and R. Zellner, *Ber. Bunsenges. Phys. Chem.* **87**, 629–633 (1983).

61. D. S. Coe, B. S. Haynes, J. I. Steinfeld, and A. F. Sarofim, unpublished work.

62. K. F. Lang, H. Buffleb, and M. Zander, *Erdoel Kohle, Erdgas, Petrochem.* **16**, 944–946 (1963).

63. H. Bockhorn, F. Fetting, G. Wannemacher, and H. W. Wenz, *Symp. (Int.) Combust.* [*Proc.*] **19**, 1413–1420 (1982).

64. A. D'Alessio, A. DiLorenzo, A. Borghese, F. Beretta, and S. Masi, *Symp. (Int.) Combust.* [*Proc.*] **16**, 695–708 (1977).

65. B. S. Haynes, H. Jander, and H. G. Wagner, *Ber. Bunsenges. Phys. Chem.* **84**, 585–592 (1980).

66. B. S. Haynes and H. G. Wagner, *Ber. Bunsenges. Phys. Chem.* **84**, 499–506 (1980).

67. J. H. Miller, W. G. Mallard, and K. C. Smyth, *Combust. Flame* **47**, 205–214 (1982).

68. A. D'Alessio, in *Particulate Carbon: Formation during Combustion* (D. C. Siegla and G. W. Smith, eds.), pp. 207–259. Plenum, New York, 1981.

69. D. S. Coe, B. S. Haynes, and J. Steinfeld, *Combust. Flame* **43**, 211–214 (1981).

70. F. Beretta, A. Cavaliere, and A. D'Alessio, *Symp. (Int.) Combust. [Proc.]* **19**, 1359–1367 (1982).

71. G. Prado, A. Garo, A. Ko, and A. F. Sarofim, *Symp. (Int.) Combust. [Proc.]* **20**, 989–996 (1985).

72. D. S. Coe and J. Steinfeld, *A. C. S. Symp. Ser.* **134**, 247–255 (1980).

73. D. S. Coe and J. Steinfeld, *Chem. Phys. Lett.* **76**, 485–488 (1980).

74. G. Spengler and J. Kern, *Brennst.-Chem.* **50**, 321–324 (1969).

75. J. Kern and G. Spengler, *Erdoel Kohle, Erdgas, Petrochem.* **23**, 813–817 (1970).

76. S. R. Smith and A. S. Gordon, *J. Phys. Chem.* **60**, 759–763 (1956).

77. B. B. Chakkaborty and R. Long, *Combust. Flame* **12**, 226–236 (1968).

78. G. Prado, M. L. Lee, R. A. Hites, D. P. Hoult, and J. B. Howard, *Symp. (Int.) Combust. [Proc.]* **16**, 649–661 (1977).

79. A. Ciajolo, R. Barbella, M. Mattiello, and A. D'Alessio, *Symp. (Int.) Combust. [Proc.]* **19**, 1369–1377 (1982).

80. M. Toqan, W. F. Farmayan, J. M. Beer, J. B. Howard, and J. D. Teare, *Symp. (Int.) Combust. [Proc.]* **20**, 1075–1081 (1985).

81. A. Herlan, *Combust. Flame* **31**, 297–307 (1978).

82. M. L. Lee, G. Prado, J. B. Howard, and R. A. Hites, *Biomed. Mass Spectrom.* **4**, 182–185 (1977).

83. J. Lahaye and G. Prado, in *Particulate Carbon: Formation during Combustion* (D. C. Siegla and G. W. Smith, eds.), pp. 33–55. Plenum, New York, 1981.

84. J. C. Street and A. Thomas, *Fuel* **34**, 4–36 (1955).

85. H. B. Palmer and C. F. Cullis, *Chem. Phys. Carbon* **1**, 265–325 (1965).

86. J. Lahaye and G. Prado, *Chem. Phys. Carbon* **14**, 168–294 (1978).

87. B. S. Haynes and H. G. Wagner, *Prog. Energy Combust. Sci.* **7**, 229–273 (1981).

88. O. I. Smith, *Prog. Energy Combust. Sci.* **7**, 275–291 (1981).

89. P. A. Tesner, *Comb. Exp. Shockwaves* **15**, 111–120 (1979).

90. B. L. Wersborg, J. B. Howard, and G. C. Williams, *Symp. (Int.) Combust. [Proc.]* **14**, 929–940 (1973).

91. K. C. Smyth and W. G. Mallard, *Combust. Sci. Technol.* **26**, 35–41 (1981).

92. S. J. Harris and A. M. Weiner, *Combust. Sci. Technol.* **38**, 75–87 (1984).

93. H. F. Calcote, in *Ion-Molecule Reactions* (J. L. Franklin, ed.), Vol. 2, pp. 673–706. Plenum, New York, 1972.

94. R. S. Tse, P. Michaud, and J. L. Delfau, *Nature (London)* **272**, 153–155 (1978).

95. A. N. Hayhurst and H. R. N. Jones, *Nature (London)* **296**, 61–63 (1982).

96. B. L. Wersborg, A. L. Yeung, and J. B. Howard, *Symp. (Int.) Combust. [Proc.]* **15**, 1439–1448 (1975).

97. K. H. Homann and E. Stroefer, in *Soot in Combustion Systems and Its Toxic Properties* (J. Lahaye and G. Prado, eds.), pp. 217–242. Plenum, New York, 1983.

98. K. H. Homann and H. Wolf, *Ber. Bunsenges. Phys. Chem.* **87**, 1073–1077 (1983).

99. K. H. Homann, E. Stroefer, and H. Wolf, *AGARD Conf. Proc.* **353**, Pap. 19 (1984).

100. G. Prado and J. B. Howard, *Adv. Chem. Ser.* **166**, 153–166 (1978).

101. D. R. Hardesty and F. J. Weinberg, *Symp. (Int.) Combust.* [*Proc.*] **14**, 907–918 (1973).

102. H. F. Calcote, in *Soot in Combustion Systems and Its Toxic Properties* (J. Lahaye and G. Prado, eds.), pp. 197–215. Plenum, New York, 1983.

103. G. P. Prado, J. Jagoda, K. Neoh, and J. Lahaye, *Symp. (Int.) Combust.* [*Proc.*] **18**, 1127–1136 (1981).

104. S. C. Graham, J. B. Homer, and J. L. J. Rosenfeld, *Mod. Dev. Shock Tube Res., Proc. Int. Shock Tube Symp., 10th, 1975*, pp. 621–631 (1975).

105. S. C. Graham, J. B. Homer, and J. L. J. Rosenfeld, in *Deuxieme symposium Européen sur la combustion*, pp. 374–379. Combustion Institute, Lyons, 1975.

106. S. C. Graham, *Symp. (Int.) Combust.* [*Proc.*] **16**, 663–669 (1977).

107. S. C. Graham, J. B. Homer, and J. L. J. Rosenfeld, *Proc. R. Soc. London, Ser. A.* **344**, 259–285 (1975).

108. S. C. Graham and A. Robinson, *J. Aerosol Sci.* **7**, 261–273 (1976).

109. G. D. Ulrich, *Combust. Sci. Technol.* **4**, 47–57 (1971).

110. F. S. Lai, S. K. Friedlander, J. Pich, and G. M. Hidy, *J. Colloid Interface Sci.* **39**, 395–405 (1972).

111. H. Bockhorn, F. Fetting, U. Meyer, R. Reck, and G. Wannemacher, *Symp. (Int.) Combust.* [*Proc.*] **18**, 1137–1147 (1981).

112. P. H. McMurray and S. K. Friedlander, *J. Colloid Interface Sci.* **64**, 248–257 (1978).

113. B. S. Haynes, H. Jander, and H. G. Wagner, *Symp. (Int.) Combust.* [*Proc.*] **17**, 1365–1374 (1979).

114. B. S. Haynes, H. Jander, H. Maetzing, and H. G. Wagner, *Combust. Flame* **40**, 101–103 (1981).

115. E. R. Place and F. J. Weinberg, *Symp. (Int.) Combust.* [*Proc.*] **11**, 245–255 (1967).

116. P. A. Tesner, *Symp. (Int.) Combust.* [*Proc.*] **7**, 546–553 (1959).

117. K. C. Hou and H. B. Palmer, *J. Phys. Chem.* **69**, 863–868 (1965).

118. V. G. Knorre, M. S. Kopylev, and P. A. Tesner, *Combust. Explos. Shock Waves* (*Engl. Transl.*) **13**, 732–736 (1977).

119. D. R. Dugwell and P. J. Foster, *Carbon* **11**, 455–467 (1973).

120. K. S. Narasimhan and P. J. Foster, *Symp. (Int.) Combust.* [*Proc.*] **10**, 253–257 (1965).

121. S. J. Harris and A. M. Weiner, *Combust. Sci. Technol.* **31**, 155–167 (1983).

122. S. J. Harris and A. M. Weiner, *Combust. Sci. Technol.* **32**, 267–275 (1983).

123. H. G. Wagner, in *Soot in Combustion Systems and Its Toxic Properties* (J. Lahaye and G. Prado, eds.), pp. 171–195. Plenum, New York, 1983.

124. B. S. Haynes and H. G. Wagner, *Z. Phys. Chem.* [N.S.] **133**, 201–213 (1983).

125. L. Baumgaertner, D. Hesse, H. Jander, and H. G. Wagner, *Symp. (Int.) Combust.* [*Proc.*] **20**, 959–967 (1985).

126. H. Bockhorn, F. Fetting, and G. Wannemacher, *Symp. (Int.) Combust. [Proc.]* **20**, 979–988 (1985).

127. C. J. Dasch, *Symp. (Int.) Combust.* **20**, Poster Sess. Abstr. PS24 (1985).

128. G. Prado, B. S. Haynes, and J. Lahaye, in *Soot in Combustion Systems and Its Toxic Properties* (J. Lahaye and G. Prado, eds.), pp. 145–170. Plenum, New York, 1983.

129. S. W. Radcliffe and J. P. Appleton, *Combust. Sci. Technol.* **4**, 171–175 (1971).

130. J. Nagle and R. F. Strickland-Constable, *Proc. Conf. Carbon, 5th, 1961* Vol. 1, pp. 154–164 (1962).

131. C. Park and J. P. Appleton, *Combust. Flame* **20**, 369–379 (1973).

132. C. P. Fenimore and G. W. Jones, *J. Phys. Chem.* **71**, 593–597 (1967).

133. F. M. Page and F. Ates, *Adv. Chem. Ser.* **166**, 190–197 (1978).

134. K. G. Neoh, J. B. Howard, and A. F. Sarofim, in *Particulate Carbon: Formation during Combustion* (D. C. Siegla and G. W. Smith, eds.), pp. 261–282. Plenum, New York, 1981.

135. K. G. Neoh, J. B. Howard, and A. F. Sarofim, *Symp. (Int.) Combust. [Proc.]* **20**, 951–957 (1985).

136. F. Takahashi and I. Glassman, *Combust. Sci. Technol.* **37**, 1–13 (1984).

137. H. F. Calcote and P. M. Manos, *Combust. Flame* **49**, 289–304 (1983).

138. R. J. Gill and D. B. Olson, *Combust. Sci. Technol.* **40**, 307–315 (1984).

139. D. B. Olson and J. C. Pickens, *Combust. Flame* **57**, 199–208 (1984).

140. R. C. Millikan and W. I. Foss, *Combust. Flame* **6**, 210–211 (1962).

141. B. S. Haynes, H. Jander, H. Maetzing, and H. G. Wagner, *Symp. (Int.) Combust. [Proc.]* **19**, 1379–1385 (1982).

142. B. S. Haynes, unpublished data.

143. J. Flossdorf and H. G. Wagner, *Z. Phys. Chem. [N.S.]* **54**, 113–128 (1967).

144. F. G. Roper and C. Smith, *Combust. Flame* **36**, 125–138 (1979).

145. I. J. Jagoda, G. Prado, and J. Lahaye, *Combust. Flame* **37**, 261–274 (1980).

146. J. H. Kent, H. Jander, and H. G. Wagner, *Symp. (Int.) Combust. [Proc.]* **18**, 1117–1126 (1981).

147. J. H. Kent and H. G. Wagner, *Combust. Flame* **47**, 53–65 (1982).

148. R. J. Santoro, H. G. Semerjian, and R. A. Dobbins, *Combust. Flame* **51**, 203–218 (1983).

149. A. Melvin, J. B. Moss, and J. F. Clarke, *Combust. Sci. Technol.* **6**, 135–142 (1972).

150. J. H. Kent and H. G. Wagner, *Combust. Sci. Technol.* **41**, 245–269 (1984).

151. J. H. Kent and H. G. Wagner, *Symp. (Int.) Combust. [Proc.]* **20**, 1007–1015 (1985).

152. W. L. Flower and C. T. Bowman, *Symp. (Int.) Combust. [Proc.]* **20**, 1035–1044 (1985).

153. R. L. Schalla and G. E. McDonald, *Symp. (Int.) Combust. [Proc.]* **5**, 316–324 (1955).

154. I. Glassman and P. Yaccarino, *Symp. (Int.) Combust. [Proc.]* **18**, 1175–1183 (1981).

155. I. Glassman and P. Yaccarino, *Combust. Sci. Technol.* **24**, 107–114 (1980).

156. A. Gomez, G. Sidebotham, and I. Glassman, *Combust. Flame* **58**, 45–57 (1984).

157. C. Wey, E. A. Powell, and J. I. Jagoda, *Combust. Sci. Technol.* **41**, 173–190 (1984).

158. R. J. Santoro and H. G. Semetjian, *Symp. (Int.) Combust. [Proc.]* **20**, 997–1006 (1985).

159. G. H. Markstein and J. de Ris, *Symp. (Int.) Combust. [Proc.]* **20**, 1055–1061 (1985).

160. J. H. Kent, *Combust. Flame* **63**, 349–358 (1986).

161. B. F. Magnussen, *Symp. (Int.) Combust. [Proc.]* **15**, 1415–1425 (1975).

162. B. F. Magnussen, B. H. Hjertager, J. G. Olsen, and D. Bhaduri, *Symp. (Int.) Combust. [Proc.]* **17**, 1383–1393 (1979).

163. H. A. Becker and S. Yamazaki, *Symp. (Int.) Combust. [Proc.]* **16**, 681–691 (1977).

164. O. Nishida and S. Mukohara, *Combust. Flame* **47**, 269–279 (1982).

165. J. H. Kent and S. J. Bastin, *Combust. Flame* **56**, 29–42 (1984).

166. K. Hein, *Combust. Sci. Technol.* **5**, 195–206 (1972).

167. J. B. Howard and W. J. Kausch, *Prog. Energy Combust. Sci.* **6**, 263–276 (1980).

168. P. A. Bonczyk, *Combust. Flame* **51**, 219–229 (1983).

169. K. P. Schug, Y. Manheimer-Timnat, P. Yaccarino, and I. Glassman, *Combust. Sci. Technol.* **22**, 235–250 (1980).

PART II
Flame Phenomena, Diffusional Processes, and Turbulent Reactive Flow

6 Premixed Flames and Detonations: Combustion Safety and Explosions

ROGER A. STREHLOW

Department of Aeronautical and Astronautical Engineering
University of Illinois
Urbana, Illinois

1. Introduction
2. Steady, inviscid, one-dimensional flow with chemical reactions
   2.1. General equations
   2.2. Rayleigh and Hugoniot solutions
   2.3. Working fluid–heat addition model
   2.4. Solution with full equilibrium chemistry
   2.5. Physically viable solutions
3. Laminar flame propagation
   3.1. Physical observations
   3.2. Laminar flame theory
   3.3. Comments on the theoretical models and real flame behavior
   3.4. Aerodynamics of the laminar flame, flame holding
   3.5. Flammability limits and extinction
   3.6. Interchangeability of fuels for laminar premixed flame burners
4. Detonation propagation: gas phase
   4.1. Gross behavior
   4.2. The Zel'dovich, von Neumann, and Döring theory
   4.3. Inherent instability of a one-dimensional detonation
   4.4. Frontal structure
   4.5. Mechanism of propagation (summary)
   4.6. Direct initiation of detonation
   4.7. Detonation failure
5. Combustion safety
   5.1. Introduction
   5.2. Flash point
   5.3. Auto ignition temperature
   5.4. Minimum ignition energy and quenching distance
   5.5. Maximum experimental safe gap (MESG)
   5.6. Inductive spark ignition
6. Explosion dynamics
   6.1. Fundamental processes
   6.2. Combustion explosions in enclosures

6.3. Nominally unconfined explosions
Problems
Nomenclature
References

1. INTRODUCTION

The purpose of this section is to introduce the reader to the subjects of premixed flames, detonations, and explosion phenomena. The rationale of the presentation is to start with fundamentals and introduce progressively more complex material which will bear on these phenomena. Finally, this section will close with a discussion of how basic understanding of premixed flames and detonations can be applied to combustion safety concepts and testing procedures and explosion dynamics under conditions of heavy and partial confinements.

2. STEADY, INVISCID, ONE-DIMENSIONAL FLOW WITH CHEMICAL REACTIONS*

2.1. General Equations

The equations of motion for strictly one-dimensional steady flow of an inviscid fluid are

$$\rho_1 u_1 = \rho_2 u_2 \tag{1}$$

$$P_1 + \rho_1 u_1^2 = P_2 + \rho_2 u_2^2 \tag{2}$$

$$h_1 + \tfrac{1}{2} u_1^2 = h_2 + \tfrac{1}{2} u_2^2 \tag{3}$$

where u is the flow velocity, P is the pressure, ρ is the density, and h is the specific enthalpy. In addition to these equations, one needs the equation of state of the substance in question. In this section we will be dealing exclusively with gases that have a low enough density, such that they obey the ideal gas equation of state:

$$P = \frac{\rho R T}{m} \tag{4}$$

---

*For a derivation of the equations used in this section the reader is referred to a general gas dynamics text such as H. W. Liepman and A. Roshko, *Elements of Gas Dynamics*, Wiley, New York, 1957, or A. H. Shapiro, *The Dynamics and Thermodynamics of Compressible Flow*, Ronald Press, New York, 1958.

where m is the molecular weight of the mixture and $R = 8.31417$ J/mol K. It can be shown thermodynamically that for a nonreactive gas, the enthalpy is a function of temperature only and is independent of the pressure. However, when the products are a reactive ideal gas, one must express the enthalpy as a function of temperature and extent of reaction,

$$h_2 = h_2(T, \lambda_j; \ i = 1, 2, \ldots, p) \tag{5}$$

where the λ_j's are the reaction coordinates for the p possible forward reactions. This means that if the gas is in chemical equilibrium, the enthalpy becomes effectively a function of temperature and pressure:

$$h_2 = h_2(T, P) \tag{6}$$

2.2. Rayleigh and Hugoniot Solutions

If we solve the mass and momentum conservation equations (Eqs. (1) and (2)) to eliminate either u_1 or u_2, we obtain the equation

$$(\rho_1 u_1)^2 = (\rho_2 u_2)^2 = \frac{P_2 - P_1}{V_1 - V_2} \tag{7}$$

where V is the specific volume, m³/kg. The right-hand side of this equation is the negative slope of a line connecting states 1 and 2 in the (P, V) plane. Both the left-hand side and the middle part of this equation represent the square of mass flow. Thus, to be physically real, the left-hand side must always be positive. This restricts all end states for a steady one-dimensional inviscid flow to the regions illustrated in Figure 1. Note that for such a flow to be real, either the pressure must increase when the volume decreases, or the pressure must decrease when the volume increases. Furthermore all thermodynamic

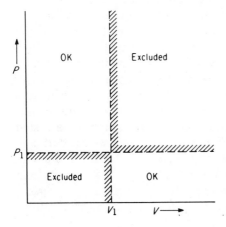

FIGURE 1. The (P, V) plane showing regions where solutions are excluded or allowed according to Eq. (7). (From *Fundamentals of Combustion*. International Textbook Co., Scranton, Pennsylvania, 1968, with permission of R. A. Strehlow.)

states of the system in such a flow must lie on a single straight line through the point (P_1, V_1). In English-language literature this line is called the *Rayleigh line*. It should be pointed out that the Rayleigh line does not represent a complete solution to the problem because the energy equation has not been included. It should also be pointed out that the Rayleigh line requirement was derived independent of any equation of state considerations. It therefore represents a fundamental constraint on one-dimensional, inviscid, steady flows irrespective of the substance that is being considered.

A solution to all equations of motion, written in such a way that both u_1 and u_2 are eliminated from the equations, yields the Hugoniot relationships. These may be written either as

$$e_2 - e_1 = \tfrac{1}{2}(P_2 + P_1)(V_1 - V_2) \tag{8}$$

or as

$$h_2 - h_1 = \tfrac{1}{2}(P_2 - P_1)(V_1 + V_2) \tag{9}$$

Notice that the Hugoniot equations were derived without any consideration of an equation of state. Thus, as written, they are applicable to a chemically reactive system. In this case, of course, the enthalpies or energies used in the equations for states 1 and 2 must be compatible; that is, they must both refer to the same standard reference state as a common zero. Thus enthalpies of reaction will appear in the equation for the Hugoniot when chemical transformations are occurring in a steady, inviscid, one-dimensional flow.

2.3. Working Fluid–Heat Addition Model

Combustion processes involve, in general, the breaking of molecular bonds in the fuel and oxidizer and the formation of stronger molecular bonds in the products of combustion. The energy that is thus released is absorbed by the system as thermal energy and causes the temperature of the system to rise. The calculation of this temperature, for example, for constant pressure combustion conditions for a gaseous fuel-oxidizer mixture such as methane-air, is performed by requiring that h_2 equal h_1. Since the temperature has risen considerably during the combustion process, it is obvious that the enthalpy temperature relationship for the products cannot simply intersect the enthalpy temperature relationship for reactants. Furthermore, in combustion with air as the oxidizer, dilution with nitrogen causes the enthalpy temperature relationship to be almost a straight line. Also, in such a dilute combustion system, even though there is a slight change in the average molecular weight during the combustion process, this change is usually unimportant. Thus, for simplified combustion considerations, one can replace the actual gas that is undergoing the combustion reactions with an inert working fluid of constant heat capacity. However, the enthalpy temperature relationship for the product fluid must be

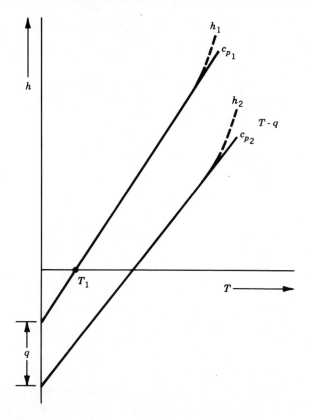

FIGURE 2. Enthalpy-temperature relationships. Dashed lines are real gas enthalpies; solid lines are the working fluid–heat addition model approximations.

displaced from that of the reactants because of the highly exothermic reactions that are occurring during the combustion process. Figure 2 is a plot showing how the real enthalpy relationships are approximated in the working fluid, heat addition model. In general, we can have

$$h_1 = c_{p_1} T \tag{10}$$

and

$$h_2 = c_{p_2} T - q \tag{11}$$

where $c_{p_1} \neq c_{p_2}$ but both are constants. However, it is common to let $c_{p_2} = c_{p_1}$.
The working fluid–heat addition model, when substituted into the Hugoniot relationship yields the following equation:

$$\frac{\gamma}{\gamma - 1}(P_2 V_2 - P_1 V_1) - q = \tfrac{1}{2}(P_2 - P_1)(V_2 + V_1) \tag{12}$$

where $\gamma = c_p/c_v =$ constant. This equation exhibits both a pressure and volume asymptote. Substituting $P_2 \to \infty$ or $V_2 \to \infty$ yields the following two asymptotes:

$$\frac{V_2}{V_1} \to \frac{\gamma - 1}{\gamma + 1}; \qquad \frac{P_2}{P_1} \to -\frac{\gamma - 1}{\gamma + 1} \tag{13}$$

We therefore can rewrite this simplified Hugoniot relationship in terms of a new coordinate system in which the variables are

$$\eta = \frac{V_2}{V_1} - \frac{\gamma - 1}{\gamma + 1} \tag{14a}$$

$$\xi = \frac{P_2}{P_1} + \frac{\gamma - 1}{\gamma + 1} \tag{14b}$$

After some algebraic manipulation, we find that the Hugoniot relationship expressed in these variables reduces to the expression

$$\eta\xi = \frac{4\gamma}{(\gamma + 1)^2} + \frac{2q}{P_1V_1}\left(\frac{\gamma - 1}{\gamma + 1}\right) \tag{15}$$

This is the equation for a rectangular hyperbola in the (η, ξ) plane, and therefore for a displaced rectangular hyperbola in the (P, V) plane. This is illustrated in Figure 3. First, note that the asymptotes are independent of the value of q, and second, note from Eq. (15) that a positive value of q displaces the Hugoniot toward the right and upward. It can be shown that when q is equal to zero, the Hugoniot passes through the point P_1, V_1. It can also be shown that at the point P_1, V_1 the tangent to the $q = 0$ Hugoniot is the equation for a sonic line. That is, such a Raleigh line represents an approach flow velocity just equal to the velocity of sound. Furthermore it can be shown that the Hugoniot for $q = 0$ is the normal shock Hugoniot when the pressure rises and the volume decreases and is physically untenable in the branch where the pressure is lower than the initial pressure.

All Hugoniot's that are displaced to the right and upward represent Hugoniot's for combustion processes. The lower branch, which lies between P_1 and 0, is called the *deflagrative branch*. A Rayleigh line of very small slope connecting the point P_1, V_1 with a point on this branch represents the subsonic flames that are observed in the laboratory. This is because the flow velocity associated with ordinary flames is so low that the pressure drop is very small. The upper branch solutions that are represented by the condition that the solutions lie between the $P = \infty$ asymptote and the $V_2 = V_1$ line are called the *detonation branch solutions* (note that all Rayleigh slopes above the slope $u = a$ represent supersonic approach flows). In this region there is a

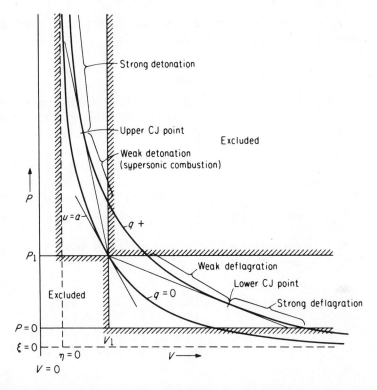

FIGURE 3. Pressure-volume plot of end states for a one-dimensional steady process, with heat addition indicating excluded regions and upper and lower Chapman-Jouguet states. (From *Fundamentals of Combustion.* International Textbook Co., Scranton, Pennsylvania, 1968, with permission of R. A. Strehlow.)

minimum supersonic velocity for steady, one-dimensional inviscid flows. This is called the *upper Chapman-Jouguet point* (CJ) and is physically observed as a self-sustaining detonation. Intersections of the Rayleigh line with the Hugoniot line above this Chapman-Jouguet point are called *strong* or *over driven detonations* and are observed physically when a shock wave in an exothermic reactive system is supported by a high-velocity piston. The weak detonation branch, or supersonic combustion branch, is observed physically when one can somehow trigger the chemistry at a high-velocity supersonic flow without having a shock wave in the system. This can be done, for example, by having a supersonic stream of air heated above the auto-ignition temperature of the fuel and injecting fuel into this very hot supersonic flow. The transition then takes place directly from the point (P_1, V_1) to the intersection of the Rayleigh line with the Hugoniot, and there is no shock in the flow.

Finally, a vertical Rayleigh line on Figure 3 from the point (P_1, V_1) represents constant volume combustion, and a horizontal Rayleigh line from the point (P_1, V_1) represents constant pressure combustion. There are no waves in the system in either of these cases.

2.4. Solution with Full Equilibrium Chemistry

The actual Hugoniot that is calculated for a fully reactive system is the equilibrium composition Hugoniot. As one travels along the Hugoniot in a direction of decreasing final volume, both the temperature and pressure increase monotonically; thus, in a real chemical system, the composition of the equilibrium gases changes continuously as one travels along the Hugoniot. In this section we will first present an interesting proof concerning the upper and lower Chapman-Jouguet points on an equilibrium Hugoniot and then discuss a simplified technique for calculating the properties along any equilibrium Hugoniot.

If we differentiate the Hugoniot relationship given in terms of energy, Eq. (8), along the Hugoniot, we obtain the expression

$$de_2 = -\frac{1}{2}(P_2 + P_1)d\left(\frac{1}{\rho_2}\right) + \frac{1}{2}\left(\frac{1}{\rho_1} - \frac{1}{\rho_2}\right)dP_2 \qquad (16)$$

However, in differential form, the first law of thermodynamics may be written as

$$de_2 = -P_2 d\left(\frac{1}{\rho_2}\right) + T_2\, ds_2 \qquad (17)$$

Therefore

$$T_2\, ds_2 = -\frac{1}{2}(P_1 - P_2)d\left(\frac{1}{\rho_2}\right) + \frac{1}{2}\left(\frac{1}{\rho_1} - \frac{1}{\rho_2}\right)dP_2 \qquad (18)$$

Now, Eq. (18) may be written in this form along the Hugoniot:

$$T_2\frac{ds_2}{d(1/\rho_2)} = \frac{1}{2}\left(\frac{1}{\rho_1} - \frac{1}{\rho_2}\right)\left[-\frac{(P_1 - P_2)}{(1/\rho_1) - (1/\rho_2)} + \frac{dP_2}{d(1/\rho_2)}\right] \qquad (19)$$

Notice that, if the quantity $ds_2/d(1/\rho_2) = 0$ at any point along the Hugoniot, the Hugoniot curve must osculate an isentrope at that point. Furthermore, at those points where this happens on the Hugoniot, one may write

$$\left(\frac{dP_2}{d(1/\rho_2)}\right)_s = -\frac{P_2 - P_1}{(1/\rho_1) - (1/\rho_2)} \qquad (20)$$

The right-hand side of this equation is the negative slope of a Rayleigh line to that point on the Hugoniot. One can therefore substitute the equation for the Rayleigh line to yield the expression

$$\left(\frac{dP_2}{d(1/\rho_2)} \right)_s = -u_2^2 \rho_2^2 \qquad (21)$$

However,

$$a_2^2 = \left(\frac{dP_2}{d\rho_2} \right)_s = -\frac{1}{\rho_2^2} \left(\frac{dP_2}{d(1/\rho_2)} \right)_s \qquad (22)$$

Therefore we find that at the point where the equilibrium Hugoniot is osculated by an equilibrium isentrope, it is tangent to a Rayleigh line, and the flow velocity in state 2 is exactly equal to the equilibrium velocity of sound. This is, of course, the condition for thermal choking in a one-dimensional flow, and it applies equally well to both the upper and lower Chapman-Jouguet points, even though the upper Chapman-Jouguet point is the only one that is observed physically.

The actual calculation of the properties of an equilibrium Hugoniot requires that one have available an equilibrium subroutine that is capable of calculating the composition and thermodynamic properties of an equilibrium mixture of appropriate species over a large range of elevated temperatures and pressures. To perform such a calculation, one also needs to know the atomic composition of the reactant mixture, as well as the heat formation of all species involved relative to the standard state elements. Once this information is available, the conservation equations along with the equation of state and caloric equation state for the mixture can be solved iteratively to find end states on the Hugoniot. We first define the density ratio

$$\lambda = \frac{\rho_2}{\rho_1} \qquad (23)$$

Substituting the ideal gas law into the momentum equation yields

$$R_1 T_1 + u_1^2 = \lambda \left(R_2 T_2 + u_2^2 \right) \qquad (24)$$

Here, R_1 and R_2 are the gas constants per unit mass for the reactants and products, respectively. This takes care of any change in molecular weight due to the chemistry that has occurred. The above equation can be rearranged to yield the expression

$$\lambda R_2 T_2 - R_1 T_1 = u_2^2 (\lambda - 1) \lambda \qquad (25)$$

Next, combining the continuity and energy equation leads to the expression

$$2(h_2 - h_1) = u_2^2 (\lambda^2 - 1) \qquad (26)$$

Multiplying the first of these by quantity $(\lambda + 1)$ and the second by λ allows one to eliminate the right-hand side of both equations, thus producing a quadratic in λ:

$$R_2T_2\lambda^2 + \left[(R_2T_2 - R_1T_1) - 2(h_2 - h_1)\right]\lambda - R_1T_1 = 0 \qquad (27)$$

The solution to this quadratic is

$$\lambda = \psi + \left(\psi^2 + \frac{T_1R_1}{T_2R_2}\right)^{1/2} \qquad (28)$$

with

$$\psi = \frac{2(h_2 - h_1) - (R_2T_2 - R_1T_1)}{2R_2T_2} \qquad (29)$$

$$h_2 = h_2(T_2, P_2) \qquad (30)$$

Equations (28), (29), and (30) can now be used to solve for the location of the complete Hugoniot, as well as the approach velocity u_1, since h_1 is known. It can also be used to solve for the three points of physical interest on the Hugoniot, namely, the point that represents constant pressure combustion, constant volume combustion, and the point where the Hugoniot and Rayleigh line are tangent on the supersonic branch (the end state for a Chapman-Jouguet detonation).

If the flow is not reactive, any assumed shock temperature will immediately yield a solution. When $h_2 = h_2(T, P)$ because state 2 is at full chemical equilibrium, the final pressure for any assumed temperature must be adjusted until both the Hugoniot and the ideal gas law (with full equilibrium dissociation) are satisfied. The solution for constant pressure explosions (which closely approximates flame temperatures) and constant volume explosions are rather straightforward and only require that $h_2 = h_1$ and $e_2 = e_1$, respectively. The solution for a Chapman-Jouguet detonation is more difficult. This is because one must first determine the locus of the Hugoniot, using an iteration technique, and then determine the location of the upper CJ point, using another overall iteration technique.

Furthermore there are two possible solutions at the upper CJ point, the frozen composition and equilibrium composition CJ solutions. The frozen composition CJ point is calculated by dividing u_2 by a_2, where a_2 is the frozen composition velocity of sound at that point on the Hugoniot. When this ratio equals unity, the frozen CJ point has been found. The equilibrium composition CJ point is determined by finding the minimum Rayleigh slope, that is, the tangent to the Hugoniot. This point is always slightly above and to the left of the frozen point on the Hugoniot. It is now usually accepted that the equilibrium CJ point is the correct one to calculate.

The application of the heat addition model to the calculation of detonation conditions will be discussed in detail in Section 4.1.

2.5. Physically Viable Solutions

The following solutions to the Hugoniot are physically viable. Overdriven detonations that are piston supported have been observed experimentally. Chapman-Jouguet detonations that are self-sustaining at a velocity very close to the calculated Chapman-Jouguet velocity have been observed in tubes under condition where there is no external support for the detonation. Constant volume combustion and constant pressure combustion are also viable solutions. There is moreover the flame solution in which ΔP is very much less than P_1 and is slightly negative. Additionally, plug flow reactors at low subsonic velocities obey the Rayleigh line relationships and therefore represent a low-velocity subsonic solution to the equations of motion.

It must be pointed out, however, that all flow properties that have been discussed up to this point are based on a Hugoniot equation in which the flow was assumed to be one-dimensional steady. Therefore no details of wave structure or of inherent stability or existence can be quantitatively given in this section. However, if a solution is found that can be shown to be one-dimensional steady in the light of stability arguments or in a one-dimensional steady wave is observed experimentally, the equations that have been introduced in this section are appropriate to that system and can be used to discuss the flow associated with that wave.

3. LAMINAR FLAME PROPAGATION

3.1. Physical Observations

Many gas-phase mixtures or pure substances that react or decompose exothermically are capable of supporting a low-velocity subsonic decomposition wave, which is called a *flame* and has the following unique properties:

1. For any initial composition, pressure, and temperature in a one-dimensional laminar flow situation, the wave exhibits a unique propagation velocity, S_u, that is generally much less than one-tenth the velocity of sound.
2. The flame exhibits a unique flame temperature that depends upon the initial conditions of composition, pressure and temperature.
3. The pressure drop across these combustion waves is very small. This last point can be illustrated by solving the Rayleigh equation (Eq. (7)) for a typical flame,

$$P_2 - P_1 = P_1\gamma(1 - \varepsilon)M_1^2 \qquad (31)$$

where $\varepsilon = V_2/V_1$ and $M_1 = S_u/a_1$ with a_1 equal to the velocity of sound in the gas ahead of the flame. Since for all observable flames $M_1 < 0.02$, $\varepsilon < 15$, and $\gamma \leq 1.667$, this equation predicts the maximum pressure change $P_1 - P_2$ should be $\leq +0.01P_1$. We thus note that flame propagation is always accompanied by a slight pressure drop and that ordinary one-dimensional flames with $M_1 \cong 0.001$ may be considered as essentially isobaric processes.

These observations justify the approximation mentioned in the previous section in which one equates the flame temperature to the isenthalpic explosion temperature for the mixture. Also, for any specific system, the isobaric assumption means that the flame temperature is relatively independent of the observed burning velocity and may be determined with reasonable accuracy without knowledge of the burning velocity. Measurements of flame temperatures have shown that the translational gas temperature rises monotonically through the flame and eventually reaches a value comparable to the isenthalpic values calculated using thermodynamics. Ordinarily, the accepted value of flame temperature is the calculated isenthalpic value, because difficulties with the experimental measurement of flame temperature and the relatively high accuracy of present-day thermodynamic calculations cause one to place more confidence in the calculated values.

Figure 4 illustrates flame temperatures for a number of different fuel-air combinations as a function of the equivalence ratio. The equivalence ratio is defined in the usual sense, that is,

$$\Phi = \frac{f/a}{(f/a)_s} \tag{32}$$

where f/a is the fuel-air ratio (on either a mass or mole basis) and the subscript s refers to a stoichiometric mixture (in which complete combustion of the fuel would occur to produce a product that contains only carbon dioxide, water, nitrogen, and sulfur dioxide). Incidentally, it is common practice to plot combustion properties versus the equivalence ratio because most combustion properties, such as flame temperature and flame velocity, have simple maxima or minima in the neighborhood of $\Phi = 1$.

It has been observed that the flame temperature of hotter flames ($T_f > 2300$ K) are somewhat insensitive to the initial temperature of the mixture because dissociation reactions cause the flame gases to have a large effective heat capacity. Low-temperature flames are relatively more sensitive to changes in initial temperature because dissociation reactions are not as important at these temperatures. On the other hand, the flame temperature of a high-temperature flame is more sensitive to changes in the ambient pressure. This is because an increase in pressure shifts the dissociation equilibria in the direction of lower dissociation at high temperatures. This causes the flame temperature of a

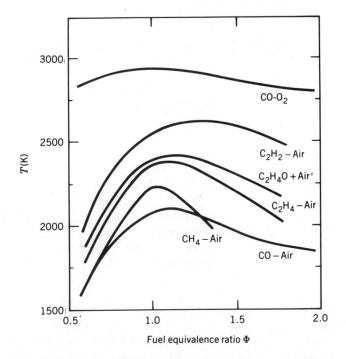

FIGURE 4. Flame temperature for various fuel-air and fuel-oxygen mixtures. (From *Fundamentals of Combustion.* International Textbook Co., Scranton, Pennsylvania, 1968, with permission of R. A. Strehlow.)

high-temperature flame to increase with ambient pressure as the ambient pressure is increased. Low-temperature flames show almost no flame temperature change as the pressure is changed. This is because there is hardly any dissociation, and thus pressure can have only a small effect on the flame temperature.

We now turn our attention to the question of burning velocity. First, we note that our assumption that a flame is a steady-flow process in no way legislates the behavior of real flames. On the contrary, the justification for this assumption arises solely from our observations of flames under well-controlled conditions in the laboratory. It has been observed that in a variety of laminar-flow situations, the flame behaves in a way best explained by assuming that it is a steady three-dimensional phenomenon. The extrapolation of this three-dimensional real behavior to the strictly one-dimensional behavior required for the definition of a burning velocity is primarily a problem of careful experimental measurement and properly assessing the effects of curvature on flame structure. Because of the difficulty of correcting for flame curvature in flows where the flame is positioned at an oblique angle to the flow, the most accurate burning velocities are obtained in situations where

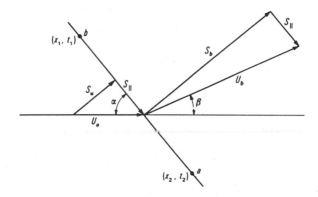

FIGURE 5. The oblique one-dimensional flame. The flame appears as an oblique flame because of a simple velocity transformation at the velocity S along itself. The stream tube area parallel to the flame front does not change in this diagram. (With permission from R. A. Strehlow, *Combustion Fundamentals*. McGraw-Hill, New York, 1984.)

either the flame is strictly flat, though oriented obliquely to the flow, or the flame, though curved, is oriented strictly normal to the flow-velocity vector. There are three measurement techniques that satisfy these requirements and that may be considered as a the primary experimental techniques for determining burning velocity: (1) the slot-burner technique, (2) the soap-bubble technique, and (3) the flat-flame burner technique.

In the slot-burner technique a flame is stabilized above a rectangular nozzle (ratio of side lengths > 3) that has either a flat velocity profile generated by a two-dimensional Mache-Hebra nozzle or a fully developed laminar pipe-flow profile generated by using a long entrance pipe of the same dimensions as the port. In this situation the flame is attached to the rim of the burner and completely burns the combustible mixture issuing from the port. The flame takes the appearance of a tent above the port and is therefore held at an oblique angle to the primary flow direction. It has been observed that a relatively flat oblique laminar flame may be transformed to a strictly one-dimensional flame by imposing a constant velocity transformation as shown in Figure 5. Therefore the advantage of the slot burner is simply that the flame has rather large flat areas (both sides of the tent) in which curvature effects are minimized. Furthermore, since the flame completely encloses the incoming flow, the incoming streamlines are not deflected until they enter the heat conduction zone of the flame. Thus the angle that the flat-flame surfaces make to the incoming flow can be measured accurately and used to evaluate the normal burning velocity by applying Eq. (33).

$$S_u = u \sin \alpha \tag{33}$$

Experiments have shown that for this geometry the measurement of the local flow velocity and local flame angle at a point in the flame yields a burning velocity that is relatively constant over a large portion of the burner width.

The soap-bubble technique is another method that is known to yield accurate burning velocities without the application of an excessive number of corrections. In this technique a soap bubble is blown with a combustible mixture and ignited centrally by means of a capacitance spark. The flame burns as an outwardly propagating spherical flame and since the bubble expands as combustion proceeds, the process occurs at constant pressure. In this case, after an initial propagation period, the flame is only slightly curved and is oriented strictly normal to the flow direction. Its apparent propagation velocity is not S_b, however, but actually $S_b = u_2$ since the gas behind the flame is quiescent. This measured velocity may be converted to the normal burning velocity by noting that the continuity equation requires that $\rho_1 S_u = \rho_2 S_b$ and therefore that

$$S_u = \left(\frac{V_1}{V_2} \right) S_b \qquad (34a)$$

$$= \left(\frac{r_1}{r_2} \right)^3 S_b \qquad (34b)$$

where r_1 and r_2 are the initial bubble radius and final flameball radius, respectively. Measurement of burning velocity using this technique thus requires the measurement of both the apparent flame velocity and the ratio of the initial to final radius of the combustible gases. Since a relatively exact final radius is required, the best measurements are obtained when the initial bubble is surrounded by an inert atmosphere to prevent afterburning. This technique is somewhat more difficult and less generally applicable than the slot-burner technique. It has the advantage, however, of requiring only small samples of combustible mixture.

The third precise technique, using the flat-flame burner, is strictly limited to very low-burning-velocity flames. In fact flames observed using this technique usually cannot be observed as stable flames in other apparatus because of their low-burning velocity. In this measurement a rather large-diameter, low-velocity combustible stream is produced by passing the flow through appropriate series of screens and honeycomb filters. The flame is stabilized as a flat flame by using a capping screen and a surrounding low-velocity, inert-gas flow. Experimentally, it is observed that a flame will be stabilized in this burner over only a very small flow range. Once the flame is stabilized at a reasonable distance from the port, the burning velocity may be determined by simply dividing the input volumetric flow rate by the total flame area. However, because of the closeness of the flame to the burner head, the velocity that is measured is usually corrected for heat loss by stabilizing the flame at different distances from the burner head and then extrapolating the zero heat loss. This apparatus has also been used to stabilize flames for structure studies.

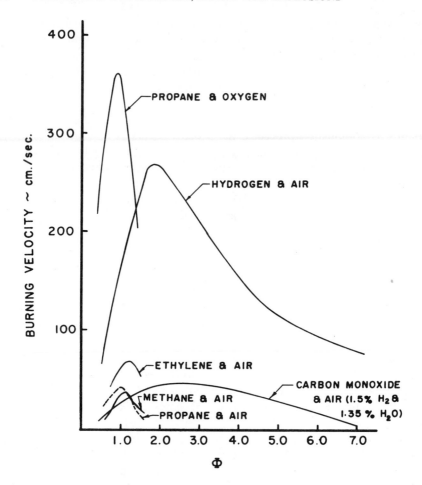

FIGURE 6. Normal burning velocity, S_u, versus equivalence ratio for a few representative systems. (From *Fundamentals of Combustion*. International Textbook Co., Scranton, Pennsylvania, 1968, with permission of R. A. Strehlow.)

The effect of mixture composition on the burning velocity for a few typical flames is shown in Figure 6. By comparing Figures 4 and 6, it is interesting to note that flame temperature and flame velocity roughly parallel each other for any specific flame system and that the maximum flame temperature ordinarily occurs in a mixture that also exhibits the maximum flame velocity.

3.2. Laminar Flame Theory

Soon after laminar premixed flames were stabilized in the laboratory, it was recognized that they represent a unique case of wave propagation; the steady self-sustaining propagation of an "exothermic reaction wave" at a low sub-

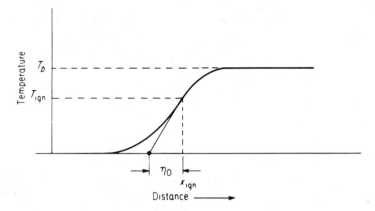

FIGURE 7. The structure of a thermal theory flame. The distance η_0 is called the *preheat zone thickness*. (From *Fundamentals of Combustion*. International Textbook Co., Scranton, Pennsylvania, 1968, with permission of R. A. Strehlow.)

sonic velocity in a direction normal to itself. The question of the propagation mechanism of this unique wave has been investigated extensively over the past hundred years. Early in this period, heat conduction was postulated as a primary mechanism of propagation. This "thermal" concept for flame propagation was first introduced by Mallard and Le Chatelier and culminated in the Zel'dovich treatment of the 1940s. In its simplest form the theory assumes that the flame

1. Has a unique propagation velocity, S_u.
2. Does not loose heat to the sides or downstream but does transmit heat to the upstream gas by thermal conduction.
3. Contains a relatively thin high-temperature reaction zone separated from a nonreactive and cooler conduction zone at a plane where "ignition" occurs. Thus this flame has a unique "ignition temperature," T_{ig}, that is closer to the flame temperature, T_b, than to the unburned gas temperature, T_u.

For simplicity, one must also assume that the heat of reaction, $q(J/kg)$, the heat capacity, $c_p(J/kg\ K)$, the thermal conductivity, κ (J/msK), and the rate combustion, $\omega(kg/m^3)$, are constants. The structure of such a flame is shown in Figure 7.

Within the framework of the assumptions we can immediately define the preheat zone thickness of this flame:

$$\eta_0 = \frac{T_{ig} - T_u}{(dT/dx)_{ig}} \tag{35}$$

If we now apply energy conservation at the ignition plane, we obtain the relationship

$$\kappa\left(\frac{dT}{dx}\right)_{ig} = \frac{\kappa(T_{ig} - T_u)}{\eta_0} = q\omega\eta_0$$

$$= c_p(T_b - T_u)\omega\eta_0 \tag{36}$$

which means (because $T_{ig} - T_u \cong T_b - T_u$) that the preheat zone thickness in such a flame is given approximately by the expression

$$\eta_0 \cong \sqrt{\frac{\kappa}{c_p\omega}} \tag{37}$$

Mass conservation states that $S_u\rho_1 = \omega\eta_0$, and therefore the thermal equation for the flame velocity becomes

$$S_u \cong \frac{1}{\rho_1}\sqrt{\left(\frac{\kappa}{c_p}\right)\omega} \tag{38}$$

Notice that we can eliminate the reaction rate from Eqs. (37) and (38) to obtain a relationship between the burning velocity and the preheat zone thickness of the flame:

$$\eta_0 \cong \frac{\kappa}{\rho_1 c_p S_u} \cong \frac{\alpha}{S_u} \tag{39}$$

where α is the thermal diffusivity of the gas.

Both the concept that a flame has a unique normal burning velocity and that it has a unique preheat zone thickness are important in industrial applications and in developing combustion safety criteria. It is important to note here that the normal burning velocity is essentially proportional to the square root of a thermal conductivity multiplied by an effective rate of reaction and that the preheat zone thickness is uniquely related to the burning velocity by the thermal diffusivity of the gas. More will be said about this later. We now turn our attention to more modern flame theory.

The development of the thermal theory of flame propagation led to many vessel studies of reaction kinetics in exothermic systems. After these studies showed the prevalence of exothermic chain-branching reactions in combustion systems, forward diffusion of highly reactive chain carriers was postulated as an alternate mechanism of propagation. The subsequent proliferation of "simple" propagation theories greatly stimulated experimental flame research, and in particular led to many studies of burning velocity as a gross property which could possibly be correlated to the various theoretical predictions.

However, it is only in the last two decades that precise experimental probing techniques in conjunction with a correct and complete steady-flow theory have yielded a good descriptive model of the highly coupled processes that cause stable flame propagation.

Accumulated experimental evidence indicates that an ordinary one-dimensional flame propagates as a steady wave and that heat conduction, diffusion, and a coupled set of rapid exothermic chemical reactions all play important roles in the propagation mechanism. In the following development, we follow the approach of Hirschfelder and Curtiss[1] and include these three processes but neglect the effect of radiative heat transfer, body forces, and second-order (i.e., coupled) transport effects, such as thermal diffusion. We also ignore nonstationary effects, such as flame acceleration or the possible occurrence of three-dimensional instabilities. That is to say, we will not consider the question of the inherent stability of a combustion wave at this time but will simply assume that the steady model represents a real flame. One important type of observed flame instability will be discussed in the next section.

Since the flame is essentially isobaric, we dispense with the momentum equation and discuss only mass and energy balances. To be complete, we must discuss the mass balance of each species that appears in the flame. The velocity of each species at any plane in the flame is simply $S + V_i$, where V_i is the average velocity due to diffusion of the ith species at that location and S is the average flow velocity at that point. Since the total mass flow is constant at each plane, and since each species by definition has the same convective velocity S, we find that the global mass-balance equation may be written as

$$\sum_{i=1}^{s} \rho_i (S + V_i) = \rho S = M \qquad (40)$$

which yields an equation relating the diffusion velocities of the individual species

$$\sum_{i=1}^{s} [I] m_i V_i = 0 \qquad (41)$$

where the symbol $[I]$ means concentration of the ith species and a lowercase m is used for molecular weight to distinguish it from the mass flow M. We may also define the diffusion velocity in terms of a diffusion coefficient,

$$[I] V_i = -N D_i \frac{DX_i}{dx} \qquad (42)$$

where N is moles of mixture per m^3, X_i is the mole fraction of the ith species, and D_i is the diffusion coefficient of the ith species in the local mixture subject to the constraint of Eq. (41). In general, these coefficients are not simply related to the binary diffusion coefficients. The negative sign appears in Eq.

(42) because a positive concentration gradient yields a negative diffusion velocity.

If we now consider a balance between two neighboring planes, x and $x + dx$, we see that the net rate of production of the ith species in the distance dx per square meter of flame area is simply

$$[I](S + V_i)|_{x+dx} - [I](S + V_i)|_x \qquad (43)$$

However,

$$[I](S + V_i)|_{x+dx} = [I](S + V_i)|_x + \frac{d}{dx}\{[I](S + V_i)\}\, dx \qquad (44)$$

and we may write

$$\left(\frac{d[I]}{dt}\right)_{\text{chem}} = \rho\left(\frac{dn_i}{dt}\right)_{\text{chem}} = \frac{d}{dx}\{[I](S + V_i)\} \qquad (45)$$

where $(dn_i/dt)_{\text{chem}}$ is defined as the net rate of production of the ith species at a station in the flame due to all the chemical reactions that are occurring at that location. Note that if $(dn_i/dt)_{\text{chem}}$ is not identically zero throughout the flame, concentration gradients will exist and cause diffusion of individual species to be superimposed on the net motion of the flame. Thus the i mass-balance equations become

$$\rho \sum_{j=1}^{p} \nu_{ij} \frac{d\lambda}{dt} j = \frac{d}{dx}\{[I]S\} - \frac{d}{dx}\left(ND_i \frac{dX_i}{dx}\right) \qquad (46)$$

where there are p chemical reactions occurring, each with their reaction coordinate λ_j and rate $d\lambda_j/dt$. The quantity λ_j has units of moles of reaction per kilogram of mixture.

We now construct an energy balance in the flame. Since the pressure is essentially constant and the flame is a low-speed subsonic wave, we write an enthalpy balance and therefore implicitly include the external work performed at constant pressure and exclude the flow energy. The enthalpy of the ith species is

$$H_i = H_{0_i}^0 + \int_0^T C_{p_i}\, dT \qquad (47)$$

where $H_{0_i}^0$ is the heat of formation at 0 K and C_{p_i} the heat capacity per mole. The total enthalpy flux through a plane located in the flame is therefore

$$\sum_{i=1}^{s} [I](S + V_i)H_i \qquad (48)$$

whereas the flux through a plane located in the equilibrium region downstream of the flame zone is given by the equation

$$\sum_{i=1}^{s} [I]_{eq} S_b H_{i_{eq}} \tag{49}$$

since no diffusion is occurring in this region. The enthalpy gain of this volume element due to the mass flow is therefore

$$\sum_{i=1}^{s} [I](S + V_i) H_i - \sum_{i=1}^{s} [I]_{eq} S_b H_{i_{eq}} \tag{50}$$

Since the flow is steady, heat must be transported out at the rate of accumulation indicated in the above equation, and since heat is transported only in the flame zone, we obtain the equation

$$\kappa \frac{dT}{dx} = \sum_{i=1}^{s} [I](S + V_i) H_i - \sum_{i=1}^{s} [I]_{eq} S_b H_{i_{eq}} \tag{51}$$

where κ is the thermal conductivity of the mixture at x and the first term on the right is evaluated at x. This equation may also be written as

$$\sum_{i=1}^{s} [I]_{eq} S_b H_{i_{eq}} = \sum_{i=1}^{s} [I] S H_i - N \sum_{i=1}^{s} H_i D_i \frac{dX_i}{dx} - \kappa \frac{dT}{dx} \tag{52}$$

The three terms on the right in Eq. (52) are energy fluxes due, respectively, to convection, diffusion, and conduction in the flame zone. Equation (52) may also be written in the differential form

$$\frac{d}{dx} \sum_{i=1}^{s} ([I] S H_i) - \frac{d}{dx}\left[N \sum_{i=1}^{s} H_i D_i \frac{dX_i}{dx}\right] - \frac{d}{dx}\left(\kappa \frac{dT}{dx}\right) \tag{53}$$

Equation (52), the mass-flow Eqs. (46) and relationships defining reaction rates, diffusion coefficients, thermal conductivities, and enthalpies may in principle be used to determine the burning velocity and detailed structure of any premixed gas flame. Integrating this equation set to obtain concentrations, state properties, and fluxes in the flame zone requires a knowledge of the burning velocity, S_u, commensurate with specific assumptions for the reaction rates and transport properties and the upstream and downstream boundary conditions. Since this information is not available, we are left with the solution of a problem in which the burning velocity is determined by a characteristic or eigenvalue that satisfies the boundary conditions for the assumed properties.

We now wish to briefly present a solution of these flame equations for a particularly simple model to illustrate the coupled nature of the processes in an ordinary flame. We follow the approach first suggested by Friedman and Burke.[2] Consider the flame reaction to be an irreversible first-order decomposition of pure A yielding only B as product:

$$A \rightarrow B \tag{54}$$

with the molecular weights $m_A = m_B = m$ and the enthalpies $H_A = C_p(T - T_u)$, $H_B - C_p(T - T_u) - Q$, where C_p is taken to be a constant and $Q = C_p(T_b - T_u)$; therefore $H_B = C_p(T - T_b)$. Thus the convected enthalpy is zero at both stations 1 and 2 in this flame. We write the first-order rate equation as

$$\frac{d(NX_B)}{dt} = k_1 NX_A \exp\left(-\frac{E}{RT_B\tau}\right) \tag{55}$$

where

$$\tau = \frac{T - T_u}{T_b - T_u} \tag{56}$$

and X_A and X_B are mole fractions.

The introduction of τ in the rate equation forces the chemical rate to be zero at T_u without significantly altering it in the hotter flame regions. This is reasonable since most real flames propagate at a steady velocity into mixtures that have a vanishingly small decomposition rate at their initial temperature.

We now define a simplified binary diffusion coefficient $D = cT^2$ and thermal conductivity $\kappa = fT$, and thus find that the Lewis number, Le $= \rho c_p D/\lambda$, is a constant throughout the flame. With this set of assumptions, we are ready to transform the general flame equations to a dimensionless form. We first note that in this binary gas mixture Eq. (41) yields

$$V_A = -V_B\left(\frac{X_B}{1 - X_B}\right) \tag{57}$$

and Eq. (52) therefore becomes

$$\frac{d\tau}{dx} = \frac{NC_p S}{\kappa}(\tau - z) \tag{58}$$

where

$$z = \frac{(S + V_B)X_B \rho}{S_u \rho_u} \tag{59}$$

Equation (40) for the species B may be written as

$$\frac{dz}{dx} = \frac{k_1\rho}{M}(1 - X_B)\exp\left(-\frac{\varepsilon}{\tau}\right) \tag{60}$$

where $\varepsilon = E/RT_b$ is a dimensionless activation energy. Furthermore the definition of a diffusion coefficient yields the equation

$$\frac{dX_B}{dx} = -\frac{X_B V_B}{D} \tag{61}$$

Since the flow is steady, these equations are parametric in x and may be written in terms of τ, z, and X_B by dividing Eq. (58) into Eqs. (60) and (61) to eliminate x as a variable. This yields the equations

$$\frac{dz}{d\tau} = \beta_c\left(\frac{1 - X_B}{\tau - z}\right)\exp\left(-\frac{\varepsilon}{\tau}\right) \tag{62}$$

and

$$\frac{dX_B}{d\tau} = \frac{1}{\text{Le}}\left(\frac{X_B - z}{\tau - z}\right) \tag{63}$$

with the boundary conditions $X_B = z = \tau = 0$ at station 1 (the cold boundary) and $X_B = z = \tau = 1$ at station 2 (the hot boundary). The constant β_c is the dimensionless eigenvalue of the physical realizable solution to this equation set and is given by the relation

$$\beta_c = \frac{k_1\kappa\rho m}{S_u^2\rho_u^2 C_p} \tag{64}$$

The value of β_c must be determined by solving Eqs. (62) and (63) simultaneously. Notice that β_c defines the burning velocity in terms of the reaction kinetic and transport properties of the combustible mixture. Also note that, for our assumptions, the temperature dependence of ρ and κ cancel so that $\rho\kappa = \rho_u\kappa_u$. Therefore we find that β_c is truly a constant throughout this flame. Thus we obtain the equation

$$S_u = \sqrt{\frac{mk_1\kappa_u}{\rho_u\beta_c C_p}} \tag{65}$$

which defines the burning velocity.

Again, following Friedman and Burke's treatment, we note that Eq. (63) is directly integrable for two cases: Le = 1 and Le = 0, yielding, respectively, $X_B = \tau$ or $X_B = z$. In the case where we assume Le = 0, we do so by setting

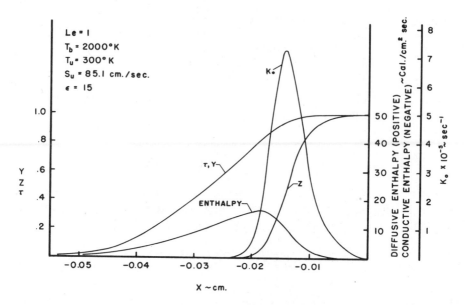

FIGURE 8. Flame structure for a flame with Lewis number = 1, supported by a first-order reaction. Rate is the rate of disappearance of species A, and equals $k_1(1 - X_B)\exp(-\varepsilon/\tau)$. (From *Fundamentals of Combustion*. International Textbook Co., Scranton, Pennsylvania, 1968, with permission of R. A. Strehlow.)

the diffusion coefficient to zero, thereby producing a flame with $V_A = V_B = 0$. With these assumptions, we may rewrite Eq. (56) to yield

$$\frac{dz}{d\rho} = \beta_c \frac{1 - \tau}{\tau - z} \exp\left(-\frac{\varepsilon}{\tau}\right) \tag{66}$$

for Le = 1, or

$$\frac{dz}{d\tau} = \beta_c \frac{1 - z}{\tau - z} \exp\left(-\frac{\varepsilon}{\tau}\right) \tag{67}$$

for Le = 0.

Figures 8 and 9 illustrate the structure of simple first-order decomposition flames for Le = 1 and Le = 0. These figures were obtained by integrating Eqs. (66) and (67) using the following assumptions: $m = 0.032$, $P = 101325$ Pa, $T_u = 300$ K (which means that $\rho_u = 1.348$ kg/m³), $C_p = 14.65$ J/mol K, $\kappa_u = 19.88 \times 10^{-3}$ J/m sec K, $k_1 = 10^{13}$, $E/R = 30,000$, and $T_b = 2000$ K. The calculated burning velocity using these assumptions are $S_u = 0.851$ m/sec for Le = 1 and $S_u = 2.546$ m/sec for Le = 0. Friedman and Burke found that for flames having Le = 1 or Le = 0, these assumed properties β_c may be quite accurately represented by the equation

$$\beta_c = \left[0.514\text{Le}\varepsilon^2 + (1.0 + 0.35\text{Le})\varepsilon + 0.20 + 0.05\text{Le}\right]\exp(\varepsilon) \tag{68}$$

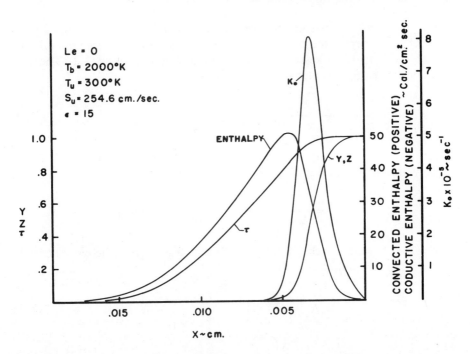

FIGURE 9. Flame structure for a flame with a Lewis number $= 0$ because the diffusion coefficient is assumed to be zero. Reaction is assumed to be first-order in the species A. Rate is the local rate of disappearance of the species A and equals $k_1(1 - X_B)\exp(-\varepsilon/\tau)$. (From *Fundamentals of Combustion*. International Textbook Co., Scranton, Pennsylvania, 1968, with permission of R. A. Strehlow.)

Note that β_c, as given by Eq. (68) has a dependence on $\varepsilon = E/RT_b$ which is primarily exponential. The significance of this will be discussed when a flame with a general kinetic expression is introduced.

The flame structure of Figures 8 and 9 must be compared to that predicted by the simple thermal theory. Note that in the Le $= 1$ flame the temperature and mole fraction curves for the product gas are coincident, whereas in the Le $= 0$ flame (with no diffusion) the net flux and mole fraction curves for the product gases are coincident. This simply means that when Le $= 1$, which is close to physical reality, the thermal conduction gradient is supported by the diffusion of the enthalpy of reaction carried with the reactive species, and there is very little convected enthalpy change throughout this flame. This means that thermal theory not withstanding, the diffusion of species in the preheat zone of a flame is important. This is particularly true because reactive radicals and atoms diffusing from the hot reaction zone can trigger the chemistry and therefore augment the propagation rate. Also note that reaction does not start to be significant until the temperature has risen by a significant amount. Thus the assumption of a high ignition temperature in the thermal theory appears to be justified.

We now wish to extend this simple theoretical model to investigate the effect of the order of the primary chemical reaction on the flame parameters. In this treatment we restrict ourselves to a flame with Le = 1. The rate equation therefore becomes

$$\frac{d(NX_B)}{dt} = k_\alpha (NX_A)^\alpha \exp\left(-\frac{E}{RT_b\tau}\right) \tag{69}$$

where α is the reaction order and k_α has appropriate units. This yields a new conservation equation (replacing Eq. (60)).

$$\frac{dz}{dx} = \frac{k_\alpha \rho^\alpha}{Mm(\alpha - 1)} (1 - X_A)^\alpha \exp\left(-\frac{\varepsilon}{\tau}\right) \tag{70}$$

Since Le = 1, we again find from Eq. (63) that $X_B = \tau$ and therefore arrive at the final flame equation:

$$\frac{dz}{dx} = \Lambda_c \left(\frac{1}{1 + \tau\phi}\right)^{\alpha - 1} \frac{(1 - \tau)^\alpha}{\tau - z} \exp\left(-\frac{\varepsilon}{\tau}\right) \tag{71}$$

where

$$\Lambda_c = \frac{k_\alpha \kappa_u \rho_u^\alpha}{M^2 C_p m^{(\alpha - 2)}} \tag{72}$$

and

$$\phi = \frac{T_b - T_u}{T_u} \tag{73}$$

Therefore

$$S_u = \rho^{[(\alpha/2) - 1]} \sqrt{\frac{k_\alpha \kappa_u}{\Lambda_c C_p m^{\alpha - 2}}} \tag{74}$$

In this flame of order α, Λ_c also shows a primary dependence on the activation energy and may be assumed to be roughly proportional to $\exp(\varepsilon)$ (see Eq. (68)). Therefore the theoretical model predicts that to a first approximation

$$S_u \propto P^{[(\alpha/2) - 1]} \exp\left(\frac{E}{RT_b}\right) \tag{75}$$

Equation (75) could have been obtained from Eq. (38) of the thermal theory if one postulated that the reaction in the flame had a rate given by $\omega \propto P^\alpha \exp(E/RT_b)$.

3.3. Comments on the Theoretical Models and Real Flame Behavior

Figure 10 is an experimental evaluation of the pressure exponent of the burning velocity, $n = \alpha/2 - 1$, for a number of hydrocarbon flames. Note that over the burning velocity range of about 0.4 to 1.0 m/sec the exponent n is 0, indicating that a second-order reaction is controlling the rate of propagation of these flames. Also note that above a burning velocity of about 1.0 m/sec, the exponent is positive and is increasing with burning velocity. Also note that for burning velocities less than about 0.4 m/sec, the exponent becomes increasingly negative as the burning velocity decreases. Both of these trends are explainable in terms of the complex kinetic processes that are occurring in these flames.

The higher value of the exponent for high-velocity flames can be explained by noting that these flames all have very high flame temperatures. Under these conditions an increase in the ambient pressure causes a marked decrease in the degree of dissociation of the hot flame gases, and therefore causes a marked increase in the rate of the chemical reactions in the flame, and this causes the pressure exponent to be positive in the high burning velocity regime.

The situation is more complex in the low burning velocity regime. It has been shown recently that in all hydrocarbon air flames, the diffusion of

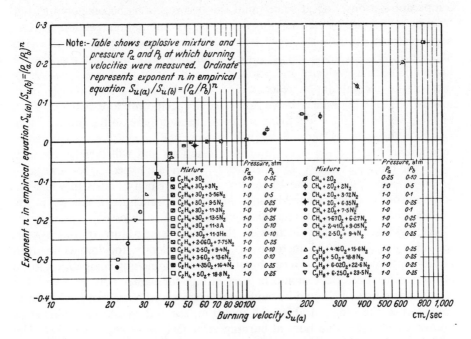

FIGURE 10. The pressure exponent n for a variety of hydrocarbon flames. (Courtesy of Dr. Bernard Lewis, Combustion and Explosions Research, Inc., Pittsburgh, Pennsylvania; originally appeared in AGARD, *Selected Combustion Problems*, p. 177. Butterworth, London, 1954.) Reprinted with permission.

hydrogen atoms into the preheat zone is the primary triggering mechanism for the chain-branching reactions that support flame propagation. It is well known that the hydrogen atom undergoes two primary competing reactions during any hydrocarbon oxidation process. These are the chain-branching reaction

$$H + O_2 \rightarrow OH + O \tag{76}$$

and the recombination reaction

$$H + O_2 + M \rightarrow HO_2 + M \tag{77}$$

Furthermore the chain-branching reaction has a first-order pressure sensitivity (because it is a second-order reaction) and has a large activation energy, which means that its rate increases very rapidly with increased temperature. On the other hand, the recombination reaction has a second-order pressure sensitivity (because it is a third-order reaction) and has essentially no temperature sensitivity at all. In the preheat zone of the flame, hydrogen atoms are diffusing toward the cold gas and at the same time reacting. In the higher-temperature regions of the flame, the chain-branching reaction dominates, and recombination is slow. However, as one travels toward the incoming gas, the temperature drops, and at some point the two reactions become competitive. In regions that are colder than this point, the recombination reaction becomes dominant and destroys hydrogen atoms. If we now consider any flame system and we lower the flame temperature by dilution with an inert gas, we will lower the burning velocity and also move the point in the flame at which the rate of the recombination reaction and chain-branching reaction are equal toward the hot boundary. If we now take a specific mixture that has a low burning velocity and a low flame temperature and increase the pressure level at which the flame is burning, we will increase the rate of the recombination reactions relative to the chain-branching reactions and will therefore tend to lower the burning velocity. In other words, we should expect to see a negative pressure exponent in Figure 10 for low burning velocity flames in which the competition between recombination and chain-branching reactions in the preheat zone is important, and this is exactly what is observed.

As was stated earlier, increasing the initial temperature of the mixture should cause a marked increase in burning velocity for low-temperature flames, and because of this recombination/chain-branching competition in the preheat zone, this effect should be very strong for flames that have very low burning velocities. It should be moderately strong for flames that have burning velocities in the range of about 0.4 to 1.0 m/sec and should have almost no influence on flames that burn at burning velocities above about 1 m/sec. This is because the high effective heat capacities of the dissociated product gases at these high temperatures will cause the flame temperature, and therefore the burning velocity, to change only very slightly as the initial temperature of the mixture is changed.

Experimental studies of real flames and theoretical modeling with relatively complete kinetic schemes have shown that in addition to the structural features of a flame illustrated by the simpler thermal and $A \to B$ theories, real flames contain a high-temperature region where the reactions slowly reach equilibrium. This has already been discussed in some detail in Chapter 3. Simply stated, the final oxidation of carbon monoxide to carbon dioxide occurs primarily through the exchange reaction

$$CO + OH \to CO_2 + H \tag{78}$$

which is relatively slow. Therefore in most flame systems the carbon monoxide concentration goes through a maximum and finally decays to its equilibrium value somewhat downstream of the rapid reaction region of the flame. Since the heat of the reaction (Eq. (78)) is about 80 KJ/mol at 2500 K, this final equilibration causes the temperature to rise downstream of the fast reaction zone.

Another point that must be made is that the theory of flame propagation developed in the previous section, is based on the a priori assumption that the flame is indeed a one-dimensional steady and self-propagating reaction wave of low subsonic velocity. Although many observed flames are known to behave in this manner, experiments have shown that there is a class of flames that prefer to propagate with a characteristic two- or three-dimensional structure rather than as a strictly one-dimensional phenomenon. Markstein has pioneered in this field and has published an extensive review[3] of the experimental and theoretical work on this so-called cellular instability.

Figure 11 contains a set of photographs of typical three-dimensional cellular flames in fuel-rich hydrocarbon-air-nitrogen mixtures propagating downward against a nearly balancing flow in a tube of 100 mm diameter. The photographs were taken using the light emitted by the flame. Clearly defined cellular structure may also be observed and studied in a rectangular slot burner, which has been specially designed to hold a flame on only one of its long edges. In this case flame curvature occurs primarily along the length of the slot, and the cellular structure may become strictly two-dimensional. Figure 12 is an edge-on photograph of a flame front for this case. Spontaneous cellular instability has also been observed in premixed gas flames having other geometries. For example, certain spherically expanding flames show the spontaneous formation of cellular structure, and certain mixtures when burned as attached flames on circular ports are known to exhibit a fluted appearance due to the occurrence of cellular instability.

Cellular structure is usually in a constant state of flux during flame propagation. Time-resolved photographs of three-dimensional flames have shown that large cells grow larger and then divide, while small cells are absorbed. Figure 13 is a streak photograph of the behavior of a typical two-dimensional cellular flame obtained on a slot burner. The dark regions are the troughs. Notice (1) that large cells split to form new cells and sometimes

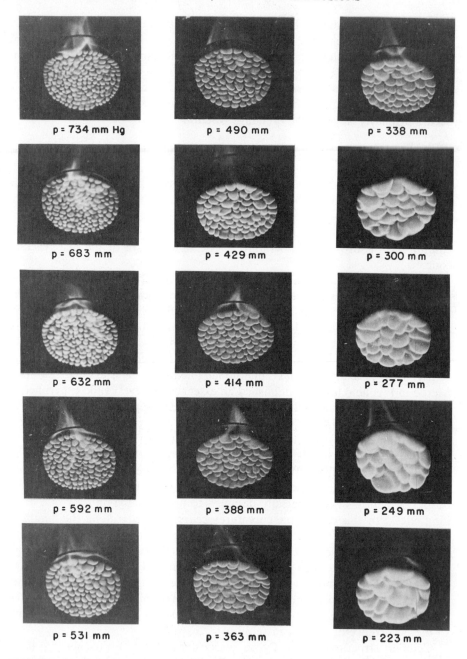

FIGURE 11. Cellular flames in fuel-rich hydrocarbon-air-nitrogen mixtures at atmospheric pressure. Tube I.D. = 10.0 cm. [Courtesy of G. H. Markstein, Factory Mutual Research, Norwood, Massachusetts; originally appeared in *J. Aeronaut. Sci.* **18**, 199 (1951). © American Institute of Aeronautics and Astronautics; reprinted with permission.]

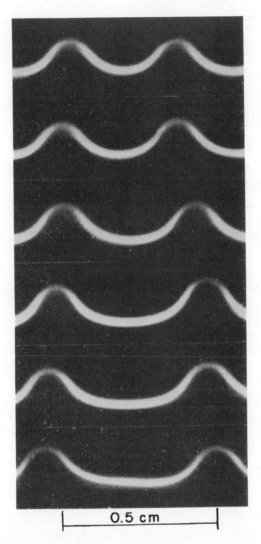

FIGURE 12. Profiles of steady cellular slot-burner flames, *n*-butane-air mixture, equivalence ratio = 1.47. Flow is upward. (Courtesy G. H. Markstein, Factory Mutual Research, Norwood, Massachusetts; originally appeared in G. H. Markstein, *Non-Steady Flame Propagation*. Macmillan, New York, 1964.)

these disappear entirely after their formation; (2) that the cells can propagate along the burner direction; and (3) that a fluctuation in one cellular region can propagate to other neighboring regions at a well-defined velocity. Other observations have shown that there is a characteristic minimum cell size for either two- or three-dimensional cells in any mixture, and that for two-dimensional cells a single cell spontaneously breaks into two cells when it reaches approximately twice the minimum size.

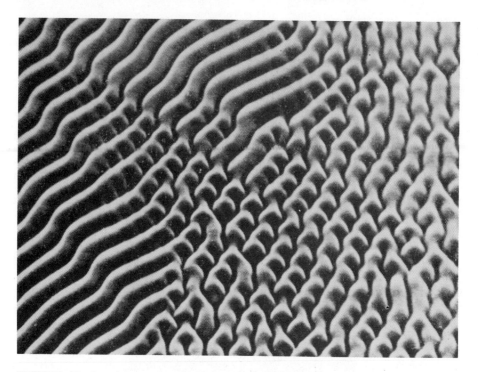

FIGURE 13. Streak-camera record of nonsteady cellular slot-burner flame. Time increases from top to bottom. *n*-butane-air mixture, $\Phi = 1.56$ (Courtesy of G. H. Markstein, Cornell Aeronautics Laboratory, Buffalo, New York; originally in G. H. Markstein *Non-Steady Flame Propogation*. Macmillan, New York, 1964.)

The most interesting and significant piece of information concerning spontaneous cellular structure is its dependence on the equivalence ratio of the combustible mixture. Experimentally it has been found that all the fuels illustrated in Figure 11 propagate spontaneously cellular flames in rich mixtures. In fact the appearance equivalence ratio has been found to be 1.00 ± 0.04 for all these hydrocarbons. It is also known that methane and hydrogen exhibit spontaneous cellular instability on the fuel lean side of stoichiometric. These facts have led to the postulation that cellular instability is due to the more rapid (preferential) diffusion of the lighter reactive species (fuel or oxygen) toward the hot reaction zone. If a trough develops in the flame, the lighter species are depleted in this region because they now diffuse more rapidly to the hot reaction zone on either side of the trough. If the initial equivalence ratio is correct, this depletion lowers the local burning velocity, and the trough deepens. This simple phenomenological theory agrees with our observations and has been verified by more sophisticated theory. Also the preferred cell size has been found to be proportional to the preheat zone thickness of the flame (i.e., $D \propto \eta_0$).

The cellular stability of flames, which is of interest in a fundamental sense, has little practical import except in lean systems in which a large percentage of the fuel is hydrogen.

3.4. Aerodynamics of the Laminar Flame, Flame Holding

Premixed gas flames may be observed as steady waves in a variety of laminar-flow situations, and except for the case of the flat flame on a flat-flame burner, they will exist as a steady-wave phenomenon only if the flow velocity of the main stream is well above the normal burning velocity of the mixture. This prevents the flame from propagating downstream into the apparatus. Thus all attached laminar flames are oriented obliquely to the flow, as indicated in Figure 5. We see from this figure that even though a flame of this local geometry will appear steady to an observer, an element of this flame is, in reality, propagating along the flame at a velocity $S_{\parallel}$. Thus, as shown in Figure 5, the flame can exist only as an apparently steady flame at some time t_2 at the point a, if at some earlier time t_1 it appeared steady at a point b whose distance from point a (along the flame in the upstream direction) is given by the expression $x_a - x_b = (t_2 - t_1)S_{\parallel}$. This implies that all steady flames must have an attachment region that continually reignites the oblique flame sheet, and that attachment must itself be steady if the flame is to appear steady.

From the vector relations in Figure 5, we note that any holding region must necessarily always be at the farthest upstream point of the steady flame, and we further see that at this point the local flow velocity vector and the local flame velocity must be equal in magnitude, coincident in direction, and of opposite sign. Furthermore, as we travel away from the attachment point into the bulk of the combustible mixture, the local flow velocity must always exceed the local burning velocity. If for any particular geometry and flow velocity these conditions are not met, the flame will either "flash back" (i.e., travel upstream), "blow off" (i.e., be washed downstream by the flow), or exist as some type of nonsteady repetitive phenomena. All three of these nonsteady behaviors have been observed.

The mechanism of steady attachment may be discussed by considering the flow in the neighborhood of an idealized burner rim as illustrated in Figure 14. Experimentally, it is observed that laminar flame attachment usually occurs within approximately 1 mm of the rim lip. The attachment point is therefore in the viscous boundary layer region for these low subsonic flows. For simplicity, we assume that the velocity gradient of the flow near the attachment point is a constant for any bulk-flow velocity and increases with the bulk-flow velocity. Typical velocity profiles just inside the rim are plotted as the straight lines 1 through 5 in Figure 14b. The normal component of the burning velocity for three arbitrarily chosen flame positions and quenching behaviors are also plotted as the curves 2', 3', and 4'. The differences in these curves are due to both the presence of the cold wall and the effect of diffusive mixing with the surrounding gas after the combustible jet issues from the tube. These are typical burning velocity curves for the flame positions shown in

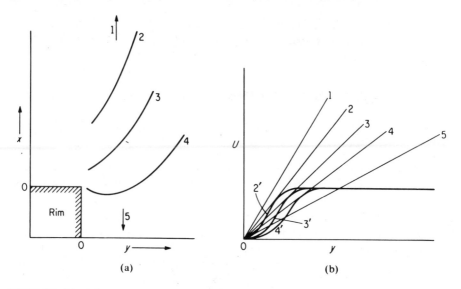

FIGURE 14. Schematic diagram of holding behavior (*a*) physical plane showing various stable flame positions, (*b*) *u-y* plane showing the relationship between the local flow and flame velocity. (From *Fundamentals of Combustion*. International Textbook Co., Scranton, Pennsylvania, 1968, with permission of R. A. Strehlow.)

Figure 14*a*. Curve 4′ represents the burning velocity for a laminar flame that is positioned entirely inside the rim of the burner. This curve is drawn tangent to the straight-line curve 4, which therefore represents a critical velocity gradient at the rim for flashback. Note that a lower-velocity gradient, such as 5, allows the local burning velocity to exceed the local flow velocity, and therefore will allow flashback to occur in the neighborhood of the rim. The velocity gradient represented by curve 3 will cause a flame in position 4 to blow off, but this flame will be stabilized in a new position because, above the burner, the effect of rim quenching is diminished before the effect of diffusion causes excessive dilution of the peripheral combustible gases. Thus as the flow velocity is increased, the flame will "seat" at progressively higher positions. However, its local burning velocity will eventually fall below the local flow velocity at all points in the flow, and blow off will occur. This blow off limit behavior is illustrated with curve 2′ in Figure 14.

This rather simple conceptual approach predicts that blow off and flashback should correlate well with the velocity gradient in the neighborhood of the holder. This correlation has been verified experimentally and Figure 15 is a typical summary of flashback and blow off behavior (in this case for a natural gas flame) plotted against the velocity gradient du/dy with units of sec^{-1}. Of these two critical behaviors, blow off is the more complex. This is because at flashback the flame is always attached in the immediate neighborhood of the rim and its attachment is therefore controlled primarily by the nature of

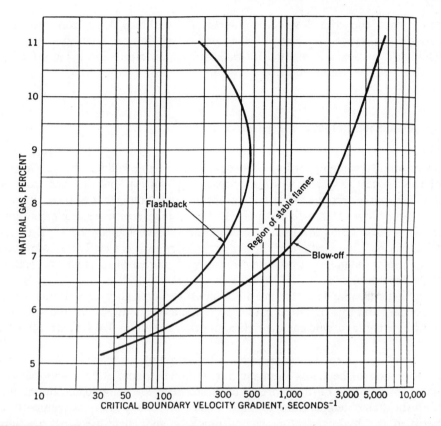

FIGURE 15. Flashback and blow off critical velocity gradient for a natural-gas air flame. (Reprinted with permission from B. Lewis and G. von Elbe, *Combustion Flame and Explosions in Gases*, 3rd ed. Academic Press, New York, 1987.)

quenching at the rim or holding body. Blow off, on the other hand, may occur when the flame is a considerable distance above the attachment body, and may therefore be controlled by other factors than simple quenching.

It has been shown experimentally[4] that the critical velocity gradient for blow off when nondimensionalized with the normal burning velocity and the preheat zone thickness of any lean flame yields a constant that is close to unity:

$$K = \left(\frac{\partial u}{\partial y}\right)_{rim} \cdot \frac{\eta_0}{S_u} \cong 1 \qquad (79)$$

The theory for this behavior is not well developed at the present time. Nevertheless, this correlation can be used to estimate the critical velocity for blow off for any lean fuel-air mixture when S_u and η_0 are known. Furthermore since the critical velocity gradient for flashback in a lean system is proportional to that for blow off, the same correlation can be used to estimate flashback gradients. A method for making such an estimation for any fuel-air mixture will be presented in Section 3.6.

3.5. Flammability Limits and Extinction

For every fuel-air combination there is a range of fuel compositions from some lower limit to some upper limit between which a flame will propagate. Outside of this range the system will not propagate a flame a long distance from an ignition source. From a safety standpoint, the most important of these limits is the lean or lower flammability limit (LFL) or the lean or lower explosion limit (LEL)—these terms are used interchangeably. Even though the UFL or the UEL can be important under certain circumstances, it is usually not important from a safety standpoint because dilution with more air can cause the mixture to be flammable.

In general, the lean flammability limit for typical hydrocarbon gas or vapor-air mixtures occurs at a fuel percentage of about 55% of the stoichiometric percentage, whereas the rich limit occurs at about 330% of stoichiometric.[5]

The flammability limit is thought to be caused by flame extinction due to a combination of heat loss from the flame, flame stretch, and/or spontaneous flame instabilities. Furthermore all of these effects are complicated by the fact that the flame temperature of limit flames is so low that the competition between the hydrogen atom chain-branching and chain-breaking reactions mentioned earlier occurs at a location very close to the hot reaction zone.

Unfortunately, there are no adequate theories for flammability limits. In essence, all the extant theories have so simplified the problem that the actual physical processes that occur at extinction are no longer modeled properly. As an example, take the heat loss theory of Spalding.[6] He models the flame as a one-dimensional flame and removes heat from the hot gases while retaining the one-dimensionality of the flame. He removes heat at a fixed rate per unit length of gas column downstream of the flame front and finds that there is a maximum value for this quantity above which one can no longer find a solution to the equations that were formulated. This critical maximum quantity of heat removal per unit length is then considered to be the condition for flame extinction. In contrast, experimentally one finds that for an upward propagating flame in a standard flame tube, extinction starts at the top center of the flame, and heat loss to the wall does not occur until the flame skirt actually touches the wall. Thus, in this case, heat loss has no effect on the extinction process. It is true that for a downward propagating flame in the same tube, heat transfer to the wall probably plays an important role in flame extinction. However, observations have shown that in this case the flame extinguishes by a complex three-dimensional nonsteady process.

One empirical observation[7] is that for many hydrocarbon-air mixtures, the lean limit volume percent of fuel multiplied by its heating value in kJ/g mol is approximately 4.35×10^3. This means, of course, that the flame temperature is approximately constant at extinction for all hydrocarbon fuels. If a lean limit is unknown for any specific mixed fuel, this rule can be used to estimate the lean limit.

The U.S. Bureau of Mines in Pittsburgh, Pennsylvania, has identified one particular technique for determining flammability limits as being their "standard" technique.[8] In this technique a 51-mm internal diameter tube 1.8 m long is mounted vertically and closed at the upper end with the bottom end opened to the atmosphere. The gaseous mixture to be tested for flammability is placed in the tube and ignited at the lower (open) end. If a flame propagates the entire length of the tube to the upper end, the mixture is said to be flammable. If the flame extinguishes somewhere in the tube during propagation, the mixture is said to be nonflammable. The choice of this technique as a standard is the result of a considerable amount of research. Specifically, it is known that upward propagation in a tube of this type has wider limits of flammability than downward propagation. In fact, if one ignites a mixture whose composition is between that of the upward and downward propagation limit in this tube, at the center of a large vessel, one finds that the flame propagates to the top of the vessel and burns only a portion of the material in the vessel before extinguishing; leaving a fair portion of the fuel-air mixture unburned. In other research it has been found that as the tube becomes smaller, the combustible range becomes narrower until one reaches the quenching diameter. At that point there is no mixture of that fuel with air that will propagate a flame through the tube. Fifty-one millimeters was chosen as the diameter for the standard tube because this is the diameter at which a further increase in tube diameter causes only a slight widening of the limits.

The initial pressure and temperature of the mixture affects the flammability limits somewhat. Increasing the temperature always widens the limits because it causes the flame temperature to increase. Increasing the pressure has little effect on the lower or lean limit but causes the upper limit to increase.

The addition of inert gases or inhibitors to the mixture causes the limits to narrow and eventually with sufficient added inert all fuel-air mixtures are nonflammable.[8] In the case of inert gases such as argon, nitrogen, or carbon dioxide, the effect is definitely due to the inert gas lowering the flame temperature. Chemical inhibitors such as the halons have a chemical inerting effect. They decompose in the hot combustion gases, act as radical scavengers, and thus markedly reduce the rates of the chain-branching reactions. It is interesting to note that at least for the inert additions, the fuel-oxygen ratio at the point where the lean and rich limit lines approach each other represents more closely the equivalence ratio where the products are CO and H_2O, not CO_2 and H_2O. This probably due to the fact, already mentioned earlier, that the one reaction that produces CO_2 in the flame, namely,

$$CO + OH \rightarrow CO_2 + H$$

TABLE 1. Scope of Existing Prediction Methods[a]

Method	Combustion Phenomena					
	Thermal Input	Lifting	Flashback	Yellow-Tipping	Sooting	Incomplete Combustion
Group I						
Wobbe	×					
AGA	×					
Willien	×					
Knoy	×					
Schuster						
Group II						
AGA		×	×	×		
Weaver	×	×	×	×		×
Group III						
Delbourg	×	×	×	×	×	×
Gilbert and Prigg	×	×	×		×	×
Holmqvist	×	×	×			
Harris and Lovelace	×	×	×			×
Harris and Wilson	×	×	×			×
France	×		×			
Grumer, Harris, and Rowe		×	×	×		
Van Krevelen and Chermin		×	×			
Van der Linden	×	×	×			×
Dutton	×	×	×	×	×	×
Group IV						
Sommers	×					

[a]Group I

C. Wobbe, *La definizione della qualità del gas*. L'industria del gas e degli acquedotta, 1926.

American Gas Association, *Mixed Gas Research Investigation*, Res. Rep. 689. AGA, New York, 1933.

L. J. Willien, "Variations in Interchangeability Formulae in High Btu Cases." *Am. Gas. Assoc., Proc.* (1938).

N. F. Knoy, "Combustion experiments with Liquified Petroleum Gases." *Gas* (*Houston*) 17, 14 (1941).

F. Schuster, "Über die Abhängigkeit der Wobbe-Zahl für Gasgemische von deren Zusammensetzung." *GWF, Gas-Wasserfach* 98, 630 (1957).

Group II

American Gas Association, *Interchangeability of Other Fuel Gases with Natural Gases*, Res. Bull. 36. AGA, New York, 1946.

E. R. Weaver, "Formulas and Graphs for Representing the Interchangeability of Fuel Gases." *J. Res. Nat. Bur. Stand.* 46, 213 (1951).

Group III

P. Delbourg, *Le contrôle de la qualitié et l'interchangeabilitié gaz*. Congrès du Gaz en France, 1951.

P. Delbourg, *Indices d'interchangeabilitié et charactéristiques de'utilisation des combustibles gazeux.* Congrés du Gaz en France, 1953.

M. G. Gilbert and J. A. Prigg, "Prediction of the Combustion Characteristics of Town Gas." *Trans. Inst. Gas Eng.* **106**, 530 (1956).

R. Holmqvist, *Internationell Bearbeting av Frågor angående Gasers Utbytbarhrt.* Svenska Gasverksföreningens, Arsbok, 1957.

J. A. Harris and D. E. Lovelace, "Combustion Characteristics of natural Gas and Manufactured Substitutes." *Inst. Gas Eng. J.* **8**, 169 (1969).

J. A. Harris and J. R. Wilson, Utilization—the burning issue. *Commun.—Inst. Gas Eng.* **949** (1974).

D. H. France, "Combustion Interchangeability Studies of Second-Family Gases. Ph.D. Thesis, University of Salford, Salford, England, 1976.

J. Grummer, M. E. Harris, and V. R. Rowe, "Fundamental Flashback, Blow-Off and Yellow-Tip Limits of Fuel Gas-Air Mixtures." *Rep. Invest.—U.S., Bur. Mines* **BI-5225** (1956).

D. W. Van Krevelen and H. A. G. Chermin, "A Generalized Flame Stability Diagram for the Prediction of the Interchangeability of Gases." *Int. Combust. [Proc.]* **7**, 358.

A. Van der Linden, "A New Combustion Diagram and Its Use in Modern Gas Technology." *Int. Gas Conf. [Proc.],* 11*th*, 1970, IGU/E15-70 (1970).

B. C. Dutton, "Interchangeability Prediction—The Framework for a New Approach." *J. Inst. Fuel* **51**, 225 (1978).

Group IV

H. Sommers and L. Joos, "SRG Method—a Contribution to the Interchangeability of Different Natural Gases." *World Gas Conf. [Proc.],* 12*th*, 1973, IGU/E29-73 (1973).

occurs late in the flame, after the CO concentration has peaked. This means that the steps that lead up to CO production are the important steps that drive flame propagation and that CO_2 production occurs so late that the heat generated in this step cannot contribute significantly to heat transfer to the preheat zone and therefore to self-sustanence of the flame.

In 1898 Le Châtelier[9] proposed a rule for determining the lean limit of a mixture of combustible gases from the known lean limits of the constituent species in the mixture. If we call L_m the volume percent of the fuel mixture at the lean limit, $L_1, L_2, \ldots, L_n$ the volume percent lean flammability limits of the n constituent fuels, and $c_1, c_2, \ldots, c_n$ the volume percent of each constituent fuel in the fuel mixture, le Châtelier's rule is given by the formula

$$L_m = \frac{100}{(c_1/L_1) + (c_2/L_2) + \cdots + (c_n/L_n)} \tag{80}$$

If the fuel contains an inert diluent such as nitrogen or carbon dioxide, the lean limit can still be calculated by not including the inert in Eq. (80). For example, the lean limit for methane air is 5%. If the methane is diluted with an inert and used as a fuel, this lean limit becomes

$$L_m = \frac{100}{(c_1/5) + 0} = \frac{500}{c_1}\% \tag{81}$$

where c_1 is the vol % CH_4 in the mixture.

The Le Châtelier formula works quite well if the fuel mixture contains hydrogen, carbon monoxide, or ordinary hydrocarbons. It does not work well for unusual compounds such as carbon disulphide, or when inhibiters are present.

3.6. Interchangeability of Fuels for Laminar Premixed Flame Burners

Because of the possible shortage of conventional gaseous fuels and the advent of new sources of gaseous fuels, the question of interchangeability of fuels on different burners has again become important. The properties that are considered important for proper burner behavior are (1) the proper thermal input, (2) lifting or blow off behavior, (3) flashback, (4) yellow tipping, (5) sooting, and (6) incomplete combustion and/or pollutant production. There have been many attempts to find a universal rule or set of rules that allow one to predict interchangeability. Recently Harsha et al.[10] have prepared an extensive review of the interchangeability prediction problem and have developed a rather complete catalog of existing prediction techniques. Table 1 from their catalog is reproduced here as Table 1 to give the reader a feeling for the amount of work that has been performed in this area. The different groups of Table 1 are as follows: group I, single index used; group II, multiple indexes used; group III, graphical techniques; group IV, other. The original references for the entries in this table are listed in the table's footnote. Relative to the problems currently associated with these various techniques for predicting interchangeability, Harsha et al.[10] state:

> Failures of existing interchangeability methods have been reported both formally and informally. The review of the published studies, combined with interface with gas companies, has identified the limitations and failure of existing prediction methods. Discussions with utilities have indicated that no common prediction technique has been employed; however, a preference for the AGA indices has been established. In general, results indicate that the AGA indices for lift, flashback, and yellow-tipping provide conservative predictions, that is, the indices predict non-interchangeability for fuel gases that laboratory experiments demonstrate to be interchangeable.

> Sixteen gas companies were surveyed and specific instances of the failure of existing procedures to reliably predict the interchangeability for residential systems have been documented.

> In addition, prediction methods are unable to reflect the needs of certain industrial consumers of fuel gases. Difficulties have been reported in operating protective atmosphere generators for heat treatment activities with propane-air, and in controlling flame size in wool and glass industries with LNG.

> Prediction procedures encompass single index, multiple indices, and diagrammatic methods. Although modern techniques have attempted to provide a more

representative description of the combustion processes by the introduction of more fundamental parameters, they do not provide a direct link between the controllable parameters and flame behavior. At present all existing prediction methods are empirical, and restricted to the fuel gases and residential burners from which the correlations are derived. In addition, the scope of existing methods is limited to flame stability and incomplete combustion criteria, and ignores other important phenomena including ignition, NO_x emissions, pilot performance, and flashback-on-ignition.

These statements are evidently true because the flames on these laminar burners are really quite small and therefore their limiting behaviors (the headings of Table 1) are very sensitive to detail of the burner construction (e.g., the shapes and sharpness of the nozzle) than to the usually measured gross properties of the flame such as the burning velocity. Also in these small curved flames preferential diffusion mechanisms can be expected to modify the local combustion process markedly. The interested reader is referred to Harsha et al.[10] for a more complete discussion of the problem.

4. DETONATION PROPAGATION: GAS PHASE

4.1. Gross Behavior

When a well-established detonation propagates down a straight channel of sufficient cross-sectional area, its gross, overall behavior can be calculated quite easily. It is well known that under these conditions the detonation propagates at a very constant supersonic velocity and that this velocity is very close to the velocity predicted by the upper Chapman-Jouguet point of Chapman-Jouguet theory. In other words, in coordinates that move with the detonation at its velocity, the wave appears, overall, to be a "steady" one-dimensional wave and the equations for one D steady inviscid flow, which were introduced in Section 2 of this chapter can be used to calculate both the average detonation velocity and the maximum average pressure at the front. The procedure for performing such a calculation was, in fact, discussed in detail in Section 2.

For a simplified treatment of detonation propagation in a fuel-air mixture initially at 1 atm pressure, one can determine the equilibrium Hugoniot over the pressure range of 1–20 atm and curve fit it to the rectangular hyperbola

$$\left(\frac{V_2}{V_1} - \beta_1 \right)\left(\frac{P_2}{P_1} + \beta_1 \right) = \beta_2 \tag{82}$$

to determine the constants β_1 and β_2, where

$$\beta_1 = \frac{\gamma - 1}{\gamma + 1} \tag{83}$$

TABLE 2. Hugoniot Curve Fit Data^a

Fuel	h_c Low Value MJ/kg Fuel	h MJ/kg Mixture	h/P_1V_1	q/P_1V_1	γ
H$_2$	120.00	3.40	28.86	33.89	1.173
CH$_4$	50.01	2.74	30.90	39.27	1.202
C$_2$H$_2$	48.22	3.38	39.12	44.79	1.195
C$_2$H$_4$	47.16	2.99	34.91	43.29	1.199
C$_2$H$_4$O	28.69	3.24	40.78	47.27	1.203
C$_3$H$_8$	46.35	2.78	35.68	47.42	1.208

(Curve Fit Data spans the last three columns: h/P_1V_1, q/P_1V_1, γ)

$^a\phi = 1.0$ (stoichiometric mixture).

and

$$\beta_2 = \frac{4\gamma}{(\gamma + 1)^2} + \frac{2q}{P_1V_1}\left(\frac{\gamma - 1}{\gamma + 1}\right) \tag{84}$$

from Eqs. (14) and (15). For a number of common fuel-air mixtures it has been found that this curve fit is surprisingly good (to within 0.25% on pressure at any specified final volume over the range 1–20 atm). This means that the overall behavior of a gas phase detonation can be modeled very adequately using the working fluid-heat addition model. Curve fit constants for six common fuels are given in Table 2.

Notice that the values of h/P_1V_1, as determined from the actual heat of combustion, is always lower than q/P_1V_1, the dimensionless heat addition needed to model the behavior of the real Hugoniot. This is simply an artifact of the model. It must be pointed out that this working fluid-heat addition model does not include molecular weight changes (which are small); nor does it correctly model the temperature or velocity of sound of the product gases.

Using the values of the last two columns of Table 2 the Chapman-Jouguet Mach number of the detonation may be calculated from the equation

$$M_{CJ} = \sqrt{\left[\frac{\gamma^2 - 1}{\gamma}\frac{q}{P_1V_1} + 1\right] \pm \sqrt{\left[\frac{\gamma^2 - 1}{\gamma}\frac{q}{P_1V_1} + 1\right]^2 - 1}} \tag{85}$$

using the positive sign solution (a negative sign on the inner square root sign yields the Chapman-Jouguet deflagration Mach number). The Chapman-Jouguet pressure may then be calculated using the equation

$$\frac{P_{CJ}}{P_1} = \frac{1 + \gamma M_{CJ}^2}{\gamma + 1} \tag{86}$$

It must be pointed out that the shock pressure is considerably higher than the Chapman-Jouguet pressure calculated using Eq. (86). In a practical sense P_{CJ} is more important than the shock pressure because the reaction zone in a propagating detonation passes over a point in space very quickly. In many cases even the Chapman-Jouguet pressure has a very small duration, and the simpler constant volume explosion pressure given by the equation

$$\left(\frac{P}{P_1}\right)_V = \frac{(\gamma - 1)q}{P_1 V_1} + 1 \tag{87}$$

should be used to estimate the effective pressurization of structures by a propagating detonation.

4.2. The Zel'dovich, von Neumann, and Döring Theory

During the period 1940–45, these three investigators independently constructed a model for a one-dimensional detonation wave which assumes a nonreactive shock transition followed by a one-dimensional reactive flow in which the chemistry, triggered by shock heating, goes to full chemical equilibrium, while the flow follows a Raleigh line to the CJ point.

The procedure for constructing such a ZND flow is as follows: For the mixture and initial state conditions of interest calculate the Chapman-Jouguet Mach number of the detonation. Use this Mach number to determine the nonreactive state (P, T) just behind the shock, and use these conditions and the prescribed mass flow to numerically integrate the reactive flow equations for steady inviscid flow until chemical equilibrium is reached. Since, mathematically, the CJ point is a strong singularity, forward numerical integration to that point is almost impossible. However, it is usually possible to integrate to within 95% of complete reaction before the singularity starts to cause trouble. Then since the CJ properties are known analytically, extrapolation is possible.

The trajectory of the calculation is indicated schematically on Figure 3. In this figure the Rayleigh line tangent to the upper CJ point, when extended to intersect the $q = 0$ shock Hugoniot, yields the conditions just downstream of the shock wave. Then the calculation follows the Rayleigh line to the CJ point.

One can also use the working fluid–heat addition model to represent the chemistry. In this case one defines an extent of heat addition

$$\phi = \frac{q}{q_{CJ}} \tag{88}$$

where q_{CJ} is the value of q tabulated in Table 2. Then, for any value of ϕ, one can calculate all the state and flow variables using the equations for volume ratio

$$\frac{V_2}{V_1} = \frac{(1 + \gamma M^2) - (M^2 - 1)\sqrt{1 - \phi}}{(\gamma + 1)M^2} \tag{89}$$

pressure ratio

$$\frac{P_2}{P_1} = \frac{(1 + \gamma M^2) + \gamma(M^2 - 1)\sqrt{1 - \phi}}{\gamma + 1} \tag{90}$$

and flow velocity

$$u_2 = \frac{V_2}{V_1} M a_1 \tag{91}$$

In order to determine the flow structure from these relations, one must of course know an effective reaction rate. In this steady flow we may write

$$\frac{d\phi}{dx} = \frac{1}{u_2 q_{\mathrm{CJ}}} \frac{Dq}{Dt} \tag{92}$$

whee Dq/Dt is the effective rate of heat release due to the chemical reactions and u_2 is the local flow velocity. Calculations of this type have been performed by Dove and Tribbeck[11] and by Strehlow and Rubins.[12]

4.3. Inherent Instability of a One-Dimensional Detonation

Early investigations of the stability of Chapman-Jouguet detonations were concerned primarily with a comparison of the properties of this flow with the properties of neighboring steady flows that also satisfy the Hugoniot and Rayleigh equations. However, transient stability cannot be fruitfully investigated using this technique, since this type of analysis does not allow for the occurrence of the truly nonsteady flow which is a possible solution to the equations of motion. In other words, any analysis of stability, to be realistic must specifically allow for nonsteady behavior.

Such realistic analyses have been made[13-15] and show that a planar ZND detonation that has been modeled to have reasonable overall reaction kinetics is linearly unstable to transverse acoustic wave propagation. This is because the region between the shock wave and the CJ plane in a ZND detonation is a region of subsonic flow and therefore can transmit acoustic signals for long distances transverse to the direction of propagation. Furthermore these propagating pressure perturbations are amplified because the exothermic reaction occurring in this region has a strong positive temperature dependence and Rayleigh's acoustic instability criterion yields a net growth of amplitude with time. Finally, the acoustic treatments show that the resulting flow should be three-dimensional and locally nonsteady and consist primarily of finite amplitude pressure waves (shocks) propagating across the main detonation front, transverse to the propagation of the front. Such frontal structure is indeed observed and will be discussed in the next section.

4.4. Frontal Structure

Detonation propagation in the gas phase was first investigated by Dixon[16] in the period 1880–1910. He used time-resolved photography to photograph detonations and measure their velocity. Even though his photographs show details that are caused by the complex frontal structure of the wave, he does not comment on this in his papers.

About 45 years after the start of Dixon's investigations, Campbell and Woodhead[17] in 1926 first observed frontal instability on a marginal detonation (i.e., one near the limit composition or pressure) in a round tube. The very violent phenomena that they observed is called *single spin detonation*. They spent considerable effort to document the fact that a marginal detonation, as it propagates down the tube, contains a single large amplitude frontal feature that travels in a helical path near the wall. They could not identify details of the structure but were able to show that the phenomena occurred in the gas and was not an interaction with an oscillation of the tube itself.

For over 25 years after its discovery by Campbell and Woodhead, it was thought that single spin was a result of a detonation being marginal and that well-established, self-sustaining detonations had a one-dimensional structure. This led to a proliferation of theories for stability of the CJ state based on entropy and velocity of sound considerations. It also led to the construction of the ZND model for detonation structure. The ZND model predicts that the shock pressure should be much larger than the pressure at the CJ plane where all the reactions are at equilibrium. This peak pressure is called the *von Neumann spike*, and the prediction of its existence in a detonation has led to many detailed experimental investigations of detonation structure. These studies led to the discovery that all self-sustaining detonations have a frontal structure that is three dimensional nonsteady. This structure is characterized by a nonplanar leading shock wave that at every instant consists of many curved shock sections that are convex toward the incoming flow. The lines of intersection of these curved shock segments are propagating in various directions at high velocity across the front and actually consist of triple-shock interactions (i.e., Mach-stem configurations). The third shock of these interactions extends back into the reactive flow regime and is required for the flow to be balanced at the intersection of the two convex leading shock waves. In general, the flow in the neighborhood of the shock front is therefore quite complex.[18,19]

One interesting feature of the triple-point Mach-stem intersections at the front is its ability to write a line on a smoked surface. These thin lines were first observed by Antolik[20] in 1875 on soot-coated plates held near a spark discharge. Mach later interpreted them as being caused by the intersection of "sound waves," and in 1877 he observed the same type of writing when a bullet passed by obstacles placed near a coated surface. This work preceded by eight years his use of spark schlieren photography to observe the shock waves associated with a traveling bullet. In 1960 Denisov and Troshin[21] revived the technique and first applied it to the observation of transverse waves in detonations. Since that time it has been extensively applied to the study of detonation structure, mainly because of its simplicity. The current consensus is that wood smoke deposited in an almost opaque layer on the surface produces the best smoke-foil records. The foils may be "fixed" after firing by spraying with a clear lacquer. A few examples of smoke-foil records obtained with propagating detonations are shown in Figure 16.

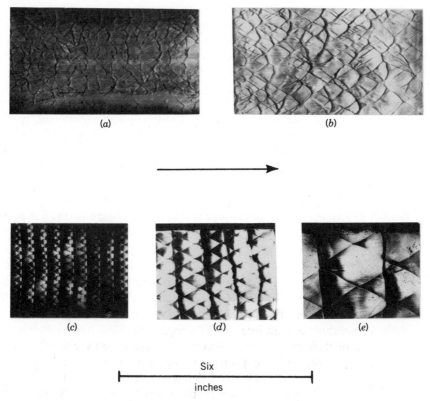

FIGURE 16. Smoked foil inscriptions obtained under the following conditions: (a) $2H_2 + O_2$ in $3\frac{1}{4} \times \frac{1}{4}$ tube, $P_0 = 150$ torr. (b) $2H_2 + O_2$ in 2-in.-diameter tube, $P_0 = 120$ torr. (c) $\frac{2}{5}C_2H_2 + O_2 + 75\%$ argon, $3\frac{1}{4} \times 1\frac{1}{2}$ in. tube, $P_0 = 100$ torr. (d) $2H_2 + O_2 + 70\%$ helium, $P_0 = 350$ torr. (e) $2H_2 + O_2 + 77.5\%$ argon, $P_0 = 100$ torr. (From *Fundamentals of Combustion*. International Textbook Co., Scranton, Pennsylvania, 1968, with permission of R. A. Strehlow.)

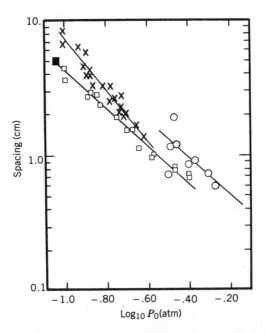

FIGURE 17. Effect of tube geometry on preferred spacing $2H_2 + O_2 + 70\%$ argon: O, round tube 2-in. diameter; □, $3\frac{1}{4} \times 1\frac{1}{2}$-in. rectangular tube; ×, $3\frac{1}{4} \times \frac{1}{4}$ in. planar tube; ■, single planar mode data point in $3\frac{1}{4} \times 1\frac{1}{2}$ in rectangular tube. (From *Fundamentals of Combustion*. International Textbook Co., Scranton, Pennsylvania, 1968, with permission of R. A. Strehlow.)

As can be seen from Figure 16, there is a wide variation in the regularity of the transverse wave structure in detonations. This regularity, or lack thereof, has been shown to be determined by some combination of the chemical kinetic and aerodynamic properties of the system. For example, stoichiometric hydrogen-, acetylene-, and ethylene-oxygen mixtures have poor structure, as shown in Figure 16, when pure but have very regular structure when highly diluted with argon or helium (Figures 16c, 16d, and 16e). Methane has highly irregular structure in all cases, and the higher hydrocarbons, C_2H_6, C_3H_8, etc., propagate detonations that have an intermediate regularity.

Irrespective of their regularity, all transverse wave systems, except the very irregular, show a propensity to propagate with a preferred transverse wave spacing that is a function of pressure and is somewhat dependent on the tube geometry and size. Figure 17 illustrates the effect of tube geometry on the spacing, and Figure 18 shows that there is a pressure-dependent preferred spacing but that a detonation with very regular spacing such as this one "locks in" to a mode number of the tube. Note in this figure that sometimes two modes can occur at the same pressure. These different modes appear on different detonations and are evidently caused by slight differences during initiation.

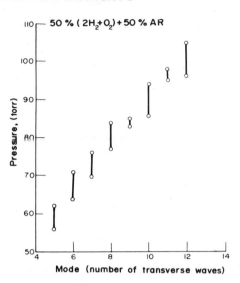

FIGURE 18. Mode number (number of transverse waves) observed in the $3\frac{1}{4}$ direction of a $3\frac{1}{4} \times 1\frac{1}{2}$-in. rectangular detonation tube as a function of initial pressure. [Originally appeared in R. A. Strehlow, A. A. Adamczyk, and R. J. Stiles, *Astronaut. Acta* **17**, 509–527 (1972), with permission of Pergamon Press PLC].

There are two independent theories that help explain the regularity of the structure and one other theory[22] that predicts the observed structure rather well, at least for the hydrogen-oxygen-argon system. The two that are concerned with structural regularity are actually concerned with the production of new transverse waves on the detonation front[23] or the destruction of existing transverse waves.[24]

In 1972 Barthel[22] presented a theory for transverse wave spacing that is based primarily on the observations that (1) equilibrium structure contains rather weak transverse waves ($M_r \cong 1.2$) and (2) acoustic ray tracing theory predicts that a random acoustic source in a ZND detonation originating between the shock and the plane where $a^2 - u^2$ is a maximum will always generate an acoustic caustic some distance away from itself at some other position behind the shock. A caustic is of course a region where acoustic energy is focused. Thus one should expect each caustic to generate a new acoustic wave propagating in both directions, and thus two new caustics should appear at the original source and twice that distance from the original source, again causing caustics at these locations. This caustic spacing was calculated by Barthel for CJ and slightly overdriven detonations and compared to experimental measurement of spacings in $[X(2H_2 + O_2) + (1 - \overline{X})Ar]$ mixtures for detonations at initial pressures of from 50–400 torr. Figure 19 is a comparison of measured and predicted spacings for various assumed overdrives from M_{CJ} to $1.2 M_{CJ}$. The agreement is excellent and is all the more striking because the structural regularity varies markedly over this composition range. Incidentally, Barthel used full H_2-O_2 kinetics and JANAF tabulated enthalpies to construct the background ZND flow.

The implications of these three theories are that caustic formation is the primary source of the preferred spacing and that the stability of the cell

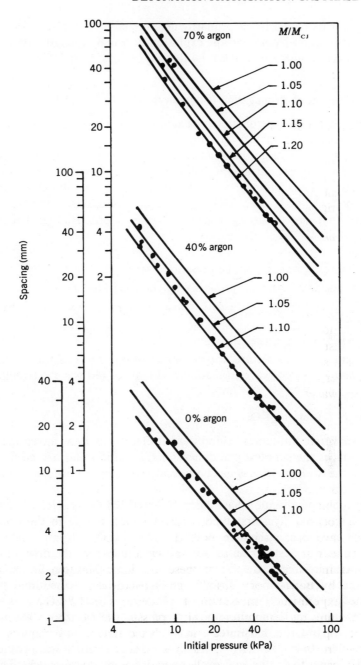

FIGURE 19. Transverse cell spacing as a function of pressure with degree of overdrive and argon dilution as parameters. Initial temperature is 25°C for calculated curves. [Originally appeared in H. O. Barthel, *Phys. Fluids* **17**, 1547 (1974), with permission of the American Institute of Physics.]

structure (or its regularity) is primarily determined by the relative values of the acoustic formation length and the explosion-destruction length, both of which are based on an analysis of the fully developed structures.

4.5. Mechanism of Propagation (Summary)

A Chapman-Jouguet detonation is a unique phenomenon in combustion, in that an inherent instability generates a finite amplitude three-dimensional nonsteady microstructure that leads to the self-sustaining time average steady propagation of a wave at the minimum velocity predicted by one-dimensional steady theory. Thus, although we still have no justification for either self-sustanence—or "why CJ?"—we do have a good understanding of the details of the processes that are occurring when the wave is propagating. These are as follows:

1. Transverse waves must be present.
2. Spacing is determined primarily by chemical kinetics.
3. The intersection of waves gives a chemical "kick" that sustains the waves.
4. Aerodynamics causes the waves to decay.
5. Regularity of structure is determined by the ratio of the characteristic spacing for new wave formation and that for old wave destruction.

4.6. Direct Initiation of Detonation

Planar Initiation. Initiation behind the reflected shock in a conventional shock tube produces the simplest one-dimensional initiation that has been observed experimentally. This process has been analyzed rather thoroughly by Strehlow and co-workers.[25]

Four types of initiation have been observed behind the reflected shock in these experiments. The first type is spurious and is due to the presence of bumps or crevasses on the tube walls or in the end plate. If large enough, these irregularities will destroy the one-dimensional nature of the flow and lead to local shock reinforcement and non-one-dimensional initiation. The second and third types of initiation are strictly one-dimensional processes and strikingly illustrate the nonsteady wave nature of initiation. Figure 20 contains schlieren and schlieren-interferometric photographs illustrating these two forms of initiation. Figure 20a illustrates one type of initiation, and Figures 20b and 20c the other. In both cases the initiation starts as a simple wave phenomenon traveling away from the end wall behind the reflected shock. This wave overtakes the reflected shock some distance from the end wall to produce a detonation in the incident gas flow. The nature of this strictly one-dimensional initiation process has been rather extensively analyzed by assuming that the development of a detonation under these conditions is primarily controlled by

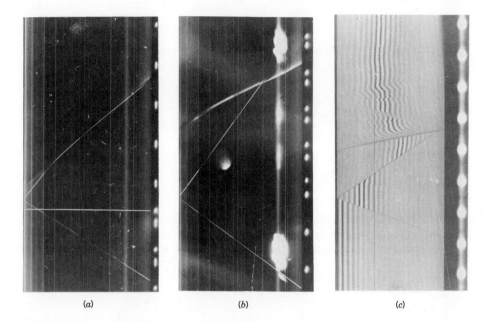

(a)   (b)   (c)

FIGURE 20. (x, t) photographs obtained behind the reflected shock in a conventional shock tube containing $2H_2 + O_2 + 70\%$ argon mixtures. Time increases upward, back wall to the right; timing dots are 100 μsec apart on left. (a) Schlieren photograph, (b) Schlieren photograph, (c) Schlieren interferometric photograph. (From *Fundamentals of Combustion*. International Textbook Co., Scranton, Pennsylvania, 1968, with permission of R. A. Strehlow.)

auto ignition due to the high temperature behind the reflected shock. This auto ignition takes on the appearance of a "reaction wave" triggered by reflected shock passage. This nonsteady "reaction wave" is fundamentally different from the "flames" discussed in Section 3 because its velocity is dictated by (1) the reflected shock velocity, and (2) a subtle interaction of local gas dynamics and reaction kinetics. Transport by diffusion or thermal conduction is essentially absent in this wave.

The fourth type of reflected shock-initiation behavior that has been observed in the hydrogen-oxygen system is illustrated in Figure 21. In this case, even though the system is apparently homogeneous, hot spots of weak ignition occur, and these each propagate spherically growing flames. When either of these flames or their pressure waves coalesce, they produce a local detonation, and the initiation process is no longer one dimensional. This type of initiation has been observed to occur in hydrogen-oxygen systems at high pressure and low temperature where the reaction

$$M + H + O_2 \rightarrow HO_2 + M \qquad (93)$$

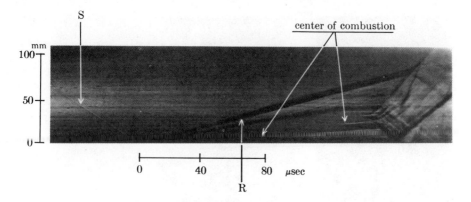

FIGURE 21. Reflected shock-initiation photographs similar to those shown in Figure 20, illustrating the occurrence of hot-spot initiation at low temperatures. Hydrogen-oxygen system, time increasing to the right back wall on the bottom. S is the incident shock. R is the reflected shock. (Courtesy of R. I. Soloukhin, Novosibirsk State University, Novosibirsk, U.S.S.R.)

becomes important during the chain-branching portion of the reaction because it competes with the rate-controlling chain-branching step. In the pressure and temperature region where this competition becomes adverse, the delay to explosion becomes very insensitive to temperature and pressure, and compression waves caused by the occurrence of random ignition centers do not augment the induction reactions in neighboring regions. Thus local flamelets propagate as spherically growing flame balls in a gas that is on the threshold of exploding, and the eventual initiation of detonation is a delayed three-dimensional event.

Planar initiation by heating the gas to the auto ignition temperature has also been observed behind an accelerating incident shock wave[26] and behind a constant velocity incident shock.[27] In all of these causes an initiation is thermal in nature and produces a detonation wave because the shock-heating process is wavelike, at a very high subsonic Mach number.

Spherical or Cylindrical Shock Wave Initiation. The rapid deposition of sufficient energy in a small volume or along a line can lead to the direct initiation of spherical or cylindrical detonation, respectively. Lee and Moen[28] have done most of the recent work on spark initiation. Bach et al.[29] first reported that there is a critical power density for initiation by spark, which is defined as the energy deposited divided by the spark kernel volume and the time of deposition. This quantity is invariant when different spark sources are used, even though the total spark energy ranges over a factor of 10^3. The spark energy was varied by such a large factor by using 20 nano second laser sparks, 1–10 μsec conventional capacitance sparks, and 100 μsec exploding wires.

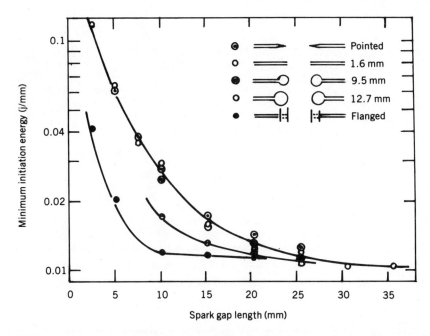

FIGURE 22. Critical energy for direct initiation with respect to electrode spacing for four electrode geometries in stoichiometric oxy-acetylene mixture at 100 torr initial pressure. [Originally appeared in H. Matsui and J. H. S. Lee, *Combust. Flame* **27**, 217–220 (1976), with permission of the Combustion Institute, Pittsburgh, Pennsylvania.]

Matsui and Lee[30] next showed that electrode shape had a profound effect on the initiation energy per centimeter of gap length for line sparks, particularly for short gaps. This is shown in Figure 22. Finally, Knystautas and Lee[31] showed that the first half cycle of a ringing spark discharge was the only phase of the discharge that determined the critical energy for direct initiation. In that paper they also determined that for very short sparks there is a critical minimum energy for initiation, even though for sparks of longer duration there is a limiting power. These two effects are shown in Figures 22 and 23.

These experimental observations show that spark initiation of detonation occurs when the spark generates a sufficiently strong shock that has a sufficient duration to heat the exothermic mixture to above its auto ignition temperature and hold it there for a time that is longer than the induction time. This is the reason why for long duration sparks the power density is important. Power density determines shock strength when the shock duration is long, and as Figure 23 shows, the power for initiation increases with total energy because the kernel size also increases with energy. Also note that there is a threshold critical energy below which initiation will not occur. This minimum energy reflects the fact that a critical minimum shock duration at a critical minimum

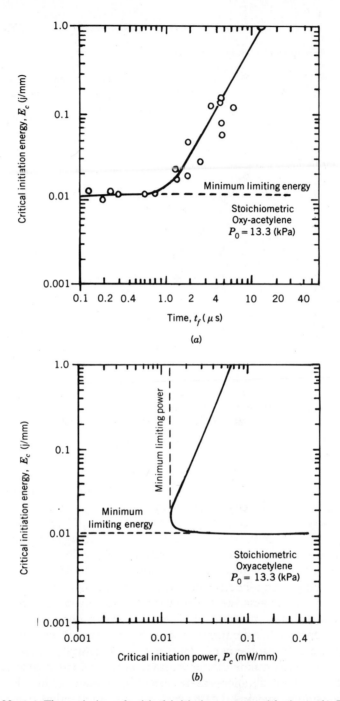

FIGURE 23. (*a*) The variation of critical initiation energy with time. (*b*) The dependence of critical initiation energy on the peak average power of the source. [Originally appeared in R. Knystautas and J. H. S. Lee, *Combust. Flame* **27**, 221–228 (1976), with permission of the Combustion Institute, Pittsburgh, Pennsylvania.]

strength is required for initiation. Finally, the fact that only the first half cycle of the power pulse from the spark is important to initiation reinforces the view that it is the initial shock wave generated by the spark that causes direct initiation.

Recently Lee and co-workers[32] have shown that the critical energy for direct initiation of detonation is directly related to the characteristic transverse spacing in a well-developed propagating detonation. Specifically, they find that for a large variety of fuel-oxidizer mixtures,

$$E_c = \frac{290\rho_0 u_{CJ}^2}{\gamma(\gamma - 1)} \cdot Z^3 \tag{94}$$

where E_c is the critical energy for direct initiation, ρ_0 and γ are the initial density and heat capacity ratio of the fuel-oxidizer mixture, and u_{CJ} and Z are the Chapman-Jouguet detonation velocity and characteristic spacing, respectively.

Free Radical Initiation. Direct initiation of detonation under conditions where the auto ignition temperature is never exceeded has been observed and studied by Lee et al.[33] They irradiated acetylene-oxygen, hydrogen-oxygen, and hydrogen-chlorine mixtures in a chamber with ultraviolet light, using quartz windows. The UV light is absorbed either by the oxygen or chlorine molecule and causes these molecules to disassociate, to form the free radicals O and Cl, respectively. These radicals, if in sufficiently high concentration, will trigger chain reactions and thereby cause a local "explosion" in the system. Three different explosion behaviors are observed. If the light intensity is low, the UV light penetrates only a very thin layer near the window, and the radicals in this layer ignite a flame or deflagration wave, which propagates away from the wall. For very high intensity irradiation, absorption is rather uniform throughout the volume, and the uniform high radical concentration causes a constant volume explosion of the vessel contents. In this case there are no waves of any consequence in the system during the explosion.

In the case of intermediate levels of radiation, a detonation is observed to form immediately and propagate away from the irradiated window. This behavior can be explained in the following way: The irradiation produces a high concentration of radicals at the wall and, more important, a gradient of radical concentration away from the wall. When this gradient has the proper shape, the localized explosion of the layer nearest the wall generates a pressure wave that pressurizes the neighboring layer and shortens the explosion delay time in that layer. This augmentation continues until a fully developed CJ detonation wave is formed. Lee et al.[33] call this *shock wave amplification by coherent energy release* (or the SWACER effect). They make an analogy to the behavior of a LASER because the simple explosion of one layer would not cause a significant reaction in the next layer if it had not been preconditioned to be ready to explode.

The significance of this mechanism of direct initiation is that the process is not triggered by the presence of a strong shock wave but instead offers a chemical mechanism for *generating* an accelerating shock wave that quickly reaches the CJ detonation velocity.

4.7. Detonation Failure

It has been observed that a well-established detonation propagating in a straight pipe will fail to form a spherical detonation when the pipe opens quickly into a large volume, unless the detonation in the pipe has 13 or more transverse cells across its diameter.[34,35] This statement is also true for a single circular orifice placed at the end of the pipe. In this case it is the orifice diameter that controls transmission or failure. Lee and co-workers[36] have found that for odd-shaped orifices, the rule of $d_c = 13Z$ also holds if d_c is calculated as the average diameter of an inscribed and circumscribed circle. This holds well up to length to width (L/W) ratios of 10. Also in the limit of very large L/W ratios the relationship $W_c = 3Z$ yields the critical width for transmission of a detonation. Finally, Lee and co-workers have found that detonation limits are single-spin limits and that, even though a detonation can propagate in the single-spin mode somewhat outside of its actual spin limit, small perturbations in the tubes structure will cause such a detonation to fail. The limit for single spin was also found to be related to the cell spacing. Specifically, $D_c = Z/2$ when D_c is the critical tube diameter below which the detonation will fail for that set of specified critical conditions. Minimum cell sizes for a few fuel-air mixtures are tabulated in Table 3. Since $ZP \cong$ constant when P is the initial pressure, these values will allow the evaluation of the minimum initiation energy and the various failure lengths at pressures in the range of 0.1 to 10 atm.

Recently Westbrook[37] and Westbrook and Urtiew[38] have performed a straightforward calculation of the induction delay time using a full kinetic mechanism for a constant volume explosion for initial conditions that correspond to nonreactive shock conditions at the computed CJ velocity. This delay time, τ, when multiplied by the flow velocity just behind the shock, yields a

TABLE 3. Minimum Transverse Wave Spacings for a Few Fuels in Air for 1 Atm Initial Pressure

Fuel	(mm)	Equivalence Ratio
Hydrogen	15	1.0
Ethylene	22	1.2
Ethane	50	1.0
Propane	50	1.0
Butane	55	1.0

length, Δ, which they find to be simply related to the properties E_c, d_c, W_c, and D_c. They also find that $E_c \propto \Delta^3$ and that all the failure lengths are simple multiples of Δ. Indications from their work is that $Z \cong 60\Delta$ when recombination is sufficiently rapid that the length of the recombination zone is immaterial. This is exactly what Strehlow and co-workers reported many years ago[39] for the hydrogen-oxygen, acetylene-oxygen, and ethylene-oxygen systems.

5. COMBUSTION SAFETY

5.1. Introduction

This section discusses flame properties related to combustion safety. The values that are determined for most of these properties are, to some extent, apparatus dependent. Historically, for each particular property, after some experimentation, a number of specific and apparatus dependent test techniques are developed, and quite often one of these techniques is declared the "standard" technique for that property. Unfortunately, in some cases, different groups of people accept different test techniques as their "standard" technique. This problem, which has led to some confusion in the marketplace, will be mentioned when appropriate.

One very important combustion safety property of any gaseous fuel is its flammability in air. The determination of flammability limits and their estimation was discussed in Section 3.5 and will not be discussed in this section.

5.2. Flash Point

When one is dealing with liquid fuels, one must also define a flash point for that fuel. Although it is true, as for flammability, that the flash point temperature of the fuel is dependent upon the technique used to measure it, the most reliable technique for pure fuels is the closed cup, equilibrium technique. In this technique the liquid in question is placed in a cup in a thermostat in contact with air, and the system is allowed to come to complete equilibrium at 1 atm pressure and that temperature. When equilibrium is reached, a portion of the top of the chamber is opened, and a small premixed flame is placed in contact with the vapor-air mixture. If the vapor-air mixture burns (or flashes), the thermostat temperature is said to be above the flash point of the liquid. If a flash does not occur, the temperature is said to be below the flash point of the liquid. With repeated tests, one can usually determine the flash point within a few degrees centigrade. A liquid is defined to be "flammable" if its flash point is less than 100°F (37.8°C). It is called "combustible" if its flash point is equal to or greater than 100°F.[40]

A rich limit flash point can also be measured because at some temperature the vapor pressure of the liquid will be so large that an equilibrium mixture of fuel with air will be above the upper flammability limit of the fuel. One should be cautious about upper flash points, however, because dilution of such a mixture with air will always cause the mixture to become flammable, even if only momentarily. It is true that with luck one can extinguish a match by dropping it into a partially filled gasoline can. This is because at room temperature the vapor pressure of gasoline is such that an equilibrium gasoline-air mixture has a composition above the ordinary rich limit. However, if any dilution with air had occurred prior to the introduction of the match, the can would undoubtedly explode. Thus lean limits are the important limit, from a safety standpoint.

Unfortunately, there are many "standard" techniques for measuring the flash point of a pure liquid or liquid mixture. In addition to the closed cup equilibrium method which is most accurate, there are also "open cup" methods in which there is no top on the cup and the vapor density of the fluid is supposed to keep the partial pressure of the vapor near the liquid close to its vapor pressure. The American Society for Testing and Materials (ASTM) has certified at least six different types of apparatus and procedures for flash point determination. The problem is not simple because viscous liquids can be heated only slowly, some liquids tend to "skin," and mixtures of liquids always change their composition as they evaporate. The problem is further complicated by the fact that different regulatory agencies at the local, state, and federal level have adapted different ASTM apparatus and procedures as standards.

Recently there has been some new research directed toward the understanding of the relation between the flash points and the flammability behavior of flammable liquid-air mixtures. Specifically, Affens and co-workers[41] have looked at the flammability properties of hydrocarbon solutions in air and have developed a "flammability index" for such mixtures based on the application of Raoult's law, Dalton's law, and the principle of Le Châtelier's rule governing the flammability of vapor mixtures. Additionally, Gerstein and Stine[42] have discussed an interesting anomaly for the case when an inert inhibitor is present. Essentially, they determined the reason why such a mixture can exhibit no flash point using the standard test procedure but still be flammable. This particular anomaly can be important when one is attempting to inert a hydrocarbon-air mixture with an inhibitor.

5.3. Auto Ignition Temperature

Vapor- and gas-air mixtures whose composition lie inside the flammable range can also be ignited by simply being heated to a high temperature. This type of ignition is usually studied in a heated vessel by an introduction technique. Alternatively, as described by Strehlow,[18] a shock-tube technique can be used to study this type of ignition. Physically it is observed that when the temperature is very high, the delay time to explosion in any particular configuration may be represented by an equation of the type

$$\log\{\tau[F]^{n}[O]^{m}\} = \frac{E}{RT} + A \tag{95}$$

where τ is a time delay before ignition in units of seconds, E is an effective activation energy, J/mol, and the concentration of fuel and oxidizer are represented by $[F]$ and $[O]$. The exponents n and m and the activation energy E were normally determined by a regression analysis using a number of data points. A good example of such a result is the curve for the initiation of methane taken from Seery and Bowman[43] (Figure 24). At a slightly lower temperature than those exemplified by the data of Seery and Bowman, the ignition to explosion cannot occur even though slow reactions may be occurring in the vessel or chamber. The temperature at which this happens for any fuel (using a standard test apparatus) is called the *auto ignition temperature* (AIT) for that fuel in air. A typical example of a measurement of the delay to ignition to determine an auto ignition temperature (AIT) is given in Figure 25 for normal propylnitrate. In this case the auto ignition temperature is 170°C.

In practice, most techniques that are used to measure the auto ignition temperature of a flammable liquid involve the addition of a small specified

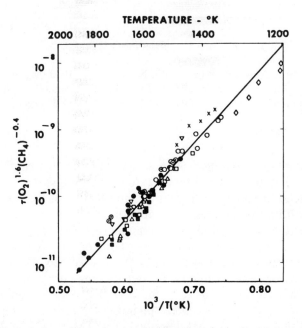

FIGURE 24. Correlation of measured induction time (τ, sec) with initial O_2 and CH_4 concentration (mol/m^3) and reflected shock temperature (K). (From Ref. 44, with permission of the Combustion Institute, Pittsburgh, Pennsylvania.)

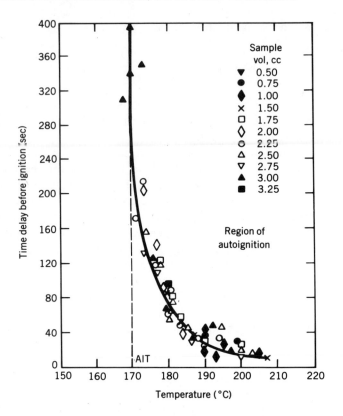

FIGURE 25. Time delay before ignition of NPN in air at 1000 Psig in the temperature range from 150° to 210°C. [From M. G. Zabetakis, "Flammability Characteristics of Combustible Gases and Vapors." *Bull.—U.S., Bur. Mines* **627** (1965).]

quantity of the liquid to a flask of some specified size that contains air and is held at some high temperature by an external heating source. In general, the operator watches for a flash of light after introduction of the sample. If a flash is observed within five minutes, auto ignition is said to have occurred. As one can imagine, the auto ignition temperature is a very apparatus-dependent quantity. A comprehensive review of the status of auto ignition temperature measurements was presented by Setchkin.[44] More recently, Hilado and Clark[45] have again reviewed the techniques for determining auto ignition temperature and present a compilation of data for over 300 organic compounds. They report that the most widely accepted apparatus and procedure for determining auto ignition temperature is that described in ASTM 2155.[46] In this apparatus a small amount of the sample is injected into a heated 200-mm Erlenmeyer flask, and the interior of the flask is watched for five minutes after injection of

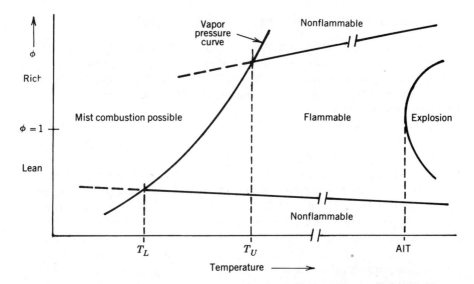

FIGURE 26. Effect of temperature on limits of flammability of a pure liquid fuel in air. T_L = lean (or lower) flash point; T_r = rich (or upper) flash point; AIT = auto ignition temperature. (From M. G. Zabetakis, "Flammability Characteristics of Combustible Gases and Vapors." *Bull.—U.S., Bur. Mines* **627** (1965).]

the sample. If no flash is observed, the temperature is said to be above the auto ignition temperature for that particular compound.

It appears that the determination of auto ignition temperature will always be rather imprecise and apparatus dependent. Nevertheless, the relative values of the auto ignition temperature of various compounds yield an indication of allowable surface temperatures of equipment and apparatus when exposed to air-vapor mixtures of these compounds. Thus these rather imprecise and apparatus-dependent auto ignition temperatures are useful from a practical standpoint. Also, because of the difficulty of determining a precise auto ignition temperature, there is little need for new research in this particular area of combustion safety.

The relationship between flash points, flammability limits, and the auto ignition temperature for a typical liquid hydrocarbon fuel in air is shown schematically in Figure 26. Notice that typically the lean flammability limit does not change very much as temperature increases, whereas the upper flammability limit increases markedly with increased temperature. The intersection of the vapor pressure curve of the fuel with the flammability limit curves defines the equilibrium flash points for lean and rich limits, as discussed earlier. Below this temperature a mist could be formed, which would also be flammable. At some high temperature the region of auto ignition occurs. This region usually has a minimum somewhere near the stoichiometric composition.

5.4. Minimum Ignition Energy and Quenching Distance

These two flame properties are quite useful for characterizing practical combustion systems. The minimum ignition energy is the smallest quantity of energy that needs to be added to a system to start flame propagation. Its value is quite dependent on the local rate and method of heat addition, and on the geometry of the heat source. Quenching distance is defined as the largest channel dimension that will just keep a flame from propagating throughout the channel. Values of the quenching distance are also quite dependent on experimental geometry. When a spark is passed through a combustible gas mixture, successful ignition leads to the propagation of either a flame or a detonation. As the total quantity of energy is reduced, direct detonation ceases to occur, if it did at all (see Section 4.6 for a discussion of detonation initiation), and flames are observed to propagate from the source. Further reduction of the total quantity of energy that is added to the fluid continues to produce ordinary flame propagation until the minimum ignition energy for that source configuration is reached. At this point only a small quantity of the combustible gas is converted to products, and the energy is dissipated harmlessly by thermal conduction.

A multitude of different energy sources and source configurations have been used to study flame ignition experimentally. However, an ordinary capacitor discharge spark has consistently yielded the lowest ignition energy for any specific combustible mixture and has a geometry and temporal history that are quite amenable to mathematical analysis.

For an experimental determination of an ignition energy, high-performance (usually air-gap) condensers are used so that the majority of the stored energy will appear in the spark gap. The stored energy may be calculated using the equation $E = CV^2/2$, where C is the capacity of the condenser and V the voltage just before the spark is passed through the gas.The spark must always be produced by a spontaneous breakdown of the gap because an electronic firing circuit or a trigger electrode would either obviate the measurement of spark energy or grossly change the geometry of the ignition source. It has been found experimentally that for this type of spontaneous spark, up to 95% of the stored energy appears in the hot kernel of gas in less than 10^{-5} s. The losses are thought to be due primarily to heat conduction to the electrodes. Since the total stored energy can be varied by changing either the capacity or the voltage, the electrode spacing (which is proportional to the voltage at breakdown) may be varied as an independent parameter. Two problems now arise: if the electrode spacing is too small, the electrodes will interfere with the propagation of the incipient flame and the apparent ignition energy will increase. If, however, the spacing is too large, the source geometry will become essentially cylindrical, and the ignition energy will again increase because the area of the incipient flame is greatly increased in this geometry. This condition is shown schematically in Figure 27a. The fact that the increase in minimum ignition energy is due to quenching at small electrode spacings has been

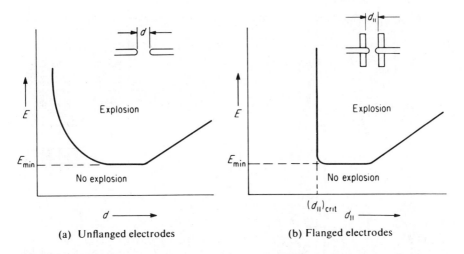

(a) Unflanged electrodes   (b) Flanged electrodes

FIGURE 27. Quenching distance and minimum-ignition-energy measurement with a capacitor spark. (From *Fundamentals of Combustion*. International Textbook Co., Scranton, Pennsylvania, 1968, with permission of R. A. Strehlow.)

confirmed by using electrodes flanged with electrically insulating material. Figure 27b shows the effect of electrode flanging on the ignition energy at small separations. Note that below a certain spacing, as the spacing is reduced, the spark energy required to ignite the bulk of the gas sample rises very much more rapidly with flanged tips. This critical flange spacing is defined as the flat-plate quenching distance of the flame. Therefore this simple experiment may be used to determine the flat-plate quenching distance and the minimum ignition energy for spherical geometry, and may also be used to evaluate whether the ignition source is essentially spherical or cylindrical. The latter piece of information is obtained by determining the effect of electrode separation on E_{min} at large separations.

Quenching distances may also be measured by quickly stopping the flow through a tube of the desired geometry when a flame is seated on the exit of the tube. If the flame flashes down the tube, the minimum dimensions are above the quenching distance for that mixture. Quenching distances that have been measured between two flat plates using this technique agree quite well with the quenching distances measured using the flanged electrodes in the spark-ignition experiment described above. Parallel-plate quenching distances and minimum ignition energies determined by the spark method for propane, oxygen, and nitrogen mixtures at various initial pressures are given in Figure 28. It should be noted that, in general, minimum ignition energies are extremely small—in some cases, corresponding to the passage of a barely audible spark through the mixtures. In fact it is well known that static electric sparks can cause the ignition of many explosive mixtures.

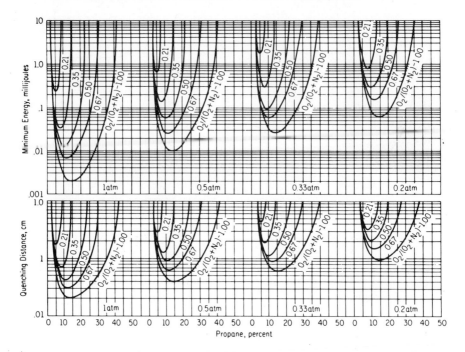

FIGURE 28. Effect of pressure, nitrogen dilution, and stoichiometry on propane-oxygen flat-plate quenching distances and spark minimum-ignition energies. (Reprinted with permission from B. Lewis and G. von Elbe, *Combustion Flames and Explosions of Gases*, 1st ed. Academic Press, New York, 1951.)

The relation between the characteristic quenching distance and the geometry of the quenching tube has been developed theoretically and confirmed experimentally in a number of instances. The theory can be based on the assumption that (1) wall capture of reactive species, or (2) heat transfer to the wall controls the quenching. Both these approaches yield the same theoretical relations relative to geometry effects, because of the similarity between diffusion and thermal conduction. Quenching measurements for a variety of geometries show quite good relative agreement ($\pm 10\%$) with the theoretical predictions for a number of hydrocarbon systems. The theory predicts that the circular tube quenching diameter should be related to the parallel plate quenching distance by the equation

$$d_{\parallel} = 0.65 d_o \tag{96}$$

Experimentally, one may show that the quenching distance is simply related to the pressure (for any one flame [Pd = constant]) and also that it is related to the minimum ignition energy of the system for virtually all hydrocarbon-air

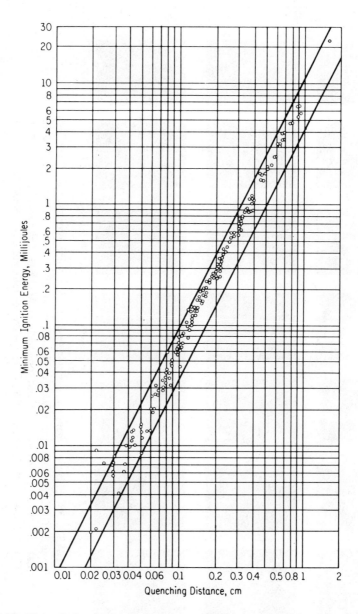

FIGURE 29. The relation between flat-plate quenching and spark minimum ignition energies for a number of hydrocarbon-air mixtures. (Reprinted with permission from B. Lewis and G. von Elbe, *Combustion Flames and Explosion of Gases*, 1st ed. Academic Press, New York, 1951.)

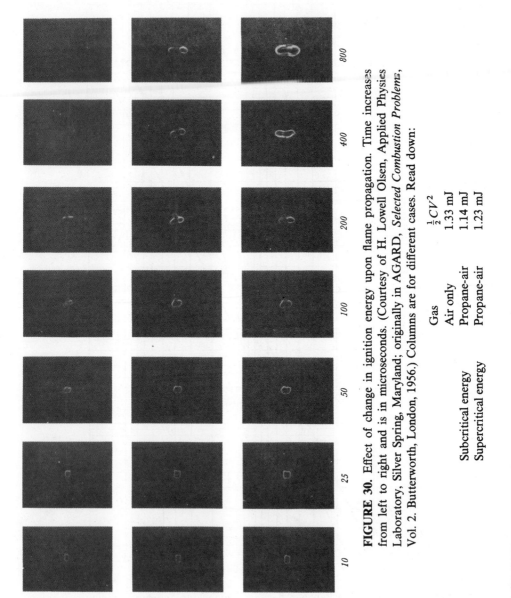

FIGURE 30. Effect of change in ignition energy upon flame propagation. Time increases from left to right and is in microseconds. (Courtesy of H. Lowell Olsen, Applied Physics Laboratory, Silver Spring, Maryland; originally in AGARD, *Selected Combustion Problems*, Vol. 2. Butterworth, London, 1956.) Columns are for different cases. Read down:

	Gas	$\frac{1}{2}CV^2$
	Air only	1.33 mJ
Subcritical energy	Propane-air	1.14 mJ
Supercritical energy	Propane-air	1.23 mJ

flames, through the equation

$$E_{\min} = 0.06 d_{\parallel}^2 \quad (\text{units} = \text{J}/\text{mm}^2) \tag{97}$$

This is shown in Figure 29. It has been found empirically that the parallel plate quenching distance is approximately three to four times the calculated preheat thickness of the flame. Thus once the normal burning velocity is known for any specific mixed fuel–inert-air mixture, Eq. (39) can be used to determine η_o and thus $d_{\parallel}$. Thus Eq. (96) can be used to determine the minimum ignition energy for that specific mixture.

We now examine the details of incipient flame growth from a small spark. Figure 30 contains repetitive schlieren photographs illustrating the appearance of the hot kernel of gas for an inert mixture and for a reactive mixture with both a subcritical and supercritical energy addition. In the reactive case the spark kernel initially contains temperatures well above the ignition temperature of the mixture, and an incipient flame is seen to form. The behavior should be contrasted with the inert case where the initially high gradient simply flattens with time and eventually disappears. However, as the two reactive cases show, the formation of an incipient flame does not insure subsequent propagation. At some point, well after spark passage (ca. 100 μsec in Figure 30), the incipient flame becomes a true flame in the supercritical case and starts to decay by thermal conduction in the subcritical case. This behavior is shown schematically in Figure 31. Thus there is evidently a

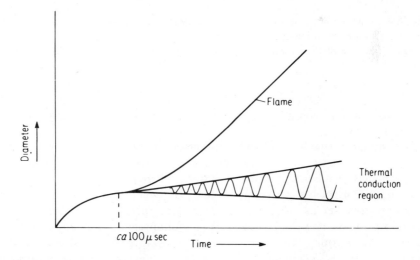

FIGURE 31. Schematic diagram of the decay or growth behavior of an ignition kernel (incipient flame ball) near the threshold value of energy for ignition of a propagating flame. (From *Fundamentals of Combustion*. International Textbook Co., Scranton, Pennsylvania, 1968, with permission of R. A. Strehlow.)

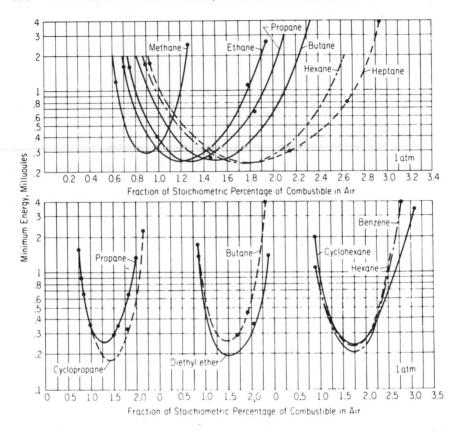

FIGURE 32. Minimum ignition vs. stoichiometry for various fuel-air mixtures showing the effects of preferential diffusion. (Reprinted with permission from B. Lewis and G. von Elbe, *Combustion Flames and Explosions in Gases*, 1st ed. Academic Press, New York, 1951.)

minimum ignition volume that will lead to stable flame propagation for spherical geometry.

One difficulty that has not been discussed is the problem of preferential diffusion associated with curved flame fronts. Evidence that preferential diffusion influences flame development in systems where the fuel and oxidizer have different diffusivities is shown in Figure 32, which is a plot of minimum ignition energy versus composition for various fuels in air. Since from the elementary theory one would expect the minimum ignition energy to occur at the maximum burning velocity of these mixtures, these curves should all show a minimum in the neighborhood of $\phi = 1.1$ (the approximate equivalence ratio for the maximum laminar burning velocity for all these systems). Notice that for methane, which is lighter than oxygen ϕ_{min} is 0.8, whereas for those fuels whose diffusivity is less than oxygen, ϕ_{min} is always greater than one and increases as the fuel molecular weight increases. Evidently, preferential diffusion of the lighter species in these highly curved flames changes the composition in the flame region markedly during the ignition process. When this composition approximates the stoichiometric composition, the minimum ignition energy is observed because the incipient flame has the maximum burning velocity. The larger the difference in relative diffusivity, the larger the deviation of ϕ_{min} from stoichiometric.

Another indication that preferential diffusion plays an important role in incipient flame behavior is obtained from the measured burning velocities of outwardly propagating flames. For example, as shown in Figure 33 in propane-air mixtures (i.e., for a heavy fuel and light oxidizer mixture), rich mixtures exhibit a higher space velocity at small radii, whereas lean mixtures

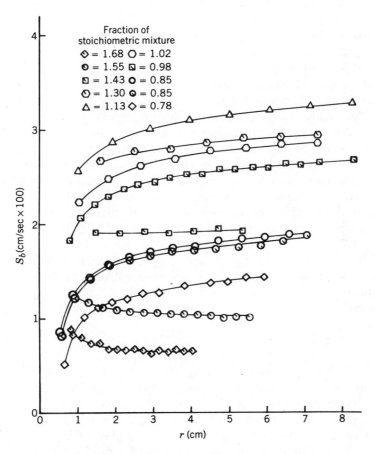

FIGURE 33. Space velocity S_b vs. radius for a laminar flame ball in propane-air mixtures. (Originally appeared in A. Palm-Leis and R. A. Strehlow, *Combust. Flame* **13**, 111 (1969), with permission of the Combustion Institute, Pittsburgh, Pennsylvania.)

show a lower space velocity at small radii. In all cases the space velocity of the flame relaxes to the theoretically expected laminar space velocity after a few centimeters of flame travel. The effect is somewhat masked by two additional effects that cannot be easily discussed qualitatively: (1) The slow relaxation of the burned products to the equilibrium composition adds a slow initial increase in velocity to the experimental curves, and (2) any excess ignition energy at the spark source can cause a higher propagation velocity of the early flame. However, the preferential diffusion effect is the only effect that shifts sign as the equivalence ratio passes through unity and is therefore quite definitely operating in this system.

5.5. Maximum Experimental Safe Gap (MESG)

The need to determine the quantity called the *maximum experimental safe gap* (MESG) is predicated on a philosophy relative to the design of electrical equipment that is to be operated in a hazardous atmosphere containing combustible vapors or gases. Specifically, the philosophy of all regulations relative to such equipment is that one should assume, unless pressurized purged enclosures are used, that the combustible gases or vapors will be able to enter the enclosure that contains the electrical apparatus. Since the sparks that are produced by the equipment will normally have sufficient energy to cause ignition of the combustible mixtures, the enclosures must be designed such that they are strong enough to contain an internal explosion without rupturing and also must have clearance gaps to the outside that are small enough such that venting of the hot products inside the chamber will not cause ignition of the combustible mixture outside the chamber. By venting hot product gases under pressure, this phenomenon (i.e., ignition) is uniquely different than the flame-quenching process that was described in an earlier section. Here the hot product gases are being forced through the gap at high velocity, and the critical gap widths at which propagation in the surrounding combustible mixture will just not occur are found to be approximately one-half the normal parallel plate quenching distance for most hydrocarbon-air mixtures.

The measurement of a maximum experimental safe gap therefore, involves the construction of a specific apparatus for this type of test. As it turns out, at the present time, three uniquely different designs have been used rather extensively for MESG testing. These are the 20-mL bomb shown in Figure 34, the 8-L bomb shown in Figure 35, and the Underwriter's Laboratories Westerburg apparatus, which is described in Dufour and Westerberg.[47] In principle, each of these apparatuses contain a chamber, which is the ignition chamber, and it contains an opening to the outside that consists of a gap of fixed length and width with an adjustable thickness. In each case the apparatus is constructed sturdily enough so that the gap thickness will not change due to internal pressure during an experiment.

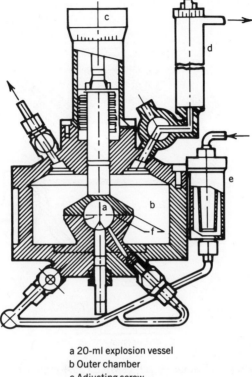

a 20-ml explosion vessel
b Outer chamber
c Adjusting screw
d Pressure adjustment
e Filling filter
f Flange gap

FIGURE 34. Twenty-milliliter explosion vessel. (Courtesy of Physikalisch-Technische Bundesanstalt, Postfach 33 45, W-3300 Braunschweig.)

Because of the large differences in the details of the construction of these three pieces of apparatus, the dynamic response during an experiment varies markedly from apparatus to apparatus.

The 20-mL apparatus is so small that the total time for exhausting the system with a nominal burning velocity of 40 cm/sec is of the order of 20 msec, and the pressure in the apparatus never rises above approximately 2 atm because of the large gap area relative to the volume of the chamber. In the 8-L vessel the absolute size and the ratios of gap size to the volume of the chamber are such that exhausting time is approximately 100 msec and the internal pressure rises to 3 to 4 atm. However, in the Westerberg apparatus the chamber is very large and the gap has a very narrow width (only 4 in.). This means that in this chamber the internal pressure almost reaches that which one would expect from a closed bomb, and the exhausting time approaches 1 s.

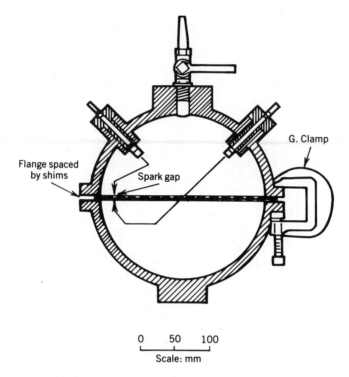

FIGURE 35. Eight-liter explosions vessel. (SMRE) 25.4-mm (1-in.) flanges, assembly: four or six G-clamps. (Reprinted with permission from Health and Safety Executive, Sheffield, England.)

The effect of chamber volume of the measured MESG for a number of compounds was studied extensively in Britain and Germany over the past 15 years.[48] Based on this work, the international standard apparatus for MESG is now the 20-mL bomb. In the United States the Westerberg apparatus is used as the standard. Additionally, in the United States, because electrical systems are housed in conduit, there is the requirement that one must also test the compound by igniting it some distance away from the test volume in a 1-in. pipe. This produces a runup to detonation condition, which in the industry is called *pressure piling*. This pressure piling can lead to much higher dynamic pressures at the gap than one would observe in a typical bomb situation. The U.S. testing technique also allows for tests in which the combustible gas in the enclosure is turbulent, thus causing a more rapid rate of pressure rise. In designing U.S. electrical proof equipment, these effects are normally taken into account.

The differences in behavior of the three different apparatuses have been investigated rather extensively by Strehlow et al.[49] Their investigation addressed reasons why certain compounds showed markedly different MESG's from apparatus to apparatus. In a number of cases, retesting in the UL apparatus reduced the anomalous behavior. Additionally, Phillips and co-workers[50] have constructed a theoretical model for the safe gap ignition process, which quite faithfully reproduces the experiment observations. Experimentally, it is observed that the hot gases escaping through the gap do not carry a flame with them, but instead, mix with the cold combustible gases external to the gap in a highly intense turbulent mixing zone some distance from the exit of the gap. If conditions are right (i.e., if the gap is large enough), after some delay time this region "explodes," and a flame propagates away from it through the rest of the combustible mixture in the surrounding enclosure. If the gap is small enough, the temperature in this region simply drops monotonically with time, and no ignition is observed. The Phillips theory predicts this behavior quite well.

The National Electric Code[51] defines four classifications for gases or vapors. These are A, B, C, and D. Class A contains only the compound acetylene. The standard compound for class B is hydrogen, and that for class C is diether ether. Gasoline is the standard compound that defines the safe gap for class D compounds. For example, if a measured safe gap for a particular compound lies between that of diether ether and that of gasoline, it is considered to be a class C compound and can be used only in locations where the equipment has been classified as class C equipment. In order to obtain a classification of this type, the equipment itself must be tested in an explosive environment, using diether ether as the fuel. If the compound has a safe gap that is larger than that of gasoline, it will be considered to be class D, and any equipment must be tested with gasoline as a fuel before it can be used in an atmosphere that contains that particular fuel.

Measured minimum experimental safe gaps have been found to be approximately half of the flat plate quenching distance for most fuels.

5.6. Inductive Spark Ignition

In ordinary industrial environments, most frequently a spark is obtained by breaking a circuit that contains a 110, 220, or 440 volt, 60 cycle, electric current. Testing to determine ignition currents required under such conditions has been done rather extensively in Germany[48] and has led to the concept of minimum igniting current. It has been found experimentally that the minimum ignition current produced during a break circuit spark correlates well with the maximum experimental safe gap for that particular compound. There are essentially no tests of this type performed in the United States.

6. EXPLOSION DYNAMICS

6.1. Fundamental Processes

Central Ignition, Spherical Vessel, with or without Venting. Transient propagation of a flame from a point ignition source through the entire contents of a closed vessel of arbitrary shape represents an interesting nonsteady propagation problem that has been studied rather extensively. By far the simplest closed vessel explosion is that following central ignition in a spherical bomb. In this case, if buoyancy may be neglected, gross aerodynamic effects are absent, and the course of the entire process may be discussed in some detail. This simplification occurs for two reasons. First, at all times during its propagation the flame front is orientated strictly normal to the local-flow velocity vector, and second, there is never any gas motion along the vessel wall and therefore never a boundary layer with which the flame may interact.

Lewis and von Elbe[52] have presented an extensive analysis of this system and have reached the following conclusions:

1. Throughout the propagation process the local flame may be assumed to be burning as though in a constant-pressure environment. This assumption is valid, since pressure changes due to the formation of hot products are relatively slow during the combustion of a gas layer of thickness equivalent to the preheat-zone thickness of the flame.
2. Because of spherical geometry the early burning takes place at essentially constant pressure (when the flame radius is one-half of the bomb radius, the hot gas ball occupies only one-eighth of the bomb's volume).
3. The early pressure rise is a cubic function of time.
4. Burning velocity is not constant during the entire propagation process. This is the result of adiabatic compression that significantly raises the temperature and pressure of the outer shell of gas before it is reached by the combustion wave.
5. The combustion process produces a gas sample that is significantly hotter at the center of the vessel. We note that the thermodynamic path for a sample at the center is different than that for a sample in the outer shell because the center sample is first burned isobarically and then compressed isentropically, whereas the outer shell sample is first compressed isentropically and then converted to products by the isobaric combustion process. Thus the specific entropy at the center will be significantly higher than that near the wall, and temperature will be higher in the center of the bomb after combustion is complete. This result has been verified experimentally.

The above results mean that with sufficient information concerning the burning velocity and the thermodynamic properties of the gas, one may follow, in detail, the combustion process as it occurs in a spherical vessel with central ignition.

Additionally, it has been shown[53] that for any particular fuel-air mixture in an essentially spherical vessel (i.e., one with a low length to diameter ratio $(L/D \cong 1)$, the maximum rate of pressure rise is related to vessel size by the equation:

$$\left(\frac{dP}{dt}\right)_{max} V^{1/3} = \text{constant} \tag{98}$$

Recently Bradley and Mitchenson[54] have performed a rather complete analysis of spherical vessel explosion dynamics, including venting. They used a very idealized model in which venting occurred uniformly over the vessel surface such that spherical geometry was always preserved. However, their idealized spherical vent allowed gas to escape as though it were a simple hole of area A_v and discharge coefficient C_d. This allowed them to define a dimensionless vent ratio where A_s is the total surface area of the sphere:

$$\bar{A} = \frac{C_d A_v}{A_s} \tag{99}$$

They found that the other dimensionless ratio that is important is the velocity of the gas ahead of growing spherical flame ball, defined for the initial conditions and nondimensionalized using the velocity of sound in the unburned gas:

$$\bar{S} = \frac{S_u}{a_o}\left(\frac{\rho_u}{\rho_b} - 1\right) \tag{100}$$

They found, in fact that the maximum overpressure generated by an explosion in a fully vented vessel (vent always open) was directly related to the quantity $\bar{A}/\bar{S}$. They also found that this quantity determined the maximum overpressure in the vessel for the case of delayed vent opening. In this case they started uninhibited venting at a number of prespecified pressures to determine the effect of vent opening delay on the maximum vessel pressure. These effects are shown in Figure 41 and values of $\bar{S}$ for various fuels are given in Table 4. In Figure 36 the line labeled "open" represents the results when the vent is continuously open. The other lines that terminate with a star represent maximum pressures generated when the vent area opens at the pressure indicated by the star. The curves that extend back to the left from each star shows that the maximum pressure is reached *after* the vent is open. For vent areas to the right of the star, the pressure drops continuously after the vent is opened.

Bradley and Mitcheson[54] point out that comparison of their predicted maximum overpressures to measured overpressures in laboratory-scale vented vessels show that their predictions are conservative. Their theory, however, does not predict the pressure time curve that one observes experimentally with a single vent. This is because, in the case of a single vent, the first gases that are vented will be the combustible mixture, but later, when the flame reaches the vent, combustion products are vented. It has been observed experimentally

TABLE 4. Properties of Selected Gas-Air Mixtures At Initial Conditions of 1 Atmosphere and 298°K[a]

Gas	Mol % in Air	ϕ^b	S_o m sec^{-1}	ρ_u/ρ_b	P_e atm^c	$\bar{S}$
CH_4	9.48	1.0	0.43	7.52	8.83	8.5×10^{-3}
C_2H_2	7.75	1.0	1.44	8.41	9.78	3.2×10^{-2}
C_2H_2	9.17	1.2	1.54	8.80	10.28	3.7×10^{-2}
C_2H_4	6.53	1.0	0.68	8.06	9.39	1.4×10^{-2}
C_3H_8	4.02	1.0	0.45	7.98	9.31	9.6×10^{-3}
C_3H_8	4.3	1.07	0.46	8.09	9.48	9.9×10^{-3}
C_3H_8	5.0	1.26	0.38	7.97	9.55	9.2×10^{-3}
C_3H_8	−6.0	1.52	0.15	7.65	9.30	3.0×10^{-3}
C_5H_{12}	2.55	1.0	0.43	8.07	9.42	9.0×10^{-3}
C_5H_{12}	2.7	1.06	0.43	8.18	9.76	9.3×10^{-3}
C_5H_{12}	3.0	1.18	0.40	8.16	9.77	8.4×10^{-3}
C_5H_{12}	3.5	1.39	0.29	7.92	9.80	6.3×10^{-3}
$C_{16}H_{34}$			0.39	7.82	9.4	7.1×10^{-3}
H_2	29.5	1.0	2.70	6.89	8.04	4.4×10^{-2}
H_2	40	1.6	3.45	6.50	7.78	5.3×10^{-2}
Town gas	25	1.4	1.22	6.64	8.03	1.9×10^{-2}

[a] From Bradley and Mitcheson.[54] Reprinted with permission from the Combustion Institute.

[b] ϕ = equivalence ratio = $\dfrac{\text{actual fuel/air volume ratio}}{\text{stoichiometric fuel/air volume ratio}}$

[c] P_e = theoretical closed vessel maximum explosion pressure.

that the pressure in the vessel reaches a peak value at the moment that the flame reaches the vent. It then drops but eventually can reach a second peak value before it finally drops back to atmospheric as combustion goes to completion.

Complications to Vessel Explosion Processes. There are four major dynamic effects that can complicate the vessel explosion process. These are (1) turbulence or large eddy production by flame-generated flow past obstacles located ahead of the flame, (2) Taylor instability of the flame front when the flame surface is accelerated toward the hot product gases, (3) Kelvin-Helmholtz instabilities of the flame due to impulsive flow accelerations perpendicular to the direction of flame propagation, and (4) combustion instability that arises because the heat release at the flame couples naturally with an acoustic mode of the chamber. Each of these will be discussed in detail in the following paragraphs.

When a flame passes over a obstacle the impulsive flow generated ahead of the flame causes eddy shedding. Eventually, the flame reaches this eddy, and then the effective flame area is increased, thus increasing the rate of conversion of reactants to products.[55] This is equivalent to an effective increase in the burning velocity, and it tends to make the explosion more violent. In the limit

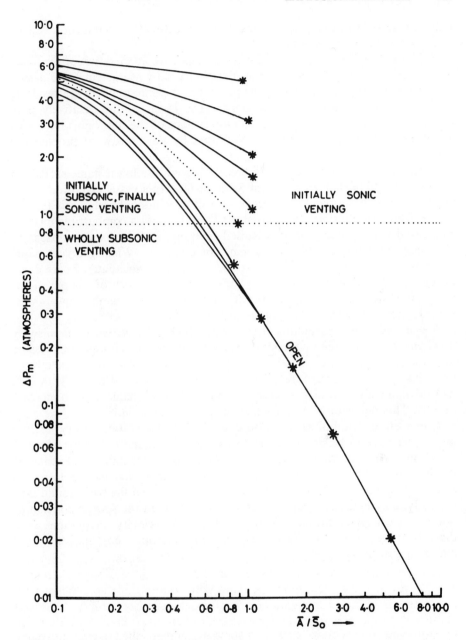

FIGURE 36. Variation of maximum pressure rise above atmospheric pressure with $\overline{A}/\overline{S}$ for present computer solutions. [Originally appeared in D. Bradley and A. Mitcheson, *Combust. Flame* **32**, 221-255 (1978), with permission of the Combustion Institute, Pittsburgh, Pennsylvania.]

this type of behavior can lead to transition to detonation. Transition will be discussed in the next section.

Taylor instability of the flame during an impulsive acceleration of the flame in a direction toward the hot gas was first observed by Markstein.[56] He passed a shock wave over a developing flame ball and photographed the resulting dynamic response using schlieren optics. The result of this relatively pure interaction is shown in Figure 37. The top surface of the hot (light) bubble of gas is unstable to the acceleration produced by the downward propagating shock in this photo, and the sphere very quickly becomes a torus. Eventually the upper flame surface completely breaks through the lower flame surface to produce a highly extended turbulent-flame surface. In a vessel explosion that is vented (either intentionally or because a wall or window bursts), this type of Taylor instability has also been observed.[57] In this case a combustible mixture in a vessel with a vent panel was ignited centrally, and the flame propagation was observed using a schlieren system movie camera. When the vent panel burst, the side of the flame opposite to the vent panel immediately broke down into a very convoluted surface due to the fact that the rush of gas out of the vent caused an impulsive motion of the bulk gases in the chamber towards the vent. Figure 38 is a schematic representation of this behavior.

Kelvin-Helmholtz instability occurs when the flame system is impulsively accelerated in a direction parallel to the flame surface. In this case the pressure drop or rarification wave accelerates the hot product gases more effectively, and this causes a shear layer to appear at the flame surface. This is a hydrodynamically unstable situation, and it causes the flame area to increase rapidly. This behavior is shown schematically in Figure 38.

Combustion instabilities can also cause flame accelerations and therefore larger pressures to be produced. Here the flame spontaneously propagates in such a manner that fundamental acoustic modes of the chamber are excited by the heat released by the flame. For many years the phenomenon has been known to occur in stationary combustion chambers when the background flow is steady. Under these circumstances, if the flame is in the proper position or interacts with either pressure fluctuations or flow velocity fluctuations such that the integral of the fluctuation in heat addition, δq, and the local fluctuation in pressure, δP,

$$\oint \delta q \, \delta P \, dt$$

is positive for one acoustic cycle of a standing mode of the system, the system is linearly unstable at that frequency, and that acoustic mode will increase in amplitude exponentially with time.

Recently, it has been shown that this phenomena occurs in larger vessels that contain a combustion explosion and are vented. Unfortunately, the combination of Taylor instability Kelvin-Helmholtz instability (Figure 39) plus a later combustion instability in large vessels can generate overpressures

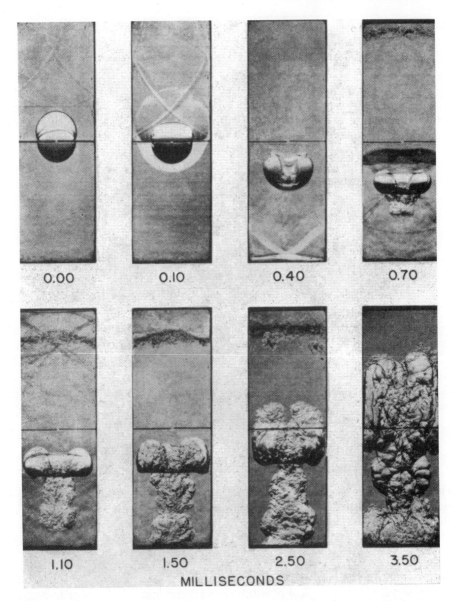

FIGURE 37. Interaction of a shock wave and a flame of initially roughly spherical shape. Pressure ratio of incident shock wave is 1.3, and stoichiometric butane-air mixture ignited at center of combustion chamber 8.70 msec before origin of the time scale. (Courtesy of G. H. Markstein, Factory Mutual Research, Norwood, Massachusetts originally in *Combust. Propuls., AGARD Colloq., 3rd*, 1958, p. 153. Reprinted with permission from Pergamon Press PLC.)

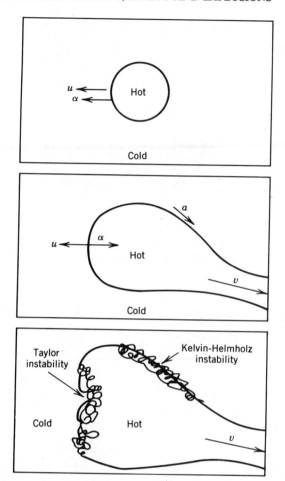

FIGURE 38. Schematic illustration of the development of a Taylor instability in a centrally ignited explosion. Here u is the flow velocity, and a is the direction of acceleration. (Courtesy of D. M. Solberg and Det Norske Veritas.)

during venting that are larger than predicted by the Bradley Mitchenson[54] theory (Figure 36). A typical pressure time record that shows acoustic oscillations and the measured overpressure for a 30-m^3 vessel are shown in Figures 39 and 40, respectively. Because of the uncertainty of extrapolation experiments using a 475 = m^3 vessel are being planned in Norway.[58]

Deflagration to Detonation Transition. We now that there are two distinct processes that can cause a deflagration wave (flame) to make a transition to a detonation wave under conditions of partial or strong confinement. They are related to the direct initiation processes that were discussed in Chapter 4.

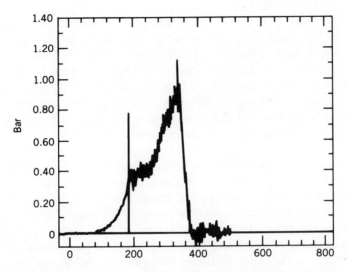

FIGURE 39. Explosion where rate of volume production exceeds venting capacity at the first peak. (Courtesy of D. M. Solberg and Det Norske Veritas.)

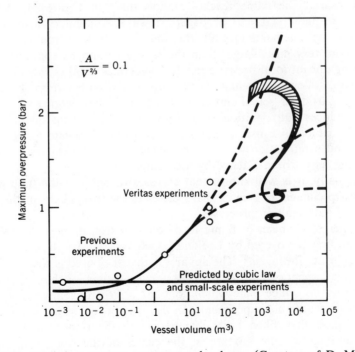

FIGURE 40. Explosion pressure versus vessel volume. (Courtesy of D. M. Solberg and Det Norske Veritas.)

Specifically, in one case a portion of this gas is heated to the auto ignition temperature by shock compression, but in the other case, hot gas–cold gas mixing or rapid flame folding trigger the SWACER mechanism.

An example of auto ignition transition is observed when a relatively high velocity flame propagates away from the closed end of a long tube. Urtiew and co-workers have studied the details of the process for the hydrogen + oxygen system.[59,60] A description of their observations follows. Ignition by a spark at the center of the closed end of the tube causes a hemispherical flame to grow until it touches the walls of the tube. At this moment the rate of flame surface area increase lessens, and the gas motion induced by flame growth slows. This impulsive deceleration of the rate of flame growth causes the flame front to become Taylor unstable and therefore causes a subsequent rapid increase of flame surface that generates more flow ahead of the flame. See station 1–2 in Figure 41. Subsequent to this, if the normal burning velocity is sufficiently high and heat transfer to the wall sufficiently slow, the gas motion ahead of the flame generates a shock of moderate strength and also generates a growing turbulent boundary layer ahead of the flame. As soon as the flame reaches this turbulent boundary layer the effective local burning velocity near the wall increases markedly because the flame becomes turbulent there. This happens some time after the flame reaches station three in Figure 41. After this happens, the flame starts to propagate down the boundary layer and pull behind it an oblique flame that fills the tube. The effective frontal area of this flame is now very much larger than the tube area and its continued growth causes a significant local pressure rise that generates another shock of moderate strength, following the first shock. Urtiew and co-workers[59] have shown that for a particular set of initial circumstances, the merging of these two shocks produces auto ignition conditions and then a subsequent localized volumetric explosion causes a detonation wave to propagate away from the end wall and a retonation wave to propagate toward the end wall in the gas that has not as yet been reached by the flame.

Additionally, it is well-known that crevases or holes in the tube walls, as well as sharp changes in area or sharp bends in the tube, augment the process of DDT by auto ignition processes.

The other mechanism of flame to detonation transition is the SWACER mechanism first discovered by Lee and co-workers at McGill, in conjunction with Wagner at Göttingen.[61] The apparatus that they used is shown in Figure 42. The small spherical and larger cylindrical chamber of this apparatus are connected by an orifice plate which contains orifices of many different sizes and shapes for different experiments. In the experiment, with a specific orifice plate in place, they filled both chambers with the same acetylene-oxygen mixture and ignited the mixture in the center of the small chamber. The pressure rise in this chamber caused a jet of gas to enter the larger chamber at the orifice place. First this turbulent jet was unreactive, but later the flame reached the orifice place and hot combustion products entered the jet. In all of these experiments a single-hole orifice failed to cause DDT. However, multiple

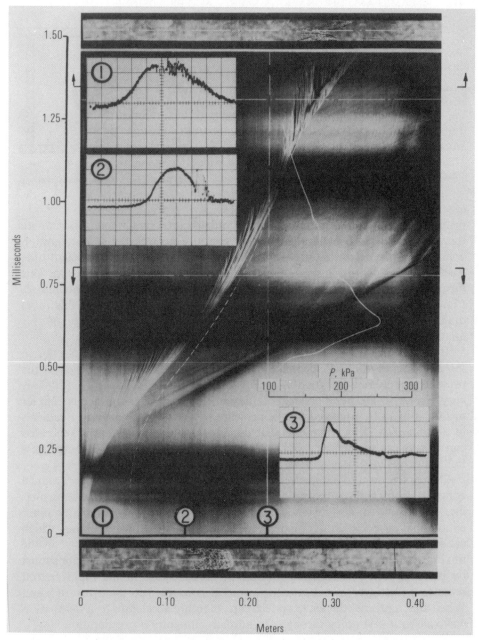

FIGURE 41. Flame propagation from the closed end of a long rectangular duct. Note that the flame is becoming highly turbulent near the end of the record and that compression waves ahead of the flame are coalescing to produce shock waves. Pressure records taken from stations 1, 2, and 3 are shown, as well as two pulsed laser photographs of the flame and shock wave development. (Courtesy of A. K. Oppenheim, University of California, Berkeley; photograph originally in paper presented at the OAR Research Applications Conference, 1966. Reprinted with permission.)

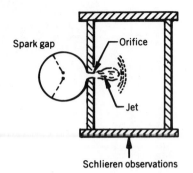

FIGURE 42. Schematic of the experimental apparatus described in Ref. 61. The larger chamber is about 0.3 m diameter and 0.4 m long. (Originally appeared in W. E. Baker, P. A. Cox, P. S. Westine, J. T. Kulesz, and R. A. Strehlow, *Explosion Hazards and Evaluation*, Fig. 1.11, p. 41, North Holland Publ., Amsterdam, 1983. Reprinted with permission.)

orifices did cause a very rapid transition to detonation under conditions where the temperature and pressure in the jet region were well below autoignition conditions. The delay to initiation was so short that there was insufficient time for a sound signal to travel from the jet to the vessel wall and back again. This meant that during the transition the jet region was essentially unconfined. Figure 43 shows the jet tip trajectory for two cases. The line on this (x,t) diagram of slope 3160 m/s represents the CJ detonation velocity of this mixture and is not a curve fit to the data points.

It is interesting to note that Meyer et al.[62] first observed this behavior in 1970 but could offer no adequate explanation for its occurrence. Even at the present time there is no good unambiguous theory for this type of DDT. It is very similar to the photochemical initiation discussed in the previous chapter in that it must be triggered by the presence of a sufficient number of radicals in a sufficiently large volume so that the adiabatic explosion that is generated can trigger an augmented reaction in neighboring layers. This is the process that quickly accelerates to a detonation.

There are two likely processes that could produce the proper radical concentrations and concentration gradient regions: (1) hot gas–cold gas mixing, which causes a sufficiently high radical concentration to trigger rapid exothermic processes in the mixing region, and (2) severe flame folding, which causes such a rapid increase in flame area that the rate of heat release in the mixing volume becomes very rapid. In both cases the neighboring regions must be preconditioned so that an increase in pressure due to the rapid central reaction will trigger a heat release wave that is coupled to the pressure wave. When this happens, the wave system can smoothly accelerate to a detonation wave without first going through the normal autoignition preconditioning.

6.2. Combustion Explosions in Enclosures

Enclosures with a Low Length to Diameter Ratio (Low L / D). Explosions in enclosures that have a low L/D are usually simple overpressure explosions. In a building the pressure rises very slowly at first, and the building will start to relieve at a very low pressure. Generally, the roof will start to rise and all the

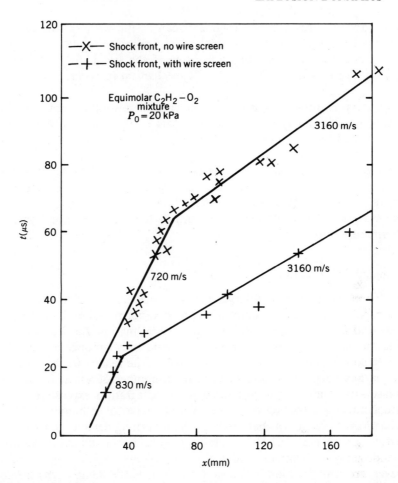

FIGURE 43. Shock and detonation front trajectories downstream of an orifice plate with four rectangular slots w = 3 mm, 1 = 38 mm, s = 2 mm with and without wire screen. (Originally appeared in R. Knystautas, J. H. Lee, I. Moen, and H. G. Wagner, *Symp.* (*Int.*) *Combust.* [*Proc.*] **17**, by permission of the Combustion Institute, Pittsburgh, PA)

walls will fall out at the same time. There will be insufficient overpressure to generate a significant blast wave and primary missiles will not be thrown long distances.

Enclosures with Obstacles or a Large L / D. If the enclosure has a large L/D ratio or contains many large obstacles, wave action similar to that which occurs during DDT can occur. In this case the damage patterns are markedly different than those observed for a low L/D explosion. The damage is quite often localized and severe. Because of the rapid rise of pressure due to wave propagation processes, primary fragments can be thrown large distances, and the blast wave can be severe. The acceleration processes can lead to transition to detonation by either heating a portion of the gas to the auto ignition temperature or by causing the SWACER mechanism to occur.

Venting. As was discussed earlier, the dimensionless venting area ratios $\bar{A}_0$ is related to the maximum overpressure in a low L/D vessel (see Figure 36). It was also stated that the current predictions in the National Fire Code (NFPA 68)[53] may not be conservative for vessels larger than 20 m³. Also it is recognized in the National Fire Code that large L/D ducting requires special venting if it is to survive a combustion explosion. There are of course, certain structures, such as high-pressure piping systems, that cannot be protected by vents and must be protected by ensuring that they do not contain an exothermic mixture.

6.3. Nominally Unconfined Explosions

There are many industrial and transportation accidents that involve an open spill of a large quantity of combustible, high vapor pressure liquid or gas. If ignition is delayed, a large premixed vapor cloud may be formed. When, and if, ignition occurs, the combustion of the cloud may produce a large fire or both a large fire and a dangerous blast wave.[63] The reason for these different behaviors has been under investigation for a number of years. In particular, the blast wave that can be generated by a deflagration explosion has been investigated using a number of numerical and analytical techniques.[64, 65] These investigations have shown that relatively high burning velocity flames $s_u \sim 50$ m sec^{-1} can generate a damaging blast wave if the flame is completely surrounded by a flammable mixture and becomes a growing flame ball. However, recently Strehlow[66] has shown that once the flame burns through a small part of the vapor cloud, the possibility of producing a damaging blast wave from even a highly subsonic deflagration wave is greatly lessened.

In addition to this fact, we now have a viable mechanism for generating transition to detonation in a nominally unconfined cloud that indicates that transition to detonation must be occurring when a damaging blast wave is observed. The most probable mechanism for this transition is the SWACER mechanism described above. In the case of hydrocarbon-air mixtures, the characteristic kinetic times are much longer than in the acetylene-oxygen systems that were studied in the laboratory. However, the possible eddy-folding regions are at least two orders of magnitude larger (such as eddy shedding by a large building ahead of the flame). John Lee[67] estimates that propane air has a characteristic eddy size of about 3 m for transition to detonation.

At the present time there is a considerable research effort to determine the effect of obstacles of different sizes, shapes, and spacing on flame acceleration processes. The hope is to find generic shapes and optimum spacings for flame acceleration.[55, 68] This information will be useful in designing chemical plants that have a smaller propensity to cause transition and therefore produce a damaging blast wave after ignition of an accidental spill of combustible material.

Recently a text on explosive hazards and evaluation[69] has appeared. It presents a comprehensive discussion of all aspects of combustion safety.

PROBLEMS

1. A room-sized enclosure $10' \times 10' \times 10'$ contains no significant obstacles to cause flame accelerations. Assume that you wish this room to survive a 1.5 psig overpressure due to a low overpressure explosion of a stoichiometric propane-air mixture. What vent area would be required if the vents released at essentially zero overpressure?

2. Determine the Chapman-Jouquet detonation velocity and the CJ and constant volume overpressures for a stoichiometric propane air mixture using the heat addition model.

3. A piping system with receivers, etc., could get into a hazardous state because it contains a stoichiometric butane-air mixture. Determine the minimum pipe diameter that will just allow a detonation to propagate and the maximum diameter that will just not allow a detonation to be transmitted to a large receiver from the pipe.

4. Consider a low L/D vessel of 90 m³ capacity (a building) that may be accidentally filled with a stoichiometric propane-air mixture. Using the data of Table 4 and Figure 36 determine the "open" vent area needed to prevent the overpressure from reaching 0.02 atm. If the building were cubic, what fraction of the wall area would this be?

NOMENCLATURE

Roman Letters

a	velocity of sound, m/sec
A	constant, dimensionless
A	species A
$\overline{A}$	vent ratio, dimensionless
A_s	surface area of equivalent sphere, m²
A_v	vent hole area, m²
B	species B
c	constant, m²/sK²
c_n	vol % of fuel in mixture
c_p	heat capacity at constant pressure, J/kg K
c_v	heat capacity at constant volume, J/kg K

C	capacity of condenser, microfarads
C_D	discharge coefficient, dimensionless
C_p	molar heat capacity, J/mol K
d_c	critical diameter for transmission of detonation, m
d	quenching distance, m
D_i	diffusion coefficient, m²/sec
D	average cell size of cellular flame, m
D_c	critical tube diameter for detonation failure, m
e	specific internal energy, J/kg
E	activation energy, J/mol
E	capacitor energy, J
E_c	critical energy detonation initiation, J
f/a	fuel/air ratio on mass or mole basis, dimensionless
f	constant, J/msec K²
F	a generic symbol for fuel
h	specific enthalpy, J/kg
h_c	heat of combustion, J/kg
H	molar enthalpy, J/mol
H_c	heat of combustion of fuel J/mol
I	species I, a generic symbol
k_1	first-order rate constant, sec⁻¹
k_α	reaction rate constant of order α(cm³/mol)$^{\alpha-1}$sec⁻¹
K	Karlovitz number, dimensionless
L	lean limit of flammability, vol % in air
L/D	length to diameter ratio of an enclosure, dimensionless
Le	Lewis number = $C_p D\rho/\kappa$, dimensionless
m	molecular weight, Kg/mol (one mole = 6.023 × 10²³ molecules)
M	Mach number, u/a, dimensionless
M	mass flow, kg/m² sec
n	pressure dependence of burning velocity
n_i	molar density of a species, mol/kg
N	mol/m³ (one mol = 6.023 × 10²³ molecules)
O	a generic term for oxidizer
p	on summation sign, total number of reactions occurring
P	pressure, pascals
P_e	theoretical closed vessel maximum explosion pressure, pascals
q	heat added to system in heat addition model, J/kg
Q	molar heat addition, J/mol
r	radius, m
R	universal gas constant, 8.31417 J/mol K (1 mole = 6.023 × 10²³ molecules)
R_i	(with subscript) specific gas content, J/kg K
s	specific entropy, J/kg K
s	on summation sign, total number of species

$\bar{S}$	dimensionless burning velocity
S	convective velocity in flame theory (no subscript), m/sec
S_b	space velocity, m/sec
S_L	burning velocity at the lean limit m/sec
S_m	burning velocity maximum, m/sec
S_r	burning velocity at the rich limit, m/sec
S_u	normal burning velocity, m/sec
$S_{\parallel}$	wave velocity parallel to flame zone, m/sec
t	time, sec
T	temperature, K
u	velocity component in x-direction, m/sec
V	specific volume, m³/kg
V_i	diffusion velocity, m/sec
V	voltage across condenser gap, volts
W_c	critical slot width for transmission of detonation, m
x	cartesian coordinate, m
X_i	mole fraction, mol/mol of mixture, dimensionless
y	cartesian coordinate, m
Y_D	mole fraction of inert in fuel at extinction
Z	detonation cell width, m
z	mass flux, dimensionless
z	mole fraction of H_2 in combustible mixture

Greek Letters

α	angle between flame front and flow direction, degrees
α	thermal diffusivity, m²/sec
β_c	Eigenvalue, dimensionless
β_1	constant, dimensionless,
β_2	constant, dimensionless
γ	heat capacity ratio, c_p/c_v, dimensionless
δ	incremental quantity
Δ	induction delay distance in a detonation, m
ε	volume ratio, V_2/V_1, dimensionless
ε	dimensionless activation energy $= E/RT_b$
η	displaced mole fraction of fuel
η	displaced volume ratio, dimensionless
η_0	preheat zone thickness, m
κ	thermal conductivity, J/msec K
λ	density ratio, dimensionless
λ_j	moles of reaction per kg of mixture
Λ_c	generalized eigenvalue, dimensionless
μ	micro, 10^{-6}
ν_{ij}	stoichiometric coefficient of the high ith species in the jth reaction, moles per unit of reaction

ξ	displaced pressure ratio, dimensionless
ξ_0	nondimensionalized distance
ρ	density, kg/m^3
τ	dimensionless temperature
τ	characteristic time to explosion or recombination, s
ϕ	temperature ratio, dimensionless
ϕ	heat addition parameter, dimensionless
Φ	equivalence ratio, dimensionless
ψ	ratio, dimensionless
ω	rate of reaction, kg/m^3

Symbols

d	differential operator
exp	exponent natural base, exp $(x) = e^x$
D	differential operator for total derivative
ln	natural logarithm
log	logarithm to the base 10
$\propto$	proportional to
[ ]	concentration of a species mol/m^3
$\sqrt{}$	square root

Acronymns or Abbreviations

AIT	auto ignition temperature, K
ASTM	American Society for Testing and Materials
CJ	Chapman-Jouquet
DDT	deflagration to detonation transition
JANAF	Joint Army Navy Air Force
LFL	lower flammability limit (vol %)
LEL	lower explosion limit (vol %)
MESG	maximum experimental safe gap
NFPA	National Fire Protection Association
NPN	normal propyl nitrate
SMRE	Safety in Mines Research Establishment (England)
SWACER	Shock Wave Amplification by Coherent Energy Release
UFL	upper explosion limit
UEL	upper explosion limit
ZND	the Zel'dovich-von Neumann-Döring of detonation structure
atm	1 atm of pressure = 101,325 Pascals

Subscripts

A	species A
b	burned gas, for flames

B	species B
BO	blow off
chem	due to chemical reaction
crit	critical value
CJ	Chapman-Jouquet state
eq	equilibrium state
f	flame
FB	flash back
i	species number; initial conditions
ig	state at which mixture will ignite
j	reaction number
L	lean flammability limit
m	maximum value
max	maximum value
min	minimum value
o	circular tube
r	rich limit
r	reflected shock
rim	at the rim of a burner
s	isentropic
s	stoichiometric mixture
u	unburned gas, for flames
1	first state; state before shock
2	second state, state after shock
$\parallel$	parallel to, parallel plate

Superscripts

m	exponent, dimensionless
n	exponent, dimensionless
o	thermodynamic standard state
α	order of a chemical reaction

REFERENCES

1. J. O. Hirschfelder and C. F. Curtiss, *J. Chem. Phys.* **17,** 1076 (1949).
2. R. Friedman and E. Burke, *J. Chem. Phys.* **21,** 710 (1953).
3. G. H. Markstein, *Non-Steady Flame Propagation*. Macmillan, New York, 1964.
4. H. Edmondson and M. P. Heap, *Combust. Flame* **15,** 179–187 (1970).
5. B. P. Mullins and S. S. Penner *Explosions, Detonations, Flammability and Ignition Agardograph*, p. 31. Pergamon Oxford (1959).
6. D. B. Spalding, *Proc. Soc. London, Ser.* **A 240,** 83 (1957).

7. F. T. Bodurtha, *Industrial Explosion Prevention and Protection.* McGraw-Hill, New York, (1980).

8. H. F. Coward and G. W Jones, *Bull.—U.S., Bur. Mines* **503,** 155 (1952); M. G. Zabetakis, *ibid.,* p. 627 (1965).

9. H. LeChâtelier and O. Boudouard, *C. R. Hebd. Seances Acad. Sci.* **126,** 1344–1347 (1898).

10. P. T. Harsha, R. B. Edelman, and D. H. France *Efficient Use of Alternate Fuels—Development of Models for the Prediction of Interchangeability, Design, and Performance of Gas Burner/Combustion Systems,* Final Rep. Phase I, Contract 5011-345-0100, GRI 79/0034. Gas Research Institute Chicago, Illinois, 1980; also *Catalog of Existing Interchangeability Prediction Methods,* SAI-80-024-CP. Science Applications, Inc., Canoga Park, Calif, (undated).

11. J. E. Dove and T. D. Tribbeck, *Astronaut. Acta* **15,** 387–397 (1970).

12. R. A. Strehlow and P. M. Rubins, *AIAA J.* **7,** 1335–1344 (1969).

13. R. A. Strehlow and F. D. Fernandes, *Combust. Flame* **9,** 109–119 (1965).

14. J. J. Erpenbeck, *Phys. Fluids* **7,** 684 (1964); **8,** 1192 (1965).

15. T. Y. Toong, *Combust. Flame* **45,** 67 (1982).

16. H. B. Dixon, *Philos. Trans. R. Soc. London, Ser. A* **200,** 328–358 (1903).

17. C. Campbell and D. W. Woodhead, *J. Chem. Soc.,* p. 30105 (1926); p. 1572 (1927).

18. R. A. Strehlow, *Fundamentals of Combustion,* Chapter 9. Krieger Press, Milbourn, Florida 1972.

19. W. Fickett and W. C. Davis, *Detonation.* University of California Press, Berkeley, 1979.

20. K. Antolik, *Pogg. Ann. Phys. Chem.* [2] **230,** 14, 154 (1875).

21. Y. H. Denisov and Y. K. Troshin, *Dokl. Akad. Sci. USSR* **125,** 217 (1960).

22. H. O. Barthel, *Phys. Fluids* **15,** 51–55 (1972).

23. R. A. Strehlow, *Astronaut. Acta* **15,** 345–358 (1970).

24. H. O. Barthel, *Phys. Fluids* **17,** 1547–1553 (1974).

25. R. A. Strehlow and A. Cohen, *Phys. Fluids* **5,** 97 (1962); R. B. Gilbert and R. A. Strehlow, *AIAA J.* **4,** 1777 (1966).

26. R. A. Strehlow, A. J. Crooker, and R. E. Cusey, *Combust. Flame* **11,** 339 (1967).

27. D. H. Edwards, *Combust. Flame* **43,** 187 (1981).

28. J. H. S. Lee and I. O. Moen, *Prog. Energy Combust. Sci.* **6,** 359–389 (1980).

29. G. G. Bach, R. Knystautas and J. H. Lee *Symp. (Int.) on Combust.* 1097 (1971).

30. H. Matsui and J. H. Lee *Combust. Flame* **27,** 217–220 (1976).

31. R. Knystautas and J. H. Lee, *Combust. Flame* **27,** 221–228 (1976).

32. J. H. S. Lee, R. Knystautas, and C. M. Guirao, in *The Link between Cell Energy Critical Tube Diameter, Initiation Energy and Detonability Limits in Fuel Air Explosions,* (J. H. S. Lee and C. M. Guirao, eds.) SM study No. 16, pp. 157–188. University of Waterloo Press, Waterloo, Canada, 1982.

33. J. H. Lee, R. Knystautas, and N. Yoshikawa, *Acta Astronaut.* **5,** 971–982 (1978).

34. I. O. Moen, M. Donato, R. Knystautas, and J. H. Lee, *Symp. (Int.) Combust.* [*Prac.*] **18,** 1695 (1981).

35. Y. B. Zeldovich, S. M. Kogarko, and N. N. Semenov, *Sov. Phys.—Tech. Phys.* (*Engl. Transl.*) **1**, 1689 (1965).

36. J. H. S. Lee, private communication, currently unpublished results, Montreal, Canada, October (1982).

37. C. K. Westbrook, *Combust. Flame* **46**, 191–210 (1982).

38. C. K. Westbrook and P. A. Urtiew, *Symp.* (*Int.*) *Combust.* [*Prac.*] **19**, (1983).

39. R. A. Strehlow, R. E. Maurer, and S. Rajan, *AIAA J.* **7**, No. 2 (1969); R. A. Strehlow and C. D. Engel, *ibid.*, No. 3 (1969).

40. *National Fire Codes*, NFPA-30-1977. Boston, Massachusetts.

41. W. A. Affens, *J. Chem. Eng. Data* **11**, 197, (1966); W. A. Affens, H. W. Carhart, and G. W. McLaren, *J. Fire Flammability* **8**, 141, 152 (1977); W. A. Affens and McLaren, *J. Chem. Eng. Data* **17**, 482–488 (1972).

42. M. Gerstein and W. B. Stein, *Ind. Eng. Chem. Prod. Res. Dev.* **12**, 253 (1973).

43. D. J. Seery and C. T. Bowman, *Combust. Flame* 37–47 (1970).

44. J. J. Setchkin, *J. Res. Nat. Bur. Stand.* **53**, 49–66 (1954).

45. C. J. Hilado and S. W. Clark, *Chem. Eng. Eng.* (*N.Y.*) Sept 4, 1972, p. 72–80.

46. American Society for Testing and Materials, *N.Y. Annu. Book ASIM Stand.* **D 2155-69**, 724–727 (1970).

47. R. E. Dufour, and U. L. Westerberg, *Bull. Res.* **58**, April 2, 1970.

48. PTB, Comments on IEC Document 31 (U.K.) 22 "Correlation of Maximum Experimental Safe Gaps (M.E.S.G.) with Minimum Ignition Current (M.E.C.)." Physikalisch-Technische Bundesanstalt, Braunschweig, W. Germany (1967); also K. Nabert "The M.E.S.G. as a function of the Bounding Volumes and the Obstacles in These Volumes." Appendix to Paper IEC/SC, 31A/WGE PTB (1967); also G. A. Lunor and H. Phillips. "Summary of experimental data on M.E.S.G." SMRE Report R2 Dept of Trade and Industry, Sheffield, England (1973).

49. R. A. Strehlow, J. A. Nicholls, P. Schram and E. Magison, *J. Hazardous Materials* **3**, 1–15 (1979).

50. H. Phillips, *Combust. Flame* **19**, 181, 187 (1972); **20**, 121 (1973).

51. *National Electrical Code*, NFPA 500, Boston, Massachusetts (1980).

52. B. Lewis and G. von Elbe, *Combustion Flames and Explosions of Gases*, 2d ed., pp. 367–381, Academic Press, New York, 1961.

53. *National Fire Codes*, NFPA 68, "Explosion Venting" (1980).

54. D. Bradley, and A. Mitcheson, *Combust. Flame* **32**, 221, 237 (1978).

55. I. O. Moen, M. Donato, R. Knystautas, and J. H. Lee *Combust. Flame* **39**, 21 (1980).

56. H. G. Markstein, Non-steady Flame Propagation, Macmillian, New York (1964).

57. H. J. Heinrich, Private Communication BAM, Berlin (Summer 1977).

58. D. Solberg, Private Communication, Oslo (1980).

59. P. A. Urtiew, A. J. Laderman, and A. K. Oppenhim, (*Int.*) *Combust.* [*Proc.*] **10**, 797–804 (1965).

60. P. A. Urtiew and A. K. Oppenheim, *Symp.* (*Int.*) *Combust.* [*Proc.*] **11**, 665–676 (1967).

61. R. Knystautas, J. H. Lee, C. Guirao, M. Frenklach, and H. G. Wagner, *Symp. (Int.) Combust.* [*Proc.*] **17**, 1235 (1979).

62. J. W. Meyer, P. A. Urtiew and A. K. Oppenheim, *Combust. Flame* **14**, 13–20 (1970).

63. R. A. Strehlow, *Symp. (Int.) Combust.* [*Proc.*], **14**, 1189–1200 (1973).

64. A. L. Kuhl, M. M. Kamel, and A. K. Oppenheim, *Symp. (Int.) Combust.* [*Proc.*] **14**, 1201–1214 (1973).

65. R. A. Strehlow, R. T. Luckritz, A. A. Adamczyk, and S. A. Shimpe, *Combust. Flame* **35**, 297-310 (1979).

66. R. A. Strehlow, *Loss Prev.* **14**, 145–153 (1980).

67. J. H. Lee private communication, McGill University Montreal, Canada September (1980).

68. R. A. Strehlow, *Flame Acceleration by Obstacles*, Gas Research Insti. Grant. University of Illinois, Urbana, 1981–1982.

69. W. E. Baker, P. A. Cox, P. S. Westine, J. J. Kulesz, and R. A. Strehlow, *Explosion Hazards and Evaluation*. Elsevier North-Holland, Amsterdam and New York, 1983.

7 Diffusion Flames

MELVIN GERSTEIN

Department of Mechanical Engineering
University of Southern California
Los Angeles, California

1. Introduction
2. Laminar diffusion flames
   2.1. Gaseous diffusion flame theory
   2.2. Single-drop evaporation and combustion
   2.3. Flame height of laminar diffusion flames
   2.4. Flame chemistry
   2.5. Smoke
3. Turbulent diffusion flames
   3.1. Introduction
   3.2. Theory
4. Flame stability
   Symbols
   References

1. INTRODUCTION

A diffusion flame is generally considered to be one in which most of the fuel and air (or oxidant) are originally unmixed. Although diffusion processes may be important in many flames, diffusion—either molecular or turbulent or both—is required to bring the reactants in contact. This requirement does not exist for premixed flames. The domain of diffusion flames may be extended to include conditions where some oxidant is included with the fuel or some fuel with the oxidant. It will be assumed here that both the fuel and oxidant streams are outside the normal flammability limits, if such premixing is considered.

Most frequently, diffusion flames consist of a fuel jet issuing into an oxidant stream, although the opposite configuration is also possible. Many different configurations of diffusion flames exist in practice. Some examples of co-linear diffusion flames are illustrated in Figure 1.[1] The fuel jet, in many

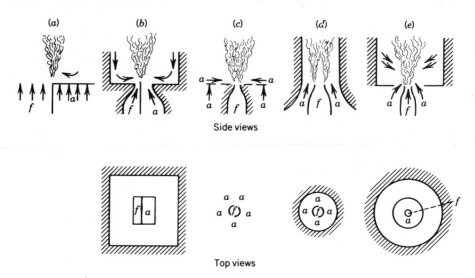

Side views

Top views

FIGURE 1. Various configurations used to obtain gaseous diffusion flames. Reprinted with permission from The Combustion Institute: *Fourth Symposium on Combustion* (1953), H. C. Hottel, pp. 97–113.

applications, is introduced at an angle to the oxidant flow as illustrated in Figure 2.[2] Opposed jet diffusion flames have also been studied. The flame appears as a "button" between the two jets.

The appearance of a gas jet diffusion flame depends on the relative quantities of fuel and air supplied. If air in excess of the stoichiometric requirement is supplied, the flame is a closed elongated figure. Such flames are typical of gas jets issuing into a large volume of quiescent air or concentric cylindrical tubes with an excess of air flowing in the outer tube. If the air flow in the outer tube is below the stoichiometric requirement, a fan-shaped flame results.

Under certain limiting flow conditions, the flame can assume a more complex structure. Flame shapes obtained in concentric tube, hydrocarbon diffusion flames, with fuel in the inner tube, for various fuel and air flows are classified in Barr[3] and illustrated in Figure 3. Zone 1 indicates a normal, overventilated diffusion flame, and zone 2 indicates the underventilated flame. The dashed line separating zones 1 and 2 denotes the smoke point, with smoke appearing as the fuel flow is increased. Zone 3 illustrates overventilated flames having a meniscus shape without the yellow glow usually associated with diffusion flames. The shaded areas are zones of unstable flames. In the lower part of zone 4, there appear flames that oscillate from side to side, called *lambent flames* in Barr,[3] and in the upper part of zone 4 rich, tilted flames appear. In the lower part of zone 5 are found toroidal-shaped flames, called *vortex flames*, and in the upper part of zone 5 are lean, tilted flames. The flame

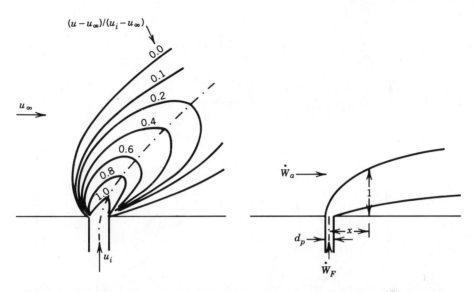

FIGURE 2. Configuration of transverse diffusion flame: (left) free transverse jet, (right) confined jet flame in a transverse air flow. Reprinted with permission from *Introduction to Combustion Phenomena* by A. M. Kanury, Gordon and Breach, Science Publishers, Inc. Copyright 1975.

in zone 6 has begun to move away from the burner port and is approaching a blow off limit.

Besides these typical gaseous diffusion flames, many condensed phase diffusion flames are known. These include such commonplace occurrences as the burning of a match, a log, or a candle and the various flames associated with the burning of liquid sprays or solid fuels such as coal. Many of the same physical processes are involved in the burning of these condensed phase fuels as are involved in the burning of gaseous fuels. This will be illustrated in a later section by a brief discussion of the burning of single liquid drops.

Diffusion flames are of obvious importance in most condensed phase fuel combustion processes. Since the fuel and air must come in contact before combustion proceeds; it is not as obvious why gaseous diffusion flames are important compared with flames of premixed fuel–oxidant combustion systems. Many combustion systems involve diffusion flames for a number of practical reasons.

The avoidance of a mixing chamber to produce a premixed flame not only eliminates the hardware and flow losses associated with the mixing process but also eliminates the hazard of an explosion in the mixing chamber if a malfunction should occur. The diffusion flame is also more tolerant to changes in flame conditions than is a premixed flame. Within a wide range of operation, the size of the flame adjusts to accommodate changes in fuel and air flow whereas for premixed flames, a control system is often required to adjust

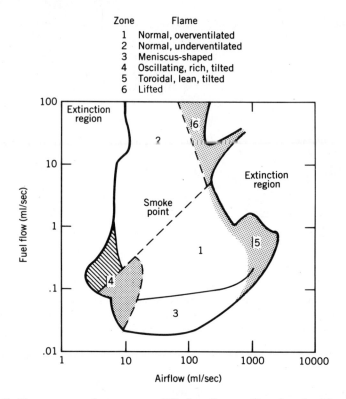

Zone	Flame
1	Normal, overventilated
2	Normal, underventilated
3	Meniscus-shaped
4	Oscillating, rich, tilted
5	Toroidal, lean, tilted
6	Lifted

FIGURE 3. Flame zones for gaseous diffusion flames. Reprinted with permission from The Combustion Institute: *Fourth Symposium on Combustion* (1953), J. Barr, pp. 765–771.

the relative flow of one stream in conjunction with changes in the other. The inherent simplicity, reliability, and efficiency of gaseous diffusion flames has resulted in their extensive use in practical systems such as furnaces and boilers.

The simplest diffusion flames involve laminar jets of fuel and air. These are the best understood theoretically and have been studied extensively in the laboratory, although a complete description of both the transport and chemical processes does not yet exist. This chapter begins with a discussion of laminar diffusion flame theory. Because of the similarity to condensed phase combustion, a brief discussion of single-drop diffusion flame theory is given following the discussion of gaseous laminar diffusion flame theory. This is not a comprehensive treatment of the burning of liquid drops of fuel but will serve as an introduction to the discussion in Chapter 11 and also shows the similarity between liquid and gaseous diffusion flame theory.

The theoretical treatment of laminar gaseous diffusion flames focuses on the calculation of flame height as the major objective. Experimental data on flame height are presented and compared with theoretical predictions.

Laminar flame theory and experiment is followed by a brief discussion of diffusion flame chemistry. Only gross mechanisms are treated since kinetics are treated in detail in Chapters 2 and 3. The special problems of smoke formation are then discussed as one important result of diffusion flame chemistry.

Turbulent flames represent the most practical application of diffusion flames but are the least understood because of the complexity of the problem of turbulent flow and mixing. Turbulent flame theory is reviewed briefly; more rigorous theory is covered in Chapters 8, 9, and 10. Experimental data are also presented at this point to show the nature of the state-of-the-art of turbulent flame theory and experiment.

Flame stability is the final section. Since laminar and turbulent flames have been treated by similar arguments, these are discussed together.

2. LAMINAR DIFFUSION FLAMES

2.1. Gaseous Diffusion Flame Theory

Burke and Schumann Theory. Burke and Schumann, in 1928,[4] developed a simplified model of diffusion flames that is still widely used, although more recent developments have relaxed many of their assumptions. Their assumptions are as follows:

1. At the port position the velocities of the air and the fuel are assumed constant, equal, and uniform across their respective tubes so that the molar ratio of fuel flow to air flow is given by $r_j^2/(r_s^2 - r_j^2)$.
2. The velocity of the fuel and air up the tube in the region of the flame is the same as the velocity at the port. (This is not actually true since heat addition causes an increase in velocity.)
3. The coefficient of interdiffusion of the two gas streams is constant. (Since the diffusion coefficient increases with temperature, as does the gas velocity treated as constant in assumption 2, Burke and Schumann suggest that these two effects may counteract each other and minimize the errors introduced by the assumptions.)
4. The interdiffusion of the two gas streams is entirely radial. (This is a reasonable assumption for tall flames.)
5. The mixing of the two streams occurs by diffusion only. In other words, recirculation eddies and radial flow components are assumed absent.
6. The flame front is a geometric surface where fuel and oxygen meet in stoichiometric proportions and react to give reaction products.
7. The oxygen is considered as negative fuel. In the reaction, fuel + iO_2 = products, each mole of fuel is equivalent to i moles of oxygen. The

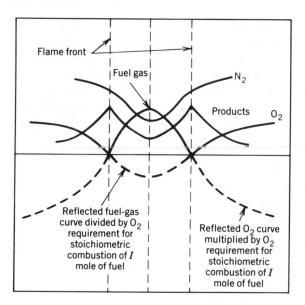

FIGURE 4. Simplified diagram of concentration profiles used in diffusion flame theory.

oxygen concentration C_{O_2} may then be replaced by its stoichiometric equivalent in terms of fuel, C_{O_2}/i, which is the number of moles of fuel that would be completely burned by C_{O_2} moles of oxygen. Since zero concentration of the reactants at the flame front is desired, the oxygen is treated as negative fuel so that the fuel concentration ranges from its value in the unreacted jet, C_j, to the equivalent value of oxygen in the unreacted air, $-C_{O_2}/i$. This is shown in Figure 4 by reflecting the oxygen concentration curve in the horizontal axis and plotting C_{O_2}/i. Burke and Schumann consider the concentration in terms of partial pressures, which are proportional to the molal concentrations.

8. The tube containing the fuel has a negligible thickness.

These assumptions reduce the problem to one of diffusion of a single gas having a certain initial distribution and subject to certain boundary conditions. Fick's law of diffusion gives in cylindrical coordinates the equation describing the concentration as a function of time and coordinates:

$$\frac{\partial P}{\partial t} = D\left(\frac{\partial^2 P}{\partial r^2} + \frac{1}{r}\frac{\partial P}{\partial r}\right) \tag{1}$$

For assumptions 1 and 2, t can be replaced by vertical distance since

$$t = \frac{y}{U_j} \qquad (2)$$

and there results

$$\frac{\partial C}{\partial y} = \frac{D}{U_j}\left(\frac{\partial^2 C}{\partial r^2} + \frac{1}{r}\frac{\partial C}{\partial r}\right) \qquad (3)$$

which now contains the independent variables of height and radius. The boundary conditions can be stated on the basis of the burner port conditions at $y = 0$.

At this point,

$$P = P_f \qquad 0 \leq r \leq r_j - \delta_{1/2}$$

$$P = \frac{P_{O_2}}{i} \qquad r_j + \delta_{1/2} \leq r \leq r_s \qquad (4)$$

The half thickness of the port $\delta_{1/2}$ is assumed negligibly small. Also

$$\frac{\partial P}{\partial r} = 0 \begin{cases} r = 0 \\ r = r_s \end{cases} \qquad (5)$$

The solution of Eq. (3) with the boundary conditions (4) and (5) is

$$P = P_T \frac{r_j^2}{r_s^2} - \frac{P_{O_2}}{i} + \frac{2r_j P_T}{r_s^2}\frac{1}{k}\frac{J_1(kr_j)J_0(kr)}{(J_0(kr_s))^2}\exp\left(\frac{Dk^2 y}{U_j}\right) \qquad (6)$$

where

J_0, J_1 = Bessel functions

k = a constant that assumes all positive roots of the equation $J_1(kr_s) = 0$

$P_T = P_f + (P_{O_2}/i)$

The equation for the flame front is obtained by setting $P = 0$ at $r = r_F$. At the flame front Eq. (6) becomes

$$\sum \frac{1}{k}\frac{J_1(kr_j)J_0(kr_F)}{(J_0(kr_s))^2}\exp\left(-\frac{Dk^2 y}{U_j}\right) = \frac{r_s^2 P_{O_2}}{2r_s i P_T} - \frac{r_j}{2} \qquad (7)$$

The shape of the flame front can be obtained by plotting the values of r_F and y that satisfy Eq. (7). The height of the flame h_F is given by the value of y when $r_F = 0$ for an overventilated flame and $r = r_s$ for an underventilated flame. The results of a typical calculation are illustrated in Figure 5 (Burke and Schumann[4]) for both types of flames. The calculated shapes agree well with observed flame shapes. The cylindrical flame resembles the configuration of Figure 1d, and the flat flame resembles that of Figure 1a.

The properties of diffusion flames are also calculated in Barr and Mullins,[5] in which the analysis was begun with Eq. (1) and essentially the same assumptions, except assumption 1, were used. The fuel and air velocities were not required to be equal. But air velocity had to be constant over the annulus and the fuel velocity over the inner tube, and no momentum transfer could occur during diffusion (assumption 2). Solution of Barr and Mullins' equation gives

$$\sum \frac{1}{k} \frac{J_1(kr_j) J_0(kr_F)}{(J_0(kr_s))^2} \exp\left(-\frac{Dk^2 y}{U_j}\right) + \frac{iU_j}{P_{O_2}U_s} \exp\left(-\frac{Dk^2 y}{U_j}\right)$$

$$= \frac{r_s^2 - r^2}{2r_j} - \frac{iU_j}{P_{O_2}U_s} \frac{r_j}{2} \tag{8}$$

Comparison of Eqs. (7) and (8) shows the similarity. The principal difference is the addition of two terms containing the ratio U_j/U_s, which was assumed equal to 1 by Burke and Schumann.

Displacement Distance Theory. Although the assumptions in the Burke and Schumann theory may seem restrictive, this model gives a good description of diffusion flame behavior. Since the structure of the flame is probably only qualitatively correct, Jost[6] has suggested that the flame height, usually the desired parameter, can be obtained in a much simpler manner through the use of the displacement length, and the assumption that the time it takes the fuel to reach the flame tip is equal to the time it takes oxygen to diffuse from the jet boundary to the jet axis. Jost suggests the relation

$$h_F = \frac{\Omega}{2\pi D} \tag{9}$$

where

h_F = flame height
Ω = volumetric flow rate
D = diffusion coefficient

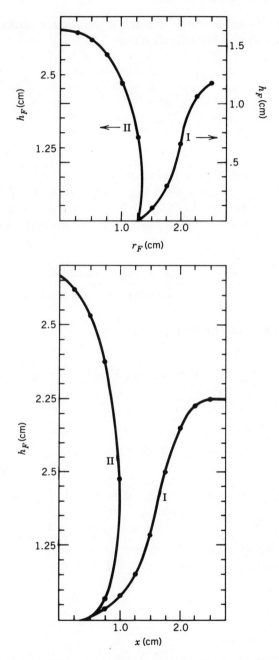

FIGURE 5. Calculated shapes of laminar diffusion flames: (top) cylindrical burner, (bottom) flat flame. Reprinted with permission from The Combustion Institute: *First Symposium on Combustion* (1928), S. P. Burke and T. E. W. Schumann, pp. 2–11.

Equation (9) is derived from Einstein's diffusion equation, which is a solution of Fick's second law of diffusion:[7]

$$\frac{\partial C}{\partial t} = -D\frac{\partial^2 C}{\partial x^2} \tag{10}$$

$$C = \frac{\alpha}{\sqrt{t}}\exp\left(-\frac{x^2}{4Dt}\right) \tag{11}$$

where α is determined by the boundary conditions. Assume now that the concentration of a species at a given location is a measure of the probability that the species reaches that location. If $P(x) = 0$, then the concentration is zero. If $P(x) = 1$, then the concentration is C_0, or the origin of the diffusion process.

$$p(x)\,dx = \frac{(\alpha/t)\exp(x^2/4\,Dt)\,dx}{\int_{-\infty}^{\infty}(\alpha/\sqrt{t})\exp(-x^2/4Dt)\,dx} \tag{12}$$

The mean value of a variable is given by

$$\overline{V} = \int_{-\infty}^{\infty} V_P(X)\,dX \tag{13}$$

so that

$$\overline{X^2} = \int_{-\infty}^{\infty} X^2 P(X)\,dX = 2Dt \tag{14}$$

If $\overline{X^2}$ is defined in terms of the radius of the jet, then t represents the time necessary for oxygen to diffuse from X_0 to 0. From the Burke and Schumann assumption of constant jet velocity the time necessary for the fuel jet to reach the flame tip (h_F) is given by

$$t = \frac{h_F}{U_j} \tag{15}$$

$$\overline{X_j^2} = 2D\frac{h_F}{U_j} \quad \text{or} \quad h_F = \frac{\overline{X_j^2}U_j}{2D} \tag{16}$$

$$\Omega = \pi X^2 U_j \tag{17}$$

$$h_F = \frac{\Omega}{2\pi D} \tag{18}$$

The flame shape can be approximated by setting

$$t = \frac{y}{U_j} \qquad (19)$$

$$\overline{X^2} = \frac{2Dy}{U_j} \qquad (20)$$

where

X = radial distance from center
y = axial distance from burner port
U_j = constant fuel jet velocity
D = diffusion coefficient.

Comprehensive Theories. A comprehensive laminar diffusion flame theory can be written using the equations of conservation of species, energy, and momentum, including diffusion, heat transfer, and chemical reaction.

The species conservation equation takes the form:[8]

$$\rho \frac{\partial X_i}{\partial t} + \rho \vec{U} \cdot \nabla X_i - \nabla \cdot (\rho D \nabla X_1) = R_i \qquad (21)$$

where X_i are the mole fractions of species i.

In the usual form of reaction rates

$$R_i = f(X_i, T) \qquad (22)$$

The enthalpy equation is given by

$$\rho \frac{\partial g}{\partial t} + \rho \vec{U} \cdot \nabla g - \nabla \cdot (\rho D \nabla g) = -\sum R_i g_f^0, i \qquad (23)$$

Coupled with the equations of motion, these serve to define laminar diffusion flames.

These equations are usually simplified by assuming a Lewis number of unity (equality of mass and energy transport) and a single global reaction for R_i. For a cylindrically symmetrical flame, the following simplified equations result:

Continuity of species

$$\frac{1}{r} \frac{\partial}{\partial r} \left(Dpr \frac{\partial C_i}{\partial r} \right) - \rho U \frac{\partial C_i}{\partial r} = R_i \qquad (24)$$

where

$$R_F = A \rho_f \rho_{O_2} \exp\left(\frac{-E}{RT} \right) \qquad (25)$$

and

$$R_{O_2} = \frac{A}{i} \rho_f \rho_{O_2} \exp\left(\frac{-E}{RT}\right) \qquad (26)$$

Conservation of energy

$$\rho C_p U \frac{\partial T}{\partial X} - \frac{k}{r}\frac{\partial}{\partial r}\left(r\frac{\partial T}{\partial r}\right) - \mu\left(\frac{\partial T}{\partial r}\right)^2 = Q A \rho_f \rho_{O_2} \exp\left(\frac{-E}{RT}\right) \qquad (27)$$

Conservation of momentum

$$\mu\frac{\partial U}{\partial X} = \frac{1}{\rho r}\frac{\partial}{\partial r}\left(\mu r \frac{\partial U}{\partial r}\right) \qquad (28)$$

Two solutions result. Since R_i is small at ambient temperature, a nonignition pure diffusion solution results unless a flame temperature is assumed at some value of r and iterations are performed until the energy equation is satisfied. Various assumptions concerning the reaction kinetics provide different approximations to the actual flame. The assumption of instantaneous reaction gives a flame sheet similar to that of Burke and Schumann. The assumption of local equilibrium provides for species in both the fuel and oxidant jets not present in the flame sheet model.

2.2. Single-Drop Evaporation and Combustion

The fact that many condensed phases evaporate and burn as diffusion flames was mentioned in the introduction. Of particular interest in furnace and propulsion applications is the evaporation and burning of liquid drops in fuel sprays. The combustion of sprays is discussed in Chapter 11, but a brief presentation of simplified drop diffusion flames is presented here for comparison with gaseous diffusion flames. Part of the Burke and Schumann approach has been used for drop burning. The assumptions of instantaneous reaction, fuel and oxidant flowing into the flame in stoichiometric proportions, and simple molecular diffusion are carried over to the burning of drops. In the simplest theory no free or forced convection is considered, and obviously, there is no gas jet or oxidant velocity. The fuel flow is replaced by the rate of consumption of the drop and is not an independent parameter.

A simplified model of single-drop evaporation can be given by considering diffusion of the fuel vapor into the surrounding atmosphere.

The equation for diffusion is

$$\frac{\dot{m}_F}{M_F} = 4\pi r^2 D \frac{dC_F}{dr} \qquad (29)$$

Treat $\dot{m}$ as a constant in steady-state evaporation.

For $r_\infty = \infty$, a burning drop in an infinite oxidant atmosphere.

$$\dot{m}_F = 4\pi r_D M_F D (C_{F,D} - C_{F,\infty})$$ (30)

For the case where the vapor behaves as a perfect gas,

$$C_F = \frac{P_F}{RT}$$ (31)

and Eq. (30) becomes

$$\dot{m}_F = -\frac{4\pi r_D M_F D}{RT}(P_{F,D} - P_{F,\infty})$$ (32)

For a spherical drop, m_F is given by

$$m_F = \tfrac{4}{3}(\pi r_D^3 \rho_F)$$ (33a)

and

$$\dot{m}_F = (4\pi r_D^2 \rho_F)\frac{dr_D}{dt}$$ (33b)

By combining Eqs. (32) and (33), we get

$$r_{D,0}^2 - r_D^2 = \frac{2M_F D}{\rho_F RT}(P_{F,D} - P_{F,\infty})t$$

$$r_{D,0}^2 - r_D^2 = \beta t$$ (34)

which is the well-known r^2 law for evaporation with

$$\beta = \frac{2M_F D}{\rho_F RT}(P_{F,D} - P_{F,\infty})$$ (35)

Equations (32) and (34) can easily be related to a burning drop by considering the burning process as a special case of evaporation. A flame surrounds the drop and the fuel and oxygen diffuse to each other arriving at the flame front in stoichiometric proportions as assumed by Burke and Schumann.

At the flame front $r = r_F$ and $P_F = 0$. Since r is not infinite, eq. (32) becomes

$$\dot{m}_F \left(\frac{1}{r_0} - \frac{1}{r_F} \right) = -\frac{4\pi M_F D}{RT_D} P_{F,D}$$ (36)

The flame radius r_F is not known and neither is the drop temperature T_D. The partial pressure of fuel at the drop surface can be assumed to be a function of the drop temperature for equilibrium evaporation. Unlike the gaseous jet where $\dot{m}_F$ is an independent variable, $\dot{m}_F$ for an evaporating or burning drop is unknown.

An additional equation can be obtained by considering oxygen diffusion to the flame front:

$$\dot{m}_{O_2}\left(\frac{1}{r_F} - \frac{1}{r_\infty}\right) = -\frac{4\pi M_{O_2} D}{RT_\infty} P_{O_2,\infty} \tag{37}$$

Since $\dot{m}_{O_2} = i\dot{m}_F$, Eqs. (36) and (37) can be combined to eliminate r_F. T_D is still unknown, however. Assuming that T_D is at the boiling temperature of the liquid fuel is a good approximation, but actually the drop temperature is less than the boiling point due to evaporative cooling.

The drop temperature can be found by the introduction of two heat transfer equations:

At the drop surface

$$\dot{m}_F(g_{vap} + \Delta g) = -4\pi r^2 \lambda \frac{dT}{dr} \tag{38}$$

$$\dot{m}_F\left(\frac{1}{r_D} - \frac{1}{r_F}\right) = -4\pi k \int_{T_D}^{T_F} \frac{dT}{g_v + \Delta g} \tag{39}$$

Outside the drop

$$\dot{m}_F g_{comb} - (g_{P,F} - g_{P,\infty}) + (g_{O_2,F} - g_{O_2,\infty}) = 4\pi r^2 \lambda \frac{dT}{dr} \tag{40}$$

$$\dot{m}_F\left(\frac{1}{r_F} - \frac{1}{r_\infty}\right) = -4\pi_r^2 \lambda \int_{T_F}^{T_\infty} \frac{dT}{g_{comb} - (g_{P,F} - g_{P,\infty}) + (g_{O_2,F} - g_{O_2,\infty})} \tag{41}$$

Equations (36), (37), (39), and (41) can be used to solve for T_D, T_F, r_F, and $\dot{m}_F$.

As in the case of gaseous diffusion flames, more comprehensive equations can be written for numerical solution. Spalding[9] defines a transfer number B in terms of a parameter

$$b = \frac{W_A}{(W_{AS} - 1)} \tag{42}$$

TABLE 1. Value of the Transfer Number, B, in Air

Combustible	B	Combustible	B
iso-Octane	6.41	Kerosene	3.4
Benzene	5.97	Gas oil	2.5
n-Heptane	5.82	Light fuel oil	2.0
Toluene	5.69	Heavy fuel oil	1.7
Gasoline	5.4	Carbon	0.12

where

W_A = mass fraction of component A

W_{AS} = mass fraction of component A at surface

and

$$B = b_\infty - b_s \tag{43}$$

For a variety of liquid fuels burning in air, values of B are given in Table 1.[10] Values of the constant β can be calculated from B:

$$\beta = \frac{8\lambda}{\rho_L C_P} \ln(1 + B) \tag{44}$$

2.3. Flame Height of Laminar Diffusion Flames

Simplified diffusion flame theory predicts that the flame height will be a linear function of the volumetric flow rate, or in other terms, a linear function of the fuel jet velocity and the square of the diameter of the port, and inversely proportional to the initial oxygen concentration

$$h_F \propto \frac{\pi U_j d^2}{4 D i C_{O_2,0}} \tag{45}$$

A comparison of calculated (using Eq. (7.8)) and experimental flame heights varying with fuel velocity is given in Figure 6.[5] The agreement is generally satisfactory. In the region of low fuel velocity, the flame height is independent of the air velocity, whereas the air velocity has an effect at higher fuel velocities.

Burke and Schumann theory, mainly because of its assumptions, predicts that flame height would be independent of temperature and pressure. This conclusion is consistent with the assumption that the temperature effect on diffusion and flow velocity counteract each other so that the ratio V/D is

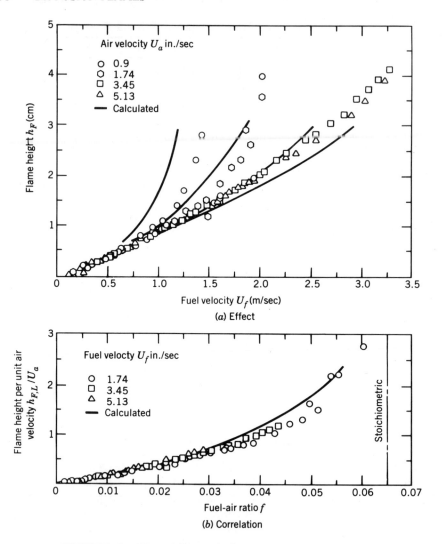

FIGURE 6. Effect of fuel and air velocity on flame height.

roughly independent of temperature and pressure. Data from Barr[11] illustrating this conclusion is presented in Table 2. The flame height is seen to be independent of pressure from 1 atm to 0.074 atm. Below 0.070 atm, lifted flames occurred under the conditions of the experiment.

Experimentally the linear dependence of flame height on volumetric flow rate is verified by some experimental data, but it is suggested in Wohl et al.[12] that the square root of the volumetric flow rate provides a better correlation. It is suggested that the dependence of diffusion coefficient on temperature cannot

TABLE 2. Variation of Laminar Flame Height with Pressure

| P, atm | h_F, cm | Lifted Flames | |
		h_{base}	h_F
1.0	1.2 (interpolated)	—	—
0.202	1.6	—	—
0.088	1.4	—	—
0.081	1.4	—	—
0.074	1.4	—	—
0.070	—	2.0	2.8
0.061	—	4.2	5.2

be ignored. If the diffusion coefficient is written as

$$D = D^\circ + kh_F \tag{46}$$

where D° is the ambient temperature diffusion coefficient. It is also stated in Wohl et al.[12] that since $kh_F \gg D^\circ$, Eq. (18) can be written as

$$h_F = \frac{\Omega}{\sqrt{(2\pi k)}} \tag{47a}$$

or including effects of jet fuel concentration (dilution by air)

$$h_F = K_1\sqrt{\Omega Y_F} \tag{47b}$$

where Y_F is the mol fraction of fuel in the jet and K_1 is an empirical constant of the order of 9.5. It changes with port diameter as shown in Table 3.

The agreement with flame height dependence on the square root of volumetric flow rate is illustrated in Figure 7.[12] The effects of fuel and oxygen concentration have been treated by various investigators.[13,14] The Burke and Schumann theory gives reasonable agreement with experimental data and there is little to recommend among the empirical correlations in the literature.

TABLE 3

Port Diameter, in.	K_1
0.180	10.4
0.290	9.9
0.40	9.7
1.03	9.1
1.99	8.3

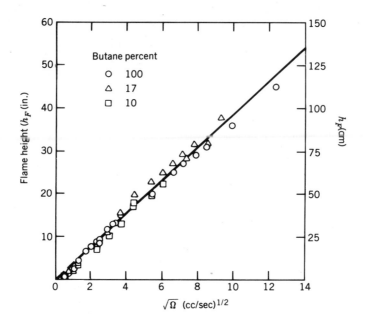

FIGURE 7. Height of butane-air flame in free air, illustrating dependence of flame height (h_F) on square root of volumetric flow rate ($\sqrt{\Omega}$). Reprinted with permission from The Combustion Institute: *Third Symposium on Combustion* (1949), K. Wohl, C. Gazely, and N. Knapp, pp. 228–300.

2.4. Flame Chemistry

The Burke and Schumann model assumes that no reactions occur anywhere except at the flame surface and that, at that location, fuel and oxygen are instantaneously converted to products of combustion. The success of the model suggests that the flame chemistry is rapid compared with other processes. This result is further confirmed by the failure to find oxygen on the fuel side of the flame and fuel on the oxidant side of the flame. In fact, however, the chemistry of combustion in a diffusion flame is not as simple as assumed, although the oxidation reactions are quite rapid.

Evidence obtained by absorption and emission spectroscopy[15] and by sampling[16,17] shows that hydrocarbon fuels undergo appreciable pyrolysis in the fuel jet before oxidation occurs. Figure 8[15] shows the temperature profile and species concentration obtained spectroscopically. It is noteworthy that carbon, C_2 and CH, are found near the centerline of the fuel side of the flame and that the fuel fragments and oxygen appear to approach zero at a location that can be considered as the flame front. The luminous carbon zone exists well on the fuel side of the flame front. The OH radical, which undoubtedly accounts for some of the fuel fragment and carbon oxidation, also extends to the fuel side of the flame front. The peak temperature appears on the oxygen

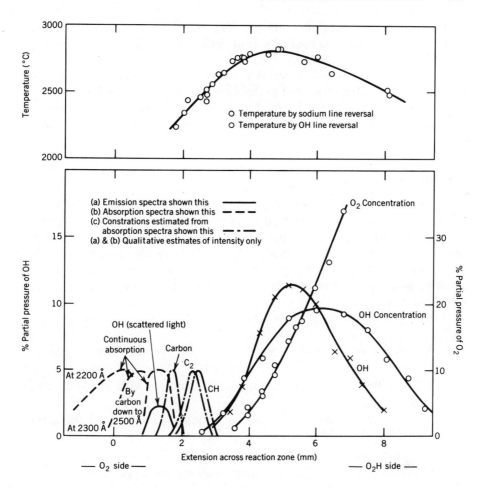

FIGURE 8. Absorption and emission spectra of an ethylene flat diffusion flame. Reprinted with permission from *Flames* by A. G. Gaydon and H. G. Wolfhard, Chapman and Hall, Publishers. Copyright 1960.

side of the assumed flame front, which may be due to equilibration processes (reassociation of free radicals and atoms) or to an effect of the measurement technique.

Sampling of a diffusion flame produces similar results with some variation. A methane flame was studied in Gordon et al.[16] The original fuel disappears very rapidly on the fuel side of the flame front, as is expected. A large number of pyrolysis products are found on the fuel side of the jet, including ethylene and acetylene. In disagreement with the Burke and Schumann assumption, some oxygen and oxygen containing molecules were found on the fuel side. The presence of O_2 and OH may produce oxidative cracking and permit the high rates of reaction required at the relatively low temperatures and residence times near the core of the fuel jet. The presence of oxygen containing molecules and radicals on the fuel side of the jet may be the result of finite rate reaction, equilibration, or both.

In terms of the gross features of the flame, particularly flame height, the preflame pyrolysis reactions are probably not very important. Fuel pyrolysis, however, probably accounts for the formation of carbon, while the presence of OH may provide a path for NO_x formation, particularly on the oxidant side of the flame.

2.5. Smoke

In the previous section the chemistry of the diffusion flame was discussed. Of primary importance in understanding the mechanism of smoke formation is the fact that fuel pyrolysis occurs extensively in the fuel jet and that olefins and acetylene are formed. Although O_2 and OH are also found on the fuel side, their concentration is very low. They may exert an influence on the cracking process, but it is unlikely that they account for appreciable oxidation of carbon particles.

Several characteristics of smoking diffusion flames are noteworthy:

1. The first appearance of smoke is generally at the flame tip, the width of the smoke trail increasing as the amount of smoke increases.
2. The appearance and quantity of smoke increases with increasing flame height.
3. The smoking tendency changes with oxidant flow rate and the oxygen content of the oxidant stream.
4. Smoking tendency is a function of fuel type.
5. Smoking tendency tends to increase with increasing pressure.

It is difficult to model the smoke formation process in a diffusion flame. Not only is the chemistry of carbon formation complicated and incompletely known, but the diffusion and heat transfer processes that determine concentration, residence time, and temperature within the fuel jet are not treated rigorously in most diffusion flame models. In an effort to model smoke formation, the diffusion processes and the pyrolysis process were decoupled in a study reported in Shoji and Gerstein.[13] The flame boundary was determined by the Burke and Schumann model. The flame temperature of the flame front was assumed to be the adiabatic, equilibrium flame temperature, and only radial heat transfer was assumed from the flame into the fuel jet and into the air stream. The combination of the diffusion and heat transfer processes provides a concentration and temperature profile within the jet. Carbon was assumed to form by the first-order pyrolysis of the fuel.

Since the pyrolysis chemistry was not known and the carbon concentration necessary to produce visible smoke was not known, these parameters were

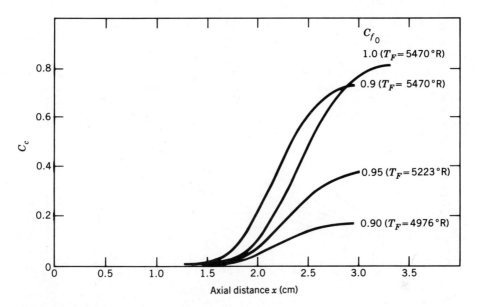

FIGURE 9. Carbon concentration profile versus axial distance for various values of initial fuel concentration ($C_{f,0}$).

treated empirically. In keeping with observation 1, the model showed that the concentration of carbon was highest along the centerline of the fuel jet and increased from the burner port to the flame tip. This centerline corresponds to the location at which the fuel is exposed to the highest temperature for the longest time.

Reducing the concentration of fuel in the jet by dilution with an inert reduces the flame height and also the flame temperature. The effect of fuel concentration reduction on carbon formation along the centerline of the fuel jet is illustrated in Figure 9.[13] The effects of flame height reduction and dilution on carbon formation are consistent with experiment. Measuring the flame height at which smoke appears is a standard test for smoking tendency of fuels. The effect appears to be the same whether measured in a gaseous diffusion flame or in a wick lamp.[18]

It has already been discussed (see Table 2) that pressure has little effect on flame height and only a modest effect on flame temperature. Pressure would have an important effect on the rate of pyrolysis, however. Assuming a first-order reaction, one would expect that carbon formation would increase linearly with pressure. The smoke-free fuel flow that governs flame height would thus vary as the reciprocal of the pressure. The data in Figure 10[18] are consistent with this prediction. Flame temperature and molecular type, which control the rate of pyrolysis, are the basis for determining smoking tendency according to Glassman and Yaccarino.[19]

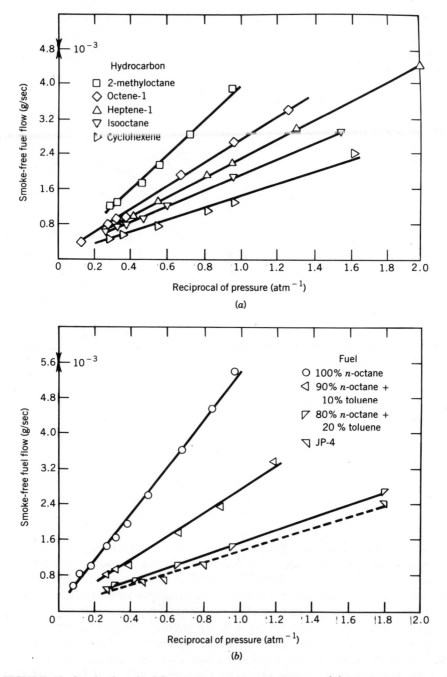

FIGURE 10. Smoke-free fuel flow versus reciprocal pressure: (*a*) five pure hydrocarbon fuels, (*b*) various fuels and blends. Reprinted with permission from The Combustion Institute: *Fifth Symposium on Combustion* (1955), R. L. Schalla and G. E. McDonald, pp. 316–324.

Much research is needed to understand the mechanism of the formation of smoke in diffusion flames. There is at the present time no accepted model for this process.

3. TURBULENT DIFFUSION FLAMES

3.1. Introduction

It is shown in Wohl et al.[12] that the independence of flame height on jet diameter and the simple relationship between flame height and volumetric flow rate ceases as the tube diameter decreases, apparently at a critical flow velocity and at a Reynolds number of about 2000. The behavior of the diffusion flame height and other characteristics are illustrated in Figure 11.[1] Illustrated is the almost linear increase of flame height with velocity in the low velocity, laminar region, a transition region and the turbulent region with flame height almost independent of velocity.

3.2. Theory

The turbulent approach flow of both fuel and oxidant consists of fluctuating velocity components. These flows may be considered to be composed of a mean velocity component, $\bar{U}$, and a fluctuating component, U', so that at any location, at any instant, the velocity, U, is given by

$$U = \bar{U} + U' \tag{48}$$

The fluctuating components of the velocity do not create a directed mass flow when averaged over time so that $\bar{U'} = 0$.

The root-mean-square velocity is not zero

$$\sqrt{\overline{U'^2}} \neq 0 \tag{49}$$

which is analogous to molecular velocities in the kinetic theory of gases. The analogy of turbulent fluctuations to molecular motion represents one approach to the description of the properties of turbulent diffusion flames.

In addition to the analogy between the fluctuating velocity to molecular velocity, it is convenient to define a quantity analogous to the mean free path — such a quantity is called the scale of turbulence. From any location in the flow, it is possible to measure the degree of correlation of the fluctuations at that point with other locations. When the distance from that point to the second point is zero, perfect correlation exists, and the correlation function has the value of 1. At some other location the correlation function becomes zero. This separation can be defined as the scale of turbulence, although other definitions have also been proposed.

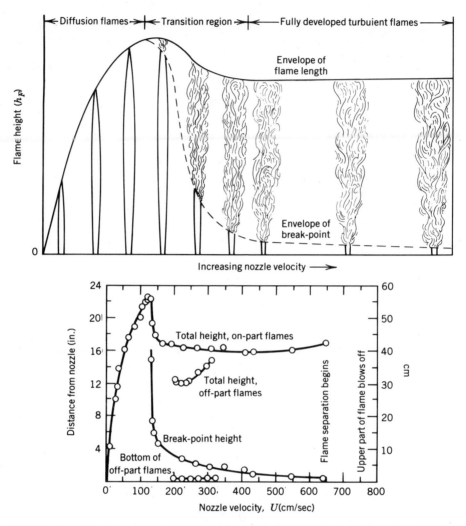

FIGURE 11. Effect of nozzle velocity on flame appearance in laminar and turbulent flow: (top) flame appearance, (bottom) flame height and break point height. Reprinted with permission from The Combustion Institute: *Third Symposium on Combustion* (1948), H. C. Hottel, pp. 254–266.

The fluctuating velocities in the approach flow of both fuel and oxidant generate fluctuations in the concentration.

In a manner analogous to the velocity description, the concentration can be written as the sum of an average component and a fluctuating component:

$$C_i = \overline{C_i} + C_i' \tag{50}$$

As with velocity

$$\overline{C_i'} = 0 \tag{51}$$

A very simplified model of turbulent diffusion flames can be developed by replacing the diffusion coefficient in the laminar diffusion flame models by an eddy or turbulent diffusivity.

Again, by analogy with kinetic theory of gases where the diffusion coefficient can be written as

$$D = l\sqrt{\overline{c^2}}$$
(52)

where

    c = molecular velocity
    l = mean free path

we can define an eddy diffusivity by the relation

$$\varepsilon = U''l'$$
(53)

where

    $U'' = \sqrt{\overline{U'^2}}$ intensity of turbulence
    l' = scale of turbulence

Equation (16) for the height of a laminar diffusion flame can now be written as

$$h_{F,t} = \frac{U_j R^2}{2\varepsilon}$$
(54)

where $h_{F,t}$ is the height of the turbulent flame.

Since U' is a function of the velocity, and l' is a function of tube radius, we can define ε by

$$\varepsilon = fRU_j$$
(55)

where f is an empirical constant.

Substituting for ε in Eq. (54), we get

$$h_{F,t} = \frac{fR}{2}$$
(56)

Note that Eq. (56) relates the turbulent flame height to the first power of the tube radius and is independent of velocity. It is shown in Figure 12 that this velocity has little, if any, effect on turbulent flame height. The ratio of flame height to tube diameter is plotted against velocity to emphasize the linear dependence on tube diameter.

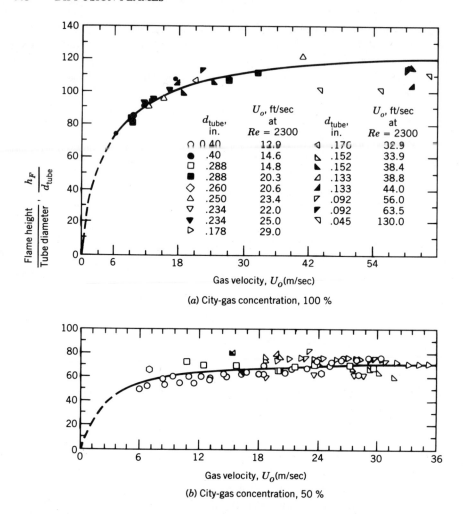

FIGURE 12. Effect of velocity on ratio of flame height to tube diameter in turbulent flow. Reprinted with permission from The Combustion Institute: *Third Symposium on Combustion* (1949), K. Wohl, C. Gazely, and N. Kapp, Figures 3 and 6.

Equation (56) is oversimplified and also does not reduce to the laminar flame height at low velocities. The latter difficulty can be removed by defining a diffusivity in the form

$$D_T = D + \varepsilon \tag{57}$$

and

$$D_T = D + fU_JR \tag{58}$$

using Eq. (58), we get

$$\frac{h}{R} = \frac{U_j R}{D + fRU_j} = \frac{1}{(D/U_j R) + f}$$ (59)

Equation (59) can be used as a correlation for both the laminar and turbulent flame regimes. Note that $f = 0$ for laminar flames, and Eq. (59) assumes the form of Eq. (45).

The advent of advanced computer techniques to solve problems of mixing with chemical reaction has largely superseded efforts to correlate turbulent diffusion flame data with highly empirical models. The details of modeling turbulent, chemically reacting flows are discussed in Chapter 8. Some of the theoretical arguments relative to diffusion flames will be summarized in this chapter.

The equations of conservation of mass and energy, the equations of motion, and descriptions of the chemical processes are the basis of turbulent diffusion flame theory as they are of all reacting flows. It has been common practice to introduce the temperature, composition, and velocity in terms of a mean value and a fluctuating component as discussed earlier. This representation is particularly useful if small perturbation techniques are used to solve the equations.

These small perturbation methods have been replaced by methods using probability density functions. The use of a probability function to obtain average values was discussed in connection with molecular diffusion processes. The use of probability density functions has its own complications. They are not easily measured, are often not symmetrical in form, and are not yet capable of being calculated from the flow parameters.

The reaction term, in the case of finite rate reactions, can also produce complications since the average reaction rate is not the rate obtained through the use of average concentrations but also contains products of fluctuating concentrations.

$$\frac{dC}{dT} = f(C_1 C_2)$$ (60)

$$C_1 = \overline{C_1} + C_1' \quad \text{and} \quad C_2 = \overline{C_2} + C_2'$$ (61)

so that

$$\frac{\overline{dC}}{dT} = f(\overline{C_1 C_2} + \overline{C_1' C_2'})$$ (62)

Bilger[8] describes the Reynolds decomposition for a global second-order reaction in terms of the double correlations. The global reaction rate,

$$R_i = \rho^2 \nu_i M_J^{-1} Y_i Y_j A \exp\left(-\frac{E}{RT}\right)$$ (63)

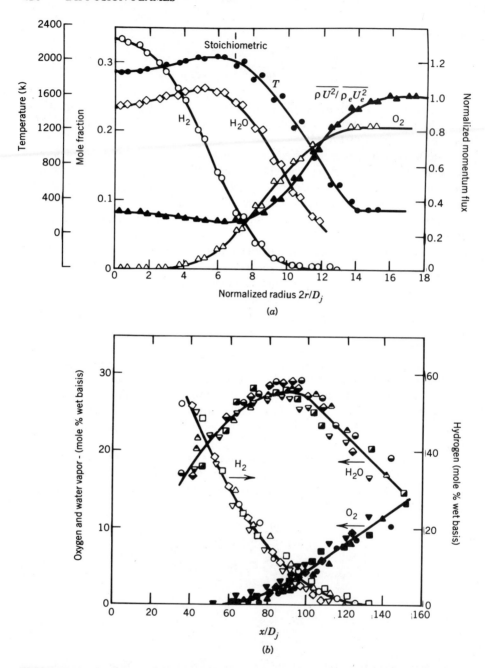

FIGURE 13. Composition profiles for a H_2/O_2 flame, illustrating crossover of fuel and oxidant species in turbulent diffusion flames. Reprinted with permission from *Energy and Combustion Science* (Student Edition), Vol. 1, R. W. Bilger, "Turbulent Jet Diffusion Flames," Copyright 1979, Pergamon Press PLC.

must be expressed as

$$\overline{R}_i = \nu M^{-1} P^{-2} \overline{Y}_i \overline{Y}_j A \exp\left(-\frac{E}{R_0 T}\right)$$

$$\times \left\{ 1 + \frac{\overline{P_1^2}}{P^2} + \frac{\overline{Y_i' Y_j'}}{\overline{Y}_i \overline{Y}_i} + \frac{\overline{2P' Y_j'}}{\overline{P} \overline{Y}_i} + \frac{2P' Y_j'}{\overline{P} \overline{Y}_j} \right.$$

$$\left. + \frac{E}{R\overline{T}} \left[\frac{\overline{Y_i T'}}{\overline{Y}_i \overline{T}} + \frac{\overline{Y_j' T'}}{\overline{Y}_j \overline{T}} + \left(\frac{E}{2R\overline{T}} - 1 \right) \frac{\overline{T'^2}}{\overline{T}^2} \right] + \cdots \right\} \qquad (64)$$

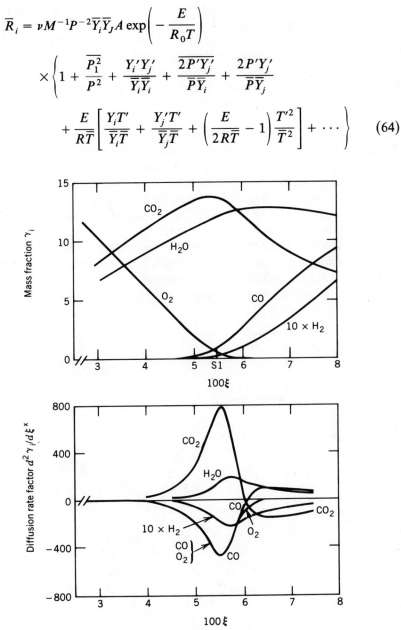

FIGURE 14. Composition profiles and reaction rate within a turbulent methane-air flame. Calculated using shifting equilibrium. Reprinted with permission from *Energy and Combustion Science* (Student Edition), Vol. 1, R. W. Bilger, "Turbulent Jet Diffusion Flames," Copyright 1979, Pergamon Press PLC.

Donaldson[20] continues the expansion to include the triple correlations and has developed a model involving two reactants and a product to simplify the reaction complexity.

The simplest approach, of course, is to adopt the Burke and Schumann assumption of instantaneous reaction, but this does not permit the calculation of reaction intermediates. Another simplified chemical model assumes that thermodynamic equilibrium applies, but this, too, does not permit the calculation of nonequilibrium products such as smoke and various pollutants.

Measurements of species concentration in turbulent diffusion flames show that fuel and oxidant components coexist as shown in Figure 13 for H_2—O_2 flames[8] and for propane-air flames.[21] Similar data for H_2—O_2 flames have been reported in Hawthorne et al.[22] The effect of the equilibrium assumption in calculating composition, which produces overlap of fuel and oxidant species, is shown for CH_4-air in Figure 14.[8] The absence of methane on the fuel side is the result of the reaction of carbon dioxide with methane to produce carbon monoxide and hydrogen.

The probability density functions are not exactly Gaussian,[23] and Favre (density) averaging is not always successful.[24] Nevertheless, reasonable predictions of turbulent diffusion flame structure can be made using truncated probability density functions as shown in Figure 15[25] and discussed in Drake et al.[26]

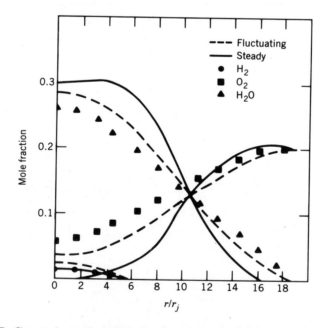

FIGURE 15. Comparison of predicted and measured species concentration using an integral method and assuming equilibrium chemistry. Reprinted with permission from *Acta Astronautica*, Vol. 1, R. P. Rhodes, P. T. Harsha, and C. E. Peters, "Turbulent Kinetic Energy and Analysis of Hydrogen/Air Diffusion Flames," Copyright 1974, Pergamon Press PLC.

Unfortunately these computer methods do not reduce to simple correlation parameters. If only an approximate flame height is desired, the simplified models are probably adequate. For more detail, the use of probability density functions for conserved scalars can provide useful information. The extent of reaction, or the "reactedness," can be used to find surfaces that define a desired level of combustion efficiency or a desired flame boundary.[27] The calculation of nonequilibrium product concentrations and distributions requires a refined model.

4. FLAME STABILITY

The diffusion flame, like the premixed flame, has certain regions of stability beyond which burning cannot occur. These regions are quite similar to those of premixed flames (except for flashback), and in fact, stabilization at the burner port may involve a region where mixing is so rapid that the behavior involves a premixed flame. Both laminar and turbulent flames appear to behave in a similar manner since the stability zone is often in the laminar sublayer of the flow. Although flashback cannot occur with a diffusion flame, a lower limit of fuel flow exists that is probably associated with quenching of the small flame by heat transfer to the burner.

Some data on the lifting and ultimate blow off of a butane diffusion flame[3] are illustrated in Figure 16. The solid line shows the location of the flame base as the flow velocity is increased and lift occurs. The dashed line, on which typical flames are shown, illustrates the path of the base of the flame as the

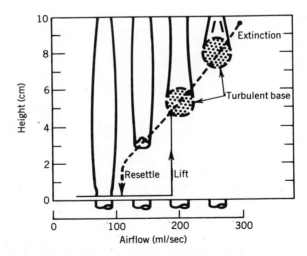

FIGURE 16. Illustration of flame lift and resettle hysteresis for a gaseous diffusion flame. Reprinted with permission from The Combustion Institute: *Fourth Symposium on Combustion* (1953), J. Barr, pp. 765–771.

butane flow velocity is reduced and the flame resettles. The ultimate point, at a flow rate of about 300 milliliters per second, corresponds to complete blow off. The phenomenon of lift is illustrated by the dashed curves of Figure 11, which show the location of the bottom of lifted flames and their corresponding total flame height.

The stability range of diffusion flames of ethylene issuing from carburetor-type jets into still air is reported in Scholfield and Garside.[28] The results showing the stability region as a function of jet diameter and Reynolds number are shown. The range of gas flow over which the flame is stable increases only slowly with increasing jet diameter. Above a certain diameter, the range over which the lifted flame persists widens as the jet size increases, because of the rapid decrease in blow off tendency of larger jets. On small jets, the lift and blow off limits coincide.

The phenomenon of lift is explained in Scholfield and Garside[28] by a comparison of the break points of ignited and unignited streams and the height of the base of the lifted flame. In Figure 17, the height to turbulence or the break point or the height to the base of the lifted flame is plotted against Reynolds number, essentially flow velocity. Curves A and B connect data for the height to turbulence in the unignited and ignited streams, respectively. Note that the height to turbulence occurs at higher Reynolds numbers for the ignited stream. The stabilization is attributed to heat addition from the flame.[28] Curve C represents the height between the top of the burner and the base of the lifted flame. Curve D represents the height to turbulence in the fuel jet of the lifted flame. At a flow rate immediately before lift, the break point in the gas stream lies well within the flame envelope. At the actual lift point, the flame is stabilized by eddies formed at the break point, and the base

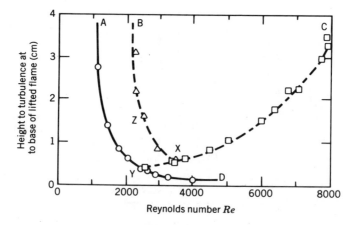

FIGURE 17. Stability diagram showing height to turbulence or to lifted flame as a function of Reynolds number. Reprinted with permission from The Combustion Institute: *Third Symposium on Combustion* (1949) pp. 102–110.

of the flame is at the same height as the height to turbulence (point x in Figure 17). After the flame has lifted, the heating effect of the flame on the gas stream near the port is reduced, so that the break point returns to that of the unignited jet, shown by the dash-dot line connecting curve D and point x. As the flow rate is reduced, the break point follows curve D, while the flame base follows curve C. When they meet at point y, the flame suddenly resettles on the port, and the break point rises to point z on curve B. This description accounts for the hysteresis in the lift curves.

Opposed jet diffusion flames exhibit a somewhat different phenomenon, but one that may be related to blow off. Both seem to be related to the extinction of the flame due to the inability of the reaction rate and the concomitant heat generation to maintain itself in the presence of the heat losses and dilution produced above a critical velocity.

In the case of an opposed jet diffusion flame, a hole appears when the mass flow exceeds a critical velocity. It is reported[29,30] that the maximum mass flow which a diffusion flame can support is directly proportional to the mass of fuel consumed per second by a unit area of flame front propagating through a fuel-air mixture, the stabilization zone behaving as a premixed flame. The laminar burning velocity can be used to generate such a quantity since $\rho_0 S_L$ is

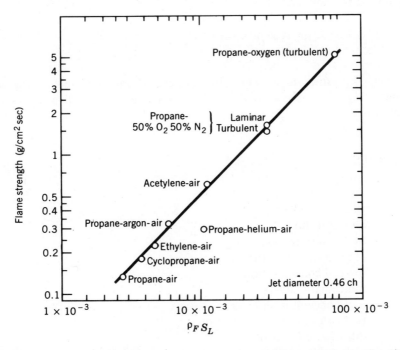

FIGURE 18. Flame strength (mass flow rate at point of flame rupture as a function of ρS_L. Reprinted with permission from The Combustion Institute: *Eighth Symposium on Combustion* (1962), A. E. Potter, S. Heimel, and J. N. Butler, pp. 1027–1034.

the mass flow of mixture per second per unit area consumed by the flame. If m_f is the mass fraction of fuel in the mixture, then the quantity $m_f \rho_0 S_L$, called the *flame strength*, measures this maximum flow. If the fuel flow into the diffusion flame exceeds this critical value, blow off occurs.

A plot of flame strength (mass flow into the opposed jet flame) versus $m_f \rho_0 S_L$ is shown in Figure 18.[29] A good correlation is apparent. This correlation with flame velocity is consistent with the model of blow off of cylindrical premixed flames where blow off is assumed to occur when the local stream velocity exceeds the laminar flame velocity near the burner port.

These same considerations may be applied to cross flows that represent the transition from coaxial diffusion flames and opposed jet diffusion flames. In each case, a critical mass flow will cause the flow rate to exceed the flame strength, and blow off will occur, usually preceded by lifted or partially lifted flames. A fuel jet in a cross flow is illustrated in Figure 2. The upper edge of the jet, l, as a function of the downstream distance, x, is given by

$$\left(\frac{1}{d_P}\right)^{1.65} = 2.91\left(\frac{W_a}{W_F}\right)\left(\frac{x}{d_P}\right)^{0.5} \tag{65}$$

or

$$W_F = m_f \rho_0 S_L = \left(\frac{1}{d_P}\right)^{1.65}\left(\frac{W_a}{2.91}\right)\left(\frac{d_P}{x}\right)^{0.5} \tag{66}$$

When W_a grows to a value that exceeds a critical value of the flame strength, W_F, the flame will blow off. Similarly, blow off will occur as d_P is reduced below some critical value.

SYMBOLS

A	Arrhenius constant
B	transfer coefficient
C	concentration
C_F	fuel concentration
C_j	fuel concentration in unreacted jet
C_{O_2}	oxygen concentration
c	molecular velocity
$C_{F,D}$	fuel concentration at drop surface
$C_{F,\infty}$	fuel concentration at infinity
D	diffusion coefficient
E	activation energy
g_c	enthalpy of combustion
g_P	enthalpy of products

g_f	enthalpy of formation
h_F	flame height
i	stoichiometric coefficient
l	mean free path
l'	scale of turbulence
m	mass
$\dot{m}$	mass flow
P	pressure
P_T	$P_F + P_{O_2}/i$
Q	heat of reaction
r	radius
r_j	radius of fuel jet
r_D	drop radius
r_F	flame radius
r_s	radius of air tube
R	universal gas constant
R_i	rate of production of species i
S_L	burning velocity
T	temperature
t	time
U	velocity
V	volume
X	distance
X_i	mol fractions
Y_i	mass fractions
λ	thermal conductivity
ε	eddy diffusivity
Ω	volumetric flow rate

REFERENCES

1. H. C. Hottel, *Symp. (Int.) Combust. [Proc.]* **4**, 97–113 (1953).
2. A. M. Kanury, *Introduction to Combustion Phenomena.* Gordon & Breach, New York, 1975.
3. J. Barr, *Symp. (Int.) Combust. [Proc.]* **4**, 765–771 (1953).
4. S. P. Burke and T. E. W. Schumann, *Ind. Eng. Chem.* **20** (10), 998–1004 (1928).
5. J. Barr and B. P. Mullins, Rep. No. R-44, British National Gas Turbine Establishment, June, 1949.
6. W. Jost, *Explosion and Combustion Processes in Gases.* McGraw-Hill, New York, 1946.
7. W. Jost, *Diffusion.* Steinkopff, Darmstadt, 1957.
8. R. W. Bilger, *Energy and Combustion Science*, Student Ed. 1, pp. 109–131, Pergamon, Oxford, 1979.

9. D. B. Spalding, *Some Fundamentals of Combustion.* Butterworth, London, 1955.

10. I. Glassman, *Combustion.* Academic Press, New York, 1977.

11. J. Barr, *Fuel* **28**, 200–205 (1949).

12. K. Wohl, C. Gazely, and N. Knapp, *Symp. (Int.) Combust. [Proc.]* **3**, 228–300 (1949).

13. J. M. Shoji and M. Gerstein, E.M.E. Thesis, University of Southern California, Los Angeles (to be published).

14. National Advisory Committee for Aeronautics, *Natl. Advis. Comm. Aeronaut. Rep.* **1300**, 207–228 (1957).

15. A. G. Gaydon and H. G. Wolfhard, *Flames*, 2nd ed. Macmillan, New York, 1960.

16. A. S. Gordon, S. R. Smith, and J. R. McNesby, *Symp. (Int.) Combust. [Proc.]* **7**, 317–324 (1959).

17. V. C. Bernez-Cambot and R. Delbourgo, *Symp. (Int.) Combust. [Proc.]* **18**, 777–783 (1981).

18. R. L. Schalla and G. E. McDonald, *Symp. (Int.) Combust. [Proc.]* **5**, 316–324 (1955).

19. I. Glassman and P. Yaccarino, *Symp. (Int.) Combust. [Proc.]* **18**, 1174–1183 (1981).

20. C. duP. Donaldson, G. R. Hilst, and A. K. Varma, *Second-Order Closure Model. Turbulent Combust., Proj. Squid Workshop Proc., and Environ. Sci. Technol.* **6**, 812 (1972).

21. J. A. Senecal and C. W. Shipman, *Symp. (Int.) Combust. [Proc.]* **17**, 355–362 (1979).

22. W. R. Hawthorne, D. S. Weddell, and H. C. Hottel, *Symp. (Int.) Combust. [Proc.]* **3**, 266–288 (1949).

23. J. H. Kent and R. W. Bilger, *Symp. (Int.) Combust. [Proc.]* **16**, 1643–1656 (1977).

24. I. M. Kennedy and J. H. Kent, *Symp. (Int.) Combust. [Proc.]* **17**, 279–287 (1979).

25. R. P. Rhodes, P. T. Harsha, and C. E. Peters, *Acta Astronaut.* **1**, 443 (1974).

26. M. C. Drake, M. Lapp, C. M. Penny, S. Warshaw, and B. W. Gerhold, *Symp. (Int.) Combust. [Proc.]* **18**, 1521–531 (1981).

27. F. C. Lockwood and A. S. Naguib, *Combust. Flame* **24**, 109–124 (1975).

28. D. A. Scholfield and J. E. Garside, *Symp. (Int.) Combust. [Proc.]* **3**, 102–110 (1949).

29. A. E. Potter, Jr. and J. N. Butler, *Am. Rocket Soc.* **29**(1), 54–55 (1959); A. E. Potter, Jr., S. Heimel, and J. N. Butler, *Symp. (Int.) Combust. [Proc.]* **8**, 1027–1034 (1962).

30. D. B. Spalding, *Fuel* **33**(3), 255–273 (1954).

8 Turbulent Reacting Flows

F. A. WILLIAMS

Department of Applied Mechanics and Engineering Sciences
University of California, San Diego
La Jolla, California

1. Perspective and Background
   1.1. Motivation and terminology for studying turbulent reacting flows
   1.2. Classification of approaches to the subject
   1.3. Basics of probability density functions
   1.4. Basics of asymptotic expansions
2. Use of moment methods
   2.1. Moments based on ordinary averages
   2.2. Moments based on Favre averages
3. Use of probability density functions
   3.1. The conserved scalar in turbulent diffusion flames
   3.2. Structures of turbulent diffusion flames
   3.3. Lift-off and blow off of turbulent diffusion flames
   3.4. Probability-density-function approaches to description of premixed turbulent
        flames
4. Use of perturbation methods for premixed combustion
   4.1. Premixed flames in small-gradient turbulence
   4.2. Regimes of turbulent flame propagation
5. Coherent structures in turbulent combustion
   5.1. Experimental observations of coherent structures
   5.2. The ESCIMO approach of spalding
   5.3. The random vortex method
   5.4. Significance of coherent structures in turbulent combustion
   References

1. PERSPECTIVE AND BACKGROUND

1.1. Motivation and Terminology for Studying Turbulent Reacting Flows

Practical Problems. In the vast majority of industrial devices, the characteristic
dimensions and flow velocities are sufficiently large for the flow to be turbulent. Therefore turbulent reacting flows may be said to be of central concern in

practical equipment for chemical processing and combustion. Design of equipment, diagnosis of malfunction problems, and development of modifications for improvement of performance rest on concepts of turbulent reacting flows. In addition analysis of hazards from accidents involving combustion utilize these concepts. Thus the topic to be considered here evidently is important.

If study of a subject is to be useful, there must be a capability of obtaining predictions that have reasonable confidence. In this respect turbulent combustion is at a disadvantage in comparison with other topics. Methods for modeling turbulent, reacting flows have been developed and are used because they are needed. However, the degree of confidence that can be placed in predictions obtained by existing modeling techniques is very low. The troubles lie at the foundations of the modeling hypotheses that have been introduced. An objective of the present exposition is to exhibit these modeling deficiencies so that more realistic estimates of confidence can be obtained. An effort will be made to provide a sufficiently fundamental coverage to enable identifications to be made of problems for which predictions can be obtained with reasonable confidence. To view this chapter in its proper historical perspective, the reader must realize that it was written before 1984. Numerous advances have occurred since that time.

It is of interest at the outset to list some specific problems that can and cannot be treated satisfactorily by currently available methods.

Tractable Problems

1. Calculate profiles of mean temperature, mean concentrations, root-mean-square (rms) temperature, and rms concentrations in a reactor without recirculation, with well-characterized inlet conditions, with single-phase flow at low Mach number, with negligible effects of radiation and buoyancy, with relatively high temperature levels but small amounts of chemical heat release and with low intensities of turbulent fluctuations, under conditions such that molecular rates of chemical reactions are known and are small. Also calculate overall rates of chemical conversion and efficiencies of conversion in such a reactor. The reactants may or may not be premixed.

2. Calculate the same quantities as in 1 for a reactor operating with nonpremixed reactants that have chemical reaction rates sufficiently rapid to maintain local chemical equilibrium and that have molecular diffusion coefficients that differ little from the molecular diffusivity of heat. The conditions of 1 concerning single-phase flow, Mach number, radiation, and buoyancy are retained, but those concerning temperature levels, heat release, and turbulence intensities are not. Also calculate flame lengths or flame heights under these same conditions.

3. Calculate the speed of propagation of a premixed turbulent flame in a well-characterized turbulent flow having turbulence scales large compared with the laminar flame thickness, under conditions such that the

corresponding laminar flame is stable and has a known flame speed. The conditions of 1 concerning a single-phase flow, Mach number, radiation, and bouyancy apply, but those concerning temperature levels and heat release do not. The characterization of the turbulence, given ahead of the flame, must include any influences related to hydrodynamic instabilities.

Intractable Problems That Require the Use of Empiricism

1. Calculate turbulent flame speeds under the transient conditions that prevail in spark-ignition engines.
2. Calculate conditions of exit velocity and exit-gas composition for liftoff and blow off of a turbulent diffusion flame produced when fuel issues from a duct into an oxidizing atmosphere.
3. Calculate critical conditions of body temperature and body length for ignition of a combustible mixture flowing turbulently in a well-characterized manner over a heated body. Calculate the time to ignition for conditions under which ignition occurs.
4. Calculate conditions of gas velocity, temperature, and composition for blow off of a rod-stabilized, premixed turbulent flame in a turbojet afterburner or conditions for flameout of a gas-turbine combustor.
5. Calculate the critical flow rate of water needed to produce extinction of a turbulent fire that is burning a vertical combustible wall.
6. Calculate rates of emission of soot and of oxides of nitrogen from a gas-turbine combustor.
7. Calculate the heat-release rate and combustion efficiency of a well-characterized burner fire by pulverized coal. Find the rate of radiant energy transfer to the walls of such a burner.
8. Calculate the combustion efficiency and limiting conditions for operation of a well-characterized supersonic combustor.
9. Calculate accurately the flame shape of a gas-fired boiler with a horizontal inlet stream.

From these lists it may be seen that most of the practical problems confronting the engineer cannot be solved in a straightforward and reliable manner by application of available methods for calculating turbulent reacting flows. However, knowledge of the extent of development of the subject may be used in selecting suitable avenues of investigation and may contribute ideas toward solutions of practical problems. Further perspective on practical problems may be gleaned from Mellor and Ferguson.[1]

Nondimensional Parameters of Turbulence and Some Basics of Turbulent Motion. The most fundamental fact that the student must realize is that there are many different regimes of turbulent combustion; the useful approaches and the uncertainties differ in the different regimes. The regimes may be investigated on the basis of relative values of various representative scales of length and time. These scales may be combined into dimensionless numbers that help to identify combustion regimes. Definitions of some of these numbers are needed at the beginning as a basis for discussion. Various definitions stem from certain basic concepts of turbulent motion, which will also be introduced here.

In terms of a characteristic mean velocity U and a characteristic size of L of a combustion chamber, an ordinary Reynolds number may be defined as

$$\text{Re} = \frac{UL}{\nu} \tag{1}$$

where ν denotes a representative value for the kinematic viscosity of the gas. Turbulent flow occurs for sufficiently large values of Re. However, Re *per se* is not the most relevant quantity in turbulent dynamics. A Reynolds number based on properties of the turbulent fluctuations is more closely associated with regimes of turbulence. Such a Reynolds number, defined in the terms of the representative magnitude u' of velocity fluctuations and an intergral scale l of the turbulence, is the turbulence Reynolds number:

$$R_l = \frac{u'l}{\nu} \tag{2}$$

A precise definition of u' may be taken as the root-mean-square value of the difference between the local, instantaneous velocity vector and the corresponding vector value of U. Typically l is on the order of the size of the large, energy-containing eddies and is less than L but may be of the same order of magnitude as L. Clearly $R_l = (l/L)(u'/U)\text{Re}$, in which u'/U often is termed the turbulence intensity. It is important to realize that the scale l as well as the intensity u'/U must be known to characterize turbulence properly.

The term *integral scale* arises from the definition of l in terms of a space integral of the mean value of the product of two fluctuating velocity components measured at different points. The space-time velocity correlation tensor, $c_{ij}(\mathbf{x}_1, t_1; \mathbf{x}_2, t_2)$, is the average of the product of two fluctuating velocities, one being the ith component of velocity at position $\mathbf{x}_1$ and at time t_1 minus its mean value, and the other being the jth component of velocity at position $\mathbf{x}_2$ and at time t_2 minus its mean value. Let v_1' and v_2', respectively, represent the root-mean-square values of each of these fluctuating velocities, and let r be the distance between points $\mathbf{x}_1$ and $\mathbf{x}_2$. Then, with attention restricted to a particular diagonal component c_{ii} (no summation intended) of the tensor c_{ij}, an integral scale may be defined explicitly as

$$l = \int_0^\infty \left[c_{ii} \left(\frac{\mathbf{x}_1, t_1; \mathbf{x}_2, t_1}{v_1' v_2'} \right) \right] dr$$

where the times t_1 and t_2 have been set equal and where, in performing the integration, the point $\mathbf{x}_1$ is held fixed while the point $\mathbf{x}_2$ varies from $\mathbf{x}_1$ to a boundary of the fluid (indicated by ∞ in the integral) in such a way that the unit vector from $\mathbf{x}_1$ to $\mathbf{x}_2$ maintains a fixed direction. In general, for a turbulent flow in any given device, the integral scale so defined depends on the component i selected, on the unit direction vector from $\mathbf{x}_1$ to $\mathbf{x}_2$ and on $(\mathbf{x}_1, t_1)$, but for locally homogeneous and isotropic turbulence (see Section 1.3 for definitions) it depends only on $(\mathbf{x}_1, t_1)$, the same variables on which u' depends, so that a unique integral scale l, termed "the" integral scale, is obtained. More generally, one might define a unique l by averaging over the three components i and over all directions between $\mathbf{x}_1$ and $\mathbf{x}_2$. From its definition it is seen that physically l measures the distance over which velocities tend to fluctuate in a relatively fixed phase (i.e., to exhibit a correlation).

Aspects of the evolution of turbulent motion associated with length scales less than l often are discussed within the context of approximations of full local homogeneity and isotropy. Although, in general, these approximations tend to be better at small scales than at large scales, their accuracies even at the smallest scales have not yet been established precisely for nonreacting turbulent flows and may be diminished appreciably by combustion. Nevertheless, physical concepts of turbulent motions, derived by use of these approximations, have been verified reasonably well. Therefore discussions of smaller-scale aspects of turbulent dynamics conveniently may be presented by introduction of these approximations. Results then are simplified when expressed by use of the spectrum of the velocity correlation tensor, the three-dimensional Fourier transform, defined as $\psi_{ij}(\mathbf{x}, t; k) = \int_0^\infty c_{ij}(\mathbf{x}, t; \mathbf{x}_2, t)e^{ikr}4\pi r^2\, dr$, where $r = |\mathbf{x}_2 - \mathbf{x}|$ and where c_{ij} depends on $\mathbf{x}_2$ only through r, under the stated approximations. Here the transform variable k is a wave number, the reciprocal of a wavelength. The inverse transform is $c_{ij} = \int_0^\infty \psi_{ij}e^{-ikr}4\pi k^2\, dk/(2\pi)^3$. At position x and time t, the average kinetic energy per unit mass associated with turbulent fluctuations is $q(x, t) = (3/2)c_{11}(\mathbf{x}, t; \mathbf{x}, t)$ (since $c_{11} = c_{22} = c_{33}$ under conditions of isotropy), whence from the inversion formula it is seen that $q(x, t) = \int_0^\infty(x, t; k)\, dk$, where $e(\mathbf{x}, t; k) = 3\psi_{11}(\mathbf{x}, t; k)k^2/(2\pi)^2$. This development provides the formal basis needed for defining the energy spectrum $e(\mathbf{x}, t; k)$, often identified simply as $e(k)$ and having the property that its integral over all wave numbers gives the turbulent kinetic energy per unit mass. With $1/k$ interpreted as an eddy dimension, $e(k)\, dk$ represents the contribution to the turbulent kinetic energy from eddies in the wave-number range dk about the length $1/k$.

In applications often $R_l \gg 1$, and under this condition the energy spectrum $e(k)$ has been found experimentally to exhibit distinguished attributes. As k approaches zero, $e(k)$ goes to zero, reflecting the fact that eddies large compared with apparatus dimensions are not present. A maximum in $e(k)$ occurs in the vicinity of $k = 1/l$. For $k > 1/l$ but less than a critical value, there is a range of eddy sizes in which only the inertial terms in the equation for momentum conservation are significant. The functional dependence of $e(k)$ on k in this "inertial subrange" was deduced by Kolmogrov through dimensional analysis. It is reasoned that only the value of k and the average rate of dissipation of velocity fluctuations, ε, can affect $e(k)$ in this subrange, since no other dimensional physical parameters occur. The representative dissipation rate is of the order of the ratio of the kinetic energy per unit mass to the characteristic time (l/u') for the large eddies; thus $\varepsilon = u'^3/l$, by definition, and to have the dimensions of the quantities agree, $e(k)$ must equal a universal dimensionless constant time $\varepsilon^{2/3} k^{-5/3}$. This Kolmogorov subrange, in which $e(k)$ decreases with increasing k in proportion to the $-5/3$ power of k, is a range of cascade of turbulent kinetic energy from larger eddies to smaller eddies through purely inertial interactions; the turbulence is generated at scales on the order of l and cannot be dissipated by viscosity until much smaller scales are achieved, where gradients will be larger, at values of k larger than those of the Kolmogorov subrange. Because of the decrease in the $e(k)$ with increasing k in the Kolmogorov subrange, most of the contribution of $e(k)$ to q comes from the values of k of order $1/l$, the major "energy-containing" eddies.

At sufficiently large values of k, viscosity becomes important and causes the spectrum $e(k)$ to depart from the Kolmogorov functional form. The characteristic length scale at which this occurs is the Kolmogorov length l_k, which is readily defined by dimensional analysis as the length that may be constructed from the two parameters ν and ε, viz., $l_k = (\nu^3/\varepsilon)^{1/4}$. From (2) it is seen that

$$l_k = \frac{l}{R_l^{3/4}} \tag{3}$$

which demonstrates that $l_k < l$ for $R_l > 1$. If k is large compared with $1/l_k$, then the viscous terms are dominant, the inertial terms may be neglected, and the problem of evolution of turbulent motion becomes linear, where it is straightforward to demonstrate that $e(k)$ becomes proportional to $e^{-\nu k^2 t}$. Thus, when $k > 1/l_k$, an exponential decrease of $e(k)$ with increasing k is expected to occur through viscous dissipation. Since this decrease is much more rapid than the $-5/3$ power variation, in a first approximation for large R_l the Kolmogorov scale is identified as a cut-off length below which it may be assumed that no turbulent eddies exist. The cascade is terminated by viscous dissipation at $k = 1/l_k$.

At this stage it must be stated that most of the recent research on turbulence dynamics runs counter to the cascade concept just developed. Instead of the "frequency-doubling" effects, evident in the nonlinearity of the inertial terms, wavelength-doubling phenomena ("frequency-halving"), associated, for example, with vortex pairing, have been predicted and observed. These "countercascade" phenomena, related in some ways, for example, to the onset of turbulence, to growths of turbulent shear layers and to the development of "spottiness" of turbulence at large Reynolds numbers through vortex stretching mechanisms that tend to concentrate fine structures in localized regions of space, are significant aspects of turbulent motions. However, these effects occur mostly at scales on the order of l or larger; they may be responsible, for example, for increases in l with increasing t as the turbulence evolves. It seems unlikely that further understanding of these interesting and important phenomena will necessitate revision of the fundamental cascade ideas of the finer structure of turbulence that have been introduced here.

It has been seen that for large values of R_l a wide range of eddy sizes, roughly from l_k to l, occurs in turbulent motion. An intermediate length scale, between l_k and l, that arises in estimates of the magnitude of the average rate of viscous dissipation, is the Taylor scale, $\lambda = l/R_l^{1/2}$. As R_l approaches unity, all of these characteristic lengths approach each other, and the distinctive features of the turbulence spectrum disappear. Turbulent flows with $R_l < 1$ are artificial; if $R_l < 1$, then the turbulence is dissipated rapidly, and the flow becomes laminar. Various time scales associated with these lengths may be defined. One is the time for convection to occur over the distance of a Kolmogorov scale, $\tau_u = l_k/U$, and another is the characteristic time associated with dissipation, $\tau_c = \lambda^2/(6\nu)$, in which the factor 6 appears through convention, based on reasoning from ideas of isotropic turbulence. Although τ_u is not of fundamental significant to turbulence dynamics, it is relevant to experiments because it usually defines the time resolution needed for instrumentation to explore the entire spectrum of turbulence.

Nondimensional Parameters of Chemistry, Regimes of Turbulent Combustion and Fundamental Questions. By use of the definitions given in the preceding subsection, a dimensional graph may be developed showing relationships among these various lengths and times for air-breathing combustors.[2] Such a graph[1,2] is shown in Figure 1, from which the times, lengths and R_l may be obtained, given values of L and U. Two sets of numbers are given, corresponding to two different chamber pressures, one atmosphere (without parentheses), and ten atmospheres (in parentheses). The parameter $U^2/(gL)$ is the Froude number, g being the acceleration of gravity, and effects of buoyancy tend to become more important below the line labeled "buoyancy limit." It is seen from the circled areas that different practical applications tend to occur in different ranges of parameters.

The parameters introduced thus far refer to fluid-mechanical aspects of the flows. There are chemical-kinetic aspects that exert an equally strong influence on the determination of combustion regimes. The most fundamental parameter of the chemical kinetics is the characteristic time τ_r required for the chemical reaction to occur. In reality there are many such times, since there are many reactions, and moreover each one of them varies strongly with local conditions. However, for the purpose of a manageable discussion, only one representative τ_r is introduced at present. It may be expected that the relative magnitudes of τ_r and of a flow time, such as τ_u or τ_c of Figure 1, will have a

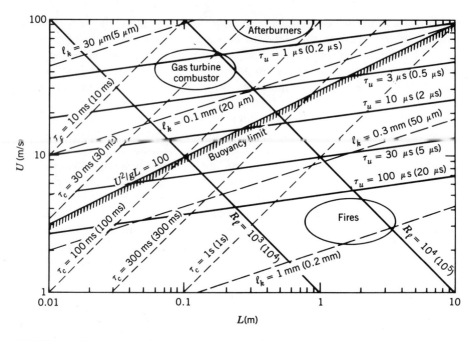

FIGURE 1. Various lengths and times in a plane of combustor size L and flow velocity U.

bearing on the regime of combustion. The ratio of a characteristic flow time to τ_r was introduced by Damköhler in 1936 as his first similarity group.[3] A conventional definition of such a Damköhler number for turbulent flows is

$$\text{Da} = \frac{l}{u'\tau_r} \qquad (4)$$

since l/u' is on the order of a characteristic turnover time for large eddies.

Qualitatively, large values of Da correspond to conditions of fast chemistry, under which reactions tend to occur in thin, distorted sheets moving in the turbulent flow. On the other hand, for sufficiently small Da, the reactions are relatively slow, and appreciable turbulent mixing occurs prior to reaction, so that conditions of a "well-stirred reactor" tend to be approached. Although there is little doubt concerning this qualitative behavior, there remain questions concerning the value of Da that defines a border between fast and slow chemical behavior in the flow, as well as questions concerning the best chemical and flow times to use in defining a Damköhler number that will best divide combustion regimes. The Kolmogorov time, $\tau_k = (\nu/\varepsilon)^{1/2}$, the characteristic time for evolution of the smallest eddies in the flow, has been employed to define an alternative Damköhler number,

$$D_k = \frac{\left(\nu l / u'^3\right)^{1/2}}{\tau_r} \tag{5}$$

that may be relevant for classification of regimes.[4] Since $D_k = Da/R_l^{1/2} < Da$, it has been suggested that in many flows chemistry should be fast for $D_k \gg 1$ and slow for $Da \ll 1$, whereas $D_k < 1 < Da$ may define an intermediate regime.[4] However, at the present state of knowledge, specific classifications are speculative.

There are flow systems in which chemical-kinetic phenomena are responsible for the existence of a characteristic length as well as a characteristic time. The most fundamental distinction in combustion is that between premixed and nonpremixed systems; for premixed systems, there exists a laminar flame speed, S_L, that together with an appropriate τ_r defines a laminar flame thickness $\delta_L = S_L \tau_r$. The definition (4) therefore may be written as $Da = (l/\delta_L)(S_L/u')$ for premixed systems and may be shown to be given by $Da = R_l(S_L/u')^2$ if use is made of the very rough formula $S_L = \sqrt{\nu/\tau_r}$. Here u'/S_L may be viewed as a relevant nondimensional measure of turbulence intensity in premixed systems. Low intensities and large scales of turbulence favor large Da, a limit that often has been called one of "fast chemistry" for premixed systems even though portions of the flow, such as those that contain the nonreacting mixture approaching the flame, clearly have slow chemistry and would have a very small value of Da if the local τ_r were used in the definition. A graph showing values of the various characteristic parameters that have been introduced here, in a plane of Reynolds number and nondimensional intensity, has been prepared[5] and is shown in Figure 2. In this figure, R_λ and R_{l_k} are Reynolds numbers based on λ and l_k, respectively, defined by (2) with l replaced by these smaller lengths.

Values of ratios of length scales of turbulence to δ_L may define different regimes of turbulent combustion in premixed systems. For example, a nondimensional flame-stretch factor $K = (\delta_L/l_k)^2$ may be identified such that wrinkled laminar flames may exist if K is below a critical value of order unity, but local disruptions of wrinkled flames by stretch would occur if K exceeds this critical value.[5] This wrinkled-flame limit is shown by the heavy line in Figure 2; disruptions occur above the line.

There is a certain degree of arbitrariness in the use of l_k rather than λ, l, or another length scale in the definition of K. The adoption of l_k is conservative in that, for $K < 1$, disruptions of laminar flame sheets are expected to be negligible even in the smallest eddies. For $\delta_L/\lambda < 1$ and $K > 1$, disruptions are anticipated in Kolmogorov-scale eddies but not in Taylor-scale eddies. For $\delta_L/l < 1$ but $\delta_L/\lambda > 1$, disruptions may occur in integral-scale eddies but not in Taylor-scale eddies. For $\delta_L/l > 1$, disruptions should be evident in all eddies, and it is unlikely that premixed laminar flames exist in the turbulent

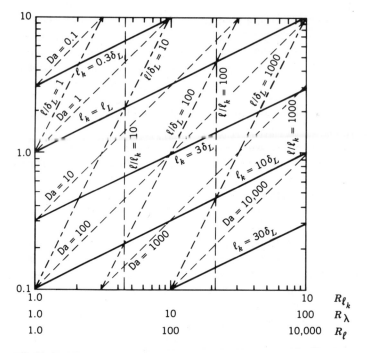

FIGURE 2. Characteristic parameters of premixed turbulent combustion in a plane of Reynolds number and nondimensional intensity.

flow. The conservative definition of K employed here seems to be most desirable at present because the dynamics of flame evolution following a disruption currently are unknown; a disruption in a Kolmogorov-scale eddy could propagate and thereby affect structures of eddies of all sizes. Continuous sheets of wrinkled laminar flames certainly may exist if K is sufficiently small compared with unity, irrespective of the magnitude of the turbulence intensity, but if $K \gtrsim 1$, then the situation is unclear.

Additional nondimensional parameters may be introduced that may be relevant to definitions of regimes of turbulent combustion.[6] Examples of such parameters are the ratio of a representative heat release by chemical reactions to a representative thermal enthalpy of the gas and the ratio of an overall energy of activation for rates of chemical reactions to thermal enthalpy. One of the main fundamental questions remaining unanswered concerns identification of the best nondimensional groups to use as a basis for separating different regimes. Additional fundamental questions concern optimum methods for analyzing turbulent reacting flows in each regime. Since it has been indicated that these methods are likely to differ in different regimes, it is helpful to be familiar with a variety of approaches to the subject.

Difficulties with Multiphase Systems, Radiation, Buoyancy, and Supersonic Combustion. A number of the practical problems identified in Section 1.1 involve two-phase flows, flows with appreciable radiative interactions, buoyancy-dominated flows, or supersonic flows. Each of these phenomena introduces complexities that prevent fundamental analyses from being developed without empiricism at present. Approaches may be found in the literature for analyzing in detail turbulent reacting flows with these effects included. For example, there are papers on modeling of turbulent combustion of pulverized coal,[7] on prediction schemes for buoyant, turbulent wall fires with significant effects of radiation,[8] and on calculation of turbulent flow properties with supersonic combustion.[9] However, the degree of confidence that can be placed in results of such ambitious analyses is low. Simple, empirical correlations, such as those presented for high-intensity, turbulent combustors burning heavy fuel sprays,[10] seem to be more useful at present.

Often these complexities need not be studied in applications. For example, fuel sprays often vaporize nearly completely before burning, and particle loadings in two-phase turbulent flows may be low enough for their influences on the dynamics of turbulence to be negligible. Under such conditions influences of these complexities may be ignored in making predictions. Proper engineering approaches should involve first making rough estimates of the importance of such complicating influences, then adopting a procedure that seems best suited to the problem in the light of such estimates. For certain turbulent flows in which these complications are absent, soundly based approaches currently are available.

Successes with Equilibrium Diffusion Flames and with Weakly Turbulent Premixed Flames. Besides the flows characterized in example 1 at the start of this section, there are two types of turbulent reacting flows for which methods of analysis now rest on firm ground, namely, nonpremixed flows with large values D_k (example 2) and premixed flames with small values of K (example 3).

If separate streams of fuel and oxidizer, each having negligible fluctuations of temperature and composition, enter a combustor with adiabatic, impermeable walls, then existing methods for analysis of nonreacting turbulent mixing may be applied to describe the combustion behavior,[4] provided that the rates of chemical reactions are sufficiently rapid to maintain chemical equilibrium and the molecular diffusivities of reactants are sufficiently close to that of heat. This well-founded description necessitates introduction of the probability density function (pdf) for developing a correspondence between the reacting and nonreacting flows. To provide the knowledge necessary for use of such an approach, basics of pdf's will be presented in the following material.

If a flame propagates in a premixed turbulent stream having $l_k \gg \delta_L$, so that conditions fall in the lower right-hand portion of Figure 2, then the turbulent flame is known to consist of an irregular and moving but continuous sheet of a wrinkled laminar flame, and methods are available for describing the structure and propagation velocity of the turbulent flame, in terms of properties of the turbulent flow in the fresh mixture.[11,12] These methods do not necessitate introducing extraneous modeling hypotheses, which always are subject to question, and therefore they are free from the usual uncertainties involved in turbulent modeling. Perturbation methods are needed to describe the turbulent flame in this regime. Basics of these methods will be presented in an effort to enable the results to be used effectively.

Limitations of Coverage. There are many techniques available in the literature that will not be covered in this presentation. These include the methods referenced in A.1.c, as well as the methods expounded in the previous sections VIII and IX. Empirical methods will not be considered, nor will the more esoteric, fundamental methods based on equations of evolution of pdf's. An effort will be made to focus on concepts that are both useful and subject to a high degree of confidence. Before embarking on detailed developments of methods, we shall present a categorization of types of approaches to the description of turbulent reacting flows.

1.2. Classification of Approaches to the Subject

Moment Methods Disregarding Probability Density Functions—Fundamental Difficulty When Applied to Combustion. For improved understanding of relationships among analytical techniques, it is useful to subdivide the types of methods currently under consideration into five separate classes.[13,14] The first of these classes, the oldest, is comprised of methods in which no special attention is paid to pdf's of fluid and thermochemical variables. Equations are written for mean values of the variables, for mean-square values, and sometimes for higher moments (the nth moment being defined as the mean value of the nth power of the variable). These equations are derived by taking suitable averages of the fundamental conservation equations. The resulting sets of equations always contain more unknowns than equations, thereby exhibiting one of the numerous diverse forms of the closure problem of turbulence. In the older moment methods, closure is achieved by approximating higher moments in terms of lower moments, or sometimes by ignoring certain higher moments outright, without regard for implications of such approximations concerning shapes of pdf's. In general, approximations introduced to achieve closure are called *turbulence modeling approximations*.

It is known[15] that chemistry introduces severe questions concerning the range of applicability of these moment methods. To illustrate the problem in its simplest form, consider an Arrhenius rate factor, $k = Ae^{-E/RT}$, in a chemical production term of a species conservation equation. Here T is temperature, R the universal gas constant, E an activation energy, and A a constant prefactor. Since moment methods involve averaging the species conservation equations, the average $\bar{k}$ appears. In terms of the average $\bar{T}$ and

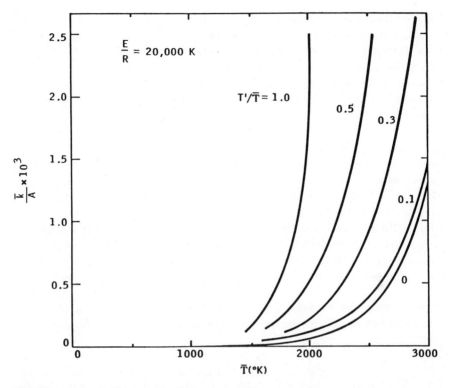

FIGURE 3. Influence of temperature fluctuations on two-term expansion of average value of specific reaction-rate constant.

the rms fluctuation $T' = [\overline{(T - \overline{T})^2}]^{1/2}$, this average is

$$\overline{k} = A e^{-E/R\overline{T}}\left[1 + \left(\frac{E}{2R\overline{T}} - 1\right)\frac{ET'^2}{R\overline{T}^3} + \cdots\right] \qquad (6)$$

where the additional terms, $+ \cdots$, involve averages of higher moments. It is seen that the coefficient multiplying the second moment involves $(E/R\overline{T})^2$, which typically is a large number. Similarly, the coefficient of the nth moment involves $(E/R\overline{T})^n$. The complete series always is convergent, but subsequent terms do not begin to decrease in magnitude until a high moment, typically on the order of the 20th in combustion problems. Therefore approximations introduced for higher moments are of critical importance, even for obtaining the simplest average $\overline{k}$, and higher moments are not represented well enough to produce reasonable accuracy for $\overline{k}$. An indication of this may obtained from Figure 3, which is a graph of (6).

There exist reacting flows in which this difficulty is not severe. Examples include flows in certain chemical reactors. If temperature levels are sufficiently high and intensities of temperature fluctuations sufficiently low for $ET'/(R\overline{T}^2)$ to be a small number, then the expansions introduced in using moment methods converge rapidly, and accuracy may be expected from application of such techniques. Whenever methods of this type are used, it is advisable to check to make sure that quantities such as $ET'/(R\overline{T}^2)$ are not large, since otherwise the system lies in a regime well beyond the range of validity of the method.

For values of $ET'/(R\overline{T}^2)$ that are too large for the full moment methods to be useful, the difficulty illustrated by (6) has been by-passed in an ad hoc manner by modeling the chemical production term without the use of moments, such as by relating it to rates of eddy breakup. This avoids the severe difficulty indicated above but not a more subtle difficulty associated with the possibility of a fluid element being transported by turbulence to a location, reacting there, then being transported back. This process affects the modeling of turbulent transport in a manner that we do not yet know how to handle, and it has been found to be a significant effect. Therefore, unless $ET'/(R\overline{T})^2$ is small, appreciable uncertainties remain in the use of moment methods.

Approximations of Probability Density Functions Using Moments. A category of approaches circumventing the difficulty identified by (6) involves physically motivated approximation of pdf shapes in terms of a small number of parameters that are related to moments. For example, if the pdf for T were known, then the average $\bar{k}$ could be evaluated directly, without introducing the expansion (6). There are flows for which the general shape of the pdf for T is reasonably evident physically. If chemistry is fast in a premixed system, it may be expected that T will be at the unburnt temperature much of the time, at the burnt temperature much of the time, and in between only seldom. Therefore the shape of the pdf in between may not be very important; a specific interior distribution such as the uniform distribution might be selected so that as few as two parameters may characterize the pdf for T, namely, the fraction of time spent at the unburnt temperature and the fraction at the burnt temperature. Considerations of pdf's enable these two parameters to be related, for example, to the mean $\overline{T}$ and to the rms fluctuation T'. It then becomes possible to express $\bar{k}$ more acurately in terms of just $\overline{T}$ and T'.

Methods of this type have been used successfully for nonpremixed systems in the equilibrium limit. These methods do not entirely eliminate the need for modeling higher moments. Second or third moments associated with turbulent transport and dissipation notably require modeling. However, fewer modeling approximations arise than those noted earlier in this section. By introducing improvements in justifications for modeling, methods in this second category hold a promise of accuracy with relative ease computation for certain regimes in which methods of the first category are invalid.

Calculation of Evolution of Probability Density Functions. A more ambitious use of pdf's is to derive and solve partial differential equations for their evolution, without approximating their shapes.[16,17] Since the closure problems that arise in these methods occur at a more fundamental level, the foundations of the modeling employed in principle can rest on somewhat firmer ground. Nevertheless, uncertainties remain in the modeling, and solution of the closed equations obtained can be challenging and usually necessitates development of numerical methods for solving relatively large numbers of partial differential equations by electronic computers. Although such methods may turn out to be the most satisfactory approaches in a number of regimes, they cannot yet be said to be very useful in engineering problems, despite some impressive recent accomplishments.

Perturbation Methods Beginning with Known Structures. In some turbulent flows it is possible to avoid modeling hypotheses entirely through the use of formal expansion procedures for limiting values of certain parameters that characterize the flow. Such expansions often are perturbations of laminar flows and therefore may apply only for relatively limited ranges of turbulent conditions. Examples are expansions for low turbulence intensity or large turbulence scale in premixed systems, such as that discussed in Section 1.1. Expansions about equilibrium in nonpremixed systems also have been pursued[4]; these are perturbations of reaction zones in turbulent structures. Within the ranges of validities of the expansions, the predictions possess very high degrees of confidence, higher than those achievable by other techniques. Therefore results obtained by perturbation methods may provide standards by which accuracies of other types of techniques may be calibrated. Carrying out analyses by perturbation methods can be challenging to analytical abilities but seldom necessitates extensive computer computations. Results often are available in the form of simple formulas that readily may be used by the engineer.

Methods Identifying Coherent Structures. The final class of methods defined here encompasses those based on introduction of coherent structures in turbulent flows. One reason for current interest in such methods is renewed study of properties of such structures by fluid dynamicists.[18] In one approach along these lines,[19,20] the chemistry is viewed as occurring in flow structures, convected and deformed by turbulence but maintaining their identity over their lifetime. Statistical approaches are employed to describe the random behavior of these structures. Methods in this category are relatively new and currently in a state of development.[6] They are not readily used by engineers, and their accuracies and regimes of utility remain to be established.

After the first version of this chapter was written, there was rapid development in new areas of analysis of turbulent reacting flows, prompted mainly by the availability of faster computers. In particular, various methods for numerical simulation of turbulent reacting flows have been developed and exercised to some extent, and there are novel approaches to statistical modeling of flame propagation in high-intensity turbulence. Although their ultimate utility remains uncertain, these methods seem likely to enjoy further advancement in the near future. More recent views[21-23] are beginning to discuss these techniques.

Subsequent Coverage. The moment methods discussed earlier in this section have been considered previously to some extent, and therefore they will be treated here only briefly. The methods of calculating evolution of the pdf's are deemed to be still relatively far from being useful in practice and therefore will not be discussed. Subsequent parts of this chapter cover approximation methods, perturbation methods, methods identifying coherent structures. Basic concepts needed for studies in approximations and perturbation are presented first. Then, in Sections 3 and 4, respectively, amplifications of approximation and perturbation methods are given. The methods of coherent structures are discussed in Section 5.

1.3. Basics of Probability Density Functions

Random Variables, Stochastic Processes, and Strange Attractors. A proper understanding of pdf's requires some background in the subject of stochastic processes. There are numerous texts on this subject, such as Papoulis[24] which is reasonably thorough. The subject begins with the concept of a random variable—a real number, whose value depends on the outcome of an experiment that contains an element of randomness. Such an experiment is not deterministic, in that repetition of the experiment does not always result in the same outcome. Instead, there is a probability of occurrence of different outcomes. Therefore there is a probability associated with obtaining different values of a random variable each time the experiment is run.

Turbulence involves randomness of this type. In a spark-ignition engine, a component of velocity at a given point in the cylinder, at a given time after firing of the spark, is not the same in every cycle. Conditions in the cylinder cannot be controlled well enough to prevent this randomness; the flow in the cylinder is turbulent. Therefore the stated velocity component is a random variable. If each engine cycle can be considered as a separate experiment, with an outcome that is not dependent upon previous cycles, then each firing of the spark may be viewed as initiating a new experiment.

For a continuous random variable, such as a velocity component at a given point at a given time, the probability structure may be defined by a probability density function (pdf) $P(v)$, having the property that for any given experiment, the probability that the outcome corresponds to the random variable lying in the range dv about a given v is $P(v)\, dv$. Such probabilities are subject to various interpretations, the simplest being the frequency interpretation, in which it is said that if the experiment is repeated many times, then the fraction of all experiments in which the random variable lies in the range dv about v is $P(v)\, dv$. The probability distribution function is properly defined as the probability that the random variable takes on a value less than, or equal to, v, that is, $\int_{-\infty}^{v} P(v)\, dv$. But colloquially this quantity often is not introduced, and instead the terms "probability distribution function" and "probability density function" are treated as synonymous. By definition, $\int_{-\infty}^{\infty} P(v)\, dv = 1$, since the probability of there being any outcome at all from the experiment is unity.

Turbulence is a process occurring in space $\mathbf{x}$ and time t, both of which are continuous variables. Thus a velocity component is a function of $\mathbf{x}$ and t. Turbulence is a random process in that, with $\mathbf{x}$ and t given, only a probability can be stated for the velocity component lying in the range dv about v. Such a random process is termed a stochastic process, and the velocity component is a random or stochastic function. A complete statistical description of the velocity component for the random process would be provided by the probability density functional, $P[v(\mathbf{x}, t)]\delta v(\mathbf{x}, t)$, giving the probability that the random function lies in the range $\delta v(x, t)$ about the function $v(\mathbf{x}, t)$. In practice, such a quantity is too complex to be specified, and therefore only limited information is obtained, which does not constitute a complete statistical description. An example of the limited information is $P(v; \mathbf{x}, t)$, where $P(v: \mathbf{x}, t)\, dv$ is the probability that at fixed $\mathbf{x}$ and t, the random function (which becomes a random variable under this restriction) lies in the range dv about v.

Complete statistical descriptions of stochastic processes are difficult to develop even in the simplest cases. Turbulent reacting flows are relatively complex cases. Turbulence arises as a result of various fluid mechanical instabilities. It can be maintained that if there were absolutely no uncertain fluctuations in then initial or boundary conditions, then stochastic aspects of turbulence would not occur, for after all, the Navier-Stokes equations are deterministic. It follows that the statistical aspects of turbulence must depend on those of the initial and boundary fluctuations. However, this input information is modified greatly by the many fluid instabilities. It is possible that these instabilities dominate the process in such a way that the statistical properties of the turbulence are virtually independent of those of initial and boundary fluctuations. Under such conditions a reasonably complete statistical description of turbulence would not require reference to initial or boundary fluctuations.

A term describing a situation of this type is "strange attractor." It may be said that turbulent processes possess strange attractors that draw the statistical properties of the stochastic process into regimes independent of the starting points. Research on strange attractors has been intensifying in recent years. For our present purposes, the concept of strange attractors may be considered as an explanation of the philosophy behind attempts to describe probabilistic structures of turbulent reacting flows without reference to external perturbations.

Calculation of Moments and of the Probability Density of a Function of a Random Variable. In applying approximation methods, it is necessary to calculate various quantities of interest from a pdf such as $P(v)$. The simplest quantities to calculate are moments of the random variable. The mean value or average of the random variable is

$$\bar{v} = \int_{-\infty}^{\infty} vP(v) \, dv \tag{7}$$

The mean-square value, or the second moment about zero, is

$$\overline{v^2} = \int_{-\infty}^{\infty} v^2 P(v) \, dv \tag{8}$$

The variance, the second moment about the mean, or the second central moment, is

$$\sigma^2 = \int_{-\infty}^{\infty} (v - \bar{v})^2 P(v) \, dv = \overline{v^2} - \bar{v}^2. \tag{9}$$

This measures the width of the distribution; its positive square root σ is the standard deviation, the rms intensity of the fluctuation which would be denoted by v' according to our earlier notation. Often the notation $v' = v - \bar{v}$ is employed instead, the prime denoting the deviation of v from the mean. The third central moment,

$$s\sigma^3 = \int_{-\infty}^{\infty} (v - \bar{v})^3 P(v) \, dv \tag{10}$$

measures the skewness of the distribution about the mean; the nondimensional quantity s may be called the *skewness*. The fourth central moment,

$$\kappa\sigma^4 = \int_{-\infty}^{\infty} (v - \bar{v})^4 P(v) \, dv \tag{11}$$

tends to measure how highly curved the distribution is; the nondimensional parameter κ may be termed the *curvature* or the *kurtosis*. These various properties of pdf's are illustrated in Figure 4. Higher moments measure finer details of $P(v)$. There is a theorem to the effect that subject to certain smoothness and convergence requirements, the specification of all moments is equivalent to the specification of $P(v)$; this theorem does not apply to many pdf's of practical interest.

The powers v^n are special functions. In applications it is often desired to calculate averages of more complicated functions. This can be done in terms of moments if the function possesses a power-series expansion that remains valid wherever $P(v)$ does not vanish. A more general and more direct approach is to first use $P(v)$ to calculate a pdf for a new random variable, defined by the function of interest, then use formulas like (7) for this new pdf. Denote the

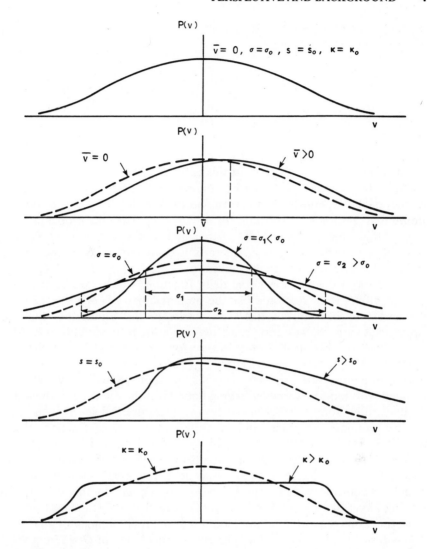

FIGURE 4. Illustrations of mean, variance, skewness, and kurtosis of probability density functions.

function by $g(v)$, and assume at first that $g(v)$ is monotonic and differentiable with $dg/dv \neq 0$. Then it becomes clear that in each outcome for which the original random variable lies in the range dv about v, the new random variable lies in the range $dg = |dg/dv|\, dv$ about $g(v)$. Hence the pdf $Q(g)$ for the new random variable must obey $Q(g)|dg| = P(v)|dv|$, whence $Q(g) = P(v)/|dg/dv|$. If $g(v)$ is not monotonic, so that $g(v_1) = g(v_2) = \cdots = g(v_n)$, then outcomes having the original random variable in the vicinity of

$v_1, v_2, \ldots, v_n$ all correspond to the same g, and

$$Q(g) = \sum_{i=1}^{n} \left[\frac{P(v_i)}{\left| \dfrac{dg}{dv} \right|_{v=v_i}} \right] \tag{12}$$

Special considerations are needed in attempting to use (12) if g remains constant over a finite range of v. In this case, as in all cases that are unclear from formulas, correct results can be obtained by considering the fraction of outcomes of the experiment that correspond to the values of g or v of interest. Suppose that $g(v) = g_0 =$ constant for $v_1 < v < v_2$. Then the probability that $g = g_0$ is the same as the probability that $v_1 < v < v_2$, which is $\int_{v_1}^{v_2} P(v)\, dv$. Hence, in the vicinity of $g = g_0$, there is a contribution to the pdf of the function of the random variable given by $Q(g) = [\int_{v_1}^{v_2} P(v)\, dv]\delta(g - g_0)$, where $\delta(g - g_0)$ denotes the Dirac delta function, which vanishes when its argument differs from zero, and has the property that $\int_{-\infty}^{\infty} \delta(g - g_0)\, dg = 1$. An example of behavior of this type occurs in the equilibrium limit for nonpremixed systems, with v representing a mixture-ratio variable and g fuel concentration; in this illustration g is zero for a range $v \leq v_0$, and therefore $Q(g)$ has a delta function of strength $\int_{-\infty}^{v_0} P(v)\, dv$ at $g = 0$.

Delta functions often appear in pdf's in applications. They are often associated with random variables having bounds on the range over which their pdf's may differ from zero. Scalars such as temperature and species concentrations are examples of such random variables; the mole fraction of any chemical species can neither be negative nor exceed unity, for example. The delta functions tend to occur at the bounds of the permissible range. There is no difficulty in using (12) if $P(v)$ has a delta function, unless its location happens to coincide with a zero or infinity of dg/dv. In this case special reasoning along the lines indicated above is needed.

The function $g(v)$ may be nonmonotonic or discontinuous or may have $dg/dv = 0$ or $dg/dv = \infty$, without preventing calculation of $Q(g)$ from $P(v)$. These various behaviors are illustrated schematically in Figure 5, where the vertical arrow signifies a delta function. A restriction is that $g(v)$ must be single valued if $Q(g)$ is to be obtained from $P(v)$, even in principle. If $g(v)$ is not single valued, then outcomes with the same value of v may have different values of g, and knowledge of only $P(v)$ provides no information concerning the relative frequencies with which each of the different possible values of g may occur. An example for nonpremixed systems with fast chemistry is that in which g is the mixture ratio and v the temperature. For temperatures above the highest inlet temperature and below flame temperature, there are two mixture ratios that correspond to a given temperature, one rich and the other lean. Therefore establishment of a pdf for temperature does not enable a pdf for mixture ratio to be calculated, even in principle, it being impossible in the

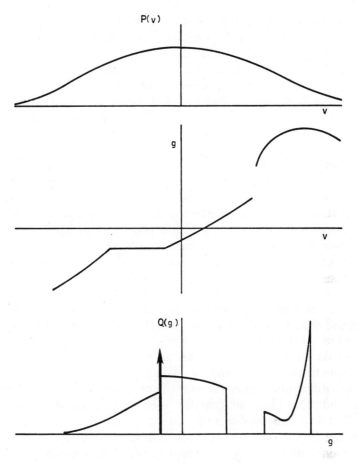

FIGURE 5. Schematic illustration of the probability density function of a function of a random variable.

absence of additional information to ascertain the frequency with which the mixture is rich or lean.

Joint, Marginal, and Conditioned Probability Density Functions. It has been indicated earlier in this section that specification of a pdf does not provide complete statistical information concerning a stochastic process. For a random function dependent on x and t, the pdf $P(v; x, t)$ for fixed x and t gives no knowledge of relationships obeyed by the random variable at adjacent values of x and t and therefore offers very limited information. The extent of information may be increased by defining a joint pdf. The value of the random function at a particular position and time, (x_1, t_1), is a random variable. The value of this same random function at a different position and time, (x_2, t_2), is another random variable. The probability that the first random variable lies in the range dv_1 about v_1 and the second in the range dv_2 about v_2 may be denoted by $P(v_1, v_2; \mathbf{x}_1, t_1, \mathbf{x}_2, t_2)\, dv_1\, dv_2$ and may be thought of as being determined by monitoring frequencies of outcomes of a large number of repetitions of the random process, counting the number of outcomes for which the random variables lie in their respective ranges, dv_1 and dv_2, and dividing by the total number of experiments. Sometimes the abbreviated notation $P(v_1, v_2)$ will be used for this joint pdf. From its definition, clearly $\int_{-\infty}^{\infty}\int_{-\infty}^{\infty} P(v_1, v_2)\, dv_1\, dv_2 = 1$.

It is clear that definitions of joint pdf's may be extended to any number of random variables; $P(v_1, v_2, \ldots, v_n)$ may be defined, with $\int_{-\infty}^{\infty}\int_{-\infty}^{\infty} \cdots$ $\int_{-\infty}^{\infty} P(v_1, v_2, \ldots, v_n)\, dv_1\, dv_2 \ldots dv_n = 1$. Allowing n to approach infinity suitably may result in a complete statistical description of a stochastic process. In practice, such information never is available for turbulence. Descriptions have been restricted to $P(v_1, v_2, \ldots, v_n)$, with n equal to 2 or 3 at most. If $t_1 = t_2$ but $\mathbf{x}_1 \neq \mathbf{x}_2$, then $P(v_1, v_2)$ is a one-time, two-point joint pdf, which is relevant to turbulent dissipation, for example. If $\mathbf{x}_1 = \mathbf{x}_2$, but $t_1 \neq t_2$, then $P(v_1, v_2)$ is a one-point, two-time joint pdf, which is relevant to aspects of the temporal behavior of the stochastic process at a fixed point. Often more than one random function is involved in a stochastic process, and joint pdf's involving different functions may be defined; for example, $P(v_1, v_2)\, dv_1\, dv_2$ may be the probability that temperature lies in the range dv_1 at $(\mathbf{x}_1, t_1)$ while fuel concentration lies in the range dv_2 at $(\mathbf{x}_2, t_2)$. The qualifier "auto" often is used when the variables v_1 and v_2 relate to the same random function, and "cross" for different random functions. For different random functions, a one-point, one-time joint pdf contains important statistical information not obtainable from a one-variable pdf.

The pdf $P(v_1)$ always can be recovered from $P(v_1, v_2)$ since, from the underlying definition, $\int_{-\infty}^{\infty} P(v_1, v_2)\, dv_2 = P(v_1)$. When viewed this way, as a projection of $P(v_1, v_2)$, the one-variable pdf $P(v_1)$ is termed a marginal pdf. It is possible to define a one-variable pdf for a random variable by considering only those experiments in which another random variable takes on a specified value or lies in a specified range. Such a pdf would be conditioned by the constraint placed on the second random variable and is termed a conditional pdf. When dealing with conditioned pdf's, it is conventional to place a vertical bar inside the parentheses of P and to write the conditioning restrictions after the bar. For example, $P(v_1|v_2)\, dv_1$ denotes the probability that a random variable lies in the range dv_1 about v_1 while another random variable (the identity of which must be known from the context since it is not indicated in the present notation) takes on the value v_2. From the fundamental meanings of the different probabilities, $P(v_1, v_2) = P(v_1|v_2)P(v_2)$, which implies Bayes's theorem, $P(v_1|v_2)Q(v_2) = Q(v_2|v_1)P(v_1)$, where the different symbols P and Q have been introduced for pdf's of v_1 and v_2, respectively, in an effort to avoid confusion.

Conditioned pdf's are arising increasingly often in applications. Examples are turbulent flows with intermittency, that is, with portions of the flow being nonturbulent; pdf's often are conditioned on turbulent fluid being present to aid in mechanistic interpretations. Another example in nonpremixed combustion would be to condition a pdf with a requirement that fuel be present. Clearly conditioned means and conditioned moments can be defined in ways entirely analogous to (7) through (11).

Statistical Independence. A random variable may be said to be statistically independent from another random variable if its conditioned pdf $P(v_1|v_2)$, conditioned on the second random variable having a particular value of v_2, is independent of v_2 for all v_2. In this case there is no need to specify the value of v_2 in the conditioned pdf, and $P(v_1|v_2) = P(v_1)$ may be written. From the general formula $P(v_1, v_2) = (P(v_1|v_2)P(v_2)$, normalization conditions show that the $P(v_1)$ defined here is simply the marginal pdf for the first random variable. Therefore the joint pdf is the product of the two marginal pdf's if the random variables are statistically independent. We may write this as $P(v_1, v_2) = P(v_1)P(v_2)$, realizing of course that $P(v_1)$ and $P(v_2)$ may be different functions since they refer to different random variables.

Statistical independence is a significant property that is assumed in the theoretical developments by researchers too often without sufficient justification. Dynamics of turbulence cause the velocity at a point in a turbulent flow to be statistically dependent on the velocity at a nearby point. The value of temperature at a point may be expected to be related statistically to the fuel concentration at that point. An indication of statistical independence is provided by the covariance, the second cross central moment, sometimes called the *cross correlation*,

$$c_{12} = \int_{-\infty}^{\infty} \int_{-\infty}^{\infty} (v_1 - \bar{v}_1)(v_2 - \bar{v}_2)P(v_1, v_2)\, dv_1\, dv_2 \tag{13}$$

The nondimensional quantity $c_{12}/(v_1' v_2')$, where v_i' denotes the rms fluctuation about the mean, is called the *correlation coefficient*. If $c_{12} = 0$, then the two random variables are said to be uncorrelated. Writing $P(v_1, v_2) = P(v_1)P(v_2)$ in (13) readily shows that if the two random variables are statistically independent, they are uncorrelated. However, uncorrelated random variables are not necessarily statistically independent. Statistical independence is a strong condition.

Ensemble Averages, Stationarity, and Homogeneity. The averages about which we have been speaking are viewed most basically as ensemble averages. This term means that a large number, an ensemble, of identical systems is imagined to exist, an experiment is run for each system, and averages are taken by first adding the values of the random variable obtained in each experiment then dividing by the total number of experiments. The ensemble view makes it unnecessary to think of averages as being taken over time or space or to think of a pdf as describing the fraction of time that a random variable lies in a given range of values. It enables probabilistic structures to be introduced without reference to space or time evolution.

The stochastic process of turbulence is defined over the space and time variables x and t. A deep question in stochastic processes concerns the establishment of conditions under which ensemble averages may be replaced by space or time averages. Theorems establishing the equivalence of ensemble and time averages are called *ergodic theorems*. In general, some form of stationarity is needed to prove ergodic theorems. There are many different definitions of stationarity, but the clearest (and the one adopted here) is stationarity in the strict sense, for which all statistical properties of the stochastic process are required to be independent of the selection of the origin of time. It is often evident physically that particular turbulent flows are statistically stationary. For example, "steady" turbulent flow through a continuous combustor such as a gas turbine or a ramjet is stationary. For such flows, pdf's and averages are readily viewed in terms of time averages. Use may be made of stationarity in these flows to simplify the descriptions of their properties; for example, two-time correlations will depend only on the time interval, not on the specific values of each of the times.

The term homogeneity refers to invariance of statistical properties under translation of the spatial coordinates. Homogeneous turbulence is turbulence with statistics independent of the selection of the origin x and having, for example, two-point correlations dependent only on the difference in the coordinates of the two points.[25] Complete homogeneity is an idealization approached only approximately in reality. Homogeneity in particular coordinate directions is achievable more accurately. For example, the turbulent flow downstream from a uniform grid in a suitably designed wind tunnel is essentially homogeneous in the two transverse directions parallel to the grid. Homogeneity should not be confused with isotropy, which refers to invariance of statistical properties under rotation of the coordinate system. For example, the statistics of streamwise and transverse velocity fluctuations must be the same in isotropic turbulence. Isotropy generally is not achieved as accurately as is homogeneity in particular directions. Under conditions of homogeneity, ensemble averages may be calculated by taking space averages over a spatial coordinate in which the process is homogeneous.

The most useful types of averages to consider depend on the stationarity and homogeneity of the flow. In stationary or "steady" turbulent flows, time averages are most useful. Motors operating on repetitive cycles, such as spark-ignition or diesel engines, satisfy neither homogeneity nor stationarity but seem well suited to definition in terms of ensemble averages. Attempts have been made to employ time or space averages over sufficiently small scales in such engines. Such approaches have the advantage of providing clearer mechanistic relationships among various averages but are difficult to justify from a fundamental viewpoint since clean separations between turbulence scales and scales for overall space or time variations of the process are needed. In general, suitable statistical descriptions of stationary flows are easier to develop.

1.4. Basics of Asymptotic Expansions

Practical Utility of Limit Processes. Perturbation methods are used in the approaches identified in Section 1.2. These methods are based on mathematical concepts of asymptotic expansions, which involve mathematical considerations of limit processes. Knowledge of the mathematics of limit processes is needed for adeptness in applying perturbation methods, especially when these methods involve matched asymptotic expansions, as they do in nearly all applications to turbulent reacting flows. Full development of the mathematics of limit processes and of the bases of methods of matched asymptotic expansions is beyond the scope of this presentation. A variety of textbooks currently are available on the subject. Notable among them are Cole[26] and Van Dyke.[27]

Methods of matched asymptotic expansions are encountering increasing use in engineering. They provide solutions in simpler forms than those obtained by numerical methods and sometimes are capable of being applied to problems that resist attack by numerical methods. The central question in matched asymptotic expansions is that of matching. Expansions are obtained in two different variables that represent the same coordinate, each expansion containing undetermined coefficients. The unknown coefficients are determined by matching the two expansions. There are different approaches and rules for matching that sometimes lead to uncertainty concerning proper matching, concerning whether matching has been achieved, or concerning whether matching is achievable. When doubts of this type arise, the best procedure is to introduce an intermediate variable and pursue a formal limit process in the intermediate variable.[26] This procedure generally removes uncertainties but, of course, requires knowledge of limit processes for its application. Presentation of this knowledge here would require too much space. Therefore a physically motivated description of matched asymptotic expansions is given.

Physical Concept of Matched Asymptotic Expansions. In the perturbation methods of Section 1.2, an expansion of the solution to the flow problem is sought for large or small values of a parameter that appears in the problem. It often occurs that as the limiting value of the parameter is approached, the flow field separates into two or more separate regions or zones, with different physical processes being dominant in each zone. The philosophy of the method of matched asymptotic expansions is to develop asymptotic expansions of the solution separately for each zone, using the independent variables that are appropriate to the zone under investigation, then match these expansions at the boundaries of the zones in a self-consistent way.

As applied to turbulent flows, this method usually necessitates consideration of zones whose locations fluctuate in time. For example, chemical reactions may occur in thin reaction sheets that move about. One zone would be the reaction zone, in which a moving coordinate normal to the sheet is stretched for the purpose of analyzing the reaction processes occurring within the sheet. Another would be a moving transport zone on one side of the sheet.

After the expansions are developed for each zone and matching is achieved, statistics of the turbulence are used to calculate desired averages of flow properties from results of the expansions.

The development here will be restricted to laminar flows for the purpose of exhibiting concepts and methods more easily. The applications to turbulent flows will be made in Section 4.

Expansion Parameters Damköhler-Number and Activation-Energy Asymptotics. A number of parameters are good candidates for use in asymptotic expansions. An example is a nondimensional heat release, which typically is large in combustion. This parameter has not yet been employed very much as a basis for asymptotic expansions. Relatively few of the attractive parameters have been exploited. Those that have been investigated are the most obviously useful parameters.

As was indicated in Section 1.1, nondimensional scales and intensities of turbulence have been employed as expansion parameters. Within the context of these expansions, additional expansions in other parameters have been employed. Two principal expansion parameters that have been investigated are a Damköhler number, such as the Da defined in (4) and a nondimensional activation energy,

$$\beta = \frac{E(T_b - T_u)}{(RT_b^2)} \tag{14}$$

where T_b is the adiabatic temperature of burnt gas and T_u the temperature of unburnt gas. Recently, the β of (14) has been termed the Zel'dovich number[22] in honor of the numerous fundamental contributions of Zel'dovich to the theory.

The first parameter to be studied in laminar combustion problems was a laminar version of Da. Frozen and equilibrium flows were obtained in the limits of small and large Da, respectively. While this parameter provided a considerable amount of information on topics such as structures of diffusion flames, it shed no light on certain important processes, such as ignition and extinction. More recently, β has been employed to obtain a large amount of information on processes of this type. A large body of research has now been completed on matched asymptotic expansions with β as the expansion parameter, and a book has been published that is devoted to this subject.[28] Expansions in β play a central role in Section 4.

Two-Zone Structure of Premixed Flames. To illustrate matched asymptotic expansion in β, consider a steady, adiabatic, premixed laminar flame in an approximation in which all species concentrations can be related to the temperature T as the single dependent variable. Write the equation for energy

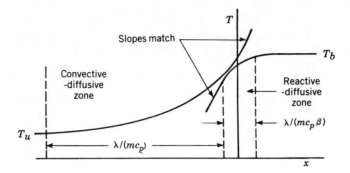

FIGURE 6. Schematic illustration of premixed-flame structure for large nondimensional activation energy.

conservation as

$$mc_p \frac{dT}{dx} - \frac{d}{dx}\left(\lambda \frac{dT}{dx}\right) = B(T_b - T)e^{-E/RT} \qquad (15)$$

where c_p is the specific heat at constant pressure, λ the thermal conductivity, m the constant mass flow rate, and B a constant related to the heat release and to the preexponential factor of the reaction rate. Introduce nondimensional variables $\xi = \int (mc_p/\lambda)\, dx$, $\tau = (T - T_u)/(T_b - T_u)$, and $\Lambda = [\lambda B/(m^2 c_p^2)]e^{-E/RT_b}$. Then (15) becomes

$$\frac{d\tau}{d\xi} - \frac{d^2\tau}{d\xi^2} = \Lambda(1 - \tau)e^{-\beta(i-\tau)/[1-\alpha(1-\tau)]} \qquad (16)$$

where $\alpha = (T_b - T_u)/T_b$.

Next consider what happens as the parameter β becomes large. The reaction term becomes very small unless τ is near unity. Hence there is a convective-diffusive zone for $\tau < 1$, described by $d\tau/d\xi = d\xi^2$ and therefore having the structure $\tau = e^\xi$, where the origin of ξ has been selected so that $\tau = 1$ at $\xi = 0$. The convective-diffusive zone is illustrated in Figure 6.

This solution breaks down when τ gets too close to unity because the reaction term becomes important there. To consider the region near $\tau = 1$, introduce the stretched variables $\eta = \beta\xi$ and $y = \beta(1 - \tau)$. In terms of these variables (16) is

$$-\frac{dy}{d\eta} + \beta \frac{d^2y}{d\eta^2} = \left(\frac{\Lambda}{\beta}\right) y e^{-y/(1-\alpha y/\beta)} \qquad (17)$$

When β becomes large, the conduction term can be balanced only if Λ

becomes proportional to β^2. Writing $\Lambda = \beta^2\Lambda_0 + \beta\Lambda_1 + \Lambda_2 + \beta^{-1}\Lambda_3 + \cdots$ and $y = y_0 + \beta^{-1}y_1 + \cdots$, we find by collecting like powers of β^{-1} in (17) that $d^2y_0/d\eta^2 = \Lambda_0 y_0 e^{-y_0}$. This describes the structure of a reaction zone that in lowest order possesses a diffusive-reactive balance. The reactive-diffusive zone is illustrated in Figure 6.

The differential equation for y_0 can be integrated twice with the boundary condition $y_0 \to 0$ as $\eta \to \infty$ and the matching condition $dy/d\eta \to -1$ as $\eta \to -\infty$ applied to obtain the reaction-zone structure $y_0(\eta)$. Fulfillment of this matching condition is achieved only if $\Lambda_0 = \frac{1}{2}$. Hence, in the first approximation,

$$m = \left[\frac{2\lambda B}{c_p^2 \beta^2} \right]^{1/2} e^{-E/(2RT_b)} \tag{18}$$

emerges as the only mass flow rate for which an internally consistent structure can be obtained. The flame speed S_L is m divided by the density of the unburnt gas.

This short derivation is subject to embellishment in many ways.[28] Both more careful development of the technique and generalization in formulation, for example, to study extinction by inclusion of heat loss, can be pursued with successful and instructive results. However, the presentation that has been given should serve to illustrate the nature of the method and the types of predictions that are obtained. Expansions in β, termed *activation-energy asymptotics*, formally involve manipulations somewhat different from those encountered in other matched asymptotic expansions and therefore call for algebra not closely resembling that presented in the standard references.

Three-Zone Structure of Diffusion Flames. Two zones were found for the premixed flame: a narrow reactive-diffusive zone and a broader convective-diffusive zone upstream. There is of course a large region downstream from the flame as well, but in the problem considered properties were uniform in that region, and therefore it did not have to be studied. If there were downstream heat losses, then a third, larger, downstream zone would have to be included as well.

In diffusion flames of interest there are temperature gradients on both sides of the reaction sheet. Therefore a three-zone is encountered, with a narrow reactive-diffusive zone separating two broader convective-diffusive zones.

Diffusion-flame analyses with Da as a parameter often are helpful in studying near-equilibrium conditions and departures therefrom.[4] Diffusion-flame studies with β as the parameter of expansion give ignition and extinction conditions as well.[29] Details of the analyses will not be pursued here, since the premixed-flame example should be sufficiently illustrative.

2. USE OF MOMENT METHODS

2.1. Moments Based on Ordinary Averages

Equations for Lower Moments. As indicated in Section 1.2, the equations for moments, used in exercising moment methods, are derived by taking suitable averages of conservation equations. For the lowest moments, such as $\overline{Y}_j$, the average mass fraction of species j, equations of the form

$$\frac{\partial}{\partial t}\left(\overline{\rho}\,\overline{Y}_j\right) + \nabla \cdot \left(\overline{\rho}\,\overline{v}\,\overline{Y}_j\right) = \nabla \cdot \left(\overline{\rho}D_T\nabla\,\overline{Y}_j\right)\overline{w}_j \tag{19}$$

eventually are derived, where $\overline{\rho}$ is an average density, $\overline{\rho}\overline{v}$ is an average mass flux, D_T is a turbulent diffusion coefficient, and $\overline{w}_j$ is the average value of the mass rate of production of species j per unit volume by chemical reactions. The term involving D_T represents an approximation to the sum of many terms that arise containing averages of fluctuating quantities, such as molecular diffusion and Reynolds transport. Usually the average product $\overline{\rho v}$ is written as the product of the averages, $\overline{\rho}\overline{v}$, it being assumed that the average of the fluctuation $\overline{(\rho - \overline{\rho})(v - \overline{v})}$ either is negligible or is included in the term involving D_T. In the lowest-order closure schemes, this average usually is neglected in the average of the continuity equation, which then is written as $\partial\overline{\rho}/\partial t + \nabla \cdot (\overline{\rho}\overline{v}) = 0$. Averaged equations for momentum conservation, for example

$$\overline{\rho}\frac{\partial\overline{v}}{\partial t} + \overline{\rho}\overline{v} \cdot \nabla\overline{v} = -\nabla\overline{p} + \nabla \cdot \overline{T} \tag{20}$$

for energy conservation and for state are then considered along with (19). In (20), $\overline{p}$ is the average pressure and $\overline{T}$ is the average value of the turbulent shear-stress tensor, which is expressed in terms of a turbulent viscosity coefficient in a manner resembling that in the third term in (19). For purposes of discussion, attention here will be focused on (19), whose four terms represent averages of accumulation, convection by mean flow, diffusion (both laminar and turbulent), and chemical production, respectively. These four effects, or sometimes only two or three of them, occur very commonly in equations for means of turbulent quantities.

In the form that the equations for the means have now been written, the closure problem manifests itself in uncertainty concerning how to evaluate quantities of the type of D_T and $\overline{w}_j$. A popular approach to obtaining D_T is to write $D_T = Cq^2/\varepsilon$, where C is a nondimensional constant, q is the turbulent kinetic energy and ε the average dissipation rate; these last two quantities have been defined in Section 1.1. The system of differential equations then is augmented by writing differential equations for q and for ε (the so-called $k - \varepsilon$ modeling, k being an alternative notation for q). These equations are derived in various ways, by introducing approximations in conservation equations for higher moments, for example, for the mean-square velocity fluctuation and for a representative length scale of turbulence. The equations turn out to be of the form[1]

$$\bar{\rho}\frac{\partial q}{\partial t} + \overline{\rho\mathbf{v}}\nabla \cdot q = \nabla \cdot (\bar{\rho}D_T\mathrm{Sc}_T\nabla q) + P_q - \bar{\rho}\varepsilon \tag{21}$$

and

$$\bar{\rho}\frac{\partial \varepsilon}{\partial t} + \overline{\rho\mathbf{v}}\nabla \cdot \varepsilon = \nabla \cdot (\bar{\rho}D_T\mathrm{Sc}_\varepsilon\nabla \varepsilon) + C_1\frac{\varepsilon}{q}P_q - C_2\bar{\rho}\frac{\varepsilon^2}{q} \tag{22}$$

where Sc_T is a turbulent Schmidt number, Sc_ε is a corresponding quantity for the dissipation rate, P_q is a rate of production of turbulent kinetic energy by deformation of the mean flow, and C_1 and C_2 are nondimensional constants. A representative expression for P_q is $P_q = \bar{\rho}D_T\mathrm{Sc}_T(\partial \bar{v}_1/\partial x_2 + \partial \bar{v}_2/\partial x_1)^2$, where the subscripts 1 and 2 identify cartesian components, and representative values of constants are $\mathrm{Sc}_T = 1$, $\mathrm{Sc}_\varepsilon = 0.7$, $C_1 = 1.4$, $C_2 = 2$, and (in the expression for D_T) $C = 0.09$. The terms in (21) and (22) are qualitatively of the same form as those in (19), with the net production rate $\bar{w}_j$ being analogous to the difference between the production and dissipation terms, the last two terms in (21) and (22). If $\bar{w}_j$ can be modeled in (19), then the equations that have been given here, along with the appropriate boundary conditions, provide a well-posed mathematical problem soluble by available computer routines.

The main advantage afforded by the introduction of (21) and (22) is to allow the turbulence the possibility of having a memory, in a sense to exert influences on evolutions of mean quantities through its own evolution. A simpler approximation that avoids additional differential equations like (21) and (22) is $D_T = l_Tq$, $q = c|\bar{v}|$, where the mixing length l_T and the constant c (perhaps $c = 0.1$) are treated as known constants for a given flow. This simpler approximation has been found to be less successful in correlating data than formulations that include differential equations for turbulence; an evolution of D_T with some memory in it seems to be wanted. More complex formulations often have been employed, with a number of additional equations like (21) and (22). Full second-order closure schemes formally write averaged differential equations for all second moments and then introduce modeling approximations only at the level of the third moments. These methods have at least as great a capability of fitting data as do those outlined here, and with present-day computers their routines are not significantly more complex to exercise, but of course they also have more constants whose values may be adjusted to provide agreement. If moment methods are adopted, then it would appear to be desirable to strike a judicious compromise concerning the number of differential equations retained. For many purposes that compromise seems likely to lie in the vicinity of $k - \varepsilon$ modeling.

Modeling Approximations for the Chemical Production Term. Thus far in the present discussion of the use of moment methods, it has been assumed that $\overline{w}_j$ in (19) can be obtained. In Section 1.2 it was indicated that the usual moment procedures reasonably can be applied for this purpose only if the rates are not too rapid and not too strongly dependent on local, instantaneous properties. An approach that has been found to be better suited to situations having strongly sensitive rates that are not too slow is to use the so-called eddy breakup model for $\overline{w}_j$. In this model chemical reaction is assumed to occur at a rate proportional to the rate of dissipation of turbulence, and a formula of the type

$$\overline{w}_j = \overline{\rho} C_j \left[\overline{\left(Y_j - \overline{Y}_j \right)^2} \right]^{1/2} \frac{\varepsilon}{q} \tag{23}$$

therefore is introduced, where C_j is a constant.[4,5] Equation (23) readily may be used in conjunction with (19) through (22) by appending to the set an equation for the mean-square concentration fluctuation that resembles (21). Usually the equation is applied only to a reactant, fuel or oxidizer, since its meaning is less clear for reaction products; the product concentrations are obtained from reactant concentrations through atom balances. Often there is a value of C_j somewhere between 10^{-1} and 10^2 that results in good agreement between calculated mean concentration profiles and measurement. To the extent that agreement can be obtained, the average chemical rate is dominated by turbulent mixing and independent of chemical kinetics.

Neither the ordinary moment procedures nor the eddy-breakup models circumvent the difficulty of interaction between chemical reaction and turbulent transport, mentioned in Section 1.2. These difficulties will be discussed further in Section 3.4. The conclusion that must be reached therefore is that the fundamental basis of (23) or of other formulas like it is uncertain.

2.2. Moments Based on Favre Averages

Equations for Lower Moments. In Section 2.1 it was found that difficulties arose even in the continuity equation in taking averages for variable-density flows. By use of mass-weighted or Favre averaging these difficulties can be minimized. It is becoming increasingly conventional to use Favre averaging in variable-density flows. The mass-weighted average of a variable v is $\tilde{v} = (\overline{\rho v})/\overline{\rho}$, where the bar denotes averages of the type defined in (7) and (8). The departure from $\tilde{v}$ is denoted by $v'' = v - \tilde{v}$ and has $\overline{\rho v''} = 0$ but $\overline{v''} \neq 0$. It is always possible in principle to relate mass-weighted averages to ordinary averages, but complications associated with averages involving density fluctuations arise.

With Favre averages, the mean of the continuity equation is simply $\partial \overline{\rho} / \partial t + \nabla \cdot (\overline{\rho} \tilde{\mathbf{v}}) = 0$, and questions of what to do with the average of the product of density and velocity fluctuations do not arise. Similar simplifications occur in the averaged equation for momentum conservation.[6] The simplifications are especially evident in inertial and convective terms. Some complications arise in molecular transport terms, but turbulence modeling is applied to these terms anyway, so these complications disappear.

It is illogical to claim that Favre averaging is inappropriate on the grounds that the long history of turbulence modeling in constant-density flows has not used it. If the density is constant, then the Favre average reduces to the ordinary average, and therefore the available constant-density modeling may be applied with Favre averages. From a fundamental viewpoint it is unclear whether in extending results of modeling of constant-density flows to variable-density turbulence, the ordinary or Favre averages are most appropriate for the generalization. On theoretical grounds this question remains open, and it has not been fully resolved by experiments in variable-density turbulence. Even if the ordinary averages were to be found appropriate for the generalization, working with Favre averages would still be reasonable as a means for combining terms to reduce the number of them. Therefore the mass-weighted averages really are unobjectionable and deserve wider use in the exercising of moment methods than they have received. Most of the averaging employed in subsequent sections (e.g., Section 3.1) will be Favre averaging.

Bilger[4] has cited additional, finer points favoring Favre averaging. In addition he has defined a Favre or mass-weighted probability density function, which in many applications produces useful simplifications. Thus, it is often helpful to think of weighting turbulence quantities in proportion to the mass.

Modeling Approximations. Use of Favre averages does not alleviate the fundamental difficulties associated with modeling chemical production terms within the context of moment methods. These difficulties, discussed in Section 2.1 and in sections cited therein, remain the major obstacles to use of classical moment methods in turbulent combustion. These methods apply best to problems typified in example 1 in Section 1.1. Other types of problems are best addressed by other methods, such as the use of probability density functions, as described in the following section.

3. USE OF PROBABILITY DENSITY FUNCTIONS

3.1. The Conserved Scalar in Turbulent Diffusion Flames

Atom Conservation and Mixture Fraction. With the background material that has been presented in Section 1, notably with the information in Section 1.3, the use of pdf's in describing turbulent diffusion flames can be understood. The first step is to introduce the conserved scalar, used in numerous descriptions of turbulent diffusion flames.[4] The existence of such a quantity rests ultimately on equations for atom conservation. Since atoms are neither created nor destroyed in chemical reactions, equations for conservation of the total number of atoms of any given kind in the system, irrespective of the molecules in which these atoms may appear, do not contain chemical source terms. It is appropriate to formulate this statement with symbols.

Consider a mixture containing N different kinds of molecules, composed of M different chemical elements. Let ν_{kj} denote the number of atoms of element k in molecule j so that k runs from 1 to M and j from 1 to N. If X_j denotes the mole fraction of species j in the mixture at any given position and time in the flow, than the fraction of elements of kind k at that position and time is

$$\hat{Z}_k = \sum_{j=1}^{N} \nu_{kj} X_j, \qquad k = 1, \dots, M \qquad (24)$$

In flow systems for which momentum conservation needs to be considered, it is often more convenient to work with mass fractions Y_j instead of mole fractions. An element mass fraction Z_k, corresponding to $\hat{Z}_k$, can be defined in terms of the alternative coefficients $\mu_{kj} = (W_k/W_j)\nu_{kj}$, the mass of the k the element in a unit mass of the jth molecule; here W_k and W_j are the molecular weights of element k and molecule j, respectively. By definition,

$$Z_k = \sum_{j=1}^{N} \mu_{kj} Y_j, \qquad k = 1, \dots, M \qquad (25)$$

The relationship between Z_k and $\hat{Z}_k$ is $Z_k = \hat{Z}_k W_k/\overline{W}$, where $\overline{W} = \sum_{j=1}^{N} X_j W_j = 1/\sum_{j=1}^{N}(Y_j/W_j)$ denotes the mean molecular weight of the mixture.

Let ω_j represent the number of moles of chemical species j produced by chemical reactions per unit volume per unit time at the position and time of interest. Then, since the reactions conserve elements, $\sum_{j=1}^{N} \nu_{kj}\omega_j = 0$ for $k = 1, \dots, M$. The mass of species j produced by chemical reactions per unit volume per unit time is $w_j = W_j\omega_j$. An alternative statement of element conservation therefore is

$$\sum_{j=1}^{N} \mu_{kj} w_j = 0. \qquad k = 1, \dots, M \qquad (26)$$

Equation (26) states that the chemical source term does not appear in the conservation equation for Z_k. The general equation for conservation of a chemical species j may be written as[30]

$$\frac{\partial(\rho Y_j)}{\partial t} + \nabla \cdot [\rho Y_j(v + V_j)] = w_j, \qquad j = 1, \dots, N \qquad (27)$$

where ρ is density, $\mathbf{v}$ flow velocity and $\mathbf{V}_j$ the diffusion velocity of species j

(measured with respect to the mass-average velocity **v** of the mixture). Multiplication of (27) by μ_{kj} and summation over j shows by virtue of (26) that

$$\frac{\partial(\rho Z_k)}{\partial t} + \nabla \cdot (\rho \mathbf{v} Z_k) = -\nabla \cdot \left(\sum_{j=1}^{N} \rho \mu_{kj} Y_j \mathbf{V}_j \right), \qquad k = 1, \ldots, M \quad (28)$$

Equation (28) is a general statement of the fact that Z_k is a conserved scalar, namely, a scalar field function whose conservation equation does not contain chemical sources. Such functions previously have been termed *coupling functions*.[30] The basis of employing conserved scalars in describing turbulent diffusion flames rests on (28).

When there are many different elements in the system, there are many different Z_k, but not all are equally useful. In idealized two-reactant problems, a particularly useful Z_k will be that for an element contained in one of the reactants but not in the other. The element may be taken to be oxygen, hydrogen, or carbon, for example, for reactions of hydrocarbons with oxygen; the best choice would depend on subsidiary considerations of preferential diffusion, condensation, etc. If $k = 1$ identifies such an element and $j = 1$ the reactant in which it is contained, then Z_1/μ_{11} is a variable whose value is unity for one of the pure reactants and zero for the other. Under these conditions, Z_1/μ_{11} is a measure of the mixture fraction of the first reactant and aside from nonreactive influences such as diffusional separation gives the mass fraction of material coming from the first reactant, irrespective of the complexity of the chemistry or of the degree of chemical conversion.

Most diffusion-flame problems are two-stream problems in which identification may be made of separate entry streams of fuel and of oxidizer. There may be more than one entry port for each of these reactants, but properties must be uniform and the same over each port for a given reactant. Under these conditions, a mixture-fraction variable may be defined in terms of Z_k for any element k having different values of Z_k in each of the two streams. Specifically, with subscripts F and O identifying conditions in fuel and oxidizer streams, respectively, (see Figure 7) introduce

$$\xi_k \equiv \frac{Z_k - Z_{kO}}{Z_{kF} - Z_{kO}} \tag{29}$$

the mass-based mixture fraction of fuel, based on measurement by element k. By definition, each ξ_k is zero in the oxidizer stream and unity in the fuel stream. Simply for purposes of normalization, it is often more convenient to work with ξ_k in place of Z_k if $Z_{kF} \neq Z_{kO}$.

Approximations of the Lewis Number of Unity and Neglect of Pressure Fluctuations in Enthalpy.

In general, effects of separation of species by molecular diffusion prevent the right-hand side of (28) from being written in terms of Z_k

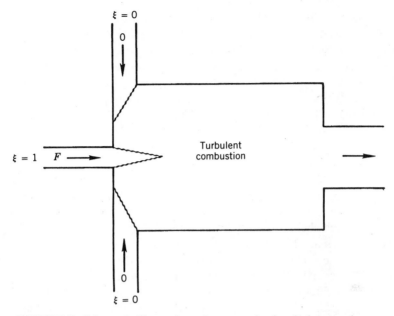

$\xi = 0$

0

$\xi = 1$ $F \longrightarrow$

Turbulent
combustion

$\longrightarrow$

0

$\xi = 0$

FIGURE 7. Schematic illustration of nonpremixed turbulent combustor.

without involving Y_j. Such effects may have a number of consequences, such as causing ξ_k to exceed unity or to be less than zero at times. If the binary diffusion coefficients for all pairs of species in the mixture are equal, and if pressure-gradient diffusion, body-force diffusion, and thermal diffusion are negligible, then Fick's law holds and may be written in the form $Y_j \mathbf{V}_j = -D\nabla Y_j$, $j = 1, \ldots, N$, where D denotes the common binary diffusion coefficient. Use of this, and (25) and (29) in (28), gives

$$\frac{\partial(\rho\xi_k)}{\partial t} + \nabla \cdot (\rho\mathbf{v}\xi_k) = \nabla \cdot (\rho D\nabla\xi_k) \tag{30}$$

Subtraction of (30) for one value of k from (30) for any other values of k results in a homogeneous equation with homogeneous initial and boundary conditions, having a unique solution of zero according to a uniqueness theorem. Therefore all ξ_k are equal under these conditions, and the subscript k may be omitted from ξ. Knowledge of ξ on a point-time resolved basis would then enable all Z_k to be calculated on a point-time resolved basis. This remarkable simplification, reducing the number of partial differential equations for Z_k from M to 1, rests only on the stated assumptions concerning molecular diffusion, on the two-stream hypothesis, and on the presumed absence of any spatial or temporal variations of Z_k in any approach stream. Typically the approximations concerning diffusion are the most severe of these, and continuing studies[4] appear to be indicating that often the diffusion approximations are reasonable.

The same simplification extends to energy conservation if additional restrictions are introduced. Attention is focused on flows at low Mach number with negligible effects of radiant energy transfer, work associated with body forces, or Dufour diffusion. Then the equation for conservation of the total (thermal plus chemical) enthalpy per unit mass of the mixture, h, is

$$\frac{\partial \rho h}{\partial t} + \nabla \cdot (\rho v h) = \frac{\partial p}{\partial t} + \nabla \cdot \left[\frac{\lambda}{c_p} \nabla h + \left(\rho D - \frac{\lambda}{c_p} \right) \sum_{j=1}^{N} h_j \nabla Y_j \right] \quad (31)$$

Here λ and c_p are the thermal conductivity and specific heat at constant pressure for the mixture, respectively, p denotes pressure, and h_j represents the (thermal plus chemical) enthalpy of chemical species j.

It is evident that (31) for h will look like (30) for ξ if $\partial p/\partial t$ is negligible and if $\rho D = \lambda/c_p$. The first of these restriction rules out transient processes such as those occurring in piston engines and requires pressure fluctuations to affect enthalpy fluctuations negligibly. The second states that the molecular Lewis number, Le $= \lambda/\rho D c_p$), is unity. These approximations seem reasonable for many flows. If, in addition, temperature fluctuations are negligible in the inlet streams and the walls are adiabatic, then the same reasoning that showed ξ to be unique results in

$$h = h_O + (h_F - h_O)\xi \quad (32)$$

entirely analogously to the earlier relationships

$$Z_k = Z_{KO} + (Z_{kF} - Z_{kO})\xi, \qquad k = 1, \dots, M \quad (33)$$

When (32) and (33) apply, a single equation of the type (30) suffices for h and all Z_k.

Conversation Equations for Moments of Conserved Scalar. From (30) may be derived equations for conservation of various moments of ξ. Because of proliferation of terms that otherwise occur in such equations for variable-density flows, it is becoming increasingly conventional to use mass-weighted or Favre averaging in these equations, as indicated in Section 2.2.

For statistically stationary flows, taking the average of (30) yields

$$\nabla \cdot (\bar{\rho} \tilde{v} \tilde{\xi}) = \nabla \cdot (\overline{\rho D \nabla \xi} - \overline{\rho v'' \xi''}) \quad (34)$$

Multiplication of (30) by ξ'' and averaging eventually produces

$$\bar{\rho} \tilde{v} \cdot \nabla \tilde{\xi}''^2 = -2\overline{\rho v'' \xi''} \cdot \nabla \tilde{\xi} - \nabla \cdot (\overline{\rho v'' \xi''^2}) - 2\overline{\rho D \nabla \xi'' \cdot \nabla \xi''}$$
$$+ \nabla \cdot \left[\overline{\rho D \nabla (\xi''^2)} \right] + 2\overline{\xi'' \nabla \cdot (\rho D \nabla \tilde{\xi})} \quad (35)$$

Derivation of (35) involves use of continuity, $\partial\rho/\partial t + \nabla \cdot (\rho v) = 0$, the average of which is $\nabla \cdot (\bar{\rho}\tilde{v}) = 0$ for stationary flow. Similar equations for changes in quantities such as $\overline{\rho v''\xi''}$ may be derived. These equations are not always closed, in that more averages occur than equations.

The conservation equations for moments may be closed by modeling hypotheses applied to higher moments. Such higher moments include the turbulent or Reynolds transport, $\overline{\rho v''\xi''}$, the production and diffusion of turbulent fluctuations (the first and second terms, respectively, on the right-hand side of (35)), and the dissipation (the first of the terms in (35) involving D). Since chemical production terms do not arise here, closure hypotheses become less questionable than in conservation equations for Y_j.

The simplest procedure having justification for high Reynolds number is first to neglect $\overline{\rho D \nabla \xi}$ in (34) and the last two terms involving D in (35) on the grounds that turbulent transport dominates molecular transport and that $|\overline{\nabla\xi''}| \gg |\nabla\tilde{\xi}|$. Next, employ a gradient approximation for turbulent transport, $\overline{\rho v''\xi''} = -\bar{\rho}D_T\nabla\tilde{\xi}$, where D_T is a turbulent diffusion coefficient, and similarly $\overline{\rho v''\xi''^2} = -\bar{\rho}D_T\nabla\overline{\xi''^2}$. Finally, relate the dissipation to $\overline{\xi''^2}$ by introducing a length scale for the turbulence; specifically, with $\chi = 2D\overline{\nabla\xi'' \cdot \nabla\xi''}$ denoting the instantaneous scalar dissipation, write $\tilde{\chi} = CD_R\overline{\xi''^2}/\lambda_\chi^2$, where C is a constant, D_R a reference value of D, and λ_ξ a Taylor length for the ξ field.

With this set of approximations, the unknowns in (34) and (35) are $\tilde{\xi}$ and $\overline{\xi''^2}$. The quantities $\bar{\rho}$ and $\tilde{v}$ are to be obtained from correspondingly averaged (and modeled) equations for mass and momentum conservation. The constants C and D_R are fixed, while D_T and λ_ξ are estimated or calculated from auxiliary equations, such as an equation for the evolution of the dissipative length scale of turbulence, by procedures of the types that have been developed for nonreactive turbulent flows (see, e.g., (21) and (22) and the corresponding discussion in Section 2.1). Thus techniques are available for rendering (34) and (35) soluble for the first two moments of ξ.

Approximations of Equilibrium and of Irreversible Chemistry. These results do not yet enable averages involving mass fractions Y_j or temperature T to be obtained. Information of ξ yields only information of h and Z_k through (32) and (33). If the chemistry is assumed to be rapid enough for equilibrium always to be attained, then knowledge of Z_k, h, and p, along with the equations of chemical equilibrium, enable all Y_j and T to be calculated. The procedure is identical to that[30] for calculating the adiabatic flame temperature and the corresponding chemical composition of the system. Consistent with the approximations introduced in obtaining (32) from (31), the local average pressure is to be used for p in these calculations; the pressure fluctuations have a small influence on the thermodynamic and chemical behavior.

Known complexities of calculations of adiabatic flame temperatures prevent simple formulas like (32) and (33) from being written for Y_k and T. To obtain illustrative simple equations, we may consider irreversible chemistry without dissociation and with constant and equal specific heats at constant pressure for each species. Assume that one gram of F combines with r grams of O to produce $1 + r$ grams of product with the release of an amount of energy q. Then $rY_F - Y_O$ is conserved in the reaction, and the analog of (33) is

$$rY_F - Y_O = (rY_{FF} + Y_{OO})\xi - Y_{OO} \tag{36}$$

where Y_{FF} is the fuel mass fraction in the fuel stream, and Y_{OO} is the oxidizer mass fraction in the oxidizer stream. Fuel but no oxidizer is present if $\xi_r < \xi < 1$, and oxidizer but no fuel is present if $0 < \xi < \xi_r$, where $\xi_r = (1 + rY_{FF}/Y_{OO})^{-1}$ is the value of ξ that corresponds to the presence of a flame sheet. The irreversible approximation thus leads to the occurrence of an infinitesimally thick flame sheet at which all of the heat release occurs.

From these results formulas for Y_F and Y_O in terms of ξ are readily derived. The fuel mass fraction is

$$Y_F = \begin{cases} Y_{FF}\dfrac{(\xi - \xi_r)}{1 - \xi_r}, & \xi > \xi_r \\ 0, & \xi < \xi_r \end{cases} \tag{37}$$

The oxidizer mass fraction is

$$Y_O = \begin{cases} 0, & \xi > \xi_r \\ Y_{OO}\left(\dfrac{1 - \xi}{\xi_r}\right), & \xi < \xi_r \end{cases} \tag{38}$$

Fuel and oxidizer diffuse into the flame sheet from opposite in stoichiometric proportions.

The temperature may be obtained as a function of ξ by the use of these results and (32). For convenience, select the zero of enthalpy so that $h = c_p T + qY_F$. Then $h_O = c_p T_O$ and $h_F = c_p T_F + qY_{FF}$. Equations (32) and (37) then give

$$T = \begin{cases} T_F + \left[T_O - T_F + \left(\dfrac{Y_{OO}}{r}\right)\left(\dfrac{q}{c_p}\right)\right](1 - \xi), & \xi > \xi_r \\ T_O + \left(T_F - T_O + \dfrac{Y_{FF}q}{cp}\right)\xi, & \xi < \xi_r \end{cases} \tag{39}$$

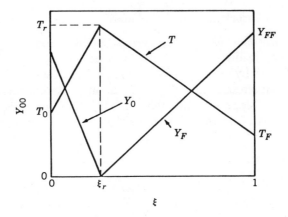

FIGURE 8. Dependences of fuel and oxidizer concentrations and of temperature on conserved scalar for irreversible chemistry.

Both of the expressions in (39) yield $T = T_r$ at $\xi = \xi_r$, where the flame temperature is

$$T_r = \frac{T_F + T_O r Y_{FF}/Y_{OO} + Y_{FF} q/c_p}{1 + r Y_{FF}/Y_{OO}} \qquad (40)$$

The results in (37), (38), and (39) are plotted in Figure 8.

Reduction to Problem of Nonreacting Turbulence. With the previous results in hand, the problem of describing turbulent diffusion flames has been reduced to that of describing nonreactive turbulent mixing. It is seen that a number of assumptions are needed to achieve this reduction. All of these assumptions have been described in the preceding subsections. The reduction is obtained only for statistically stationary, two stream problems. The most important assumption is chemical equilibrium, and the second most important is equal diffusivities. Ascertaining how good these assumptions are currently is a topic of active research.

The thermodynamics of the nonreactive mixing problem derived here are not simple. Since T and $\overline{W}$ are complicated functions of ξ, obtained through calculations of adiabatic flame temperatures, the density $\rho = p\overline{W}/RT$ is a complicated function of the concentration ξ of the nonreactive scalar that is mixing. This complication appears to be relatively small price to pay for the ability to obtain all properties of interest in terms of ξ without considering chemical rates. Modeling of nonreactive mixing with complex thermodynamics is more firmly based than modeling of reactive mixing.

3.2. Structures of Turbulent Diffusion Flames

Probability Density Functions for Conserved Scalars. From the preceding development, it may readily be understood why pdf's arise in describing turbulent diffusion flames by use of a conserved scalar. The relationships between ξ and quantities of interest, such as Y_j and T, are highly nonlinear, as is seen, for example, from the slope discontinuities in (37), (38), and (39) or Figure 8. Therefore moments of Y_j and T are not simply related to moments of ξ. The shape of the pdf for ξ must be known to calculate moments of Y_j and T with

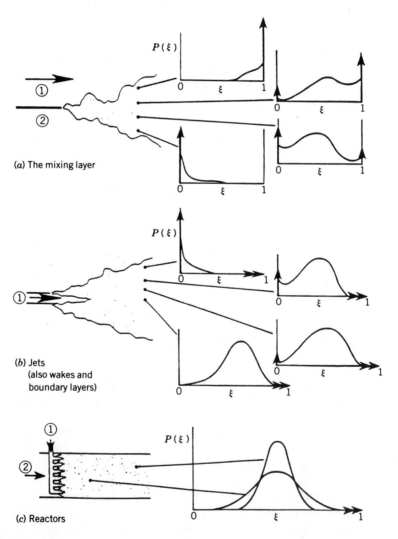

(a) The mixing layer

(b) Jets
(also wakes and
boundary layers)

(c) Reactors

FIGURE 9. Schematic illustration of probability density functions for conserved scalar in various flows.

reasonable accuracy. Thus the background material in Section 1.3 is relevant for obtaining the properties of Y_j and T that are of interest.

Under the assumptions that led to (32) and (33), the bounds $0 \leq \xi \leq 1$ apply. Therefore the pdf for ξ must vanish if $\xi < 0$ or $\xi > 1$. Quite a variety of different shapes for the pdf $p(\xi)$ are encountered under this restriction, depending on the type of flow and the position in the flow field. Many of these pdf's are shown in Figure 9. There are flows, especially mixing layers, that exhibit intermittency, in the sense that at certain points pure oxidizer and/or pure fuel are present for part of the time while turbulently mixed fluid is present of the time. Such flows will have delta functions in $P(\xi)$ at $\xi = 0$ or at $\xi = 1$, illustrated in Figure 9. Other flows, such as well-stirred reactors tend to have more nearly Gaussian pdf's without delta functions, as illustrated.

To facilitate analysis, there has been interest in obtaining functional forms that simulate the shapes shown in Figure 9. One approach has been to employ renormalized, truncated Gaussian Functions with delta functions added at $\xi = 0$ and at $\xi = 1$:

$$P(\xi) + \alpha\delta(\xi) + \beta\delta(\xi - 1) + \gamma e^{-(\xi - \bar{\xi})^2/2\sigma^2} \tag{41}$$

where γ vanishes for $\xi < 0$ or $\xi > 1$ and in between equals a constant that depends on the constants, α, β, σ, and $\bar{\xi}$ and that is obtained from the normalization condition $\int_{-\infty}^{\infty} P(\xi)\, d\xi = 1$. The value in $0 < \xi < 1$ is

$$\gamma = \frac{1 - \alpha - \beta}{\sqrt{2}\sigma\left[\mathrm{erf}\left(\dfrac{1 - \bar{\xi}}{\sqrt{2}\sigma}\right) - \mathrm{erf}\left(\dfrac{-\bar{\xi}}{\sqrt{2}\sigma}\right)\right]} \tag{42}$$

where erf denotes the error function. This versatile form has four parameters, or two if the delta functions are excluded.

A simpler form with two parameters is the beta function,

$$P(\xi) = \frac{\xi^{\alpha-1}(1 - \xi)^{\beta-1}\Gamma(\gamma)}{\Gamma(\alpha)\Gamma(\beta)} \tag{43}$$

where $\gamma \equiv \alpha + \beta$ and where Γ denotes the gamma function. Equation (43) is restricted to $0 < \xi < 1$ and for certain values of α and β produces infinities at $\xi = 0$ and at $\xi = 1$ that may simulate delta functions, as shown in Figure 10. For the beat function, the mean is $\bar{\xi} = \alpha/\gamma$ and the variance is $\overline{\xi'^2} = \bar{\xi}(1 - \bar{\xi})/(1 + \gamma)$, restrictions being $\alpha > 0$ and $\beta > 0$. Beta functions have found appreciable use in analyses involving conserved scalars.[31]

Adopting a two-parameter representation for $P(\xi)$ enables $P(\xi)$ to be obtained at any point in the flow by solving conservation equations for $\bar{\xi}$ and $\overline{\xi'^2}$, like those given in Section 3.1. If more parameters are retained, then in

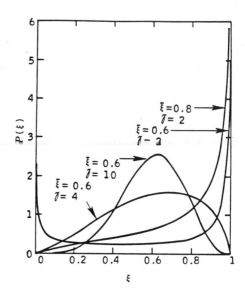

FIGURE 10. Various shapes of beta-function densities for conserved scalar.

predictive methods that do not calculate the evolution of $P(\xi)$ directly, either additional hypotheses must be introduced relating these parameters or additional conservation equations must be included. The use of (43) along with the method of Section 3.1 appears to give adequate accuracy for many purposes.

Probability Density Functions for Temperature and Concentrations. When α and β are known, use of (43) and (12) enables the pdf's for temperatures and concentrations to be calculated on the basis of the information developed in Section 3.1. With general equilibrium chemistry, the calculation of course requires the use of electronic computers. For the irreversible chemistry of Section 2.1, (37), (38), and (39) may be employed with (12) to enable calculations to be done by hand. An illustration of the results of such a computation is shown in Figure 11.

Reference to Figure 11 indicates that for fuel the $P(\xi)$ for $0 < \xi < \xi_r$ collapses into a delta function at $Y_F = 0$. For oxidizer, the $P(\xi)$ for $\xi_r < \xi < 1$ collapses into a delta function at $Y_O = 0$, thereby resulting in two delta function in $P(Y_O)$ for the illustrated $P(\xi)$. For temperature, since ξ is a double-valued function of T, there are two additive contributions to $P(T)$ in the range $T_O < T < T_r$; each of these is indicated by a dashed line in Figure 11.

It is seen that the resulting shapes for $P(Y_j)$ and $P(T)$ may be more complex than those of $P(\xi)$ and may not be readily characterized by simple formulas. However, subject to the approximations12 in Section 3.1 they always may be deduced from $P(\xi)$. This property helps to make ξ a useful variable.

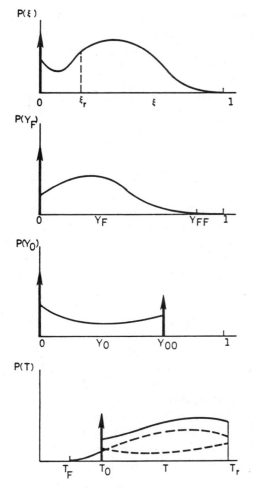

FIGURE 11. Illustration of deduction of probability density functions for reactant concentrations and for temperature from that for the conserved scalar.

Perturbation Approaches for Finite-Rate Chemistry and Production of Trace Species. The most critical approximation underlying these results is the assumption of chemical equilibrium. Criteria for the validity of this approximation can be stated.[14] Perturbation methods may be pursued to account for departures from equilibrium. Formalisms for the perturbations have been developed and in some cases employed to check the accuracies of equilibrium approximations.[14] The checks indicated that equilibrium often affords a good starting point for analyses of diffusion flames.

Trace species, those present in small concentrations, often exhibit greater departures from equilibrium than do the principal products. Nitric oxide (NO) and soot are examples of trace species that usually are far from equilibrium. Appreciable simplifications are possible in accounting for nonequilibrium of trace species since these species may not contribute significantly to the overall behavior of the flow. The presence of trace species may be neglected at first, and an analysis for equilibrium flow performed. Then the trace species may be studied by a finite-rate analysis with a known flow. Formalisms for doing this have been published.[14]

As an example, consider NO production according to the Zel'dovich mechanism, $O + N_2 \rightarrow NO + N, N + O_2 \rightarrow NO + O$.[32] Since the second step is fast, the production rate of NO may be written as $\omega_{NO} = 2k[N_2][O]$, where k is the specific reaction-rate constant for the first step, and square brackets identify concentrations. This rate is strongly dependent on T through K and also depends on the concentration of O atoms, which are not major species. If equilibrium is maintained for O_2 dissociation, then $[O]^2 = K[O_2]$, where K is and equilibrium constant. Under this condition, $\omega_{NO} = 2k\sqrt{K}[N_2][O_2]^{1/2}$ expresses the local instantaneous rate of production of NO explicitly in terms of the major variables T, $[N_2]$ and $[O_2]$. The equilibrium calculations give these three variables in terms of ξ. Hence the function $\omega_{NO}(\xi)$ may be evaluated. The average rate of production of NO at any given position then may be calculated readily from

$$\bar{\omega}_{NO} = \int_0^1 P(\xi)\omega_{NO}(\xi)d\xi \qquad (44)$$

after $P(\xi)$ has been obtained. The values $k = 7 \times 10^{13}\exp(-37,750/T)$ cm³/mol s and $\sqrt{K} = 4.1\exp(-29,150/T)$ (mol/cm³)$^{1/2}$ may be employed here, with T in K.[32]

The dependence of ω_{NO} on T is so strong that major contribution to the integral in (44) comes from a narrow range near the maximum temperature. This enables the simplification $\bar{\omega}_{NO} = CP(\xi_r)$ to be obtained from (44), where the constant C may be calculated by an asymptotic expansion.[33] Thus the average production rate is closely proportional to the value of the pdf at the stoichiometric value of ξ.

It should be emphasized that the procedure gives only the average production rate of NO, not the average concentration. It would be appreciably more difficult to calculate the average concentration because this will depend on a conservation equation for a species with a nonnegligible chemical production term. Fortunately the production rate is of the greater practical interest.

The calculation is inaccurate in that equilibrium for $2O \rightleftarrows O_2$ has been employed. It is known that superequilibrium O atom concentrations occur in turbulent diffusion flames and may lead to appreciably larger values of ω_{NO}. Attempts to treat O in a framework similar to that of NO, on the grounds that O also is a trace species, meet with complication because reactions involving collisions of O with other trace species are important. Therefore the best approach may be to estimate the extent of superequilibrium from considerations of laminar diffusion flames and to insert the resulting estimate of [O] as a function of ξ directly into the original formula for ω_{NO}, and then use this modified $\omega_{NO}(\xi)$ in (44).

g to calculate blow
region between the du
cone of fuel and oxidiz
exit ct premixed. It is thu
occ nonreactive mixing i
ass his mixture-ratio profil
est btained by equating
is elocity locally. At suffi-
pr d is less that the local
cie
ve lift-off region certainly
sults presumes that the
oc ocally the mixture is of
m that this full degree of
er some success has been
m ame-speed balance, there
a e use of the approach.
a n conditions the turbulent
fluctuating, laminar flame
1.4 it was indicated that
developed[29] that enable
lyses may be employed in

er number for extinction.[29]
hler number. For turbulent
e fluctuating laminar flame
gradients or dissipation $\hat{\chi}$ at
sheet, thereby increasing the
for local extinction of the
al extinction should begin to

blow off. Locally quenched
in a premixed fashion. The
the burner in the presence of
sion flamelets extends to the
s may be expected to disrupt

spatial distribution of $\hat{\chi}$ in the
off. Since values of $\hat{\chi}$ at the
l pdf $P(\hat{\chi}|\xi_r)$ is relevant; from
). Spatial distributions of this
lifted flames and on blow off.
n in lifted flames seems desir-

Additional complications arise when NO concentrations become large enough to approach their equilibrium values or it additional reactions involving nitrogen in fuel molecules are important. The former difficulty prevents (44) from being employed since it causes production rates to depend on the concentration of NO. The latter difficulty necessitates much more involved considerations of chemical kinetics.

This extended discussion of NO production is intended to illustrate the methods, advantages, and limitations in use of $P(\xi)$ for calculating production rates of trace species. Further development, evaluations, and discussion may be found in the literature.[14]

Average Rates of Chemical Heat Release. If we attempt to calculate average overall rates of chemical heat release from reactions of major species by use of $P(\xi)$ alone, we find that it cannot be done. As equilibrium is approached, the chemical production terms in the species conservation equations become indeterminate, involving differences of large numbers that cancel. A somewhat more circuitous route therefore is needed.[4,14]

In the Fick's law approximation, the equation for conservation of the jth chemical species in the mixture is

$$\frac{\partial(\rho Y_j)}{\partial t} + \nabla \cdot (\rho v Y_j) - \nabla \cdot (\rho D \nabla Y_j) = w_j, \qquad j = 1, \dots, N \quad (45)$$

where w_j denotes the mass per unit volume per unit time of species j produced by chemical reactions. Averages involving w_j are best computed from the left-hand side of (45) under near-equilibrium conditions.

Use of (32), (33), and equilibrium equations provides equilibrium functions $Y_j^e(\xi)$ and $T^e(\xi)$ for the species mass fractions and temperature. Substitution of these functions into the left-hand side of (45) gives

$$w_j = \frac{dY_j^e}{d\xi}\left[\frac{\partial(\rho\xi)}{\partial t} + \nabla \cdot (\rho v \xi) - \nabla \cdot (\rho D \nabla \xi)\right] - \rho D |\nabla \xi|^2 \frac{d^2 Y_j^3}{d\xi^2} \quad (46)$$

The quantity in the square brackets vanishes by virtue of (30). Therefore at equilibrium

$$w_j = -\rho D \left(\frac{d^2 Y_j^e}{d\xi^2}\right)(\nabla \xi) \cdot (\nabla \xi), \qquad j = 1, \dots, N \quad (47)$$

In (47), $d^2 Y_j^e/d\xi^2$ is a function of ξ. However, the nonnegative quantity $\hat{\chi} = 2D\nabla\xi \cdot \nabla\xi$ is not. Here $\hat{\chi}$ is closely related to the instantaneous scalar dissipation χ, introduced in Section 3.1; for high turbulence Reynolds numbers these two dissipations are practically equal. The occurrence of $\hat{\chi}$ in (47) emphasizes the importance of a joint pdf for the conserved scalar ξ and its dissipation $\hat{\chi}$. In terms of this pdf, denoted by $P(\hat{\chi}, \xi)$, the average of (47) is

$$\bar{w}_j = -\frac{1}{2}\int_0^1\int_0^\infty \rho\left(\frac{d^2Y_j^e}{d\xi^2}\right)\hat{\chi}P\,(\hat{\chi},\xi)\,d\hat{\chi}\,d\xi, \qquad j=1,\dots,N \qquad (48)$$

If h_j° is the heat of formation of species j, then $\dot{q} = -\Sigma_{j=1}^N h_j^\circ w_j$ is the energy per unit volume per unit time released in chemical reactions. Thus, the mean rate of heat release is

$$\bar{q} + \frac{1}{2}\int_0^1\int_0^\infty \rho\left[\sum_{j=1}^N h_j^\circ \frac{d^2Y_j^e}{d\xi}^2\right]\hat{\chi}P(\hat{\chi},\xi)\,d\hat{\chi}\,d\xi \qquad (49)$$

These formulas show that knowledge of the mean production rates of species or mean rate of heat release relies on knowledge of the joint pdf for ξ and the magnitude of its gradient. This can be understood on the basis of rates of diffusion into thin flame sheets.[33] At chemical equilibrium, turbulent diffusion flames may be viewed as collection of moving, wrinkled, laminar diffusion flames.

The derivatives $d^2Y_j^e/d\xi^2$ are readily calculated through thermochemistry. With the approximations in (37), (38), and (39), $d^2Y_j^e/d\xi^2$ are delta functions situated at $\xi = \xi_r$, the flame sheet. Under this approximation the averages w_j and $\dot{q}$ may be obtained if the joint pdf is known as a function of $\hat{\chi}$ at the stoichiometric value ξ_r only. The difficulties in use of (48) or (49) stem from difficulties in finding the function $P(\hat{\chi}, \xi_r)$ of the single variable $\hat{\chi}$ or the function $P(\hat{\chi}, \xi)$ of the two variables, ξ and $\hat{\chi}$.

Calculations of $\bar{w}_j$ and $\bar{q}$ have been made under the assumption that ξ and $\hat{\chi}$ are statistically independent random variables.[34] The results thereby obtained seem likely to have the proper order of magnitude. However, $\hat{\chi}$ is relatively difficult to measure; pdf data on $\hat{\chi}$ are recent and sparse, and essentially no experimental information is available yet on the joint pdf. Therefore at present the utility of (48) and (49) appears to be restricted to obtaining order-of-magnitude estimates.

With the simplification of (37), it is seen that

$$\frac{d^2Y_F^e}{d\xi^2} = \frac{Y_{FF}}{1-\xi_r}\delta(\xi-\xi_r),$$

and (48) becomes, approximately,

$$\bar{w}_F = -\frac{1}{2}\bar{\rho}\tilde{\chi}P(\xi_r)\left(Y_{FF}+\frac{Y_{OO}}{r}\right) \qquad (50)$$

when use is made of the expression for ξ_r. The mean rate of heat release

3.4. Probability-Density-Function Approaches to Description of Premixed Turbulent Flames

Probability Density Function for Reaction Progress Variable. In premixed systems the mixture ratio is a fixed quantity, established by the initial condition of mixing. Therefore, in a sense, $P(\xi) = \delta(\xi - \xi_i)$, where ξ_i is the initial value of ξ. The final diagram in Figure 9 may be used to illustrate the situation. View the "reactor" as a mixing chamber in which no reaction occurs. As the fuel-oxidizer mixture proceeds through the mixing chamber, the variance of ξ decreases, and $P(\xi)$ approaches a delta function. A premixed system may be defined as one in which this mixing has proceeded to completion prior to the beginning of the process to be studied. Thus every premixed system has a fixed value of ξ.

It is evident that the pdf $P(\xi)$ is of no use or relevance in premixed systems. One might think that the absence of variation of ξ provides simplifi-

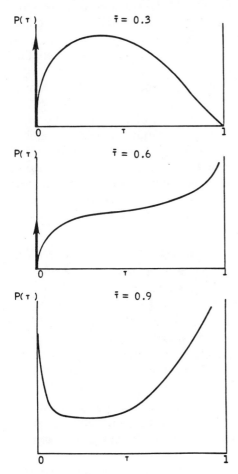

FIGURE 12. Schematic illustration of probability density functions for the reaction progress variable in premixed turbulent flames.

cation. It certainly does if the chemical equilibrium hypothesis, useful in nonpremixed systems, is retained. Unfortunately there is little interest in premixed systems that maintain complete equilibrium. In combustion such systems involve only completely reacted combustion products. Interest lies instead in systems involving conversion of reactants to products. Such systems cannot be described on the basis of equilibrium or near-equilibrium hypotheses.

Because of the necessity of considering nonequilibrium, the premixed combustion problems of interest are more complex than many of the non-premixed combustion problems discussed previously. The pdf procedures discussed thus far are largely irrelevant. However, different pdf's may be introduced in describing premixed turbulent combustion. In adiabatic systems the variable $\tau = (T\text{-}T_u)/(T_b - T_u)$ introduced in (16) becomes an important random variable. This nondimensional temperature change goes from zero in the unburnt gas to unity in the fully reacted mixture. Therefore in some sense it provides a measure of the extent of progress of the reaction. In studies of premixed turbulent combustion, this variable has been called a reaction progress variable and has been assigned various symbols, such as θ and very often c. For the sake of consistency with our earlier notation we retain here the symbol τ.

The pdf for the reaction progress variable may be denoted by $P(\tau)$. Many studies of structures of premixed turbulent flames have focused on $P(\tau)$. Approaches have involved parameterizing the shape of $P(\tau)$ in terms of a small number of variables and calculating the evolution of these variables from various averaged conservation equations.[5] The assumption of "fast chemistry" often motivates the selection of the shape of $P(\tau)$. In premixed systems this term has come to mean not equilibrium but instead a pdf dominated by delta functions at $\tau = 0$ and at $\tau = 1$. Some representative pdf's $P(\tau)$ for premixed turbulent flames are shown in Figure 12.

Modeling Assumptions and Difficulties with Countergradient Diffusion. With a two-parameter representation of $P(\tau)$, conservation equations for $\bar{\xi}$ and $\overline{\xi'^2}$ were seen to provide the information needed for calculating the evolution of $P(\xi)$. Similarly, with a two-parameter representation of $P(\tau)$, conservation equations for $\bar{\tau}$ and $\overline{\tau'^2}$ would give the history of $P(\tau)$. However, there is an important difference between ξ and τ as variables. The basic conservation equation for the former does not contain a chemical source term, but that for the latter does. This difference leads to numerous problems in attempting to write suitable equations for $\bar{\tau}$ and $\overline{\tau'^2}$.

There are not an unduly large number of terms in the equations for $\bar{\tau}$ and $\overline{\tau'^2}$ that need to be modeled. Turbulent transport and dissipation appear, as with ξ, and in addition averages of the chemical production term occur (compare (19)). The shape of $P(\tau)$ may be used in obtaining suitable modeling for the chemical production averages.[5] However, the existence of the chemical source term introduces significant uncertainties in the validity of the conventional modeling of turbulent transport.

Qualitatively, the fundamental difference is that the turbulence may move the reacting fluid to a different position, where it reacts, and then move the reacted fluid back. This phenomenon interferes with the modeling of the turbulent transport. For the reacting fluid, it is entirely possible that with the approximation $\overline{v'\tau'} = -D_T\nabla\bar{\tau}$, the turbulent diffusion coefficient may be negative. Such an occurrence is termed *countergradient diffusion*. When there is a possibility of countergradient diffusion, it becomes entirely unclear how to select an appropriate modeling for D_T.

The occurrence of countergradient diffusion is an established fact for premixed turbulent flames.[14] The theoretical work on the basis of $P(\tau)$ that led to the discovery involved retaining a conservation equation for $\overline{v'\tau'}$, eliminating the gradient modeling of this term and effecting closure at a higher order with modeling assumptions for higher-order terms. Countergradient diffusion has also been predicted by the perturbation methods to be discussed in Section 4[36] and often is observed experimentally. Therefore its existence is a fact that must be dealt with in developing predictive methods.

There are a variety of possible causes of countergradient diffusion. One cause may be understood on the basis of the action of the pressure drop across the flame on the turbulent transport.[37] Much recent work has been devoted toward reconciling the phenomenon with pdf approaches to the description of premixed turbulent combustion. Useful computational techniques that might be employed in design studies are emerging from these investigations but are not yet fully available. For this reason use of pdf's in premixed systems is not pursued in detail here.

4. USE OF PERTURBATION METHODS FOR PREMIXED COMBUSTION

4.1. Premixed Flames in Small-Gradient Turbulence

The Wrinkled-Laminar-Flame Structure of Premixed Turbulent Flames. In Section 1.1 it was stated that for premixed turbulent flames with $l_k \gg \delta_L$, the turbulent flame consists of a moving, continuous, wrinkled laminar-flame sheet. This is illustrated schematically in Figure 13, which shows temperature profiles associated with two different instantaneous positions of the wrinkled laminar flame, as well as profiles of the mean temperature and its variance. Perturbation methods have been employed to describe the turbulent flame structure and turbulent flame speed in this limit.[14,36]

Because of the mathematical complexities involved, none of the details of the perturbation approach will be exposed here. Most engineers will not wish to follow through all of the intermediate equations. Instead, only a general discussion will be given, and the results will be presented. The most useful

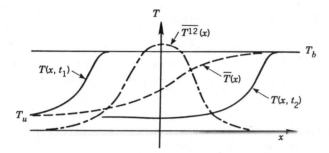

FIGURE 13. Schematic illustration of wrinkled-laminar-flame structure of premixed turbulent flame.

aspects of the work are the results obtained. These results, notably flame-spread formulas, may be used without knowledge of the details of the analysis.

The approach has been to recognize first that the structure is that of a wrinkled laminar flame. Because of this, the time-dependent conservation equations are expressed in a coordinate system tied to the moving laminar flame. A double expansion is performed in this coordinate system for large values of the parameters β and l_k/δ_L. By methods like that exhibited in Section 1.4, the expansion in β is employed to provide jump conditions across the thin reaction zone. The expansion in l_k/δ_L is then employed to solve the chemistry-free problem that results. The analyses that have been performed make full of stationarity and homogeneity in transverse directions in carrying out the expansion in l_k/δ_L. Therefore the results are best applicable to flames propagating normal to the oncoming turbulent flow, such as would be encountered for flames in ducts downstream from grids that produce turbulence.

The analyses initially were restricted to one-reactant systems with a Lewis number of unity and with small heat release. However, they have recently been generalized to account for Lewis numbers differing from unity and for arbitrarily large heat release.[11,12,38] The results will be quoted here for these more general conditions.

Premixed-Flame Instabilities. There are a number of sources of instability in premixed combustion systems.[39] Since instabilities have not yet been taken into account in analyzing the wrinkled-laminar-flame propagation, it is of interest to summarize the instabilities before quoting flame-speed results.

At the largest scales there are system instabilities that involve interactions of flows in different parts of a reacting system. These can modify turbulence but for analysis necessitate consideration of the entire system. From the viewpoint of the methods discussed here, their effects are to be accounted for by appropriate specification of inlet and outlet conditions.

There are acoustic instabilities that involve the interactions of acoustic waves with the combustion processes. These instabilities usually are of high frequency and have not been considered here at all, although there certainly are situations in which they can be quite significant. Generation of sound by turbulent combustion is an example of acoustic effects. By suitable design it is usually possible to avoid acoustic instabilities, if desired. The discussion here presumes that these instabilities are negligible in the system to be considered.

Taylor instabilities involve effects of buoyancy or acceleration in fluids with variable density. They often can be important in turbulent combustion but are not considered here. A light fluid beneath a heavy fluid is unstable by the Taylor mechanism. Upward propagation of premixed flames in tubes is subject to Taylor instability. In a configuration of downward propagation the buoyancy effects are stabilizing. When these effects are stabilizing, they can be useful in producing a more regular type of flame propagation, effectively insulating the flame from Landau instabilities (see below), for example. In stable configurations the Taylor mechanism may effectively inhibit other instabilities but exhibit no other significant influence on flame structure, thereby extending the regime of applicability of the turbulent flame analysis outlined here.

Landau instabilities are hydrodynamic instabilities of flame sheets that are associated neither with acoustics nor with buoyancy but instead involve only the density decrease produced by combustion in incompressible flow. The mechanism of Landau instability is purely hydrodynamic. In principle, Landau instability always is present in premixed flames, but in practice apparently it is seldom observed. It may be stabilized by combinations of buoyancy effects, diffusive-thermal effects (see below), and finite dimensions of chambers in which combustion occurs (walls, flame holders, etc.). In comparison with diffusive-thermal instabilities, Landau instabilities are relatively far-field phenomena; the Landau analysis for premixed laminar flames treats the entire flame as a discontinuity propagating at a flame speed that is independent of stretch or curvature of the sheet. Theoretical analyses are beginning to take into account effects of Landau instability on turbulent flame propagation in the wrinkled-laminar-flame regime.[12] In most of the available results on turbulent flame propagation, any influences of Landau phenomena must be include in specifying the upstream turbulence when the formulas are applied.

The smallest-scale flame instability is the diffusive-thermal instability. This instability involves relative diffusion of reactants and of heat within a laminar flame and depends on the nondimensional activation energy β, defined in (14) and on the Lewis number, $Le = \lambda/(\rho D c_p)$. Planar laminar flames are known to be unstable to transverse disturbances and to develop cellular shapes if $\beta(Le - 1) < l^*$, a critical value.[40, 41] This critical value has been found in a representative analysis [11] to be

$$l^* = -\frac{[2/(1 - \alpha)]\ln[1/(1 - \alpha)]}{\int_0^{\alpha/(1-\alpha)} x^{-1} \ln(1 + x)\, dx} \tag{52}$$

which is less than -2 but approaches -2 as α approaches zero. Here a heat-release parameter is $\alpha = (T_b - T_u)/T_b$. When $\beta(\text{Le} - 1) < l^*$, results of available analyses of turbulent flame speeds in the wrinkled-laminar-flame regime cannot be meaningful because the cellular instability invalidates the perturbation theory on which they are based. When $\beta(\text{Le} - 1) > l^*$, the results may be employed, and it is found that in this regime the stretch and curvature effects cause S_T/S_L, the ratio of turbulent to laminar flame speeds, to be greater than the unmodified wrinkled flame value if $\text{Le} < 1$ and less if $\text{Le} > 1$. At reasonable values of α, the effect of the heat release on l^* is a significant stabilizing influence. Thus the available results for turbulent flame speeds in the wrinkled-laminar-flame regime may have an appreciably wide range of applicability.

Prediction of Turbulent Flame Speed. The general formula obtained for the ratio of the turbulent flame speed S_T to the laminar flame speed S_L is

$$\frac{S_T}{S_L} = \overline{\left(1 + |\nabla a|^2\right)^{1/2}} - \delta_L^2 \overline{\left(\frac{1}{S_L}\frac{\partial u}{\partial x} - \nabla^2 a\right)^2} F(\alpha, \beta, \text{Le}) \qquad (53)$$

Here x is the longitudinal coordinate, perpendicular to the turbulent flame, and ∇ is a two dimensional gradient involving only the derivatives in the transverse directions, parallel to the turbulent flame. The longitudinal component of the velocity fluctuation upstream from the flame (but downstream from the Landau region of hydrodynamic adjustment), measured in a coordinate system fixed with respect to the turbulent flame (i.e., propagating into the unburnt mixture at the velocity S_T), is denoted by u here, while $a = \int'u\,dt$, which is the streamwise displacement associated with u. Since u is measured in an Eulerian frame, a has been called the *streamwise Eulerian displacement of the incoming turbulence.*

The function $F(\alpha, \beta, \text{Le})$ depends on the heat-release parameter, $\alpha = (T_b - T_u)/T_b$, the nondimensional activation energy, β, defined in (14) and the Lewis number, $\text{Le} = \lambda/(\rho D c_p)$. As α approaches zero, $F(\alpha, \beta, \text{Le})$ approaches

$$F(0, \beta, \text{Le}) = \beta(\text{Le} - 1)\left[1 + \frac{\beta(\text{Le} - 1)}{8}\right] \qquad (54)$$

The dependence of F on α has not yet been determined precisely, but it appears that the magnitude of F will decrease slowly as α increases, eventually approaching zero as α approaches unity. The change may be slight enough to enable (54) to be used with reasonable accuracy for values of α of practical interest.

From (54) it is seen that the term involving F in (53) vanishes if $\text{Le} = 1$. This term arises from physical effects operative only if the molecular diffusivi-

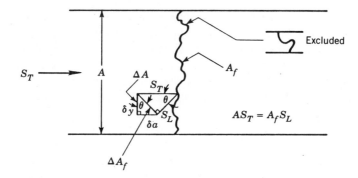

FIGURE 14. Effect of area increase of wrinkled laminar flame on turbulent flame speed.

ties of reactants and heat differ. Discussion of these physical effects is postponed to the following subsection. Attention will be focused here on the first term, which often represents the dominant contribution.

The first term in (53) described in influence of the area increase associated with flame wrinkling on the turbulent flame speed. This area-increase effect is illustrated schematically in Figure 14. It involves only wrinkling of the flame by the turbulence, with no change whatever in the internal structure of the wrinkled laminar flame. Therefore each element of the wrinkled flame propagates normal to itself at velocity S_L. Since all of the fluid flowing through the duct passes through the wrinkled flame, on the average $AS_T = A_f S_L$, where A and A_f are the cross-sectional area of the duct and the average area of the wrinkled flame sheet, respectively. The first term in (54) represents the ratio A_f/A,

That A_f/A involves an average of the transverse gradient of the streamwise Eulerian displacement is understandable. The Eulerian displacement defines the extent of motion of the flame sheet, and the transverse gradient of this provides the differences in the extent of longitudinal motion at adjacent transverse locations of points on the sheet. This difference is the tangent of the angle θ between the local normal to the sheet and the average stream direction. Thus, $A_f/A = \overline{1/\cos \theta} = (1 + \tan^2 \theta)^{1/2}$. The first term in (54) expresses this average in terms of properties of the incoming turbulence. The form given in (54) breaks down if wrinkling occurs to such an extent that the flame sheet turns back on itself in a manner identified in Figure 14 as to be excluded; this turn-back does not occur unless the relative intensity, u'/S_L, becomes too large, and in recent work[12] progress has been made toward removing the turn-back restriction, without yet uncovering any modification to (53).

The general form of the first term in (53) may be difficult to use because statistical properties of $|\nabla a|^2$, needed for finding the average, often are unavailable. Therefore it is of interest to introduce further approximations, to relate A_f/A to properties of the turbulence that are more readily available. If the upstream turbulence is isotropic, then statistical properties of a transverse derivative of the longitudinal velocity fluctuation are the same as those of the longitudinal derivative of a transverse component of the fluctuating velocity. The time integral of the longitudinal derivative of a fluctuating quantity is approximately this fluctuating quantity itself divided by S_T (in turbulence, a statement of this type is called a Taylor hypothesis). Hence in isotropic turbulence the statistical properties of ∇a are approximately the same as those of a transverse velocity fluctuation divided by S_T, which are the same as those of the longitudinal velocity fluctuation divided by S_T in view of the isotropy. Thus, approximately, in the present notation,

$$A_f/A = \overline{\left[1 + 2(u/S_T)^2\right]^{1/2}},$$

where the factor 2 occurs because there are two components of ∇a.

This simplification enables A_f/A to be evaluated from a pdf for u if S_T is known. For turbulence of low intensity, the relevant average may be expressed explicitly in terms of the rms velocity fluctuation u' introduced in Section 1.1. Since the mean-square value of the longitudinal velocity fluctuation is $u'^2/3$ under isotropy, and since S_T differs from S_L by an amount that is small if u'/S_T is small, it is found that in the simplest approximation $A_f/A = 1 + (u'/S_L)^2/3$. Hence in its simplest form the first term of (53) gives approximately

$$\frac{S_T}{S_L} = 1 + \frac{(u'/S_L)^2}{3} \tag{55}$$

This indicates that $S_T/S_L - 1$ is proportional to u'^2 and independent of turbulence scale at small values of u'/S_L.

Because of the occurrence of a derivative and an integral in $\nabla \int u\, dt$, insensitivity to turbulence scale seems to be a general property of the first term in (54); S_T/S_L may depend on ratios of scales that define anisotropy but not directly on magnitudes of scales. Anisotropy effects and inaccuracy of a Taylor hypothesis may be included in an unknown factor c in the expression $A_f/A = \left[1 + 2c(u/S_T)^2\right]^{1/2}$. To obtain a feeling for the influences of departures from the low-intensity approximation, the average may be taken inside the square root in this last expression. Equation (53) with only the first term retained then becomes $S_T/S_L = [1 + 2c(u'/S_T)^2/3]^{1/2}$, which after squaring may be written as $(S_T/S_L)^2 = 1 + (2/3)c(u'/S_L)^2/(S_T/S_L)^2$. From this expression a quadratic equation for $(S_T/S_L)^2$ is obtained, the solution to which gives

$$\frac{S_T}{S_L} = \left\{ \frac{1}{2}\left[1 + \sqrt{1 + \frac{8}{3}c\left(\frac{u'}{S_L}\right)^2}\right] \right\}^{1/2} \tag{56}$$

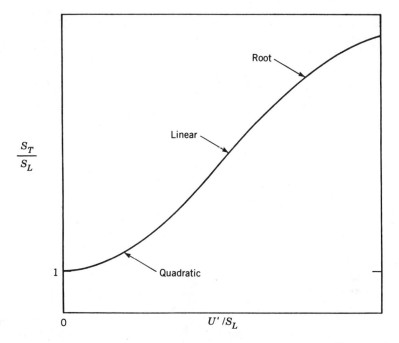

FIGURE 15. Illustration of dependence of ratio of turbulent to laminar flame speeds on ratio of rms velocity fluctuation upstream to laminar flame speed according to flame-wrinkling effect in wrinkled-laminar-flame regime.

This result reduces to (55) if u'/S_L is small and if $c = 1$. When u'/S_L is large, (56) gives $S_T/S_L = (2c/3)^{1/4}\sqrt{u'/S_L}$, but the basis of (56) is not well justified for large u'/S_L. Use of a pdf for ∇a is a better way to calculate S_T/S_L when u'/S_L is of order unity of larger. Nevertheless, the result given here may well be qualitatively correct. The results seem to indicate that as u'/S_L increases, $S_T/S_L - 1$ tends to progress from a quadratic to a square-root dependence on u'/S_L, as indicated schematically in Figure 15 This prediction agrees qualitatively with much of the experimental data that have been reported.[42]

It is worthwhile to reemphasize that these results involve nothing more than flame wrinkling. In the literature may be found a number of wrinkled-laminar-flame theories that postulate particular shapes of flame wrinkling, and these typically result in $S_T = S_L + \text{const} \cdot u'$, usually with the constant equal to unity. It is clear from (56) that such theories are inaccurate but may represent the correct predictions approximately over the intermediate portion of Figure 15.

The problem of calculating S_T/S_L when u'/S_L is of order unity or larger has recently been approached from a number of different viewpoints giving results that, although somewhat contradictory, nevertheless tend to be illuminating. Reviews of much of this work are available.[22, 43] However, new and

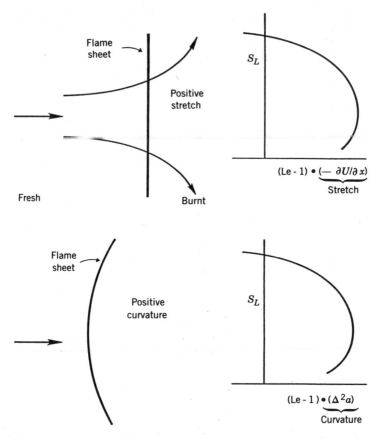

FIGURE 16. Illustration of effects of stretch and curvature of the wrinkled-laminar-flame sheet.

interesting ideas, not yet discussed in reviews, continue to be put forward on this difficult problem.

Effects of Stretch and Curvature on Turbulent Flame Speeds. The last term in (53) involves the average of the square of the difference of the first longitudinal derivative of u and the second transverse derivative of a. Each of these derivatives describes a different physical effect. By continuity, the longitudinal derivative of u is the negative of the transverse divergence of the transverse velocity, which has been described as effectively stretching the laminar flame. The second transverse derivative of a defines the curvature of the flame sheet. Therefore flame stretch and flame curvature both occur in the second term in (53). These phenomena are illustrated schematically in Figure 16.

These stretch and curvature terms have been written in the form indicated in (53) to emphasize their joint occurrence. The reason that they appear together is that in fact they are different aspects of the same phenomenon; by adopting different coordinate systems one may define a flame-surface stretch produced by the flow with respect to the moving, curved flame or a curvature of the flame with respect to the flow, both of which are equivalent and include both of the effects identified here. A general rate-of-stretch expression for a moving, curved flame sheet in a flow, the time rate of increase of the logarithm of the area, may be shown to be[12, 23]

$$\kappa = \nabla \cdot \mathbf{v} + \frac{S_L \nabla^2 a + \dfrac{D}{Dt}\left[(1 + |\nabla a|^2)^{1/2}\right]}{(1 + |\nabla a|^2)^{1/2}} \tag{57}$$

where $D/Dt = \partial/\partial t + \mathbf{v} \cdot \nabla$, where ∇ is the two-dimensional gradient that appears in (53), and where $\mathbf{v}$ is the two-dimensional, transverse velocity vector of velocity fluctuations just upstream from the flame. The general quantity that properly appears under the average sign in the second term of (53) is $(-\kappa/S_L)^2$. This general quantity reduces to that shown if $|\nabla a|^2 \ll 1$. The less general form has been written in (53) only because of its greater transparency.

Stretch and curvature are known to influence the laminar flame speed when Le $\neq$ 1.[40] These influences are indicated in Figure 16. The second term in (53) represents the average of these combined influences on S_T. Unlike the first term, the second term depends on the scale as well as the intensity of the upstream turbulence. For example, $(\partial u/\partial x)^2$ is proportional to u'^2/λ^2, where λ is the Taylor scale defined in Section 1.1. Thus influences of turbulence scales on S_T arise through terms involving modifications of structures of the wrinkled laminar flames. This may be expected, since the ratio l_k/δ_L governs the extent of such modifications.

Results like those shown qualitatively in Figure 16 originally were derived neglecting the density changes associated with combustion. In more recent work (e. g., Libby and Williams[44] and subsequent papers) effects of density changes are taken into account. These effects are substantial in that they tend to reduce significantly the sensitivity of the flame sheet to fluctuations in the external velocity field. The density decrease makes the gradient locally more shallow and thereby tends to "insulate" the flame. Consequently the sensitivity of S_L to stretch becomes much weaker than indicated in Figure 16. This also underlies the reduction in l^* with increasing α in (52).

In extreme cases, β(Le $-$ 1) < l^*, and the flames are unstable. Flame-speed formulas for turbulent flames in cellularly unstable regimes are unavailable today. Proper theoretical consideration has not been give to the influence of cellular instability on turbulent flame propagation. Moreover experimental correlations do not distinguish between cellularly stable and unstable conditions. The ability to achieve rough correlations without addressing questions of cellular instability suggests that when l_k/δ_L is large the first term of (53) provides reasonable results irrespective of the occurrence of cellular instability.

Figure 16 indicates that for Le $\neq$ 1, if the stretch or curvature becomes too large, then there is no value of S_L. The maximum stretch or curvature value defines a condition of laminar flame extinction. Thus turbulence with sufficiently large stretch or curvature will produce local disruptions of the continuous wrinkled flame sheet. Such disruptions may be expected to occur for turbulence of sufficiently small scale, that is, if l_k/δ_L is not sufficiently large. The basis of (53) fails under these conditions. Comparably accurate analyses of turbulent flame propagation are not available for these higher-gradient conditions. It is known only that the wrinkled-laminar-flame structure breaks down when the regions of disruption become too extensive. Increasing α tends to delay disruption appreciably; extinctions by stretch alone would appear to be very rare for most real turbulent flames, and small heat losses may play an important role.

Prediction of Mean Reactant Flux and Effects of Flame on Turbulence. The perturbation methods that led to (53) also provide predictions of the mean turbulent transport of the reacting species[36] and of the modifications of the turbulence velocities produced by the flame[11] in the wrinkled-laminar-flame regime. Since this information is not needed as often as are turbulent flame speeds, details of results will not be given here. We merely state that even in the absence of the heat release that causes a pressure drop across the flame, countergradient diffusion is found to occur,[36] and that in the presence of heat release the flame enhances transverse velocity fluctuations.[11] Equations for evolution of a wrinkled-laminar-flame front in a turbulent flow can be derived[12] and employed to study instabilities, statistics of velocity fluctuations, etc.

4.2. Regimes of Turbulent Flame Propagation

Intensity and Scale Parameters. Regimes of turbulent flame propagation can be considered in a plane of turbulence intensity and turbulence scale. A related diagram, with turbulent Reynolds number employed instead of scale as the horizontal coordinate, was discussed in Section 1.1 and shown in Figure 2, where it was indicated that the wrinkled-laminar-flame regime, discussed above, occurs in the lower right-hand portion of the diagram. If instead the coordinates had been simply u'/S_L and l_k/δ_L, then the wrinkled-laminar-flame regime would occur on the right-hand side of the diagram. Some experimental measurements of turbulent flame speeds have been reported that extend into this regime, as well as into others.[45] Some such results are shown in Figure 17.

From Figure 17 it is seen that at the largest turbulence scales, S_T/S_L tends to become independent of scale and dependent only upon u'/S_L. This behavior is consistent with (56), as is to be expected since the wrinkled-laminar-flame regime occurs for large scales.

The existence of other regimes may be observed in Figure 17. For the smaller values of u'/S_L, the ratio S_T/S_L is seen to increase with increasing scale. This behavior appears to characterize a regime of low-intensity turbulence that would occur near the lower left-hand boundary of Figure 2.

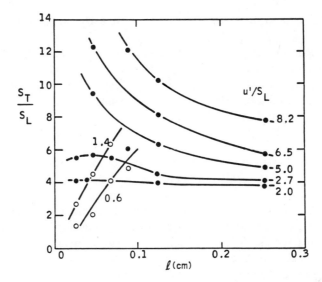

FIGURE 17. Experimentally measured influences of turbulence scale and intensity on the turbulent flame speed of propane-air mixtures.

At the higher intensities and smaller scales, Figure 17 shows a tendency for S_T/S_L to decrease with increasing scale. This behavior is consistent with (53) if Le < 1. However, it appears likely that extensive flame-sheet disruptions would occur in the upper left-hand portion of Figure 17 and that therefore these conditions lie not in the wrinkled-laminar-flame regime but rather in a third regime of high intensity and intermediate or small scale that is situated in the upper left-hand portion of Figure 2.

Thus at least three different regimes of turbulent flame propagation can be located in Figure 2. The large-scale regime occurs at the lower right, the low-intensity, small-scale regime at the lower left, and the high-intensity, small-scale regime at the upper left. Additional regimes may occur, for example with δ_L intermediate between two different turbulence scales. Flame-speed behaviors differ markedly in each of the three regimes indicated, as is seen from Figure 17, for example. To seek common formulas that will correlate flame speeds over all regimes would seen to be folly, since the physical processes of turbulent flame propagation clearly differ greatly in each regime. The engineer seems best advised to ascertain first, from Figure 2, for example, the regime of propagation as well as possible, and then to look for a flame-speed formula for the regime that has been identified.[43]

Flame Speeds in Different Regimes. Flame-speed formulas, such as (53) and (56), for the wrinkled-laminar-flame regime have been discussed in Section 4.1. Flame-speed formulas are not available for the low-intensity, small-scale regime; this regime appears to be, perhaps, of the least practical importance. There are theoretical indications that the flame speed may increase with both intensity and scale in this regime.[33] The high-intensity, small-scale regime is of appreciable practical importance. It has not yet received proper theoretical elucidation. Rudimentary concepts are available for this regime,[5,23] such as the idea that the ratio of the turbulent to laminar flame speed should be proportional to the square root of a turbulent diffusivity, but these concepts are unsubstantiated and most likely incorrect. Various correlations of experimental data may be found in the literature.[5]

Prospects for Prediction by Perturbation Methods. The low-intensity, small-scale regime can be studied by perturbation methods in which u'/S_L, instead of δ_L/l_k, is treated as a small parameter, with departures from the wrinkled-laminar-flame structure being sought. The possibility of investigating the high-intensity, small-scale regime by perturbation methods remains distant; structures about which perturbations might be developed are entirely unknown for this regime. The main regime in which further advances may be expected to be seen by accounting for additional phenomena through perturbation methods is the wrinkled-laminar-flame regime.

5. COHERENT STRUCTURES IN TURBULENT COMBUSTION

5.1. Experimental Observations of Coherent Structures

Observation in Nonpremixed Combustion. Only introductory statements and literature references on coherent structures will be given here since the subject has not progressed to a stage at which it is generally useful to engineers.

The coherent structures observed in nonreacting flows typically are large-scale vortices that occur in turbulent flows with shear, such as turbulent mixing layers.[18] Nonpremixed combustion involves such shear layers where the fuel and oxidizer mix. Therefore it is not surprising that vortex-type structures are found in nonpremixed combustion. A number of observations have been reported of these and other types of coherent structures in nonpremixed combustion.[46] They may take various forms, such as billowing flames or tonguelike flamelets protruding periodically from a turbulent flame brush. It is likely that different underlying mechanisms are responsible for the occurrence of the different forms. Hence the definition of what is to be called a *coherent structure* may vary with the phenomenon considered.

Observations in Premixed Combustion. Recently there has also been firm documentation of coherent structures in premixed combustion.[47] A premixed combustible enters a combustion chamber by flowing over a sharp downward step. In the shear layer downstream form the step large-scale vortices develop and may be seen by high-speed photography of the premixed turbulent flame. The flame induces modifications of the properties of these structures from those present in nonreacting flow. Causes for these modifications currently are understood only qualitatively.

Relevance of Observations to Predictive Methods. The observations of coherent structures are largely qualitative and merely provide an impetus for attempting to develop new predictive methods based on coherent-structure ideas of one kind or another. The specific structures that have been put into such predictive techniques are not the structures that have been observed. A successful technique would predict observed structures of the types seen. Thus the experimental observations provide only a qualitative background against which results of predictive methods can be judged.

5.2. The ESCIMO Approach of Spalding

Elements of the Approach. Only two predictive methods that may be viewed as being based on coherent-structure ideas will be mentioned here. The first is an approach, developed by Spalding,[19] which he terms ESCIMO, an acronym for engulfment, stretching, coherence, interdiffusion, and moving observer. In this method he identifies parcels of fluid which he calls "folds" and attributes to them both "demographic" and "biographic" aspects.

In the biographic aspect, each fold is followed by a moving observer, remains coherent and experiences stretching, chemical reaction, and molecular diffusion. Within each fold, time is termed "age," and the biographic aspect of the approach is to calculate the history of the fold as a function of its age. The demographic aspect of the method concerns probalistic conservation equations for the fraction of fluid in each age range, with account taken of turbulent engulfment of folds. By carrying out the biographic and demographic calculations with feedback between the two, it is hoped to obtain a reasonable description of the turbulent reacting flow. There is some relationship here to the methods of more traditional "age" theories.[48]

Advantages and Disadvantages of the Approach. The ESCIMO approach, along with others like it, has certain advantages, such as the ability to include complex chemical kinetics. Some successes have been obtained in applications to simple flows, such as well-stirred reactors, but problems of intermittency affect results for more complex flows, such as diffusion flames.[20] These problems need further study, and the technique requires judgment in its application and is not readily available to the engineer. Moreover, the description is not formally derived from the underlying Navier-Stokes equations and therefore is not subject to accuracy tests from fundamentals. The nature of the approach makes it difficult to see just what has been left out.

5.3. The Random Vortex Method

Basis of the Approach. The second predictive method to be mentioned, the random vortex method, is very different from the first. It is derived, with stated assumptions, from the Navier-Stokes equations. It may be classified as a coherent-structure method in that it employs vortices as fundamental structures for describing the flow. The vortices are randomly distributed, and each moves under the influences of all other vortices. In a sense the random vortex method may better be viewed as a technique for numerical modeling of flows, like finite-difference methods, since it employs large numbers of vortices and requires a great deal of computer capacity and computer time in application.

Successes Achieved by the Method. A notable accomplishment[49] of the random vortex method has been calculation of premixed turbulent combustion in a two-dimensional chamber like that studied experimentally.[47] This involved extending the technique through addition of an algorithm for propagation of heat-release sites at the flame speed. The flow simulated by the computational approach bears a striking qualitative resemblance to the experimentally observed flow. Large-scale vortex structures are predicted to occur, as observed experimentally.

Although this success is promising, a number of difficulties remain with the method. It is better suited to describing flow in the central part of a chamber than near walls; there are complexities associated with applying boundary conditions. Proper inclusion of three-dimensional vortices with reasonable accuracy is a challenge replete with fundamental questions and with the potential of requiring computers larger than any currently available; the combustion computations employed two-dimensional vortices.[49] Further detailed comparisons of computational efficiencies relative to finite-difference methods are needed. At the present stage of development, the random vortex method can be applied only by a specialist in the technique. The method does not constitute a design tool for the engineer.

5.4. Significance of Coherent Structures in Turbulent Combustion

Relevance to Flame Holding and Blow out. Coherent structures appear to play a central role in the stabilization and blow off of flames.[46] It is well known that standing vortices behind bluff bodies may be stabilized by the viscosity increase associated with the heat release of combustion and in turn may serve to stabilize a premixed turbulent flame in a duct. Large-scale vortices are of significance for holding premixed flames in many geometries. Diffusion flames also are sensitive to large structures. A lifted diffusion flame may remain stable for a long time, until an unusually large vortex passes and causes blow off. Statistics associated with large-scale structures thus appear to be important in connection with flame holding. Unfortunately, very little information is available that can be used to calculate these influences. Nevertheless, in recent years there have been many new experimental observations of these structures in various flames, and these observations can be useful in formulating ideas for quantitative descriptions.

Relevance to Other Flame Properties. Sizes of the larger coherent structures tend to establish the maximum length and time scales of the turbulence. In that way they may influence combustion regimes, reaction-zone structures, and overall rates of heat release. Large structures in heated fuel-rich regions may provide increased residence times needed for enhanced soot production. They may thereby influence radiative as well as convective heat transfer. The characteristic scales given in Section 1.1 can be helpful in estimating some of these effects. However, the recognition of many types of coherent structures is too recent for establishment of engineering methods of calculation based on them.

Prospects for Impact on Combustor Design. Recirculation zones behind bluff bodies already are taken into consideration in certain combustor designs. Knowledge, gained from experience, of types of coherent structures that may be expected to occur can greatly aid the engineer in efficiently designing combustors. Although general prescriptions cannot be given for the use of these concepts, it may be anticipated that coherent structures will play an increasing role in combustor design in the future.

REFERENCES

1. A. M. Mellor and C. R. Ferguson "Practical Problems in Turbulent Reacting Flows."In *Turbulent Reacting Flows* (P. A. Libby and F. A. Williams, eds.), pp. 45–64. Springer-Verlag, Berlin and New York, 1980.

2. R. Goulard, A. M. Mellor, and R. W. Bilger, "Combustion Measurements in Air Breathing Propulsion Engines. Survey and Research Needs." *Combust. Sci. Technol.* **14**, 195–219 (1976).

3. G. Damköhler, "Einflüsse der Strömung, usw. auf die Leistung von Reaktionsöfen I." *Z. Elektrochem.* **42**, 846–882 (1936).

4. R. W. Bilger, "Turbulent Flows with Nonpremixed Reactants." In *Turbulent Reacting Flows* (P. A. Libby and F. A. Williams, eds.), pp. 65–113. Springer-Verlag, Berlin and New York, 1980.

5. K. N. C. Bray, "Turbulent Flows with Premixed Reactant." In *Turbulent Reacting Flows* (P. A. Libby and F. A. Williams, eds.), pp. 115–183. Springer-Verlag, Berlin and New York, 1980.

6. P. A. Libby and F. A. Williams, eds., *Turbulent Reacting Flows*, pp. 1-43, 219–236. Springer-Verlag, Berlin and New York, 1980.

7. P. J. Smith, T. H. Fletcher, and L. D. Smoot, "Model for Pulverized Coal-Fired Reactors." *Symp. (Int.) Combust. [Proc.]* **18**, 1285–1293 (1981).

8. F. Tamanini, "A Numerical Model for the Prediciton of Radiation-Controlled Turbulent Wall Fires." *Sym. (Int.) Combust. [Proc.]* **17**, 1075–1085 (1979).

9. J. S. Evans and C. J. Schexnayder, Jr., "Influence of Chemical Kinetics and Unmixedness on Burning in Supersonic Hydrogen Flames." *AIAA J.* **18**, 188–193 (1980).

10. A. M. Mellor, "Turbulent-Combustion Interaction Models for Practical High Intensity Combustors." *Symp. (Int.) Combust. [Proc.]* **17**, 377–387 (1979).

11. P. Clavin, and F. A. Williams, "Effects of Molecular Diffusion and of Thermal Expansion on the Structure and Dynamics of Premixed Flames in Turbulent Flows of Large Scale and Low Intensity." *J. Fluid Mech.* **116**, 251–282 (1982).

12. P. Clavin, "Dynamical Behavior of Premixed Flame Fronts in Laminar and Turbulent Flows." *Prog. Energy Combust. Sci.* **11**, 1–59 (1985).

13. F. A. Williams, "Current Problems in Combustion Research." *Dynamics and Modeling of Reactive Systems* (W. E. Stewart, W. H. Ray, and C. C. Conley, eds.), pp. 293–314. Academic Press, New York, 1980.

14. P. A. Libby and F. A. Williams, "Some Implications of Recent Theoretical Studies in Turbulent Combustion." *AIAA J.* **19**, 261–274 (1981).

15. P. A. Libby, and F. A. Williams, "Turbulent Flows Involving Chemical Reactions." *Annu. Rev. Fluid Mech.* **8**, 351–379 (1976).

16. E. E. O'Brien, "The Probability Density Function Approach to Reacting Turbulent Flows." In *Turbulent Reacting Flows* (P. A. Libby and F. A. Williams, eds.), pp. 185–218. Springer-Verlag, Berlin and New York, 1980.

17. S. B. Pope, "PDF Methods for Turbulent Reactive Flows." *Prog. Energy Combust. Sci.* **11**, 119–192 (1985).

18. A. Roshko, "Structure of Turbulent Shear Flows: A New Look." *AIAA J.* **14**, 1349–1357 (1976).

19. D. B. Spalding, "Mathematical Models of Turbulent Flames: A Review." *Combust. Sci. Technol.* **13**, 3–25 (1976).

20. A. S. C. Ma, D. B. Spalding, and R. L. T. Sun, "Application of 'ESCIMO' to the Turbulent Hydrogen-Air Diffusion Flame." *Symp. (Int.) Combust.* [Proc.] **19**, 393–402 (1983).

21. G. I. Sivashinsky, "Instabilities, Pattern Formation and Turbulence in Flames." *Annu. Rev. Fluid Mech.* **15**, 179–199 (1983).

22. F. A. Williams, "Theory of Turbulent Flames." *Combustion Theory*, 2nd ed., pp. 373–445. Benjamin/Cummings, Menlo Park, California, 1985.

23. F. A. Williams, "Turbulent Combustion." In *The Mathematics of Combustion* (J. D. Buckmaster, ed.), pp. 97–131. SIAM, Philadelphia, Pennsylvania, 1985.

24. A. Papoulis, *Probability, Random Variables and Stochastic Processes*. McGraw-Hill, New York, 1965.

25. G. K. Batchelor, *Homogeneous Turbulence*. Cambridge University Press, London and New York, 1953.

26. J. D. Cole, *Perturbation Methods in Applied Mathematics*. Ginn/Blaisdell, Waltham, Massachusetts, 1968.

27. M. Van Dyke, *Perturbation Methods in Fluid Mechanics*. Academic Press, New York, 1964.

28. J. Buckmaster, and G. S. S. Ludford, *Theory of Laminar Flames*. Cambridge University Press, London and New York, 1982.

29. A. Liñán, "The Asymptotic Structure of Counterflow Diffusion Flames for Large Activation Energies." *Acta Astronaut.* **1**, 1007–1039 (1974).

30. F. A. Williams, *Combustion Theory*, 2nd ed. Benjamin/Cummings, Menlo Park, California, 1985.

31. J. Janicka, and W. Kollman, "A Two-Variable Formalism for the Treatment of Chemical Reactions in Turbulent H_2-Air Diffusion Flames." *Symp. (Int.) Combust.* [*Proc.*] **17**, 421–430 (1979).

32. N. Peters, "An Asymptotic Analysis of Nitric Oxide Formation in Turbulent Diffusion Flames." *Combust. Sci. Technol.* **19**, 39–49 (1978).

33. F. A. Williams, "A Review of Some Theoretical Consideration of Turbulent Flame Structure." *AGARD Conf. Proc.* **164**, III-1–III-25 (1975).

34. R. W. Bilger, "The Structure of Diffusion Flames." *Combust. Sci. Technol.* **13**, 155–170 (1976).

35. N. Peters, and F. A. Williams, "Lift-off Characteristics of Turbulent Jet Diffusion Flames." *AIAA J.* **21**, 423–429 (1983).

36. P. Clavin, and F. A. Williams, "Theory of Premixed-Flame Propagation in Large-Scale Turbulence." *J. Fluid Mech.* **90**, 589–604 (1979).

37. P. A. Libby, and K. N. C. Bray, "Countergradient Diffusion in Premixed Turbulent Flames." *AIAA J.* **19**, 205–213 (1981).

38. P. Clavin, and F. A. Williams, "Effects of Lewis Number on Propagation of Wrinkled Flames in Turbulent Flow." In *Combustion in Reactive Systems* (J. R. Bowen, N. Manson, A. K. Openheim, and R. I. Soloukhin, eds.), pp. 403–411. American Institute of Aeronautics and Astronautics, New York, 1981.

39. F. A. Williams, "Laminar Flame Instability and Turbulent Flame Propagation." In *Fuel-Air Explosions* (J. H. S. Lee and C. M. Guirao, eds.), pp. 69–76. University of Waterloo Press, Waterloo, Ontario, Canada, 1982.

40. G. I. Sivashinsky, "On a Distorted Flame as a Hydrodynamic Discontinuity." *Acta Astronaut.* **3**, 889–916 (1976).

41. T. Mitani and F. A. Williams, "Studies of Cellular Flames in Hydrogen-Oxygen-Nitrogen Mixtures." *Combust. Flame* **39**, 169–190 (1980).

42. R. G. Abdel-Gayed and D. Bradley, "A Two-Eddy Theory of Premixed Turbulent Flame Propagation." *Philos. Trans. R. Soc. London, Ser. A* **301**, 1–25 (1981).

43. J. Abraham, F. A. Williams, and F. V. Bracco, "A Discussion of Turbulent Flame Structure in Premixed Charges." In *Engine Combustion Analysis: New Approaches* (S. M. Shahad, ed.), pp. 27–42. Society of Automotive Engineers, Warrendale, Pennsylvania, 1985.

44. P. A. Libby, and F. A. Williams, "Structure of Laminar Flamelets in Premixed Turbulent Flames." *Combust. Flame* **44**, 287–303 (1982).

45. D. R. Ballal, and A. H. Lefebvre, "Turbulence Effects on Enclosed Flames." *Proc. R. Soc. London, Ser. A* **344**, 217–234 (1975).

46. N. Peters, and F. A. Williams, "Coherent Structures in Turbulent Combustion." In *The Role of Coherent Structures in Modeling Turbulence and Mixing* (J. Jimenez, ed.), pp. 364–393. Springer-Verlag, Berlin and New York, 1981.

47. A. R. Ganji, and R. E. Sawyer, "Experimental Study of the Flow Field of a Two-Dimensional Premixed Turbulent Flame." *AIAA J.* **18**, 817–824 (1980).

48. D. T. Pratt, "Mixing and Chemical Reaction in Continuous Combustion." *Prog. Energy Combust. Sci.* **1**, 73–86 (1976).

49. A. F. Ghoniem. A. J. Chorin, and A. K. Oppenheim, "Numerical Modeling of Turbulent Combustion in Premixed Gases." *Symp. (Int.) Combust.* [*Proc.*] **18**, 1375–1383 (1981).

PART III
Heterogeneous Combustion

9 Fuel Atomization, Droplet Evaporation, and Spray Combustion

ARTHUR H. LEFEBVRE

School of Mechanical Engineering
Purdue University
West Lafayette, Indiana

1. Atomization
   1.1. Introduction
   1.2. The atomization process
   1.3. Spray characteristics
   1.4. Spray measurement
   1.5. Plain-orifice atomizer
   1.6. Simplex atomizer
   1.7. Wide-range atomizers
   1.8. Fan-spray atomizers
   1.9. Rotary atomizers
   1.10. Air-assist atomizers
   1.11. Airblast atomizers
   1.12. Gas turbines
   1.13. Diesel injection
   1.14. Power systems
   1.15. Miscellaneous types of atomizers
2. Droplet Evaporation
   2.1. Introduction
   2.2. Steady-state evaporation
   2.3. Drop lifetime
   2.4. Evaporation at high temperature
   2.5. Heat-up period
   2.6. Convective effects on evaporation
   2.7. Effective evaporation constant
   2.8. Droplet burning
   2.9. Multicomponent fuel drops
   2.10. Drop transport in sprays
3. Spray combustion
   3.1. Droplet arrays, groups, and sprays
   3.2. Mathematical models of spray combustion
   3.3. Dense sprays

3.4. Experiments on spray flames
3.5. Combustion performance
Nomenclature
References

1. ATOMIZATION

1.1. Introduction

Atomization is the process whereby a volume of liquid is converted into a multiplicity of small drops. Its principal aim is to produce a high ratio of surface to mass in the liquid phase, resulting in very high evaporation rates. Fortunately, atomization is fairly simple to accomplish since, for most liquids, all that is needed is the existence of a high relative velocity between the liquid to be atomized and the surrounding air or gas. With some atomizers this is accomplished by discharging the liquid at high velocity into a relatively slow moving stream of air or gas. Notable examples include the various forms of pressure atomizer, and also rotary atomizers that eject the liquid at high velocity from the periphery of a rotating cup or disc. An alternative approach is to expose the relatively slow moving liquid to a high velocity air stream. The latter method is generally known as "twin fluid," "air-assist," or "airblast atomization."

The fuel injection process plays a major role in many key aspects of combustion performance. Its influence seems likely to assume even greater importance in the future, since most types of combustion devices will probably be subjected to increasingly severe emissions regulations and will be called upon to burn a larger proportion of heavy distillate and synthetic fuels. Fuel injectors having multifuel capability will also be in increasing demand. To meet these changing needs, the combustion engineer should have some understanding of the basic atomization process and also be fully conversant with the capabilities and limitations of all the relevant fuel-injection devices. All of these aspects will be discussed in some detail. The early sections are mainly concerned with the mechanisms of atomization and with the various means employed to measure and characterize the drop sizes produced. The remainder is devoted to practical methods of achieving a well-atomized spray and to the influence of fuel properties, atomizer geometry, and chamber-operating conditions on spray characteristics.

1.2. The Atomization Process

Diesel engines employ a plain-orifice atomizer in which fuel is forced under pressure through a small hole and is discharged in the form of a jet that rapidly disintegrates into a well-atomized spray of narrow cone angle.

In gas turbine combustion chambers atomization is normally accomplished by transforming the bulk fuel into fine jets or thin conical sheets in order to induce instability and promote disintegration into drops. Thin sheets may be obtained by discharging the fuel through orifices with specially shaped approach passages, by forcing it through narrow slots, by spreading it over a metal surface, or by feeding it to the center of a rotating disc or cup. Disintegration of the fuel jets or sheets into ligaments and then drops is induced by interaction with the surrounding air or gas. An alternative method of atomization—and one that is generally preferred for high-pressure ratio engines—is to expose the fuel jets or sheets to high-velocity air, which also serves to convey the resulting drops into the combustion zone.

Oil-fired furnaces use both pressure and air (or steam) atomization, and sometimes a combination of the two. However, regardless of whether atomization is achieved by hydraulic, pneumatic, or mechanical means, the function of the atomizer is always to attenuate the fuel into fine jets or thin sheets from which ligaments and ultimately drops will be produced, and then to distribute the resulting drops throughout the combustion zone in a controlled pattern and direction.

Jet Breakup. Several modes of jet disintegration have been identified, but in all cases the final mechanism involves the breakup of unstable threads of liquid into rows of drops conforming to the classical mechanism postulated by Rayleigh.[1] According to this theory a circular column of liquid becomes unstable and breaks up into drops when the amplitude of a small disturbance, symmetrical about the axis of the jet, grows to one-half the diameter of the undisturbed liquid jet. This occurs when $\lambda/d_0 = 4.5$, where λ is the wavelength of the disturbance and d_0 is the initial jet diameter. For this value of $\lambda/d_0 = 4.5$, the average drop diameter at breakup is calculated to be $1.89d_0$, that is, almost twice the diameter of the initial jet. Tyler[2] subsequently measured the frequency of formation of drops as a jet disintegrates and found that $\lambda/d_0 = 4.69$, thus confirming the mathematical predictions of Rayleigh.

Rayleigh's analysis assumed that surface tension was the only force resisting breakup and took no account of liquid viscosity. This analysis was later extended by Weber[3] to include the effect of viscous forces on jet breakup. According to Weber the ratio of λ/d_0 required to produce maximum instability for viscous jets is given by

$$\frac{\lambda}{d_0} = \pi\sqrt{2}\left[1 + \frac{3\mu_L}{\sqrt{\rho_L \sigma d_0}}\right]^{0.5} \tag{1}$$

When $\mu_L = 0$, the value of λ/d_0 reduces to 4.44, which is close to the value of 4.5 predicted by Rayleigh for this case.

Haenlein[4] studied experimentally the breakup of liquid jets and found, for a liquid viscosity of 0.86 kg/ms, that the ratio of λ/d_0 for maximum instability ranged from 30 to 40, which is clearly very different from Rayleigh's value of 4.5. Haenlein identified four distinct regimes of breakup, namely, drop formation with the influence of air, drop formation without the influence of air, the formation of waves due to air friction, and complete disintegration of the jet.

For the case of liquid jet disintegration occurring due to the influence of the surrounding air, the drop sizes obtained are governed by the ratio of the disruptive aerodynamic force ($\rho_A U_R^2$) to the consolidating surface tension force (σ/d_0). This dimensionless ratio is known as the Weber number, We:

$$We = \frac{\rho_A U_R^2 d_0}{\sigma} \tag{2}$$

For the case of liquid jet breakup occurring without the influence of surrounding air, dimensional analysis suggests that atomization quality is dependent on jet diameter and on the liquid properties of density, surface tension, and viscosity. The breakup mechanism of the jet is found to be dependent on the so-called "Z number,"[5] which is obtained as the ratio of the square root of the Weber number to the Reynolds number:

$$Z = \frac{We^{0.5}}{Re} = \left(\frac{\mu_L}{\sqrt{\sigma \rho_L d_0}} \right) \tag{3}$$

For any given atomizer, where d_0 and the relevant air and liquid properties are known, the relevant values of Weber number and Z number provide a useful indication of both the quality and the nature of the atomization process. The application of these numbers to the correlation of experimental data on mean drop size, for both pressure-swirl and airblast atomizers, is discussed later in this chapter.

Sheet Breakup. Fraser and Eisenklam[6] have defined three modes of sheet disintegration which are described as "rim," "wave," and "perforated sheet." In the first mode forces created by surface tension cause the free edge of a liquid sheet to contract into a thick rim that then breaks up by a mechanism corresponding to the disintegration of a free jet. When this occurs, the resulting drops continue to move in the original flow direction but remain attached to the receding surface by thin threads that also rapidly break up into rows of drops. This mode of disintegration is most prominent under conditions where the viscosity and surface tension of the liquid are both high. It tends to produce large drops together with numerous small satellite droplets.

In "perforated sheet" disintegration holes appear in the sheet at a certain distance from the orifice, these holes are delineated by rims formed from the liquid initially included inside. They grow rapidly in size until the rims of adjacent holes coalesce to produce ligaments of irregular shape that finally break up into drops of varying size.

Disintegration can also occur in the absence of perforations by the generation of a wave motion on the sheet, whereby areas of the sheet, corresponding to a half or full wavelength of the oscillation, are torn away before the leading edge is reached. These areas rapidly contract under the action of surface tension but may suffer disintegration by air action or liquid turbulence before a regular network of threads can be formed. As Fraser[7] has pointed out, the orderliness of the disintegration process and the uniformity of production of threads has a large influence on drop-size distribution. Perforations occurring in the sheet at the same distance from the orifice have a similar existence, and thus thread diameters tend to be uniform and drop sizes fairly constant. However, wavy-sheet disintegration is highly irregular, and consequently drop sizes are much more varied.

Atomizers that discharge the fuel in the form of a sheet are usually capable of exhibiting all three modes of sheet disintegration. Sometimes two different modes occur simultaneously and their relative importance can greatly influence both the mean drop size and the drop-size distribution.

The breakup of liquid sheets under the action of strong aerodynamic forces is discussed later in the section on airblast atomization.

More detailed descriptions of the various modes of jet and sheet disintegration are to be found in the various publications of Fraser[7,8] and Dombrowski,[8,9] some of which contain useful reviews of other work in this field.

1.3. Spray Characteristics

The spray characteristics of importance to combustion performance include mean drop size, drop-size distribution, patternation, cone angle, and penetration. Special importance is attached to mean drop size, drop-size distribution, and patternation because they are almost solely dependent on atomizer design, whereas cone angle and penetration are governed partly by atomizer design and partly by the aerodynamic influences to which the spray is subjected after atomization is complete.

Mean Drop Size. In order to facilitate the calculation of evaporation rates, and to compare the atomization quality of different sprays, the term "mean diameter" has been introduced. Its definition varies according to the particular purpose for which it is used, the general idea being to replace a given spray by a fictitious one in which all the drops have the same diameter and yet retain certain characteristics of the original spray. The Sauter mean diameter (SMD) is the one most widely used. It is defined as the diameter of a drop having the same volume to surface ratio as the entire spray:

$$\text{SMD} = \frac{\sum nD^3}{\sum nD^2} \tag{4}$$

TABLE 1. Some Mathematical Definitions of Mean Drop Size

Parameter	Description	Symbol for Mean Diameter	Mathematical Definition of Mean Size
Mean diameter	Linear mean diameter of drops in spray	D_{10}	$\Sigma n_i D_i / \Sigma n_i$
Area-length mean diameter	Diameter of a drop having the same surface/diameter ratio as the entire spray	D_{21}	$\Sigma n_i D_i^2 / \Sigma n_i D_i$
Volume-surface mean diameter	Diameter of a drop having the same volume/surface ratio as the entire spray	D_{32} (SMD)	$\Sigma n_i D_i^3 / \Sigma n_i D_i^2$
Area mean diameter	Diameter of a drop whose surface area is equal to the mean surface area of all the drops in the spray	D_{20}	$(\Sigma n_i D_i^2 / \Sigma n_i)^{0.5}$
Volume (mass) mean diameter	Diameter of a drop whose volume (mass) is equal to the mean volume (mass) of all the drops in the spray	D_{30}	$(\Sigma n_i D_i^3 / \Sigma n_i)^{0.33}$
Mass median diameter	Diameter to a drop below or above which lies 50% of the mass of the drops	MMD	—

Also frequently used is the mass median diameter (MMD), which is the diameter of a drop below or above which lies 50% of the mass of the drops. Table 1 gives the meaning and mathematical definitions of this and other mean diameters that are sometimes quoted.

Drop-Size Distribution. Owing to the heterogeneous nature of the atomization process, the threads and ligaments formed by the various mechanisms of jet and sheet disintegration vary widely in diameter, and the resulting main drops and satellite drops produced from them vary in size correspondingly. Only under certain special conditions, such as are obtained with a rotary cup atomizer when operating within a limited range of fuel-flow rates and rotational speeds, can a fairly homogeneous spray be produced. Thus, in addition to mean drop size, another parameter of importance to the definition of a spray is the distribution of drop sizes it contains. The drop sizes normally encountered in modern gas turbines range in diameter from 10 to 400 microns.

Drop-size distribution is a property that is most difficult to predict theoretically and to determine experimentally. An instructive picture may be obtained

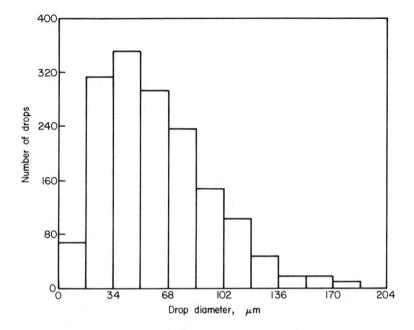

FIGURE 1. Typical drop-size histogram.

by plotting a histogram of drop size, each ordinate representing the number of drops whose dimensions fall between the limits $(x - \Delta x/2)$ and $(x + \Delta x/2)$. A typical histogram of this type is shown in Figure 1 in which $\Delta x = 17 \ \mu$m. As Δx is made smaller, the histogram assumes the form of a frequency curve which, provided it is based on sufficiently large samples, may be regarded as a characteristic of the spray. Such a plot is shown in the sketch in Figure 2, and is usually referred to as a frequency-distribution curve. If the surface area or volume of the spray corresponding to a given diameter were plotted versus diameter, the distribution curve would be skewed as shown in the figure, owing to the weighting effect in the large-diameter range.

The ordinate values of the frequency-distribution curve may indicate the number of drops with a given diameter, or the relative number of fraction of the total, or the fraction of the total number per size class. If the ordinate is expressed in the latter manner, the area under the frequency-distribution curve must be equal to 1.0.

In addition to representing the drop-size distribution by a frequency plot, it is also informative to use a so-called cumulative distribution representation. This type of plot is esentially the integral curve of the frequency plot and may represent the percentage of the total number of drops in the spray less than a given size, or it may express the percentage of the total surface or volume of a spray contained by drops less than a given size. Cumulative distribution curves when plotted on arithmetic coordinates have the general shape of the curve

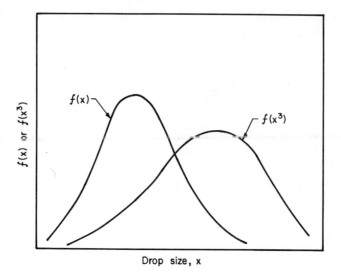

FIGURE 2. Drop-size frequency on number and volume basis.

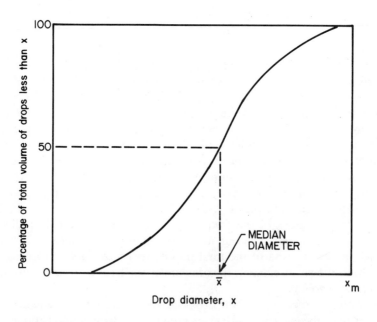

FIGURE 3. Typical shape of cumulative drop-size distribution curve.

shown in Figure 3, the ordinate of which may be the percentage of drops by number, surface, or volume less than a given drop diameter, x. For a spray it might be expected that a finite maximum drop size would occur, and hence the curve in Figure 3 would intersect the 100% ordinate at this maximum size, x_m. Furthermore drops of diameter less than zero could not be a physical reality, and hence the distribution curves would not be expected to extend below $x = 0$.

Because graphical representation of drop-size distribution is laborious, and not easily related to experimental results, many workers have attempted to replace graphical representation of drop-size distribution by mathematical expressions whose parameters can be obtained from a limited number of drop-size measurements. A relatively simple mathematical function that adequately describes the actual distribution is

$$\frac{dv}{dx} = ax^P \exp(-bx)^q \tag{5}$$

which has four independent constants, namely, a, p, b, and q. Most of the commonly used size distribution functions represent either simplifications or modifications of this function. One example is the Nukiyama-Tanasawa equation[10]

$$\frac{dv}{dx} = 1.5\pi x^5 \exp(-bx)^q \tag{6}$$

that is, $p = 5$.

At the present time the most widely used expression for drop-size distribution is one that was originally developed for powders by Rosin and Rammler.[11] It may be expressed in the form

$$1 - v = \exp(-bx)^q \tag{7}$$

where v is the fraction of the total volume contained in drops of diameter less than x, and b and q are constants. Thus by applying the Rosin-Rammler relationship to sprays, it is possible to describe the drop-size distribution in terms of two parameters, b and q. The exponent q provides a measure of the spread of drop sizes. The higher the value of q, the more uniform is the spray. If q is infinite, the drops in the spray are all the same size. For most fuel sprays the value of q lies between 2 and 4. However, for rotary atomizers q can be as high as 7.

Although it assumes an infinite range of drop sizes, the Rosin-Rammler expression has the virtue of simplicity, and it permits the data curves to be extrapolated into the range of very fine droplets where measurements are most difficult and least accurate.

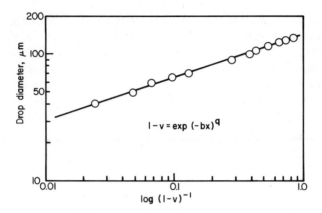

FIGURE 4. Typical Rosin-Rammler plot.

A typical Rosin-Rammler plot is shown in Figure 4. From such a plot the value of q is obtained as the slope of the line, while b, which represents a mean diameter of some kind, is given by the value of x for which $1 - v = \exp^{-1}$.

Mugele and Evans[12] analyzed the various functions used to represent drop-size distribution data by computing mean diameters for the experimental data and comparing them with various means calculated from the distribution functions given above. The empirical constants used in these functions were determined from the experimental distributions. From their analysis, Mugele and Evans proposed a so-called "upper limit function" as being the best method for representing the drop-size distribution of sprays. This is a modified form of the log-probability equation which is based on the normal distribution function. The volume distribution equation is given by

$$\frac{dv}{dy} = \frac{\delta \exp - \left(\delta^2 y^2\right)}{\sqrt{\pi}} \qquad (8)$$

where

$$y = \log\left[\frac{ax}{x_m - x}\right]$$

As y goes from $-\infty$ to $+\infty$, x goes from x_0 (minimum drop size) to x_m (maximum drop size), while δ relates to the standard deviation of y, and hence of x also. The a is a dimensionless constant. The Sauter mean diameter is given by

$$\text{SMD} = \frac{x_m}{1 + a \exp\left(1/4\delta^2\right)} \qquad (9)$$

It follows that a reduction in δ implies a more uniform distribution.

The upper limit distribution function assumes a realistic spray of finite minimum and maximum drop sizes, but it involves difficult integration, which requires the use of log-probability paper. The value of x_m must be assumed, and usually it takes many trials to find the most suitable value.

Mugele and Evans[12] have summarized the various statistical formulae for computing the different mean diameters and the variances. The principal conclusion appears to be that drop-size distribution should be presented in each case by the best empirical representation obtainable. Until atomization mechanisms can be suitably related to one or more distribution functions, there seems to be no theoretical justification for a belief, or expectation, that one function should be generally superior to another for representing drop-size distribution. Probably the best reasons for selecting a given distribution function would be (1) mathematical simplicity, (2) ease of manipulation in computations, and (3) consistency with the physical phenomena involved.

Chin and Lefebvre[13] have argued in support of the Rosin–Rammler distribution function because it is simple to use and can be readily obtained using a standard drop-size analyzer. A further useful advantage is that all the representative diameters in the spray are uniquely related to each other via the distribution parameter q. For example, we have

$$\frac{\text{MMD}}{\text{SMD}} = (0.693)^{1/q} \Gamma\left(1 - \frac{1}{q}\right) \tag{10}$$

where Γ denotes the gamma function. Also

$$\frac{D_{0.1}}{\text{MMD}} = (0.152)^{1/q} \tag{11}$$

$$\frac{D_{0.9}}{\text{MMD}} = (3.322)^{1/q} \tag{12}$$

$$\frac{D_{0.999}}{\text{MMD}} = (9.968)^{1/q} \tag{13}$$

where

$D_{0.1}$ = drop diameter such that 10% of total liquid volume is in drops of smaller diameter

$D_{0.9}$ = drop diameter such that 90% of total liquid volume is in drops of smaller diameter

$D_{0.999}$ = drop diameter such that 99.9% of total liquid volume is in drops of smaller diameter (i.e., the maximum drop size in the spray)

Patternation. The uniformity of the circumferential distribution of fuel in a conical spray is generally referred to as its "patternation." Poor patternation adversely affects many important aspects of combustion performance by creating local pockets of mixture in the combustion zone that are either appreciably richer or weaker than the design fuel/air ratio.

Cone Angle. With plain-orifice atomizers the cone angle is narrow, and the fuel drops are fairly evenly dispersed throughout the entire spray volume. Sprays of this type are often described as "solid." It is possible to produce solid cones with swirl atomizers but, for combustion applications, the spray is usually in the form of a hollow cone of wide angle, with most of the drops concentrated at the periphery.

In both types of atomizers the fuel jet or sheet rapidly disintegrates into drops that tend to maintain the general direction of motion of the original jet or cone. However, because of air resistance, the leading drops and the drops formed on the outside of the spray rapidly lose their momentum and form a cloud of finely atomized drops suspended around the main body of the spray. Their subsequent dispersion is governed mainly by the general movement of air and gaseous products in the combustion zone.

Because plain-orifice atomizers produce a narrow, compact spray in which only a small proportion of drops are subjected to the effects of air resistance, the distribution of the spray as a whole is dictated mainly by the magnitude and direction of the velocity imparted to it at exit from the atomizer orifice. With swirl atomizers, however, the hollow conical structure of the spray incurs appreciable exposure to the influence of the surrounding air. Normally any increase in spray cone angle will increase the extent of this exposure, leading to improved atomization and an increase in the proportion of drops whose distribution is dictated by the aerodynamics of the primary zone. This is one reason why spray angle is such an important characteristic of a swirl atomizer.

With airblast atomizers the fuel drops are airborne right from their inception, and thus their subsequent trajectory is dictated by the air movements created by the interaction of the atomizer air with the overall combustor airflow pattern.

Dispersion. The dispersion of a spray may be expressed quantitatively if at any given instant the volume of fuel within the combustion space is known. According to one definition the degree of dispersion may be stated as the ratio of the volume of the spray to the volume of the fuel contained within it. Other ways of expressing the degree of dispersion are discussed by Giffen and Muraszew.[14]

The advantage of good dispersion is that the fuel mixes rapidly with the surrounding gas, and the subsequent rates of evaporation and heat release are high. With plain-orifice atomizers dispersion is governed mainly by other spray characteristics, such as cone angle, mean drop size, and drop-size distribution, and to a lesser extent by the physical properties of the fuel and the surrounding medium. In general, all the factors that increase the spray cone angle also tend to improve the spray dispersion.

Penetration. The penetration of a spray may be defined as the maximum distance it reaches when injected into stagnant air. It is governed by the relative magnitude of two opposing forces: (1) the kinetic energy of the initial fuel jet, and (2) the aerodynamic resistance of the surrounding gas. The initial jet velocity is high, but as atomization proceeds and the surface area of the spray increases, the kinetic energy of the fuel is gradually dissipated by frictional losses to the gas. When the drops have finally exhausted their kinetic energy, their subsequent trajectory is dictated mainly by the effects of gravity and surrounding gas movements.

In general, a compact narrow angle spray will have a high penetration, whereas a well-atomized spray of a high cone angle, incurring more air resistance, will tend to have a low penetration. In all cases the penetration of a spray is much greater than that of a single drop because the first drops to be formed impart their energy to the surrounding gas which begins to move with the spray, thereby offering less resistance to the following drops which consequently penetrate further.

1.4. Spray Measurement

The spray properties of most interest to the designer are mean drop size, drop-size distribution, spray cone angle, and patternation. The methods most commonly used to measure these properties are described below.

Drop Size. Although a spray contains a far larger proportion of small drops than large ones, it is the relatively few large drops that predominate in determining the average drop diameter of the spray. Thus, for a sample of drops to be truly representative of the spray as a whole, it is vitally important that the large drops be included. As Lewis et al.[15] have pointed out, the presence or absence of one large drop in a sample of about 1000 drops may affect the average diameter of the sample by as much as 100%. Thus, in order to achieve an accuracy of within $\pm 5\%$, it is necessary to measure about 5000 drops.

Methods of measuring drop size fall into three main groups: direct methods, indirect methods, and simulation methods in which the fuel is replaced by a liquid that solidifies after atomization so that the solid drops can be easily measured and counted. A simple example of the direct method is one that involves the impaction of drops on a magnesium oxide-coated slide. The slide is prepared by moving it to and fro over a length of burning magnesium ribbon, until an oxide layer is obtained of thickness equal to the diameter of the largest anticipated drop. After exposing the slide to the spray, the sizes of the droplet impressions are measured, usually by means of a microscope fitted with a traversing scale, and these sizes are then converted to actual droplet sizes using a correction factor derived by May.[16] Soot-coated slides have also been used to measure both the radial and the overall drop-size distribution in a spray.[17]

One problem associated with slides is that of determining what fraction of the slide area should be covered by drops. If too many drops are collected, the probability of error due to droplet overlap is high, and also drop counting becomes tedious. On the other hand, if too few drops are collected the sample may not be sufficiently representative of the spray. From the standpoint of ease of measurement and counting, a 0.2% coverage is satisfactory, but a coverage as high as 1.0% can be tolerated before the problem of droplet overlap becomes significant.

Other important considerations are droplet evaporation and collection efficiency. The lifetime of small drops is extremely short. For example, a water droplet 10 μm in diameter has a lifetime of about one second in a 90% relative humidity atmosphere.[18] Thus evaporation effects are very significant in the measurement of fine sprays. Collection efficiency is especially important with airblast atomizers due to the flowfield created around the collecting surface. The large drops have enough inertia to impact on the surface, but the small drops tend to follow the streamlines. For these reasons the drop-size data for airblast atomizers obtained by direct drop collection methods tend to be larger than the actual sizes.

Yet another problem that arises with the collection of droplets on coated slides is the determination of the correction factor by which the diameters of the flattened droplets must be multiplied to obtain the original diameter of the spherical volume. Its value depends on liquid properties, notably surface tension, and on the nature of the coating employed. For example, with oil drops the correction factor is around 0.5 for a clean glass surface and 0.86 for a magnesium-coated slide.

Thus it is very desirable to determine the correction factor for the specific test conditions employed.

Another interesting method of drop-size measurement is the "wax droplet" technique which was first used successfully by Joyce.[19] The basic idea is to replace the fuel used in a normal spray by a molten paraffin wax that solidifies after atomization. Joyce employed a paraffin wax that, when molten, had physical properties close to aviation kerosine (density, 780 kg/m^3; surface tension, 0.027 kg/sec^2; kinematic viscosity, 1.5×10^{-6} m^2/sec). The attraction of the wax method is that direct readings are obtained of the volume (or mass) fractions in each size range by microscope counting, or by sifting the solidified drops of the entire spray. Molten wax, and melts of wax-polyethylene mixtures of various composition to obtain variations in viscosity, were also used successfully by Kim and Marshall.[18] The main disadvantages of the method are practical problems associated with preheating the wax, and the errors incurred due to changes in the physical properties of the wax droplet as it rapidly cools after leaving the atomizer, such that the processes of formation and secondary recombination may not be accurately reproduced. For this reason the air near the nozzle (i.e., in the region where the key atomization processes are taking place) should be heated to the same temperature as the molten wax.

A natural extension of the wax droplet technique is to attempt to solidify the liquids drops as they emerge from the nozzle. In 1957 Choudhury, Lamb, and Stevens[20] described a method for freezing the entire spray in a bath of liquid nitrogen. The atomizer was mounted with its discharge nozzle pointing downward at a height of approximately 0.4 m above the liquid nitrogen surface. After sufficient drops had been collected, the liquid nitrogen was decanted off, and the frozen drops were passed through a series of screens, ranging in size from 53 to 5660 μm. It was found that in order to avoid any agglomeration of drops on the liquid nitrogen surface, the density of the liquid being atomized should exceed 1200 kg/m³. Thus the method is unsuitable for kerosine, fuel oils, and water; this clearly represents a serious disadvantage, especially when coupled with a minimum measurable drop size of 53 μm.

Various refinements to the nitrogen-freezing technique have been made by Nelson and Stevens,[21] Street and Danaford,[22] and Norster and his co-workers at Cranfield. The Cranfield method has been described by Rao.[23] It employs a specially designed isokinetic probe that is mounted with its inlet facing the atomizer. Gaseous or liquid nitrogen is conveyed to the tip of the probe and injected into the incoming spray through a narrow annular slot. The nitrogen, at a temperature of around 140 K, rapidly freezes the droplets, which are then collected in a nitrogen-cooled perspex pot and photographed through a microscope. The final prints, showing the drops enlarged by a factor of 50 or more, are subsequently analyzed to determine the number and size of drops in the sample. The method is simple, elegant, and convenient, but it still involves a correction factor to account for the change in drop size that occurs during freezing.

In recent years considerable effort has been applied to research on optical methods for the measurement of drop sizes and drop-size distribution. Such methods have an important inherent advantage in that they do not require the insertion of a sampling device into the flow. The drops of most interest in combustion work are those which exceed 10 μm in diameter, since smaller drops evaporate very rapidly and, in any case, represent only a very small fraction of the total liquid mass in the spray. Various optical methods have been developed to measure drop sizes in the range of practical interest. The Sauter mean diameter for the spray in a nonreacting flow can be obtained by the light-scattering technique first proposed by Dobbins, Crocco, and Glassman.[24] It is based on a direct measurement of the scattered light intensity profile after the monochromatic light beam has passed through the spray. The SMD is obtained directly from measurement of intensity versus radius in the focal plane of the receiving lens. In practice, this is accomplished by measuring the traverse distance (r) between the optical axis and a point on the profile at which the light intensity is equal to one-tenth of the normalized intensity in the scattered profile. The SMD of the spray can then be determined using the relationship between r and SMD as derived by Roberts and Webb.[25]

Lefebvre and his co-workers have made extensive measurements on fuel atomizers using the Dobbins' method.[26-34] These results, and the good agree-

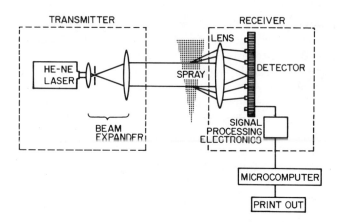

FIGURE 5. Schematic of Malvern particle size analyzer.

ment found with other independent measurements, indicate the practical usefulness of the method. One version of the technique, as described by Lorenzetto[30] is capable of measuring SMD values with good accuracy down to 20 μm.

A commercially available instrument, the Malvern particle analyzer, is based on measurement of the "scattered energy" distribution, rather than the "light intensity" distribution. It works on the principle of Fraunhofer diffraction from particles illuminated by a laser beam, as described by Swithenbank et al.[35] The main components of the instrument are shown in Figure 5. The light source is a 2-mW, He-Ne laser, with a spatial filter and beam expander to provide a 9-mm diameter collimated beam. The standard detector arrangement employs a 300-mm focal length collecting lens that focuses the light scattered from particles in the beam onto a photodiode detector having 30 concentric rings. The signal from each ring is sampled and processed to yield the drop-size distribution of the spray. Specific size distribution functions, such as the Rosin-Rammler distribution discussed above (Eq. (7)), can be assumed as part of the normal data processing, but the processing can also be carried out in a model-independent mode that yields the mass percentage of the spray in 15 size ranges. The accuracy and limitations of the Malvern particle analyzer have been discussed by Hirleman et al.[36] and Dodge.[37]

The need to avoid making drop-size measurements close to the nozzle has been stressed by Wittig et al.[38] The rapid deceleration of the smallest drops in the spray in this region gives rise to readings of SMD that are appreciably lower than the true value. Chin et al.[39] recommend a downstream distance of around 25 cm as being the ideal plane in which to make drop-size measurements. However, their calculations take no account of evaporation. To minimize errors arising from this effect, it is desirable to keep the distance between the nozzle and the plane of measurement as short as possible. For fairly volatile fuels, such as gasoline, a downstream distance of around 15 cm would appear to present the best compromise.

Many photographic techniques have been used with varying degrees of success to measure the drop sizes in a spray. Some of these rely on a short light pulse to "freeze" the spray image so that direct measurements of drop size can be made.[40, 41] In the case of double-flash photography, two closely spaced light pulses are used to obtain double images of each droplet so that velocity can also be determined.

The growing interest in fuel atomization has led in recent years to the development of several new laser-scattering and intensity-ratioing techniques. The latter offers considerable promise for the accurate measurement of drop sizes in the very small size range, say from 0.3 to 3 μm, while laser scattering should prove useful in the range above 10 μm. So far these new techniques have been tested only under "cold" conditions, but they are now being applied to the more hostile combustion situation. For further information on these and other developments in optical sizing methods, reference should be made to the review by Chigier.[41]

Cone Angle. A major difficulty in the definition and measurement of cone angle is that for swirl atomizers the spray cone has curved boundaries due to the effects of air entrainment by the spray. To overcome this problem, the cone angle is often quoted as the angle formed by two straight lines drawn from the discharge orifice to cut the spray contours at some specified distance from the atomizer face. One method of observing the spray cone angle is to project a silhouette of the spray onto a ground glass screen at two or three magnifications. The drawback to this technique is that it only defines the outer boundary of the spray and provides no information on the radial distribution of fuel within this boundary.

Radial Fuel Distribution. The radial fuel distribution within a spray is measured using a "patternator," as illustrated in Figure 6. This device collects fuel from the spray in 29 tubes, the level of flow collected in each tube being a measure of the flow rate at the corresponding radial location of the fuel spray cone. Tests are normally conducted with the nozzle spraying downward. The juxtaposition of nozzle and patternator is such that the nozzle discharge orifice is located at the center of curvature of the patternator. Spray samples are collected by rotating the inverted patternator to the upright position after the nozzle has stabilized at the correct flow rate and fuel pressure conditions. After a sufficient quantity of fuel has been collected, the fuel is turned off and a screen is moved over the top of the tubes to shelter them from any residual fuel dribbling from the nozzle.

Some typical radial fuel distributions obtained by this technique are shown in Figure 13.

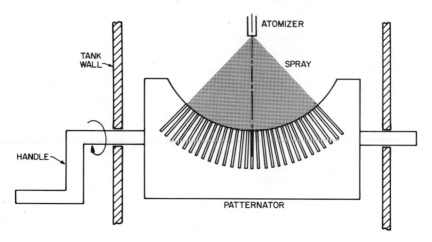

FIGURE 6. Schematic diagram of patternator for measuring radial fuel distribution.

Circumferential Fuel Distribution. The circumferential distribution, or patternation, of a conical spray can be determined by mounting the atomizer vertically in a chamber with the spray discharging downward into a collection tray that is partitioned off into a number of radial segments, each segment draining into a separate sampling tube. The flow is turned off when one sampling tube is full, and the lowest level observed in any one of the other tubes is taken as the patternation reading. A patternation of 80% is generally acceptable for most applications. This corresponds to a 20% variation in circumferential fuel distribution.

1.5. Plain-Orifice Atomizer

The plain-orifice type of nozzle is widely used in Diesel engines and in the afterburners of turbojet engines.

The flow rate of a plain-orifice atomizer is given by the expression

$$\dot{m}_L = 35.1 C_D d_0^2 (SG \Delta P_L)^{0.5} \tag{14}$$

The discharge coefficient, C_D, of a plain hole is dependent on many factors, but especially on the diameter of the hole and its length to diameter ratio, l/d_0. Haenlein[14] provides several graphs illustrating the variation of discharge coefficient with Reynolds number, l/d_0 ratio, orifice diameter, and fuel injection pressure, based mainly on the experimental findings of Gelalles.[42] They show that discharge coefficients reach a maximum at a value of l/d_0 between 4 and 6. This range of l/d_0 ratio also corresponds to the maximum penetration of the fuel jet into stagnant air.

The sprays produced by plain-orifice atomizers have a cone angle that usually lies between 5 and 15 degrees. The cone angle is only slightly affected by the diameter and l/d_0 ratio of the orifice, and is mainly dependent on the viscosity and surface tension of the fuel and the turbulence of the issuing jet. Any increase in turbulence will increase the ratio of radial to axial components of velocity in the jet and thereby increase the cone angle.

In plain-orifice atomizers the mechanism of atomization is one in which the liquid jet is first converted into ligaments and then into drops. Disintegration of the jet is promoted by an increase in flow velocity, which increases both the level of turbulence in the issuing jet and the aerodynamic drag forces exerted by the surrounding medium, and is opposed by an increase in liquid viscosity that resists breakup of the ligaments. These considerations are embodied in the following relationship for the mean drop size of sprays when injected from a plain circular orifice into stagnant air:[43]

$$\text{SMD} = \frac{3.15 \times 10^4 (\nu_L)^{0.2}}{U_L} \tag{15}$$

A more comprehensive equation, due to Tanasawa,[44] is the following:

$$\text{SMD} = 83.2 \frac{d_0}{U_L} \left(\frac{\sigma}{\rho_A} \right)^{0.25} \left(1 + 3.3 \frac{\mu_L}{(\sigma \rho_L d_0)^{0.5}} \right) \tag{16}$$

It should be noted that the above expressions for SMD apply only to situations where liquid is injected into stagnant air. There are two other cases of practical importance:

1. Injection into a co-flowing or contra-flowing stream of air.
2. Transverse injection across a flowing stream of air.

The influence of air or gas velocity is important because the atomization process is not complete immediately the jet leaves the orifice but continues in the surrounding medium until the drop size falls to a critical value below which no further disintegration can occur. For any given fuel, this critical value of drop size depends not on the absolute velocity of the liquid jet but on its velocity relative to that of the surrounding medium. If both are moving in the same direction, penetration is augmented, atomization is retarded, and mean drop diameter is increased. When the movements are in opposite directions penetration is decreased, the cone angle widens, and the quality of atomization is improved. Thus in so far as gaseous flow can affect the formation and development of the spray, and the degree of atomization achieved, it is the relative velocity that should be taken into consideration.

When a plain-orifice atomizer is oriented normal to the air flow, the larger drops penetrate farther into the air stream, and the spectrum of drop sizes produced in the flowing air stream is skewed radially. This distortion of the spray is not necessarily disadvantageous; for example, ignition performance could be improved by locating the igniter in a region of small droplets.

The above discussion on the effects of air motion on the spray characteristics of plain-orifice atomizers is relevant only to situations where the air velocity is not sufficiently high to change the basic nature of the atomization process. If, however, the issuing liquid jet is subjected to a high-velocity air stream, the mechanism of jet disintegration is different and corresponds to airblast atomization. Plain-jet airblast atomizers of this type are used in gas turbines, and their features and spray characteristics are discussed later in this chapter.

1.6. Simplex Atomizer

The narrow spray cone angle exhibited by plain-orifice atomizers makes it unsuitable for most practical applications. Much wider cone angles are achieved in the pressure-swirl atomizer in which a swirling motion is imparted to the fuel so that, under the action of centrifugal force, it spreads out in the form of a hollow cone as soon as it leaves the orifice.

The simplest form of pressure-swirl atomizer is the so-called "simplex" atomizer, as illustrated in Figure 7. Fuel is fed into a swirl chamber through tangential ports that give the fuel a high angular velocity, thereby creating an air-cored vortex. The outlet from the swirl chamber is the final orifice, and the rotating liquid flows through this orifice under both axial and radial forces to emerge from the atomizer in the form of a hollow conical sheet, the actual cone angle being determined by the relative magnitude of the tangential and axial components of velocity at exit.

The development of the spray passes through several stages as the fuel injection pressure is increased from zero:

1. Fuel dribbles from the orifice.
2. Fuel leaves as a thin distorted pencil.
3. A cone forms at the orifice but is contracted by surface tension forces into a closed bubble.
4. The bubble opens into a hollow tulip shape, terminating in a ragged edge where the fuel disintegrates into fairly large drops.
5. The curves surface straightens to form a conical sheet. As the sheet expands its thickness diminishes, and it soon becomes unstable and disintegrates into ligaments and then drops in the form of a well-defined hollow-cone spray.

These five stages of spray development are illustrated in Figure 8. Various types of simplex atomizer have been designed that differ mainly in the method used to impart swirl to the issuing jet.[19] They include swirl chambers with

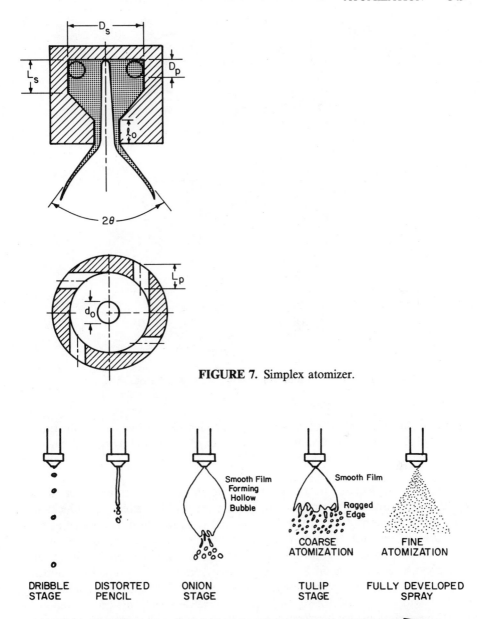

FIGURE 7. Simplex atomizer.

FIGURE 8. Illustrations of the five stages of spray development observed for pressure-swirl atomizers with increase in fuel injection pressure.

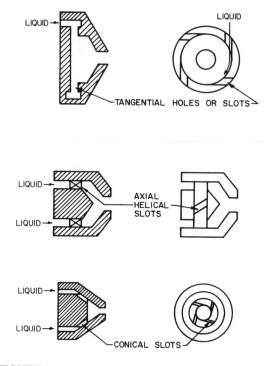

FIGURE 9. Various designs of simplex swirl atomizers.

tangential slots or drilled holes, swirling by helical slots or vanes, as illustrated in Figure 9, and the use of thin swirl plates out of which the swirl chamber and entry ports are cut or stamped.

Discharge Coefficient. The discharge coefficient of a swirl atomizer is inevitably low, due to the presence of the air core, which effectively blocks off the central portion of the orifice. The air core is almost fully developed at fuel pressures of around 400 kPa (60 psi), although the experimental results of Kutty et al.[45] suggest that air-core diameter continues to increase slightly with fuel pressure up to around 3.4 MPa (500 psi). Radcliffe[46] studied the performance of a family of injectors based on common design rules, using fluids that covered wide ranges of density and viscosity. He demonstrated the existence of a unique relationship between the discharge coefficient and the Reynolds number based on orifice diameter. At low Reynolds numbers the effect of viscosity is to thicken the fluid film in the final orifice, and thereby increase the discharge coefficient. With nozzles of small flow number, and for kerosine fuel on cold days at low flow rates, this effect can be significant over the low fuel-flow range associated with light-up and acceleration. However, for Reynolds numbers larger than 3000 (i.e., over most of the normal working range), the discharge coefficient is practically independent of Reynolds number. Thus, for fuels of low viscosity, the convention is to disregard conditions at low Reynolds number and assume that any given atomizer has a constant discharge coefficient.

The first attempt to analyze the hydrodynamics of swirl atomizers was made by Taylor,[47] who used inviscid flow theory to derive the following expression for discharge coefficient

$$C_D^2 = 0.225 \left[\frac{A_p}{D_s d_0} \right] \tag{17}$$

Subsequently, Giffen and Muraszew[14] proposed the following equation for discharge coefficient:

$$C_D^2 = \frac{(1 - X)^3}{(1 + X)} \tag{18a}$$

where X is the ratio of the air-core area to the orifice area.

The values of C_D calculated from the above equation were found to be higher than the experimental values, so a constant K_V was introduced to take account of various losses in the nozzle. Equation (18a) thus becomes

$$C_D = K_V \left[\frac{(1 - X)^3}{(1 + X)} \right]^{0.5} \tag{18b}$$

Eisenklam[48] and Dombrowski and Hasson[49] have used the following dimensionally correct group to correlate their experimental data on discharge coefficients:

$$\frac{A_p}{D_s d_0} \left[\frac{D_s}{d_0} \right]^{1-n}$$

Although Eisenklam states that the value of n varies from 0.1 to 0.5, Dombrowski and Hassan claim that a unique correlation is obtained for $n = 0.5$.

According to Carlisle,[50] Taylor's equation, which is based on inviscid flow theory, can be used to predict experimental data on C_D, provided it is modified to include the effects of D_s/d_0 and L_s/D_s on discharge coefficient. The influence of D_s/d_0 is readily accommodated by the correction factor 0.55 $(D_s/d_0)^{0.5}$. In regard to the L_s/D_s ratio, Carlisle recommends a correction factor of 0.95. Incorporating these correction terms into Eq. (16) gives the relationship

$$C_D^2 = 0.616 \frac{D_s}{d_0} \frac{A_p}{D_s d_0} \tag{19}$$

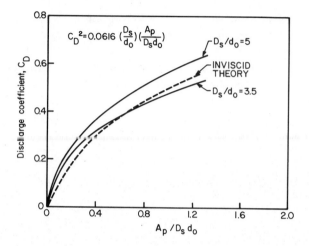

FIGURE 10. Relationship between discharge coefficient and atomizer geometry. Reproduced with permission from Hemisphere Publishing Corp., N. Y. *Atomization and Sprays*, edited by A. H. Lefebvre.[213]

This equation is shown plotted in Figure 10 for two values of D_s/d_0, namely, 5.0 and 3.5. For comparison, also plotted in the figure is the inviscid relationship due to Taylor.[47] More recently, Rizk and Lefebvre[51] have derived the following equation for C_D:

$$C_D = 0.35 \left[\frac{A_p}{D_s d_0} \right]^{0.5} \left[\frac{D_s}{d_0} \right]^{0.25} \tag{20}$$

This relationship is illustrated in Figure 11.

Flow Number. The effective flow area of an atomizer is usually described in terms of a "flow number," which is expressed as the ratio of atomizer throughput to the square root of the fuel-injection pressure differential. Two definitions of flow number are in common use, one based on the mass flow rate and the other on the volume flow rate:

$$FN_{USA} = \frac{\text{flow rate, lb/h}}{(\text{fuel injection pressure differential, psi})^{0.5}} \tag{21}$$

$$FN_{UK} = \frac{\text{flow rate, U.K. gal/h}}{(\text{fuel injection pressure differential, psi})^{0.5}} \tag{22}$$

Note that 1 U.K. gal = 1.2 U.S. gal.

A basic defect of Eqs. (21) and (22) is that, for any given nozzle, the flow number is not constant but varies with the density of the fuel. Thus, although

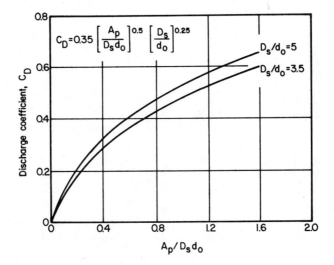

FIGURE 11. Relationship between discharge coefficient and atomizer geometry. Reproduced with permission from Hemisphere Publishing Corp., N. Y. *Atomization and Sprays*, edited by A. H. Lefebvre.[213]

it is customary to stamp or engrave a value of flow number on the body of a simplex atomizer, this value is correct only when the nozzle is flowing a standard calibrating fluid of density 765 kg/m³. In the past this has posed no problems with aircraft gas turbines because 765 kg/m³ roughly corresponds to the density of aviation kerosine. However, for the alternative fuels now being actively studied for gas turbine applications, some of which have densities considerably higher than 765 kg/m³, the use of the above equations to calculate fuel flow rates or fuel injection pressures could result in appreciable error.

The basic deficiency of Eqs. (21) and (22) is the omission of fuel density. Inclusion of this property would not only allow these equations to be rewritten in dimensionally correct form but would also enable the flow number to be defined in a much more positive and useful manner than at present, namely, as the effective flow area of the nozzle. Thus the flow number of any given nozzle would have a fixed and constant value for all types of fuels. If SI units are employed, the flow number in m² is obtained as

$$FN_{SI} = \frac{\text{flow rate, kg/sec}}{(\text{pressure differential, Pa})^{0.5}(\text{liquid density, kg/m}^3)^{0.5}} \quad (23)$$

The standard U.S. and U.K. flow numbers may be obtained from the SI flow number using the formulae

$$FN_{U.S.} = 0.66 \times 10^6 \times \rho_L^{0.5} \times FN_{SI}$$
$$FN_{U.K.} = 0.66 \times 10^8 \times \rho_L^{-0.5} \times FN_{SI}$$

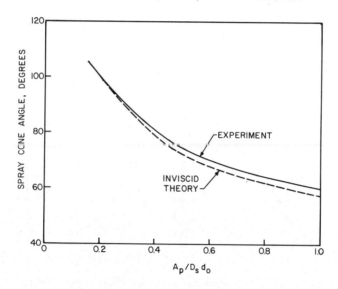

FIGURE 12. Practical relationship between spray cone angle and atomizer geometry.

Spray Cone Angle. The spray cone angle is governed mainly by the swirl port area and the diameters of the swirl chamber and the final orifice. The theoretical relationship between spray cone angle and the geometrical parameter $A_p/D_s d_0$ is shown in Figure 12. This relationship is unique only for nonviscous fluids. It is modified in practice by viscous effects that depend on the form and area of the wetted surface as expressed in the ratios D_s/d_0, L_s/D_s, and l_0/d_0. The predictions of Taylor's inviscid theory are represented by the dotted line in Figure 12, whereas the full line corresponds to experimental data obtained by Carlisle[50] and Giffen.[52] Agreement between theory and experiment is clearly satisfactory at high cone angles, but the theoretical predictions are about 3 degrees too low at a cone angle of 60 degrees.

The influence of fuel viscosity on spray cone angle was examined by Giffen and Massey[52] over a range of viscosities from 2×10^{-6} to 50×10^{-6} m^2/sec. They found the following relationship between cone angle and viscosity for a swirl atomizer:

$$\tan \phi \propto \nu_L^{-0.131} \tag{24}$$

The effects of fuel-injection pressure and ambient pressure on spray angle for large capacity simplex nozzles have been investigated by De Corso and Kemeny.[53] Their results showed that, over a range of injection pressures from 1.7 to 2700 kPa, and a range of gas pressures from 10 to 800 kPa, the spray cone angle is an inverse function of $\Delta P_L(P_A)^{1.6}$. A similar study carried out by Ortman and Lefebvre[54] on kerosine sprayed into stagnant air, produced a similar result, as illustrated in Figure 13. This "collapse" of spray cone angle

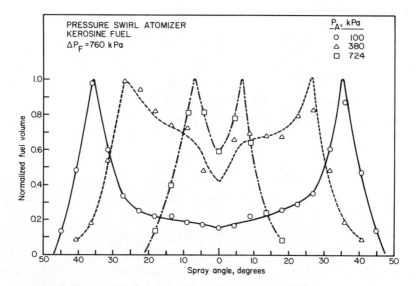

FIGURE 13. Influence of ambient air pressure on spray cone angle.[62]

at high fuel-flow rates and high ambient pressures is a primary cause of the soot formation and smoke associated with this type of nozzle.

As stated above, the cone angle is determined mainly by the value of $A_p/D_s d_0$, but it can also be modified within wide limits by using different orifice edge configurations—radiusing, chamfers, lips, steps, etc.—and this is frequently done to reduce the effects of fuel viscosity.

Effect of Variables on Mean Drop Size. The main factors governing atomization quality are fuel properties, air or gas properties, fuel-injection pressure, and the size and geometry of the atomizer. These various influences are discussed below in turn.

Liquid Properties. The fuel properties of importance are surface tension, viscosity, and density. In practice, the significance of surface tension is diminished by the fact that gas turbine fuels exhibit only minor differences in this property. This is also true for density. However, viscosity varies by almost two orders of magnitude in some applications, so its effect on mean drop size can be appreciable. Very little experimental work has been done on the effect of surface tension on SMD, whereas the influence on density does not appear to have been examined at all. The results obtained on the effect of surface tension are usually expressed by a relationship of the form

$$SMD \propto \sigma^a \qquad (25)$$

Jasuja[55] observed a value for a of 0.6, but this result is not considered significant because the total range of variation of surface tension was only 20%, and this was accompanied by a more than fiftyfold variation in viscosity. Simmons and Harding[56] studied the difference in atomizing performance between water and kerosine for six simplex fuel nozzles. These two liquids have virtually the same viscosity and a 30% difference in density, but the surface tension of water is higher by a factor of 3. Thus, according to Simmons and Harding, any significant differences in SMD could be attributed to differences in surface tension rather than density. For the important practical case of Weber number less than unity, it was found that $a = 0.16$ or 0.19, depending on whether ΔP_L or $\dot{m}_L$ was held constant.

More consistency is observed in the published data on the effect of viscosity on mean drop size. In all cases the experimental data are expressed in the form

$$\text{SMD} \propto \mu_L^b \tag{26}$$

Values for b of 0.16, 0.20, and 0.215 have been reported by Jasuja,[55] Radcliffe,[57] and Knight,[58] respectively. An average of these three values is about 0.2, which gives

$$\text{SMD} \propto \mu_L^{0.2} \tag{27}$$

Air Properties. The two properties of interest are air pressure and temperature. From a fundamental standpoint they are usually regarded only as components of air density. However, this represents a gross oversimplification of the practical situation which, as discussed later, could result in serious error.

Experimental studies on the effect of ambient pressure on mean drop size have produced conflicting results. De Corso[59] found that SMD increased with P_A, and this was confirmed by Neya and Sato[60] whose experimental data conform to the relationship SMD $\propto P_A^{0.27}$. However, the study by Abou-Ellail et al.[61] gave the result SMD $\propto P_A^{-0.25}$, a relationship that subsequent work by Rizk and Lefebvre[62] generally confirmed. For air pressures up to around 0.4 MPa (4 atm), they found that increasing the air pressure reduced the mean drop size according to the relationship SMD $\propto P_A^{-0.1}$, whereas at pressures greater than 0.4 MPa, the effect of air pressure was stronger (SMD $\propto P_A^{-0.28}$). Rizk and Lefebvre suggest that a pressure exponent of -0.25 be used overall. Lefebvre's[63] dimensional analysis of published experimental data on pressure-swirl atomization also led to a value for the air pressure exponent of -0.25.

Air temperature can influence atomization for several reasons. At constant pressure, it affects the air density, which is readily accommodated by plotting the data in terms of density rather than pressure. For constant air density, the Weber number is unaffected by changes in air temperature. However, if a change in air temperature causes the fuel temperature to change also, Weber number will be affected by the change in surface tension. Increases in fuel temperature cause a large decrease in viscosity and a small decrease in surface tension, the result being an improvement in atomization quality.

Abou-Ellail et al.[61] studied the effect of ambient air temperature on mean drop size. They found that drop sizes diminished with increase in air temperature according to the relationship

$$\text{SMD} \propto T_A^{-0.56} \tag{28}$$

This somewhat surprising result is attributed to the fact that "as the fuel is injected in the hot air, it is heated by convection and radiation which lowers its surface tension and hence increases its Weber number. Therefore, to keep the Weber number below a certain critical value, the large droplets break up into smaller droplets to balance the effect of decreasing surface tension as T_A increases." Although not mentioned by these workers, these same mechanisms of heat transfer to the fuel drops should also assist atomization by lowering the fuel viscosity.

In their experiments on simplex swirl atomizers, Dodge and Biaglow[64] found that air temperature had no effect on the initial mean drop size close to the nozzle if density rather than pressure was used to correlate the data, and if viscosity changes due to fuel heating were accounted for. Further downstream, increasing the air temperature led to higher SMD's due to preferential evaporation of the smaller drops.

Fuel-Injection Pressure. Increase in the fuel-injection pressure differential causes the fuel to be discharged from the atomizer at a higher level of velocity. This raises the Weber number, thereby promoting finer atomization. The effect may be expressed quantitatively as

$$\text{SMD} \propto \Delta P_L^{-c} \tag{29}$$

Reported values of c include 0.275, 0.35, 0.4, and 0.43, due to Simmons,[65] Abou-Ellail et al.,[61] Radcliffe,[57] and Jasuja,[55] respectively.

Atomizer Dimensions. The work of Lefebvre and his colleagues[27,28,31] has demonstrated that the dimension of most importance to atomization is the thickness of the liquid sheet as it leaves the final orifice. SMD is roughly proportional to the square root of liquid sheet thickness so, provided the other key parameters that are known to affect atomization are maintained constant, an increase in atomizer size will reduce atomization quality according to the relationship SMD $\propto d_0^{0.5}$. However, although this relationship is generally supported by the results of dimensional analysis, to the author's knowledge the influence of atomizer size has not yet been subjected to systematic investigation.

Drop-Size Relationships. Empirical drop-size relationships have been derived by many workers.[9,57,58] They are usually of the form

$$\text{SMD} \propto \sigma^a \nu_L^b \dot{m}_L^c \Delta P_L^d \tag{30}$$

One of the earliest and most widely quoted expressions is that of Radcliffe,[57]

$$\text{SMD} = 7.3 \sigma^{0.6} \nu_L^{0.2} \dot{m}_L^{0.25} \Delta P_L^{-0.4} \tag{31}$$

More recent work by Jasuja[55] yielded the following expression:

$$\text{SMD} = 4.4 \sigma^{0.6} \nu_L^{0.16} \dot{m}_L^{0.22} \Delta P_L^{-0.43} \tag{32}$$

This investigation covered a range of liquid viscosities from 1.0×10^{-6} to 93.0×10^{-6} m^2/sec, which gives substance to the exponent for ν_L of 0.16. However, as noted earlier, the variation of surface tension in these experimentals was very small and was accompanied by wide variations in viscosity. Thus the exponent of 0.6, which presumably was chosen to be consistent with the previously reported value of 0.6 by Radcliffe,[57] has no special significance in Eq. (32).

Drop-Size Analysis. The atomization process in simplex fuel nozzles is governed by the relative magnitude of two opposing forces: (1) the aerodynamic forces that tend to break up the liquid sheet formed at the nozzle exit into ligaments and drops, and (2) the surface tension forces that oppose sheet disintegration. The relative magnitude of these opposing forces may be expressed quantitatively as the Weber number. As mentioned earlier, the fuel thickness at the final orifice is of prime importance and should be designated as the characteristic dimension in the Weber number. Thus we have

$$\frac{\text{SMD}}{t} \propto \text{We}^{-x} \propto \left(\frac{U_L^2 \rho_A t}{\sigma} \right)^{-x} \tag{33}$$

This equation suggests that SMD is independent of liquid viscosity. However, Rizk and Lefebvre[31] have shown that t varies with viscosity according to the relationship

$$\frac{t}{d_0} \propto \left(\frac{U_L \rho_L d_0}{\mu_L} \right)^{-y} \tag{34}$$

Substituting for t from Eq. (34) into Eq. (31) gives

$$\text{SMD} \propto \left(\frac{\sigma}{U_L^2 \rho_A} \right)^x \left(\frac{\mu_L}{U_L \rho_L} \right)^{y(1-x)} d_0^{(1-x)(1-y)} \tag{35}$$

Analysis of the available experimental data, in which added weight is given to the results that covered the widest range of any particular variable or were obtained by the most reliable measuring technique, suggests "best" values for x and y of 0.25 and 0.333, respectively. Substitution of these values into Eq. (35) and also for $U_L^2 = 2\Delta P_L/\rho_L$ gives

$$\text{SMD} = A\sigma^{0.25}\mu_L^{0.25}\rho_L^{0.125}d_0^{0.5}\Delta P_L^{-0.375}\rho_A^{-0.25} \tag{36}$$

The value of 0.25 for the surface tension exponent is reasonably close to Simmons' estimates (0.16 to 0.19). The exponent of 0.25 for viscosity agrees well with all reported values. No reliable data exist to check the exponent for liquid density, but it is generally agreed that the effect of this property on SMD is quite small. The d_0 exponent of 0.5 confirms the theoretical value of 0.5. The exponents for injection pressure differential and air density of -0.375 and -0.25, respectively, are identical to the experimental values obtained by Abou-Ellail et al.[61] and Rizk and Lefebvre.[62] Substituting in Eq. (36) for $d_0 \propto \dot{m}_L^{0.5}/(\Delta P_L\rho_L)^{0.25}$, and using Jasuja's data[55] to determine the value of A, gives

$$\text{SMD} = 2.25\sigma^{0.25}\mu_L^{0.25}\dot{m}_L^{0.25}\Delta P_L^{-0.5}\rho_A^{-0.25} \tag{37}$$

Spray Penetration. As discussed earlier, the penetration of a spray is much higher than that of a single drop due to the favorable gas currents induced by the leading drops, which facilitate the penetration of the remaining drops in the spray. However, in the absence of any detailed information on spray penetration, some idea of the relative importance of the variables involved can be gained by considering the penetration of a single drop. Lefebvre[66] has shown that the penetration of a liquid drop of diameter D, when injected into stagnant gas, may be expressed as

$$S \propto \Delta P_L^{0.42}D^{1.84}P_A^{-0.16}\rho_L^{-0.42} \tag{38}$$

The terms of most practical interest in this expression are fuel pressure and gas pressure. The effect of an increase in fuel-injection pressure is twofold. By discharging the fuel at higher velocity, it tends to increase spray penetration. However, the finer atomization produced by higher fuel-injection pressure increases the surface area, and hence the aerodynamic drag of the spray, and thereby tends to reduce spray penetration.

For any given simplex atomizer we have, from Eq. (36)

$$\text{SMD} \propto \Delta P_L^{-0.375} \tag{39}$$

Substitution in Eq. (38) gives

$$S \propto \Delta P_L^{-0.22} \tag{40}$$

namely, the net effect of an increase in fuel-injection pressure is to reduce the penetration of the spray.

The influence of air or gas pressure on penetration was also examined by Lefebvre[66] who showed that, for any given gas turbine engine, the penetration distance of the spray is inversely proportional to the cube root of the air pressure:

$$S \propto P_A^{-0.33} \tag{41}$$

This reduction in spray penetration with increase in air pressure, coupled with the corresponding decrease in spray cone angle, can lead to excessive concentrations of fuel in the region adjacent to the atomizer. As discussed elsewhere, the most serious consequence of this high fuel concentration at high chamber pressures is an increase in soot formation, leading to higher flame radiation and a smokier exhaust.

1.7. Wide-Range Atomizers

A major drawback of the simplex atomizer is that its flow rate varies as the square root of the injection pressure differential. Thus doubling the flow rate demands a fourfold increase in fuel-injection pressure. For kerosine the minimum injection pressure for satisfactory atomization quality is about 100 kPa. This means that an increase in fuel-flow rate to some 20 times the minimum value would require an injection pressure of 40 MPa. This basic drawback of the simple atomizer has led to the development of various "wide-range" atomizers such as duplex, dual-orifice, and spill atomizers, in which ratios of maximum to minimum fuel output in excess of 20 can readily be achieved with fuel-injection pressures not exceeding 7 MPa. These various designs have been compared by Mock and Ganger,[67] Carey,[68] Radcliffe,[46,57] and Dombrowski and Munday.[9]

Duplex. The main feature that distinguishes the duplex nozzle from the simplex nozzle is that its swirl chamber employes two sets of tangential swirl ports, one set being the primary ports for low flows, while the other set comprises the secondary passage for the large flows met with under normal operating conditions. The primary swirl ports are the first to be supplied with fuel from the primary manifold, while a spring-loaded pressurizing valve prevents fuel from entering the secondary fuel manifold. Only when a predetermined injection pressure has been reached does the valve open, and fuel is then supplied through primary and secondary swirl ports simultaneously. The essential features of one type of duplex nozzle are shown in Figure 14. Its flow characteristics are illustrated in Figure 15. This figure serves to illustrate the superior performance of the duplex nozzle, especially at low fuel flows. Consider, for example, a condition where the fuel-flow rate is 10% of the maximum value. Inspection of Figure 15 shows that the fuel-injection pressure

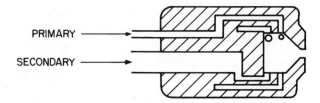

FIGURE 14. Duplex atomizer.

for the duplex nozzle is about eight times higher than that of the simplex nozzle. According to Eq. (36) this should improve atomization quality almost threefold. This same advantage of better atomization at low fuel flows applies equally well to the dual-orifice atomizer which is described in the next section.

A drawback of the duplex atomizer is a tendency for the spray cone angle to be smaller in the combined flow range than in the primary flow range by about 20 degrees. This is because in going from the primary flow to the combined flow, the ratio $(A_p/D_s d_0)$ is increased and the spray cone angle thereby reduced, (see Figure 12). In some designs this problem is overcome by setting the primary swirl ports on a smaller tangent circle than the secondary ports. The effect of this is to reduce the swirl component and hence the spray cone angle at low fuel output.

The design procedure for duplex atomizers is the same as for simplex atomizers, except that special treatment is required if the primary ports are to be set on a smaller tangent circle than the main ports to give a constant spray cone angle.

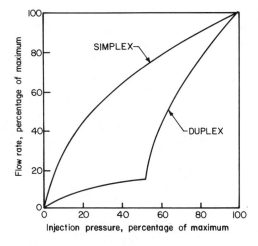

FIGURE 15. Flow characteristic of simplex and duplex atomizers.

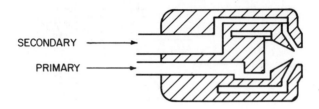

FIGURE 16. Dual-orifice atomizer.

Dual-Orifice. The dual-orifice atomizer, sometimes known as the duple atomizer, is shown diagrammatically in Figure 16. Essentially it comprises two simplex nozzles that are fitted concentrically, one inside the other. The primary nozzle is mounted on the inside and the juxtaposition of primary and secondary is such that the primary spray does not interfere with either the secondary orifice or the secondary fuel within the orifice. At low fuel flows all the fuel is supplied from the primary nozzle, which has a low flow number. However, when the fuel pressure has reached a certain predetermined value, a pressurizing valve opens and admits fuels to the secondary nozzle, which has a high flow number. The pressurizing valve opening pressure varies with different engines between 1.4 and 3 MPa. When the fuel delivery is low, it all flows through the primary nozzle, and atomization quality tends to be high because a high fuel pressure is needed to force the fuel through the small ports in the primary swirl chamber. As the fuel flow is increased, a fuel pressure is eventually reached at which the pressurizing valve opens and admits fuel to the secondary. At this point atomization quality is poor because the secondary fuel pressure is low. With further increase in fuel flow, the secondary fuel pressure increases, and atomization quality starts to improve. However, there is inevitably a range of fuel flows, starting from the point at which the pressurizing valve opens, over which atomization quality is relatively poor. In order to alleviate this problem, it is customary to arrange for the primary spray cone angle to be slightly wider than the secondary spray cone angle so that the two sprays coalesce and share their energy within a short distance from the atomizer. This helps the situation to some extent, but the atomization quality is still unsatisfactory. Thus the designer must make sure that opening of the pressurizing valve does not coincide with an engine operating point at which high combustion efficiency and low pollutant emissions are prime requirements. Carlisle[69] has analyzed the atomizing conditions in this critical flow range, starting from the assumption that the primary and secondary sprays coalesce and share momenta. The result of this analysis shows that

$$\Delta P_e = \left(\frac{4\Delta P_p}{R^2}\right)\left((1 + R)^{0.5} - 1\right)^2 \tag{42}$$

where

ΔP_e = the equivalent fuel injection pressure of combined spray
ΔP_p = primary injection pressure
R = ratio of secondary to primary flow number

Thus for any combination of primary and secondary flow numbers, and any given pressurizing valve-opening pressure, a minimum value ΔP_e, the equivalent injection pressure of the combined spray, can be calculated and used in Eq. (37) for estimating mean drop size. (For kerosine, the value of ΔP_e should not be allowed to fall below 140 kPa if high combustion efficiency is to be maintained.) The fuel flow corresponding to worst atomization occurs when the ratio of total fuel flow to primary fuel flow is equal to $(1 + R)^{0.5}$.

Spill. The spill atomizer is basically a simplex atomizer, except that the rear wall of the swirl chamber, instead of being solid, contains an annular passage through which fuel can be "spilled" away from the atomizer, as shown in Figure 17. Its basic features have been described by Carey[68] and Rizk and Lefebvre.[70] Fuel is supplied to the swirl chamber at high pressure and high flow rate. The fraction of fuel injected into the combustion zone is varied and controlled by a valve located in the spill-return line. As the fuel demand decreases with increase in altitude, or reduction in engine speed, more fuel is spilled away from the swirl chamber, leaving less to pass through the atomizing orifice. The spill atomizer's constant use of a relatively high pressure means that even at extremely low fuel flows, there is adequate swirl to provide efficient atomization of the fuel. According to Carey[68] satisfactory atomization can be achieved even when the fuel flow rate is as low as 1% of its maximum value, and in general, the tendency is for atomization quality to improve as the fuel flow is reduced.

Radcliffe[57] gives the following empirical relationship for the mean drop diameter of kerosine sprays with the spill line closed:

$$\text{SMD} = 0.0157 \dot{m}_L^{0.318} \Delta P_L^{-0.53} \tag{43}$$

The wide range of fuel flows over which atomization quality is high is a most useful characteristic of the spill-return atomizer. Other attractive features include an absence of moving parts and, because the flow passages are designed to handle large flows all the time, freedom from blockage by contaminants in the fuel.

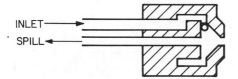

INLET
SPILL

FIGURE 17. Schematic diagram of spill-return atomizer.

A disadvantage of the spill-return atomizer is the large variation in spray angle with change in fuel flow. The effect of a reduction in flow rate is to lower the axial component of velocity without affecting the swirl component; consequently the spray angle widens. The spray angle at minimum flow can be up to 50 degrees wider than at maximum fuel flow. From a combustion viewpoint this is very unsatisfactory; a spray angle that narrows with reduction in fuel flow would be much more desirable.

Another disadvantage of the spill system is that problems of metering the fuel are more complicated than with other types of atomizers, and a larger capacity pump is needed to handle the large recirculating flows. For these reasons interest in the spill-return atomizer has declined in recent years. However, if the aromatic content of gas turbine fuels continues to rise, it could pose serious problems due to blockage, by gum formation, of the fine passages of conventional pressure atomizers. The spill-return atomizer, having no small passages, is virtually free from this defect. This factor, combined with its excellent atomizing capability, makes it an attractive proposition for dealing with the various alternative fuels now being considered for gas turbine applications, most of which have high aromaticity, high viscosity and low volatility.

1.8. Fan-Spray Atomizers

Flat sprays are usually obtained by cutting slots into plane or cylindrical surfaces and by arranging for the fuel to flow into the slot from two opposite directions. In the single-hole fan-spray injector, there is only a single shaped orifice that forces the liquid into opposing streams within itself so that a flat spray issues from the orifice and spreads out in the shape of a sector of a circle of about 75 degree angle. An air shroud is usually fitted around the nozzle to provide both air assistance in atomization and air scavenging of the nozzle tip, using the pressure differential across the liner wall.

The behavior of flat sprays has been investigated by Dombrowski and Fraser[8] and Lewis.[71] The results obtained by Lewis suggest that the mean drop diameter for fan-spray injectors should correlate in the following form:

$$\text{SMD} = f\left[\text{FN}^{0.25}\Delta P_L^{-0.18}\sigma^{0.25}\rho_L^{0.18}\rho_A^{0.2}\right] \tag{44}$$

The larger drop sizes obtained at higher air pressures are attributed by Lewis to sheet disintegration occurring closer to the injector face so that drops are produced from a thicker sheet initially. Lewis also comments that ambient pressure has little effect on the spray angle obtained with fan-spray injectors.

Fan-spray atomizers lend themselves ideally to small annular combustors because they provide a good lateral spread of fuel and allow the number of injection points to be minimized. The AVCO Lycoming Company has manu-

AIR

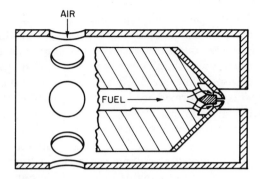

FUEL ⟶

FIGURE 18. Avco-Lycoming fan-spray injector.

factured flat-spray atomizers in several versions. One type is illustrated, in a much simplified form, in Figure 18.

1.9. Rotary Atomizers

In the rotary atomizer, liquid is fed onto a rotating surface where it spreads out fairly uniformly under the action of centrifugal force. The rotating surface may take the form of a flat disc, top, vaned disc, cup, or slotted wheel. Diameters vary from 25 to 450 mm, the small discs rotating up to 60,000 rpm, whereas the larger discs rotate up to 12,000 rpm, with atomizing capacities up to 1.4 kg/sec.[9] Where a coaxial air blast is used to assist atomization, lower speeds of the order of 3000 rpm may be used. The system has extreme versatility and has been shown to atomize successfully liquids varying widely in viscosity. An important asset is that the thickness and uniformity of the liquid sheet can be readily controlled by regulating the liquid flow rate and the rotational speed.

The process of centrifugal atomization has been studied by several workers, including Hinze and Milborn,[72] Dombrowski and Fraser,[8] and Nukiyama and Tanasawa,[10] and has been described in detail in a review of atomization methods by Dombrowski and Munday.[9]

Several mechanisms of atomization are observed with a rotating flat disc, depending on the flow rate of the liquid and the rotational speed of the disc. At low flow rates the liquid spreads out across the surface and is centrifuged off as discrete drops of uniform size, each drop drawing behind it a fine ligament. The drops finally separate from the ligaments, which are themselves converted into a series of fine drops of fairly uniform size. This process is essentially a discontinuous one and occurs from place to place at the periphery of the rotating cup or disc. If the flow rate is increased, the atomization process remains basically the same, except that ligaments are formed along the entire periphery and are larger in diameter. With further increase in flow rate the condition is eventually reached when the ligaments can no longer accommodate the flow of liquid, and a thin continuous sheet is formed that extends from the lip until an equilibrium condition is achieved—at which the contraction force at the free edge due to surface tension is just equal to the kinetic energy of the advancing sheet. A thick rim is produced that again disintegrates into ligaments and drops. However, because the rim has no controlling solid surface, the ligaments are formed in an irregular manner, which results in appreciable variation in drop size. When the peripheral speed of a rotating disc becomes very high, more than 50 m/sec, say, the liquid appearing at the edge of the disc is immediately atomized by the surrounding air.

Fraser et al.[73] have demonstrated that, except for a limited range of flow conditions, a spinning cup is not capable of smoothing out the flow of liquid over its surface under the action of centrifugal force, and that sheet uniformity is critically dependent on the method of feed distribution. It was also found that under normal conditions the sheet thickness at the point of free disintegration is independent of rotary speed or cup diameter and is a function of flow rate only.

From the mechanism for rotary atomization as outlined above, one would expect the mean drop size to diminish with increase in rotational speed and increase with flow rate and fuel viscosity. This view is confirmed by the various theoretical and experimental investigations that have been carried out on the factors governing the atomization properties of rotary atomizers and which have resulted in a number of drop-size relationships.[9] One that appears to encompass the widest range of parameters is the following due to Kayano and Kamiya:[74]

$$\text{SMD} = 0.016 Q_F^{0.32} N^{-0.79} D_d^{-0.69} \rho_F^{-0.29} \sigma^{0.26} \left(1 + \mu_F^{0.65}\right) \qquad (45)$$

1.10. Air-Assist Atomizers

In this chapter reference has been made several times to the beneficial effect of flowing air in assisting the disintegration of a liquid jet or sheet. Examples include the use of shroud air on fan-spray and simplex nozzles. As discussed earlier, a basic drawback of the simplex nozzle is that if the swirl ports are sized to pass the maximum fuel flow at the maximum fuel injection pressure, then the fuel pressure differential will be too low to give good atomization at the lowest fuel flow. This problem can be overcome by using dual-orifice or duplex nozzles, but an alternative approach is to retain the simplex swirl atomizer, which can always give good atomization at high fuel flows, and then

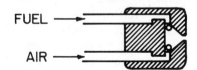

FIGURE 19. Internal mixing air-assist atomizer.

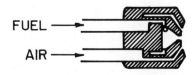

FUEL ⟶

AIR ⟶

FIGURE 20. External mixing air-assist atomizer.

use air or steam to augment the atomization process at low fuel flows. A wide variety of designs have been produced for use in industrial gas turbines and oil-fired furnaces. Useful descriptions of these have been provided by Gretzinger and Marshall,[75] Yeager and Coffin,[76] and Mullinger and Chigier.[77] In all designs a high-velocity gas stream impinges on a relatively low-velocity fuel stream, either internally, as shown in Figure 19, or externally, as shown in Figure 20.

With the internal-mixing type the spray cone angle is a minimum for maximum air flow; the spray opens up as the air flow is reduced. This type of atomizer is very suitable for highly viscous fuels, and good atomization can be obtained right down to zero fuel flow. External-mixing types can be designed to give a constant spray angle at all fuel flows and have the advantage that there is no danger of fuel finding its way into the air line. However, their utilization of air is less efficient, and consequently their power requirements are higher.

Several drop-size equations have been derived to correlate experimental data on various fuel nozzle types that fall into the category of "air assist". These expressions are included in Table 2.

The main drawback to air-assist atomizers from a gas turbine standpoint is the need for an external supply of high-pressure air. This virtually rules it out for aircraft applications. For large industrial engines it is a much more attractive proposition, especially since the high-pressure air is needed only during engine light-up and acceleration.

1.11. Airblast Atomizers

In principle airblast atomizers function in exactly the same manner as air-assist atomizers because both types employ the kinetic energy of a flowing air stream to shatter the fuel jet or sheet into ligaments and then drops. The main difference between the two systems lies in the quantity of air employed and its atomizing velocity. With the air-assist nozzle, where the air is supplied from a compressor or a high-pressure cylinder, it is important to keep the air flow rate down to a minimum. However, since there is no special restriction on air pressure, the atomizing air velocity can be made very high. Thus air-assist atomizers are characterized by their use of a relatively small quantity of very high-velocity air. However, because the air velocity through an airblast atomizer is limited to a maximum value (usually around 120 m/sec), corresponding to the pressure differential across the liner wall, a larger amount of air is required to achieve good atomization. However, this air is not wasted. After

TABLE 2. Drop-Size Equations for Miscellaneous Airblast Atomizers

Investigators	Atomizer Type	Equation	Remarks
Wigg[82]	Thin sheet, external airblast	$\mathrm{MMD} = 20\left(\dfrac{\nu_L^{0.5}\,\dot{m}_L^{0.1}h^{0.2}\sigma^{0.2}}{\rho_A^{0.3}U_R}\right)\left(1+\dfrac{\dot{m}_L}{m_A}\right)^{0.5}$	h = height of air annulus
Wetzel and Marshall[80]		$\mathrm{MMD} = 2.0\left(\dfrac{d_0^{0.35}}{U_R^{1.68}}\right)$	
Gretzinger and Marshall[75]	Impingent nozzle	$\mathrm{MMD} = 1.22\times10^{-4}\left(\dfrac{\mu_A}{\rho_A U_A L}\right)^{0.15}\left(\dfrac{\dot{m}_L}{\dot{m}_A}\right)^{0.6}$	L = diameter of wetted periphery. Drop size independent of liquid viscosity but dependent on air viscosity
Kim and Marshall[18]	Double concentric airblast	$\mathrm{MMD} = 2.62\left[\dfrac{\sigma^{0.41}\mu_L^{0.32}}{(\rho_A U_A^2)^{0.72}\rho_L^{0.16}}\right] + 0.00106\left(\dfrac{\mu_L}{\sigma\rho_L}\right)^{0.17}\left(\dfrac{\dot{m}_A}{\dot{m}_L}\right)^{m}\dfrac{1}{U_R^{0.54}}$	$m = -1$ at $\dot{m}_A/\dot{m}_L < 3$; $m = -0.5$ at $\dot{m}_A/\dot{m}_L > 3$
Sakai et al.[130]	Internal mixing air-assist	$\mathrm{SMD} = 14\times10^{-6}D_h^{0.75}\left(\dfrac{\dot{m}_L}{\dot{m}_A}\right)^{0.75}$	Water flow rates up to 100 kg/hr; $(\dot{m}_L/m_A)$ from 5 to 100

atomizing the fuel, it flows into the primary zone where it provides part of the air required for primary combustion.

Airblast atomizers have many advantages over pressure atomizers, especially in their application to gas turbine engines of high-pressure ratio. They require lower fuel pressures and produce a finer spray. Moreover, because the airblast atomization process ensures thorough mixing of air and fuel, the ensuing combustion process is characterized by very low soot formation and a blue flame of low luminosity, resulting in relatively cool liner walls and a minimum of exhaust smoke. A further asset of the airblast atomizer is that it provides a sensibly constant fuel distribution over the entire range of fuel flows. This offers an important practical advantage in that the temperature distribution in the chamber efflux gases, which is determining to turbine blade life at higher pressures, may be adequately predicted from temperature surveys carried out at lower and more convenient levels of pressure.

The merits of the airblast atomizer have led is recent years to its installation in a wide variety of industrial and aircraft engines. Most of the systems now in service are of the "prefilming" type, in which the fuel is first spread out into a thin, continuous sheet and then subjected to the atomizing action of high-velocity air. In other designs the fuel is injected into the high-velocity airstream in the form of one or more discrete jets. In all cases the basic objective is the same, namely, to deploy the available air in the most effective manner to achieve the best possible level of atomization.

During the past few decades numerous experimental studies have been carried out on many different types of airblast atomizers. In the following sections the result of these investigations will be reviewed and the main conclusions present in a form that may be useful to combustion engineers and particularly to those involved in atomizer performance and design.

Experimental Studies. The first major study of airblast atomization was conducted over 40 years ago by Nukiyama and Tanasawa[78] on plain-jet airblast atomizers of the type illustrated in Figure 21. The drop sizes were measured by collecting samples of the spray on oil-coated glass slides. Liquid properties were investigated over the following ranges:

Viscosity	0.001–0.050 kg/msec
Surface tension	0.019–0.073 kg/sec^2
Density	700–1200 kg/m^3

Drop-size data were correlated by the following experimental equation for SMD:

$$\text{SMD} = \frac{0.585}{U_R}\left(\frac{\sigma}{\rho_L}\right)^{0.5} + 53\left(\frac{\mu_L^2}{\sigma\rho_L}\right)^{0.225}\left(\frac{Q_L}{Q_A}\right)^{1.5} \tag{46}$$

It is of interest to note that the right-hand side of Eq. (46) is expressed as the sum of two separate terms, one of which is dominated by air velocity and

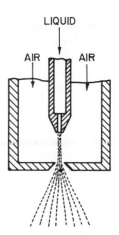

FIGURE 21. Nukiyama-Tanasawa atomizer.[78]

the other by viscosity. More recent work by Lefebvre and co-workers[27-34] has confirmed the validity of this form of equation for SMD.

Although Eq. (46) is not dimensionally correct, it nevertheless allows some useful conclusions to be drawn. For example, it shows that for liquids of low viscosity the mean drop size is inversely proportional to the relative velocity between the air and liquid, whereas for large values of air/liquid ratio, the influence of viscosity on SMD becomes negligibly small.

From dimensional analysis it appears that in order to make Eq. (46) dimensionally correct, all that is needed is the introduction of a term to denote length, raised to the power of 0.5. One obvious choice for this length would be the diameter of the liquid orifice or air nozzle. However, from tests carried out with different sizes and shapes of nozzles and orifices, Nukiyama and Tanasawa[78] concluded that these factors have virtually no effect on mean drop size. Thus the absence of atomizer dimensions is a notable feature of Eq. (46).

Another significant omission is air density, which was kept constant (at the normal atmospheric value) in all experiments. This represents a serious limitation of Eq. (46) since it prohibits its application to the many types of combustion system that are required to operate over wide ranges of air pressure and temperature.

During the past 40 years many investigations have been conducted on the process of airblast atomization. The most important results of these early studies are contained in the papers of Kim and Marshall,[18] Fraser et al.,[73] Gretzinger and Marshall,[75] Weiss and Worsham,[79] Wetzel and Marshall,[80] Wigg,[81,82] Mayer,[83] and Ingebo and Foster.[84]

Since the early 1960s a series of experimental studies on airblast atomization have been conducted at Cranfield.[27-34,55,85] This continuing research program was prompted by a recognition of the deficiencies of conventional pressure atomizers, in terms of excessive flame radiation and exhaust smoke, in their application to gas turbine engines of high compression ratio.

In an attempt to elucidate the key factors involved in the design of airblast atomizers, Lefebvre and Miller[85] carried out a large number of tests, using both water and kerosine on several different atomizer configurations. Drop sizes were measured by collecting the drops on magnesium-oxide-coated slides. Their experimental data confirmed the effect of increasing air velocity in reducing mean drop size and also showed that variation in air/liquid ratio over the range from 3.0 to 9.0 had little effect on atomization quality. However, the main conclusion from this study was that:

> Minimum drop sizes are obtained by using atomizers designed to provide maximum physical contact between the air and the liquid. In particular it is important to arrange that the liquid sheet formed at the atomizing lip is subjected to high velocity air on *both* sides. This also ensures that the droplets remain airborne and avoids deposition of liquid on solid surfaces.

Lefebvre and Miller also stressed the importance of spreading the liquid into the thinnest possible sheet before subjecting it to airblast action, on the grounds that "any increase in the thickness of the liquid sheet flowing over the atomizing lip will tend to increase the thickness of the ligaments which, upon disintegration, will then yield drops of larger size."

The main value of this study stems not so much from the experimental data obtained, which are necessarily suspect due to the relatively crude measuring techniques employed, but in its clear identification of the essential features of successful nozzle design, which may be summarized as follows:

1. The liquid should first be spread into a thin continuous sheet of uniform thickness.
2. Finest atomization is obtained by producing a liquid sheet of minimum thickness. In practice, this means that the atomizing lip should have the largest possible diameter.
3. The annular liquid sheet formed at the atomizing lip should be exposed on both sides to air at the highest possible velocity. Thus the atomizer should be designed to achieve minimum loss of total pressure in the airflow passages and maximum air velocity at the atomizing lip.

Rizkalla and Lefebvre[27,28] used the light-scattering technique in a detailed investigation of airblast atomization. A cross-sectional drawing of the atomizer employed is shown in Figure 22. In this design the liquid flows through six equispaced tangential ports into a weir, from which it spills over the prefilming surface before being discharged at the atomizing lip. In order to subject both sides of the liquid to high-velocity air, two separate airflow paths are provided. One airstream flows through a central circular passage and is deflected radially outward by a pintle before striking the inner surface of the liquid sheet. The other airstream flows through an annular passage surrounding the main body of the atomizer. This passage has its minimum flow area in the plane of the

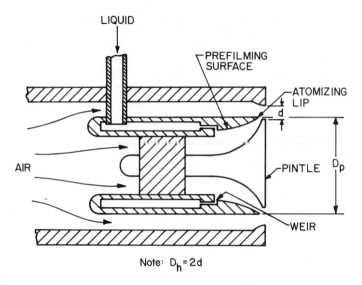

FIGURE 22. Prefilming type of airblast atomizer.[28, 34]

atomizing lip in order to impart a high velocity to the air where it meets the outer surface of the liquid sheet.

In the first phase of their program, Rizkalla and Lefebvre[27] examined the effects of changes in liquid properties, namely, viscosity, surface tension, and density, on mean drop size. In the second phase of their work they investigated the effect of changing airflow properties by varying the air temperature using a kerosine-fired preheater located upstream of the atomizer. The influence of air pressure on SMD was also examined.

From inspection of all the data obtained on the effects of air and liquid properties on atomization quality,[27,28] Rizkalla and Lefebvre drew certain general conclusions concerning the main factors governing mean drop size. For liquids of low viscosity, the key factors are surface tension, air velocity, and air density. From results obtained over a wide test-range, it was concluded that liquid viscosity has an effect that is quite separate and independent from that of surface tension. This suggested a form of equation in which size is expressed as the sum of two terms, the first term being dominated by surface tension and air momentum, and the second term by liquid viscosity. Dimensional analysis was used to derive the following equation, the various constants and indices being deduced from the experimental data:

$$\text{SMD} = 3.33 \times 10^{-3} \left(\frac{\sigma \rho_L D_p}{\rho_A^2 U_A^2} \right)^{0.5} \left(1 + \frac{\dot{m}_L}{\dot{m}_A} \right)$$

$$+ (13.0 \times 10^{-3}) \left(\frac{\mu_L^2}{\sigma \rho_L} \right)^{0.425} D_p^{0.575} \left(1 + \frac{\dot{m}_L}{\dot{m}_A} \right)^2 \qquad (47)$$

where L_c, the characteristic dimension of the atomizer, is made equal to D_p.

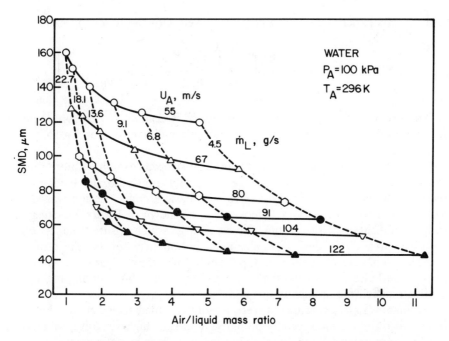

FIGURE 23. Influence of air/liquid ratio on mean drop size.[28]

For liquids of low viscosity, such as water and kerosine, the first term predominates, and SMD thus increases with increase in liquid surface tension, liquid density, atomizer dimensions, and liquid/air ratio, and declines with increase in air velocity and air density. With liquids of high viscosity, the second term acquires greater significance, and in consequence SMD becomes less sensitive to variations in air velocity and density. The deleterious effect of a decrease in air/liquid ratio is clearly brought out in Figure 23. This figure has direct relevance to the design of airblast atomizers because it highlights two main points:

1. Atomization quality starts to decline when the air/liquid ratio falls below about 4, and deteriorates quite rapidly at air/liquid ratios below about 2.
2. Once the air/liquid ratio exceeds about 5, only very slight improvement in atomization quality is gained by the addition of more air.

The range of variables tested by Rizkalla and Lefebvre[27,28] was; air velocity from 55 to 125 m/sec, air temperature from 296 to 424 K, air pressure from 100 to 850 kPa, air/liquid mass ratio from 1 to 10, liquid viscosity from 0.001 to 0.044 kg/ms, and surface tension from 0.024 to 0.074 kg/sec².

Jasuja[55] carried out an experimental study of the airblast atomization characteristics of kerosene, gas oil, and various blends of gas oil with residual fuel oil. His experimental data correlated well with the following expression:

$$\text{SMD} = 10^{-3}\left[1 + \frac{\dot{m}_L}{\dot{m}_A}\right]^{0.5} \times \left[\frac{(\sigma\rho_L)^{0.5}}{\rho_A U_A} + 0.06\left(\frac{\mu_L^2}{\sigma\rho_A}\right)^{0.425}\right] \qquad (48)$$

This equation is almost identical to that of Rizkalla and Lefebvre,[28] except for the absence of a term to represent atomizer dimensions and a somewhat lower dependence of SMD on air/liquid ratio.

The effect of atomizer scale on mean drop size was examined more directly by El-Shanawany.[33] He used three geometrically similar nozzles having cross-sectional areas in the ratio of 1 : 4 : 16. The basic atomizer design was similar to that used by Rizkalla and Lefebvre,[27,28] as shown in Figure 22, except for some modifications to the inner airflow passage which ensured that its cross-sectional area diminished gradually and continuously toward the atomizing lip. These changes eliminated any possibility of airflow separation within the passage and also encouraged the liquid film to adhere to the prefilming surface until it reached the atomizing lip.

El-Shanawany's experiments were mainly confined to water and kerosine, but he also used some specially prepared liquids of high viscosity. Mean drop sizes were measured by the light-scattering technique in its improved form, as developed by Lorenzetto.[30] His results for water showed that SMD $\propto D_h^{0.42}$. From analysis of all the experimental data, El-Shanawany and Lefebvre[34] concluded that the mean drop sizes produced by prefilming airblast atomizers could be predicted by the following dimensionally correct equation:

$$\frac{\text{SMD}}{D_h} = \left[1 + \frac{\dot{m}_L}{\dot{m}_A}\right]$$

$$\times \left[0.33\left(\frac{\sigma}{\rho_A U_A^2 D_p}\right)^{0.6}\left(\frac{\rho_L}{\rho_A}\right)^{0.1} + 0.068\left(\frac{\mu_L^2}{\rho_L \sigma D_p}\right)^{0.5}\right] \qquad (49)$$

An essential feature of all the Cranfield atomizers described above is the use of a prefilming surface to produce a thin, uniform sheet of liquid at the atomizing edge. To be fully effective, this system requires both sides of the liquid sheet to be exposed to the air. This requirement introduces a complication in design, since it usually necessitates two separate airflows through the atomizer. For this reason the "plain-jet" airblast atomizer is sometimes preferred in which the fuel is not transformed into a thin sheet but instead is injected into the high-velocity airstream in the form of discrete jets, as illustrated in Figure 21. This is the type of atomizer on which the pioneering work of Nukiyama and Tanasawa[78] was carried out almost 50 years ago.

The performance of plain-jet atomizers was investigated in detail by Lorenzetto and Lefebvre,[29,30] using a specially designed system in which

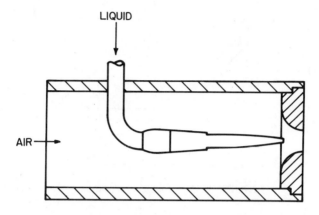

FIGURE 24. Plain-jet airblast atomizer.[29]

liquid physical properties, air/liquid ratio, air velocity, and atomizer dimensions could be varied independently over wide ranges and examined for their separate effects on spray quality. A cross-sectional drawing of the atomizer employed is shown in Figure 24.

The results of this study are generally similar to those obtained with prefilming atomizers, except that atomization quality improves slightly with increase in liquid density. The influence of air/liquid ratio on SMD is also more pronounced. In a separate series of tests the effect of fuel orifice diameter, d_0, was found to be quite small except for liquids of high viscosity, where SMD varied roughly in proportion with $d_0^{0.5}$.

The atomizing performance of prefilming and plain-jet airblast nozzles was compared by plotting some of the measured values of SMD alongside the experimental data obtained by Rizkalla and Lefebvre[28] under comparable operating conditions. The results of this comparison showed that the plain-jet atomizer performs less satisfactorily than the prefilming type, especially under the adverse conditions of low air/liquid ratio and/or low air velocity.

The following dimensionally correct expression for SMD was derived from analysis of the experimental data

$$
\mathrm{SMD} = 0.95 \left[\frac{(\sigma \dot{m}_L)^{0.33}}{U_R \rho_L^{0.37} \rho_A^{0.30}} \right] \left[1 + \frac{\dot{m}_L}{\dot{m}_A} \right]^{1.70}
$$

$$
+ 0.13 \mu_L \left[\frac{d_0}{\sigma \rho_L} \right]^{0.5} \left[1 + \frac{\dot{m}_L}{\dot{m}_A} \right]^{1.70} \tag{50}
$$

Further studies at Cranfield on plain-jet airblast atomization were made by Jasuja,[55] using a single-nozzle configuration, as illustrated in Figure 25. With this nozzle the fuel flows through a number of radially drilled, plain circular

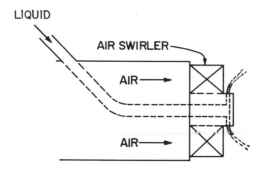

FIGURE 25. Plain-jet airblast atomizer.[55]

holes, from which it emerges in the form of discrete jets that enter a swirling airstream. These jets then undergo in-flight disintegration without any further preparation such as prefilming. Jasuja derived the following equation for drop-size correlations:

$$\text{SMD} = \frac{0.19}{U_A} \left(\frac{\sigma}{\rho_L \rho_A} \right)^{0.35} \left(1 + \frac{\dot{m}_L}{\dot{m}_A} \right)^{0.25}$$
$$+ 0.127 \left(\frac{\mu_L^2 d_0}{\sigma \rho_L} \right) \left(1 + \frac{\dot{m}_L}{\dot{m}_A} \right) \tag{51}$$

This equation is virtually the same as that proposed by Lorenzetto and Lefebvre,[29] except for a much lower dependence of SMD on air/liquid mass ratio.

Many expressions for predicting the mean drop sizes of airblast atomizers have been proposed. Those believed to be the most significant are listed in Tables 3, 4, and 2 for prefilming, plain-jet, and other types, respectively.

Effect of Variables on Mean Drop Size. The main factors influencing the mean drop size of the spray are liquid properties, air properties, and atomizer geometry.

Liquid Properties. The liquid properties of importance in airblast atomization are viscosity, surface tension, and density. High-speed photographs indicate that viscosity forces tend to suppress the formation of waves on the liquid surface which normally precede atomization. Instead, the liquid is drawn out from the atomizing lip in the form of long ligaments. Viscosity forces also resist deformation of these ligaments into drops, so that when atomization occurs, it does so well downstream of the atomizing lip, in regions of relatively low velocity. Thus, increase in viscosity would be expected to yield larger drop sizes, and this is borne out by experimental observations.[27,29,34,55]

TABLE 3. Drop-Size Equations for Prefilming Airblast Atomizers

Investigators	Equation	Remarks
Rizkalla and Lefebvre[28]	$$\text{SMD} = 3.33 \times 10^{-3}\,\frac{(\sigma \rho_L t)^{0.5}}{\rho_A U_A}\left(1 + \frac{\dot{m}_L}{\dot{m}_A}\right) + 13.0 \times 10^{-3}\left(\frac{\mu_L^2}{\sigma \rho_L}\right)^{0.425} t^{0.575}\left(1 + \frac{\dot{m}_L}{\dot{m}_A}\right)^2$$	L_c equated to t
Jasuja[55]	$$\text{SMD} = 10^{-3}\,\frac{(\sigma \rho_L)^{0.5}}{\rho_A U_A}\left(1 + \frac{\dot{m}_L}{\dot{m}_A}\right)^{0.5} + 0.6 \times 10^{-4}\left(\frac{\mu_L^2}{\sigma \rho_A}\right)^{0.425}\left(1 + \frac{\dot{m}_L}{\dot{m}_A}\right)^{0.5}$$	No effect of nozzle dimensions
Rizk and Lefebvre[31]	$$\text{SMD} = 0.5\left(\frac{\sigma^{0.6}\rho_L^{0.25}}{\rho_A^{0.85}U_A^{1.2}}\right)t^{0.4}\left(1 + \frac{\dot{m}_L}{\dot{m}_A}\right)^{0.85} + 0.107\left(\frac{\mu_L^2}{\sigma \rho_L}\right)^{0.45} t^{0.55}\left(1 + \frac{\dot{m}_L}{\dot{m}_A}\right)$$	t = liquid film thickness
Lefebvre[87]	$$\frac{\text{SMD}}{L_c} = A\left(\frac{\sigma}{\rho_A U_A^2 D_p}\right)^{0.5}\left(1 + \frac{\dot{m}_L}{\dot{m}_A}\right) + B\left(\frac{\mu_L^2}{\sigma \rho_L D_p}\right)^{0.5}\left(1 + \frac{\dot{m}_L}{\dot{m}_A}\right)$$	"Basic" equation for prefilming airblast atomizers L_c = characteristic dimension D_p = prefilmer diameter A, B = constants whose values depend on atomizer design
El-Shanawany and Lefebvre[34]	$$\frac{\text{SMD}}{D_h} = \left[1 + \frac{\dot{m}_L}{\dot{m}_A}\right]\left[0.33\left(\frac{\sigma}{\rho_A U_A^2 D_p}\right)^{0.6}\left(\frac{\rho_L}{\rho_A}\right)^{0.1} + 0.068\left(\frac{\mu_L^2}{\sigma \rho_L D_p}\right)^{0.5}\right]$$	L_c equated to D_h
Fraser et al.[73]	$$\text{SMD} = 6 \times 10^{-6} + 0.019\left(\frac{\sigma^{0.5}\nu_r^{0.21}}{\rho_A^{0.5}(aD_L + a^2)^{0.25}}\right)\left(1 + 0.065\left(\frac{\dot{m}_L}{\dot{m}_A}\right)^{1.5}\right)$$ $$\times \left(\frac{Q_L}{U_p^2(0.5U_r^2 - U_r + 1)}\right)^{0.5}$$	a = radial distance from cup lip U_p = cup peripheral velocity U_r = air/liquid velocity ratio

TABLE 4. Drop-Size Equations for Plain-Jet Airblast Atomizers

Investigators	Equation	Remarks
Nukiyama and Tanasawa[78]	$$\text{SMD} = 0.585\left(\frac{\sigma}{\rho_L U_R^2}\right)^{0.5} + 53\left(\frac{\mu_L^2}{\sigma\rho_L}\right)^{0.225}\left(\frac{Q_L}{Q_A}\right)^{1.5}$$	No effect of nozzle dimensions or air density
Gretzinger and Marshall[75]	$$\text{MMD} = 2.6\times10^{-3}\left(\frac{\mu_A}{\rho_A U_A L}\right)^{0.4}\left(\frac{\dot{m}_L}{\dot{m}_A}\right)^{0.4}$$	L = diameter of wetted periphery Drop size independent of liquid viscosity but dependent on air viscosity
Kim and Marshall[18]	$$\text{MMD} = 5.36\times10^{-3}\left[\frac{\sigma^{0.41}\mu_L^{0.32}}{\left(\rho_A U_R^2\right)^{0.57}A^{0.36}\rho_L^{0.16}}\right] \\ + 3.44\times10^{-3}\left[\left(\frac{\mu_L^2}{\sigma\rho_L}\right)^{0.17}\left(\frac{\dot{m}_A}{\dot{m}_L}\right)^m\frac{1}{U_R}\right]^{0.54}$$	A = flow area of atomizing air stream $m = -1$ at $\dot{m}_A/\dot{m}_L < 3$ $m = 0.5$ at $\dot{m}_A/\dot{m}_L > 3$
Lorenzetto and Lefebvre[29]	$$\text{SMD} = 0.95\left[\frac{\left(\sigma\dot{m}_L\right)^{0.33}}{\rho_L^{0.37}\rho_A^{0.30}U_R}\right]\left[1 + \frac{\dot{m}_L}{\dot{m}_A}\right]^{1.70} + 0.13\left(\frac{\mu_L^2 d_0}{\sigma\rho_L}\right)^{0.5}\left(1 + \frac{\dot{m}_L}{\dot{m}_A}\right)^{1.70}$$	For low viscosity liquids SMD is independent of initial jet diameter, d_0
Jasuja[55]	$$\text{SMD} = 0.19\left(\frac{\sigma}{\rho_L\rho_A}\right)^{0.35}\left(\frac{1}{U_A}\right)^{0.25}\left(1 + \frac{\dot{m}_L}{\dot{m}_A}\right)^{0.25} + 0.127\left(\frac{\mu_L^2 d_0}{\sigma\rho_L}\right)^{0.5}\left(1 + \frac{\dot{m}_L}{\dot{m}_A}\right)^{0.5}$$	For low viscosity liquids SMD is independent of initial jet diameter, d_0
Ingebo and Foster[84]	$$\text{MMD} = 3.9\left(\frac{\sigma\mu_L d_0^2}{\rho_A U_A^3 \rho_L}\right)^{0.25}$$	For cross-current breakup
Weiss and Worsham[79]	$$\text{SMD} \propto \left[\frac{U_L^{0.08}d_0^{0.16}\mu_L^{0.34}}{\rho_A^{0.30}U_R^{1.33}}\right]$$	Experiments covered wide range of variables

Surface tension forces tend to impede atomization by resisting any disturbance or distortion of the liquid surface, thereby opposing the creation of surface waves and delaying the onset of ligament formation.

Liquid density affects droplet size in a fairly complex manner. For example, with prefilming atomizers, the distance to which the coherent liquid sheet extends downstream of the atomizing lip increases with density, so that ligament formation occurs under conditions of lower relative velocity between the air and the liquid. Moreover, for any given flow rate, an increase in liquid density will produce a more compact spray that is less exposed to the atomizing action of the high-velocity air. Both these effects combine to increase the mean drop size. However, an increase in liquid density can also improve atomization by reducing the thickness of the sheet produced at the atomizing lip of prefilming systems ($t \propto \rho_L^{-0.4}$), and by increasing the relative velocity, U_R, for plain-jet nozzles. The net effect of these conflicting factors is that the influence of liquid density on SMD is quite small.

Air Properties. Of all the various factors influencing mean drop size, air velocity is undoubtedly the most important. This underlines the importance in airblast atomizer design of arranging for the liquid to be exposed to the highest possible air velocity consistent with the available pressure drop.

Another important consideration in atomizer design is the air/liquid mass ratio. For all types of airblast atomizers, it is found that atomization quality starts to decline when the air/liquid ratio drops below about 4, and deteriorates quite rapidly at air/liquid ratios below about 2. At air/liquid ratios higher than about 5, only marginal reductions in SMD are gained by the utilization of more air in atomization, as illustrated in Figure 23. The observed effects of air/liquid ratio on drop size are consistent with the view that a deficiency of air results in failure to overcome the viscous and surface tension forces that act together to oppose drop formation, whereas an excess of air gives no advantage since it is physically too remote from the liquid jet or sheet to affect its disintegration.

The effects of air temperature and pressure on mean drop size are manifested through their effects on air density. The measurements carried out by Rizkalla[28] suggest that, for prefilming atomizers, SMD is proportional to air density to the power -0.7. For plain-jet airblast atomizers, the dependence of SMD on air density is somewhat less, with reported exponents varying between -0.25 and -0.5.[29, 55, 84, 86]

For prefilming atomizers it is found that SMD increases with increase in atomizer scale (size), according to the relationship SMD $\propto L_c^{0.4}$.[34]

Analysis of Drop-Size Relationships. From inspection of all the experimental data obtained on prefilming types of airblast atomizer, some general conclusions concerning the effects of air and liquid properties on mean drop size can be drawn. For liquids of low viscosity, such as water and kerosine, the main factors governing SMD are liquid surface tension, air density, and air velocity, whereas for liquids of high viscosity the effects of air properties are less significant, and SMD becomes more dependent on the liquid properties, especially viscosity.

The fact that viscosity has an effect that is quite separate from that of air velocity, as observed experimentally by Nukiyama and Tanasawa,[78] Kim and Marshall,[18] Lefebvre et al.,[27,29,31,34] and Jasuja,[55] suggests a form of equation in which SMD is expressed as the sum of two terms, one being dominated by air velocity and air density, and the other by liquid viscosity. For prefilming types of atomizer, it can be shown[87] that

$$
\frac{\text{SMD}}{L_c} = \left[A' \left(\frac{\sigma}{\rho_A U_A^2 D_p} \right)^{0.5} + B' \left(\frac{\mu_L^2}{\sigma \rho_L D_p} \right)^{0.5} \right] \left[1 + \frac{\dot{m}_L}{\dot{m}_A} \right] \tag{52}
$$

where A' and B' are constants whose values depend on the atomizer design and must be determined experimentally.

Equation (52) is the basic equation for the mean drop size of prefilming airblast atomizers. It shows that SMD is proportional to the square root of the characteristic dimension, L_c, which represents the linear scale of the atomizer. Equation (52) also shows that mean drop size decreases with increase in prefilmer lip diameter, D_p. This is not surprising since, if other parameters are kept constant, any increase in D_p will reduce the liquid film thickness at the atomizing lip and hence also the SMD.[31] The implication of this to atomizer design is clear—for any given atomizer size (i.e., for any given value of L_c) the ratio of D_p/L_c should be made as large as possible.

In practice, some secondary factors, such as Reynolds number and Mach number effects in the liquid and airstreams, respectively, affect the atomization process in a manner that is not yet fully understood. Thus it is found that the ability of Eq. (52) to predict mean drop diameters can be improved by raising the exponent of the term $(\sigma_L/\rho_A U_A^2 D_p)$ from 0.5 to 0.6. Moreover high-speed photographs of the atomization process have shown that for prefilming atomizers, the effect of an increase in ρ_L/ρ_A is to extend the distance downstream of the atomizing lip at which sheet disintegration occurs so that it takes place in a region of lower relative velocity.[32] From analysis of the experimental data, it is found that this effect may be accommodated by introducing into Eq. (52) the dimensionless term $(\rho_L/\rho_A)^{0.1}$. With these modifications Eq. (52) becomes

$$
\frac{\text{SMD}}{L_c} = A \left(\frac{\sigma_L}{\rho_A U_A^2 D_p} \right)^{0.6} \left(\frac{\rho_L}{\rho_A} \right)^{0.1} \left(1 + \frac{\dot{m}_L}{\dot{m}_A} \right)
$$

$$
+ B \left(\frac{\mu_L^2}{\sigma_L \rho_L D_p} \right)^{0.5} \left(1 + \frac{\dot{m}_L}{\dot{m}_A} \right) \tag{53}
$$

A logical choice for the characteristic dimension, L_c, is the hydraulic mean diameter of the atomizer air duct at its exit plane (see Fig. 22). Equating $L_c = D_h$ and substituting into Eq. (53) the values of A and B, derived from the experimental data of El-Shanawany and Lefebvre,[34] yields Eq. (49).

Summary of Main Points. From analysis of the experimental data obtained by many workers on many different types of airblast atomizers, the following conclusions are drawn:

1. The mean drop size of the spray increases with an increase in liquid viscosity and surface tension and with a reduction in the air/liquid ratio. Ideally, the air/liquid mass ratio should exceed a value of 3, but little improvement in atomization quality is gained by raising this ratio above a value of around 5.

2. Liquid density appears to have little effect on mean drop size. The data indicate that for prefilming nozzles, the mean drop size increases slightly with increase in liquid density, whereas for plain-jet nozzles, the opposite effect is observed.

3. The air properties of importance in airblast atomization are density and velocity. In general, it is found that mean drop size is inversely proportional to air velocity. The effect of air density may be expressed as SMD $\propto P_A^{-n}$, where n is about 0.4 for plain-jet atomizers, and between 0.6 and 0.7 for prefilming types.

4. For plain-jet nozzles the initial liquid jet diameter has little effect on mean drop size for liquids of low viscosity, but for high-viscosity liquids, atomization quality deteriorates with increase in jet size.

5. For prefilming atomizers, it is found that mean drop size increases with increase in atomizer scale (size) according to the relationship SMD $\propto L_c^{0.4}$.

6. For any given size of prefilming atomizer (i.e., for any fixed value of L_c), the finest atomization is obtained by making the prefilmer lip diameter, D_p, as large as possible. This is because increase in D_p reduces the liquid film thickness, t, which in turn reduces the mean drop size (SMD $\propto t^{0.4}$).

7. With prefilming nozzles the atomization process is sensibly complete at a downstream distance of around $1.5D_p$. Beyond this distance the mean drop size increases gradually due to evaporation and, possibly, droplet coalescence.

8. Minimum drop sizes are obtained by using atomizers that are designed to provide maximum physical contact between the air and the liquid. With prefilming systems the best atomization is obtained by producing the thinnest possible liquid sheet of uniform thickness. In practice, this means that the atomizing lip should have the largest possible diameter.

It is also important to arrange that the liquid sheet formed at the atomizing lip is subjected to high-velocity air on both sides. This not only provides the finest atomization, but it also eliminates droplet deposition on adjacent solid surfaces.

9. The performance of prefilming atomizers is superior to that of plain-jet types, especially under adverse conditions of low air/liquid ratio and/or low air velocity.

10. There are at least two different mechanisms involved in airblast atomization, the relative importance of each being mainly dependent on the level of liquid viscosity. For low-viscosity liquids injected into a low-velocity airstream, waves are produced on the liquid surface, which becomes unstable and disintegrates into fragments. These fragments then contract into ligaments, which, in turn, break down into drops. With increase in air velocity, the liquid sheet disintegrates earlier so that ligaments are formed nearer the lip. These ligaments tend to be thinner and shorter and disintegrate into smaller drops. With liquids of high viscosity, the "wavy surface" mechanism is no longer present. Instead, the liquid is drawn out from the atomizing lip in the form of long ligaments. When atomization occurs, it does so well downstream of the atomizing lip in regions of relatively low velocity. In consequence drop sizes tend to be larger.

More detailed information on airblast atomization is contained in the review paper of Lefebvre.[87]

1.12. Gas Turbines

The processes of liquid atomization and evaporation are of fundamental importance to the behavior of a gas turbine combustion system. Normal fuels are not sufficiently volatile to produce vapor in the amounts required for ignition and combustion unless they are atomized into a large number of drops with corresponding vastly increased surface area. The smaller the drop size, the faster the rate of evaporation. The influence of drop size on ignition performance is of special importance, since large increases in ignition energy are needed to overcome even a slight deterioration in atomization quality. Spray quality also affects stability limits, idle combustion efficiency, and the emission levels of smoke, carbon monoxide, and unburned hydrocarbons.

The earliest practical form of fuel injector used on gas turbines was the "simplex" pressure-swirl atomizer. One problem that soon arose with this atomizer was that burning was possible only over a fairly narrow range of fuel-flow rates. Also, on aircraft engines, several limitations on performance were experienced—notably low combustion efficiency and inadequate relight capability at high altitudes. To eliminate these shortcomings, various "wide-range" atomizers were introduced, of which the most outstanding examples are probably the "dual-orifice" and "duplex" spray atomizers. These developments, along with parallel advances in the performance of fuel pumps and control systems, provided a period of many years in which fuel injection was one of the most satisfactory and trouble-free aspects of combustion chamber operation.

In due course, and as a direct result of the continuing trend toward engines of higher compression ratio, combustors featuring pressure atomizers became increasingly afflicted by excessive exhaust smoke and poor pattern factor. Consideration of these problems evoked new interest in the airblast atomizer. This type of fuel injector is ideally suited for gas turbine applications because high-velocity air is always readily available, owing to the pressure differential across the liner. An important feature of the airblast atomizer is that it provides some premixing of fuel and air prior to combustion, which leads to worthwhile reductions in exhaust smoke and in the emission of nitric oxides.

Airblast atomizers are now in widespread use in both aircraft and industrial gas turbines. In comparison with pressure atomizers they offer significant advantages in terms of good temperature traverse quality, freedom from exhaust smoke, low fuel-pressure requirements, good mechanical reliability, and low sensitivity to variations in fuel viscosity. Their narrow stability limits can be overcome by using a "hybrid" or "piloted" system that employs both pressure and airblast atomization. At start-up all the fuel is supplied from the pressure atomizer, which is usually a plain-orifice or simplex nozzle. For normal operation, some fuel is still supplied to the pressure nozzle, but most is supplied to the airblast atomizer. By this means the merits of pressure atomization at low fuel flows—namely, easy light-up, satisfactory combustion efficiency, and a high weak-extinction AFR—are realized, while the defects of the pressure system at high fuel flows (corresponding to high air pressures)—a highly luminous flame, copious amounts of exhaust smoke, and sensitivity of pattern factor to variations in fuel-flow rate—are eliminated.

Vaporizers. Apart from the various atomization methods already discussed, an alternative method of preparing a liquid fuel for combustion is by heating it above the boiling point of its heaviest hydrocarbon ingredient, so that it is entirely converted to vapor prior to combustion. This method is, of course, applicable only to such high-grade fuels as can be completely vaporized, leaving no solid residue. When vaporizers are employed to handle grades of fuel that are appreciably heavier than kerosine, it is necessary to scrape clean the vaporizing elements at fairly frequent intervals.

Historically, vaporizing systems were developed before atomizers; in fact, in the early Whittle chambers, the fuel was heated in tubes located in the flame zone and was maintained at high pressure so that vaporization could not occur until the fuel had been injected through a nozzle and its pressure reduced to combustion-chamber pressure. This method of vaporization, called *flash vaporization*, is seldom used nowadays because of the problems of thermal cracking and solid deposition referred to above, and because of difficulties that arise in controlling the flow of fuel.

An alternative and much simpler method of vaporization is to inject the fuel, along with some air, into tubes that are immersed in the flame. The injected fuel-air mixture is heated by the tube walls and, under ideal conditions, emerges as a mixture of vaporized fuel and air. The remainder of the combustion air is admitted through apertures in the liner dome and reacts with the fuel-air mixture issuing from the tubes. During the starting cycle, when the tubes are too cold to effect vaporization, a simple form of torch igniter is used to initiate combustion.

It is now recognized that the term "vaporizer" is largely a misnomer, since insufficient heat can be transferred to the tubes to vaporize more than a small fraction of the fuel. Thus only at the lowest fuel flow rates can the system be regarded as a true vaporizer. Where vaporizers are used on modern engines, their main function appears to be that of providing a satisfactory distribution of fuel throughout the primary combustion zone. The types of vaporizers used by Rolls Royce have been described by Sotheran.[88]

Premix-Prevaporize Combustor. This concept is especially suitable where very low emissions is a prime requirement. Fuel is injected in a finely atomized form into high-velocity air which then flows into the combustion zone. The design objective is to attain complete evaporation of the fuel and thorough mixing of fuel and air prior to combustion. By avoiding droplet burning, and by operating the reaction zone at a lean fuel/air ratio, nitric oxide emissions are drastically reduced, owing to the low flame temperature and the elimination of "hot spots" within the combustion zone. Problems with this system include incomplete fuel vaporization and mixing, the risk of autoignition and/or flashback in the premixing passages, poor lean blowout characteristics, and difficult light-up. Some of these problems can be overcome, at the expense of additional cost and complexity, through the use of staged combustion or variable geometry.

Gas Injection. Gaseous fuels, especially those of high calorific value, present few problems from a combustion viewpoint. With low-heat-content gases, however, the fuel-flow rate may comprise about one-fifth of the total combustor mass flow; this may lead to a mismatch between the compressor and the turbine, especially if the engine is intended for a multifuel application. Another problem with low-heat-content gases is their low reaction rate, which may necessitate additional combustion-zone volume, over and above the extra volume needed to accommodate the large volumetric flow of gaseous fuel. The methods used to inject gaseous fuels include plain orifices and slots, swirlers, and venturi nozzles. It is sometimes difficult to achieve the optimum mixing rates in the combustion zone. Too high a mixing rate results in poor lean blowout characteristics, while too low a mixing rate could give rise to rough combustion.

Slinger System. In France the engine company Turbomeca has successfully developed a drum atomizer for gas turbine engines that is employed in conjunction with a radial-annular combustor chamber, as illustrated in Figure 26. Fuel is supplied at low pressure along the hollow main shaft and is discharged radially outward through holes drilled in the shaft. These injection holes vary in number from 9 to 18 and in diameter from 2.0 to 3.2 mm. The holes may be drilled in the same plane as a single row, but some installations feature a double row of holes. The holes never run full; they have a capacity that is many times greater than the required flow rate. The reason they are made large is to obviate blockage. However, it is very important that the holes be accurately machined and finished, since experience has shown that uniformity of flow between one injection hole and another is very dependent on their dimensional accuracy and surface finish. Clearly, if one injection hole supplies more fuel than the others, this will produce a rotating "hot spot" in the exhaust gases, with disastrous consequences for the particular turbine blade on which the hot spot happens to impinge.

Flow uniformity is also critically dependent on the flow path provided for the fuel inside the shaft, especially in the region near the holes. Where there are two rows of holes, it is very important to achieve the correct flow division between the two rows. Again the internal geometry of the shaft in the vicinity of the holes is of prime importance.

The main advantages of the "slinger" system lie in its cheapness and simplicity. Only a low-pressure fuel pump is needed, and the quality of atomization depends only on engine speed. The equivalent injection pressures are very high, of the order of 34 MPa at full speed, and satisfactory atomization is claimed at speeds as low as 10% of the rated maximum. The influence of fuel viscosity is small so that the system has a potential multifuel capability.

The system seems ideally suited for small engines of low compression ratio, and this has been its main application to date. As the success of the system

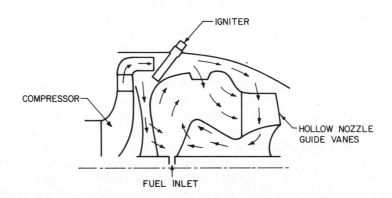

FIGURE 26. Turbomeca system.

depends on high rotational speeds, usually greater than 20,000 rpm, it is clearly less suitable for large engines where shaft speeds are much lower. In the United States the slinger system has been used successfully on several engines produced by the Williams Research Corporation. For further information on this system, see Lefebvre.[63]

1.13. Diesel Injection

The combustion process in a diesel engine may be described as a complex sequence of overlapping and mutually interacting events. Characteristics of the process are a complicated function of the size and velocity distribution of the atomized droplets, the physical and chemical properties of the fuel and its environment, and the characteristics of the mixing process of the fuel with air and with combustion products. Atomization is accomplished by injecting compressed liquid fuel from a nozzle into air at high pressures and temperatures. The high velocity of discharge of the fuel into closely packed air molecules causes the liquid jet to disintegrate rapidly into fine drops.

Knowledge of diesel fuel spray characteristics is quite meager. The few attempts that have been made to measure drop sizes in diesel sprays show a low degree of accuracy, and there is conflicting evidence as to the effect of pressure on drop size. It is well known that atomization, drop size, and vaporization are radically different at typical diesel engine pressures (3 MPa) and temperatures (800 K) so that measurements made under atmospheric pressure and temperature cannot be related to sprays in diesel chambers.

The first experimental studies on drop sizes in a diesel spray were carried out by Woltjen[89] and Sass.[90] Sass obtained the drop-size distribution for three different nozzle diameters at pressures from 0 to 1 MPa. He showed that mean drop diameter increases linearly with nozzle diameter and decreases with increase in ambient air pressure. The effect of ambient pressure on drop-size distribution has been studied by Hiroyasu and Kadota.[91] They measured the drop sizes in a spray injected into a high-pressure bomb and derived generalized forms of empirical correlations between effective injection pressure, ambient pressure, the rate of fuel delivery, and the Sauter mean diameter of the spray. Their results showed that Sauter mean diameter increases with increasing ambient pressure.

One of the main complicating factors in the study of diesel sprays is that in diesel engines the fuel injection is both intermittent and unsteady. The injection pressure, for example, exhibits rapid variation with time, which causes a rapid fluctuation of the initial velocity of the fuel jet issuing from the nozzle. The unsteadiness and intermittency of the spray makes it extremely difficult to analyze the combustion process in a diesel engine.

Knowledge of fuel spray penetration in diesel engines is important both from the viewpoint of establishing mathematical models to represent the combustion process and for preventing overpenetration of the fuel spray which could lead to fuel impingement on a cold surface and consequent fuel wastage.

Many investigators, using different techniques, have attempted to measure the penetration of a single spray injecting into quiescent air.[92-98] Hay and Jones[98] reviewed the various penetration equations deduced from these experiments and concluded that for low air densities, best results are obtained using the following expression due to Dent:[95]

$$S = 0.0865 \left[\left(\frac{\Delta P_F}{\rho_A} \right)^{0.5} t d_0 \right]^{0.5} \left[\frac{530}{T_A} \right] \tag{54}$$

The following conclusions may be drawn from this equation:

1. Spray penetration increases with increase in fuel injection pressure. This is due to the corresponding increase in droplet momentum.
2. Spray penetration increases with increase in injector nozzle diameter. This is because larger nozzles produce larger drops of higher momentum.
3. Spray penetration is reduced by an increase in ambient air pressure. This is accounted for by the finer atomization produced by increase in air back pressure. The smaller drops increase the surface area and aerodynamic drag of the spray so that its penetration power is diminished.

Elkotb et al.[99-101] have measured drop sizes and drop-size distributions of diesel fuel sprays over wide ranges of operating conditions, using a Bosch DND SD 211 pintle nozzle, as illustrated in Figure 27. During the investigation fuel-injection pressure and ambient air pressure were varied from 7.8 to

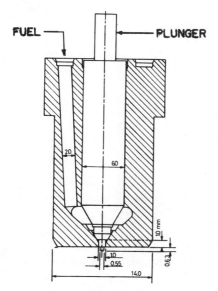

FIGURE 27. Bosch pintle diesel injector, type DND SD 211.[100]

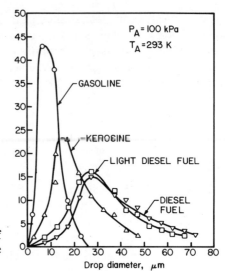

FIGURE 28. Droplet size distribution of several fuel sprays for an injection pressure of 118 bar.[100]

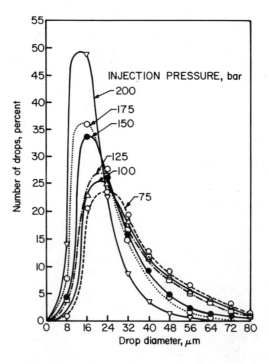

FIGURE 29. Droplet size distribution for diesel fuel sprays inside a cold engine model at various injection pressures.[100]

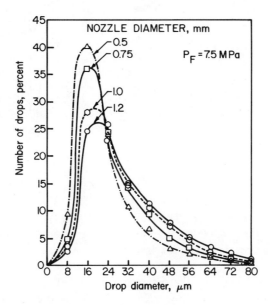

FIGURE 30. Droplet size distribution for diesel fuel sprays inside a cold engine model for various nozzle diameters.[100]

11.8 MPa and from 1.2 to 8.2 kg/m³, respectively. The slide-sampling technique was used to measure the drop-size distribution. Some examples of the results obtained are shown in Figures 28 to 30.

Based on these and other experimental data Elkotb[101] derived the following empirical equation for mean drop size:

$$\text{SMD} = 0.074\nu_F^{0.385}\sigma^{0.737}\rho_F^{0.737}\rho_A^{0.06}\Delta P_F^{-0.54} \tag{55}$$

Thus, in common with most other expressions for the drop sizes produced by pressure atomizers, Eq. (55) shows that atomization quality is improved by increase in fuel-injection pressure and deteriorates with increases in any of the three main fuel properties of relevance to atomization, namely, viscosity, surface tension, and density.

1.14. Power Systems

A problem of continuing concern with oil-fired boiler plants is the emission of particulate material that is known to be linked to the quality of atomization.[102,103] The relationship is difficult to define, but generally a 10% reduction in mean drop size yields around 25% reduction in solids emission.[103]

Many different types of atomizer are employed in utility boilers and industrial furnaces, ranging from conventional pressure-swirl nozzles to fairly complex twin-fluid atomizers. A typical commercial design of large pressure-

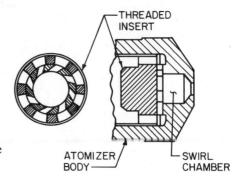

FIGURE 31. Commercial design of large pressure atomizer.[103]

swirl nozzle is shown in Figure 31. Atomizers of this type are designed for oil flow rates up to around 4000 kg/hr. Figure 32 shows a more advanced version that embodies several features for improving the quality of atomization.[103] These include (1) minimum area of wetted surface to reduce frictional losses, (2) improved mixing between the discrete jets emanating from each inlet slot, and (3) reduced losses due to flow separations at sharp corners. According to Jones[103] these design improvements yield a 12% reduction in mean drop size, corresponding to a 30% reduction in the emission of particulates.

One of the most widely used types of twin-fluid atomizer is the Babcock Y-jet design, shown schematically in Figure 33.[104] It is a multi-ported system, and in each individual Y oil is injected at an angle into the exit port, where it mixes with the atomizing fluid (air or steam) admitted through the air port. The exit ports are uniformly spaced around the atomizer body at an angle to the nozzle axis so that the individual jets of two-phase mixture issuing from the exit ports rapidly merge to form a hollow conical spray.

According to Bryce et al.[102] the most common method of operating Y-jet atomizers is to maintain the pressure of the atomizing fluid constant over the full range of oil flows, with oil flow controlled by variation of the oil pressure.

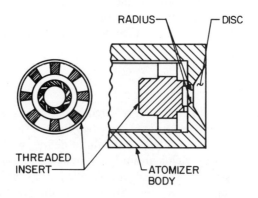

FIGURE 32. Improved design of pressure atomizer.[103]

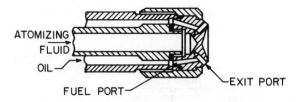

FIGURE 33. Typical Y-jet trip arrangement.[104]

This control mode results in the atomizing fluid/oil mass flow ratio increasing as the fuel flow decreases.

An alternative method of operation is to maintain the ratio of fluid/oil pressure constant over the fuel-flow range. This results in higher atomizing fluid flows being available at the upper end of the oil flow range, and lower atomizing fluid flows at the lower oil flows, which gives a saving in atomizing fluid at low loads without significantly reducing the quality of atomization.

A typical throughput of heavy residual fuel oil for a large utility boiler burner is around 7500 kg/hr. Steam is generally employed as the atomizing fluid. An obvious advantage to using steam is that any heat transferred from the steam to the fuel in the mixing ports will enhance atomization by reducing the fuel's viscosity and surface tension. However, comparative tests carried out by Bryce et al.[102] showed that compressed air produced a much finer spray than steam.

A typical furnace installation for a Y-jet nozzle is shown in Figure 34. It features a convergent-divergent (venturi) register and a Y-jet atomizer mounted at the end of a fuel feed arm. Most of the air passing through the venturi register flows axially into the combustion zone, but the central portion flows through an air swirler that surrounds the Y-jet fuel nozzle. The function of this swirler is to provide flame stabilization by inducing the recirculation of burned gases into the flame initiation zone.

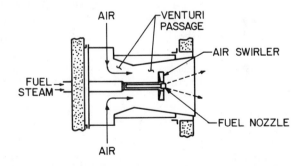

FIGURE 34. Typical installation of Y-jet fuel nozzle.

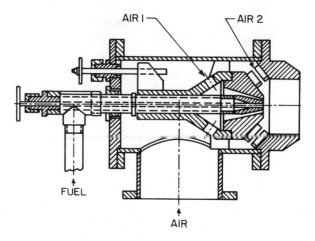

FIGURE 35. Combined pressure and twin-fluid atomizer.[105]

Another type of twin-fluid atomizer is illustrated in Figure 35. The fuel is first pressure-atomized and then subjected to two stages of air (or steam) atomization. In the second stage swirl vanes or slots are used to impart a helical motion to the air flow.

Additional information of fuel nozzles for industrial applications may be found in Stambuleanu.[105]

1.15. Miscellaneous Types of Atomizers

Electrostatic. A basic requirement for atomizing any liquid is to make some area of its surface unstable. The surface will then rupture into ligaments that subsequently disintegrate into drops. In electrical atomization the energy causing the surface to disrupt comes from the mutual repulsion of like charges that have accumulated on the surface. An electrical pressure is created that tends to expand the surface area. This pressure is opposed by surface tension forces that tend to contract or minimize surface area. When the electrical pressure exceeds the surface tension, the surface becomes unstable, and droplet formation begins. If the electrical pressure is maintained above the critical value consistent with the liquid flow rate, then atomization is continuous.

The electrical pressure, P_e, has been derived by Graf[106] as

$$P_e = \frac{FV^2}{2\pi D^2} \tag{56}$$

where V is the applied voltage, and D is the drop diameter. F is a charging factor that represents the fraction of the applied potential attained on the drop

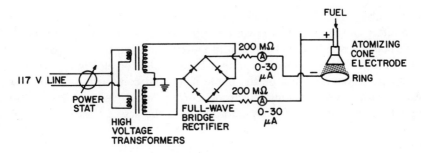

FIGURE 36. Cone and ring electrode system for electrostatic atomization.[107]

surface. It is found[106] that F decreases with increasing liquid conductivity and increasing electrode spacing.

Many configurations of atomizing electrode have been tested, including hypodermic needles, sintered bronze filters, and cones.[107] Both direct and alternating current electrical systems have been used to generate the high voltage needed for fine atomization. A typical DC circuit, of the type used successfully by Luther,[107] is illustrated in Figure 36.

The results of many experimental studies on electrostatic atomization[106–114] show that, in general, the size of the drops produced depends on the applied voltage, surface tension, electrode size and configuration, liquid flow rate, and the electrical properties of the liquid, such as dielectric constant and electrical conductivity. The smallest drops are produced by the highest voltage that can be used without encountering excessive corona losses.

The very low liquid flow rates that are generally associated with electrostatic atomization have tended to restrict its practical application to electrostatic painting and nonimpact printing. However, an invention by Kelly,[114] called the "Spray Triode," shown schematically in Figure 37, appears to have great promise for the development of electrostatic atomizers capable of handling the high fuel-flow rates required by most practical combustion devices. This figure shows that a voltage differential is impressed between a central, submerged, emitter electrode and a blunt-orifice electrode. The two electrodes

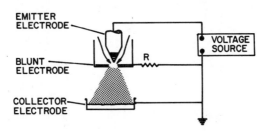

FIGURE 37. Schematic diagram of Spray Triode atomizer.[114]

form a submerged electron gun and serve to charge the fluid that flows around the emitter electrode and exits via the orifice. Once free of the confines of the interelectrode region, the charged fluid undergoes disruption and spray formation. Charge is returned to the circuit by a "collector electrode" which in a combustion system would be the flame front and combustion chamber wall. The resistor serves to limit electrode current in the event of an internal breakdown in the fluid. It may be noted that the active electrodes are in an ideal location; they are submerged in an insulating fluid. This arrangement precludes the possibility of external corona breakdown and permits operation in ionizing combustion gases.

A key feature of Kelly's atomizer is the special emitter electrode which comprises a multiplicity of very small tungsten fibers embedded in a refractory material. It is claimed that submicron-diameter, continuous monocrystalline fibers can be developed at densities up to $10^7/cm^2$.[114] Because each fiber is capable of handling flow rates up to 1 mL/sec, this would imply that even a 1-mm diameter electrode could effectively handle fuel flow rates up to 30 kg/hr. This is clearly very satisfactory from the standpoint of its application to practical combustion systems.

Ultrasonic. This method of atomization is based on the principle that liquid droplets are produced when powerful, high-frequency sound waves are focused onto the liquid surface. A schematic diagram of the type of ultrasonic atomizer used by Gershenzon and Eknadiosyants[115] is shown in Figure 38. It employs a deep reservoir in conjunction with a saucer-shaped, focusing transducer. The function of the transducer is to transmit high-frequency waves to the surface. If the wave strength at the surface exceeds the surface tension, the liquid disintegrates into drops.

The drop formation mechanism for this type of atomizer was originally attributed to the creation and subsequent collapse of cavities caused by the intense wave. However, other workers[116-119] have favored a capillary wave theory which is based on the observation that the mean drop size produced from thin layers of liquid on an ultrasonically excited plate are proportional to the capillary wavelength on the liquid surface. This implies that the drop size should be related to the ripple wavelength, which in turn is controlled by the vibration frequency. The results of experiments tend to support this hypothe-

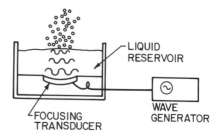

FIGURE 38. Schematic design of ultrasonic atomizer.

sis. For example, Lang[116] found that

$$D = 0.34\lambda = 0.34\left(\frac{8\pi}{\rho_F F}\right) \tag{57}$$

which is in close agreement with Lobdell's theoretical value of 0.36λ, obtained from considerations of drop formation from high-amplitude capillary waves.[117]

Another capillary wave theory was developed by Peskin and Raco[118] in terms of Taylor instability. They found that the atomization process is governed by several nondimensional parameters, including transducer amplitude a, frequency ω_0, liquid film thickness t, surface tension σ, and density ρ_L. For large film thickness, their analytical result reduces to

$$D = \left(\frac{4\pi^3\sigma}{\rho\omega_0^2}\right)^{1/3} \tag{58}$$

The ranges of drop sizes obtained by various workers are summarized in Table 5 from Lee et al.[120] The operating frequency used in these studies varied from 10 to 2000 kHz.

Unfortunately, ultrasonic atomizers have a low maximum flow rate capacity, typically around 4 L/hr at a frequency of 40 kHz. This is due primarily to the low amplitude of oscillations that ultrasonic transducers generate. One method of increasing the flow rate is to combine the principle of ultrasonic atomization with that of whistle-type atomization. In this manner the amplitude of the signals produced by the ultrasonic vibrator is further amplified by the resonant effects created within the hollow space of the horn.[120] However, a

TABLE 5. Range of Drop Sizes Covered by Various Studies Employing an Ultrasonic Atomizer[a]

Investigators	Drop Size, μm			
	1	10	100	1000
Wilcox and Tate[128]			___	
Crawford[124]			_	
Antonevich[122]			___	
Lang[116]		___		
Bisa et al.[123]		_		
McCubbin[125]		_		
Topp[129]			_	
Muromstev and Nenashev[126]	___			

[a] From Lee et al.[120]

drawback to this approach is that the operating frequency of the transducer is limited to one value.

Mochida[121] has studied the horn type of ultrasonic atomizer operating at a frequency of 26 kHz, at flow rates up to 50 L/hr. Using distilled water, and solutions of water with methanol and glycerin to obtain suitable variations in the liquid properties that influence atomization, he obtained the following empirical equation for SMD:

$$\text{SMD} = 0.06(\sigma/\rho_L)^{0.354}\mu_L^{0.303}Q_L^{0.139} \tag{59}$$

Further information on the performance of ultrasonic atomizers is contained in Mochida,[121] Antonevich,[122] Bisa et al.,[123] Crawford,[124] McCubbin,[125] Muromtsev et al.,[126] Topp and Eisenklam,[127] Wilcox and Tate,[128] Topp,[129] and Sakai et al.[130]

Whistle Type. In a similar manner to an ultrasonic atomizer using a transducer, liquid can also be disintegrated into drops by directing high-pressure gas into the center of a liquid jet. Due to the strong sound waves created inside the nozzle by the focusing airflow, this type of atomizer is frequently called the *whistle*-, or *stem-cavity* atomizer. It generally operates at a sound frequency of about 10 kHz, and produces droplets around 50 microns in diameter at flow rates up to 4500 L/hr.

A drawback to whistle atomizers is that the drop size cannot be easily controlled unless the nozzle dimension is changed. Wilcox and Tate[128] studied this type of nozzle systematically and concluded that the sound field was not an important variable in the atomizing process. This led Topp and Eisenklam[127] to suspect that all whistle atomizers operate simply as twin-fluid types—that is, the liquid is disintegrated primarily by the aerodynamic interactions between the gas and the liquid. No reliable or proven theoretical analyses on the performance of whistle atomizers seem to be available.

2. DROPLET EVAPORATION

2.1. Introduction

Most practical combustion systems employ liquid fuels. Usually the combustion process includes injection and atomization of the liquid fuel, followed by evaporation, mixing of the vaporized fuel with air, ignition of the fuel-air mixture, and, finally, chemical reaction leading to the formation of combustion products. Chemical processes can be limiting to rates of combustion in low-temperature and/or low-pressure situations and are also important in the formation of pollutant emissions. However, in many practical combustion devices the rate of chemical reaction is so high that the burning rate is controlled mainly by the rate at which the fuel can evaporate and mix in combustible proportions with air. Thus a sound knowledge of the factors governing the rate of evaporation of fuel sprays is a key requisite to any examination or assessment of the main performance parameters of liquid fuel-fired combustors, such as combustion efficiency, burning range, and ignition performance. Moreover the process of fuel evaporation seems likely to assume added importance in the future, since the alternative fuels that are being investigated for various engine applications are less volatile than current conventional fuels.

Examples of combustion systems burning liquid fuels include diesel engines, spark ignition engines, gas turbines, liquid rocket engines, and industrial furnaces. Usually the fuel is injected in the form of a well-atomized spray. In some systems, such as the carburettor of a spark-ignition engine and the jet engine afterburner, the fuel is sprayed into the airstream ahead of the combustion zone. The objective is to achieve complete evaporation of the fuel and thorough mixing of fuel and air prior to combustion. In other systems, such as the diesel engine and the gas turbine combustor, the fuel is sprayed directly into the combustion zone.

The foundations of our present understanding of droplet evaporation and combustion are based largely on studies of the low-temperature evaporation of single drops. Obviously, spray evaporation constitutes rather more than the summation of the evaporation of a large number of single drops, due to the interaction between adjacent evaporating drops. However, if the average spatial distance between drops is sufficiently large, estimations based on single-drop evaporation can provide a good approximation to the evaporation of the entire spray.

It is proposed to consider first the case of single-droplet evaporation in stagnant air, without combustion. This discussion will provide an introduction to the more complicated situations involving air movement and combustion. Relationships will be derived for expressing rates of fuel evaporation in terms of the relevant air and fuel properties. These relationships will then be used to determine the burning rate of a fuel spray, the time required for evaporation, and the level of combustion efficiency.

2.2. Steady-State Evaporation

The term "steady-state" is not really applicable to fuel drop evaporation or drop combustion because it can easily be demonstrated that in high-temperature flames a fuel drop seldom attains a steady state during its lifetime. This is especially true for multicomponent fuel drops, which may contain several different petroleum compounds, each possessing its own individual physical and chemical properties. Nevertheless, for most light distillate fuel oils, it is convenient to consider a quasi-steady gas phase that embodies the main features of the mass and thermal diffusion processes and that allows mass evaporation rates and drop lifetimes to be estimated to a reasonable level of accuracy.

Measurement of Evaporation Rate. Two general approaches have been made in the experimental study of single-drop evaporation. In one, a drop is suspended from a silica fiber or thermocouple wire. (The latter method allows the liquid temperature to be measured.) The change of drop diameter with time is recorded using a cinecamera operating at a speed of around 100 frames per second. The elliptical shape of the drop is corrected to a sphere of equal volume. After an initial transient period, steady-state evaporation is soon established, and the drop diameter diminishes with time according to the relationship

$$D_0^2 - D^2 = \lambda t \tag{60}$$

This is sometimes called the "D^2 law" of droplet evaporation. λ is known as the "evaporation constant."

An alternative method of determining evaporation rates is by feeding the liquid into the inside of a hollow sphere constructed with a porous wall, as illustrated in Figure 39. The liquid supply should be adjusted to maintain a wetted surface on the outside of the sphere. With this technique the diameter

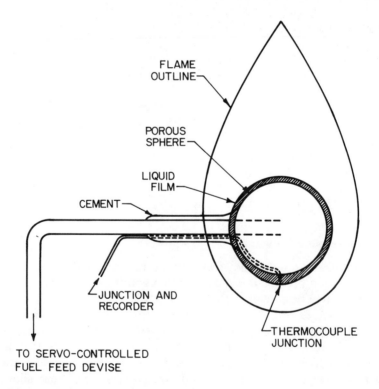

FIGURE 39. Schematic diagram of fuel-wetted porous sphere with thermocouple for measurement of liquid surface temperatures.

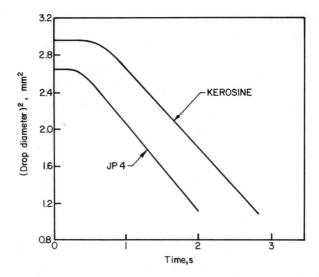

FIGURE 40. Burning rate curves for kerosine and JP4.[131]

of the vaporizing spherical surface remains constant, and the rate of vaporization is equal to the supply rate of the liquid. A typical plot of D^2 versus time is shown in Figure 40.[131] Values of λ for various fuels are listed in Table 6.

Theoretical Background. The development of drop evaporation theory has been largely motivated by the needs of the aero gas turbine and the liquid propellant rocket engine. Following Godsave[132] and Spalding[133] the approach generally used is to assume a spherical symmetrical model of a vaporizing drop in which the rate-controlling process is that of molecular diffusion. The following assumptions are usually made:

1. The droplet is spherical.
2. The fuel is a pure liquid having a well-defined boiling point.
3. Radiation heat transfer is negligible.

Except for conditions of very low pressure, or for highly luminous flames, these assumptions are considered valid.

Consider the hypothetical case of a pure fuel drop that is suddenly immersed into gas at high temperature. Following Faeth,[134] the ensuing evaporation process proceeds as follows: At normal fuel-injection temperatures the concentration of fuel vapor at the liquid surface is low, and there is little mass transfer from the drop in this initial stage, which corresponds to the first part of the plots in Figure 40, where the slope is quite small. Under these conditions the fuel drop heats up exactly like any other cold body when placed

TABLE 6. Some Values of Evaporation Constant for Various Stagnant Fuel-Air Mixtures

Fuel	Ambient Air Temperature, K	λ, m^2/sec	λ, ft^2/sec
Gasoline	971	1.06×10^{-6}	11.4×10^{-6}
Gasoline	1068	1.49×10^{-6}	16.0×10^{-6}
Kerosine	923	1.03×10^{-6}	11.1×10^{-6}
Kerosine	971	1.12×10^{-6}	12.1×10^{-6}
Kerosine	1014	1.28×10^{-6}	13.8×10^{-6}
Kerosine	1064	1.47×10^{-6}	15.8×10^{-6}
Diesel oil	473	0.79×10^{-6}	8.5×10^{-6}
Diesel oil	971	1.09×10^{-6}	11.7×10^{-6}

in a hot environment. Due to the limited heat conductivity of the fuel, the temperature inside the drop is not uniform but is cooler at the center of the drop than at the liquid surface.

As the liquid temperature rises so also does the partial pressure and concentration of fuel vapor at the surface. This increases the rate of dispersion of fuel vapor away from the droplet which has two effects: (1) a larger proportion of the heat transferred to the drop is needed to furnish the heat of vaporization of the liquid, and (2) the outward flow of fuel vapor impedes the rate of heat transfer to the drop. This tends to diminish the rate of increase of the surface temperature so that the level of temperature within the droplet becomes more uniform. Eventually, a stage is reached where all of the heat transferred to the droplet is used as heat of vaporization, and the droplet temperature stabilizes at its "wet bulb temperature." This condition corresponds to the straight lines drawn in Figure 40.

Mass Transfer Number. An expression for the rate of evaporation of a fuel drop may be derived as follows: More comprehensive and rigorous treatments are provided in the publications of Faeth,[134] Goldsmith and Penner,[135] Kanury,[136] Spalding,[137] and Williams.[138]

Neglecting thermal diffusion, and assuming that the driving force for species diffusion is a concentration gradient in the direction of the diffusion path, the following expression is obtained for an evaporating drop:

$$\dot{m}_F = 2\pi D_s \left(\frac{k}{c_p} \right)_g \ln(1 + B_M) \qquad (61)$$

This is the basic equation for the rate of evaporation of a fuel drop of diameter, D_s. Its accuracy is very dependent on the choice of values of k, c_p, and B_M. According to Hubbard et al.,[139] best results are obtained using the one-third rule of Sparrow and Gregg,[140] where average properties are evaluated at the following reference temperatures and compositions:

$$T_r = T_s + \frac{T_\infty - T_s}{3} \tag{62}$$

$$Y_{F_r} = Y_{F_s} + \frac{Y_{F_\infty} - Y_{F_s}}{3} \tag{63}$$

where T is temperature, Y_F is mass fraction of fuel vapor, and subscripts r, s, and ∞ refer to reference, surface, and ambient conditions. If the fuel concentration at an infinite distance from the drop is assumed to be zero, then Eq. (63) becomes

$$Y_{F_r} = \tfrac{2}{3} Y_{F_s} \tag{64}$$

and

$$Y_{A_r} = 1 - Y_{F_r} = 1 - \tfrac{2}{3} Y_{F_s} \tag{65}$$

Equations (62)–(65) are used to calculate the reference values of the relevant physical properties of the vapor-air mixture that comprises the environment of an evaporating drop. For example, the reference specific heat at constant pressure is obtained as

$$c_{p_g} = Y_{A_r}\left(c_{p_A} \text{ at } T_r\right) + Y_{F_r}\left(c_{p_v} \text{ at } T_r\right) \tag{66}$$

The reference value of thermal conductivity is estimated in a similar manner as

$$k_g = Y_{A_r}\left(k_A \text{ at } T_r\right) + Y_{F_r}\left(k_v \text{ at } T_r\right) \tag{67}$$

Heat Transfer Number. Similar arguments to those employed above, but based on considerations of conductive and convective heat fluxes across a thin shell surrounding the evaporating drop, lead to the following expression for heat transfer number:

$$B_T = \frac{c_{p_g}(T_\infty - T_s)}{L} \tag{68}$$

where L is the latent heat of fuel vaporization corresponding to the fuel surface temperature, T_s.

B_T denotes the ratio of the available enthalpy in the surrounding gas to the heat required to evaporate the fuel. As such it represents the driving force for the evaporation process. Where heat transfer rates are controlling to evapora-

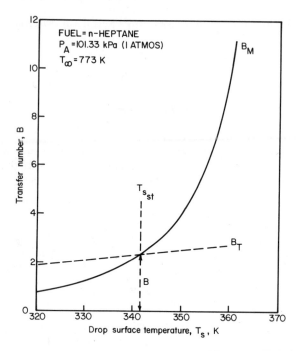

FIGURE 41. Variation of B_M and B_T with drop surface temperature.

tion, the rate of fuel evaporation for a Lewis number of unity is obtained as

$$\dot{m}_F = 2\pi D \left(\frac{k}{c_p} \right)_g \ln(1 + B_T) \tag{69}$$

At low fuel temperatures the value of B_T is higher than that of B_M. B_M and B_T both increase with increase in fuel temperature, but the rate of increase of B_M is higher than that of B_T so that plots of B_M and B_T versus T_s eventually intersect, as illustrated in Figure 41. At the point of intersection, which corresponds to the "steady-state" or "wet bulb" temperature, $B_M = B_T$, and either Eq. (61) or Eq. (69) may be used to calculate the rate of fuel evaporation. The advantage of Eq. (61) is that it applies under all conditions, including the transient state of droplet heat-up. On the other hand, Eq. (69) is usually easier to evaluate, since the magnitude of the various terms are either contained in the data of the problem or are readily available in the literature. However, its practical utility is restricted mainly to the steady-state portion of the evaporation process.

Calculation of Steady-State Evaporation Rate. The term "steady state" is used to describe the stage in the drop evaporation process where the drop surface has attained its "wet bulb temperature" and all of the heat reaching the surface is utilized in providing the latent heat of vaporization. Where $T_{s_{st}}$ is known, the transfer number B is easy to evaluate. We have

$$B_M = \frac{Y_{F_s}}{1 - Y_{F_s}} \tag{70}$$

Now

$$Y_{F_s} = \frac{P_{F_s} M_F}{P_{F_s} M_F + \left(P - P_{F_s}\right) M_A} \tag{71}$$

$$= \left[1 + \left(\frac{P}{P_{F_s}} - 1\right) \frac{M_A}{M_F}\right]^{-1} \tag{72}$$

where P_{F_s} is the fuel vapor pressure at the drop surface, P is the ambient pressure, which is the sum of the fuel vapor pressure and the air partial pressure at the droplet surface, and M_F and M_A are the molecular weights of fuel and air, respectively.

For any given value of surface temperature, the vapor pressure is readily estimated from the Clausius-Clapeyron equation:

$$P_{F_s} = \exp\left[a - \frac{b}{T_s - 43}\right] \tag{73}$$

Values of a and b for several hydrocarbon fuels are listed in Table 7.

TABLE 7. Some Relevant Thermophysical Properties

Fuel	n-Heptane	Aviation Gasoline	JP4	JP5	DF2
T_{cr}, K	540.17	548.0	612.0	684.8	725.9
$L_{T_{bn}}$, kJ/kg	317.8	346	292	266.5	254
Molecular mass	100.16	108.0	125.0	169.0	198.0
Density, kg/m³	687.8	724	773	827	846
T_{bn}, K	371.4	333	420	495.3	536.4
a in Eq. (16) for $T > T_{bn}$	14.2146	14.1964	15.2323	15.1600	15.5274
a in Eq. (16) for $T < T_{bn}$	14.3896	13.7600	15.2323	15.1600	15.5274
b in Eq. (16) for $T > T_{bn}$	3151.68 K	2777.65 K	3999.66 K	4768.77 K	5383.59 K
b in Eq. (16) for $T < T_{bn}$	3209.45 K	2651.13 K	3999.66 K	4768.77 K	5383.59 K

At the steady-state condition, $B_M = B_T = B$, and the mass rate of fuel evaporation is given by

$$\dot{m}_F = 2\pi D_s \left(\frac{k}{c_p}\right)_g \ln(1 + B) \qquad (74)$$

2.3. Drop Lifetime

In many practical combustion systems, the rates of chemical reaction are so high that the burning rate is controlled mainly by the rate of fuel vaporization. In such situations the evaluation of drop lifetime could be important since it determines the residence time needed to ensure completion of combustion. Droplet evaporation times are also important to the design of premix-pre-vaporize combustors for gas turbines, since they govern the length of the prevaporization passages.

One of the first theoretical attacks on the problem of droplet evaporation was made by Godsave.[132] Using the usual assumptions of a spherical system with a steady temperature distribution, where both the flow and temperature distribution are spherically symmetrical, he derived the rate of vaporization of a single droplet as follows:

$$m_F = \rho_F \left(\frac{\pi D^3}{6}\right)$$

$$\frac{\dot{m}_F}{\rho_F} = \frac{d}{dt}\frac{\pi D^3}{6}$$

$$= \frac{\pi D^2}{2}\frac{dD}{dt}$$

$$= -\left(\frac{\pi}{4}\right)\lambda D$$

where

$$\lambda = -\frac{d(D^2)}{dt} \qquad (75)$$

This definition of the evaporation constant, λ, by Godsave and his co-worker Probert,[141] greatly simplifies many engineering calculations. For example, the rate of evaporation of a liquid droplet is given by

$$\dot{m}_F = \left(\frac{\pi}{4}\right)\rho_F \lambda D \qquad (76)$$

Also the drop lifetime is readily obtained by assuming that λ is constant and by integrating Eq. (76) to get

$$t_e = \frac{D_0^2}{\lambda} \tag{77}$$

By comparing Eqs. (74) and (76), it is apparent that the evaporation constant, λ, can be expressed as

$$\lambda = \frac{8\ln(1 + B)}{\rho_F \left(c_p/k\right)_g} \tag{78}$$

The evaporation time of a droplet is thus equal to

$$t_e = \frac{D_0^2}{\lambda} = \frac{\rho_F D_0^2}{8\left(k/c_p\right)_g \ln(1 + B)} \tag{79}$$

It should be noted that when chemical reaction rates are so high that the rate of combustion is limited by the rate of fuel evaporation, the above expressions for $\dot{m}_F$ also represent the rate of combustion. Furthermore the droplet evaporation time, t_e, in Eq. (79) then becomes the droplet combustion time.

2.4. Evaporation at High Temperature

In many practical combustion systems the liquid fuel is sprayed directly into the flame zone so that evaporation occurs under conditions of high ambient temperature, usually around 20,000 K. At such elevated temperatures the wet bulb temperature is only slightly lower than the boiling temperature at that pressure. This observation has been made previously by Spalding,[137] Kanury,[136] and others. It can be illustrated by reference to Figure 42 which shows B_M and B_T plotted against temperature. B_M is low at low temperatures but increases rapidly with increase in temperature and becomes asymptotic to the vertical line drawn through $T_s = T_b$. The figure shows that there is a wide range of values of B_T, corresponding to a wide range of high values of ambient temperature, over which $T_{s_{st}} \simeq T_b$. This means that for high temperatures, it is usually satisfactory to substitute T_b for T_s in Eq. (68) so that the steady-state value of B becomes

$$B = B_{T_{st}} = \frac{c_{p_g}(T_\infty - T_b)}{L} \tag{80}$$

Clearly this substitution greatly simplifies the calculation of B_T and λ_{st}.

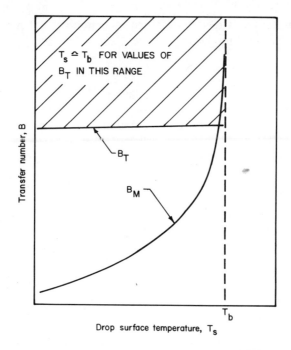

FIGURE 42. Graphs illustrating that $T_s \simeq T_b$ for high values of B_T corresponding to high ambient temperatures.

2.5. Heat-up Period

Figure 40 shows a straight line relationship between (droplet diameter)^2 and time during most of the evaporation period. However, inspection of the graphs reveals that the slope of the D^2/t line is almost zero in the first stage of evaporation and then gradually increases with time until eventually it reaches its steady-state value, which then remains fairly constant throughout the remainder of the drop lifetime. The D^2/t relationship exhibited in these curves is consistent with the notion that initially the rate of evaporation is quite low, so most of the heat transferred to the droplet from its surroundings is employed in heating up the droplet. As the fuel temperature rises, the fuel vapor concentration at the drop surface increases, and a larger proportion of the heat reaching the drop goes to supply the latent heat of vaporization than goes to sensible heat. This applies especially to high-volatility fuels where the net heat transfer from the ambient gas to the drop goes mostly to latent heat of vaporization. Eventually, the drop stabilizes at its wet bulb temperature, $T_{s_{st}}$, and during the remainder of the drop lifetime all of the heat reaching the drop surface is utilized as heat of vaporization.

Although the fuel drops reach their wet bulb temperature asymptotically with time, for the purposes of analysis and calculation, the vaporization time can roughly be separated into a "heat-up" period and a "steady-state" period. For low-volatility fuels and low ambient temperatures, the heat-up period is only a very small portion of the total evaporation time. However, for many fuels at high ambient pressures and temperatures, the heat-up period is relatively much larger in magnitude and cannot be neglected.[142]

2.6. Convective Effects on Evaporation

For drop evaporation under quiescent conditions, the principal mode of heat transfer is by thermal conduction. Where relative motion exists between the drop and the surrounding air or gas, the rate of fuel evaporation is enhanced. Surprisingly, perhaps, convective effects do not change the steady-state evaporation temperature, $T_{s_{st}}$. This is because forced convection affects both the rate of heat transfer to the fuel drop and the rate of fuel evaporation to the same extent so that under steady-state conditions the two effects cancel each other. Also the duration of the heat-up period is unaffected by convective effects.[142] However, since rates of fuel evaporation are enhanced by convection, the total drop lifetime must diminish. Thus the effect of convection is to reduce the duration of the steady-state portion of the drop evaporation process.

If the rate of heat transfer to a fuel drop per unit surface area per unit time is denoted by h, then

$$h = \frac{\dot{m}_F L}{\pi D^2} \tag{81}$$

Substituting for $\dot{m}_F$ from Eq. (69) gives

$$h = \frac{2(k/c_p)_g L \ln(1 + B_T)}{D} \tag{82}$$

For constant fluid properties, in the absence of convection, the Nusselt number is

$$Nu = \frac{hD}{k(T_\infty - T_s)} \tag{83}$$

Substituting for h from Eq. (82) gives

$$Nu = 2\left(\frac{L}{c_{p_g}(T - T_s)}\right)\ln(1 + B_T)$$

or, since

$$\frac{c_{p_g}(T_\infty - T_s)}{L} = B_T$$

$$Nu = \frac{2\ln(1 + B_T)}{B_T} \tag{84}$$

As B_T tends toward zero, $\ln(1 + B_T)$ approaches unity, and $Nu = 2$, which is the normal value for low heat transfer rates to a sphere.

In a comprehensive theoretical and experimental study, Frossling[143] showed that the effects of convection on heat and mass transfer rates can be accommodated by a correction factor which is a function of Reynolds number and Schmidt (or Prandtl) number. Where diffusion rates are controlling, the correction factor is

$$1 + 0.276Re_D^{0.5}Sc_g^{0.33} \tag{85}$$

For the more usual case where heat transfer rates are controlling, it becomes

$$1 + 0.276Re_D^{0.5}Pr_g^{0.33} \tag{86}$$

The velocity term in Re_D should be the relative velocity between the drop and the surrounding gas, that is, $Re_D = UD/\nu_g$. However, both calculations and experimental observations suggest that small droplets rapidly attain the same velocity as the surrounding gas, after which they are susceptible only to the fluctuating component of velocity, u'. The appropriate value of Reynolds number then becomes $(u'D/\nu_g)$.

Another important convective heat transfer correlation is the following due to Ranz and Marshall[144]

$$Nu = 2 + 0.6Re_D^{0.5}Pr_g^{0.33} \tag{87}$$

which corresponds to a correction factor of

$$1 + 0.3Re_D^{0.5}Pr_g^{0.33} \tag{88}$$

Note that the physical properties ρ_g, μ_g, c_{p_g}, and k_g embodied in Re_D and Pr_g should be evaluated at the reference temperature, T_r. Faeth[134] has analyzed the available data on convective effects and has proposed a synthesized correlation for Nu that approaches the correct limiting values at low and high Reynolds numbers ($Re_D < 1800$). This correlation yields the following correction factor to account for the augmentation of evaporation due to forced convection:

$$\frac{1 + 0.276\mathrm{Re}_D^{0.5}\mathrm{Pr}_g^{0.33}}{\left[1 + 1.232/\left(\mathrm{Re}_D\mathrm{Pr}_g^{1.33}\right)\right]^{0.5}} \tag{89}$$

Thus the primary factors affecting the lifetime of a drop are

1. The physical properties of the ambient air or gas, that is, pressure, temperature, thermal conductivity, specific heat, and viscosity.
2. The relative velocity between the drop and the surrounding medium.
3. The properties of the fuel and its vapor, including density, vapor pressure, thermal conductivity, and specific heat.
4. The initial condition of the drop, especially size and temperature.

The combination of Eqs. (74) and (88) yields the following equation for the rate of fuel evaporation with forced convection:

$$\dot{m}_F = 2\pi D\left(\frac{k}{c_p}\right)_g \ln(1 + B)\left(1 + 0.3\mathrm{Re}_D^{0.5}\mathrm{Pr}_g^{0.33}\right) \tag{90}$$

This equation gives the instantaneous rate of fuel evaporation for a droplet of diameter, D. To obtain the *average* rate of evaporation of the drop during its lifetime, the constant should be reduced from 2 to 1.33. Substituting also for $\overline{D} = 0.667D_0$ and $\mathrm{Pr} = 0.7$ gives

$$\overline{\dot{m}}_F = 1.33\pi D_0\left(\frac{k}{c_p}\right)_g \ln(1 + B)\left(1 + 0.22\mathrm{Re}_{D_0}^{0.5}\right) \tag{91}$$

and

$$t_e = \frac{\rho_F D_0^2}{8(k/c_p)_g \ln(1 + B)\left(1 + 0.22\mathrm{Re}_{D_0}^{0.5}\right)} \tag{92}$$

2.7. Effective Evaporation Constant

During steady-state evaporation all of the heat transferred to the drop is employed in fuel vaporization, and consequently evaporation rates are relatively high. However, during the heat-up period much of the heat transferred to the drop is absorbed in heating up the drop so that the amount of heat that is available for fuel vaporization is correspondingly less. This lower rate of vaporization, when considered in conjunction with the significant proportion of the total drop lifetime that is occupied by the heat-up period, means that overall evaporation rates can be appreciably lower, and drop lifetimes much longer, than the corresponding values calculated on the assumption that the heat-up time is zero.

From a practical viewpoint it would be very convenient if the effect of the heat-up period could be combined with that of forced convection in a manner that could lead to the derivation of an "effective" value of evaporation constant for any given fuel at any stipulated conditions of ambient pressure, temperature, velocity, and drop size. To achieve this goal, it is necessary to select a correlative fuel property that will define to a sufficient degree of accuracy the evaporation characteristics of any given fuel. It is recognized that no single chemical or physical property is completely satisfactory for this purpose. However, the average boiling point (50% recovered) has much to commend it, since it is directly related to fuel volatility and fuel vapor pressure. It also has the virtue of being easy to measure, and is usually quoted in fuel specifications. For these reasons the average boiling point was chosen by Chin and Lefebvre[145] to characterize the fuel's propensity for evaporation.

The effective evaporation constant is defined as

$$\lambda_{eff} = \frac{D_0^2}{t_e} \tag{93}$$

where t_e is the total time required to evaporate the fuel drop, including both convective and transient heat-up effects.

The drop diameter at the end of the heat-up period, D_1, is readily obtained as

$$D_1^2 = D_0^2 - \lambda_{hu}\Delta t_{hu} \tag{94}$$

and the drop lifetime is given by

$$t_e = \Delta t_{hu} + \Delta t_{st} \tag{95}$$

$$= \Delta t_{hu} + \frac{D_1^2}{\lambda_{st}} \tag{96}$$

From Eqs. (88) and (94) we have

$$t_e = \Delta t_{hu} + \frac{\left[D_0^2 - \lambda_{hu}\Delta t_{hu}\left(1 + 0.3\mathrm{Re}_{hu}^{0.5}\mathrm{Pr}_g^{0.33}\right)\right]}{\lambda_{st}\left(1 + 0.3\mathrm{Re}_{st}^{0.5}\mathrm{Pr}_g^{0.33}\right)} \tag{97}$$

in which Re_{hu} and Re_{st} are based on D_{hu} and D_{st}, respectively. (Note that $D_{st} = 0.25D_1$.) Values of λ_{eff}, calculated from Eqs. (93) and (97), are shown plotted in Figure 43. This figure shows plots of λ_{eff} versus T_{bn} for various values of UD_0, and three levels of ambient temperature, namely 500, 1200, and

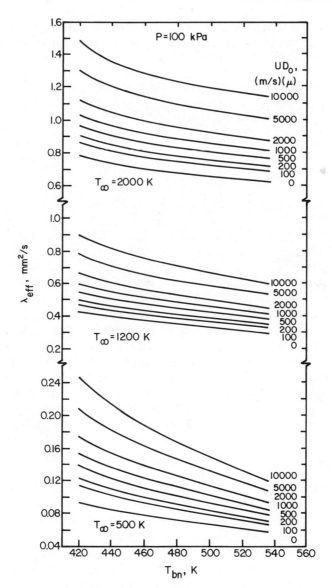

FIGURE 43. Values of effective evaporation constant for $P_A = 101$ kPa.[145] Reproduced with permission from Hemisphere Publishing Corp., N. Y. *Atomization and Sprays*, edited by A. H. Lefebvre.[213]

2000 K. They show, in general, that λ_{eff} increases with increase in ambient temperature, velocity, and drop size and diminishes with increase in normal boiling temperature. Figure 43 is drawn for a pressure of 100 kPa (1 atm). Similar charts for 1 MPa and 2 MPa are contained in Chin and Lefebvre.[145]

The concept of an effective value of evaporation constant considerably simplifies calculations on the evaporation characteristics of fuel drops. For example, for any given conditions of pressure, temperature, and relative velocity, the lifetime of a fuel drop of any given size is obtained from Eq. (93) as

$$t_e = \frac{D_0^2}{\lambda_{eff}} \tag{98}$$

whereas the average rate of fuel evaporation is determined by dividing the initial mass of the droplet, $(\pi/6)\rho_F D_0^3$, by t_e from Eq. (98) to obtain

$$\dot{m}_F = \left(\frac{\pi}{6}\right)\rho_F \lambda_{eff} D_0 \tag{99}$$

2.8. Droplet Burning

Most experimental studies on the burning of single fuel drops have either used drops suspended at the end of wires or have simulated the drop-burning process by the use of a porous sphere covered with a liquid film, as discussed earlier. Spalding's early studies[133] on the influence of relative air velocity on the combustion of liquid fuel spheres revealed a critical velocity above which the flame could not be supported at the upstream portion of the sphere. At air velocities below the critical level, the sphere is entirely enveloped in flame. But at higher relative velocities, the flame on the upstream half of the sphere is extinguished, and the burning zone is confined to a small wake region, similar to the flame behind a bluff-body flameholder. Thus within a liquid fuel-fired combustion zone, if the temperature and oxygen concentration of the ambient gas are both high enough to effect rapid ignition, and the relative air velocity is below the critical value, the fuel drop will become surrounded by a thin spherical flame. This "attached" flame is sustained by the exothermic reaction of fuel vapor and oxygen in the thin combustion zone, with oxygen diffusing radially inward from the outer regions (against the outward flow of combustion products formed at the flame), while the fuel vapor diffuses radially outward from the drop surface. The process is quite analogous to droplet evaporation without combustion, except that in the latter case the heat is derived from regions far removed from the drop, whereas in the burning case the heat is supplied from the flame.

Measurements of temperature and concentration profiles were conducted by Aldred et al.[146] at the lower stagnation point of a 9.2-mm porous sphere burning n-heptane fuel in still air. Their results show that in addition to oxidation there is significant cracking of the fuel vapor between the flame zone and the liquid surface. This cracking produces the carbon particles that are largely responsible for the characteristic yellow flames of fuel sprays.

As the burning rates of individual fuel drops are limited by evaporation rates, the equations derived in the previous sections for the mass evaporation rate, $\dot{m}_F$, and the drop lifetime, t_e, are directly applicable. Since heat transfer processes are dominant, it is appropriate to use B_T for the transfer number. At the beginning of droplet combustion, while the temperature at the center of the drop still retains its initial value, a suitable expression for B is

$$B = \frac{c_{P_g}(T_f - T_b)}{L + c_{P_F}(T_b - T_{F_0})} \tag{100}$$

where

$$T_f = \text{flame temperature}$$
$$T_b = \text{boiling point of fuel}$$
$$T_{F_0} = \text{initial fuel temperature}$$
$$L = \text{latent heat of vaporization at } T_b$$
$$c_{P_g} = \text{mean specific heat of gas between } T_b \text{ and } T_f$$
$$c_{P_F} = \text{mean specific heat of fuel between } T_b \text{ and } T_{F_0}$$

As burning proceeds, the flow recirculations induced within the drop soon create a fairly uniform temperature distribution so that, throughout its volume, the drop temperature is everywhere fairly close to the surface value, T_b. Thus under steady-state conditions the transfer number simplifies to

$$B = \frac{c_{P_g}(T_f - T_b)}{L} \tag{101}$$

Substituting this value of B into Eq. (90) gives the following expression for the burning rate of a fuel drop:

$$\dot{m}_F = 2\pi D\left(\frac{k}{c_p}\right)_g \ln\left[1 + \frac{c_{P_g}(T_f - T_b)}{L}\right]\left(1 + 0.30\mathrm{Re}_D^{0.5}\mathrm{Pr}_g^{0.33}\right) \tag{102}$$

and the lifetime of a burning droplet is obtained from Eqs. (77), (78), and (92) in conjunction with Eq. (101). For example, Eq. (92), which defines the lifetime of a droplet under conditions where the heat-up period is effectively zero, becomes

$$t_e = \frac{\rho_F D_0^2}{8\left(k/c_p\right)_g \ln\left[1 + c_{P_g}(T_f - T_b)/L\right]\left(1 + 0.22\mathrm{Re}_{D_0}^{0.5}\right)} \tag{103}$$

2.9. Multicomponent Fuel Drops

Much of the available information on drop evaporation and combustion has been gained using pure fuels, such as *n*-heptane. For such fuels, the boiling temperature is clearly defined, and the drop cross-sectional area decreases linearly with time during the steady-state evaporation (or combustion) period. Commercial fuels, however, are generally multicomponent mixtures of various petroleum compounds possessing individually different physical and chemical properties. According to Wood et al.,[131] as combustion proceeds the composition of a multicomponent fuel drop changes by a process of simple batch distillation; that is, the vapors produced are removed continuously without further contact with the residual liquid mixture. Hence, as burning proceeds, the more volatile constituents of the liquid drop vaporize first and the concentration of the higher boiling fractions in the liquid phase increases. The drop cross-sectional area decreases in a nonlinear manner with time. This characteristic, incidentally, offers a suitable means of analyzing the vaporization process occurring during the combustion of a multicomponent fuel.

There have been many experimental studies on multicomponent fuels, and numerous analyses have been made to compare the predictions of diffusion flame models of droplet combustion with measurements at moderate pressures. A review of these studies is beyond the scope of the present work, but further information is available in the various publications of El Wakil, Faeth, Law, Sirignano, and Chin and Lefebvre.[134, 147–165]

The available experimental evidence on light distillate petroleum fuels suggests that λ remains sensibly constant during steady-state evaporation, as illustrated in Figure 40 for JP4 and kerosine burning in air. However, as pointed out by Faeth,[134] this may be the case for combustion, where flame temperatures are much higher than drop temperatures, but it should not be expected for drop evaporation in a low-temperature environment. At low gas temperatures, variation of drop-surface temperature due to composition changes exerts a stronger influence on the temperature difference between the drop and its surroundings, causing significant variations in B_M. More experimental work is needed on the vaporization characteristics of multicomponent fuels in low-temperature gas streams.

The combustion of medium and heavy fuel oils is complicated by the disruptive boiling and swelling of the burning drops and the formation of a carbonaceous residue.[131, 138] The "D^2 law" no longer holds, although apparently it is still possible to define equivalent burning rate constants.

According to Williams[138] the course of combustion of a medium to heavy fuel oil drop can be summarized as follows:

1. Heating up of the droplet and vaporization of the low boiling point components.
2. Self-ignition and combustion with slight thermal decomposition and continued vaporization of the volatile components.

3. Extensive disruptive boiling and swelling of the droplet together with considerable thermal decomposition to give a heavy tar.

4. Combustion of remaining volatile liquids and gases from the decomposition of the heavy tar which collapses and forms a carbonaceous residue having a open structure called a cenosphere.

5. Slow heterogeneous combustion of the carbonaceous residue at a rate of about one-tenth that of the initial droplet (in terms of burning rate constants).

It is claimed that when disruptive swelling is augmented by emulsification of water with the fuel, extensive swelling of the droplet can produce secondary atomization due to smaller droplets being explosively ejected from the parent drop.

2.10. Drop Transport in Sprays

In order to determine the fuel distribution downstream of a fuel injector located within a gas stream, it is necessary to calculate the droplet trajectory, which is described by the droplet motion equations (two-dimensional motion) as follows:

$$\frac{dU_{D \cdot x}}{dt} = \frac{3}{4} \frac{\rho_g}{\rho_L} \frac{C_D}{D} |U_R|(U_{g \cdot x} - U_{D \cdot x}) \tag{104}$$

$$\frac{dU_{D \cdot y}}{dt} = \frac{3}{4} \frac{\rho_g}{\rho_L} \frac{C_D}{D} |U_R|(U_{g \cdot y} - U_{D \cdot y}) \tag{105}$$

where C_D is the droplet drag coefficient. Subscripts x and y represent the velocity components in the x and y directions. Clearly in any calculations concerned with fuel distribution or spray combustion that involve droplet trajectory, a most important parameter is the droplet drag coefficient. Unfortunately this is difficult to determine due to the drop deformation that occurs if there is a high relative velocity between the drop and the surrounding gas. Furthermore, in a high-temperature environment where the moving drop is evaporating, the evaporation process thickens the boundary layer, which changes the drag coefficient. Under certain conditions the moving droplet may have an attached flame, in which case its drag coefficient will vary from that of a solid sphere by an amount that depends on whether the attached flame is an "envelope" flame that completely surrounds the drop, or a "wake" flame that, as its name suggests, is confined to a small recirculation zone just downstream of the moving drop. Due to these and other factors, it is hardly surprising that wide differences exist between the results obtained by various workers.

Drag Coefficient of a Solid Sphere. Stokes[166] analyzed the steady flow of a real fluid past a solid sphere (or a solid sphere moving in a real fluid). He showed that the drag force on the sphere can be expressed as

$$F = 3\pi D \mu_L U_R$$

If we define the drag coefficient C_D as

$$C_D = \frac{F}{(\pi/4) D^2 \left(\rho_g U_g^2 / 2 \right)} \tag{106}$$

then the drag coefficient can be expressed as

$$C_D = \frac{24}{\text{Re}} \tag{107}$$

where

$$\text{Re} = \frac{U_R D \rho_g}{\mu_g}$$

Stokes' law holds only for flows that are entirely dominated by viscosity forces, that is, for flows characterized by low Reynolds number. For higher Reynolds numbers the inertia forces become significant, and the drag force may be increased due to the formation of a wake and possible detachment of energy-consuming vortices. Another departure from Stokes' law is caused by the presence of a boundary layer, since the actual volume moving through the fluid consists of the sphere together with its boundary layer. These considerations have led to various theoretical correction factors to Stokes' drag law, but most of the correlations for drop drag coefficient in common use are based on the analysis of experimental data.

There are many empirically determined functions that approximate to the experimental data obtained by different investigators. For example, Langmuir and Blodgett[167] suggested the following equation:

$$C_D \frac{\text{Re}}{24} = 1 + 0.197 \text{Re}^{0.63} + 2.6 \times 10^{-4} \text{Re}^{1.38} \tag{108}$$

whereas according to Prandtl[168]

$$C_D = \frac{24}{\text{Re}} + 1 \quad (\text{for Re} < 1000) \tag{109}$$

Morsi and Alexander[169] suggest that in order to improve accuracy and keep the deviation from the experimental data to within 2%, the experimentally determined drag coefficient curve should be divided into a number of regions and the curve in each region should be approximated by an equation of the form

$$C_D = \frac{K_1}{Re} + \frac{K_2}{Re^2} + K_3 \tag{110}$$

Other researchers have used similar drag coefficients. For example, Mellor[170] employed the following equation for low values of Re for his prediction of droplet trajectories:

$$C_D = \frac{1}{Re}\left[23 + (1 + 16Re^{1.33})^{0.5}\right] \tag{111}$$

Onuma and Ogasawara,[171] on the other hand, prefer for spray combustion predictions the expression

$$C_D = \frac{24}{Re}(1 + 0.15Re^{0.687}) \tag{112}$$

which was obtained by Schiller and Naumann.[172]

All the above drag coefficient equations approach Stokes' law (i.e., $C_D = 24/Re$) as Reynolds number tends to zero. Some of these equations do not lend themselves to the mathematical integration that is necessary for trajectory predictions. Equations (109) and (110) are both suitable for integration, but the former is too simple and the latter is too complicated. This is why Putnam[173] tried to obtain an empirical equation that is suitable for integration and also reasonably accurate. He suggested the following equation for trajectory prediction; it fits the experimental data well and also is mathematically tractable:

$$C_D = \frac{24}{Re}\left(1 + \frac{1}{6}Re^{2/3}\right) \tag{113}$$

This equation is easily integrated to obtain droplet velocities and trajectories. It is recommended for Reynolds numbers less than 1000. It is convenient to use since it covers the Reynolds number range that is normally encountered in spray analysis. Since combustion modeling and spray analysis all involve droplet trajectory prediction, an equation for droplet drag coefficient equation that can be integrated easily is clearly very desirable.

Effect of Acceleration on Drop Drag Coefficient. So far we have considered only drops moving steadily in a flow field with no acceleration or deceleration. For this situation the drag coefficients obtained theoretically or experimentally are called the "standard" drag coefficient. However, within the combustion zone the fuel drops are accelerated by the gas flow if the gas velocity is higher than the droplet velocity, or decelerated if the gas velocity is lower than the droplet velocity. Thus the effect of acceleration on drag coefficient is always present, although for low accelerations the drop drag coefficient approaches the steady-state value.

Ingebo[174] conducted experiments in which clouds of liquid and solid spheres accelerating in airstreams were studied for various values of airstream pressure, temperature, and velocity. Diameter and velocity data for individual drops and solid spheres in the clouds were obtained with a high-speed camera. From these data the linear accelerations of spheres of 20 to 120 microns in diameter were determined, and instantaneous drag coefficients for unsteady momentum transfer were calculated. The drag coefficient for drops of iso-octane, water, and tri-chloroethylene, and solid spheres (magnesium and calcium silicide) were found to correlate with Reynolds number according to the following equation, for $6 < \text{Re} < 400$:

$$C_D = \frac{27}{\text{Re}^{0.84}} \tag{114}$$

Drag Coefficients for Evaporating and Burning Drops. The influence of evaporation on droplet drag may be due to two factors:

1. The effect of mass transfer on drag, known as the "blowing" effect.
2. The temperature and concentration gradient near the drop surface (due to evaporation) affects the dependence of drag coefficient on Reynolds number. This means that the physical properties chosen (variable properties or constant properties with different averaging methods) will affect the equation $C_D = f(\text{Re})$.

Yuen and Chen[175] have shown that the drag data for evaporating drops can be correlated with the drag data for solid spheres, provided that the average viscosity in the boundary layer is calculated at the $\frac{1}{3}$ reference state (see Eqs. (62) and (63)). A possible explanation for this result is that at high Reynolds numbers the decrease in friction drag due to evaporation is offset by the increase in pressure drag from flow separation due to blowing. At low Reynolds numbers, of course, mass efflux has little effect on drag. Thus, at both high and low Reynolds numbers, the effect of mass efflux (due to evaporation) from a sphere has little effect on the drag coefficient. Clearly this is most advantageous from the standpoint of combustor modeling.

Yuen and Chen's basic conclusion is that the deviation of drag coefficient of an evaporating drop from that of a corresponding nonevaporating drop is due not to the blowing effect but to changes in the physical properties of the gas at the drop surface.

Eisenklam et al.[176] determined drag coefficients using experimental data on small, free falling, burning, and/or evaporating drops of fuel. They proposed the following correlation for evaporating or burning drops:

$$C_{D,m} = \frac{C_D}{1 + B_M} \tag{115}$$

where $C_{D,m}$ is the drag coefficient evaluated at a mean condition for intense mass transfer.

Equation (115) is considered valid over the following range of conditions:

Drop diameter, 25–500 μm
Approach Re, 0.01–15
Transfer number, 0.06–12.3

Onuma[171] used Eqs. (112) and (115) to calculate the droplet drag for spray combustion as

$$C_D = \frac{24}{Re}(1 + 0.15Re^{0.687})(1 + B_M)^{-1} \tag{116}$$

The correlation obtained by Spalding,[137] based on his experimental data on the effect of intense mass transfer on the drag coefficient of flat plates in laminar flow, gives the reduction of drag by mass transfer as

$$\frac{\bar{C}_D}{C_D} = \frac{\ln(1 + B_M)}{B_M} \tag{117}$$

where $\bar{C}_D$ is the drag coefficient under intense mass transfer.

Both Eqs. (115) and (117) result in drag coefficients that are considerably lower than the standard values.

According to Law[177,178] the droplet drag coefficient can be written as

$$C_D = \frac{23}{Sc^{0.14} Re} \cdot (1 + 0.276 \cdot Sc^{0.33} \cdot Re^{0.5}) \cdot G(B) \tag{118}$$

where $G(B) = (1 + B_M)^{-1}$, (Eq. (115)), or $G(B) = B_M^{-1}\ln(1 + B_M)$, (Eq. (117)).

This expression is valid only for Re < 200. Law argues that it is seldom necessary to investigate the behavior of the system for larger Reynolds numbers, since, for most liquids that are not too viscous, the droplet would then become unstable and tend to break up. This statement is true only for low pressures; for high pressures and relative low temperatures, Re > 1000 is quite common. As Re approaches zero, this correlation fails to asymptote to

Stokes' expression, $C_D = 24/\text{Re}$, but the difference is small and is usually unimportant, since in such situations the drops follow the gas motion almost exactly.

The droplet drag coefficient equations suggested by Lambiris and Combs[179] and used by many researchers are as follows:

$$C_D = 27\text{Re}^{-0.84} \qquad \text{(for Re} < 80) \tag{119}$$

$$= 0.271\text{Re}^{0.217} \qquad \text{(for 80} < \text{Re} < 10{,}000) \tag{120}$$

$$= 2 \qquad \text{(for Re} > 10{,}000) \tag{121}$$

Finally, it can be stated that at the present time there is no single expression for drop drag coefficient that applies to all conditions. In the meantime the following equations are recommended:

1. For low temperatures, and Re < 1000, for easy integration use

$$C_D = \frac{24}{\text{Re}}\left(1 + \frac{1}{6}\text{Re}^{2/3}\right) \tag{113}$$

2. For low temperatures, over wide ranges of Re, it is more accurate to use

$$C_D = \frac{K_1}{\text{Re}} + \frac{K_2}{\text{Re}^2} + K_3 \tag{110}$$

3. For high temperatures (with the drop evaporating), use Yuen and Chen's correlation.
4. For burning conditions, use Eq. (120).

3. SPRAY COMBUSTION

Perhaps the most distinguishing feature of spray combustion is that, unlike premixed combustible gaseous systems, it is not uniform in composition. The fuel drops within the flame zone vary widely in size, and they move with different velocities and in different directions to that of the mainstream flow. Thus spray combustion is characterized by wide variations in the physical and chemical state of the burning mixture, leading to wide ranges of temperature, composition, and burning rate within the flame region. These difficulties, along with a lack of reliable experimental data, have inhibited the development of spray combustion models, in comparison to turbulent combustion processes in general. However, some greatly oversimplified models have been used with success, and examples of these are referred to in the following sections.

One obvious method of calculating the burning rates of fuel sprays is to assume that the spray is dilute, that the drops are spaced too far apart to have any appreciable influence on each other. The main advantage of this approach is that it allows all the theoretical and experimental knowledge that has been acquired on the evaporation and combustion rates of isolated drops to be applied directly to the combustion of fuel sprays. However, in many practical sprays the average distance between adjacent drops can be as short as a few drop diameters. Under these conditions the behavior of each drop will be strongly influenced by neighboring drops and, to a lesser extent, by all the drops in the spray. Clearly the presence of droplets cools the environment, thereby reducing the rate of heat transfer to the drops. Also, by increasing the local concentration of fuel vapor, the rate of mass transfer must also decrease. These two effects combine to lower the rate of vaporization.

3.1. Droplet Arrays, Groups, and Sprays

The problem of droplet interaction has been examined theoretically in terms of droplet arrays, droplet groups, and sprays.[160] Array theory assumes an artificial well-defined geometry or a large number of droplets in a periodic configuration. The number of drops in a typical array is small enough to have little influence on the ambient gas environment. Droplet group theory differs from array theory in that a statistical description of drop spacing, rather than a precise geometrical formulation, is employed. It can also deal with a larger number of droplets than array theory. Spray theory also uses a statistical or average representation of properties to take account of large numbers of drops. It considers full coupling among ambient gas properties, local gas properties, and drop properties, so that the effects of drop spacing on Nusselt number and vaporization rates can be included.

A simple, but very useful example of array theory was provided by Twardus and Brzustwoski[180] who considered the vaporization and combustion of two adjacent drops of equal size. They found a critical value of the ratio of the distance between droplet centers to the droplet radius above which the droplets burn with two separate flames and below which the two droplets burn together with one envelope flame. This result has considerable practical significance because in typical sprays the droplet spacing is smaller than this critical value. It is also consistent with Chigier's observations that in combustors the flame envelopes many droplets rather than single drops.[181,182]

Considerations of droplet group theory have revealed four different regimes of combustion; isolated drop combustion, internal group combustion, external group combustion, and sheath combustion. Which particular regime will be dominant in any given spray flame will depend on droplet number density, overall ambient conditions, droplet diameter, and fuel volatility. Chui and co-workers[162,183] considered quasi-steady diffusion and vaporization processes and derived a group combustion number as

$$G = 1.5(1 + 0.276 \, \mathrm{Re}^{1/2} \, \mathrm{Sc}^{1/3}) \mathrm{Le} \, N^{2/3}\left(\frac{D}{s}\right) \tag{122}$$

where Re, Sc, Le, N, D, and s are, respectively, the Reynolds number, Schmidt number, Lewis number, total number of droplets in the cloud, instantaneous average droplet diameter, and average spacing between the centers of the droplets. Thus G increases with Reynolds number, Prandtl number, and the size of the cloud (measured in droplet numbers), whereas it decreases with increasing Schmidt number and spacing between drops.

For values of G less than 10^{-2}, isolated droplet combustion occurs with a separate flame surrounding each droplet. For values of G between 10^{-2} and somewhere below unity, the principal mode of combustion involves a core within the cloud where vaporization is occurring with the core totally surrounded by flame. This process is described as internal group combustion. With further increase in G, the size of the core increases, and when a single flame envelops all droplets, we have external group combustion. This regime begins when G is close to unity so that many furnaces and gas turbine burners fall within this range. At very high values of G (> 100) only the drops in a thin layer at the edge of the clouds are evaporating. This regime is described by Chui and co-workers as external sheath combustion.

Other approaches to group combustion theory, also based on the assumption of infinitely fast chemical kinetics, have been made by Labowsky and Rosner,[184] Correa and Sichel,[185] and Kerstein and Law.[186] These and other theories of spray vaporization and combustion have been reviewed by Sirignano[160] and Chigier.[182]

3.2. Mathematical Models of Spray Combustion

A satisfactory model for spray combustion should represent a reasonable balance between accuracy of representation, utility, ease of use, economy of operation, and capability for further improvement. In recent years a large number of models have been proposed to account for the behavior of evaporating and burning sprays. Their practical applications include furnaces, gas turbine combustors, and direct injection stratified charge engines. The dilute spray assumption is generally made, which implies that direct interaction between drops can be neglected. This allows single-drop formulae to be used in calculating evaporation and burning rates. Droplet collisions are ignored, and the volume occupied by liquid fuel is considered negligibly small.

Burning sprays are usually modeled by assuming that the drops simply serve as distributed sources of fuel vapor that subsequently mix with air and burn in the manner of gas flames.[150] Swithenbank and co-workers[187-189] and Gosman and Johns[190] employ a single-step reaction with rates found from the smallest of eddy breakup and global Arrhenius expressions. Mongia and Smith[191] employ a single-step reaction with rates found from the smallest of eddy breakup and global Arrhenius expressions. They use a two-step reaction involving fuel and CO as reactants. Burning rates are determined from an eddy-breakup model and global Arrenhius expressions. The global kinetic parameters for this model were obtained by fitting predictions to measurements in a gas turbine combustor. El-Banhawy and Whitelaw[192] employ a statistical formulation for turbulent reaction using a $k - \varepsilon$ turbulence model with an assumed probability density function for mixture fraction.

The results achieved to date from mathematical modeling of spray flames have been quite modest, and even the most successful models can do little more than predict trends. Major obstacles to success have been insufficient knowledge of key spray properties, such as cone angle, mean drop size, and drop-size distribution, and on drag coefficients for evaporating/burning drops. Current research in these areas, along with continuing advances in computer technology, are expected to lead to the development of satisfactory models for spray evaporation and spray combustion in the forseeable future.

3.3. Dense Sprays

A dilute spray, by definition, is one in which the void space is so large that the system behaves as a collection of individual, isolated drops. With dense sprays, however, of the type found in diesel engines and in the region close to the atomizer in high-pressure gas turbines, the void fraction is low, and the volume occupied by the liquid phase is an appreciable fraction of the total spray volume. Interactions between drops and the effects of drop spacing on heat and mass transfer rates become very significant, as also do drop collisions and the modification of turbulence properties in the spray region due to the presence of high drop concentrations.

Some of the mathematical models discussed earlier address certain aspects of dense spray phenomena, for example, Gosman and Johns[190] allow for the volume occupied by the liquid in the governing equations for the gas phase. The most detailed treatment of very dense sprays yet reported is that due to O'Rourke and Bracco.[193] Their analysis includes the effects of drop collisions, drop heat-up, and vaporization for a fully interacting two-phase flow. The validity of the model was assessed by comparison with measurements of drop size and spray penetration carried out by Hiroyasu and Kadota[91] on a diesel-type injector. As drop vaporization was negligible for the experimental conditions, evaluation of the model was limited to the dynamics of an isothermal spray within a constant pressure chamber. Fairly good agreement was obtained between the predictions of the model and the measured values of spray penetration. Drop collisions were found to be very important, and O'Rourke and Bracco invoked this phenomenon to rationalize the very small calculated values of initial drop size with the much larger sizes measured downstream.

Progress toward a better understanding of very dense sprays is inevitably slow due to the experimental difficulties involved both in probing the flow and in trying to penetrate it by optical means. Some interesting ideas for modeling the dense spray region have been proposed, for example, the suggestion by Putnam[194] and adopted by O'Rourke and Bracco[193] to adapt the results of measurements in fluidized and packed beds. For further details of this approach, and other theoretical models for the dense spray region, reference should be made to the reviews by Chigier,[195,196] Faeth,[150] and Sirignano.[160]

3.4. Experiments on Spray Flames

In a classic experiment, using monodisperse tetralin sprays, Burgoyne and Cohen[197] found that for drop sizes less than 10 μm in diameter, flame speeds are the same as for fully premixed gaseous mixtures. This shows that for very small drops of a volatile fuel, droplet evaporation is completed early in the heat-up process, and the combustion process is relatively uninfluenced by the existence of the fuel spray. For the conditions of their experiment, larger drop sizes result in a reduction of flame speed due to the time required to evaporate the fuel. For large drop sizes and nonvolatile fuels, little fuel evaporation occurs upstream of the flame zone, and an enveloping flame surrounds each drop. In this case the end of the flame zone coincides with the disappearance of the largest drops.

Onuma and Ogasawara[171] conducted an experiment to compare the structures of a spray combustion flame and a turbulent gas diffusion flame in a vertical cylindrical furnace. Spatial distributions of drop concentration and size were measured by inserting a magnesium-oxide-coated slide, covered by a shutter. They also measured temperature distribution, velocity distribution, and gas concentrations with gas chromatography. The same measurements were made in a turbulent, gaseous diffusion flame using gaseous propane as fuel instead of kerosine. They found that the spray flame could be subdivided into three main regions: (1) an initial region consisting of a two-phase mixture in which most of the drops evaporate and where soot is formed due to the very rich mixture strengths, (2) an intermediate region in which the concentration

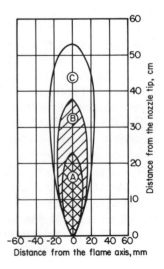

FIGURE 44. Zones in a spray flame.[171]

and size of drops are very low but concentrations of combustible gases are high, and (3) a final region where intermittent burning takes place as combustion is completed. These three regions are shown schematically in Figure 44. The direct comparisons made by Onuma and Ogasawara among temperature, velocity, and concentration profiles in gaseous diffusion flames and spray flames showed that they are very similar. It was concluded therefore that most of the drops do not burn individually, but that the vapor cloud from the evaporated drops burns like a turbulent diffusion flame. In this case the change of drop size can be calculated on the basis of single drops vaporizing in a hot gas environment.

In a later paper Onuma et al.[198] used a heavy oil to examine the extent to which the above conclusions remain valid. A heavy oil flame was compared with a kerosine flame under the same experimental conditions. The result showed that the shape and the measured profiles of various quantities were almost the same for both flames. They concluded that the heavy oil flame does not differ significantly in structure from the kerosine flame under the conditions of their experiment.

Mizutani et al.[199] studied a spray diffusion flame stabilized in a heated air stream, using a pressure atomizer to obtain a full-cone spray. Profiles of velocity, temperature, and liquid flux were determined at various positions in the flame. For this type of flame, it was found that the flame lies slightly inside the spray boundary and that drops participate in the combustion process to a certain degree. The temperature near the centerline is low, and maximum drop concentrations appear in this region at the downstream stations. This cool core region appears to have relatively low evaporation rates that are not indicative of strong combustion effects. It is likely that fuel vaporized in this region is transported to the flame zone by turbulent mixing, similar to a gaseous diffusion flame. For the conditions of the experiment conducted by Mizutani et al., the evaporated fuel does not burn near the axis, but in the outer regions of the spray. Also, in the vicinity of the flame zone, direct combustion of fuel drops may occur.

The picture of the spray combustion process that emerges from these studies on relatively simple spray configurations is shown in Figure 45, for a case where the flame is stabilized at the injector exit. The spray leaving the injector is nonuniform, with generally smaller drops at the periphery and larger drops near the centerline. The small drops readily exchange momentum with the gas, causing the spray jet to entrain surrounding air, similar to a gas jet. The rapid evaporation of the small drops feeds the turbulent diffusion flame positioned toward the outer portion of the flow. Hot combustion products are transported away from the combustion zone, providing a heated-gas surrounding to evaporate the fuel drops. Although drops are observed beyond the maximum reaction zone, small drops tend to follow the flow, and they may be largely confined to fuel-rich eddies, evaporating before significant concentrations of oxygen appear in their immediate surroundings. However, larger drops may pass through oxygen-rich regions due to their inertia. With

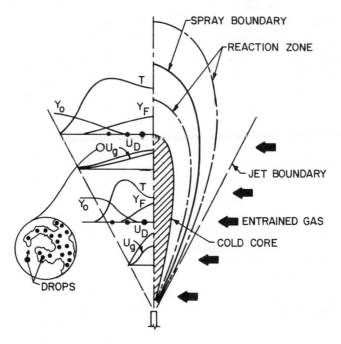

FIGURE 45. Schematic representation of a coaxial spray diffusion flame.[134]

increasing distance from the injector, centerline temperatures increase, causing more rapid evaporation of the large drops, particularly in the high-temperature reaction zone close to the centerline. The measurements also showed that the disappearance of drops is closely correlated with positions of maximum temperature, in both the radial and axial directions.

Chigier[181] proposed an idealized spray flame model that is based on the hypothesis that spray flames and gas diffusion flames are similar and that both temperature and oxygen concentrations are low within the spray. In the idealized model the flame acts as an interface, totally separating the inner fuel vapor from the outer air (see Figure 46). All the fuel vapor originates from vaporization at drop surfaces. Burning occurs as a consequence of fuel vapor diffusing outward and air diffusing inward to the flame front. The interface is convoluted by turbulence so that in the time mean fuel drops, and flame can occur at any one position but never at the same time. The drop velocity can either be greater or smaller than the fuel vapor velocity. The surrounding air velocity will vary from zero for stagnant air surroundings to values higher than drop velocities. All drops are injected into the spray through the atomizer, and vaporization of all drops is completed within the flame volume. Most of the drops are contained within the spray boundary so that there is a zone of fuel vapor, without drops, between the spray boundary and the flame. No interaction takes place between drops, and the separation between drops is suffi-

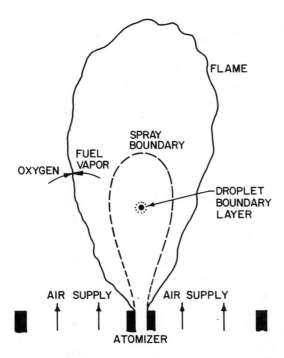

FIGURE 46. Vaporization of a typical droplet in an idealized spray flame.[182]

ciently large for each drop to be considered in isolation. A spherical boundary layer with a diameter of approximately twice the drop diameter, envelops the drop. Chigier ascribes bulk gas temperatures, velocities, and concentrations to the gas outside the boundary layer, while restricting the significant gradients of temperature, velocity, and concentration to within the boundary layer. The spray is subdivided into an initial "cool" zone, where temperatures are maintained at low levels due to strong quenching action of the drops, followed by a "hot" zone where the temperature rises due to turbulent convective transfer of heated products from the flame. Within the cool zone, heat transfer is restricted to radiation from the flame front to the drop surface. In the hot zone, heat transfer takes place both by radiation from the flame front and by turbulent convection. Predictions of the rate of vaporization of the drops within the spray can be made by using the methods described in the evaporation section.

Onuma and Ogasawara[171] have calculated vaporization rates and drop velocities in a spray flame. They showed that the rate of vaporization of drops is initially low because of the low temperature and high fuel vapor pressure of the gas in the cool zone. This rate of vaporization increases in the hot zone so that, for drops up to 150 μm in diameter, vaporization is just complete at the tip of the spray. For drops greater than 200 μm, it is possible that isolated drops will pass through the spray boundary and the flame. When these liquid drops come into contact with cooled chamber walls, deposition and coke formation can occur. Thus, in this idealized model of a spray flame, the rate of vaporization of drops may play only a minor role in the combustion process, and the rate-controlling step is the rate of turbulent diffusion of vapor and air to the flame front. This idealized model of spray combustion can be applied to situations where there is no evidence of envelope flames around individual drops, and only the flame front surrounds the spray. The experimental studies that have been made on spray flames suggest that the above model is a good approximation for the combustion of light, volatile fuels, and finely atomized sprays, but poor atomization and/or heavy fuels may enhance the importance of drop combustion, leading to a brushlike flame where the drops burn with individual flames. The factors that determine whether a drop burns individually or after it evaporates may be temperature, oxygen concentration of the surrounding gas, drop size, and relative velocity between the drop and the gas. In a poorly atomized spray, drop sizes are large, and the number of drops is correspondingly small. In this situation sufficient oxygen to allow the drops to burn individually may be available. In a well-atomized spray, drop sizes are very small and the spatial evaporation rate is very high. In consequence there is insufficient oxygen to support an envelope flame around each individual drop. The important parameter appears to be characteristic drop lifetime divided by a characteristic eddy dissipation time. It seems therefore that more information is required before a general statement can be made about the combustion conditions under which turbulent diffusion or drop evaporation respectively controls the structure of liquid spray flames.

Whether or not spray combustion is in the mode of droplet burning or in the mode of a gaseous diffusion flame, liquid sprays play a very important role in combustion performance, pollutant formation, and emissions from any kind of liquid-fueled combustor.[200-202] Local values of fuel/air ratio will determine combustion efficiency, temperature distribution, and pollutant formation, but local fuel/air ratio distribution is determined by the trajectories of individual drops, rates of vaporization, and mixing of fuel vapor and air. Drop size has a very strong influence on combustion. For instance, large drops may leave the combustion chamber unburned, causing loss of efficiency, or they may impinge on the liner walls to cause overheating or cooling at the wall. Too high a concentration of liquid drops will form a rich zone that will influence the soot formation and the radiative properties of the flame. Ignition is directly affected by the presence of liquid drops. The temperature distribution in the exit gases will be influenced by the location of the heat-release zone which, in turn, is governed by the drop trajectories and subsequent mixing. This is why it is important to know the trajectory of drops in order to calculate local and overall fuel distributions and fuel/air ratio distributions.

3.5. Combustion Performance

For most practical combustion systems, the key aspects of combustion performance are combustion efficiency, weak extinction limits, and light-up characteristics. Exhaust smoke and particulates can also be important, but the exhaust concentrations of these species are usually more dependent on fuel chemistry, especially aromatics content, and carbon/hydrogen ratio than on fuel spray characteristics.[63]

Combustion Efficiency. For a dilute spray that comprises a volume of air V, containing n fuel drops whose initial drop size is D_0, the average rate of fuel evaporation is given by (see Eq. (91))

$$\bar{\dot{m}}_F = 1.33\pi D_0 n \left(\frac{k}{c_p}\right)_g \ln(1 + B)\left(1 + 0.22\,\mathrm{Re}_{D_0}^{0.5}\right) \tag{123}$$

where B is the mass-transfer number. The Reynolds number in Eq. (123) is expressed in terms of the fluctuating velocity u' rather than the mainstream velocity U; that is, $\mathrm{Re} = u'D_0/\nu_g$. This modification is used because experience and observations suggest that most droplets rapidly assume the same velocity as the surrounding air, after which they are susceptible only to the fluctuating component of velocity.

The fuel/air ratio in the combustion zone is given by

$$q_c = \frac{n(\pi/6)D_0^3\rho_F}{V_c\rho_g} \tag{124}$$

or

$$n = \frac{6}{\pi}\frac{\rho_g}{\rho_F}\frac{V_c}{D_0^3}q_c \tag{125}$$

Substituting n from Eq. (125) into Eq. (123) yields

$$\dot{m}_F = 8\left(\frac{\rho_g}{\rho_F}\right)\left(\frac{k}{c_p}\right)_g (V_c/D_0^2)q_c \ln(1 + B)\left(1 + 0.22\,\mathrm{Re}_{D_0}^{0.5}\right) \tag{126}$$

It is assumed that the fuel instantly mixes and burns with the surrounding air as it evaporates. Thus the combustion efficiency is obtained as the ratio of the mass of fuel evaporated within the combustion zone to the mass of fuel supplied:

$$\eta_c = \frac{\dot{m}_F t_{res}}{\rho_g V_c q_c} \tag{127}$$

Substituting $\dot{m}_F$ from Eq. (126) into Eq. (127) gives

$$\eta_c = \frac{8(k/c_p)_g \ln(1 + B)\left(1 + 0.22\,\mathrm{Re}_{D_0}^{0.5}\right)t_{res}}{\rho_F D_0^2} \tag{128}$$

This equation may be used for calculating or correlating combustion efficiencies in situations where fuel evaporation is known to be the rate-controlling step. It shows, under these conditions, that combustion efficiency is improved by increases in fuel volatility, turbulence intensity, combustion volume, and gas pressure (via Re_{D_0}) and is impaired by increases in air mass flow rate and mean drop size.

Note that although it is mathematically possible for η_c to exceed unity; when this occurs, it simply means that evaporation is not limiting to combustion efficiency, and $\eta_c = 1.0$.

Unfortunately, little or no experimental data are available against which to test the validity of Eq. (128). This equation relates combustion efficiency to combustor dimensions (via t_{res}), to combustor design (via u' and ΔP_L), to combustor operating conditions (via k_g, c_{p_g}, and ρ_g), to atomizer characteristics (via D_0), and to fuel type (via ρ_F and B). However, a useful guide to the influence of fuel type on combustion efficiency may be obtained by defining a dimensionless efficiency ratio as the ratio of the combustion efficiency of any alternative fuel a to that of some baseline fuel b, when both fuels burn in the same combustor at the same operating conditions. From Eq. (128) we have, for low Re_D,

$$\frac{\eta_{c_a}}{\eta_{c_b}} = \frac{\rho_{F_b}}{\rho_{F_a}} \frac{D_b^2 \ln(1 + B_a)}{D_a^2 \ln(1 + B_b)} \tag{129}$$

This equation has considerable practical utility: it provides a means of assessing the effect on combustion efficiency of replacing aviation kerosine with some alternative fuel, without reference to any particular combustor or operating conditions. It may also be used more generally to compare the combustion-efficiency characteristics of any two liquid fuels, provided, of course, that fuel evaporation is known to be the rate-controlling step. If the level of combustion efficiency of interest is high, say, greater than 90%, it is more useful and more accurate to rewrite Eq. (129) as

$$\frac{1 - \eta_{c_b}}{1 - \eta_{c_a}} = \frac{\rho_{F_b} D_b^2 \ln(1 + B_a)}{\rho_{F_a} D_a^2 \ln(1 + B_b)} \tag{130}$$

For swirl atomizers, the mean drop size depends on the surface tension and viscosity of the fuel. However, conventional fuels exhibit only slight differences in surface-tension values, so only the influence of viscosity on SMD need be considered. From Eq. (37) we have

$$SMD \propto \mu_F^{0.25} \tag{131}$$

The substitution of SMD from Eq. (131) into Eqs. (129) and (130) gives

$$\frac{\eta_{c_a}}{\eta_{c_b}} = \frac{\rho_{F_b}\mu_{F_b}^{0.5}\ln(1 + B_a)}{\rho_{F_a}\mu_{F_a}^{0.5}\ln(1 + B_b)} \tag{132}$$

and

$$\frac{1 - \eta_{c_b}}{1 - \eta_{c_a}} = \frac{\rho_{F_b}\mu_{F_b}^{0.5}\ln(1 + B_a)}{\rho_{F_a}\mu_{F_a}^{0.5}\ln(1 + B_b)} \tag{133}$$

The practical utility of Eq. (133) for predicting the influence on combustion efficiency of a change in fuel type is well demonstrated in Figure 47; the experimental data in the figure were obtained by Moses,[203] using a T63 combustor. With Jet A designated as the baseline fuel, and using values for B and ρ_F from Table 3, the combustion efficiencies of the other fuels are readily calculated from Eq. (133). Owing to the lack of detailed information on the fuel spray characteristics, the effects of differences in mean drop size among the various fuels cannot be included. The results of the calculations are shown in Figure 47. The dashed curve describes the combustion efficiency of the baseline fuel as determined experimentally. The solid curves represent the predictions of Eq. (133) for the other fuels. The level of agreement between the predicted and measured values is clearly very satisfactory.

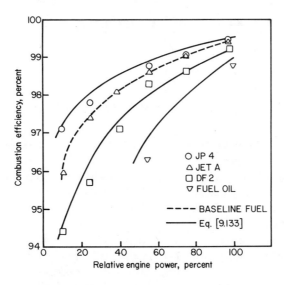

FIGURE 47. Use of Eq. (133) to predict the combustion efficiencies of various alternative fuels from data obtained with baseline fuel (experimental data from Moses[203]).

Another parameter of importance to combustion efficiency is the evaporation time of the largest drop in the spray, which is typically around three times larger in diameter than the SMD. If evaporation rates are limiting to burning rates then all that is needed to attain 100% combustion efficiency is to ensure that the residence time in the combustion zone exceeds the lifetime of the largest drop. This is obtained from Eq. (92) as

$$ t_e = \frac{\rho_F D^2}{8(k/c_p)_g \ln(1 + B)(1 + 0.22\,\mathrm{Re}_D^{0.5})} \tag{134} $$

Weak Extinction Limits. For premixed, gaseous combustible mixtures, flame blowout occurs when the rate of heat liberation in the recirculation zone becomes insufficient to heat the incoming fresh mixture up to the required reaction temperature. With heterogeneous mixtures, however, an additional factor is the time required for fuel evaporation. For fuel sprays of low volatility and large mean drop size, this time is relatively long and is often the main factor limiting the overall rate of heat release. In the analysis of weak extinction limits, it is convenient to consider homogeneous mixtures first and then to examine how the results obtained should be modified to take account of fuel evaporation.

For homogeneous mixtures, it can be shown that the maximum airflow rate corresponding to flame blowout is[204]

$$ \dot{m}_{\max} = 1.93 V_c P^{1.25} \exp\left(\frac{T_0}{150}\right)\phi^{6.25} \tag{135} $$

or

$$ \phi_{WE} = C_4 \left[\frac{\dot{m}_A}{V_c P^{1.25} \exp(T_0/150)}\right]^{0.16} \tag{136} $$

where V_c is the volume of the combustion zone, P is the gas pressure, and T_0 is the initial temperature of the fuel-air mixture. C_4 is a constant whose value is dependent on the geometry of the combustion zone and must be determined experimentally. Having derived the value of C_4 at any convenient test condition, Eq. (136) may then be used to predict the weak extinction value of ϕ at any other operating condition.

Equation (136) may also be used to predict the weak extinction limits of combustion zones supplied with heterogeneous fuel-air mixtures, provided that the rate of fuel evaporation is sufficiently high to ensure that all of the fuel is fully vaporized within the combustion zone. If the fuel does not fully vaporize, then clearly the "effective" equivalence ratio will be lower than the nominal value. However, if the fraction of fuel that is vaporized is known, or can be calculated, it can then be combined with Eq. (136) to yield the weak extinction value of equivalence ratio:

$$\phi_{WE(heterogeneous)} = \phi_{WE(homogeneous)} \times f^{-1} \tag{137}$$

where f is the fraction of fuel that is vaporized either in or upstream of the combustion zone. An expression for f may be readily derived by dividing the rate of fuel evaporation, $\dot{m}_F$, from Eq. (126) by the rate of fuel supplied $(= q\dot{m}_A)$. Hence

$$f = 8\left(\frac{\rho_g}{\rho_F}\right)\left(\frac{k}{c_p}\right)_g\left(\frac{V_c}{\dot{m}_A D_0^2}\right)\ln(1 + B)\left(1 + 0.22\,\mathrm{Re}_{D_0}^{0.5}\right) \tag{138}$$

Substitution of $\phi_{WE(hom)}$ from Eq. (136) and f from Eq. (138) into Eq. (137) gives

$$\phi_{WE(het)} = C_5\left[\frac{\dot{m}_A}{V_c P}\right]^{1.16}\left[\frac{\rho_F\left(c_p T/k\right)_g D_0^2}{\exp(T_0/940)\ln(1 + B)\left(1 + 0.22\,\mathrm{Re}_{D_0}^{0.5}\right)}\right] \tag{139}$$

The above equation has proved successful in predicting the weak extinction limits of bluff-body stabilized flames supplied with heterogeneous mixtures of air and heavy fuel oil, diesel oil, and iso-octane, as illustrated in Figures 48 and 49.[205]

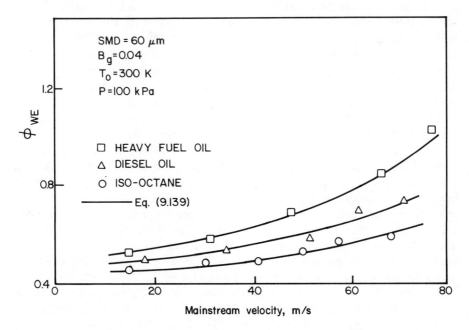

FIGURE 48. Influence of mainstream velocity on weak extinction limits.[205]

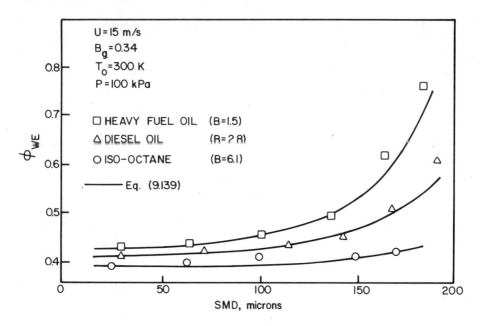

FIGURE 49. Influence of mean fuel drop size on weak extinction limits.[205]

Ignition. Ignition of a combustible mixture may be accomplished by various means—for example, by an electric spark, by contact with a hot surface or any other external source of heat—or it may occur spontaneously if the pressure and/or temperature are raised to levels at which the reaction becomes self-accelerating. In gasoline and gas turbine engines, ignition is usually effected by means of an electric spark. When flame-out occurs in turbojet engines at high altitudes, large amounts of energy are needed to ignite the heterogeneous and highly turbulent mixture flowing at velocities of the order of 25 m/sec.

In recent years a series of detailed experimental studies has been carried out on the influence of electrical and flow parameters on minimum spark energy in flowing mixtures of fuel drops and air. These studies have led to a better conceptual understanding of the basic ignition process and have provided a useful theoretical foundation for relating ignition characteristics to all of the operating variables involved. They confirm practical experience in showing that ignition is made easier by increases in pressure, temperature, and spark energy and that it is impaired by increases in velocity and turbulence intensity. They also emphasize that ignition performance is markedly affected by fuel properties through the way in which they influence the concentration of fuel vapor in the immediate vicinity of the igniter plug. These influences arise mainly through the effect of volatility on evaporation rates, but also through the effects of surface tension and viscosity on mean fuel drop size. The amount of energy required for ignition is very much larger than the values normally associated with gaseous fuels at stoichiometric fuel/air ratio. Much of this extra energy is absorbed in the evaporation of fuel droplets, the actual amount depending on the distribution of fuel throughout the primary zone and on the quality of atomization.

Ignition Theory. Ballal and Lefebvre[206] have proposed a model for the ignition of fuel sprays that is based on the assumption that chemical reaction rates are infinitely fast and that the onset of ignition is limited solely by the rate of fuel evaporation. Support for this notion may be found in the literature on the ignition of turbojet combustors. For example, the reports by Foster and Straight[207] and Wigg[208] contain ample evidence that various fuel spray characteristics, such as mean drop size and volatility, can appreciably affect the energy required for ignition. These effects are due to their influence on fuel evaporation rates that govern the mixture strength in the ignition zone. Further confirmation of the importance of fuel vapor concentration to ignition is provided in the more basic studies conducted by Rao and Lefebvre[209] on the ignition of heterogeneous, flowing kerosine/air mixtures. They found that for mixtures weaker than stoichiometric the main factor limiting ignition is a deficiency of fuel vapor in the ignition zone.

The process of ignition is envisaged to occur in the following manner. Passage of the spark creates a small, roughly spherical, volume of air (referred to henceforth as the spark kernel) whose temperature is sufficiently high to initiate rapid evaporation of the fuel drops contained within the volume. It is assumed that reaction rates and mixing times are infinitely fast so that any fuel vapor created within the spark kernel is instantly transformed into combustion products at the stoichiometric flame temperature. If the rate of heat release of combustion exceeds the rate of heat loss by thermal conduction at the surface of the inflamed volume, then the spark kernel will grow in size to fill the entire combustion volume. If, however, the rate of heat release is less than the rate of heat loss, the temperature within the spark kernel will fall steadily until fuel evaporation ceases altogether.

Thus of crucial importance to ignition is the size of the spark kernel at which the rate of heat loss at its surface is just balanced by the rate of heat release, due to the instantaneous combustion of fuel vapor, throughout its volume. This concept leads to the definition of *quenching distance* as the critical size that the inflamed volume must attain in order to propagate unaided, and the amount of energy required from an external source to attain this critical size is termed the *minimum ignition energy*.

By equating the heat generated within a spherical spark kernel to the heat lost by conduction from its surface, Ballal and Lefebvre[210] derived the following equation for quenching distance, for conditions where the rate of heat release is limited by evaporation rates:

$$d_q = \left[\frac{\rho_F D^2}{\rho_A \phi \ln(1 + B_{st})} \right]^{0.5} \tag{140}$$

It should be noted that this equation was derived through consideration of the basic mechanisms of heat generation within the kernel and heat loss from its surface, and it contains no experimental or arbitrary constants. It is valid for monodisperse sprays only. However, for polydisperse sprays of the type provided by most practical atomizing devices, it can be shown[211] that the quenching distance is given by

$$d_q = \left[\frac{C_3^3 \rho_F D_{32}^2}{C_1 \rho_A \phi \ln(1 + B_{st})} \right]^{0.5} \tag{141}$$

where C_1 and C_3 are parameters that describe the drop-size distribution in the spray;

$$C_1 = \frac{D_{20}}{D_{32}} \qquad C_3 = \frac{D_{30}}{D_{32}}$$

where

D_{20} = mean surface-area diameter
D_{30} = mean volume diameter
D_{32} = Sauter mean diameter

Equations (140) and (141) provide simple dimensionless relationships between the quenching distance and the fuel-drop size in the spray. Essentially, they state that quenching distance is directly proportional to drop size and is inversely proportional to the square root of gas pressure. An increase in ϕ and a reduction in ρ_F both reduce d_q because they promote evaporation by increasing the surface area of the fuel. Similarly, an increase in B also accelerates evaporation and thereby decreases d_q.

Values of $E_{\min}$ may be obtained for quiescent or low-turbulence mixtures by inserting the calculated values of d_q from Eq. (140) or Eq. (141) into the expression

$$E_{\min} = c_{p_A} \rho_A \Delta T_{st} \left(\frac{\pi}{6} \right) d_q^3 \tag{142}$$

The results of such calculations are shown as solid curves in Figures 50 and 51.

Although the above equations for d_q in heterogeneous fuel-air mixtures were derived for quiescent mixtures, they may be applied to flowing mixtures in combustion systems without much loss of accuracy. This is because, apart from the very largest drops, most of the fuel spray is airborne and the relative velocity between the fuel drops and the surrounding air or gas is too small to enhance appreciably either the rate of fuel evaporation or the rate of heat loss from the spark kernel.

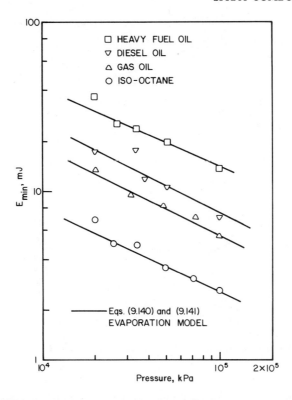

FIGURE 50. Effect of air pressure on minimum ignition energy, $\phi = 0.65$, SMD $= 60$ μm, $T_0 = 290$ K, $U = 0$.[210]

In a later study Ballal and Lefebvre[212] extended the model described above to include (1) the effects of finite chemical reaction rates, which are known to be significant for very well-atomized fuels at low pressures and low equivalence ratios, and (2) the presence of fuel vapor in the mixture flowing into the ignition zone. Thus the model has general application to both quiescent and flowing mixtures of air with gaseous, liquid, or evaporated fuel or any combination of these fuels. For example, by including chemical effects, Eq. (140) becomes

$$d_q = \left[\frac{\rho_F D^2}{\rho_A \phi \ln(1 + B_{st})} + \left(\frac{10\alpha}{S_L} \right)^2 \right]^{0.5} \tag{143}$$

From a practical viewpoint, the value of the models described above lies not so much in their ability to predict minimum ignition energy requirements, since the available spark energy is determined by the specifications of the ignition system, but in highlighting the key parameters that are controlling to

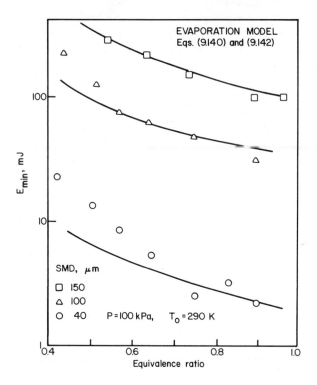

FIGURE 51. Minimum ignition energies of stagnant heavy fuel oil and air mixtures for various mean drop sizes.[210]

ignition and in providing quantitative relationships or "rates of exchange" between these parameters.

Consider, for example, a combustor that normally operates on kerosine fuel and it is required to estimate what improvement in atomization quality would be needed to achieve the same ignition performance when burning diesel oil. Now the relevant fuel properties are density and volatility, the latter being represented in the quenching distance equations by the mass transfer number, B. From Eq. (140) we have

$$d_q \propto \left[\frac{\rho_F D^2}{\ln(1 + B_{st})} \right]^{0.5} \tag{144}$$

Thus for constant ignition performance (i.e., constant d_q), we can write

$$\left(\frac{D_{do}}{D_k} \right)^2 = \frac{\{ \ln(1 + B_{st})/\rho_F \}_{do}}{\{ \ln(1 + B_{st})/\rho_F \}_k} \tag{145}$$

TABLE 8. Some Relevant Fuel Properties

Fuel	Density, kg/m^3	Mass Transfer Number, B_{st}	Viscosity, kg/msec
Gasoline (JP4)	692	6.10	0.00070
Kerosine (Jet A)	775	3.75	0.00129
Diesel oil (DF2)	900	2.80	0.0030
Light fuel oil	930	2.50	0.0037
Heavy fuel oil	970	1.50	—

TABLE 9. Fuel Properties Relevant to Calculation of Mass Transfer Number, B

Fuel	Fuel Density, ρ_F		Latent Heat of Vaporization, L		Specific Heat of Fuel, c_{p_F}		Average Boiling Temperature,
	kg/m^3	lb/ft^3	kJ/kg	Btu/lb	J/kg K	Btu/lb K	K
Iso-octane	702	43.8	328	141	2156	0.515	398
Benzene	884	55.2	432	186	1846	0.441	353
Methanol	796	49.7	1101	473	2369	0.566	338
Ethanol	794	49.6	837	360	2344	0.560	352
Gasoline	720	44.9	339	146	2051	0.49	428
Kerosine	775	51.5	291	125	1926	0.46	423
Light diesel	900	54.7	267	115	1884	0.45	523
Medium diesel	920	57.4	244	105	1800	0.43	533
Heavy diesel	960	60.0	232	100	1758	0.42	543

From Tables 8 and 9 for diesel oil, $\rho_F = 900$ and $B_{st} = 2.80$, whereas for kerosine, $\rho_F = 775$ and $B_{st} = 3.75$. Substituting these values into Eq. (145) gives

$$\frac{D_{do}}{D_k} = 0.86$$

which suggests that for diesel oil the SMD should be reduced by about 14% in order to retain the same ignition performance as for kerosine.

NOMENCLATURE

ALR	air/liquid mass ratio
A_p	swirl chamber inlet port area, m^2
B	steady-state transfer number
B_M	mass transfer number
B_T	heat transfer number

C_D	drag coefficient, or hole discharge coefficient
c_p	specific heat at constant pressure, J/kg K(Btu/lb F)
D	drop diameter, or mean drop size in spray, m
D_c	diffusion coefficient, m²/sec
D_d	disc diameter, m
D_h	hydraulic mean diameter of atomizer air duct at exit plane
D_p	prefilmer diameter, m
D_s	swirl chamber diameter, m
d	spark kernel diameter, m
d_0	liquid orifice diameter, or initial jet diameter, m
d_q	quenching distance, m
E_{min}	minimum ignition energy, J
F	charging factor
f	fraction of fuel vaporized
FN_{USA}	atomizer flow number, (lb/hr) (psi)$^{-0.5}$
FN_{UK}	atomizer flow number, (gal/hr) (psi)$^{-0.5}$
H	heat of combustion, J/kg
k	thermal conductivity, J/msk
L	latent heat of vaporization, J/kg
L_c	characteristic atomizer dimension, m
L_{T_b}	latent heat of fuel vaporization at normal temperature, T_b, J/kg
$L_{T_{bn}}$	latent heat of vaporization at normal boiling temperature, T_{bn}, J/kg
L_s	length of swirl chamber, m
l	length of final orifice, m
M	molecular weight
m	mass of fuel drop, kg
$\dot{m}$	mass flow rate, kg/sec
m_F	rate of fuel evaporation per unit surface area
$\dot{m}_F$	rate of fuel evaporation, kg/sec (lb/sec)
MMD	mass median diameter, m
N	rotational speed, rpm
n	number of drops
Nu	Nusselt number
P	pressure, kPa
P_e	electrical pressure
P_{F_s}	fuel vapor pressure at drop surface, kPa
Pr	Prandtl number
Q	volumetric flow rate, L/sec
Q_c	rate of heat loss by conduction, J/sec
Q_H	rate of heat release by combustion, J/sec
q	fuel/air ratio by mass, or drop-size distribution parameter
R	gas constant, or ratio of secondary to primary flow number
r	drop radius, m
r_s	radius at drop surface, m

Re_D	drop Reynolds number
S	spray penetration, m
Sc	Schmidt number
SG	specific gravity
SMD	Sauter mean diameter, m
T	temperature, K
T_b	boiling temperature, K
T_{bn}	boiling temperature at normal atmospheric pressure, K
T_{cr}	critical temperature, K
T_f	flame temperature, K
T_r	reference temperature, K
T_s	drop surface temperature, K
T_∞	ambient temperature, K
t	time, sec
t_e	drop evaporation time, sec
t_{res}	residence time in combustion zone, sec
Δt_{hu}	duration of heat-up period, sec
Δt_{st}	duration of steady-state period, sec
U	velocity, or relative velocity, m/sec
u'	fluctuating component of turbulence velocity, m/sec
V	volume, m^3, or applied voltage
V_C	volume of combustion zone, m^3
v	fraction of total liquid volume contained in drops of diameter less than x
We	Weber number
X	ratio of air core area to discharge orifice area
x, y	drop-size parameters
x_0	minimum drop diameter, m
x_m	maximum drop diameter, m
Y_A	mass fraction of air
Y_F	mass fraction of fuel vapor
Z	Z number
δ	standard deviation
v	kinematic viscosity, m^2/sec
λ	wavelength, m, or evaporation constant, m^2/sec
λ_{eff}	effective average value of λ during drop lifetime, m^2/sec
ϕ	equivalence ratio, or spray cone angle
ϕ_{WE}	equivalence ratio at weak extinction
μ	dynamic viscosity, kg/msec
ν	kinematic viscosity, m^2/sec
ρ	density, kg/m^3
α	thermal diffusivity, m^2/sec
σ	surface tension, kg/sec^2
η_c	combustion efficiency
ω_0	frequency

Subscripts

A	air
c	combustion zone value
F	fuel
L	liquid
R	air relative to liquid
g	gas
max	maximum value
0	initial value
r	reference value
s	value at drop surface
st	steady-state value
hu	mean or effective value during heat-up period
∞	ambient value

REFERENCES

1. Lord Rayleigh, "On the Instability of Jets." *Proc. London Math. Soc.* **10**, 4–13 (1879).
2. F. Tyler, "Instability of Liquid Jets." *Philos. Mag.* [7] **16**, 504–518 (1933).
3. C. Weber, "Disintegration of Liquid Jets." *Z. Angew. Math. Mech.* **11**, No. 2 (1931).
4. A. Haenlein, "On the Disruption of a Liquid Jet." *NACA Tech. Memo* **659** (1932).
5. W. Ohnesorge, "Formation of Drops by Nozzles and the Breakup of Liquid Jets." *Z. Angew. Math. Mech.* **16** (1936).
6. R. P. Fraser and P. Eisenklam, *J. Imp. Coll. Chem. Eng. Soc.* **7** (1953).
7. R. P. Fraser, "Liquid Fuel Atomization." *Symp. (Int.) Combust.* [*Proc.*] **6**, 687–701 (1957).
8. N. Dombrowski and R. P. Fraser, "A Photographic Investigation into the Disintegration of Liquid Sheets." *Philos. Trans. R. Soc. London, Ser. A* **247**, 101–130 (1954).
9. N. Dombrowski and G. Munday, "Spray Drying." In *Introduction to Biochemical and Biological Engineering Science* (N. Blakeborough, ed.), Vol. 2, pp. 209–320. Academic Press, New York, 1968.
10. S. Nukiyama and Y. Tanasawa, *Experiments on the Atomization of Liquids in an Air Stream*, Rep. No. 3. Defence Res. Board, Dept. Natl. Defence, Ottawa, Canada, Mar. 18, 1950 [transl. from *Trans. Soc. Mech. Eng., Tokyo* **5**(18), 62–67 (1939).]
11. P. Rosin and E. Rammler, "The Laws Governing the Fineness of Powdered Coal." *J. Inst. Fuel* **7**(31), 29–36 (1933).
12. R. Mugele and H. D. Evans, "Droplet Size Distribution in Sprays." *Ind. Eng. Chem.* **43**(6), 1317–1324 (1951).

13. J. S. Chin and A. H. Lefebvre, "Some Comments on the Characterization of Drop-Size Distribution in Sprays." *Proc. Int. Conf. Liq. Atom. Spray Syst., 1985,* pp. NA/1/1-12 (1985).

14. E. Giffen and A. Muraszew, *The Atomization of Liquid Fuels.* Chapman & Hall, London, 1953.

15. H. C. Lewis, D. G. Edwards, M. J. Goglia, R. I. Rice, and L. W. Smith, "Atomization of Liquids in High Velocity Gas Streams." *Ind. Eng. Chem.* **40**(1), 67–74 (1948).

16. K. R. May, "The Measurement of Airborne Droplets by the Magnesium Oxide Method." *J. Sci. Instrum.* **27**, 128–130 (1950).

17. M. M. Elkotb, N. M. Rafat, and M. A. Hanna, "The Influence of Swirl Atomizer Geometry on the Atomization Performance." *Proc. Int. Conf. Liq. Atomization Spray Syst. 1st 1978,* pp 109–115 (1978).

18. K. Y. Kim and W. R. Marshall, "Drop-Size Distributions from Pneumatic Atomizer." *AIChE J.* **17**(3), 575–584 (1971).

19. J. R. Joyce, "The Atomization of Liquid Fuels for Combustion." *J. Inst. Fuel* **22**(124), 150–156 (1949).

20. A. P. R. Choudhury, G. G. Lamb, and W. F. Stevens, "A New Technique for Drop-Size Distribution Determination." *Trans. Indian Inst. Chem. Eng.* **10**, 21–24 (1957).

21. P. A. Nelson and W. F. Stevens, "Size Distribution of Droplets from Centrifugal Spray Nozzles." *AIChE J.* **7**(1), 8–86 (1961).

22. P. J. Street and V. E. J. Danaford, "A Technique for Determining Drop Size Distribution Using Liquid Nitrogen." *J. Inst. Pet.* **54**(536), 241–242 (1968).

23. K. V. L. Rao, "Liquid Nitrogen Cooled Sampling Probe for the Measurement of Spray Drop Size Distribution in Moving Liquid-Air Sprays." *Proc. Int. Conf. Liq. Atomization Spray Syst. 1st, 1978,* pp. 293–300 (1978).

24. R. A. Dobbins, L. Crocco, and I. Glassman, "Measurement of Mean Particle Sizes of Sprays from Diffractively Scattered Light." *AIAA J.* **1**(8), 1882–1886 (1963).

25. J. M. Roberts and M. J. Webb, "Measurement of Droplet Size for Wide Range Particle Distribution." *AIAA J.* **2**(3), 583–585 (1964).

26. A. H. Lefebvre and E. R. Norster, *A Proposed Double-Swirler Atomizer for Gas Turbine Fuel Injection,* SME Rep. No. 1. Cranfield Institute of Technology, Cranfield, England, 1972.

27. A. A. Rizkalla and A. H. Lefebvre, "Influence of Liquid Properties on Airblast Atomizer Spray Characteristics," *J. Eng. Power,* pp 173–179 (April 1975).

28. A. Rizkalla and A. H. Lefebvre, "The Influence of Air and Liquid Properties on Air Blast Atomization." *J. Fluids Eng.* **97**(3), 316–320 (1975).

29. G. E. Lorenzetto and A. H. Lefebvre, "Measurements of Drop Size on a Plain Jet Airblast Atomizer." *AIAA J.* **15**(7), 1006–1010 (1977).

30. G. E. Lorenzetto, "Influence of Liquid Properties on Plain Jet Airblast Atomization." Ph.D. Thesis, School of Mechanical Engineering, Cranfield Institute of Technology, Cranfield, England, 1976.

31. N. K. Rizk and A. H. Lefebvre, "Influence of Liquid Film Thickness on Airblast Atomization." *J. Eng. Power* **102**, 706–710 (1980).

32. N. K. Rizk, "Studies on Liquid Sheet Disintegration in Airblast Atomizers." Ph.D. Thesis, Cranfield Institute of Technology, Cranfield, England, 1977.

33. M. S. M. R. El-Shanawany, "Airblast Atomization—The Effect of Linear Scale on Mean Drop Size." Ph.D. Thesis, Cranfield Institute of Technology, Cranfield, England, 1978.

34. M. S. M. R. El-Shanawany and A. H. Lefebvre, "Airblast Atomization: The Effect of Linear Scale on Mean Drop Size." *J. Energy* **4**(4), 184–189 (1980).

35. J. Swithenbank, J. M. Beer, D. Abbott, and C. G. McCreath, "A Laser Diagnostic Technique for the Measurement of Droplet and Particle Size Distribution." *AIAA Pap.* **76-69** (1976).

36. E. D. Hirleman, V. Oechsle, and N. A. Chigier, "Response Characteristics of Laser Diffraction Particle Size Analyzers: Optical Sample Volume Extent and Lens Effects." *Opt. Eng.* **23**(5), 610–619 (1984).

37. L. G. Dodge, "Change of Calibration of Diffraction-Based Particle Sizers in Dense Sprays." *Opt. Eng.* **23**(5), 626–630 (1984).

38. S. Wittig, M. Aigner, Kh. Sakbani, and Th. Sattelmayer, "Optical Measurements of Droplet Size Distributions: Special Considerations in the Parameter Definition for Fuel Atomizers." *Pap. AGARD Meet. Combust. Probl. Turbine Engines*, Cesme, Turkey, October, *1983*.

39. J. S. Chin, D. Nickolaus, and A. H. Lefebvre, "Influence of Downstream Distance on the Spray Characteristics of Pressure-Swirl Atomizers." *J. Eng. Gas Turbines Power* **106**(1), 219–224 (1986).

40. N. A. Chigier, "The Atomization and Burning of Liquid Fuel Sprays." *Prog. Energy Combust. Sci.* **2**, 97–114 (1976).

41. N. A. Chigier, "Drop Size and Velocity Instrumentation." *Prog. Energy Combust. Sci.* **9**, 155–177 (1983).

42. A. G. Gelalles, "Effect of Orifice Length-Diameter Ratio on Spray Characteristics," *Natl. Advis. Comm. Aeronaut., Tech. Notes* **NACA TN-352** (1930); also see "Effect of Orifice Length-Diameter Ratio on Fuel Sprays for Compression-Ignition Engines." *Natl. Advis. Comm. Aeronaut., Rep.* **402** (1931).

43. A. C. Merrington and E. G. Richardson, "The Break-up of Liquid Jets." *Proc. Phys. Soc. London* **59**(33), 1–13 (1947).

44. Y. Tanasawa, "Liquid Fuel Injection." *Diesel Engine*, Part 1. Sankaido Book Company, Tokyo, 1956.

45. P. Kutty, M. V. Narasimhan, and K. Narayanaswamy, "Measurement of Air-Core Size in Swirl Chamber Atomizers. I." *Pet. Rev.* **DDD** 357, IP-75-003 (1975).

46. A. Radcliffe, "The Performance of a Type of Swirl Atomizer." *Proc. Inst. Mech. Eng.* **169**, 93–106 (1955).

47. G. I. Taylor, "The Boundary Layer in the Converging Nozzle of a Swirl Atomizer." *Q. J. Mech. Appl. Math.* **3**, Part 2, 129–139 (1950).

48. P. Eisenklam, "Atomization of Liquid Fuels for Combustion." *J. Inst. Fuel* **34**, 130 (1961).

49. N. Dombrowski and D. Hasson, "The Flow Characteristics of Swirl (Centrifugal) Spray Pressure Nozzles with Low Viscosity Liquids. *AIChE J.* **15**, 604 (1969).

50. D. R. Carlisle, Communication on "The Performance of a Type of Swirl Atomizer," by A. Radcliffe. *Proc. Inst. Mech. Eng.* **169**, 101 (1955).

51. N. K. Rizk and A. H. Lefebvre, "Internal Flow Characteristics of Simplex Swirl Atomizers." *AIAA J. Propul. Power* **1**(3), 193–199 (1985).

52. E. Giffen and B. S. Massey, Rep. No. 1950/5. Motor Industry Research Association, England, 1960.

53. S. M. De Corso and G. A. Kemeny, "Effect of Ambient and Fuel Pressure on Nozzle Spray Angle." *Trans. ASME* **79**(3), 607–615 (1957).

54. J. Ortman and A. H. Lefebvre, "Fuel Distributions from Pressure-Swirl Atomizers." *AIAA J. Propuls. Power* **1**(1), 11–15 (1985).

55. A. K. Jasuja, "Atomization of Crude and Residual Fuel Oils." *Am. Soc. Mech. Eng.* [*Pap.*] **78-GT-83** (1978).

56. H. C. Simmons and C. F. Harding, "Some Effects of Using Water as a Test Fluid in Fuel Nozzle Spray Analysis." *Am. Soc. Mech. Eng.* [*Pap.*] **80-GT-90** (1980).

57. A. Radcliffe, *High Speed Aerodynamics and Jet Propulsion*, Fuel Inject., Sect. D, Vol. XI. Princeton University Press, Princeton, New Jersey, 1960.

58. B. E. Knight, Communication on "The Performance of a Type of Swirl Atomizer," by A. Radcliffe. *Proc. Inst. Mech. Eng.* **169**, 104 (1955).

59. S. M. Decorso, "Effect of Ambient and Fuel Pressure on Spray Drop Size." *J. Eng. Power* **82**, 10–18 (1960).

60. K. Neya and S. Sato, "Effect of Ambient Air Pressure on the Spray Characteristics of Swirl Atomizers." *Pap. Ship Res. Inst.* (*Tokyo*) **27** (1968).

61. M. M. M. Abou-Ellail, M. M. Elkotb, and N. M. Rafat, "Effect of Fuel Pressure, Air Pressure and Air Temperature on Droplet Size Distribution in Hollow-Cone Kerosene Sprays." *Proc. Int. Conf. Liq. Atomization Spray Syst., 1st, 1978*, pp. 85–92 (1978).

62. N. K. Rizk and A. H. Lefebvre, "Spray Characteristics of Simplex Swirl Atomizers." *Prog. Astronaut. Aeronaut.* **95**, 563–580 (1984).

63. A. H. Lefebvre, *Gas Turbine Combustion*. Hemisphere, New York, 1983.

64. L. G. Dodge and J. A. Biaglow, "Effect of Elevated Temperature and Pressure on Sprays from Simplex Swirl Atomizers." *Am. Soc. Mech. Eng.* [*Pap.*] **85-GT-58** (1985).

65. H. C. Simmons, "The Prediction of Sauter Mean Diameter for Gas Turbine Fuel Nozzles of Different Types." *Am. Soc. Mech. Eng.* [*Pap.*] **79-WA / GT-5** (1979).

66. A. H. Lefebvre, "Factors Controlling Gas Turbine Combustion Performance at High Pressure." *Cranfield Int. Symp. Ser.* **10**, 211–226 (1968).

67. F. C. Mock and D. R. Ganger, "Practical Conclusions on Gas Turbine Spray Nozzles." *SAE Q. Trans.* **4**(3), 357–367 (1950).

68. F. H. Carey, "The Development of the Spill Flow Burner and Its Control System for Gas Turbine Engines." *J. R. Aeronaut. Soc.* **58**(527), 737–753 (1954).

69. D. R. Carlisle, private communication.

70. N. K. Rizk and A. H. Lefebvre, "Drop-Size Distribution Characteristics of Spill-Return Atomizers." *AIAA J. Propuls. Power* **1**(1), 16–22 (1985).

71. J. D. Lewis, "Studies of Atomization and Injection Process in the Liquid Propellant Rocket Engine." *Combust. Propuls. AGARD Colloq., 5th, 1963*, pp. 141–174 (1963).

72. J. O. Hinze and H. Milborn, "Atomization of Liquids by Means of a Rotating Cup." *ASME J. Appl. Mech.* **17**(2), 145–153 (1950).

73. R. P. Fraser, N. Dombrowski, and J. H. Routley, "The Production of Uniform Liquid Sheets from Spinning Cups; the Filming of Liquids by Spinning Cups; the Atomization of a Liquid Sheet by an Impinging Air Stream." *Chem. Eng. Sci.* **18**, 315–321, 323–337, 339–353 (1963).

74. A. Kayano and T. Kamiya, "Calculation of the Mean Size of the Droplets Purged from the Rotating Disk." *Proc. Int. Conf. Liq. Atomization Spray Syst., 1st*, pp. 133–138 (1978).

75. J. Gretzinger and W. R. Marshall, Jr., "Characteristics of Pneumatic Atomization." *J. Am. Inst. Chem. Eng.* **7**(2), 312–318 (1961).

76. M. L. Yeager and C. L. Coffin, "A Survey of Components for Use with Air-Atomizing Oil-Burner Nozzles." *API Publ.* No. **1720** (1961).

77. P. J. Mullinger and N. A. Chigier, "The Design and Performance of Internal Mixing Multi-jet Twin-Fluid Atomizers." *J. Inst. Fuel* **47**, 251–261 (1974).

78. S. Nukiyama and Y. Tanasawa, "Experiments on the Atomization of Liquids in an Airstream." *Trans. Soc. Mech. Eng., Tokyo* **5**, 68–75 (1939).

79. M. A. Weiss and C. H. Worsham, "Atomization in High Velocity Air-Streams." *ARS J.* **29**(4), 252–259 (1959).

80. R. H. Wetzel and W. R. Marshall, "Venturi Atomization." *Pap. Natl. Meet. AIChE*, Washington D.C., *1954*.

81. L. E. Wigg, *The Effect of Scale on Fine Sprays Produced by Large Airblast Atomizers*, British National Gas Turbine Establishment Rep. No. 236, 1959.

82. L. D. Wigg, "Drop-Size Predictions for Twin Fluid Atomizers." *J. Inst. Fuel* **27**, 500–505 (1964).

83. E. Mayer, "Theory of Liquid Atomization in High Velocity Gas Streams." *J. Am. Rocket Soc.* **31**, 1783–1785 (1961).

84. R. D. Ingebo and H. H. Foster, "Drop-Size Distribution for Cross-Current Break-Up of Liquid Jets in Air Streams." *Natl. Advis. Comm. Aeronaut., Tech. Note* **NACA TN 4087** (1957).

85. A. H. Lefebvre and D. Miller, *The Development of an Air Blast Atomizer for Gas Turbine Application*, CoA-Rep.-AERO-193. College of Aeronautics, Cranfield, Bedford, England, June, 1966.

86. N. K. Rizk and A. H. Lefebvre, "Spray Characteristics of Plain-Jet Airblast Atomizers." *J. Eng. Gas Turbines Power* **106**, 634–638 (1984).

87. A. H. Lefebvre, "Airblast Atomization." *Prog. Energy Combust. Sci.* **6**, 233–261 (1980).

88. A. Sotheran, "The Rolls Royce Annular Vaporizer Combustor." *Am. Soc. Mech. Eng.* [*Pap.*] **83-GT-49** (1983).

89. A. Woltjen, "Uber die Feinheit der Brennstoffzerstaubung in Olmaschinen." Dissertation, Technische Hochschule Darmstadt (1925).

90. F. Sass, *Die Kompresslose Dieselmaschinen*, p. 47. Springer, Berlin, 1929.

91. H. Hiroyasu and T. Kadota, "Fuel Droplet Size Distribution in Diesel Combustion Chamber." *SAE Natl. Power Plant Meet.*, Milwaukee, Wisconsin, Pap. No. 740715 (1974).

92. P. H. Schweitzer, "Mechanism of Disintegration of Liquid Jets." *J. Appl. Phys.* **8**, 513 (1937).

93. M. Parks, C. Polonski, and R. Toye, "Penetration of Diesel Fuel Sprays in Gases." *SAE Tech. Pap. Ser.* **660747** (1966).

94. R. Burt and K. Troth, "Penetration and Vaporization of Diesel Fuel Sprays." *Proc. Inst. Mech. Eng., Part 3J* **184**, 147 (1970).

95. J. C. Dent, "A Basis for the Comparison of Various Experimental Methods for Studying Spray Penetration." *SAE Tech. Pap. Ser.* **716571** (1971).

96. D. H. Taylor and B. E. Walsham, "Combustion Processes in a Medium Speed Diesel Engine." *Proc. Inst. Mech. Eng., Part 3J* **184**, 67 (1970).

97. T. J. Williams, "Parameters for Correlation of Penetration Results for Diesel Fuel Sprays." *Proc. Inst. Mech. Eng.* **187**, 771 (1973).

98. N. Hay and P. L. Jones, "Comparison of the Various Correlations for Spray Penetration." *SAE Tech. Pap. Ser.* **720776** (1972).

99. M. M. Elkotb and N. M. Rafat, "Fuel Spray Trajectory in Diesel Engines." *J. Eng. Power* **100**, 326 (1978).

100. M. M. Elkotb and M. A. Abdalla, "Atomization of Multifuel Sprays," *Proc. Int. Conf. Liq. Atomization Spray Syst., 2nd, 1982*, pp. 237–244 (1982).

101. M. M. Elkotb, "Fuel Atomization for Spray Modelling." *Prog. Energy Combust. Sci.* **8**, 61–91 (1982).

102. W. B. Bryce, N. W. Cox, and W. I. Joyce, "Oil Droplet Production and Size Measurement from a Twin-Fluid Atomizer using Real Fluids." *Proc. Int. Conf. Liq. Atomization Spray Syst., 1st, 1978*, pp. 259–263 (1978).

103. A. R. Jones, "Design Optimization of a Large Pressure Jet Atomizer for Power Plant." *Proc. Int. Conf. Liq. Atomization Spray Syst., 2nd, 1982*, pp. 181–185 (1982).

104. M. Sargeant, "Blast Atomizer Developments in the Central Electricity Generating Board." *Proc. Int. Conf. Liq. Atomization Spray Syst., 2nd, 1982*, pp. 131–135 (1982).

105. A. Stambuleanu, *Flame Combustion Processes in Industry*. Abacus Press, Turbridge Wells, England, 1976.

106. P. E. Graf, "Breakup of Small Liquid Volume by Electrical Charging." *API Publ.* **1701**, Pap. CP62-4 (1962).

107. F. E. Luther, "Electrostatic Atomization of No. 2 Heating Oil." *API Publ.* **1701**, Pap. CP62-3 (1962).

108. R. L. Peskin, R. J. Raco, and J. Morehouse, "A Study of Parameters Governing Electrostatic Atomization of Fuel Oil." *API Publ.* **704** (1965).

109. R. Bollini, B. Sample, S. D. Seigal, and J. W. Boarman, "Production of Monodisperse Charged Metal Particles by Harmonic Electrical Spraying." *J. Colloid Interface Sci.* **51**(2), 272–277 (1975).

110. V. G. Drozin, "The Electrical Dispersion of Liquids as Aerosols." *J. Colloid Sci.* **10**(2), 158–164 (1955).

111. W. A. Macky, "Some Investigations on the Deformation and Breaking of Water Drops in Strong Electric Fields." *Proc. R. Soc. London, Ser. A* **133**, 565–587 (1931).

112. M. A. Nawab and S. C. Mason, "The Preparation of Uniform Emulsions by Electrical Dispersion." *J. Colloid Sci.* **13**, 179–187 (1958).

113. B. Vonnegut and R. L. Neubauer, "Production of Monodisperse Liquid Particles by Electrical Atomization." *J. Colloid Sci.* **7**, 616–622 (1952).

114. A. J. Kelly, "The Electrostatic Atomization of Hydrocarbons." *Proc. Int. Conf. Liq. Atomization Spray Syst., 2nd, 1982*, pp. 57–65 (1982).

115. E. L. Gershenzon and O. K. Eknadiosyants, "The Nature of Liquid Atomization in an Ultrasonic Fountain." *Sov. Phys.—Acoust. (Engl. Transl.)* **10**(2), 127–132 (1964).

116. R. J. Lang, "Ultrasonic Atomization of Liquids." *J. Acoust. Soc. Am.* **34**(1), 6–8 (1962).

117. D. D. Lobdell, "Particle Size-Amplitude Relation for the Ultrasonic Atomizer." *J. Acoust. Soc. Am.* **43**(2), 229–231, Sept. 1967.

118. R. L. Peskin and R. J. Raco, "Ultrasonic Atomization of Liquids." *J. Acoust. Soc. Am.* **35**(9), 1378–1381 (1963).

119. B. W. Hansen, "High-Speed Photographic Studies of Droplet Formation at 20 KHz Ultrasonic Atomization of Oil." *Ultrasonics* **8**(2), 97–99 (1970).

120. K. W. Lee, A. A. Putnam, J. A. Gieseke, M. N. Golovin, and J. A. Hale, "Spray Nozzle Designs for Agricultural Aviation Applications." *NASA [Contract. Rep.] CR* **NASA-CR-159702** (1979).

121. T. Mochida, "Ultrasonic Atomization of Liquids." *Proc. Int. Conf. Liq. Atomization Spray Syst., 1st, 1978*, pp. 193–200 (1978).

122. J. Antonevich, "Ultrasonic Atomization of Liquids." *IRE Trans. PGUE* **7**, 6–15 (1959).

123. K. Bisa, K. Dirnagl, and R. Esche, "Zerstaubung von Flussigkeiten mit Ultraschall." *Siemens-Z.* **28**(8), 341–347 (1954).

124. A. E. Crawford, "Production of Spray by High Power Magnetostriction Transducers." *J. Acoust. Soc. Am.* **27**, 176 (1955).

125. T. McCubbin, Jr., "The Particle Size Distribution in Fog Produced by Ultrasonic Radiation." *J. Acoust. Soc. Am.* **25**, 1013–1014 (1953).

126. S. N. Muromtsev and V. P. Nenashev, "The Study of Aerosols. III. An Ultrasonic Aerosol Atomizer." *J. Microbiol. Epidemiol. Immunol. (Engl. Transl.)* **31**(10), 1840–1846 (1960).

127. M. N. Topp and P. Eisenklam, "Industrial and Medical Uses of Ultrasonic Atomizers." *Ultrasonics* **10**(3), 127–133 (1972).

128. R. L. Wilcox and R. W. Tate, "Liquid Atomization in a High Intensity Sound Field." *AIChE J.* **11**(1), 69–72 (1965).

129. M. N. Topp, "Ultrasonic Atomization—A Photographic Study of the Mechanism of Disintegration." *Aerosol Sci.* **4**, 17–25 (1973).

130. T. Sakai, M. Kito, M. Saito and T. Kanbe, "Characteristics of Internal Mixing Twin Fluid Atomizer." *Proc. Int. Conf. Liq. Atomization Spray Syst., 1st, 1978*, pp. 235–241 (1978).

131. B. J. Wood, H. Wise, and S. H. Inami, "Heterogeneous Combustion of Multicomponent Fuels." *NASA Tech. Note* **NASA TN D-206** (1959).

132. G. A. E. Godsave, "Studies of the Combustion of Drops in a Fuel Spray—The Burning of Single Drops of Fuel." *Symp. (Int.) Combust. [Proc.]* **4**, 818–830 (1953).

133. D. B. Spalding, "The Combustion of Liquid Fuels." *Symp. (Int.) Combust. [Proc.]* **4**, 847–864 (1953).

134. G. M. Faeth, "Current Status of Droplet and Liquid Combustion." *Prog. Energy Combust. Sci.* **3**, 191–224 (1977).

135. M. Goldsmith and S. S. Penner, "On the Burning of Single Drops of Fuel in an Oxidizing Atmosphere." *Jet Propuls.* **24**, 245–251 (1954).

136. A. M. Kanury, *Introduction to Combustion Phenomena*. Gordon & Breach, New York, 1975.

137. D. B. Spalding, *Some Fundamentals of Combustion*. Academic Press, New York, 1955.

138. A. Williams, "Fundamentals of Oil Combustion." *Prog. Energy Combust. Sci.* **2**, 167–179 (1976).

139. G. L. Hubbard, V. E. Denny, and A. F. Mills, "Droplet Evaporation: Effects of Transients and Variable Properties." *Int. J. Heat Mass Transfer* **18**, 1003–1008 (1975).

140. E. M. Sparrow and J. L. Gregg, "The Variable Property Problem in Free Convection." *Trans. ASME* **80**, 879–886 (1958).

141. R. P. Probert, "The Influence of Spray Particle Size and Distribution in the Combustion of Oil Droplets." *Philos. Mag.* **37**, 94 (1946).

142. J. S. Chin and A. H. Lefebvre, "The Role of the Heat-up Period in Fuel Drop Evaporation." *Int. J. Turbo and Jet Engines* **2**, 315–325 (1986).

143. N. Frossling, "On the Evaporation of Falling Droplets." *Gerlands Beitr. Geophys.* **52**, 170–216 (1938).

144. W. E. Ranz and W. R. Marshall, Jr., "Evaporation from Drops." *Chem. Eng. Prog.* **48**, Part I, 141–146 (1952); **48**, Part II, 173–180 (1952).

145. J. S. Chin and A. H. Lefebvre, "Effective Values of Evaporation Constant for Hydrocarbon Fuel Drops," *Proc. Automot. Technol. Dev. Contract. Coord. Meet.*, *20th*, pp. 325–331 (1982).

146. J. W. Aldred, J. C. Patel and A. Williams, "The Mechanism of Combustion of Droplets and Spheres of Liquid n-Heptane." *Combust. Flame* **17**, 139–149 (1971).

147. M. M. El Wakil, O. A. Uyehara, and P. S. Myers, "A Theoretical Investigation of the Heating-up Period of Injected Fuel Drops Vaporizing in Air." *Natl. Advis. Comm. Aeronaut., Tech. Note* **NACA TN 3179** (1954).

148. M. M. El Wakin, R. J. Priem, H. J. Brikowski, P. S. Myers, and O. A. Uyehara, "Experimental and Calculated Temperature and Mass Histories of Vaporizing Fuel Drops." *Natl. Advis. Comm. Aeronaut., Tech. Note* **NACA TN 3490** (1956).

149. R. J. Priem, G. L. Borman, M. M. El Wakil, O. A. Uyehara, and P. S. Myers, "Experimental and Calculated Histories of Vaporizing Fuel Drops." *Natl. Advis. Comm. Aeronaut., Tech. Note* **NACA TN 3988** (1957).

150. G. M. Faeth, "Evaporation and Combustion of Sprays." *Prog. Energy Combust. Sci.* **9**, 1–76 (1983).

151. G. M. Faeth, "Spray Combustion Models—A Review." *AIAA Pap.* **79-0293** (1979).

152. C. K. Law, "Multicomponent Droplet Combustion with Rapid Internal Mixing." *Combust. Flame* **26**, 219–233 (1976).

153. C. K. Law, "Unsteady Droplet Combustion with Droplet Heating." *Combust. Flame* **26**, 17–22 (1976).

154. C. K. Law, "Recent Advances in Droplet Vaporization and Combustion." *Prog. Energy Combust. Sci.* **8**, 171–201 (1982).

155. C. K. Law, S. Prakash and W. A. Sirignano, "Theory of Convective, Transient, Multi-component Droplet Vaporization." *Symp. (Int.) Combust. [Proc.]* **16**, 605–617 (1977).

156. W. A. Sirignano and C. K. Law, "Transient Heating and Liquid Phase Mass Diffusion in Droplet Vaporization." *Adv. Chem. Ser.* **166**, 1–26 (1978).

157. C. K. Law and W. A. Sirignano, "Unsteady Droplet Combustion with Droplet Heating. II. Conduction Limit." *Combust. Flame* **28**, 175–186 (1977).

158. W. A. Sirignano, "Theory of Multicomponent Fuel Droplet Vaporization." *Arch. Thermodyn. Combust.* **9**, 235–251 (1979).

159. P. Lara-Urbaneja and W. A. Sirignano, "Theory of Transient Multicomponent Droplet Vaporization in a Convective Field." *Symp. (Int.) Combust. [Proc.]* **18**, 1365–1374 (1981).

160. W. A. Sirignano, *Fuel Droplet Vaporization and Spray Combustion Theory*, Rep. School of Mechanical Engineering, Carnegie-Mellon University, Pittsburgh, Pennsylvania, May 1983.

161. W. A. Sirignano, "Fuel Droplet Vaporization and Spray Combustion." *Prog. Energy Combust. Sci.* **9**, 291–322 (1983).

162. H. H. Chui, H. Y. Kim, and E. J. Croke, "Internal Group Combustion of Liquid Droplets." *Symp. (Int.) Combust. [Proc.]* **19**, 971–980 (1983).

163. S. K. Aggarwal and W. A. Sirignano, "Ignition of Fuel Sprays: Deterministic Calculations for Idealized Droplet Arrays." *Symp. (Int.) Combust. [Proc.]* **20**, 1773–1780 (1985).

164. W. A. Sirignano, "The Effect of Droplet Spraying on Spray and Group Combustion," *Am. Soc. Mech. Eng. [Pap.]* **84-WA / HT-26** (1984).

165. J. S. Chin and A. H. Lefebvre, "Steady-State Evaporation Characteristics of Hydrocarbon Fuel Drops." *AIAA J.* **21**(10), 1437–1443 (1983).

166. G. G. Stokes, *Scientific Papers*. Cambridge University Press, London and New York, 1901.

167. I. Langmuir and K. Blodgett, *A Mathematical Investigation of Water Droplet Trajectories*, A.A.F. Tech. Rep. 5418. Air Material Command, Wright Patterson Air Force Base, Ohio, 1946.

168. L. Prandtl, *Guide to the Theory of Flow*, 2nd ed., p. 173. Braunschweig, 1944.

169. S. A. Morsi and A. J. Alexander, "An Investigation of Particle Trajectories in Two-Phase Flow Systems." *J. Fluid Mech.* **55**, Part 2, 193–208 (1972).

170. R. Mellor, Ph.D. Thesis, University of Sheffield, 1969.

171. Y. Onuma and M. Ogasawara, "Studies on the Structure of a Spray Combustion Flame." *Symp. (Int.) Combust. [Proc.]* **15**, 453–465 (1975).

172. L. Schiller and A. Z. Naumann, *Ver. Dtsch. Ing.* **77**, 318 (1933).

173. A. Putnam, "Integratable Form of Droplet Drag Coefficient." *J. ARS* **31**, 1467–1468 (1961).

174. R. D. Ingebo, "Drag Coefficients for Droplets and Solid Spheres in Clouds Accelerating in Airstreams." *Natl. Advis. Comm. Aeronaut., Tech. Note* **NACA TN 3762** (1956).

175. M. C. Yuen and L. W. Chen, "On Drag of Evaporating Liquid Droplets." *Combust. Sci. Technol.* **14**, 147–154 (1976).

176. P. Eisenklam, S. A. Arunachlaman, and J. A. Weston, "Evaporation Rates and Drag Resistance of Burning Drops." *Symp. (Int.) Combust. [Proc.]* **11**, 715–728 (1967).

177. C. K. Law, "Motion of a Vaporizing Droplet in a Constant Cross Flow." *Int. J. Multiphase Flow* **3**, 299–303 (1977).

178. C. K. Law, "A Theory for Monodisperse Spray Vaporization in Adiabatic and Isothermal Systems." *Int. J. Heat Mass Transfer* **18**, 1285–1292 (1975).

179. S. Lambiris and L. P. Combs, "Steady State Combustion Measurement in a LOX RP-1 Rocket Chamber and Related Spray Burning Analysis. Detonation and Two Phase Flow." *Prog. Astronaut. Rocketry* **6**, 269–304 (1962).

180. E. M. Twardus and T. A. Brzustowski, "The Interaction between Two Burning Fuel Droplets." *Arch. Procesow Spalania* **8**, 347–358 (1977).

181. N. A. Chigier and C. G. McGreath, "Combustion of Droplets in Sprays." *Acta Astronaut.* **1**, 687–710 (1974).

182. N. A. Chigier, "Group Combustion Models and Laser Diagnostic Methods in Sprays: A Review." *Combust. Flame* **51**, 127–139 (1983).

183. H. H. Chui and T. M. Liu, "Group Combustion of Liquid Droplets." *Combust. Sci. Technol.* **17**, 127–142 (1977).

184. M. Labowsky and D. C. Rosner, "Group Combustion of Droplets in Fuel Clouds. I. Quasi-Steady Predictions." *Adv. Chem. Ser.* **166**, 63–79 (1978).

185. S. M. Correa and M. Sichel, "The Group Combustion of a Spherical Cloud of Monodisperse Fuel Droplets." *Symp. (Int.) Combust. [Proc.]* **19**, 981–991 (1983).

186. A. R. Kerstein and C. K. Law, "Percolation in Combustion Sprays. I. Transition from Cluster Combustion to Percolate Combustion in Non-Premixed Sprays." *Symp. (Int.) Combust.* **19**, 961–970 (1983).

187. F. Boyson and J. Swithenbank, "Spray Evaporation in Recirculating Flow." *Symp. (Int.) Combust. [Proc.]* **17**, 443–454 (1979).

188. J. Swithenbank, A. Turan, and P. G. Felton, "Three-dimensional Two-Phase Mathematical Modelling of Gas Turbine Combustors." In *Gas Turbine Combustor Design Problems* (A. H. Lefebvre, ed.), pp. 249–314. Hemisphere Press, Washington, D.C., 1980.

189. F. Boyson, W. H. Ayers, J. Swithenbank, and Z. Pan, "Three-Dimensional Model of Spray Combustion in Gas Turbine Combustors." *AIAA Pap.* **81-0324** (1981).

190. A. D. Gosman and R. J. R. Johns, "Computer Analysis of Fuel-Air Mixing in Direct-Injection Engines." *SAE Pap.* **800091** (1980).

191. H. C. Mongia and K. Smith, "An Empirical/Analytical Design Methodology for Gas Turbine Combustors." *AIAA Pap.* **78-998** (1978).

192. Y. El-Banhawy and J. H. Whitelaw, "Calculation of the Flow Properties of a Confined Kerosene-Spray Flame." *AIAA J.* **18**, 1503–1510 (1980).

193. P. J. O'Rourke and F. V. Bracco, *Modeling of Drop Interactions in Thick Sprays and a Comparison with Experiments*, Pap. C404/80. Institution of Mechanical Engineers, London, 1980.

194. A. A. Putnam, "Injection and Combustion of Liquid Fuels." *WADC Techn. Rep.* **56-344** (1957).

195. N. A. Chigier, *Energy, Combustion and Environment*. McGraw-Hill, New York, 1981.

196. N. A. Chigier, "Spray Combustion Processes, A Review," *Am. Soc. Mech. Eng.* [*Pap.*] **82-WA / HT-86** (1982).

197. J. H. Burgoyne and L. Cohen, "The Effect of Drop Size on Flame Propagation in Aerosols." *Proc. R. Soc. London, Ser. A* **225**, 375–392 (1954).

198. Y. Onuma, M. Ogasawara, and T. Inoue, "Further Experiments on the Structure of a Spray Combustion Flame." *Symp. (Int.) Combust. [Proc.]* **16**, 561–567 (1977).

199. Y. Mizutani, G. Yasuma, and M. Katsuki, "Stabilization of Spray Flames in a High-Temperature Stream." *Symp. (Int.) Combust. [Proc.]* **16**, 631–638 (1977).

200. R. Bolado and A. J. Yule, "The Relationship between Atomization Characteristics and Spray Flame Structures." *Proc. Int. Conf. Liq. Atomization Spray Syst., 2nd, 1982*, pp. 221–227 (1982).

201. A. M. Mellor, "Simplified Physical Model of Spray Combustion in a Gas Turbine Engine." *Combust. Sci. Technol.* **8**, 101–109 (1973).

202. J. H. Tuttle, M. B. Colket, R. W. Bilger, and A. M. Mellor, "Characteristic Times for Combustion and Pollutant Formation in Spray Combustion." *Symp. (Int.) Combust. [Proc.]* **16**, 209–222 (1977).

203. C. A. Moses, *Studies on Fuel Volatility Effects on Turbine Combustor Performance*. Western Central States Section/Combustion Institute Joint Spring Meeting, San Antonio, Texas, 1975.

204. D. R. Ballal and A. H. Lefebvre, "Weak Extinction Limits of Turbulent Flowing Mixtures." *J. Eng. Power* **101**(3), 343–348 (1979).

205. D. R. Ballal and A. H. Lefebvre, "Weak Extinction Limits of Turbulent Heterogeneous Fuel/Air Mixtures." *J. Eng. Power* **102**(2), 416–421 (1980).

206. D. R. Ballal and A. H. Lefebvre, "Ignition and Flame Quenching of Flowing Heterogeneous Mixtures." *Combust. Flame* **35**, 155–168 (1979).

207. H. H. Foster and D. M. Straight, "Effect of Fuel Volatility Characteristics on Ignition Energy Requirements in a Turbojet Combustor." *Natl. Advis. Comm. Aeronaut.* **NACA RM E52J21** (1953).

208. L. D. Wigg, *The Ignition of Flowing Gases*, Select. Combust. Probl. II, pp. 73–82. Butterworth, London, 1956.

209. K. V. L. Rao and A. H. Lefebvre, "Minimum Ignition Energies in Flowing Kerosine/Air Mixtures." *Combust. Flame* **27**(1), 1–20 (1976).

210. D. R. Ballal and A. H. Lefebvre, "Ignition and Flame Quenching of Quiescent Fuel Mists." *Proc. R. Soc. London, Ser. A* **364**, 277–294 (1978).

211. D. R. Ballal and A. H. Lefebvre, "Ignition and Flame Quenching of Flowing Heterogeneous Fuel-Air Mixtures." *Combust. Flame* **35**(2), 155–168 (1979).

212. D. R. Ballal and A. H. Lefebvre, "General Model of Spark Ignition for Gaseous and Liquid Fuel/Air Mixtures." *Symp. (Int.) Combust. [Proc.]* **18**, 1737–1746 (1981).

213. A. H. Lefebvre, ed., *Atomization and Sprays*. Hemisphere, New York, 1989.

10 Coal and Char Combustion

L. DOUGLAS SMOOT

Professor of Chemical Engineering
Brigham Young University
Provo, Utah

1. Introduction
   1.1. Objectives
   1.2. Scope
   1.3. Approach
   1.4. General references
2. Coal processes and properties
   2.1. Coal availability and uses
   2.2. Coal processes
   2.3. Issues in increasing use of coal
   2.4. Coal characteristics
3. Coal particle ignition and devolatilization
   3.1. Introduction and scope
   3.2. Elements of coal reactions
   3.3. Ignition
   3.4. Particle heat-up
   3.5. Devolatilization
   3.6. Combustion of volatiles
4. Heterogeneous char reaction processes
   4.1. Reaction variables
   4.2. Global reaction rates
   4.3. Intrinsic reaction rates
   4.4. Other aspects of particle reaction
5. Practical coal flames
   5.1. Background and scope
   5.2. Flame classification and description
   5.3. Practical combustion process characteristics
   5.4. Practical gasification process characteristics
6. Modeling of coal processes
   6.1. Background and scope
   6.2. Basic model's elements and premises
   6.3. Plug-flow model
   6.4. A multidimensional flame model (PCGC-2)
   6.5. Other coal combustion codes
   6.6. Status of coal process model development and application
Nomenclature
References

1. INTRODUCTION

1.1. Objectives

This chapter deals with reaction processes involving coal, char, and other solid fossil fuels. Properties and uses of these fossil fuels are treated; reaction processes of coal particles are also considered and modeled. Then these results are applied to the description of practical coal processes.

The key objectives of this chapter are the following:

1. Review the existing and potential uses of coal and the processes most commonly applied.
2. Identify the general chemical and physical properties of coal, emphasizing the complexity and variability of these natural materials.
3. Summarize major issues being addressed in the increasing uses of coal and other solid fossil fuels.
4. Characterize effects of key variables such as coal type, particle size, heating rate, temperature, pressure, and oxidizer type on coal particle reaction rate.
5. Outline useful existing methods for modeling of coal particle reactions, including devolatilization and heterogeneous oxidation processes.
6. Identify the nature and controlling processes of practical coal dust flames in various coal processes.
7. Outline general methods for modeling of coal reaction processes, and illustrate by application of a one-dimensional model.

1.2. Scope

The entire area of coal reaction processes is very extensive. This field of study includes or interacts with such topics as (1) the origin and geologic nature of coal; (2) the chemical and physical properties and classification of coal; (3) the relationship of coal structure and composition to coal reaction processes; (4) the relationship of coal to other solid and solid-derived fossil fuels, such as oil shale or solvent-refined coal; (5) thermal devolatilization of coal and its dependence on coal type, particle size, heating rate, temperature, etc.; (6) the nature and chemical composition of coal volatiles and their dependence on coal type, heating rate, temperature, etc.; (7) the chemical reaction of coal volatiles in the gas phase, including formation of soot and cracking of hydrocarbons; (8) the formation of char during devolatilization, including swelling, softening, cracking, and formation of internal surfaces; (9) the reaction of char particles, including oxidizer diffusional processes internal and external to the particle, effects of volatiles transpiration, surface reaction, and product diffusion; (10) formation and control of a variety of pollutant species, including oxides of nitrogen and their precursors, oxides of carbon, potentially carcinogenic hydrocarbons, carbon dioxide, volatile trace metals, and small particulates; (11) radiative processes of coal and its solid products (i.e., soot, ash, slag, and char) and gaseous products (e.g., CO_2 and H_2O); (12) formation of ash and slag particles, their change in particle sizes, and their control and removal; (13) interaction of particles with walls and surfaces, including formation of ash or slag layers; (14) particle-gas interactions including convective and radiative heat transfer, reactant and product diffusion, and particle motion in the turbulent gas media; (15) design and optimization of coal reaction processes.

In this chapter, only solid fossil fuels are considered, and emphasis is placed on finely pulverized coal reaction processes. This form of coal is dominant in existing coal processes. Less treatment is given to processing of larger coal particles. Reactions of coal processes are considered in some detail. However, the application of these reaction processes to modeling of coal reaction processes is only introduced and illustrated with a one-dimensional model. The important topic of coal slurries is only mentioned.

1.3. Approach

An effort is made to apply the foundations of combustion established in earlier parts to the treatment of coal reaction processes. An attempt is also made to identify the relationships among coal properties, basic coal particle reactions, and the behavior of coal flames in practical processes. Selected experimental data are presented to illustrate typical coal behavior and reaction rates. Then existing methods for correlating these data are developed and generalized where possible. These correlative and predictive methods are integrated into methods for describing coal reaction processes.

1.4. General References

The multivolume series *Chemistry of Coal Utilization*,[1] edited by Lowry, published over two decades ago, still provides a good description of coal origins, properties, and reactions and coal processes, including storage, handling, processing, and products. More recently, Elliott[2] has edited a second supplementary volume of *Chemistry of Coal Utilization*. The book *Pulverized Coal Combustion*, by Field et al.,[3] published in England in 1967, provides a good general description of the modeling of coal reactions. More recently, the two books, *Pulverized Coal Combustion and Gasification*, edited by Smoot and Pratt[4] and published in 1979, and *Coal Combustion and Gasification*, by Smoot and Smith,[5] provide advanced summaries of the foundations and methods for describing and modeling coal flames and reaction processes. In addition to the above books, several review articles have been prepared on various aspects of

TABLE 1. Summary of Selected Surveys in Coal Combustion in the Past Decade

Author(s)	Date	Topic	Subject Area(s)
Grumer[6]	1975	Explosions	Review of practices for extinguishment of coal dust explosions, including rock dust, water, chemicals; characteristics of explosions (63 refs.)
Anson[7]	1976	Fluidized beds	Descriptive review of the state of development of fluidized beds for generating power from coal, including aspects of combustion, emissions, cycles (50 refs.)
Anthony and Howard[8]	1976	Coal devolatilization	Basic review of thermal decomposition of coal in inert gases and in hydrogen, including product yield, pyrolysis models (180 refs.)
Beer[9]	1976	Fluidized beds	Basic and applied aspects of fluid bed coal combustion, including theory, related kinetic data, sulfur and nitrogen emissions, heat transfer (59 refs.)
Godridge and Read[10]	1976	Large furnace heat transfer	Combustion and heat transfer in large coal and oil boilers, including boiler characteristics, burner designs, heat transfer models, and furnace measurements (58 refs.)
Sarofim and Flagen[11]	1976	NO_x pollution control	Summary of NO_x pollutant formation processes and control methods for sources (95 refs.)
Breen[12]	1977	Boiler	Impact of furnace/boiler modifications (e.g., reduced xs air) on efficiency and pollutant level (15 refs.)
Essenhigh[13]	1977	Particle, dust reactions	Basic discussion of coal particle reaction; coal dust flame propagation (160 refs.)
Littlewood[14]	1977	Gasification	Theory and applications for gasification of coal and other fossil fuels, including a description of several commercial and development systems (119 refs.)
Smoot and Horton[15]	1977	Dust flames	Review of measurements and theory for propagation of premixed, laminar coal-dust flames (86 refs.)
Belt and Bissett[16]	1978	Flash and hydropyrolysis	Review of past 15 years' work in pyrolysis; eight heating methods reviewed; key variables identified; need for additional data recognized (35 refs.)
Gibson[17]	1978	Coal properties	The constitution of coal and its relationship to pyrolysis, liquefaction, gasification, and hydrogenation processes (46 refs.)
Laurendeau[18]	1978	Heterogeneous kinetics	Detailed review of char combustion and gasification, including coal characteristics, surface phenomena, kinetics (214 refs.)
Sarofim and Hottel[19]	1978	Radiation	Summary of optical properties of coal and its products: ash, char, and soot; application to solid fossil fuels; state of prediction of radiation in coal flames (77 refs.)

Author	Year	Topic	Description
Krazinski et al.[20]	1979	Coal dust flames	Review of measurements and models in dust flames; development of a model for dust flame prediction (92 refs.)
Macek[21]	1979	Pollution from combustor types	Stoker, cyclone, pulverized combustion comparison for NO_x, fly ash, SO_x, coal/oil dispersions (34 refs.)
Wall et al.[22]	1979	Coal mineral matter	Mineral matter in coal and its effect on large boiler performance, characteristics of coal ash, release during grinding, mining, heating, combustion; wall deposits, radiative properties (85 refs.)
Smoot[23]	1980	Theory, modeling	Review of foundations and submodels for pulverized coal turbulent flames; summary and comparison of p.c. models developed in past decade;
Wendt[24]	1980	NO_x pollution	A review of coal combustion mechanisms and pollutant formation in furnaces. Furnace-flame types are illustrated. Effects of staging and other variables on NO_x levels; basic coal particle reaction processes (61 refs.)
Howard[25]	1981	Coal devolatilization	Review of experimental methods, data and modeling of coal devolatilization, applied particularly to gasification (312 refs.)
Brown[26]	1982	Gasification	A review of the potential for use of coal gasification for electrical power generation, particularly with combined cycle application (36 refs.)
Smith[27]	1982	Char oxidation	Review of data correlation methods, and experimental data (non-U.S. coals) for oxygen reaction heterogeneously with chars and cokes (55 refs.)
Simons[28]	1983	Pore structure coal reactions	Review of chemistry, physics and modeling of the pore structure of coals and chars (47 refs.)
Smoot and Hill[29]	1983	Critical needs in combustion	An extensive review of combustion research work over the past five years, including coal combustion, with recommendations for further work (258 refs.)
Solomon and Hamblen[30]	1983	Devolatilization	A review of coal pyrolysis data and an evaluation of the theory of the rank insensitive nature of chemical functional group release (carboxyl, hydroxyl, etc.) during pyrolysis of coal (56 refs.)
Several Authors[31]	1984	Coal combustion and applications	A special collection of eight articles dealing with coal combustion technologies (S. S. Penner, et al.), coal structure and its relations to combustion (P. H. Given), fouling and slagging (W. T. Reid), fluid dynamics (J. M. Beer et al.), diagnostics (S. S. Penner, et al.), acoustic instrumentation (A. C. Paptis), modeling of coal combustion processes (L. D. Smoot), and coal preparation (L. G. Austin and P. T. Luckie). This collection resulted from a DOE-sponsored working group who considered the state-of-the-art and key research needs in combustion of coal (533 refs.)

coal combustion. Table 1 summarizes several of these recent reviews, which by themselves contain nearly 2000 references to studies of coal combustion and related topics.[6-31]

2. COAL PROCESSES AND PROPERTIES

2.1. Coal Availability and Uses

Coal is the most abundant fossil fuel that exists in the United States. Recoverable reserves have been estimated to be nearly 300 billion tons,[32] with potential total reserves far in excess of this amount. Deposits of coal are located in most regions in the United States.[32] Estimated production of coal by the year 2000 will be nearly two billion tons per year, with the bulk being consumed through combustion processes as illustrated in Figure 1.[33] Thus present recoverable reserves are adequate to meet the nation's coal needs for many decades and potentially much longer.

Most of the combustion of coal by the year 2000 will be for electric power generation as illustrated in Figure 1, with industrial consumption of coal for steam and heat and for metallurgical processes and other major uses. Areas for potential expanding uses of coal, though hard to project in magnitude at the present time, include the following: (1) fluidized bed combustion, (2) coal exports, (3) combined cycle power generation, (4) direct gasification for medium or high BTU gas, (5) liquefaction, (6) chemical feedstocks, (7) improved solid fuels (e.g., solvent refined coal), (8) MHD power generation, (9) fuel cells, (10) direct firing in gas turbines.

2.2. Coal Processes

Most of the coal presently being consumed is by direct combustion of finely pulverized coal in large-scale utility furnaces for generation of electric power, and this is likely to remain the way through the end of this century (see Figure 1). However, there are a large number of other processes for the conversion of coal into other products or for the direct combustion of coal.[1,34,35] These processes can be classified by process type (or end product) or by coal particle size or temperature, etc. Some examples of coal processes are the following:

Direct Combustion

Pulverized coal combustors (smallest particles)
Fluidized bed (medium-sized particles)
Fixed bed (e.g., stokers) (larger particles)

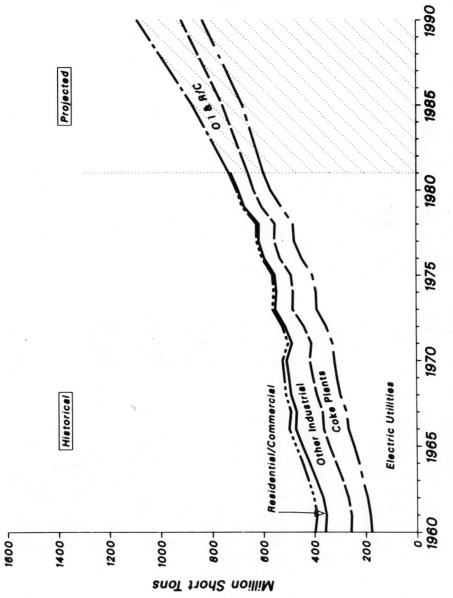

FIGURE 1. Combustion of coal and lignite by end-use section.[33]

Gasification

Entrained bed (smallest particles)
Fluidized bed (medium-sized particles)
Fixed bed (larger particles)
Others (molten bath)

Carbonization and Coking

Low temperature (750–975 K)
Medium temperature (1025–1175 K)
High temperature (1175–1325 K)

Liquefaction

Pyrolysis
Extraction

Other

MHD
Fuel cell

Although not a complete review these examples illustrate the wide variability of coal processes. Table 1a summarizes some of the characteristics of these coal processes, including extent of use, scale size, and the coal particle sizes employed.[36–41]

2.3. Issues in Increasing Use of Coal

Increasing use of coal in the United States presents many technical problems. Table 2 summarizes related technical problems areas that are presently being emphasized.[42–49] A host of other problems also have a direct bearing on the consumption of coal, including mine safety, labor availability and relations, water availability, feedstock transportation, mine equipment manufacture, capital availability, and control of solid and liquid wastes.[32] Although these issues are well beyond the scope of this part, they will continue to have a major impact on the consumption of coal.

2.4. Coal Characteristics

Formation and Variation. Coal is a black, inhomogeneous, organic fuel, formed largely from partially decomposed and metamorphosed plant materials. Formation has occurred over long time periods, often under high pressures of overburden and at elevated temperatures. Differences in plant materials and in their extent of decay influence the components present in coals, such as vitrain (from "vitro" meaning glass), clarain (from "clare" meaning clear or bright), durain (from "dur" meaning hard or tough), and fusain (meaning charcoal). Descriptions of such coal components are part of the science of petrography.[50]

Coals vary greatly in their composition. Of 1200 categorized coals, no two had the same composition.[50] Typical compositions (mass percentages) of coal include 65–95% carbon, 2–7% hydrogen, up to 25% oxygen and 10% sulfur, and 1–2% nitrogen.[13] Inorganic mineral matter (ash) as high as 50% has been observed, but 5–15% is more typical. Moisture levels commonly vary from 2 to 20%, but values as high as 70% have been observed. The process of conversion of plant materials such as peat to coal is called *coalification* and takes place in stages producing a variety of coal products. Hendrickson[50] provides the following description of some of these coal types:

> Lignite, the lowest rank of coal, was formed from peat which was compacted and altered. Its color has become brown to black and it is composed of recognizable woody materials imbedded in pulverized (macerated) and partially decomposed vegetable matter. Lignite displays jointing, banding, a high moisture content, and a low heating value when compared with the higher coals. Subbituminous coal is difficult to distinguish from bituminous and is dull, black colored, shows little woody material, is banded, and has developed bedding planes. The coal usually splits parallel to the bedding. It has lost some moisture content, but is still of relatively low heating value. Bituminous coal is dense, compacted, banded, brittle, and displays columnar cleavage and a dark black color. It is more resistant to disintegration in air than are subbituminous and lignite coals. Its moisture content is low, volatile matter content is variable from high to medium, and its heating value is high. Several varieties of bituminous coal are recognizable. Anthracite is the highly metamorphosed coal, is jet black in color, is hard and brittle, breaks with a conchoidal fracture, and displays a high luster. Its moisture content is low and its carbon content is high. Neither peat nor graphite are coal, but they are the initial and end products of the progressive coalification process.

Lowry,[1] IGT,[34] Spackman,[45] Hendrickson,[50] Given,[51] Neavel,[52] Hamblen et al.,[53] and Elliott[2] discuss further the origin and/or characteristics of coal.

Coal Classification. Efforts have been made to broadly classify coal and to relate similarities among coals to their potential behavior in conversion processes. The most common of these is the ASTM Classification, which is based upon fixed carbon level and heating value. Figure 2 illustrates the general characteristics of 12 such coal groups, ranging from soft lignite to very hard metaanthracite.

Classifications have also been based on petrographic parameters. Lowry[1] and Elliott[2] summarize several other systems for general classification of coal including: (1) English National Coal Board System, based upon percent of proximate volatiles and Gray-King coking properties; (2) International System

TABLE 1a. Summary of Selected Coal Processes

Process Type	Description	Extent of Coal Use in U.S. (% of Total Used)	Commercial Use	Scale Size (TPD)[a]	Coal Types	Coal Size
DIRECT COMBUSTION	Burning coal to produce, electricity heat, and steam					
Power Station	Commercial electrical production	78[36]–80%[37]				
Pulverized	Rapid burning of finely grained coal		Common	1000–10,000[38]	All	0.01–0.025mm[38]
Fluidized bed[b]	Well-stirred combustion		Pilot plant	2000–8000[38]	All	0.15–0.6 cm[38]
Stoker	Mechanically fed fixed bed		Small	100[38]	Noncaking	1–5 cm[38]
MHD	Combustion energy capture by magnetic fields	Laboratory		800–4000[39]		
Coal/oil mixture (COM)	Burning coal/oil mixtures in oil furnaces		Demonstration	TPC ce[c]		
Industrial Heat/Steam	Industrial plant power	8–11%			Same as above	
Pulverized			Small	1–100		
Fluidized bed			Pilot	1–100		
Stoker			Common	1–100		
COM			Demonstration	1–100		
Domestic/Commercial	Hand-stoked space heating	1%		0.005–0.05	Noncaking	3–10 cm
Transportation	Fuel for railroads	0.01–0.02%		0.1–1	Noncaking	
GASIFICATION[40]	Converting coal to low, medium, or high BTU gas for use as fuel or feedstock	0				
Fixed bed Single-stage dry ash	Reacting coal in a fixed bed		Common	60–800	Noncaking	0.3–0.5 cm[32]

Process	Description	Value	Status	TPD[a]	Coal type	Particle size
Two-stage dry ash			Common	100	Noncaking	0.3–0.5 cm[32]
Slagging			Test units	20	Noncaking	0.3–0.5 cm[32]
Fluidized bed	Reacting coal is a well-mixed reactor		Common		All	0.8 cm[32]
Entrained flow	Gasifying coal rapidly					
Single stage			Common	600	All	0.01–0.02 mm
Two stage			Pilot plant	5–60	All	0.01–0.02 mm
Tumbling bed	Mechanically mixing the bed		Pilot plant	44	All	0.3–0.2 cm
Molten bath	Gasifying coal in a hot liquid					
In situ	Reacting coal in place, underground		Pilot tests			
LIQUEFACTION[41]	Converting coal to liquid fuel					
Pyrolysis	Removal of volatile compounds by heating	0				
Gentle			Pilot plant[d]	25		1–2 cm
Flash			Commercial[e]			0.01–0.025 mm
Direct Liquefaction	Hydrogenating, dissolving, and heating the coal to derive a liquid					
Solvent extraction			Pilot plant	50–250		0.01–0.025 mm
Catalytic			Pilot plant	10–600		0.01–0.026 mm
Indirect Liquefaction	The gasification of coal followed by catalytic recombination		Commercial[f]		see gasification	
Coking	Carbonizing coal for use in metallurgical processes	10^{37}–15^{36}	Common		All	0.15–0.6 cm[3]

[a] TPD = tons/day.
[b] Very adaptable, exhibits high heat transfer in bed as well as low-level pollutant products.
[c] Coal equivalent.
[d] Was once widespread as a feedstock for chemicals.
[e] Commercial application is only found in Europe.
[f] Commercial application is only found in South Africa.

TABLE 2. Some Technical Issues and Problems Arising from the Expanding Use of Coal

Problem Area	General Description	Related References
1. Environmental concerns	Need for control of emissions from coal processes, including oxides of nitrogen, oxides of sulfur, carbon dioxide, fine particulates, soot, heavy hydrocarbons, and trace elements	11, 21, 24, 32, 42
2. Safety and explosions	Need for methods to control explosions and fires in mines, utility and industrial boilers, and in coal storage facilities	6, 43
3. Combustion/conversion efficiency	Need to maintain high combustion efficiency as process variables change for pollutant control, coal variability, etc.	44
4. Ash/slag	Need to control ash and slag deposits and removal, and associated maintenance of surfaces, corrosion, etc., particularly for the variety of coals being considered, some with very high ash levels	22
5. Process design/ optimization	Need for development and verification of more efficient methods for designing, scaling, and optimizing coal conversion processes	23
6. Basic process data	Need for relating coal characteristics (composition, structure, etc.) to conversion/combustion characteristics, including radioactive properties, pyrolysis qrates, and char oxidation rates	17, 34, 35
7. Use of alternate solid fossil fuels	Basic information on conversion and combustion on alternate solid fuels and fuel products such as oil shale, char, solvent refined coal, peat, and tar sands	46–48
8. Coal water mixtures	For retrofiting existing oil-fired furnaces and for direct coal ignition systems. Needs for coal preparation, CWM slurry characteristics, and full-scale boiler tests.	49

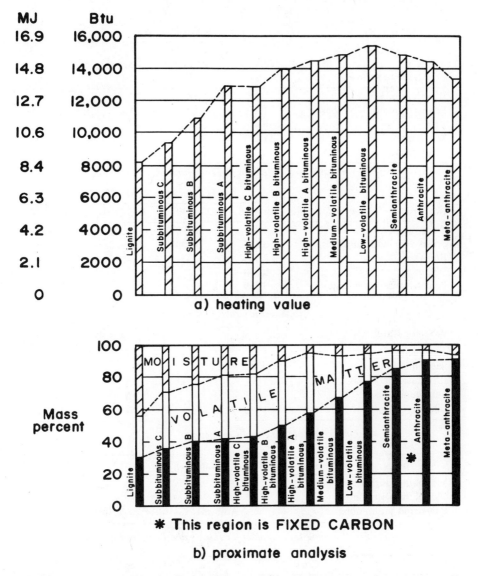

MJ	Btu
16.9	16,000
14.8	14,000
12.7	12,000
10.6	10,000
8.4	8000
6.3	6000
4.2	4000
2.1	2000
O	O

a) heating value

b) proximate analysis

❋ This region is FIXED CARBON

FIGURE 2. Energy content and composition of coals according to ASTM coal rank.[50]

for hard coals and brown coals; (3) Mott System, based upon volatile matter, heating values, and ultimate (O—H) analysis. None of these approximate systems is able to deal with the complex structural and compositional differences in coals.

Coal Physical and Chemical Properties. Modeling of coal conversion processes requires data for physical properties of coal, such as thermal conductivity, specific heat or density. Table 3 summarizes selected values of common

TABLE 3. Typical Values for Selected Coal Physical Properties[a]

(a) Specific Heats of Air-Dried Coals

| | Proximate Analysis | | | |
| | Moisture (%) | Volatile Matter (%) | Carbon (%) | Ash (%) |
Coal Sample Source				
West Virginia	1.8	20.4	72.4	5.4
Pennsylvania (bituminous)	1.2	34.5	58.4	5.9
Illinois	8.4	35.0	48.2	8.4
Wyoming	11.0	38.6	40.2	10.2
Pennsylvania (anthracite)	0.0	16.0	79.3	4.7

Mean Specific Heat for K Temperature Ranges

| | Temperature Range (K) | | | |
Coal Sample Source	301–338	298–403	298–450	298–500
West Virginia	0.261	0.288	0.301	0.314
Pennsylvania (bituminous)	0.286	0.308	0.320	0.323
Illinois	0.334			
Wyoming	0.350			
Pennsylvania (anthracite)	0.269			

(b) Specific Gravity[b] of Coal

Probable Rank	Specific Gravity
Anthracite	1.7
Semianthracite	1.6
Bituminous	1.4
Subbituminous	1.3
Lignite	1.2

(c) Thermal Conductivity[c]

Coal Type	Temperature (K)	$kJs^{-1}m^{-1}K^{-1}$
Monolithic anthracite	303	0.2–0.4
Monolithic bituminous	303	0.17–0.3
Pulverized bituminous	Ambient	0.10–0.15

(d) Average Free-Swelling Index Values for Illinois and Eastern Bituminous Coals

Rank	Coals	ASTM Free-Swelling Index
High-volatile C	Illinois No. 6	3.5
High-volatile B	Illinois No. 6	4.5
High-volatile B	Illinois No. 5	3.0
High-volatile A	Illinois No. 5	5.5
High-volatile A	Eastern	6.0–7.5
Medium-volatile	Eastern	8.5
Low-volatile	Eastern	8.5–9.0

[a] Data from Hendrickson.[50]
[b] Specific gravity has been shown by Lowry,[1] to be a function of hydrogen content.
[c] Thermal conductivity usually increases with increasing apparent density, volatile matter content, ash content, temperature, and probably with moisture content. Coal is thermally anisotropic with k greater perpendicular to bed.

physical properties of coal, including specific heat, specific gravity, thermal conductivity, and swelling index. Heating value data are shown in Figure 2. These properties will vary among coals, even of common rank, and will often be related to temperature and moisture content. Lowry,[1] IGT,[34] Spackman,[45] Hendrickson,[50] Hamblen et al.,[53] and Elliott[2] discuss several other physical, mechanical, thermal, and chemical properties of coal, including fate of trace elements, grindability, friability, compressive strength, dustiness, electrical resistivity, plasticity, optical density, indices of refraction, reflection, and absorption, magnetic susceptibility, electrical conductivity, dielectric constants, and forms of sulfur.

Properties of char are even more variable, since such properties are a function of the nature of the conversion process from which they were produced, in addition to being related to the coal from which they were

TABLE 4. Typical Proximate and Ultimate Analyses of Coals and Char[a]

Coal I.D. Rank	Utah Church Mine Bituminous	Pittsburgh[b] Bituminous	Pittsburgh[b] Bituminous (High-Volatile)	Sewell[b] Bituminous (Medium-Volatile)	Anthracite[b] (Low-Volatile)
Moisture %	2.5–2.7	2.0	1.0	1.9	1.3
Proximate %					
Volatiles	44.1–45.5	36.6	28.9	16.3	8.8
Fixed carbon	42.6–44.2	55.4	63.2	75.6	71.8
Ash	9.2–9.5	6.0	6.9	6.2	18.1
Ultimate %					
Carbon	69.8–71.5	77.5	80.6	84.2	73.2
Hydrogen	5.5–5.6	5.3	4.9	4.3	3.1
Nitrogen	1.4–1.5	1.5	1.5	1.2	0.9
Sulfur	0.4–0.7	1.2	0.7	0.7	0.9
Oxygen	11.2–13.2	8.5	5.4	3.4	3.8
Ash	9.2–9.5	6.0	6.9	6.2	18.1

Coal I.D. Rank	Illinois Coal Bituminous	Illinois Coal Char[c]	North Dakota Lignite	Wyoming Subbituminous	Kentucky Bituminous
Moisture %	10.1	0.9	29.9	27.8	5.0
Proximate %					
Volatiles	35.9	2.4	29.5	32.9	39.5
Fixed carbon	46.7	76.1	33.4	34.3	49.7
Ash	7.3	20.6	7.2	5.0	10.5
Ultimate %					
Carbon	68.3	74.0	69.7	76.3	—
Hydrogen	5.0	0.7	3.8	4.4	—
Nitrogen	1.3	1.0	1.9	1.1	—
Sulfur	3.5	3.3	1.1	0.5	3.3
Oxygen	13.8	0.2	13.2	10.8	—
Ash	8.1	20.8	10.3	6.9	11.1

[a] Data courtesy of Bureau of Mines, Brigham Young University, and ERDA.
[b] From U.S. Bureau of Mines, Pittsburgh, PA.
[c] From *COED Gasification Process*, supplied by ERDA, Pittsburgh, PA (fourth-stage product).

TABLE 5. Compositions of Typical Ashes[a]

(a) Variations in Coal Ash Compositions with Rank

Rank	SiO_2 (%)	Al_2O_3 (%)	Fe_2O_3 (%)	TiO_2 (%)	CaO (%)	MgO (%)	Na_2O (%)	K_2O (%)	SO_3 (%)	P_2O_5 (%)
Anthracite	48–68	25–44	2–10	1.0–2	0.2–4	0.2–1	—	—	0.1–1	—
Bituminous	7–68	4–39	2–44	0.5–4	0.7–36	0.1–4	0.2–3	0.2–4	0.1–32	—
Subbituminous	17–58	4–35	3–19	0.6–2	2.2–52	0.5–8	—	—	3.0–16	—
Lignite	6–40	4–26	1–34	0.0–0.8	12.4–52	2.8–14	0.2–28	0.1–13	8.3–32	—
Utah bituminous[b]	43–48	16–19	3.8–4.2	0.0–1.0	6.5–8.1	0.9–1.1	4.3–4.9	0.4–4.7	3.5–4.1	1.0

(b) Range of Amount of Trace Elements Present in Coal Ashes (ppm on Ash Basis)

Element	Anthracites	High Volatiles, Bituminous	Low Volatiles, Bituminous	Medium Volatiles, Bituminous	Lignites and Subbituminous	Utah[b] Bituminous
Ag	1	1–3	1–1.4	1	1–50	—
B	63–130	90–2800	76–180	74–780	320–1900	700–1500
Ba	540–1340	210–4660	96–2700	230–1800	550–13900	700–1500
Be	6–11	4–60	6–40	4–31	1–28	5–7
Co	10–165	12–305	26–440	10–290	11–310	7–15
Cr	210–395	74–315	120–490	36–230	11–140	70–100

Element						
Cu	96–540	30–770	76–850	130–560	58–3020	62–68
Ga	30–71	17–98	10–135	10–52	10–30	30–70
Ge	20	20–285	20	20	20–100	—
La	115–220	29–270	56–180	19–140	34–90	70–100
Mn	58–220	31–700	40–780	125–4400	310–1030	400
Ni	125–320	45–610	61–350	20–440	20–420	15–30
Pb	41–120	32–1500	23–170	52–210	20–165	35–45
Sc	50–82	7–78	15–155	7–110	2–58	15
Sn	19–4250	10–825	10–230	29–160	10–660	—
Sr	80–340	170–9600	66–2500	40–1600	230–8000	1000–1500
V	210–310	60–840	115–480	170–860	20–250	150
Y	70–120	29–285	37–460	37–340	21–120	50–70
Yb	5–12	3–15	4–23	4–13	2–10	7
Zn	155–350	50–1200	62–550	50–460	50–320	58–64
Zr	370–1200	115–1450	220–620	180–540	100–490	200–300
Cd	—	—	—	—	—	1
Li	—	—	—	—	—	133–155
Nb	—	—	—	—	—	20–30

[a] Table used with permission from Hendrickson.[50]
[b] Utah Power and Light Coal, Church Mine, Analyses by U.S. Geological Survey, Denver, Colorado.

derived. Chars are richer in carbon and leaner in hydrogen than the parent coal and are often porous and more regular in shape, having softened during the conversion process.

The composition of coal is traditionally characterized by ASTM proximate analysis or ASTM ultimate analysis. The former determines only the moisture content (by drying), percent volatiles (from inert devolatilization at about 1200 K), ash (residual after complete combustion in air), and fixed carbon (by difference). Coal rank versus proximate analysis was shown in Figure 2. The proximate analysis for a selected variety of coals is shown in Table 4. In these coals alone, the percent of volatiles varies from 8.8 to 45.5% by weight. The char shown has only 2.4% proximate volatiles.

ASTM ultimate analysis gives element percentages for carbon, hydrogen, nitrogen, sulfur, and oxygen, the latter often determined by difference. The

TABLE 6. Typical Size Distributions for Pulverized Coal

(a) Fluid-Bed Gasifier[a] (Cumulative wt.%)

Tyler Screen Mesh	Illinois Bituminous	Illinois Char
14	15.2	3.3
28	47.0	28.7
48	69.0	54.0
100	81.9	72.1
200	90.2	84.2
325	94.2	91.2
Pan	99.7	99.8

(b) Utility Boiler[b] (Utah Bituminous[c])

Increment Size (μm)	Percentage in Increment[d]	Increment Size (μm)	Percentage in Increment[d]
2.85	0.3–0.5	22.80	7.1–9.0
3.59	0.4–0.6	28.70	9.0–11.1
4.52	0.5–0.7	36.15	10.9–13.3
5.70	0.7–1.0	45.55	10.7–12.4
7.18	1.2–1.6	57.40	11.5–12.1
9.04	2.0–2.5	72.30	9.2–12.4
11.39	3.0–3.7	85.30	5.3–8.3
14.35	4.1–5.1	90 +	7.4–12.9
18.10	5.5–6.9		

[a]Courtesy of U.S. Department of Energy.
[b]Termed 70% through 200 mesh.
[c]Church Mine, Brigham Young University data.
[d]Mass mean diameter is about 50 μm (Coulter counter measurement).

residual mineral matter is shown as ash. Ultimate analyses for selected coals and a char are also shown in Table 4. Ash in coals has been shown to contain significant amounts of some elements, together with trace amounts of several elements, as shown in Table 5 for selected coals. Sarofim et al.[54] discuss particle sizes of ash from pulverized coals and note that some ash components are volatile at higher temperatures.

Particle sizes of coal dust vary greatly, depending upon grinding technique and desired application. Typical size distributions for a fluid-bed gasifier application and a utility boiler application are shown in Table 6. Size distribution for one specific char is also shown in Table 6.

Particle size distributions produced by a full-scale exhauster-type coal mill[55] have recently been measured as illustrated in Figure 3. This large variation in particle sizes (from 1 to 600 μm) has important implications on the combustion of coal. In Figure 3 the upper change in slope in the log-normal distribution is the result of the mill classifier which returns the larger sizes to the mill. The lower slope changes are due to the action of the sampling cyclone.

Structural Characteristics of Coal. It is thought that coal structure is highly planar and layered with pore volume of 8–20%. References 1, 8, 34, 45, 50, 51, 53, and 56 discuss chemical properties of coal, including details of proximate and ultimate analyses, plastic properties of coal, coal hydrogenation and hologenation, solvent extraction of coal components, and properties of minerals and coal chemical structure.

A recent "model" for the chemical organic structure of coal is illustrated in Figure 4. The model shown from Solomon[56] is similar to that prepared by other recent investigators[57,58] and is based on information from infrared measurements, nuclear magnetic resonance, ultimate and proximate analyses, and pyrolysis data. Mineral matter can exist as occlusions within the base organic structure. This is not the specific structure of any particular coal but only a model for interpreting and correlating coal conversion data.

Infrared (Fourier transform) measurements provide quantitative concentrations for the following constituents in the raw coal or its components: hydroxyl, aliphatic or hydroaromatic hydrogen, aromatic hydrogen, aliphatic carbon, and aromatic carbon. From this and other information, the infrared spectrum of the parent coal or its constituents can be synthesized from the basic organic constituents, as illustrated by Solomon.[56]

Presently, methods for design and analysis of existing coal-conversion processes do not make significant use of this kind of information on the structure of coals. However, with increased interest in the use of coal in the United States, more attention is being given on the relationship of coal structure to coal reaction and conversion processes. Gibson[17] discussed the relevance of coal constitution to coal conversion processes, including carbonization, liquefaction, gasification, pyrolysis, extraction, and hydrogenation. Given and Biswas[59] have conducted an investigation of the relation of coal

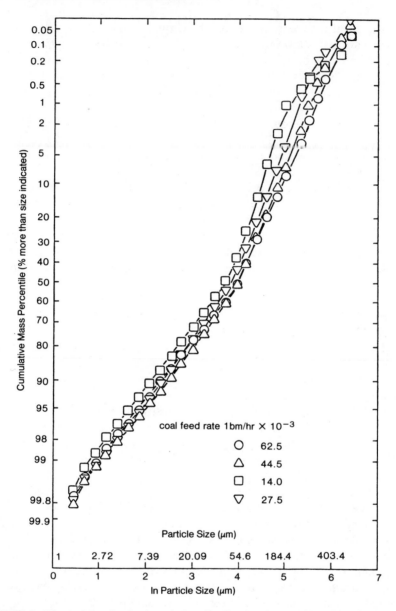

FIGURE 3. Measured pulverized coal size distribution for a full-scale, exhauster-type mill operating at a steady state.[55]

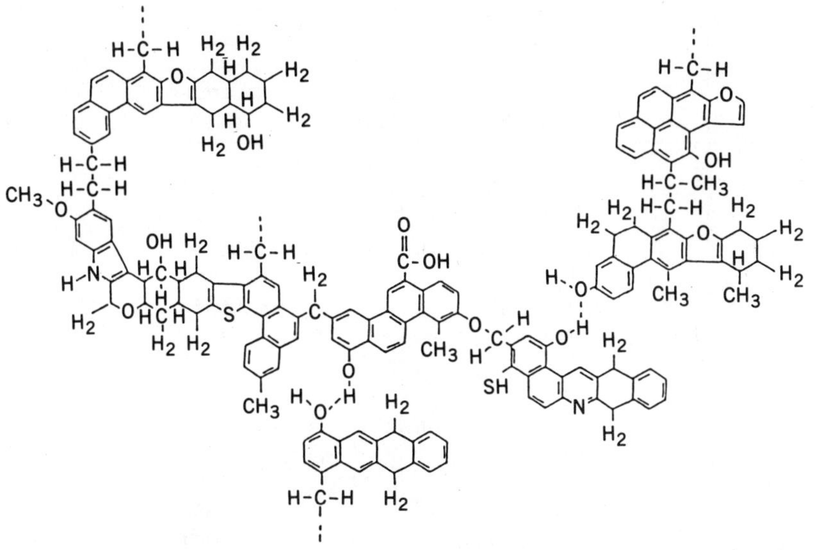

FIGURE 4. Summary of coal structure information in a hypothetical coal molecule.[8]

characteristics to liquefaction of coal, and Spackman[45] has considered this issue with respect to production of clean energy fuels.

Mineral Matter Removal. Environmental constraints, together with development of coal-based fuels for oil substitution, have increased interest in cleaning of coal. Singh et al.[60] and Liu[61] have provided recent reviews of physical and chemical methods for cleaning of coal. Reduction in mineral and sulfur content of coals is of particular concern.

Ultrafine grinding (micronization) of coal to an average particle diameter of about 10 microns separates much of the mineral matter from the organic coal. Subsequent separation of the coal and mineral matter by differences in specific gravity have led to coals with mineral matter levels of less than 1% and with sulfur levels about half the original value. Reduction in sulfur level by physical methods depends strongly upon the relative proportions of sulfur in the mineral matter (pyritic, sulfates) and that bound in the organic structure. Work is continuing on several physical and chemical coal cleaning processes[60,61] in an effort to demonstrate economically viable processes.

3. COAL PARTICLE IGNITION AND DEVOLATILIZATION

3.1. Introduction and Scope

Coal particle reactions are an essential aspect in all coal processes. Often these reactions are among the rate-limiting steps that control the nature and size of coal processes. These reactions also have a direct impact on the formation of

TABLE 7. Typical Coal Particle Reaction Rates

Process	Particle Size	Pressure (atm)	Process Temperature (K)	Typical Residence Time
Direct Combustion				
Pulverized	10–100 μm[a]	1	1750[b]	1 sec[a]
Fluidized bed	1–5 mm[a]		1150[a]	100–500 sec[a]
Stoker	1–5 cm[a]		1750[a]	3000–5000 sec[a]
MHD	10–100 μm			
Coal oil mixtures	Slurry			
Gasification				
Fixed bed	0.3–5.0 cm[c]	1–27[b]	1420[c]	> 1 hr
Fluidized bed	< 8 mm[c]	1–68[b]	1200[b]	0.5–1 hr[c]
Entrained flow	0.01–0.025 mm	1–82[b]	1640–1920[b]	5 s[c]
In situ	5–30 cm			
Liquefaction				
Pyrolysis	0.01 mm–2 cm	1–35[d]	590–1260[d]	0.1–10 sec[d]
Direct liquefaction	15–100 μm			
Solvent extraction		100–270[d]	700[d]	40–100 m[d]
Catalytic		136[d]	700[d]	30–60 m[d]
Indirect liquefaction	(see gasification)			
Coking				
	0.15–0.6 cm[e]	1	1200–1300[e]	18–16.4 hr[e]

[a] Essenhigh et al.[38]
[b] Bodle and Schora.[40]
[c] EPRI Journal.[62]
[d] Nowacki.[41]
[e] Lowry.[63]

fine particulates and on nitrogen- and sulfur-containing pollutant species. Table 7[62,63] summarizes several coal processes that were identified in Section 2, with particular emphasis on particle sizes, residence times, and particle reaction times. A wide range of conditions under which coal is processed is evident. This, coupled with the extensive variety of coals of interest, defines the enormous scope and complexity of this subject area. Sections 3 and 4 deal with the following aspects of coal particle reaction: (1) the two-component concept of coal reactions, which consists of coal devolatilization and char oxidation; (2) coal particle heat-up and ignition; (3) coal devolatilization, including illustrative experimental data for rates and products, volatiles combustion, and common methods for modeling; (4) char oxidation, with typical test data, factors influencing rates, and methods for modeling; (5) intrinsic reactivity of char.

3.2. Elements of Coal Reactions

Development of a complete description of coal processes requires incorporation of a "coal particle reaction" model. A single-model framework would not be adequate for all coal varieties and sizes. One "model" suggests that the particle, at any time in the reaction process, is composed of (1) moisture, (2) raw coal, (3) char, and (4) ash (mineral matter). Also, the particle may be surrounded by volatilized matter. A general description of experimental observations for pulverized coal particles, and assumptions used in developing this model concept, are outlined in detail by Smoot and Smith.[64] An adequate description of coal reaction processes requires the modeling of each of these particle elements.

Coal reactions are generally divided into two distinct components[65]:

1. *Devolatilization of the raw coal.* This part of the reaction cycle occurs as the raw coal is heated in an inert or oxidizing environment; the particle may soften (become plastic) and undergo internal transformation. Moisture present in the coal will evolve early as the temperature rises. As the temperature continues to increase, gases and heavy tarry substances are emitted. The extent of this "pyrolysis" can vary from a few percent up to 70–80% of the total particle weight and can take place in a few ms or several minutes depending on coal size and type, and on temperature. The residual mass, enriched in carbon and depleted in oxygen and hydrogen, and still containing some nitrogen, sulfur, and most of the mineral matter, is referred to as *char*. The char particle is often spherical, has many cracks or holes made by escaping gases, may have swelled to a larger size, and can be very porous. The nature of the char is dependent on the original coal type and size and also on the conditions of pyrolysis.

2. *Oxidation of the residual char.* The residual char particles can be oxidized or burned away by direct contact with oxygen at sufficiently high temperature. This reaction of the char and the oxygen is thought to be heterogeneous, with gaseous oxygen diffusing to and into the particle, adsorbing and reacting on the particle surface. This heterogeneous process is often much slower than the devolatilization process, requiring seconds for small particles to several minutes or more for larger particles. These rates vary with coal type, temperature, pressure, char characteristics (size, surface area, etc.), and oxidizer concentration. Other reactants, including steam, CO_2, and H_2, will also react with char, but rates with these reactants are considerably slower than for oxygen.

These two processes (i.e., devolatilization and char oxidation) may take place simultaneously, especially at very high heating rates. If devolatilization takes place in an oxidizing environment (e.g., air), then the fuel-rich gaseous and tarry products react further in the gas phase to produce higher temperatures in the vicinity of the coal particles. In Section 3 coal particle heat-up, particle ignition, devolatilization, and combustion of volatiles are considered. In Section 4 heterogeneous reaction of char is treated, with presentation of rate data for reaction in O_2 and CO_2 and with brief consideration of intrinsic reactivity.

3.3. Ignition

Ignition of coal is a particularly complex issue and is not yet adequately described or correlated. The literature contains much information relating to ignition, under such topics as induction periods, quenching distances, flame propagation, flammability limits, flashback, blow off, and extinction. Williams,[66] Kanury,[67] Glassman,[68] and Smoot and Smith[5] provide general treatments of ignition in gaseous systems. Field et al.[69] gave an earlier treatment of ignition in pulverized coal, whereas Singer[70] and Elliott[2] provide some practical information on coal ignition systems. Essenhigh[71] also includes a recent review of ignition phenomena in coal.

Ignition characteristics of coal are strongly dependent upon the way the coal particles are arranged or configured. Three arrangements can be identified: (1) single coal particles, (2) coal piles or layers, and (3) coal clouds. For single particles, no interaction among particles occurs. This arrangement provides basic data and may provide practical insight in very dilute coal flames. Coal piles or layers occur in moving beds, fixed beds, coal storage piles, and dust layers. Coal clouds exist in pulverized coal combustors, furnaces, entrained gasifiers, coal mine dust explosions, and in fluidized bed systems (more dense in latter case). Coal ignition characteristics in these various configurations differ substantially.

Results of various typical ignition or "onset" temperatures are summarized in Table 8.[72-75] Coal onset temperatures vary from that of coal pile self-heating, which occurs at near ambient temperatures, to single-particle ignition temperatures as high as 1200 K (900°C). All of these onset or ignition temperatures are functions of coal type, size, and condition. Values also depend on experimental method and environment. These representative values emphasize the wide variability in ignition or onset temperatures. Chemical reactions proceed at some rate over a wide range of temperatures. It takes several hours at low temperature for a coal pile to ignite, whereas a coal dust cloud can be ignited in milliseconds at high temperature.

No single definition of ignition for coal seems appropriate. In general terms, ignition can be described as a process of achieving a continuing reaction of fuel and oxidizer. Ignition is most often identified by a visible flame. However, reactions can proceed very slowly at low temperatures,

TABLE 8. Typical Ignition (or Onset) Temperatures for Various Coal Reaction Processes

Coal Process	Time	Typical Ignition (Onset) Temperature Range		Experimental Methods	Reference
		°C	K		
Coal pile self-heating (ambient conditions)	hr	30–105	303–378	Adiabatic calorimeter (airflow 50 cm³/min)	Kuchta et al.[72]
Coal pile auto ignition[a]	min	170–225	443–500	Samples in isothermal electrically preheated cylindrical (8.0 cm I.D.) oven	Hertzberg et al.[73]
Coal pile thermal ignition[b]	min	200–500	473–773	Coal layer on isothermal hot plate	Nagy et al.[74]
Coal devolatilization	ms	300–500	573–773	Several techniques used with various heating rates and residence times	Howard,[25] Solomon and Colket[75]
Pulverized coal cloud ignition	s	400–800	673–1073	Coal dust injected into isothermal oven that is open at bottom	Nagy et al.[74]
Single-particle ignition	s	800–900	1073–1173	"Few" particles on tip of platinum wire in isothermal laboratory furnace	Essenhigh[71]

without a visible flame. Ignition is sometimes said to occur when the rate of heat generation of a volume of combustibles exceeds the rate of heat loss. Ignition is more complex in condensed phases and particularly in heterogeneous solids such as coal where particles can react slowly by oxygen attack on the coal surface, leading to spontaneous ignition.

In either inert or reactive hot gases, the coal will react internally, softening and devolatilizing as the particle increases in temperature. In the process, gases and tars are released. Although no flame is visible, and this process is not called "ignition," the onset of this thermal decomposition is an ignitionlike process. These off-gases and tars can also be ignited in a surrounding oxidizer. This ignition process could involve only gases or also the tars. The remaining char can be ignited in oxygen by surface reaction processes. This reaction could also be CO_2, H_2O, or H_2, and a visible flame may not result, yet coal reactions continue.

Ignition is often characterized by the time required to achieve a certain temperature, rate of temperature increase, a visible flame, or a certain consumption of fuel for a specified set of conditions. However, no unique ignition time exists either. Potentially important variables that influence coal ignition temperature and time include coal type, system pressure, volatiles content, gas composition, particle size, coal moisture content, size distribution, residence time, gas temperature, coal concentration or quantity, surface temperature, gas velocity, mineral matter percentage, and coal aging since grinding. Variables that dominate the ignition process depend strongly on the configuration of the coal particles, as will the range of ignition temperatures and times. Williams[66] and Remenyi,[76] among others, treat the general theory of ignition. It is beyond the scope of this section to provide a comprehensive treatment of theories of ignition or to provide a summary of experimental data. However, Smoot and Smith[5] provide some additional information on this topic.

3.4. Particle Heat-up

Coal reaction processes are very dependent on the rate at which the particles heat and on the maximum particle temperature. Heating of the particle is complicated by coal reaction processes, with devolatilization initiated at about 600 K (300°C). As the particles continue to heat, convective heating is retarded by the flow of volatile matter from the particle. Reaction of the volatiles in the gas phase changes the surrounding temperature and heat-up rate of the particle; exothermic surface oxidation of the char with oxidizer releases energy to the particle and to the gas in the immediate vicinity of the surface. Further oxidation of the surface products (e.g., $CO + O_2 \rightarrow CO_2$) causes additional heat release in the vicinity of the particle.

Recently, optical pyrometers have been successfully used to measure particle and gas temperature histories in reacting coal systems as shown in Table 9.[77-85] Results show peak particle surface temperatures in the range of 2100–2200 K in air or similar O_2/gas mixture. The peak particle temperature

TABLE 9. Measured Temperatures of Reacting Coal Particles[77]

Reactor Type	Stoichiometric Ratio	Gas System	Coal Type	Particle Size (μm)	Particle Temperatures (K)	Estimated Heat-up Rate (K/sec)	Gas Temperature (K)	Reference
Large-scale entrained flow	0.9–1.1	Air	Gillete, WY subbituminous	49 (mass mean)	Up to 2250	10^5	—	LaFollette[77]
Small, heated entrained flow	Dilute	10–20% O_2	Semi anthracite	6, 22, 49, 78	1400–2200	—	Entering gases preheated to 1400–1800	Ayling and Smith[78]
Eight-liter explosion bomb	Close to 1.0	Air	Pittsburgh seam	7 (surface weighted mean)	1. 1400–2100 2. 700–1300	—	Gas temperature nearly always in excess of particle temperature	Cashdollar and Hertzberg[79]
Premixed laminar, flat flame burner	Close to 1.0	23% O_2	Eastern Kentucky bituminous	6.4 (number based mean)	1100–2100	10^5–10^6	Gas temperature nearly always in excess of particle temperature	Mackowski et al.[80]
Premixed transparent flat flame	Dilute	0–25% O_2 75–100% CH_4–N_2	1. Millmerran char 2. Petroleum coke	90 (mass mean)	1. 1400–1900 2. 1400–1900	10^5	1300–1800	Mitchell and McLean[81]
Drop tube furnace	Dilute	Air/oxygen	Gedling char	400–925	Up to 2100	10^4	1200	Nettleton[82]
Fixed-bed furnace	0.7–1.2	Air	UT bituminous	12,500–25,000	1400–2100	—	—	Starley et al.[83]
Heated muffle tuble	Dilute	O_2—He—Ar mixtures	1. ND lignite 2. MT lignite 3. AL Rosa 4. UT bituminous 5. IL No. 6	38–45 90–105	2300 (21% O_2) Up to 3100 With high oxygen	10^5	1250, 1700	Timothy et al.[84]
Flat flame burner	1.35 (for CH_4-air mixture)	CH_4-Air	1. MD lignite 2. MT lignite 3. Al bituminous 4. NM bituminous 5. UT bituminous 6. PA anthracite	40, 80	1500–2200	10^5–10^6	1750	Seeker et al.[85]

TABLE 10. Experimental Techniques and Conditions for Coal Devolatilization[8]

Investigator[a]	Technique	Residence time, sec	Temperature, K	Heating rate, K/sec	Pressure, atm[b]	Ambient Gas	Particle Size, μm
Captive Sample Techniques							
Standard proximate analysis (ASTM, 1974)	Crucible	420	1220	15–20	1.0	Air (lid on)	≤ 250
Wiser et al. (1967)	Crucible	300–72,000	670–770	15–20	1.0	N_2	246–417
Portal and Tan (1974)	Crucible and basket	15–1200	820–1420	0.5–250	1.0	N_2	< 45–88
Gray et al. (1974)	Crucible	420	1220–1470	0.3–20	1.0	N_2	≤ 200
Hiteshue et al. (1962a, b, 1964)	Hot rod	20–900	750–1470	10	18–400	H_2	250–600
Feldkirchner and Linden (1963); Feldkirchner and Huebler (1965)	Semiflow	10–480	980–1200	100–300 (seems high)	34–168	H_2, H_2O	841–1000
Moseley and Paterson (1965a)	Railway heater	15–165	1090–1220	25	18–95	H_2	150–300
Feldkirchner and Johnson (1968); Johnson (1971)	Thermobalance	Several to 7200	≤ 1200	≤ 100	≤ 100	H_2, N_2, H_2O, etc.	425–850
Gardner et al. (1974)	Thermobalance	25–3000	1120–1220	—	35–69	H_2	500–1000
Loison and Chauvin (1964)	Electric grid	0.7	≤ 1320	1500	Vacuum	—	50–80
Rau and Robertson (1966)	Electric grid	1–1.5	1170–1470	600	1.0	—	250–425
Juntgen and Van Heek (1968)	Electric grid	0.7	≤ 1270	1500	Vacuum	—	50–60
Koch et al. (1969)	Electric grid	7	≤ 1770	167	Vacuum	—	75–630
Mentser et al. (1970, 1974)	Electric grid	0.05–0.15	670–1470	8250	Vacuum	—	44–53
Cheong et al. (1975)	Electric grid	< 1–1800	570–1270	≤ 1000	Vacuum–3.4	—	90–355
Anthony et al. (1974, 1975, 1976)	Electric grid	0.1–20	670–1370	100–12,000	0.001–100	H_2, He, N_2	53–1000

Graff et al. (1975); Squires et al. (1975)	Ring of coal	1–6 (gas), 10–30 (solid)	870–1270	650	69	H_2	≤ 44
Suuberg et al. (1978)	Electric grid	1–15	570–1370	< 3(10⁴)	10²–10⁴	He	74–1000
Blair et al. (1976)	Heated ribbon		1070–1970	2–8(10⁴)	1	He, Ar	500–600
Solomon et al. (1978)	Electric grid		570–1570	< 10³	~ 0.1	Vacuum	> 100
Coal-Flow Techniques							
Stone et al. (1954)	Fluidized bed	10–2500	670–970	—	1.0	N_2	200–600
Peters et al. (1960, 1965)	Fluidized bed	1–15	870–1370	300	1.0	N_2	1000–3000
Pitt (1962)	Fluidized bed	10–6000	670–970	—	1.0	—	200–600
Jones et al. (1964)	Fluidized bed	~ 2400	700–1370	1000 +	1.0	N_2, H_2O	≤ 1000
Friedman (1975)	Fluidized bed	1800–3600	570–920	—	1.0	H_2O	250–710
Zielke and Gorin (1955)	Fluidized bed	—	1090–1200	—	1–30	H_2, H_2O, etc.	150–212
Birch et al. (1960, 1969)	Fluidized bed	500–9500	770–1220	—	21–42	H_2, etc.	150–710
Eddinger et al. (1966)	Entrained flow	0.008–0.04	900–1270	2500 +	1.0	He	6150
Howard and Essenhigh (1967)	Entrained flow	0–0.8	470–1820	22,000	1.0	Air	80% ≤ 74
Badzioch et al. (1967, 1970)	Entrained flow	0.03–0.11	670–1270	25,000–50,000	1.0	N_2	20
Kimber and Gray (1967a, b)	Entrained flow	0.012–0.34	1050–2270	150,000–400,000	1.0	Ar	22–50
Belt et al. (1971, 1972)	Entrained flow	≤ 1	1090–1310	—	1–28.2	H_2, N_2	70% ≤ 74
Sass (1972)	Entrained flow	A few	810–920	10,000	1.0	—	25–80
Coates et al. (1974)	Entrained flow	0.012–0.34	920–1640	—	1.0	H_2, H_2O, etc.	≤ 74
Moseley and Paterson (1965b)	Entrained flow	0.17–2.5	1060–1270	1000 +	50–520	H_2	100–150
Glenn et al. (1967)	Entrained flow	2.4–10.4	1190–1240	—	70–84	H_2, CO, etc.	≤ 44
Johnson (1975)	Entrained flow	5–14	760–1120	28	18–52	H_2, He	75–90
Shapatina et al. (1960)	Free fall	0.45–14,400	≤ 820	—	45–490	N_2	150–200
Moseley and Paterson (1967)	Free fall	A few	1110–1270	—	45–490	H_2	100–150
Feldmann et al. (1970)	Free fall	A few	920–1170	—	35–205	H_2	150–300

[a] References in this table are identified in detail in Anthony and Howard[8] or Smoot.[23]

[b] To convert to MPa, divide by 10.

can significantly exceed the surrounding gas temperature, partly because of heat-release during exothermic surface reaction. As particle burnout continues, the gas temperature approaches the particle temperature. In systems highly loaded with coal, the gas temperature may exceed the particle temperature.

Several variables influence the particle temperature. Coals with high volatile content tend to exhibit highest particle temperatures. Coal types with highest temperature are, in descending order, high-volatile bituminous, medium-volatile bituminous, lignites, and anthracite. Particles with high mineral matter content exhibit lower temperatures. Smaller particles exhibit higher peak temperatures and heat up more quickly than larger particles. Increased oxygen concentration raises particle temperature significantly. Values as high as 3100 K have been observed[84] for systems with high oxygen percentage.

3.5. Devolatilization

Experimental Studies. Anthony and Howard[8] and Howard[25] and to a less general extent, Horton,[86] Wendt,[24] and Solomon[56] provide recent reviews of coal devolatilization. Anthony and Howard[8] also note several reviews of devolatilization over the period from 1963 to 1972. Most of the recent work has emphasized finely pulverized coal particles at high temperatures. Significant devolatilization does not start until temperatures of 625–675 K.[25] The heating causes thermal rupture of bonds, and volatile fragments escape from the coal. Table 10 taken principally from Anthony and Howard,[8] with more recent entries added, provides a detailed summary of much of the experimental methods and test conditions of previous studies of devolatilization. Variables that influence devolatilization rates include temperature, residence time, pressure, particle size, and coal type. Table 10 provides a general summary of the observed effects of these key variables.

Final temperature is possibly the most important variable, as illustrated in Figure 5[87] for a lignite coal and a bituminous coal. This figure illustrates several important observations relating to these data: (1) The extent to which the coal devolatilizes varies greatly for both coals (less than 5% to over 60%) as a function of final temperature. This extent of devolatilization can obviously differ significantly from the "proximate" level. (2) The residence time required to devolatilize the pulverized coal is very short (10–200 ms) for these small particles and high temperatures. (3) Although these two coals have significantly different chemical characteristics, they have similar devolatilization rate characteristics.

These data have emphasized the rate of weight loss during devolatilization. The nature of the pyrolysis products is illustrated in Figure 6,[88, 89] again for a bituminous coal and a lignite coal. This figure correlates the results from different investigations. Results show (1) significantly different behavior between the two coal types, with more liquid/tar products for the bituminous coal and more gas products (including H_2O) from the lignite, and that (2) the proportions of various products change with changes in pyrolysis temperature.

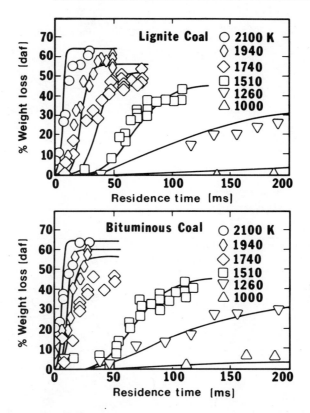

FIGURE 5. Comparison of calculated weight losses with experimental results for devolatilization of lignite and bituminous coals.[87] For Eqs. (2) and (4), $Y_1 = 0.3$, $E_1 = 104.6$ MJ/kmol (25 kcal/mol), $A_1 = 2 \times 10^5$ sec^{-1}, $Y_2 = 1.0$, $E_2 = 167.4$ MJ/kmol (30 kcal/mol), and $A_2 = 1.3 \times 10^7$ sec^{-1}.

These data are for rapid heating rates, for particles ranging from 74 to 1000 μm, and for a wide range of pressures. For the range of small particles tested, neither the product distribution nor the rates of devolatilization seems to vary significantly with the particle diameter. Several studies noted by Anthony and Howard[8] also provided observations that weight loss was not size dependent for particles up to 400 μm in diameter. However, very large particles behave differently from finely pulverized coal. Larger particles will not heat rapidly or uniformly, so that a single temperature cannot be used to characterize the entire particle. The internal char surface provides a site where secondary reactions can occur. Pyrolysis products generated near the center of a particle must migrate to the outside to escape. During this migration they may crack, condense, or polymerize, with some carbon deposition taking place. Larger particles lead to greater amounts of deposition and hence smaller volatiles' yield.

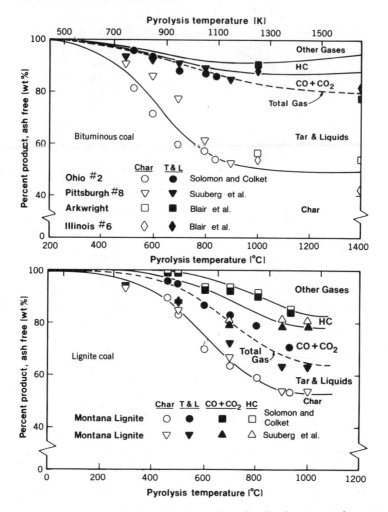

FIGURE 6. Comparison of coal pyrolysis product distributions at various temperatures from three independent investigators. d_p = coal diameter, h = heating rate, HC = hydrocarbons, T and L = tar and liquids. Data points not shown for gaseous components for bituminous coal. For Suuberg et al.,[88] d_p = 74–1000 μm, in helium, 10^2–10^4 atm, $h < 10^4$ K/s; for Blair et al.,[89] d_p = 500–600 μm, in helium/argon, 1 atm, $h = 2 - 8 \times 10^4$ K/sec; for Solomon and Colket,[75] $d_p > 100$ μm, vacuum, 1 kPa (0.01 atm), $h < 10^3$ K/s.

Devolatilization processes during gasification are thought to be quite similar to that during combustion. Major differences are elevated pressure and a more fuel-rich environment that often occur in gasification. Anthony et al.[90] reported that a lower ambient pressure favors the liberation of a greater mass of volatiles. Weight loss of a bituminous coal had declined from 50–55% (daf) for pressures under 10 kPa (0.1 atm) to below 40% at 10 MPa (100 atm).[25]

This appears to be consistent with other results, since an increase in pressure also increases the transit time of volatiles within the particle. An increase in pressure thus has an effect analogous to an increase in particle size. Howard[25] more recently reviewed the data available on effects of elevated pressure. Howard notes results of approximately 20 devolatilization studies at elevated pressures up to 50 MPa.

Devolatilization Models. Badzioch and Hawksley[91] postulated that the devolatilization was a first-order reaction process, with the reaction rate being proportional to the amount of volatile matter still remaining in the coal:

$$\frac{dv}{dt} = k(v_\infty - v) \tag{1}$$

with

$$k = A \exp\left(-\frac{E}{RT}\right) \tag{2}$$

This treatment required a method for relating the amount of "total" volatile matter (v_∞) to that obtained from proximate analyses. The correlation used was

$$v_\infty = Q(1 - v_c)v_p \tag{3}$$

The parameters Q and v_c were empirically determined, although a value of 0.15 for v_c was suitable for all of the nonswelling coals tested. This is a satisfactory single-step reaction to describe pyrolysis. However, it lacks the flexibility required to describe much of the experimental data available and may be inadequate to describe nonisothermal pyrolysis. The fact that the parameters v_p, Q, v_c, A, and E may depend upon the specific coal dust also limits the generality of this model.

Kobayashi et al.[87] suggested that pyrolysis could be modeled with the following pair of parallel, first-order, irreversible reactions:

$$C \xrightarrow{k_1} (1 - Y_1)S_1 + Y_1V_1$$
$$C \xrightarrow{k_2} (1 - Y_2)S_2 + Y_2V_2 \tag{4}$$

with the rate equations

$$\frac{dc}{dt} = -(k_1 + k_2)c \tag{5}$$

and

$$\frac{dv}{dt} = \frac{dv_1 + dv_2}{dt} = (Y_1k_1 + Y_2k_2)c \tag{6}$$

k_1 and k_2 are Arrhenius-type rate coefficients. An important feature of the model is that $E_1 < E_2$. This approach satisfactorily correlates the data of Badzioch and Hawksley[91] and Kimber and Gray,[92] and the more recent data of Kobayashi et al.[87] obtained under conditions of transient temperature as illustrated in Figure 5. In fact the agreement for weight loss is impressive for both coals and for a range of temperatures. Computations were made with a single set of parameters shown in Figure 5. The model is conceptually sound in that the variation in volatiles yield with temperature is explained by a second reaction rather than by a correlating parameter like that of Eq. (3). The general utility may be limited because the parameters Y_1, Y_2, A_1, A_2, E_1, and E_2 may depend on the specific coal dust. The stoichiometric parameters Y_1 and Y_2 can be estimated as the fraction of coal devolatilized during proximate analysis (Y_1) and the fraction that can be devolatilized at high temperatures (Y_2, often near unity).

In another treatment Anthony et al.[90] postulated that pyrolysis occurs through an infinite series of parallel reactions. A continuous Gaussian distribution of activation energies is assumed, along with a common value for the frequency factor so that

$$\frac{(v_\infty - v)}{v_\infty} = \left[\sigma(2\pi)^{1/2}\right]^{-1}\left\{\int_0^\infty \exp\left[-\left(\int_0^t kdt\right)f(E)\,dE\right]\right\} \qquad (7)$$

with

$$f(E) = \left[\sigma(2\pi)^{1/2}\right]^{-1}\exp\left\{-\frac{(E - E_0)^2}{2\sigma^2}\right\} \qquad (8)$$

This approach also provided very good correlation of the data from Anthony et al.[90] as well as those of Suuberg et al.[88] The model is attractive because it requires only four correlating constants. The utility of this model, too, may be restricted by the need to determine the parameters v_∞, k, E_0, and σ for the specific coal dust of interest.

Sprouse and Schuman[93] have compared the lignite data of Figure 5 (1000 K–2100 K) with predictions from Eqs. (7) and (8) as shown in Figure 7a. The authors note the excellent agreement with four parameters (σ, E_0, k, v_∞), compared to five for comparisons of Figure 5. A further comparison for lignite data of Anthony et al.,[90] for lower temperatures (973–1273 K) but for variable heating rates (10^2–10^4 K/sec) is shown in Figure 7b. Again, agreement is very good with the same parameters. It is also interesting to note the absence of significant effect of heating rate on the maximum extent of devolatilization.

Both methods (i.e., Eq. (6) and Eqs. (7)–(8)) give good results. Differences in the number of coefficients, (4) versus (5), are considered secondary, since it is likely that "a priori" information may be available for stoichiometric

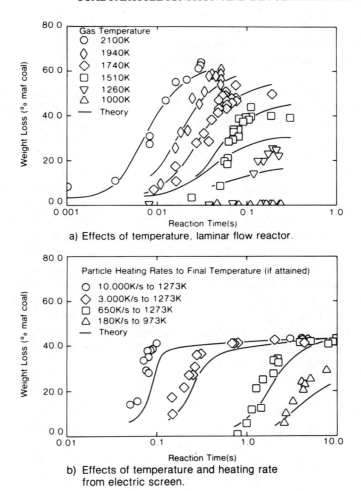

a) Effects of temperature, laminar flow reactor.

b) Effects of temperature and heating rate
from electric screen.

FIGURE 7. Comparison of measured lignite weight loss data (0.1 mpa) and predictions from Eqs. (7) and (8). $\sigma = 1 \times 10^5$ (J/mol), $v_\infty = 63.5$ (wt %), $E_0 = 3.15 \times 10^5$ (J/mol), $k = 1.67 \times 10^{13}$ sec^{-1}. Parameters were not quite optimum but were common in both figures.[93]

coefficients of Eq. (6). For example, Y_1 could be taken as the proximate volatiles level, while Y_2 may be near unity on a daf basis. More complex reaction mechanisms were reviewed by Anthony and Howard,[8] and often involved several coal reaction steps.

In addition comprehensive codes require a devolatilization model that will produce the composition of the volatile gaseous products and the residual char as well as the rate of volatiles evolution. The two-step model of Eq. (4) produces these compositions. Knowing the elemental composition of the virgin coal and specifying the char composition (mostly carbon), the elemental volatiles composition V_1 and V_2 can be calculated by mass balance. The total volatile matter evolved from the competing two-step process produces different amounts and composition of volatile matter from different temperature histories.

A more recent method for predicting both devolatilization rates and product distribution has been developed by Solomon and Colket.[56,75] Prediction of time and temperature variation in these two properties is based on coal structural groups, using general kinetic parameters for each of the coal constituents. This technique was based on two key observations: (1) While the overall rates of devolatilization of coal vary with coal type, the rates of individual functional groups in the coal (ether groups, hydroxyl groups, tar, etc.) are *independent* of coal type; (2) the chemical composition of the tars is essentially that of the virgin organic coal. Thus predictions can be made from quantitative measurements of coal composition and the set of parameters for each of the functional groups that are taken to be valid for all coals.

Figure 4 showed a model of the virgin coal organic structure, while Figure 8 illustrates a model for thermal decomposition of this same virgin coal. The virgin organic coal is thus identified as being composed of a series of chemical constituents. These include carboxyl (producing CO_2), hydroxyl (producing H_2O), ether (producing CO), aromatic hydrogen, aliphatic hydrogen, nitrogen, and nonvolatile carbon. For many coals, the general composition of these tars can be identified from the parent coal.

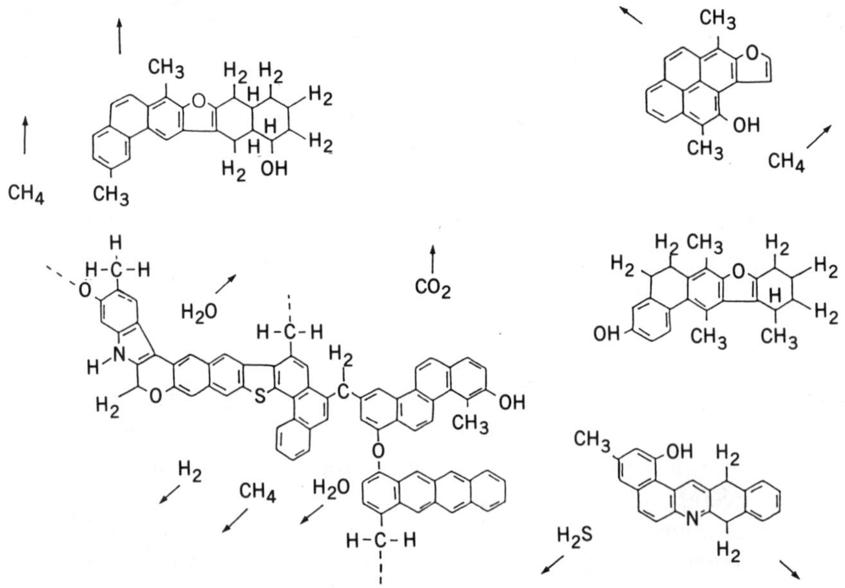

FIGURE 8. Cracking of hypothetical coal molecule during thermal decomposition. (Refer to Figure 4 for the postulated original structure before thermal decomposition.)[56]

Thermal decomposition is postulated to follow separate activated chemical reaction processes for each of these functional groups, and for the tar. A given chemical species such as CO, OH, or CH_4 then is evolved from the coal by two independent, first-order processes: one for the species directly, and the second for the fraction of that chemical species in the tar. The rates of reaction for thermal decomposition for each species including the tar are assumed to be first order:

$$W_i = W_i^0 [1 - \exp(k_i t)] \tag{9}$$

where W_i^0 is the weight fraction of the functional group evolved during devolatilization and W_i is the weight fraction of the functional group in the organic part of the raw coal; $k_i = A_i \exp(-E_i/RT)$, and A_i and E_i are determined from time-dependent functional group composition data during devolatilization. Additional details and constraints involved in this model scheme are discussed by Solomon[56] and Howard.[25] Measurements of the products and rate of tar formation are particularly important issues in this treatment. Most recent compilations of rate results for tar, together with rates of each of the functional groups are given in Solomon and Hamblen.[94, 95]

Figure 9 shows comparisons of measured and predicted levels of evolved products H_2, tar, and aliphatics, together with the residual C, H, and O in the char for a bituminous coal and a lignite coal for various devolatilization temperatures. Other comparisons for several other coals and other functional groups are shown by Solomon[56] and Solomon and Hamblen.[94, 95] Correlation of coal devolatilization rate data requires reliable values for coal particle temperature. Direct measurement of coal particle temperatures by optical methods is made more difficult by interfering escaping volatiles material. Recent measurements of devolatilizing coal particle temperatures with FT-IR emission and transmission spectroscopic methods[96] have suggested that significant errors may exist in computed values in much of the earlier work. This issue is still one of controversy.

3.6. Combustion of Volatiles

A variety of products are produced during devolatilization of coal, including tars and hydrocarbon liquids, hydrocarbon gases, CO_2, CO, H_2, H_2O, and HCN (see Figure 6). Several hundred, mostly aromatic hydrocarbon compounds have been identified in the tars produced during devolatilization.[97] These products react with oxygen in the vicinity of the char particles, increasing temperature and depleting the oxidizer (e.g., oxygen). This complex reaction process is very important to the control of nitrogen oxides, formation of soot, stability of coal flames, and ignition of char. Some aspects of this process include the following: (1) release of volatiles from the coal/char; (2) condensation and repolymerization of tars in char pores; (3) evolution of hydrocarbon cloud through small pores from the moving particle; (4) cracking

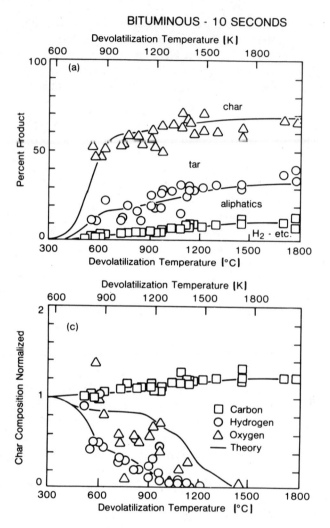

FIGURE 9. Pyrolysis product yields—experiment and theory. The char composition normalized to the composition of the parent coal.[56]

of the hydrocarbons to smaller hydrocarbon fragments, with local production of soot; (5) condensation of gaseous hydrocarbons and agglomeration of sooty particles; (6) macro-mixing of the devolatilizing coal particles with the oxidizer (e.g., air); (7) micro-mixing of the volatiles cloud and the oxygen; (8) oxidation of the gaseous species to combustion products; (9) production of nitrogen oxides and sulfur oxides by reaction of devolatilized products and oxygen; (10) heat transfer from the reacting fluids to the char particles. At very high combustion temperatures, up to 70% or more of the reactive coal mass is consumed through this process (see Figure 5).

LIGNITE - 10 SECONDS

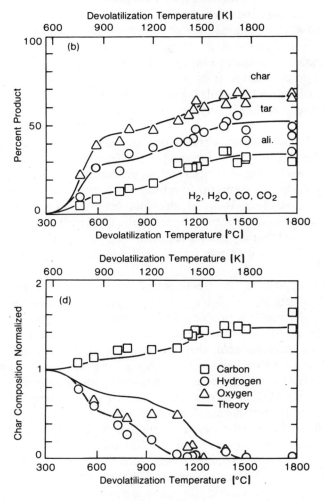

FIGURE 9. (*Continued*)

In an earlier treatment Field et al.[69] discussed combustion of coal volatiles, and Thurgood and Smoot[98] and Smoot and Smith[5] have treated this topic more recently. Also Essenhigh[71] mentions this subject briefly, noting the significant lack of specific kinetic information. Hucknall[99] has prepared a very comprehensive review of the combustion of hydrocarbons, including a brief treatment of modeling.

Recently Seeker et al.[85] photographically observed these clouds and jets of volatiles being emitted from devolatilizing coal particles. Their tests were done with five pulverized coals of various sizes in a methane-heated laboratory air flame at 1750 K with 35% excess air. They observed, with holographic methods, jetlike volatiles evolution forming a cloud of volatiles from larger (80 μm) bituminous coal particles, with subsequent soot particle (ca. 3 μm) formation and agglomeration. Each of these little clouds can form a diffusion flame with the oxidizing gas, if the gas temperature and residence time are adequate. Smaller bituminous coal particles (40 μm) and anthracite or lignite coals did not produce the soot trails.

Thurgood and Smoot[98] tabulated basic reaction rate coefficients for oxidation of hydrogen, methane, actylene, ethylene, and ethane. In the methane oxidation mechanism, 41 basic radial reactions were included. Engleman[100] has provided a more extensive review of basic gaseous reaction mechanisms, which includes 322 reactions for methane oxidation. It is apparent that a rigorous treatment of the volatiles reaction process is not a realistic expectation in a heterogeneous, turbulent flow involving hundreds of chemical species and many more reactions. However, this volatiles reaction process is rapid at high temperatures, and for some purposes such as overall coal burnout rates, volatiles reaction rates may not be required. Yet, for other purposes, such as nitrogen oxide formation or soot formation, these volatiles reaction rates may control.

Two useful approaches for treatment of volatiles reaction are considered here: local equilibrium and global reaction rates. For local equilibrium, it is assumed that the volatiles products and the oxidizing gas are in thermodynamic equilibrium locally where the gas temperature is sufficiently high. Thus, as the volatiles leave the coal, they are assumed to immediately equilibrate with the gas in that local region. From this assumption the gas temperature and gas composition can be estimated without a knowledge of the evolved chemical species. All that is required is the elemental composition of the volatiles and the heat of devolatilization (thought to be near zero). Data on elements emitted during the devolatilization process can be deduced from measurements (e.g., see Figures 6 and 9). This information is often adequate for examining many aspects of coal flames, such as exit or local gas concentration and char consumption times, even in turbulent systems.

Sometimes volatiles and oxidizing gases are not in equilibrium, such as in formation of NO_x from HCN or in formation of soot in fuel-rich regions. Use of global reaction rates provides some estimate of volatiles combustion rates. The approach is to correlate chemical reaction rates of various fuels (CH_4, H_2, C_nH_m, etc.) for the overall reactions, such as

$$C_nH_m + \left[\left(\frac{n}{2}\right) + \left(\frac{m}{4}\right)\right]O_2 \xrightarrow{k} nCO + \left(\frac{m}{2}\right)H_2O \qquad (10)$$

where

$$k = A \exp\left(-\frac{E}{RT}\right) \qquad (11)$$

Edelman and Fortune[101] and Siminski et al.[102] proposed a global reaction for consumption of hydrocarbons where reaction products are CO and H_2 rather than CO_2 and H_2O. Siminski et al.[102] have correlated kinetic rates for heavy

TABLE 11. Parameters for Global Hydrocarbon Reaction, Eq. (13)[a]

Hydrocarbon	A	E/R
Long-chain	59.8	12.20×10^3
Cyclic	2.07×10^4	9.65×10^3

[a]From Siminski et al.[102]

hydrocarbon combustion for the global reaction:

$$C_nH_m + \left(\frac{n}{2}\right)O_2 \rightarrow \left(\frac{m}{2}\right)H_2 + nCO \qquad (12)$$

Different finite rates were specified for long-chained hydrocarbons and for cyclic hydrocarbons. The rate is given as

$$\frac{dC_H}{dt} = -ATP^{0.3}(C_H)^{0.5}(C_O)\exp\left(-\frac{E}{RT}\right) \qquad (13)$$

where T is the temperature in K, P is the pressure in pascals, C_H and C_O are molar concentrations of hydrocarbon and oxygen in kmol m^{-3}, t is the time in seconds, E is the activation energy in kcal/kmol, and the constants are given in Table 11.

Once a global scheme is selected, a determination of the carbon-to-hydrogen ratio to be used for the volatiles pseudomolecule C_nH_m in Eq. (12) must be made. One possible approach is based upon ultimate analysis of the char. All coal components other than char can be grouped together, and a simple material balance can be made to determine the carbon-to-hydrogen ratio of the group. The group ratio can then be used as a rough estimate of stoichiometric coefficients m and n for the global reaction. When tars represent a major fraction of the volatiles products, this estimate should give reasonable results.

More recently, Hautman et al.[103] provided overall reaction rates for H_2, CO, C_2H_4, and for alkanes (primarily from propane data):

$$C_nH_{2n+2} \rightarrow \left(\frac{n}{2}\right)C_2H_4 + H_2 \qquad (14)$$

$$C_2H_4 + O_2 \rightarrow 2CO + 2H_2 \qquad (15)$$

$$CO + \tfrac{1}{2}O_2 \rightarrow CO_2 \qquad (16)$$

$$H_2 + \tfrac{1}{2}O_2 \rightarrow H_2O \qquad (17)$$

Rate expressions for these four reactions are summarized in Table 12 where coefficients for these rates are also shown. This sequence of reactions has been shown to be quite reliable for aliphatic hydrocarbons for a range of stoichiometric ratios (0.12–2), pressures (0.1–0.9 mpa or 1–9 atm) and temperatures (960–1540 K). If Eqs. (14) and (15) are added, an overall rate for consumption

TABLE 12. Summary of Global Reaction Rates for Various Major Species in Gaseous Hydrocarbon Flames[103] (gmol / cm³ sec)[a]

$C_nH_{2n+2} \rightarrow \left(\dfrac{n}{2}\right)C_2H_4 + H_2$ $\dfrac{d[C_nH_{2n+2}]}{dt} = -10^{17.32}\exp\left(-\dfrac{49,600}{RT}\right)[C_nH_{2n+2}]^{0.50}[O_2]^{1.07}[C_2H_4]^{0.40}$

$C_2H_4 + O_2 \rightarrow 2CO + 2H_2$ $\dfrac{d[C_2H_4]}{dt} = -10^{14.70}\exp\left(-\dfrac{50,000}{RT}\right)[C_2H_4]^{0.90}[O_2]^{1.18}[C_nH_{2n+2}]^{-0.37}$

$CO + 0.5O_2 \rightarrow CO_2$ $\dfrac{d[CO]}{dt} = \left\{-10^{14.6}\exp\left(-\dfrac{40,000}{RT}\right)[CO]^{1.0}[O_2]^{0.25}[H_2O]^{0.50}\right\}$

$\times 7.93\exp(-2.48\phi)$

$H_2 + 0.5O_2 \rightarrow H_2O$ $\dfrac{d[H_2]}{dt} = -10^{13.52}\exp\left(-\dfrac{41,000}{RT}\right)[H_2]^{0.85}[O_2]^{1.42}[C_2H_4]^{-0.56}$

[a] ϕ = initial equivalence ratio, and $7.93\exp(-2.48\phi) \leq 1$.

of the aliphatic hydrocarbon C_2H_{2n+2} results:

$$C_nH_{2n+2} + \left(\frac{n}{2}\right)O_2 \rightarrow nCO + nH_2 \qquad (18)$$

Here no knowledge of the specific intermediate hydrocarbons (i.e., C_2H_4) is required.

From these rates, reaction times required to consume such gaseous reactant species, C_nH_m, C_nH_{2n+2}, CO, and H_2 were estimated at 1500 K.[5] With these very fast rates, treatment of volatiles reaction may not generally be required. For cyclic hydrocarbons, results are generally consistent with observations noted by Field et al.[69] that the slowest, and therefore rate-controlling, step in the high-temperature combustion of hydrocarbons is the oxidation of CO. Aliphatic decomposition rates from the more recent work of Hautman et al.[103] are much faster than those suggested by Siminski et al.[102]

DeSoete[104] has reported global oxidation rates for HCN by oxygen:

$$HCN + O_2 \xrightarrow{k} NO + \dots \qquad (19)$$

where

$$k = 10^{10}\exp\left(-\frac{67,000}{RT}\right) \qquad (20)$$

From this expression, for HCN in air at 0.1 mpa (1 atm) and SR = 1.0, 50% consumption times are very long (5 sec at 1500 K, 20 msec at 2000 K), compared to oxidation times for CO, H_2, and hydrocarbons. Thus formation of NO from fuel-nitrogen will often depend on the rate of HCN reaction.

Even with some estimate of volatiles reaction rates, the influence of gas turbulence on the rates makes kinetics calculations of volatiles consumption in a practical combustor a more difficult task with more uncertain results.

4. HETEROGENEOUS CHAR REACTION PROCESSES

4.1. Reaction Variables

The time required for consumption of a char particle is a significant part of the coal reaction process; it can range from a few milliseconds to over an hour. Modeling of the heterogeneous reaction process is complicated by (1) coal structural variations, (2) diffusion of reactants, (3) reaction by various reactants (O_2, H_2O, CO_2, and H_2), (4) particle size effects, (5) pore diffusion, (6) char mineral content, (7) changes in surface area, (8) fracturing of the char, (9) variations with temperature and pressure, and (10) moisture content of the virgin coal. Because of these uncertainties modeling of this process has been highly dependent on laboratory rate data for the specific coal and test conditions used. Recent reviews of this important topic include Essenhigh,[71,105] Laurendeau,[18] Smith,[106] and Skinner and Smoot.[107]

The rate-limiting step in the oxidation of char can be chemical (adsorption of the reactant, reaction, and desorption of products) or gaseous diffusion (bulk phase or pore diffusion of reactants or products). Several investigators, such as Walker et al.[108] and Gray et al.,[65] have postulated the existence of different temperature zones or regimes which determine the controlling resistance. In zone I, which occurs at low temperatures, or for very small particles, chemical reaction is the rate-determining step. Zone II is characterized by control due to both chemical reaction and pore diffusion. Zone III, which occurs at high temperatures, is characterized by bulk mass-transfer limitations. Figure 10 illustrates these zones graphically and shows the theoretical dependence of the reaction rate on particle diameter and oxidizer concentration. Kinetic data must be interpreted in light of the conditions under which the data were obtained. To obtain intrinsic rate data, a variety of schemes have been tried to minimize the effects of diffusion. This includes efforts to conduct tests at low pressure or low temperature, by microsampling of gases very close to the solid surface, by use of high-velocity gas streams to decrease the boundary layer thickness, and by use of small particles.[69,109]

The rates of carbon oxidation by steam and carbon dioxide are of the same order of magnitude, and are much lower than the reaction of carbon with oxygen, whereas the hydrogenation reaction (i.e., H_2 + char) is several orders of magnitude slower than the steam-char and the CO_2-char reactions. Walker et al.[108] estimated the relative rates of the reactions at 1073 K and 10 kPa pressure as being 3×10^{-3} for H_2, 1 for CO_2, 3 for H_2O, and 10^5 for O_2. At higher temperatures the differences in rates will not be so extreme, but rates for H_2, CO_2, and H_2O will still remain relatively unimportant as long as comparable concentrations of oxygen are present. Often, only the heterogeneous carbon oxidation reaction with oxygen needs to be considered for combustors.[109] In coal gasifiers, however, the reactions with steam and CO_2 can be quite important, especially after the rapid depletion of oxygen.

In examining rate data from various sources, it is apparent that temperature and reactant concentration are not the only variables that influence reaction

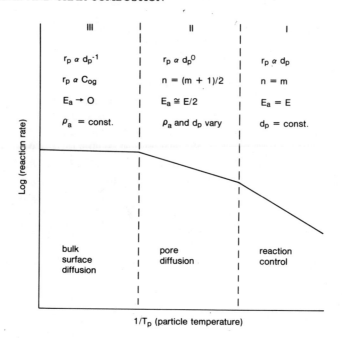

FIGURE 10. Rate-controlling regimes for heterogeneous char oxidation.

rates. The composition of the feed material, its pretreatment, and its thermal history are also important. Some experiments have used pure forms of carbon, while others have used char. In general, it has been found that the more pure forms of carbon are less reactive than the chars. Also chars from the lower-rank coals are more reactive than those from higher-rank coals. Differences in reactivity at different heating rates and temperatures have been noted in chars formed from the same parent coal, but some of the differences in reactivity noted here may be due to additional volatiles reactions from the low-temperature chars, which were not totally pyrolyzed.

Other important variables are the internal and external surface areas of the char. Under some conditions reactant gases diffuse into the particles before reacting. An observed increase in reaction rate with increasing burn-off, followed by a subsequent decline in rate, has been explained in terms of expanding of the pores to expose more and more internal area.[110] The surface area continues to increase up to a certain point; then, due to coalescence of pores and thermal annealing, the area and reaction rate decrease. Another factor that may affect the oxidation rates is the presence of catalytic impurities in the char.

4.2. Global Reaction Rates

It has been common practice[27] to relate measured char consumption rates to the external char surface area, even when pore reaction may occur. The resulting rate is termed *global*. The basic approach and illustrative data are discussed below.

Char Oxidation Data. Recently, Goetz et al.[111] reported char reaction rate parameters for four typical U.S. coals. Other data are reported in the review by Smith.[27] Results were reported for reaction with O_2 and CO_2. Measurements were made in a drop-tube furnace. Properties of the parent coals are shown in Table 13.[112] Chars were prepared from 200–400 mesh parent coal in the furnace at 1750 K in a nitrogen atmosphere prior to oxidation tests. Extent of combustion was determined from oxygen depletion, whereas extent of reaction with CO_2 was determined from CO concentration.

Char combustion was performed at 0.1 MPa (1 atm) with $P_{O_2} = 1$–5 kPa (0.01–0.05 atm), an over a temperature range of 1250–1730 K. Key assumptions in data reduction were first-order reaction, CO as the surface product of combustion and shrinking core particle based on external surface area. Figure 11 shows the reaction rate coefficients (k_p values) for three of the U.S. coals of Table 13. For these data, the specific coal reaction rate (R_{po}, grams oxidizer consumed/m² sec, based on external surface area), expressed as a first-order, heterogeneous reaction is

$$R_{po} = \frac{\rho_{og}}{(1/k_m) + (1/k_r)} \tag{21}$$

where ρ_{og} is the oxidizer concentration (g/m³) in the bulk gas, k_m is the oxidizer mass-transfer coefficient to the particle surface, and k_r is the reaction rate coefficient for O_2 with the char. The specific coal reaction rate (R_p) is computed from weight of the particle, the mean initial external surface area, and the reaction time. The consumption rate of the char is related directly to that of the oxidizer, R_{po}, once the reactants and products are specified (see Eq. (25)). The particle temperature (T_p) is backcalculated from measured gas temperature. Then, k_m, the mass transfer coefficient for oxidizer transfer from the bulk gas to the particle surface, is computed. With R_p, ρ_{og}, and k_m, the specific reaction rate coefficient (k_r) is computed for each experiment. k_r is assumed to have an Arrhenius form of temperature dependence:

$$k_r = A \exp\left(\frac{-E}{RT_p}\right) \tag{22}$$

For a family of experiments at various T_p values, reaction rate values are plotted as $\ln k$ versus $1/T_p$, as illustrated in Figure 11, with one additional variation. The authors elected to plot the parameter k_p, which is the reaction rate coefficient in the expression: $r_{po} = k_p A_p P_{O_2}$. Thus k_p is based on pressure and has units of g/cm² sec atm O_2. By comparison, k_r has units of g/cm² sec-g/cm³ or cm/sec. Thus, from k_p values of Figure 11, k_r values can be computed from $k_r = k_p T_g R / M_{O_2}$.

TABLE 13. Analyses of Parent U.S. Coals[112]

Analysis	Texas (Monticello) Lignite A		Wyoming (Jacobs Ranch Range) Subbituminous C		Illinois No. 6 (Freeman) High-Volatile Bituminous C		Pittsburgh No. 8 (Delton) High-Volatile Bituminous A	
	As Received	Dry-Ash Free	As Received	Dry-Ash Free	As Received	Dry-Ash Free	As Received	Dry-Ash Free
Proximate wt %								
Moisture (total)	29.3	—	28.0	—	19.4	—	0.8	—
Volatile matter	30.0	54.0	32.2	48.6	31.3	43.6	38.1	42.1
Fixed carbon	25.5	46.0	34.2	51.4	40.4	56.4	52.6	57.9
Ash	15.2	—	5.6	—	8.9	—	8.5	—
Total	100.0	100.0	100.0	100.0	100.0	100.0	100.0	100.0
Ultimate, wt %								
Moisture (total)	29.3	—	28.0	—	29.4	—	0.8	—
Hydrogen	3.2	5.7	3.5	5.2	3.6	5.1	5.1	5.6
Carbon	39.4	71.0	49.3	74.3	53.0	73.9	75.9	83.7
Sulfur	1.0	1.9	0.4	0.6	3.0	4.2	2.5	2.8
Nitrogen	0.7	1.3	0.8	1.1	1.0	1.4	1.4	1.5
Oxygen (diff.)	11.2	20.1	12.4	18.8	11.1	15.4	5.8	6.4
Ash	15.2	—	5.6	—	8.9	—	8.5	—
Total	100.0	100.0	100.0	100.0	100.0	100.0	100.0	100.0
Higher heating value								
kJ/kg	16,092	28,964	19,871	29,871	21,813	30,406	31,755	35,010
(Btu/lb)	(6,920)	(12,455)	(8,525)	(12,845)	(9,380)	(13,075)	(13,655)	(15,055)

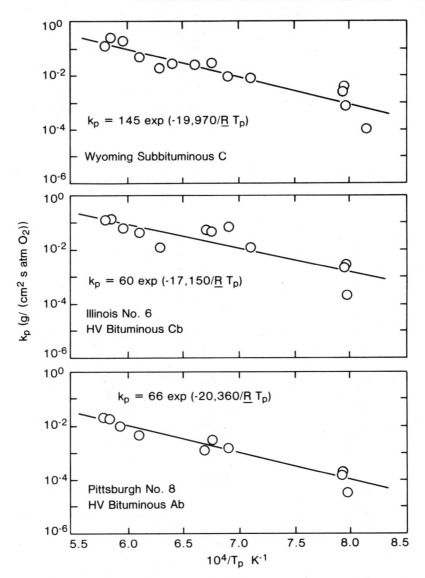

FIGURE 11. Char reaction rate coefficients in oxygen for three U.S. coals.[111]

Observations from these and other data include the following:

1. For a given char,[5] the variation with temperature is reasonably well correlated in Arrhenius form, supporting an activated reaction process.
2. For small dust particles (20–100 μm), k_p is not a function of particle size, suggesting that the external surface area adequately correlates the char oxidation rate, even though reactions may be occurring internally.

TABLE 14. Rate Parameters for Char Reaction with Oxygen for Rate[108] Equation $r = AT^N \exp(-E/RT)P_{O_2}$

References	A, kg m^{-2} sec^{-1} kPa^{-n} K^{-N}	E/R, K	N	Order n	Particle Type	Size-Graded	Sizes, μm	Temperature Range, K
Smith and Tyler[112]	1.32×10^{-1}	16,400	0	0	Brown coal char	Yes	22, 49, 89	630–1812
Field et al.[69]	8.6×10^{2}	18,000	0	1	Various fuels	Varied	Varied	950–1650
Essenhigh et al.[113]	—	20,100	0	0,1^a	Carbon	Yes	2.54×10^{4}	—
Howard and Essenhigh[114]	—	3000–6000	0	1^b	Bituminous char	No	0–200	—
Nettleton[115]	—	15,000–32,700 6500–25,000^c	1.75–3.5	0 1	Various coals	Yes	420–1000	1100–1500
Hamor et al.[116]	9.18×10^{-1}	8200	0	0.5	Brown coal char	Yes	22, 49, 89	630–2200
Smith[117]	2.013	9600	0	1	Semianthracite	Yes	6, 22, 49, 78	1400–2200
Smith and Tyler[118]	5.428	20,100	0	1	Semianthracite	Yes	6, 22, 49, 78	1400–2200
Sergeant and Smith[119]	2.903^d	10,300^d	0	1	Bitummious char	Yes	18, 35, 70	800–

^{a} for $T < 1000$ K; 1 for $T > 1000$ K.
^{b} 1 indicates adsorption control before flame front; 0 indicates desorption control in tail of flame.
^{c} Most values between 11,600 and 14,600 K.
^{d} Calculated from plot in Seargeant and Smith.[119]

3. These data are correlated with the assumption that the surface oxidation rate is first order in oxygen concentration, as specified by Eq. (22) and indicated by the units on k_p. There is still uncertainty on reaction order and activation energy as illustrated by the data of Table 14.[112-119] From this summary of char-oxygen data from several investigators, reaction orders from zero to unity have been identified, while E/R values (K) vary from 3000 to over 30,000.

4. Specific rate coefficients vary by about an order of magnitude for the three different coals. The decrease was specifically associated with a measured decrease in internal surface area.

5. Char rate coefficient decreased with increasing coal rank.

6. The range of reaction rate values is of the same order as those of the non-U.S. coals.[5]

7. Small variations in activation energy 71–84 kJ/gmol (17–20 kcal/mol) were observed among the three coals.

Specific reaction rate coefficient data for surface reaction with CO_2 are shown in Figure 12 for the four U.S. coals. Values also vary by over three orders of magnitude among the four coals. Again, rates varied directly with coal rank, and more specifically with measured char internal surface area. Reaction rate coefficients for CO_2 are also up to six or seven orders of magnitude lower than values for O_2 at a given temperature. Activation energies for CO_2-char reaction are much greater than for O_2-char reaction, ranging from 146–234 kJ/gmol (35–56 kcal/gmol). These value provide specific rate data for a family of commonly used U.S. coals.

Recent results for O_2-char reactions are also reported by Dutta and Wen,[120] Mandel,[121] Froberg and Essenhigh,[122] and Young and Smith.[123] Other specific data are also available for reaction for various chars with CO_2, and with steam and H_2. Much of these data are summarized by Skinner and Smoot[107] and more recently and more exhaustively by Essenhigh.[71] Recent data on coal and char reactions with carbon dioxide,[121,124,125] steam,[126-129] and hydrogen[130] have also been reported. Reactions of char with CO_2 and H_2O will be particularly important in gasification applications where oxygen is rapidly depleted before the char is consumed.

Effects of Mineral Matter. As many of 35 elements were identified in the ashes of typical coals (Table 5). Large quantities of highly variable mineral matter can have significant impact on the combustion and gasification of coal. Major reviews have been reported on the properties of ash,[131,132] effects of mineral matter on the combustion of coal particles,[71] and its effects on combustion systems.[22,133,134] Yet the behavior and effects of ash remain a major unresolved issue in the utilization of coal. New physical and chemical cleaning methods are also being developed to reduce the percentage of ash in coal prior to combustion.[61]

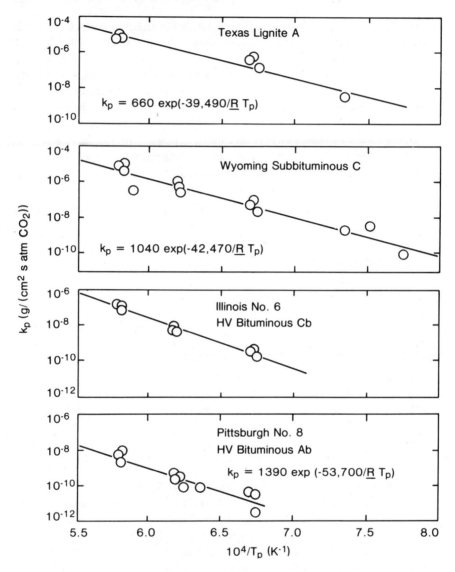

FIGURE 12. Char reaction rate coefficients in CO_2 for four U.S. coals.[111]

The presence of ash and of the specific minerals in ash can have several potential effects on the combustion of coal, including the following:

1. *Thermal effects*. Large quantities of ash change the thermal behavior of particles. Ash consumes energy as it is heated to high temperatures and changes phase.

2. *Radiative properties*. Radiative properties of ash differ from those of char or coal[22]; and the presence of the ash provides a solid medium for radiative heat transfer when the carbon is consumed.

3. *Particle size.* Char particles, toward the end of burnout tend to break into smaller fragments.[54] This breakup process is undoubtedly related to the quantity and nature of mineral matter in the char.

4. *Catalytic effects.* Various minerals in the char have been shown to cause increases in char reactivity, particularly at low temperature. For example, Essenhigh[71] reports a factor of 30 increase in lignite char reactivity at 923 K as calcium percentage in the char (achieved through demineralization and ion exchange) was varied from near zero to 13%. However, these catalytic effects are likely far less prominent at high temperature. Walker et al.[135] provides a comprehensive review of the effects on catalysts on carbon consumption and also discusses mechanisms.

5. *Hindrance effects.* Mineral matter provides a barrier through which the reactant (e.g., oxygen) must pass to reach the char. Particularly toward the end of burnout, it is possible that high quantities of mineral matter will impede combustion. This can be worsened by softening and melting of the mineral matter.[131]

The above effects deal directly with the impact of mineral matter on the consumption of char. Mineral matter also plays a vital role in the operation of practical combustion systems.[22,133] These effects include fouling and slagging on reactor walls and heat transfer tubes, sulfur pollutant formation, radiative heat transfer, corrosion, vaporization of trace metals and formation of fly ash.

Global Modeling of Char Oxidation. It is assumed that the oxidizer (e.g., O_2) must diffuse from the bulk gas, adsorb onto the surface of the particle, and react with the carbon in the char. The rate of reaction will relate to the diffusion and chemical reaction rates and to the specific products formed. For example, if CO were the product, the reaction rate may be more rapid than if CO_2 were the product, since less oxygen would be required.

If the particle is porous, then diffusion of the product into the porous volume and reaction on the larger internal surface area will influence the overall reaction rate. Because the details of this complex internal structure are not always known, it has been common to relate the char reaction rate to the external surface area. This type of char model is referred to as a *global model*.

Assuming a heterogeneous surface reaction, the reaction rate (kg/s) for a single spherical char particle is

$$r_k = \nu_s M_p k_r \xi_p A_p C_{op}^n \tag{23}$$

where C_{op} is the molar concentration of the oxidizer in the gas phase at the surface of the particle, A_p is the external surface area of the sphere (or equivalent sphere), k_r is the rate constant for heterogeneous reaction, M_p is the molecular weight of the reactant in the particle (e.g., carbon), ν_s is the stoichiometric coefficient to identify the number of moles of product gas per mole of oxidant (i.e., 2 for CO product; 1 for CO_2), ξ_p is the particle area factor to account for internal surface burning (effective burning area of the entire particle/external area of the equivalent spherical particle), and n is the apparent order of the reaction. Smith[27,106] suggests that carbon monoxide is the most likely surface product of reaction.

By analogy to Newton's law of cooling, the oxidizer diffusion rate is often expressed as

$$r_{do} = k_m M_o A_p \left(C_{og} - C_{op} \right) \tag{24}$$

This analogy is only correct for equi-molar counter diffusion in a stagnant film or for very low rates of mass transfer. More generally, the diffusion rate requires a contribution from the bulk flow motion induced by the mass transfer itself:

$$r_{do} = k_m M_o A_p \left(C_{og} - C_{op} \right) + \frac{r_d C_{og}}{C_g} \tag{25}$$

where r_d is the total diffusion rate and C_g is the total molar gas concentration. The total diffusion rate r_d includes the contribution of the oxidizer r_{do}, and thus Eq. (25) is implicit in the diffusion rate of the oxidizer.

The particle reaction rate, determined from the diffusion of oxidizer is

$$r_p = \frac{v_s r_{do} M_p}{M_o} \tag{26}$$

When the particle is burning in a quasi-steady manner, the rate of diffusion of the oxidizer must also equal the rate of consumption of the oxidizer (i.e., $r_{do} = r_{ko} = r_p$). When the order of the reaction (n) is unity, Eqs. (25) and (26) are often combined straightforwardly to eliminate r_{do} and the unknown quantity, C_{op} to give

$$r_p = \frac{A_p v_s M_p C_{og}}{\left(1/k_r \xi_p \right) + \left(1/k_m \right)} \tag{27}$$

When ξ_p = unity, this result is equivalent to Eq. (21), which was the basis for reporting the data of Figures 11 and 12. Although Eq. (27) is attractive and often used, it neglects the bulk diffusion term as shown in Eq. (25). The more complete expression is implicit in the reaction rate

$$r_p = \frac{A_p v_s M_p C_{og}}{\left(1/k_r \xi_p \right) + \left(1/k_m \right) + \left[r_{pt} / \left(A_p M_g C_g k_r \xi_p k_m \right) \right]} \tag{28}$$

where r_{pt} is the total reaction rate of the particle.[107] When the reaction order is nonunity, Eqs. (21)–(26) are still appropriate and can be solved iteratively.

For a single spherical particle, the rate of mass change is

$$\frac{dm_p}{dt} = -r_p = \frac{d \left(\rho_{ap} V_p \right)}{dt} = \frac{d \left[\rho_{ap} (\pi/6) d_p^3 \right]}{dt} \tag{29}$$

One of the two assumptions is commonly made in tracking particle burnout:

1. The particle is highly porous and burns internally as well as externally with variable density and near-constant diameter.
2. Or, the particle is a uniform solid and burns with constant density and shrinking diameter.

In the first case

$$\frac{dm_p}{dt} = \frac{V_p d\rho_{ap}}{dt} \tag{30}$$

whereas in second, if the particle is assumed to be spherical,

$$\frac{dm_p}{dt} = \frac{\rho_{ap}(\pi/2) d_p^2 d(d_p)}{dt} \tag{31}$$

Alternatively, Smith,[27,106] presents correlations of particle size versus fractional char burn-off that could be used to establish the relationship between d_p and ρ_{ap}.

To compute the time required to consume a char particle, the following is required (for the variable diameter case):

Parameter	Units	Definition	Source
ρ_{ap}	kg/m³	Bulk or apparent density of the char particle	Lab measurements (see Table 3)
A_p	m²	External surface area of equivalent sphere	Initial value specified; can vary during burnout
ν_s	—	Stoichiometric coefficient	2 and CO 1 for CO₂
M_p	kg/kmol	Molecular weight of the combustible fuel	12 for carbon
k_r	m/sec	Reaction rate	$A \exp(-E/RT)$
A, E	m/sec, kcal/kmol	Reaction rate	Lab data constants (Figures 11 and 12)
ξ_p	—	Surface area factor	From lab measurements— usually unity to correspond with basis for reported data
k_m	m/sec	Mass-transfer coefficient	~ $2 D_{om}/d_p$ with low Re_p and r_p
D_{om}	m²/sec	Oxidizer diffusion coefficient	Data tables

The internal surface area of the char can be far greater than the external area. However, if the basic char oxidation data have been based on the external area, then this factor (ξ_p) is contained in the reported value of k_r and will thus be taken as unity.

Although the use of this model and the data correlations provide a basis for estimating char oxidation rates, many questions remain. Influences of pore diffusion affect the values of k_r and are not modeled by Eqs. (21) and (22). Devolatilization influences the nature of the char, and the char also changes characteristics during burnout.[27,106] Reaction order (n) has been established over small variations in oxygen partial pressure. Much less information is available for CO_2, H_2O, or H_2.[71] Impurities in the coal apparently influence char reactivity. Ash may accumulate on the particle surface, increasing resistance to diffusion of oxidizer and products. The particles can fracture into several smaller pieces toward the end of burnout.[54] High-reaction temperatures can lead to slag droplet formation on the surface of the char.

Because of the complexity of these coal particle reaction processes, it is emphasized that a realistic model must be closely related to laboratory data for the same type of coal under similar conditions of pressure, temperature, size, and oxidizer concentration.

Diffusion Limited Combustion. For large particles or high temperatures, consumption of char can be dominated by diffusion of oxygen (or other oxidizer to the char particle). The data of Figure 13 were obtained for several chars and other carbonaceous substances in air at pressures from 0.1–0.45 MPa. Calculated burning times were for 1500 K. Computation of the burning time in Figure 13 is essentially through integration of Eqs. (27) and (31) with $k_r \to \infty$, and with constant temperature, oxygen concentration, and particle density but with negligible convective effects on oxygen diffusion rate (i.e., very small rates of mass transfer with use of Eq. (24) in a stagnant film). Thus

$$\frac{dm_p}{dt} = \rho_{ap} \frac{(\pi/2)d_p^2 d(d_p)}{dt} = -r_p = \frac{-2D_{om}\pi d_p^2 v_s M_p C_{og}}{d_p} \qquad (32)$$

or

$$\frac{d_p d(d_p)}{dt} \cong \frac{-4v_s M_p C_{og} D_{om}}{\rho_{ap}} \qquad (33)$$

Upon integration to complete burnout (i.e., particle diameter approaches zero,

$$t_b = \left(\frac{\rho_{ap}}{8v_s M_p C_{og} D_{om}}\right) d_{p_o}^2 = K_D d_{po}^2 \qquad (34)$$

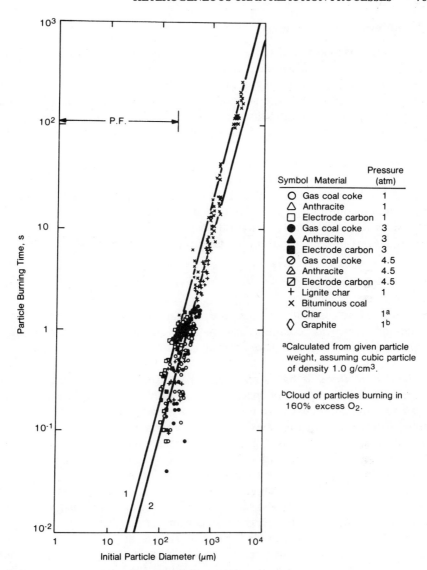

Symbol	Material	Pressure (atm)
O	Gas coal coke	1
△	Anthracite	1
□	Electrode carbon	1
●	Gas coal coke	3
▲	Anthracite	3
■	Electrode carbon	3
⊘	Gas coal coke	4.5
◮	Anthracite	4.5
▨	Electrode carbon	4.5
+	Lignite char	1
×	Bituminous coal Char	1[a]
◊	Graphite	1[b]

[a]Calculated from given particle weight, assuming cubic particle of density 1.0 g/cm³.

[b]Cloud of particles burning in 160% excess O_2.

FIGURE 13. Comparison of experimental and theoretical particle burning times in air at 1500 K. Curve 1, bulk density = 2 g cm⁻³; curve 2, bulk density = 1 g cm⁻³.[71]

The experimental data of Figure 13 confirm this predicted variation of burn time with the square of particle diameter. Measured values correlated well with predicted burning times based on the rate of diffusion of oxygen to the particle surface for particles above 100 μm. For smaller particles or lower temperatures, internal pore diffusion and chemical reaction effects become important, particularly at lower temperatures.

4.3. Intrinsic Reaction Rates

Background. Methods outlined above for describing char reaction are based on the external area of the char particle and do not formally account for (1) continued swelling and pyrolysis during char oxidation, (2) porosity of the char and changes in the pore structure as the char is consumed, (3) internal diffusion of oxidizer and products in the porous char, (4) internal reaction of oxidizer and char surface, (5) influence of carbon structure and char impurities on consumption, or (6) fracturing of particle toward the end of consumption. A more realistic treatment of char reaction would be to formally account for the internal char reaction. If the specific internal structure were described (A_t, m^2 total area/kg particles) and the true order of the reaction known (m), the instantaneous rate of consumption of char can be computed from the intrinsic reaction rate, k_i (reaction rate based on the area of the pore walls) in the absence of internal diffusion limitations (compare to Eq. (23)):

$$r_{pi} = k_i A_t C_{og}^m \tag{35}$$

Work has been conducted to measure r_{pi} and m for various carbonaceous solids and to provide computational methods for predicting A_t during char consumption and to account for internal diffusion resistance.

Intrinsic Rate Data. It has been clearly established that char oxidation can occur in the internal pores of char particles. Smith[27] notes that internal surface areas vary from about 1 to 10^3 m^2/g for chars and cokes. Smith[27] also reported intrinsic reaction rates of several porous carbonaceous materials in oxygen including chars, cokes, graphite, and highly purified carbons, as shown in Figure 14. Data are in terms of specific particle consumption rate, R_{pi}, at atmospheric pressure in pure oxygen. The correlation of Figure 14 is shown for data over 11 orders of magnitude in char reactivity for temperatures from 700–2000 K. Yet reactivity also varies by up to four orders of magnitude for a given solid fuel at fixed temperature. According to Smith,[27] the wide range of intrinsic reactivities observed for various substances at a single temperature is due to differences in both the carbon structure and to impurities in the solids and gaseous reactants. Intrinsic reactivity values for highly purified carbons show improved correlation with temperature.

   Char reactivity (g char consumed/unit time/g char present) was recently measured for the five process-development chars of Table 15 in three different temperature ranges.[136] Char reactivity in oxygen at lower pressure and lower temperature (500–750 K) have also been measured[137] for these process chars in a TGA apparatus as shown in Figure 15. Reactivity was also measured[138] at 1173 K (900°C) in a small fixed bed of char particles in CO_2 (Figure 15b) Char samples had been preheated in an inert atmosphere to temperatures above 1173 K (900°C) before performing CO_2 reactivity tests in order to eliminate volatile products. Reactivities of the various chars differ by nearly

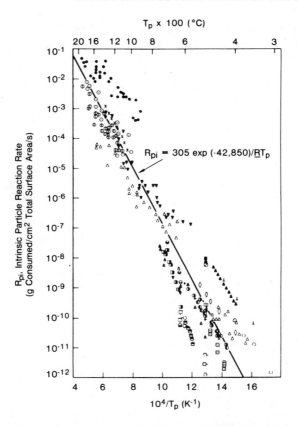

FIGURE 14. Intrinsic reactivity of several porous carbonaceous solids in oxygen (at an oxygen pressure of 0.1 MPa (1 atm)).[27]

two orders of magnitude. Reactivities of the two chars with internal surface areas (BET) of 200–220 m^2/g are an order of magnitude or more above that for the three chars whose internal surfaces areas are less than 20 m^2/g. Internal surface area is not the only factor governing char reactivity. However, from the results of Table 15 and Figure 15, it is obviously a major factor.

Strong variations of reactivity with temperature and with char type shown in Figure 16a are similar to those in Figure 15b at higher temperatures in CO_2. However, the order among char reactivities differs somewhat for the two different sets of tests. These differences are likely due in part to differences in preparation methods (1173 vs. 973 K), and to differences in grinding or classification prior to testing, as well as to the different temperature, pressure, and oxidant.

Reactivity for these same chars have also been measured at high temperature in a drop-tube furnace.[136] Test were made in air for a particle temperature range of about 1500–2100 K. Results are shown in Figure 16 for classified

TABLE 15. Elemental and Proximate Analyses of Chars and Parent Coals[136]

(a) Chars	Bi Gas	FMC COED	Occidental	Rockwell	Toscoal
		Elemental			
H	1.0	1.7	1.8	1.9	2.7
C	77.5	80.1	78.7	73.8	71.2
N	0.4	1.8	1.3	1.3	1.8
O	0.2	0.7	6.5	3.6	8.7
S	0.4	2.8	0.3	1.6	0.5
Ash	21.5	12.9	11.4	17.8	15.1
		Proximate			
Moisture	1.2	2.0	1.3	2.5	1.0
Ash	21.8	13.7	11.5	18.0	16.8
Volatile matter	5.6	9.1	15.1	7.0	17.0
Fixed carbon	71.4	75.2	72.1	72.5	65.2
Heating value, 10^3 kJ/kg	27.7	29.7	28.1	27.2	26.4
N_2-BET surface area, m^2/g	340	115	196	5.6	8.5
Apparent density,[a] g/cm^3	0.87	1.52	0.86	0.93	1.43
Porosity[a]	0.58	0.13	0.55	0.51	0.11
Char diameter,[b] μm	260	60	63	110	90

(b) Parent Coals	Montana Rosebud	Utah King	Wyodak	Pittsburgh	Kentucky #9
		Elemental			
H	5.1	6.1	5.8	5.3	5.1
C	70.1	74.2	67.5	77.5	73.5
N	1.3	1.4	1.3	1.5	1.3
O	15.4	11.0	19.8	8.5	8.5
S	0.6	0.7	0.3	1.2	2.9
Ash	7.5	6.6	5.3	6.0	8.8
		Proximate			
Moisture	2.8	4.0	6.0	2.0	5.1
Ash	9.2	6.3	6.4	6.0	9.1
Volatile matter	39.0	34.9	47.3	36.6	38.2
Fixed carbon	49.0	54.8	40.3	55.4	47.6

[a] For −200 mesh fraction.
[b] Mass mean, after size reduction in some cases.

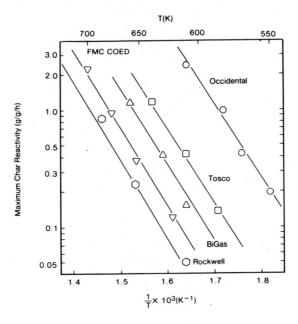

(a) TGA 0.1 mPa air

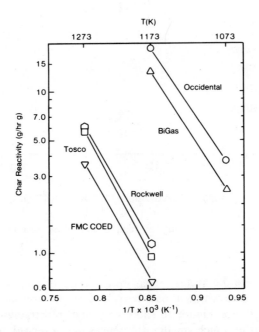

(b) Char Reactivity in CO_2 (0.25 MPa at 30% Carbon Conversion)

FIGURE 15. Effect of temperature and char type on reactivity.

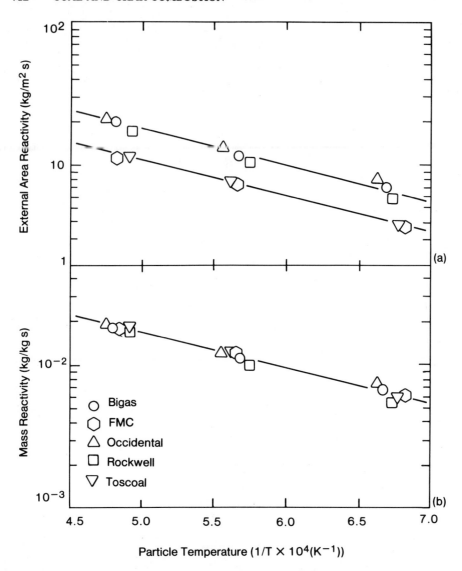

FIGURE 16. Process char reactivity at high temperature (*a*) mass basis and (*b*) external area basis.[136]

char fractions, where the average mass mean diameter was about 68 μm. The reactivity values are shown on a mass basis and on per unit of external particle surface area. Results of Figure 16 show very high reactivities at these higher temperatures. Further much smaller differences are observed among the chars, compared to the dramatic differences shown in Figure 15. Results imply that the reaction occurs far less on the internal surfaces at higher temperatures.

Thus variations in the internal surface area do not cause such marked changes in reaction rate. External diffusional rates have become an important controlling factor in consumption rate at these high temperatures, together with surface reaction. These results, when combined with those of Figure 15, dramatically illustrate the variation in char reactivity among chars over a wide temperature range. They also suggest that control of internal structure is not so important to char consumption at elevated temperatures.

Intrinsic Reactivity Modeling. Experimental results suggest that knowledge of internal surface area and internal diffusion effects are important factors in determining heterogeneous char consumption. In spite of the difficulty of developing an adequate model with the limited data available, recent articles or reviews[18, 27, 71] have stressed the importance of implementing intrinsic rate expressions. Intrinsic treatment allows the greatest potential of differentiating among the various char reactivities on the basis of fundamental kinetic parameters. This necessitates a knowledge of not only the intrinsic kinetics and reaction conditions but also of the initial pore structure and its changes during char consumption. Two different classifications of pore structure models have been defined; macroscopic and microscopic.[18] Macroscopic modeling of pore diffusion constitutes the majority of the existing char models. It utilizes an effective diffusivity throughout the particle and includes the classical shrinking core model and the progressive conversion model.

Smith and co-workers[27, 112, 139] have utilized a series of simplified theoretical equations for calculating intrinsic particle consumption rates. These equations relate the intrinsic kinetics for a given porous structure and gaseous environment to char reactivity based on external surface area or weight of unreacted char. The intrinsic reaction rate coefficient, k_i, was defined by Eq. (35) in the absence of any mass-transfer or pore diffusional limitation as the rate of particle reaction per unit of total surface area per (unit of oxidizer concentration)^m. The exponent m is the true order of reaction. The mass rate of consumption of char particle with mass-transfer or pore diffusional limitations is

$$r_{pi} = \eta k_i A_t C_{os}^m$$

$$= \eta k_i A_t C_{og}(1 - \chi)^m \tag{36}$$

where $\chi = r_{pi}/r_{p_\infty}$ and $\eta = $ the effectiveness factor, which is defined as the particle consumption rate over the maximum value in the absence of pore-diffusional limitations. The second equality, shown by Smith,[27] provides a expression for any m in terms of the known, bulk gas oxidizer concentration

TABLE 16. Comparison of Microscopic Pore Char Models

Characteristics	Simons and Finson[140]	Gavalas[141, 142]	Amundson et al.[143-145]
1. Initial pore shape	Tree or river stream	Cylindrical	Cylindrical micropores spherical macropores (optional)
2. Consideration of Knudsen and molecular diffusion	Yes	Yes	Yes
3. Statistical pore distribution function pdf	Provided for the user and is proportional to (pore radius)$^{-3}$	Based on pore intersections with a surface and the initial porosity distribution	Absorbed in terms of moments
4. Change of the pore distribution function (pdf) with conversion	Yes	No	Not directly applicable
5. Utilization of Langmuir rate type expressions	One format only	Yes	Yes
6. Computational computer time	Small	Large	Large
7. Nonisobaric nonisothermal	No/no	Yes/optional	Yes/optional
8. Capability to monitor local species concentration and pore structure parameters	No	Yes	Yes

C_{og}. Thus computation of a particle consumption rate from intrinsic reaction rate data requires a value for η (which is related to pore diffusion and reaction processes) together with the total particle surface area and intrinsic rate coefficient.

Microscopic models of pore diffusion attempt to describe diffusion through a single pore and then to predict the overall particle consumption rates by an appropriate statistical description of the pore size distribution. Microscopic models treat variation of diffusivity with position and burn-off, intrinsic rate expressions with variable reaction orders, and development of pore intersection and expansion due to reaction. Table 16[140-145] briefly summarizes features of three such models recently developed; Smoot and Smith[5] give an expanded discussion of these models.

TABLE 17. Other Important Aspects of Coal Particle Reaction

Item	Description	Key References
1. Particle fracturing	Particles may swell during devolatilization, giving rise to significant increase in porosity; then near the end of particle burnout, char particles may fracture into several smaller fragments, changing their structure, size, and shape and their motion in the gas.	54, 86
2. Larger particles	Much of the information presented herein has focused on finely pulverized coal dust ($< 150 \ \mu$m). In fluidized bed and fixed bed coal processes, particles are much larger and behave differently. Cracking, internal heat and mass transfer, product condensation and repolymerization influence the burnout.	8, 146
3. Nitrogen oxide	Most coals have about 1–2% of nitrogen, called fuel-nitrogen. Nitrogen pollutants (NO, NO_2, HCN, NH_3) form readily from this nitrogen during combustion. Typically 30% of the fuel nitrogen can be converted, largely to NO. NO levels can be controlled through control of the combustion process.	11, 24, 90, 147, 148
4. Sulfur pollutants	Sulfur levels in coal vary considerably (e.g., 0.5–5%) and exist as organic (i.e., in the coal structure) or pyritic (e.g., FeS). Some of the pyritic sulfur can be removed prior to combustion; in some processes (e.g., fluidized beds) the sulfur can be removed with limestone or other additive during combustion. Most of the sulfur forms as H_2S, SO_2, or SO_3 in the exhaust.	147
5. Trace metal pollutants	All coals have a variety of trace metals, some of which are potentially harmful (Be, Pb, As, P, etc.). Some can exist in volatile forms, whereas others can escape in very small particulates formed during the combustion process.	149–151
6. Other oxidants	Char particles can also be consumed through reaction with CO_2, steam, or H_2. Reaction rates of these reactants are much slower than with oxygen but are important and often controlling in gasifiers, or fuel-rich combustors.	107
7. Ash/slag formation	Coals contain varying amounts of ash. Some coals have over 50% ash, but 5–25% is common. These minerals form into very small particles or droplets during the combustion process. They also form layers on walls, reducing heat transfer rates and corroding surfaces. Ash ingredients may also catalyze heterogeneous reactions.	22, 134, 152, 153
8. Coal particle clouds	Much of the information reported here on coal reactions is from tests with small quantities of coal. In practical reactors, the particles form clouds, and their behavior may differ from that observed for individual particles.	4

4.4. Other Aspects of Particle Reaction

Several important aspects of coal particle reaction have not been addressed in this chapter. Table 17 provides a summary of some of these considerations.[146-153] These issues serve to further illustrate the scope and complexity of coal reaction processes.

5. PRACTICAL COAL FLAMES

5.1. Background and Scope

Sections 3 and 4 dealt with characteristics and reaction rates of small quantities of coal particles under very carefully controlled laboratory conditions. In this section the nature and characteristics of practical, coal-containing flames are considered. This section treats direct combustion and coal gasification. In principal, there is little difference in these two flame types. Direct combustion flames are those where the reactions of coal take place to near completion with CO_2 and H_2O being among the major product species and where a high-temperature gas is a principal purpose.

Direct combustion of coal has been identified in some of the earliest recorded history. According to Elliott and Yoke,[154] the Chinese used coal as early as 1000 BC, and the Greeks and Romans made use of coal before 200 BC. By AD 1215, trade in coal was started in England. Pioneering uses of coal (e.g., coke, coal tars, and gasification) have continued to be advanced since the late sixteenth century. The fixed bed stoker system was invented in 1822, the firing of pulverized coal occurred in 1831, and fluidized beds were invented in 1931.[154]

Specific purposes for use of direct combustion processes include (1) power generation, (2) industrial steam and heat, (3) kilns (cement, brick, etc.), (4) coking for steel processing, and (5) space heating and domestic consumption. Most of the world's coal is consumed in pulverized form for power generation (see Section 2). The various forms in which the coal is used also differ substantially. Coal particle diameters vary from micron size through cm size. Use is made of virginal coal, char, and coke. Pulverized coal is also slurried with water or other carriers, both for transporting and for direct combustion.

Various processes that are used to directly combust or to gasify coal can be classified in several different ways. This section presents a discussion of the general features of these practical flames, together with various classification methods. A brief description of various direct coal combustion and gasification processes is also included. This information provides a basis for mathematical description (i.e., modeling) of these practical flames, which follows in the next section. It is beyond the scope of this chapter to provide an extensive discussion of coal process hardware and operating characteristics. The reader is referred to Elliott[2] for extended treatments on these topics.

5.2. Flame Classification and Description

Flame Classification. Practical flames can be classified in several ways, including the following:

Classification	Examples
1. Flow type	Well-stirred
	Plug-flow
	Recirculating
2. Mathematical model complexity	Zero-dimensional
	One-dimensional
	Multidimensional
	Transient
3. Process type	Fixed bed
	Moving bed
	Fluidized bed
	Entrained bed (i.e., suspension firing)
4. Flame type	Well-stirred
	Premixed
	Diffusion
5. Coal particle Size	Large
	Intermediate
	Small

Table 18 summarizes these classification groups and illustrates typical relationships among the various classifications. For example, a fluidized bed combustor may be approximated as a well-stirred reactor (zero-dimensional) and may be assumed to be well-mixed and kinetically controlled. An entrained gasifier may be considered as a plug-flow reactor (one-dimensional), with solid kinetics and heat transfer controlling the reaction rate. A coal dust explosion may be treated as a transient, premixed, one-dimensional flame with kinetic and convective control. Field et al.[69] and Essenhigh[71, 105] also discuss classification of flames.

Classic Flame Types

1. *Well-stirred reactor.* An ideal fluidized bed combustor is an example of a well-stirred reactor. This kind of idealized reactor, with instantaneous, complete mixing, is simple to describe, since fluid motion need not be described in detail and properties vary only with time. Heat transfer and solids kinetics are dominant processes. When the bed is not perfectly stirred, as in practical fluidized beds, mass transfer and bubble dynamics are important issues.

TABLE 18. Classification of Practical Coal Flames

Flow Type	Process Application	Typical Flame Type/ Control	Particle Size	Typical Mathematical Complexity
Perfectly stirred reactor	Fluidized bed	Well-mixed, kinetically controlled	Intermediate 500–1000 μm	Zero-dimensional
Plug flow	Moving beds, steady fixed beds, shale retort	Solid reaction/ heat transfer control	Large 1–5 cm	One-dimensional, steady
Plug flow	Pulverized coal furnace, entrained gasifier	Mixing specified, solid reaction/ heat transfer control	Small 10–100 μm	One-dimensional, steady
Plug flow	Coal mine explosions, coal process explosions, flame ignition/stability	Premixed flame, kinetic and diffusion control	Small 10–100 μm	One-dimensional, transient
Recirculating flow	Power Generators, entrained gasifiers, industrial furnaces	Diffusion flames complex control (mixing, gas kinetics solid kinetics)	Small 10–100 μm	Multidimensional, steady or transient

2. *Premixed flame.* A premixed coal dust flame may exist in a coal mine explosion, a coal pulverizer explosion, or a recirculation zone where a diffusion flame is stabilized. This type of flame can be very small (mm to cm in thickness) and is very complex, with diffusion, solids reaction, heat transfer, and gas kinetics being important rate-controlling processes. Of particular importance to laminar flames of this type are counterdiffusion of gaseous reactants and products and convection between the hot gas and incoming cooler particles.

3. *Diffusion flame.* A diffusion flame is characterized by injection of separate streams of fuel and oxidizer into the reactor. In this flame the mixing of fuel (gaseous or solid) and oxidizer play an important role. However, gas kinetics, particle kinetics, and heat transfer are also often important. This type of flame is characterized by very complex fluid mechanics that may include recirculating and swirling flows. Interactions of the turbulence and the reactions further complicate this kind of flame. This flame type will predominate in pulverized coal combustors, furnaces, and gasifiers. These general flame classifications often form the basis for modeling of practical flames.

5.3. Practical Combustion Process Characteristics

Bed Type and Coal Size. Direct coal combustion processes are most frequently classified according to bed type: fixed or moving beds (large particles),

TABLE 19. Comparison of Various Characteristics of Practical Direct Coal Combustors

Process Type	Fixed/Moving Bed	Fluidized Bed	Suspended Bed
Coal size, μm	10,000–50,000[a]	1500–6000[a]	1–100[a]
Bed porosity, %	Low	95–99[a]	Very high
Operating temperature, K	< 2000 K	1000–1400[a]	1900–2000[b]
Residence time, sec	500–5000	10–500	< 1
Coal feed rate, kg/hr	Up to 40,000[b]	Up to 40,000[a]	Up to 120,000[b]
Distinguishing features	Countercurrent or cross-flow	Particle–particle interaction important, much of heat transfer due to conduction	Very small particles, high rates
Advantages	Established technology, low grinding, simple	Low SO_x and NO_x pollutants; less slagging; less corrosion, low-grade fuel possible[a]	High efficiency, large-scale possibilities, high capacity
Disadvantages	Emissions, especially particulates,[a] less efficient than other methods	New technology, feeding fuel[a] tube erosion	High NO_x, fly ash, pulverizing expensive
Key operational variables	Fuel feed rate, airflow rate	Bed temperature pressure, fluidizing velocity	Air/fuel ratio
Commercial operations	Stokers	Industrial boilers	Pulverized coal furnaces and boilers

[a]Elliott[2]
[b]Singer[70]

fluidized beds (intermediate sized particles), and suspended or entrained beds (small particles). Coal strokers are an example of the first, fluidized bed combustors the second and pulverized coal furnaces the third. Table 19 provides a more detailed comparison among these three distinct types of direct coal combustion processes. It is clear that particle size is a major distinguishing characteristic among these types of processes, with the coal size varying from about 10 μm to 5 cm, depending on process type. Flame temperature also varies with the fluidized bed operating at uniquely lower temperatures (1100–1200 K) compared to fixed and entrained bed systems. These predominant differences have a marked impact on residence times of the coal particle in the bed, which vary from less than a second to more than an hour. This observation was well illustrated by measurements and predictions in Figure 13, where burning times for particles from 100 μm to about 5000 μm vary from about 100 ms to 500 s.

Flame Processes. The most distinguishing feature of practical flames is that several physical and chemical processes can occur simultaneously. These processes can include convective and conductive heat transfer, radiative heat transfer, turbulent fluid motion, coal particle devolatilization, volatiles reaction with oxidizer, char reaction with oxidizer, particle dispersion, ash/slag formation, soot formation, and others. Whichever of these processes occurs—and more important, whichever controls—the coal reaction processes differ, depending upon the type of practical flame and upon the operating conditions.

Most of the world's coal consumption is presently through finely pulverized form. Practical flames with pulverized coal include premixed flames, and particularly diffusion flames. In order to provide a foundation for mathematical description of these flames, the specific features of these two flame types are outlined. Laboratory measurements are used to illustrate the nature of these flames.

Pulverized Coal Diffusion Flames

Flame Characteristics. In these flames, fuel and oxidizer, not completely premixed, are delivered to a reactor vessel. The most common practical diffusion flame is that in large-scale utility and industrial furnaces where pulverized coal (usually less than 150 μm but can be up to 400 μm; Figure 3) is pneumatically conveyed with a small percentage of air into a furnace through several ducts. The larger percentage of the total air, which is preheated to higher temperatures, enters through different ducts, placed close to the coal-carrying ducts. The specific numbers and combinations of these inlet ducts vary greatly among various furnaces.[155] In order for the coal to burn to completion, the coal-air "primary" stream and the hot air "secondary" stream must mix. This turbulent mixing process that occurs in these diffusion flames has a major impact on the burning of the coal.

Major operational variables that influence the behavior of diffusion flames include reactor size and shape, wall materials and temperature, coal type and size distribution, moisture percentage, ash percentage and composition, primary stream temperature, velocity and mass flow, coal percentage, secondary stream temperature, velocity, mass flow and swirl level, and burner configuration and location (i.e., arrangement of ducts).

Detailed characteristics of pulverized coal diffusion flames have been studied experimentally for several years. Work at the International Flame Research Foundation has been in progress for many years, but much of this work is not available in the open literature. Studies that reveal details of these coal diffusion flames most commonly use intrusive probes for measurement of velocity, temperature, and gas and solids composition. Such measurements have been made for various coals.

Combustion Data. Typical results for two very different coals are shown here to illustrate the features of these flames. Data are taken principally from

Harding et al.[156] and Asay et al.[157] from measurements made in a laboratory-scale combustor. The combustor was oriented vertically and fed from the top to facilitate particulate and ash removal and to minimize asymmetric effects due to particle settling and natural convection. The combustor consisted of a series of sections, each lined with castable, high purity aluminum oxide insulation. One of the sections contained a sampling probe for removal of gaseous and particulate material.

The tests were devised in order to determine several key rate parameters, including gas and particle mixing and coal reaction. Effects of coal type,

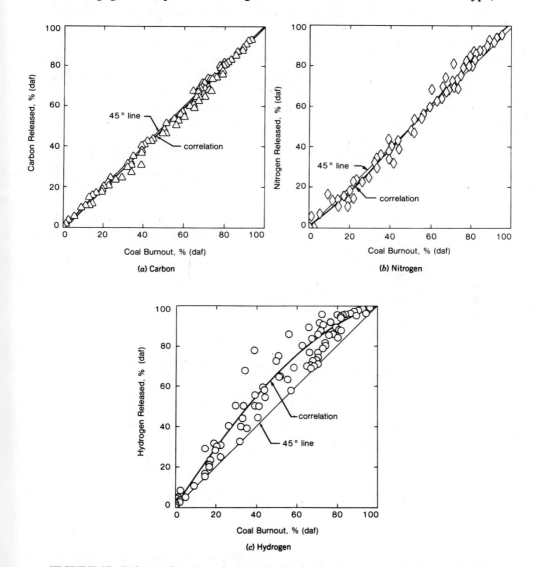

FIGURE 17. Release of various elements for Utah bituminous coal during combustion in a laboratory combustor.[159]

moisture percentage, swirl number of the secondary stream, stoichiometric ratio, coal particle size, secondary stream velocity, and injection angle of the secondary stream were investigated. Test conditions were similar to operating conditions existing in industrial furnaces. The first coal was a Utah (Desert Mine), high-volatile butiminous coal with a proximate analysis (weight percent) of 2.4% moisture, 45.4% volatiles, 43.6% fixed carbon, and 8.6% ash. Ultimate analysis (weight percent, moisture free) was 5.7% H, 70.2% C, 1.4% N, 0.5% S, 8.9% ash, and 13.4% O. The Utah coal was pulverized to 70% through 200 mesh and the mass-average coal particle diameter of the standard size coal was 50 μm. The Wyoming subbituminous coal was taken from the Belle Ayre Mine. The as-received coal contained 5.0% ash, 27.8% moisture, 32.9% proximate volatiles, and 34.3% fixed carbon. Elemental composition (daf) was 4.7% H, 72.0% C, 1.2% N, 0.56% S, and 21.6% O. The undried coal was pulverized to 80% through 200 mesh, with a 3% moisture loss occurring during grinding. The undried coal contained 25% moisture at the time of testing, whereas the dried coal contained 4.5% moisture.

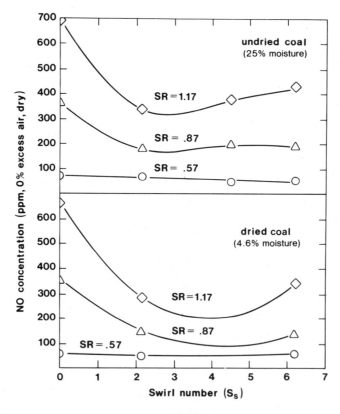

FIGURE 18. Effect of swirl number, stoichiometric ratio (SR), and coal moisture on NO level for Wyoming subbituminous coal in a laboratory combustor.[157]

Data were obtained from the ultimate analyses on the relative rates of release of elements (C, H, O, N, and S) from the coal at various points in the combustor. Results[158] of Figure 17[159] indicate that the extent of element release was largely a function of the extent to which the coal was burned and was far less dependent on either test condition or location within the combustor, results further indicated that hydrogen (see Figure 17c) and oxygen were liberated more rapidly than the coal as a whole, whereas carbon (Figure 17a), nitrogen (Figure 17b) and sulfur emitted at about the same rate as total coal mass release.

Detailed combustion characteristics of undried (25% moisture) and dried (4.5% moisture) Wyoming subbituminous coal were also determined in the combustor. Measurements were made of coal burnout and nitrogen pollutant concentrations for various secondary stream swirl numbers (S_s) and stoichiometric ratios. The stoichiometric ratio (SR = inlet air/air required for complete coal combustion) was varied by changing the secondary air mass flow

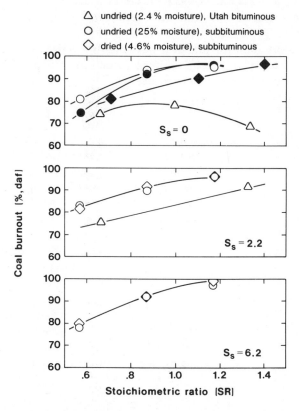

FIGURE 19. Effect of swirl number, stoichiometric ratio (SR), and coal moisture on Wyoming subbituminous coal and Utah bituminous coal burnout. (Closed symbols from carbon balance; open symbols from ash in char.)[157]

rate. Effects of coal moisture at various SR and S_s values on nitrogen oxide concentrations are shown in Figure 18. The substantial decrease in coal moisture from 25% to 4.5% caused little change in NO level, except at high SR and S, where a decrease of about 20% in NO concentration was observed. However, a change in the swirl number, which causes dramatic changes in the fluid structure of the flame, caused marked changes in NO concentration.

Figure 19 shows effects of S_s, SR, and coal moisture on percentage of coal burnout near the reactor exit. Two methods for computing burnout were used. A carbon balance was performed based on carbon-containing species in the gas, and ash was used as an inert particle tracer. Due to ash volatility at flame temperatures as well as moderate ash solubility in sample quench water, these latter values are thought to be somewhat low but were more reproducible. Residence time in the combustor was only 200–300 ms, so coal burnout was not always complete. Measured burnout values are expectedly a strong function of SR but were influenced little by variation in coal moisture. An increase was observed in burnout with increasing swirl number. Burnout was approximately 80% for $SR = 0.7$ and 90–95% burnout for $SR = 1.2$. Use of swirl thus caused small increases in total burnout, together with a significant reduction in N-pollutant level.

These laboratory data for practical but small pulverized coal flames show effects of several test variables on flame structure, gas composition, pollutant level and coal burnout, as summarized below:

Variable	Range	Effects
Coal type	Subbituminous, bituminous	Subbituminous more reactive
Particle size	20 μm, 50 μm	More rapid ignition for small particles
Moisture level	4–25%	Little effect on NO_x or burnout
Stoichiometric ratio	0.6–1.3	Major effect on NO_x, burnout
Secondary swirl number	0–6	Major effect on jet structure, NO_x; leads to rapid flame ignition
Secondary injection angle	0–30°	Increases jet mixing, little effect on ignition
Secondary velocity	35, 55 m/sec	Little impact

These variables represent only a small set of possible test conditions in complex practical flames. Data show that marked changes can be made in flame structure through control of test variables. Processes that control these diffusion flames can vary markedly depending on the test conditions. However, turbulent mixing processes and particle ignition processes are clearly important. Providing for a clear understanding of the behavior of these practical flames goes well beyond inspection of experimental data.

Combustor Profile Data. Characteristics of the combustion process are more clearly shown from measurements of gas composition, coal burnout, temperature, velocity, and other fluid-particle properties within the combustor as combustion occurs. Such measurements are quite common, having been reported by several laboratories, including International Flame Research Foun-

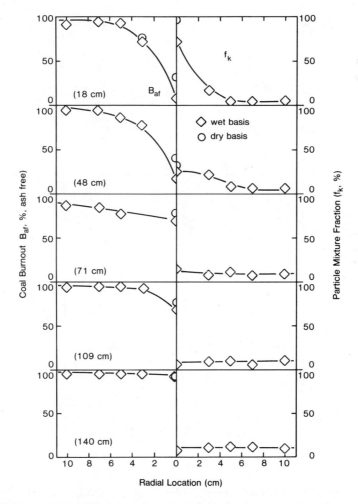

FIGURE 20. Radial profiles of coal burnout and particle mixture fraction at five axial locations for subbituminous coals.[157]

dation, Massachusetts Institute of Technology, Imperial College, and Brigham Young University. Smoot and Smith[5] provide a review of selected profile data, and a comprehensive summary of these data has recently been published.[160]

Profile data for the moist Wyoming subbituminous coal (80% through 200 mesh) are shown in Figures 20–22. Tests were performed in a down-fired, cylindrical, laboratory-scale combustor (20 cm I.D. × 1.5 m long) with pulverized coal feed rate of 10.2 kg/hr, $SR = 1.06$, and secondary air swirl number of 2. All test conditions are shown in Asay et al.[157] Gas composition, coal burnout percentage (B_{af}), coal particle fraction (f_k), and gas mixture fraction

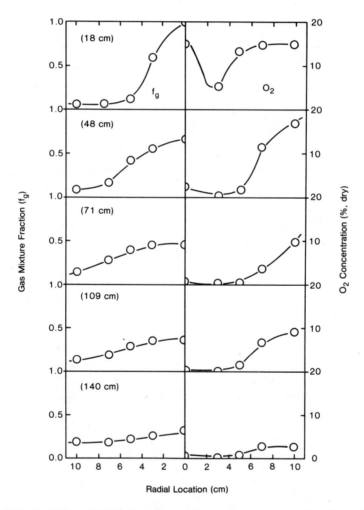

FIGURE 21. Radial profiles of O_2 concentration and gas mixture fraction (f_g) at five axial positions for subbituminous coal.[157]

(f_g) were measured. These latter two properties are defined by

$$f_k = \frac{m_k}{(m_k + m_g)} \tag{37}$$

$$f_g = \frac{m_p}{m_p + m_s} \tag{38}$$

and provide measurements on the extent of coal particle dispersion in the gas and extent of mixing of primary and secondary gas feed streams, respectively.

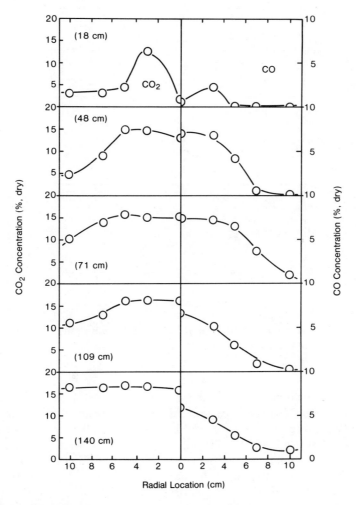

FIGURE 22. Radial profiles of CO and CO_2 concentrations at five axial positions for subbituminous coal.[157]

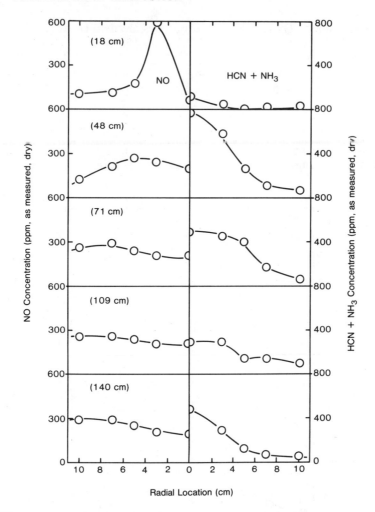

FIGURE 23. Radial profiles of NO and HCN + NH$_3$ concentrations at five axial positions for subbituminous coal.[157]

At an axial position of 18 cm and on the combustion centerline, the B_{af} values indicate that little coal reaction had occurred (Figure 20). However, all of the coal moisture had been released by the time that the coal reached this position, as suggested by the differences between the coal burnout values computed with the inlet coal on dry and wet basis (see Figure 20). At a radial position of 3 cm, the coal/air ratio was near stoichiometric and significant coal reaction had occurred, causing a drop in O$_2$ (Figure 21) and f_k (Figure 20), with increases in CO and CO$_2$ concentrations (Figure 22) and coal burnout. Also NO concentrations increased dramatically (Figure 23). Near the wall very little CO$_2$, CO, or NO were present, since most of the coal remained near the centerline. The coal distribution is shown by the f_k ($m_k/(m_k + m_g)$) value (Figure 20) which was very nearly zero at this location.

At 48 cm the flame had spread out considerably. CO_2 and CO levels were high from the centerline to 5 cm, while O_2 was low over the same range. The NO concentrations had become more uniform radially, but the f_g values indicate that at radial positions of 7 and 10 cm, the gas composition was mostly secondary air. The burnout values indicate the coal reaction had still not progressed to a great extent in regions where the coal was concentrated, but the HCN + NH_3 levels had risen dramatically. The region near the centerline was still quite fuel rich, as evidenced by the high CO levels.

At an axial position of 71 cm, the region of major coal-air reaction had approached the 7-cm radial location with a corresponding decrease in O_2 (see Figure 21). As more secondary air was mixed inward, the NH_3 + HCN levels declined (see Figure 23). Coal burnout was > 75%, while NO values had changed little. At the 109-cm axial positions all of the coal (daf) from 3 cm to the wall was virtually consumed. CO_2 levels were higher, while CO and HCN + NH_3 were declining. Particles were well mixed (f_k constant), and gases were approaching that state (lower f_g gradient).

At 140 cm (near exit) CO_2, f_k, and B_{af} gradients had essentially vanished. Small gradients remained in O_2, f_g, and NO, while CO and NH_3 + HCN still varied considerably across the reactor. The centerline value of HCN + NH_3 at the exit may be erroneously high compared to the centerline value at an axial location of 109 cm. Measured gas temperatures by suction pyrometry[161] were well below anticipated values but did show regions with temperature gradients and regions of vigorous reaction. Although these profile results are only exemplary, they do illuminate significant details about the coal combustion process. And they provide a basis for evaluation of comprehensive models.

5.4. Practical Gasification Process Characteristics

Process Applications. Gasification of coal is a process that occurs when coal or char is reacted with an oxidizer to produce a fuel-rich product. Principal reactants are coal, oxygen, steam, carbon dioxide, and hydrogen, whereas desired products are usually carbon monoxide, hydrogen, and methane. Potential uses of this product are for (1) use as a gaseous fuel for generation of electrical power, (2) use as a gaseous fuel for production of industrial steam or heat, (3) production of substitute natural gas, (4) use as a synthesis gas for subsequent production of liquids such as alcohols or gasoline, and (5) for direct production of coal liquids and char.

Gasification of coal has been practiced commercially for nearly 200 years, with the first gas produced from coal in the late eighteenth century.[162] However, the production of gas from coal gradually decreased in the midtwentieth century as abundant natural gas resources were widely distributed through general pipeline delivery. A sharply renewed interest in gasification was evident in the mid-1970s with concern over dwindling reserves of oil and gas. Several commercial gasification processes are in use in the world, particularly in Europe and South Africa.[163] In the United States major gasification demonstration projects have been conducted in North Dakota[164] and Southern California.[165] It is not the purpose of this chapter to give a detailed accounting of the basic aspects of gasification or a comprehensive description of various gasification processes. Johnson[166] provides a detailed treatment of the fundamentals of coal gasification, and Hebden and Stroud[163] give an extensive review of various commercial and developmental gasification processes Bissett[167] provides a summary of entrained coal gasification research from the Morgantown Energy Technology Center and its predecessor, the U.S. Bureau of Mines. This work dates from 1947. A summary of other U.S. and European entrained gasification research is also included. In this section general features of coal gasification are identified, and the relationship to coal combustion is discussed. From the standpoint of theoretical description, coal gasification processes are considered to be much like that of fuel-rich coal combustion. From this concept, similarities and differences in combustion and gasification of coal are identified.

Basic Process Features

Basic Reactions. A principal overall reaction during the gasification of coal and other carbonaceous materials is

$$C(s) + O_2 \rightarrow CO; \quad \Delta H_R^0 * = -110.5 \text{ MJ/kmol} \quad (39)$$

This reaction is exothermic. Further the reaction of $C(s)$ does not stop at CO, but any free oxygen rapidly reacts with CO in the gas phase to produce CO_2. Thus for a fuel-rich system, in order to consume the remaining $C(s)$, the much slower endothermic reaction

$$C(s) + CO_2 \rightarrow 2CO; \quad \Delta H_R^0 = +172.0 \text{ MJ/kmol} \quad (40)$$

must occur.

In order to control high temperatures resulting from $C(s) - O_2$ reaction, and to increase the heating value of the product gas through addition of hydrogen, steam is often added as a reactant:

$$C(s) + H_2O(g) \rightarrow CO + H_2; \quad \Delta H_R^0 = +131.4 \text{ MJ/kmol} \quad (41)$$

This reaction is also endothermic and must rely on the heat release from the $C(s) - O_2$ reaction for energy requirements. Further the rate of reaction of $C(s) + H_2O$ is also very slow compared to $C(s) + O_2$. However, the resulting product gas can have an increased heating value, as illustrated in Figure 24.

---

*All ΔH_R^0 at 298 K and 1 atm.

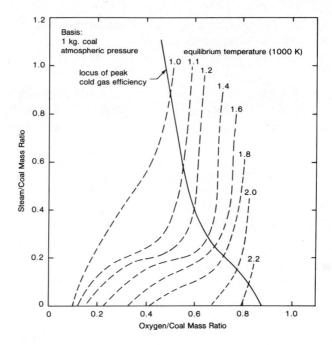

FIGURE 24. Predicted effect of coal gasification oxygen/steam/coal feed ratios on cold gas efficiency (thermal equilibrium).[169]

This figure shows the results of a series of computations for various mixtures of a typical HV bituminous coal with steam and oxygen, assuming thermodynamic equilibrium.[169] Temperature is shown to increase with increasing oxygen content and to decrease with increasing steam content. Also shown is the peak cold gas efficiency, which is the ratio of heating value of the product gas at ambient temperature to the heating value of the coal (daf).

This value increases with increasing steam and decreases with increasing oxygen. Thus, for development of a given gasification process, a balance between sufficiently high reacting temperature for flame stability and carbon conversion, and acceptably high cold gas efficiency, must be achieved.

In a coal/steam/oxygen system, gaseous reactions are also very important, particularly,

$$CO + \tfrac{1}{2}O_2 \rightarrow CO_2; \qquad \Delta H_R^0 = -283.1 \text{ MJ/kmol} \qquad (42)$$

$$CO + H_2O(g) \rightarrow CO_2 + H_2; \qquad \Delta H_R^0 = -41.0 \text{ MJ/kmol} \qquad (43)$$

The first of these reactions causes rapid consumption of oxygen, increases gas temperature, and forces the requirement for slow, heterogeneous C(s) reaction with CO_2. The second, slightly exothermic water-gas shift reaction, also produces CO_2 from CO and tends to control the final product distribution. In some gasification processes, hydrogen is added as a reactant in order to increase the quantity of methane as a product:

$$C(s) + 2H_2 \rightarrow CH_4 \qquad (44)$$

However, with the high cost of hydrogen, this is not common in practical gasification processes. Edmister et al.,[170] Batchelder and Sternberg,[171] and particularly Johnson[166] consider the thermodynamics of coal gasification in greater detail.

Gasification Process Types. Section 2 outlined key process types for gasification: fixed (or moving) bed, fluidized bed, entrained bed, and other (e.g., molten bath); Table 1 compared various features of these processes. Hebden and Stroud[163] describe the details of several specific gasifiers in each category.

TABLE 20. Comparison of Coal Gasification Process Types (Air and Oxygen Blown)

	Fixed Bed	Fluidized Bed	Entrained Bed
Residence time	1–3 h[a]	20–150 min[a]	0.4–12 sec
Coal size	6–50 mm	500–2400 μm	10–150 μm
O_2/coal	0.14–0.81	0.25–0.97	0.28–1.17
Steam/coal	0.28–3.09	0.11–1.93	0.1–1.20
Coal type	Most types, no fines	Noncaking coals	All types
Temperature range, K	1150–1300	600–1470	1150–2500
Pressure range, MPa	0.1–2	0.1–10	0.1–30
(atm)	(1–20)	(1–100)	(1–300)
Product gases, mol %			
CO + H$_2$	39–66	2–80[a]	35–91
CH$_4$	2–15	3–68	0.1–17
HHV, Btu/SCF	(250–320)	(300–800)	(115–550)
Applications	Extensive	Slight	Moderate
Commercial operations	Lurgi Wellman-Glusa	Winkler	Koppers-Totzek, Texeco
Principal advantages	High turndown ratio, mature technology, low thermal losses	Lower temperature, reduced thermal losses, variety of coal sizes, moderate residence time	Smaller, simple design, all coal types, highest capacity/ volume

[a]Information from Perry.[162] All other information from Hebden and Stroud.[163]

Further division is made between commercial and developmental gasifiers, and between those containing nitrogen (i.e., air-blown) in the product gas and those without nitrogen (i.e., oxygen-blown). Air is used when the presence of nitrogen in the product gas is not economically detrimental (e.g., industrial fuel gas), whereas oxygen is used to produce a nitrogen-free product gas (e.g., substitute natural gas).

A brief summary of the various features of coal gasification process types is given in Table 20. Key differences among the various processes relate to coal particle size and operating temperature. Large coal sizes (6–50 mm) and moderate operating temperatures lead to very long residence times (hours) in fixed or moving beds. Fluidized bed gasifiers operate with residence times of minutes and with smaller coal size (500–1000 μm) but at lower temperatures (1100–1200 K) that result from cooling by in-bed tubes. Highly pulverized coals (1–150 μm) are used in entrained systems with very high operating temperatures (up to 2200 K in oxygen-blown gasifiers) that result in very short (< 1 s) residence times. Rates and the nature of coal reactions differ dramatically among these process types. Other key process variables include operating pressure and reactant composition (i.e., stream/oxygen/coal ratio). High-pressure operation is common in all types of gasifiers and is particularly attractive for combined cycle operation when generating electrical power.

Comparison with Combustion. Gasification of coal can be compared to fuel-rich combustion. Methods of modeling coal gasification processes are thus essentially identical to those for coal combustion.[4] Many similar aspects exist for gasification and combustion, including use of the same types of processes, coal preparation and grinding, and use of varieties of coal. Table 21 summarizes some of the major differences between the direct combustion of coal and its gasification. Even with these differences, methods of computing properties of gasification and combustion processes are very similar.

TABLE 21. Differences in Direct Coal Combustion and Coal Gasification

	Direct Coal Combustion	Coal Gasification
Operating temperature	Lower	Higher
Operating pressure	Usually atmospheric	Often high pressure
Ash condition	Often dry	Often slagging
Feed gases	Air	Steam, oxygen
Product gases	CO_2, H_2O	CO, H_2, CH_4, CO_2, H_2O
Gas cleanup	Postscrubbing	Intermediate scrubbing
Polutants	SO_2, NO_x	H_2S, HCN, NH_3
Char reaction rate	Fast (with O_2)	Slow (with CO_2, H_2O)
Oxidizer	In excess	Deficient
Tar production	None	Sometimes
Purpose	High-temperature gas	Fuel-rich gas

Physical and Chemical Gasification Mechanisms. Although essentially the same physical and chemical processes occur during gasification and direct combustion, these basic processes interact in different ways with different results. Significant differences among these basic reactions also occur in fixed, fluidized, and entrained gasifiers. The following physical and chemical processes control both combustion and gasification of coal: (1) turbulent mixing of reactants (except fixed beds), (2) turbulent dispersion of particles (except fixed beds), (3) convective coal particle heat-up; (4) radiative coal particle heat-up, (5) coal devolatilization, (6) gaseous reaction of volatiles products, (7) heterogeneous reaction of char, and (8) formation of ash/slag residue. Critical differences in gasification, compared to combustion include the following, particularly in oxygen-blown fluidized and entrained beds:

1. Peak gasification temperatures are often higher in gasification due to absence of the nitrogen diluent. Thus the extent to which the coal devolatilizes, which is a strong function of peak temperature (see Section 3), can be greater during gasification, reaching as high as 80% of the total daf coal.[172]

2. These volatiles products quickly mix and react with available oxygen in the gas phase, completely depleting oxygen and producing the very high temperatures that cause the increased extent of devolatilization. High concentrations of CO_2 and increased concentrations of H_2O are produced through these gaseous reactions.

3. The residual char is reacted relatively slowly through heterogeneous attack by the CO_2 and steam. This process is faster at higher temperature but still slow compared to oxygen-char reaction that occurs in direct combustion.

This series of processes is quite different from direct combustion where oxygen is usually present in excess and dominates the consumption of residual char. In fixed bed gasifiers, a flow of coal is countercurrent to that of oxygen. Thus oxygen first encounters and reacts directly with the residual char, producing hot gases that devolatilize the coal in upper gasifier regions. Thus, unlike fluidized and entrained beds, oxygen plays a key role in the last stages of char consumption in the fixed bed gasifier systems. Devolatilization rates during gasification are treated as outlined in Section 3, with the same treatment used for devolatilization during combustion. Differences during gasification may be due principally to effects of pressure and the surrounding atmosphere of gases. Howard[173] treats the effects of pressure and hydrogen on coal devolatilization.

Rates of char reaction differ greatly at a given temperature for different reactants, generally in the order

$$O_2 \gg H_2O > CO_2 \gg H_2 \qquad\qquad (45)$$

Smoot and Pratt,[4] Essenhigh,[71] and Johnson[166] give recent reviews of rate data for char reaction with these reactants. Section 4 presented recent data for O_2 and CO_2 reaction of chars for four U.S. coals.

Reaction rates for those chars in CO_2 (Figure 15) vary by three orders of magnitude among the chars and are six to seven orders of magnitude below O_2 values for the practical temperature range of 1250–1750 K. Thus consumption of char by CO_2 at temperatures where kinetic rate controls will be dramatically slower than in oxygen and will vary substantially among coals. At higher temperatures, where diffusion of oxygen to the particle surface dominates, effects of coal type and of oxidizer type will diminish.

It is presumed in the theoretical description of coal conversion processes that if appropriate laboratory rates for coal devolatilization and char reaction are available, differences in the nature of combustion and gasification will be adequately described by the same theory.

Entrained Gasifier Data. Data for gasifier design are summarized by Johnson[166] and by Bissett[167] for entrained gasifiers. It is beyond the scope of this chapter to present or discuss these data in any detail. Gasification data are of two basic types:

1. Correlation of laboratory data on characteristics of various coals and chars such a reactivity or surface area. Data of this type were illustrated in Sections 3 and 4 (e.g., Figure 5 for coal weight loss during devolatilization and Figures 11 and 12, for char reaction rates in O_2 and CO_2).

2. Correlations of gasifier effluent data for such properties as carbon conversion, pollutant concentration, product gas composition, and temperature. These correlations are shown as functions of key gasifier variables such as reactant gas composition, coal type, pressure, or residence time. Figure 27 (shown later) shows data of this type where carbon conversion is shown as a function of steam-oxygen reactant concentrations from various investigators for entrained gasification. This figure also shows that carbon conversion during entrained gasification varies dramatically with oxygen/coal ratio.

Recently, new data on entrained coal gasification data from a laboratory gasifier (about 30 kg/hr) have been reported.[174-177] Measurements were made by analysis of char-gas samples removed from within the gasifier. Test variables that have been investigated include oxygen/coal ratio (0.5–1.1), steam/coal ratio (0–0.50), flame type—achieved by feeding the oxygen with the coal (premixed) in the primary stream, or separately from the coal in the secondary stream (diffusion flame)—pressure (1–17 atm), coal type (bituminous, subbituminous, lignite), secondary stream swirl number, coal particle size (37–68 μm), coal feed rate (24–55 kg/hr), and reactor length

TABLE 22. Effect of Variables on Carbon Conversion during Entrained Coal Gasification

Increase in	Effect on Carbon Conversion	References
Oxygen/coal ratio	Large increase	169, 174–176
Steam/coal ratio	Small decrease	169, 174, 175
Coal particle size	Large decrease	174–176
Coal type/rank	Decrease	174
Coal feed rate	Increase	174, 175
Primary oxygen percentage	Large increase	176
Secondary stream temperature	Little effect	167
Secondary stream swirl number	Little effect	177
Reactor static pressure	Increase	176
Reactor length	Increase	174
Reactor insulation	Increase	174

(1.7–2 m). Table 22 summarizes observed effects of key variables on carbon conversion during entrained gasification of coal. Variables that showed the greatest impact on carbon conversion were oxygen/coal ratio, particle diameter, and flame type.

Figure 25 shows that a 35% increase in particle diameter caused about a 15–20% increase in carbon conversion, whereas premixing of feed oxygen with coal caused an even greater increase in carbon conversion, compared to the unmixed case (diffusion flame) at atmospheric pressure as shown in Figure

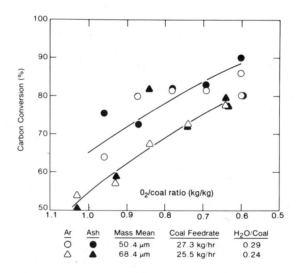

FIGURE 25. Influence of coal particle diameter on carbon conversion in a laboratory-scale entrained coal gasifier.[175]

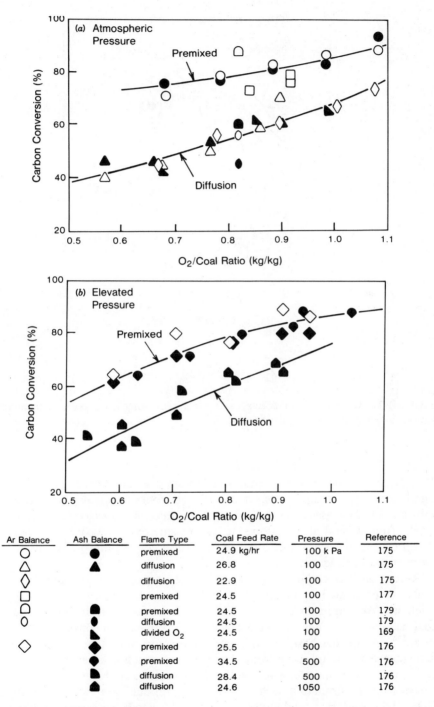

Ar Balance	Ash Balance	Flame Type	Coal Feed Rate	Pressure	Reference
○	●	premixed	24.9 kg/hr	100 k Pa	175
△	▲	diffusion	26.8	100	175
◇		diffusion	22.9	100	175
☐		premixed	24.5	100	177
⍟	◖	premixed	24.5	100	179
○	•	diffusion	24.5	100	179
	◣	divided O_2	24.5	100	169
◇	◆	premixed	25.5	500	176
	●	premixed	34.5	500	176
	◣	diffusion	28.4	500	176
	◤	diffusion	24.6	1050	176

FIGURE 26. Effect of oxygen/coal ratio and flame type on carbon conversion for Utah bituminous coal; H_2O/coal = 0.03; (*a*) atmospheric pressure; (*b*) elevated pressure.

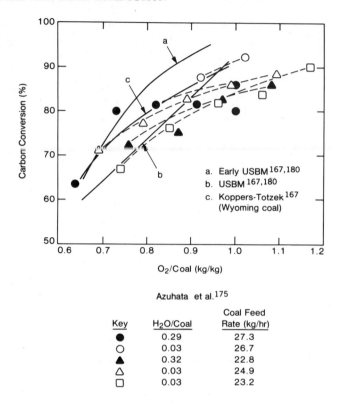

FIGURE 27. Comparison of effects of O_2/coal ratio on carbon conversion among various investigators (atmospheric pressure).[175]

26a. Effects of premixing of coal and oxygen are diminished but still very significant at elevated pressure, where residence times for similar feed rates are substantially increased. Figure 27 shows the influence of oxygen/coal ratio for results from several investigators. Here increasing oxygen/coal ratio from 0.6 to 1.0 causes at least a 30% increase in carbon conversion, but at the lower product gas heating value.

Recent gasifier carbon conversion data at atmospheric pressure for four coals are correlated in Figure 28. Results show that the oxygen/coal ratio (including oxygen in the coal), particle size, coal heat release rate (feed rate x coal heating value), and coal char reactivity govern carbon conversion. Brown et al.[174] discuss the basis for this correlation and note that char reactivity dependence seems to occur from char chemical reaction rate resistance that arises at lower temperatures in the aft of the reactor. With carbon conversion values from a correlation such as Figure 28, product gas composition (CO, CO_2, H_2O, and H_2), and other gasifier performance parameters (cold gas

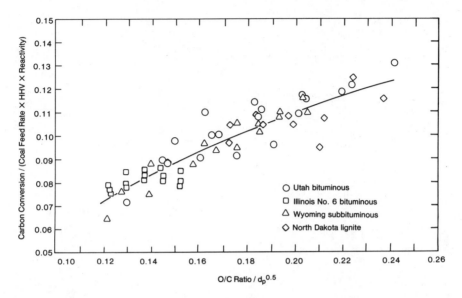

FIGURE 28. Correlation of laboratory entrained gasifier carbon conversion data for four coals.[174]

efficiency and temperature) can be estimated from thermal equilibrium considerations.

A third type of data is also vital to the development and evaluation of predictive methods for coal gasification. Here various properties, such as temperature, residue composition, gas composition, and velocity, are measured throughout a gasifier, and from these data, maps or profiles are constructed. These locally measured data provide unique insight into the gasification process. They can also be used for comparison with predictions from gasification theories. The extent of such gasification data is very limited. Skinner et al.[178] and, more recently, Soelberg et al.,[177] Highsmith et al.,[178] and Brown et al.[174] reported profiles of gas species concentration for several gases and gas mixture fraction from within a laboratory-scale entrained gasifier. Figure 29 shows data for CO_2, CO, and H_2.[178] The data show the very rapid ignition of pulverized coal in an intense reaction region near the inlet of reactants and near the reactor centerline. Up to 80% of the coal has been consumed in this region through devolatilization (according to the theory) with predicted local temperatures as high as 3000 K. Oxygen concentration declines rapidly to zero in this zone. A recirculation zone is shown in the forward region off-centerline. Downstream of the intense reaction zone, residual char is slowly consumed, through (according to the theory) reaction with CO_2 and H_2O, which were present in significant concentrations. From these data, and comparison with model computations, a more complete description of the gasification process is derived.

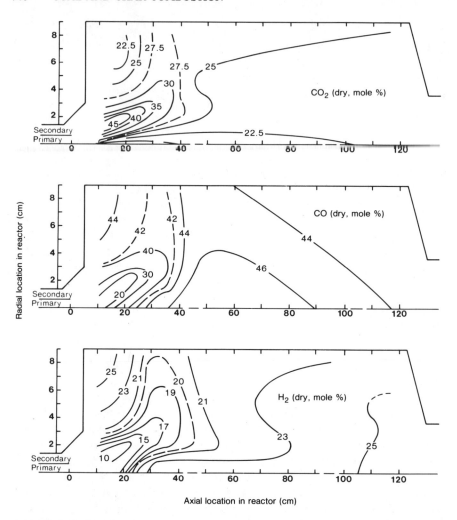

FIGURE 29. Measured species profiles for CO_2, CO, and H_2 from a laboratory-scale entrained gasifier. (Utah bituminous coal, feed rate = 25 kg/hr; mean diameter = 41 μm; oxygen/coal = 0.91; steam/coal = 0.27.)[177]

6. MODELING OF COAL PROCESSES

6.1. Background and Scope

In the past, comprehensive modeling of coal processes has been restricted by lack of computer speed and capacity and by technical difficulties in describing essential model elements. Field et al.[69] discussed many of the elements of modeling of pulverized coal processes but emphasized only the more elementary well-stirred and plug-flow models of coal processes. These authors noted that no multidimensional models existed at that time for pulverized coal systems. Smoot and Pratt[4] edited a treatment of pulverized coal models where multidimensional models were considered. More recently, Smoot[23] and Smoot and Smith[5] have reviewed the state of development of models for pulverized and other coal flames, both for combustion and gasification.

Until recently, modeling of complex coal furnaces emphasized radiative heat transfer,[181-183] with limited attempts to describe reaction and flow processes. A more advanced treatment of radiative heat transfer in large-scale coal furnaces has recently been published by Richter and Heap.[184] In a review of radiative transfer in combustion chambers, Sarofim and Hottel[19] note that ability to estimate radiative effects in furnaces is limited primarily by inadequate information on chemical and physical processes governing the concentration and size of particulates. These latter quantities are closely related to the complex flow and relation processes that take place in coal reactors. Figure 30

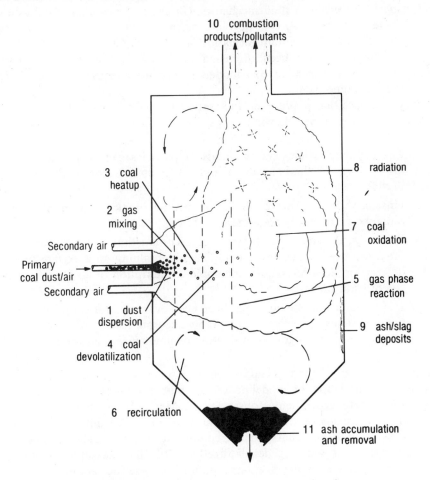

FIGURE 30. Aspects of pulverized coal flames.[5]

illustrates several aspects of a complex pulverized coal flame in industrial or utility coal boilers. Complete modeling of coal flames must account for all of these aspects. However, for a specific application, not all of these aspects will necessarily be important and may not need to be considered to develop an adequate description. The foundations and elements of modeling are presented in this section and illustrated through one- and two-dimensional models. Key references to comprehensive modeling are also noted.

6.2. Basic Model's Elements and Premises

In this section a "model" is taken to be a comprehensive, computerized code that combines several model components and can be applied to the description of complex processes. The key components used in comprehensive, coal reaction models for application to combustors and gasifiers often include the following:[5] (1) turbulent fluid mechanics, (2) gaseous, turbulent combustion, (3) particle dispersion, (4) coal devolatilization, (5) heterogeneous char reaction, (6) radiation, (7) pollutant formation, and (8) ash/slag formation. Several of these elements have been discussed in earlier sections. It is a major challenge in development of coal models to provide adequate submodels for each of these components.

Key premises that provide the foundation for development of comprehensive models include the following:

1. The behavior of clusters or clouds of particles can be predicted from information on the behavior of individual particles or small groups of particles.
2. Unsteady or quasi-steady behavior can be predicted from basic data obtained from steady-state measurements.
3. Processes as complex as those in coal combustion and gasification are often governed by key controlling steps; obtaining an adequate process description does not require a complete description of all aspects of the combustion process.
4. Use of numerical methods that have been evaluated for simple computational systems can produce acceptably accurate results in more complex systems that defy rigorous evaluation of uniqueness and accuracy.

Since the validity of these premises cannot always be demonstrated directly, the entire foundation of modeling of complex processes relies heavily on comparison with experimental observations. Many studies have emphasized fundamental aspects of this problem, such as coal pore diffusion, radiative properties of coals and chars, coal structure and its relationship to reactions and particle changes during devolatilization. Still, the development of these coal process models requires a large number of specific assumptions. Frequently these assumptions are not strongly supported by experimental data.

Thus demonstration of validity and accuracy of model predictions is a continuous challenge. Code computations are often too extensive to permit separate and complete evaluation of every component. Comparison with gross data such as outlet temperature or composition provides little confidence. Comparison with mean values of space-resolved properties is a much stronger test.

A general, comprehensive code usually includes input properties, independent and dependent variables, and model parameters. These properties typically include the following:

INDEPENDENT VARIABLES

Physical coordinates (x, y, z) Time (t)

DEPENDENT VARIABLES

Gas species composition	Particle temperature
Gas temperature	Particle velocity
Gas velocity	Turbulent energy
Pressure	of dissipation
Mean turbulent kinetic energy	Particle size distribution
Mixture fraction (mean and	Elemental composition
variance)	Extent of reaction
Bulk density	Radiant heat flux

INPUT DATA FOR EACH INLET STREAM

Gas velocity	Fuel elemental composition
Gas composition	Particle temperature(s)
Gas temperature	Particle size distribution
Gas turbulent intensity	Particle velocity(s)
Gas mass flow rate	Particle mass flow rate(s)
Pressure	Particle bulk density(s)

REACTOR PARAMETERS

Specific configuration	Dimensions
Inlet configurations	Wall materials
Inlet locations	Wall thickness

In addition key physical parameters arising in the model subcomponents must be specified. Parameters required will vary, depending on model assumptions, and are obtained principally from basic laboratory measurements.

Comprehensive models of this type are time-consuming to develop, require significant computer time to operate, and their reliability is difficult to determine. It is therefore appropriate to consider the justification for development of such codes. Reasons frequently cited are to (1) identify general reactor features (2) interpret measurements, (3) identify important test variables, (4) identify rate-controlling processes, (5) identify areas requiring additional investigation, (6) assist in scaleup, (7) assist in design and optimization of reactors, and (8) provide a basis for system control.

Given the high cost of large-scale process development, reliable predictive methods are of significant value. However, the question of model reliability is central to model utility. Are reasonable descriptions of the various processes possible? Are the models financially practical to exercise? Can reliability be demonstrated? Most of these issues are not yet resolved; yet the potential is such that development of these codes is receiving significant emphasis.

Models of coal processes follow these four basic classifications, as noted in Section 5:

Classification	Examples
1. Flow type	Well-stirred
	Plug-flow
	Recirculating
2. Mathematical complexity	One-dimensional
	Multidimensional
	Transient
3. Process type	Fixed bed
	Moving bed
	Fluidized bed
	Entrained bed
4. Flame type	Premixed
	Diffusion

In the text that follows, modeling of coal processes is illustrated through discussion of the elements of a one-dimensional, plug-flow model and briefly with a generalized multidimensional model. These models apply to pulverized coal flames and to a variety of other nonreacting and reacting gaseous and particle-laden flows. A review of models for entrained, fixed, and fluidized beds is also prevented.

6.3. Plug-Flow Model

In order to illustrate typical features of coal process models, a one-dimensional model of pulverized coal combustion and gasification (1-DICOG) is outlined. This code is particularly applicable to suspended or entrained flow systems, especially those that are essentially one dimensional (e.g., long and narrow). However, plug-flow models can also be applied to fixed beds, moving beds, and reverse-flow reactors. This model contains a rigorous treatment of coal particle reactions but avoids the complexities of multidimensional fluid motion. Such codes are relatively inexpensive to operate and sufficiently general to permit application to a wide range of reaction proceses. This particular model has been documented in detail in the literature.[5, 64, 185] Several other one-dimensional codes have also been reported (see Table 25, shown later).

Model Assumptions. 1-DICOG uses the integrated or macroscopic form of the general conservation equations[4] for a volume element inside the gasifier or combustor, as illustrated in Figure 31. The following aspects of pulverized coal combustion and gasification have been included in the model: (1) mixing of primary and secondary streams (specified as input), (2) recirculation of reacted products (supplied as input), (3) devolatilization and swelling of the coal, (4) reaction of the char by oxygen, steam, carbon dioxide, or hydrogen, (5) conductive and convective heat transfer between the coal or char particles and the gases, (6) convective losses to the reactor wall, (7) particle–particle radiation and particle–wall radiation, with transparent gases, (8) variations in composition of inlet gases and solids, (9) variation in coal or char particle sizes, (10) oxidation of the coal devolatilization products, and (11) inclusion of multiple sizes or types of coal particless, each with their own individual properties, composition, and reaction rates.

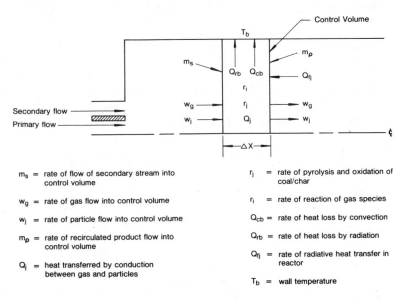

m_s = rate of flow of secondary stream into control volume

w_g = rate of gas flow into control volume

w_j = rate of particle flow into control volume

m_ρ = rate of recirculated product flow into control volume

Q_j = heat transferred by conduction between gas and particles

r_j = rate of pyrolysis and oxidation of coal/char

r_i = rate of reaction of gas species

Q_{cb} = rate of heat loss by convection

Q_{rb} = rate of heat loss by radiation

Q_{fj} = rate of radiative heat transfer in reactor

T_b = wall temperature

FIGURE 31. Schematic diagram of reactor volume element for 1-DICOG.[5]

The following are considered to be the major limitations of 1-DICOG: (1) The model does not predict local fluctuating properties within the pulverized coal reactor, and predicted mean properties are only a function of axial position. (2) The model does not predict rates of jet mixing or recirculation; rather, these values are required input. (3) The detailed behavior of coal reactions is not yet well understood, which leads to uncertainty in the kinetic description and parameters for the pulverized coal systems. (4) Some details of the pulverized coal gasification and combustion processes have been neglected in this version to reduce model complexity and computation time. These include micromixing processes, gas-phase, rate-limiting reactions and gas-phase radiation.

1-DICOG has been developed for an arbitrary number of chemical elements (K in total number). Equations are derived for elemental balances as opposed to species balances, since the gas phase is assumed to be in chemical equilibrium locally. This does not require that neighboring elements are in equilibrium with each other. Use of these elemental balances reduces the number of differential equations and simplifies the required link with the coal reaction model. Volatiles, for example, need not be specified by species composition, which is impossible to identify, but only by elemental composition, which can be defined from ultimate analyses of the coal and char. From only the element balances, the calculated energy level and pressure, the complete species composition, gas temperature, and other properties are computed.

The particle is assumed to be composed of specified amounts of raw coal, char, ash, and moisture. The dry, ash-free portion of the coal undergoes devolatilization to volatiles and char by one or more reactions (M in total number) of the form:

$$(\text{raw coal})_j \xrightarrow{k_{jm}} Y_{jm}(\text{volatiles})_{jm} + (1 - Y_{jm})(\text{char}) \qquad (46)$$

The volatiles react further in the gas phase. The char reacts heterogeneously after diffusion of the reactant (i.e., O_2, CO_2, H_2O, and H_2) to the particle surface by one or more reactions (L in total number) of the form:

$$\phi_l(\text{char}) + (\text{oxidizer})_l \xrightarrow{k_l} (\text{gaseous products})_l \qquad (47)$$

Differential Equations. The differential equations for 1-DICOG are first-order, nonlinear, highly coupled, ordinary differential equations, derived from the volume element of Figure 31. The mass rate of change of each chemical element, k, is

$$\frac{d(w_g \omega_k)}{dx} = A \sum_j r_{jk} + m_{sgk} + m_{pgk} \qquad (48)$$

One equation is solved for each of the K total elements considered. The sources or sinks of elemental mass addition or depletion for the gas phase are only three in number. Mass addition or depletion may take place through reaction with the particles (term 1, RHS). Other sources or sinks are due to secondary mixing (term 2) and recirculation (term 3). These last two terms must be specified by the user, since they represent multidimensional effects that might be important in the early regions of the reactor.

The mass rate of change of each particle phase is

$$\frac{dw_j}{dx} = -Ar_j + m_{sj} + m_{\rho j} \tag{49}$$

One of these equations is solved for each particle phase classification. The latter two terms of this equation are not included in 1-DICOG. The only source or sink is the overall particle reaction. The gas-phase energy balance is

$$\frac{d(w_g h_g)}{dx} = h_{sg} m_{sg} + h_{\rho g} m_{\rho g} + A\left[\sum_j Q_j - Q_{cb} + \sum_j (r_j h_{jg} + \xi r_j \Delta h_j)\right] \tag{50}$$

The first two terms on the right-hand side depict the energy exchange due to secondary mixing and recirculation, respectively. The final terms enclosed between parentheses represent all other gas-phase heat transfer mechanisms. Q_j accounts for the conductive and convective heat transfer from the jth particle type to the gas. Q_{cb} is the convective heat transfer from the gas to the boundary or reactor walls. The last term is the energy carried to the gas by the mass addition from the particle phase. The summation over each of the particle terms accounts for all particle classifications.

The particle phase energy equation is

$$\frac{d(w_j h_j)}{dx} = m_{sj} h_{sj} + m_j h_j + A(Q_{fj} + Q_{rbj} - Q_j - r_j h_j - \xi r_j \Delta h_j) \tag{51}$$

Energy exchange due to secondary mixing or recirculation of particles is accounted for in the first two terms. The particle energy equation is similar to the gas-phase energy equation, with the exception of two radiation terms and no convective wall term. The radiation terms include particle phase radiation, but with the assumption of no gas-phase radiation. Q_{fj} accounts for particle–particle radiation within the reactor. Q_{rbj} is the radiative heat transfer rate from the jth particle cloud to the reactor boundaries. Both Eqs. (50) and (51) include the term, $\xi r_j \Delta h_j$. This term allows for the heat of reaction for the heterogeneous reactions to be partitioned between the solid and the gas phases. Since these reactions take place at the boundary between the two phases, it is not immediately apparent which phase should be credited with the energy generated by exothermic reactions or consumed by endothermic reactions. In the approach incorporated in 1-DICOG, this partitioning is specified by the user by setting the value of ξ. If $\xi = 0$, all of the energy is given to the gas phase; if $\xi = 1.0$, all of the energy is given to the particulate phase. Any value between 0 and 1.0 can be specified. The rate of change of the gas-phase mass flow rate is

$$\frac{d(w_g)}{dx} = A \sum_j r_j + m_{sg} + m_{\rho g} \tag{52}$$

This equation results from the sum of the gas element balances (Eq. (46)) over all of the elements. From the coal particle model, the change of each of the constituents of each of the particle classifications considered is:

$$\frac{d(\alpha_{cj})}{dx} = \frac{r_{cj}}{n_j v} \tag{53}$$

$$\frac{d(\alpha_{hj})}{dx} = \frac{r_{hj}}{n_j v} \tag{54}$$

$$\frac{d(\alpha_{wj})}{dx} = \frac{r_{wj}}{n_j v} \tag{55}$$

Initially, the reacting particles are composed of specified amounts of virgin coal (α_{cjo}), char (α_{hjo}), moisture (α_{wjo}), and ash (α_{ajo}). Ash is defined as that portion of the initial particle that is inert and thus constant. This set of differential equations is sufficient to describe the changes occurring in the volume element.

Auxiliary Equations. In addition to the required differential conservation equations, a large number of algebraic, auxiliary equations is required. It is beyond the scope of this section to present these equations in detail.[185] However, they occur in the following categories:

1. *Enthalpy-Temperature Relations* Nine equations are used to express the enthalpies of the coal particle components, in terms of heat capacity and temperature, for each particle type.
2. *Physical Properties* $5i + 3j + 4$ equations are used to compute the physical properties C_p, k, μ, D_{im} as functions of composition and temperature, where i = number of gaseous chemical species in the calculation and j = number of particle types.
3. *Radiative Heat Transfer* A radiatively transparent gas phase and radiatively gray particles are assumed with negligible particle scattering. The only radiative energy transfer is thus among the particles within the reactor (Q_{fj}) and between the particle cloud and the reactor walls (Q_{rbj}).

Three different options have been coded for the radiation model (zone model, diffusion model, and flux model). Each model requires different numbers of equations. Several equations are used to describe the dependence of these terms on temperature, dust concentration and size, and radiative properties.

4. *Convective Heat Transfer* Two equations were formulated to describe convective heat transfer rates between each particle type or size and the gas and between the fluid and the walls. Reduction of convective heat transfer by gas evolution from particles is included.

5. *Particle Reactions* The devolatilization rate was treated according to Eqs. (2) and (4) with parameters shown in Table 23.[186] Char reaction with O_2, CO_2, or H_2O was treated according to Eq. (27) with parameters also shown in Table 23.[187-189] In addition the coal particle diameter was assumed to vary with the extent of devolatilization, with the swelling coefficient, γ, specified

$$d_j = d_{jo}\left[1 + \frac{\gamma(\alpha_{cjo} - \alpha_{cj})}{\alpha_{cjo}}\right] \qquad (56)$$

After devolatilization, the particle could be envisioned as a porous sphere where reaction takes place within the pores. As char combustion proceeds, the particle diameter can be taken to remain relatively constant, but with variable density until breakup or complete burnout of the fuel. Alternatively, the particle may be perceived to react mostly on the surface and thus burn out at a near constant density but as a shrinking particle. These two options were treated in Section 4 (see Eqs. (30)–(31)). In this latter case

$$d_j = \left(\frac{6\alpha_j}{\pi\rho_{aj}}\right)^{1/3} \qquad (57)$$

where ρ_{aj} is the solid density of the jth particle type after complete devolatilization. Both options have been coded in 1-DICOG.

Model Predictions and Comparisons. Computations show that the rate of initial particle heat-up is an important step in the overall reaction process. The coal or char particles receive or lose energy by radiation from other particles and from the vessel walls; they also exchange energy by conduction and convection to the gases that surround them. The rate of energy exchange with the incoming particles determines where in the reactor the particle ignition will occur.

Selected computations to investigate devolatilization were also conducted. As soon as devolatilization begins, the process proceeds rapidly to completion.

TABLE 23. Selected Kinetic Parameters for Particle Model in 1-DICOG

Process	Reaction	Rate Constant	Parameters	Reference
Devolatil-ization	Raw coal $\xrightarrow{k_1}$ Y_1(volatiles)$_1$ + (1 − Y_1)(char) simultaneously with reaction 2	$k_1 = A_1 \exp(-E_1/RT_j)$	$Y_1 = 0.39$ $A_1 = 3.7 \times 10^5$ sec^{-1} $E_1 = 17.6$ kcal mol^{-1}	Ubayaker et al.[186]
	Raw coal $\xrightarrow{k_2}$ Y_2(volatiles)$_2$ + (1 − Y_2)(char) simultaneously with reaction 1	$k_2 = A_2 \exp(-E_2/RT_j)$	$Y_2 = 0.8$ $A_2 = 1.46 \times 10^{13}$ sec^{-1} $E_2 = 60.0$ kcal mol^{-1}	
Char oxidation with O_2	$2C + O_2 \xrightarrow{k} 2CO$	$k = T_j(A + BT_j)$	$A = -1.68 \times 10^{-2}$ msec^{-1} K^{-1} $B = 1.32 \times 10^{-5}$ msec^{-1} K^{-2}	Field[a,187]
Char oxidation with CO_2	$C = CO_2 \xrightarrow{k} 2CO$	$k = AT^n \exp(-E/RT_j)$	$n = 1.0$ $A = 4.40$ msec^{-1} K^{-1} $E = 38.7$ kcal mol^{-1}	Mayers[a,188]
Char oxidation with H_2O	$C + H_2O \xrightarrow{k} CO + H_2$	$k = AT^n \exp(-E/RT_j)$	$n = 1.0$ $A = 1.33$ msec^{-1} K^{-1} $E = 35.1$ kcal mol^{-1}	Mayers[189]

[a]More recent correlations for data of Goetz et al.[111] for four U.S. coals should be considered for char reaction with O_2 and CO_2.

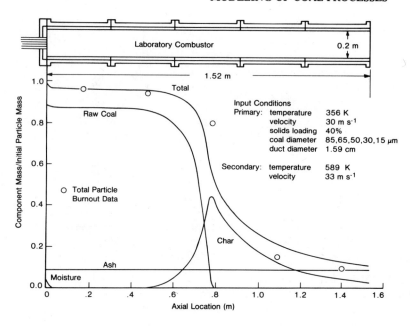

FIGURE 32. Predictions and measurements of coal particle burnout in laboratory combustor.[5]

The devolatilization rate is affected only slightly by particle size, whereas the other particle rate processes are strongly influenced by particle diameter. After devolatilization is initiated, the gaseous products react in the gas phase, and the gas temperature rises rapidly. The particle temperature subsequently rises by transfer of heat from the hot gaseous surroundings. Parametric predictions were made by selectively altering the diffusion rate to determine whether the surface reaction rate or the oxygen diffusion rate was the controlling mechanism during char burnout. Pore diffusion and reaction are accounted for in this formulation only indirectly through the magnitude of the experimental rate constants, which are based on external spherical surface areas for the coal char. It was found for small pulverized coal particles that surface reaction was rate controlling over oxidizer diffusion.

Thurgood et al.[190] have made local measurements of both gas and solid phases in a laboratory combustor. The reaction was a 20 cm I.D. axisymmetric refractory-lined combustor, 1.5 m in length. Figure 32 includes a reactor schematic, typical test conditions, and model predictions compared with laboratory combustor measurements of coal particle burnout as a function of reactor length. These calculations and measurements were performed for a high-volatile B-bituminous coal. Five size classifications were used to approximate the measured distribution and the various particle sizes are consumed at different rates. Figure 33 shows the measured and predicted gas-phase mole fraction history for the same case. One reason for the good agreement between laboratory measurements and predictions is the one-dimensional nature of the

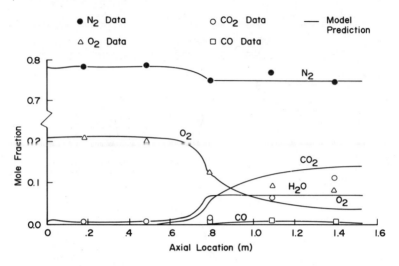

FIGURE 33. Predictions and measurements of gas mole fractions in a laboratory combustor.[5]

laboratory combustor. Mixing of primary and secondary gases is rapid, and particle ignition occurs later in a fully mixed, gas environment; particle combustion is thus not significantly affected by the gas mixing and recirculation processes in the upper regions of the reactor. During the first 70 cm the particles are heated by radiation. Vaporization of moisture from the coal is very rapid, and devolatilization begins at about the same time for all particles and is completed very rapidly. As the virgin coal devolatilizes, gas-phase products are evolved that are further reacted in bulk gas phase, and the temperature rises sharply. The devolatilization process also forms the residual char which rises to peak quantities at the point of complete devolatilization.

These few comparative results illustrate the value of the one-dimensional model in describing coal combustion processes. Other applications with this model have been reported recently by Suzuki et al.[191] for high-intensity combustion of several coals and by Kramer et al.[192] for combustion of several coal chars.

6.4. A Multidimensional Flame Model (PCGC-2)

Basis. Comprehensive, multidimensional modeling of turbulent combustion and gasification has been recognized as a difficult problem, due not only to the problems associated with solving the differential equation set but also to complexities in describing the interactions between chemical reactions and turbulence. Such a model is introduced briefly here. It is referred to as PCGC-2 (*P*ulverized *C*oal *G*asification or *C*ombustion-2 Dimensional). This description introduces the complexities of comprehensive modeling. PCGC-2 is an attempt to use currently available technology to combine knowledge of the turbulent fluid mechanics with a reasonable approach to the reaction processes. PCGC-2 is applicable to nonreactive, gaseous or particle-laden flows or to gaseous or pulverized, coal-fired combustion and to entrained flow gasification in axisymmetric, cylindrical coordinates. Variations in the predicted quantities are considered only in the axial and radial directions. Symmetryis assumed in the angular direction. The particular cylindrical reactor modeled is coaxial, with coal entering the reactor in the central (primary) stream and the majority of the oxidizer entering in the outer (secondary) annular stream. The pulverized coal particles range in size from about 1 μm to over 100 μm. The model predicts the mean gas field properties for axisymmetric, steady-state, turbulent diffusion flames (local velocity, temperature, density, species composition, etc.). Local particle properties are also computed, such as coal burnout, particle velocity, and temperature and coal component composition. Smoot and Smith[5] provide a detailed discussion of the model. At the time of publication of this book, Smith and co-workers had nearly completed work on a generalized highly efficient three-dimensional combustion code.

General Description of Subroutines

Gas Phase. The gas phase is assumed to be turbulent, reacting, continuum field that can be described locally by general conservation equations. The flow is assumed to be time steady. Gas properties (i.e., density, temperature and species composition) fluctuate randomly according to an assumed shape of a probability density function (pdf) that is characteristic of the turbulence. Turbulence is modeled by Favre averaging of gradient diffusion with the two-equation $k - \varepsilon$ model for closure. The effect of particles on the gas-phase turbulence is modeled with an empirical correlation. Gas-phase reactions are assumed to be limited by mixing rates, and not by kinetic limitations. Gaseous properties are calculated assuming local instantaneous equilibrium.

Radiation. The radiation field is a multicomponent, nonuniform, emitting, absorbing, scattering gas-particle system. The coal particles cause anisotropic and multiple scattering. The flame may be surrounded by nonuniform, emitting, reflecting, absorbing surfaces. An Eulerian framework is used to model the radiation, which facilitates incorporation of radiation properties into gas-phase equations and also specifies a radiation field for the Lagrangian particle calculations.

Particle Flow. Pulverized coal flames of interest have a void fraction of near unity and the individual particles are very dispersed. This results in few particle–particle collisions, and hence, the particle phase has not been considered a continuum like the gas phase. Different particles at the same location may exhibit completely different properties due to the different particle paths, giving rise to a particle history effect. A Lagrangian treatment of the particles is utilized, representing the particle field as a series of trajectories. Particle properties are obtained along these trajectories.

Particle Reaction. The coal reaction rates are assumed to be slow compared to the turbulence time scale. This allows the particle properties to be calculated from the mean gas properties instead of the fluctuating gas properties. PCGC-2 assumes that the off-gas from the coal is of constant species composition. Particles are defined to consist of coal, char, ash, and moisture. Ash is taken to be inert. Any volatile mineral matter is considered as part of the volatile matter of coal. Coal reaction rates are characterized by parallel reaction rates with fixed activation energies. Char particle swelling is accounted for empirically. The interior of the particle is assumed to have the same temperatures as that of the particle surface. Devolatilization often occurs more rapidly than the char combustion reactions. The efflux of off-gases from the coal reactions can affect diffusion to the particle for the heterogeneous char combustion reactions, heat conduction from gas to particle, and momentum exchange between gas and particle; these effects are included through corrections from stagnant film theory. The successful prediction of particle reactions is dependent upon the available coal reaction rate data.

Model Application and General Observations. Formulation, numerical solution, and evaluation of PCGC-2 has been completed over the past several years. Development has reached the point where the code has been applied. Table 24[193-198] provides a summary of sources of predictions and data comparisons with this comprehensive predictive method. Predictions have been compared with measurements from gaseous and particle-laden cold flows, gaseous and pulverized coal combustion, and pulverized coal gasification. Results and interpretations of these predictions are shown in at least the ten references noted in Table 24. It is well beyond the scope of this treatment to discuss all of these predictions or comparisons. Two comparisons of PCGC-2 predictions with measurements follow for illustrative purposes.

Figure 34 shows predictions and data for measurements from combustion of a pulverized Wyoming subbituminous coal in a laboratory combustor. Data are those of Figures 21 and 22. This case includes a swirl component in the secondary stream ($S = 2.0$). Results show comparisons for mixture fraction, CO_2, and O_2. Agreement for these coal combustion comparisons is not as good as for gaseous combustion, but they demonstrate that overall trends can be predicted in coal reaction processes.

Figure 35 shows PCGC-2 predictions of CO_2, CO, and H_2 concentrations for entrained gasification of Utah bituminous coal. Predictions are to be compared to corresponding data of Figure 29. The model predicted the same three zones of reaction and mixing as found experimentally. Good quantitative agreement was found in the respective trends, peaks, and concentration levels for the gas species. The model predicted that coal devolatilization was complete within the first one-third of the reactor and accounted for over 80% of the predicted carbon conversion. Gas mixing and heterogeneous reactions took place in the lower two-thirds of the reactor. Char–CO_2 and char–H_2O

TABLE 24. Summary of Predictions and Data Comparisons with
Comprehensive Model (PCGC-2)

Data Category	Description of Comparisons	References
Gaseous cold flow	Profile comparisons for nonswirling, mildly swirling, strongly swirling flows, velocity, pressure and turbulence properties; data from several investigators.	5, 159, 193–195
Particle-laden cold flow	Gaseous and particulate profiles for silicon and coal-dust containing fluids with and without swirling flows.	5, 159, 193, 195
Gaseous combustion	Profile comparisons for hydrogen- and methane-containing flames. Gas composition, temperature, and velocity (mean and RMS).	5, 64, 159, 193, 195
Coal combustion	Profile and variables comparisons for pulverized bituminous, subbituminous, and anthracite combustion. Carbon conversion, temperature and gas-composition, and NO_x comparisons. Non swirling and mildly swirling flows. Effects of swirl number, coal type, moisture percentage, and particle size.	5, 159, 193, 196
Coal gasification	Profile and variable comparisons for entrained coal gasification for coals, oxygen/steam/coal stoichiometric ratio variation, coal feed rate, coal size.	177, 197
Coal-water mixture combustion	Limited predictions for a coal-water mixture fired gas turbine combustion. No data comparisons.	198

reactions occurred but still accounted for only a small part of the total predicted outlet carbon conversion.

The average particle heating rate was $\sim 4 \times 10^5$ K sec^{-1}, requiring ~ 6 msec to reach a temperature of 2300 K. The estimated time for complete devolatilization was ~ 11 msec. The qualitative agreement between observations and calculations in the figures suggests that these reaction times are approximately correct.

Mathematical modeling of pulverized-coal systems, including coal combustion and/or gasification and combustion of coal-water mixtures, requires description of many physical and chemical processes that are not fully understood. It is principally this lack of information about fundamental processes that necessitates association of a mathematical modeling program with an experimental program. Current predictive techniques can provide a quantitative description of local behavior in the near burner-field, providing a valuable interpretive tool for experimental data. Multidimensional computer models can also provide insight into the controlling processes involved. The

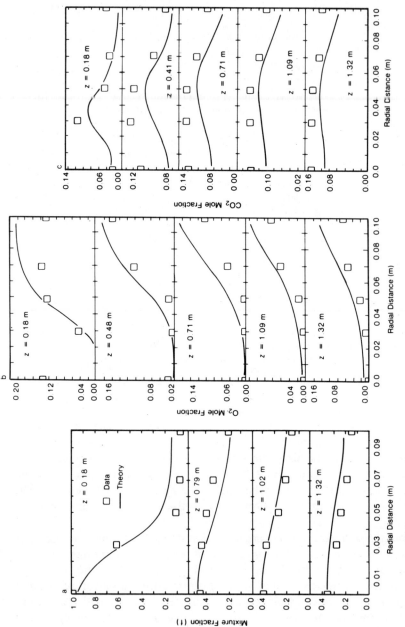

FIGURE 34. Comparisons of predicted and measured radial profiles of (*a*) mixture fraction, (*b*) O_2 concentration, (*c*) CO_2 concentration for combustion of pulverized Wyoming subbituminous coal in a laboratory combustor. SR = 1.06, swirl number = 2.0, 25% moisture in coal, coal feed rate = 10.2 kg/hr. Data profiles were shown in Figures 21 and 22.

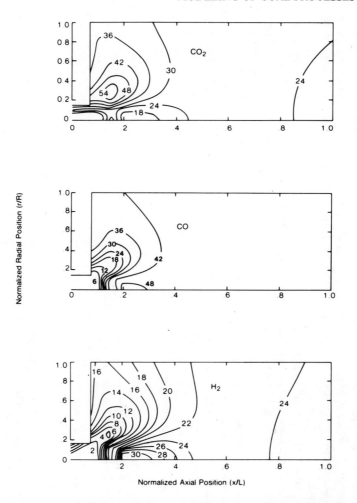

FIGURE 35. Predictions of (*a*) CO_2, (*b*) CO, and (*c*) H_2 concentrations (dry, mol %) for entrained gasification of Utah bituminous coal in a laboratory gasifier. O_2/coal mass ratio = 0.91 steam/coal mass ratio = 0.27, coal feed rate = 25 kg/hr, mass mean coal diameter = 41 μm (40% heat loss assumed in computations, compared to 25–30% heat loss subsequently estimated for data). Predictions to be compared to measurements in Figure 29.

use of these models requires not only sophisticated computer equipment but also experienced users. Currently, these models cannot be used as a "black box" calculation of coal reaction chambers. Simplified input and colored graphics output display improves the usability of these computer programs, but to interpret the output requires some knowledge of the assumptions made as well as an understanding of the numerical algorithms used. In many applications of coal combustion and gasification, the transient start-up of the

TABLE 25. Review of Models for Pulverized Coal Combustors / Gasifiers in Past Ten Years[23]

Author(s)	Year	Description	Major Results
Gibson and Morgan[199]	1970	Two-dimensional, axisymmetric, elliptic, stream function–vorticity formulation; local gas equilibrium; particle diameter change with time based on measurement.	With simplistic radiation and coal combustion model, the predicted wall heat flux, temperature, and coal burnout agree reasonably well with the measured values.
Mehta and Aris[200]	1971	Phenomenological model combination of PFR and PSR calculations, gas-phase chemical equilibrium, overall char gasification with Arrhenius-type rates.	Describes the outlet conditions for a three-stage entrained flow gasifier.
Richter and Quack[201]	1974	Two-dimensional, axisymmetric, elliptic, stream function–vorticity, two-equation turbulence model, four flux radiation	Technique similar to Ref. 199 with improved radiation and turbulence submodels. Heterogeneous reaction rates fitted to experimental furnace data.
Blake et al.[202, 203]	1977, 1979	Three-dimensional, transient, mixed finite-element, finite-difference scheme with separate Eulerian particle equations; two-equation gaseous turbulence; separate turbulent kinetic energy for particles.	Under development; illustrative predictions shown for gasification.
Lewellen et al.[204]	1977	Phenomenological model, swirling flow reactor divided into four regions of interest; dominant physical processes identified for each region.	Predicts effects of four performance parameters on certain design variables.
Sprouse[205]	1977	One-dimensional hydrogasification; free-stream equilibrium chemistry; one-step devolatilization and hydrogasification model.	Extensive study of the boundary layer around the particles; compares well with outlet measures coal gasifier tests.
Ubhayakar, et al.[206]	1977	One-dimensional steady-state; two-step devolatilization; combined diffusion and surface reaction rates; no addition; equilibrium gas-phase reactions.	Applied to gasification combustion and hydro-pyrolysis with good agreement.
Finson et al.[207]	1978	Steady, one-dimensional model, with gaseous kinetics; kinetics from data for coal pyrolysis; heterogeneous oxidation; considers char structure and reactivity.	Developed code; predictions compared with laboratory gasification measurements of axial temperature and gas composition.

758

Wen and Chuang[208]	1978.	Single-step pyrolysis, global rate for gaseous CO combustion or gaseous equilibrium; data based rate expressions for CO_2, O_2, and H_2O reactions with char.	Predictions for Texaco entrained gasifier compared with limited pilot plant temperature and composition data.
Barnhart et al.[209]	1979	Combination of PFR and PSR elements; gas phase in equilibrium with two-equation kinetics for pyrolysis, char oxidation, and CO oxidation.	Developed code; predicts carbon burnout, product gas composition, temperature; reported fair to good comparison between experiment and theory for cyclone gasifier.
Smith and Smoot,[185] Smoot[23]	1979, 1980	One-dimensional steady-state, two-step devolatilization, Combined diffusion and surface reaction rates, one-dimensional zonal radiation, equilibrium gas-phase reactions	Good agreement with one-dimensional combustor data; poor agreement with gasifier measurements.
Chan et al.[210]	1980	Two-dimensional, time-dependent, axisymmetric code, applied to high-pressure gasification, Lagrangian particle motion, global gas kinetics, devolatilization and char oxidation in coal reaction model; first-order (algebraic) turbulence closure.	Treats multistage process, no comparisons with data; potential for three dimension.
Lockwood et al.[211]	1980	Two-dimensional, axisymmetric, with gas recirculation, $k - \varepsilon$ turbulence model; single-step pyrolysis, mixed-controlled volatiles reaction, four-flux radiation with scattering, Lagrangian particle motion.	Computations for pulverized coal furnace with concentric burner.
Smith et al.[212]	1981	Two-dimensional, axisymmetric, with recirculation, $k - \varepsilon$ turbulence model, Lagrangian particles two-step devolatilization, diffusion, and kinetic heterogeneous rates, four-flux radiation with scattering.	Gas-phase components reported and compared with data; coal code recently completed, example computations.
Oberjohn et al.[213]	1982	Two-dimensional, steady, axisymmetric code, applied to pulverized coal combustion. Eulerian gas, Lagrangian particle motion	Applications to industrial systems.

759

TABLE 26. Submodel Summary for Multidimensional Pulverized Coal Reactor Models[a]

Author(s)	Date	Dimension	Turbulence	Gaseous Combustion
Gibson and Morgan[199]	1970	Two-dimensional steady-state with swirl	Algebraic equation for μ_{eff}	Burns instantaneously when fuel and oxidizer mix
Richter and Quack[201]	1974	Two-dimensional steady-state	Two-equation $k - \varepsilon$ model for μ_{eff}	Instantaneous reaction of $CO + O_2 = CO_2$
Blake et al.[202, 203]	1977, 1979	Three-dimensional transient	Two-equation $k - \varepsilon$ model for gas, plus a k equation for particles	Global kinetics with partial equilibrium
Smith et al.[212]	1981	Two-dimensional steady-state	Two-equation $k - \varepsilon$ model for μ_{eff}	Equilibrium with probability density function to account for "unmixedness"
Chan et al.[210]	1980	Two-dimensional transient	Algebraic expression for μ_{eff}	Global kinetics, seven stable species
Lockwood et al.[211]	1980	Two-dimensional steady-state	Two-equation $k - \varepsilon$ model for μ_{eff}	Reaction rate eddy time scle
Oberjohn et al.[213]	1982	Two-dimensional steady	$k - \varepsilon$ model	Various options

[a]Smoot and Pratt.[4]

reactor is of interest. However, transient computer programs require more computational time.

6.5. Other Coal Combustion Codes

Pulverized Coal Models. Attempts to calculate the detailed performance of turbulent, pulverized coal reaction processes have only been undertaken during the last decade. Modeling of turbulent reaction processes is still in a state of development. Pulverized coal conversion models developed since 1970 are reviewed in Table 25.[199-213] Three codes use the perfectly stirred reactor approach in combination with plug flow, five are one-dimensional codes (including that 1-DICOG), and six are multidimensional codes (including

TABLE 26   (*Continued*)

Devolatilization	Heterogeneous Reaction	Radia- tion	Particle Dispersion	Numerical Method
No devolatilization; volatiles admitted separately in gaseous form	Mass fraction of char solved by seprate transport equation for 5 size classes; diameter change with time based on measurement	Two-flux model (radial flux only)	Particles treated like a gas	Stream function– vorticity formulation
Considered only low-voaltile coals	Arrhenius-type reaction rate with constants obtained by fitting model to data	Four-flix model	Particles treated like a gas	Stream function– vorticity formulation
Single reaction with distribution of activation energies	Reaction with steam, CO_2, O_2, and H_2 based on Arrhenius or Langmuir expression and laboratory data	Not clarified	Separate Eulerian particle equations with different classifica- tions	Primitive variable, Eulerian for both phases; mixed finite element- finite difference
Multiple-step devolatilization model	Diffusion rate and kinetic rate included	Four-flux model, including scatter due to particles	Lagrangian particles with turbulent diffusion velocity	Primitive variable Eulerian gas- phase technique Lagrangian particle technique
Single activated step products specified	With steam, CO_2, and O_2; considers intrinsic rates	High-pressure system diffusion	Lagrangian particles, no effects of gas turbulence	Primitive variable finite difference Eulerian gas Lagrangian particle
Constant release rate	Diffusion and kinetic rates included	Grey; flux method with scattering	Lagrangian particles with with turbu- lent diffusion	Primitive variable Eulerian gas-phase technique Lagrangian particle technique
Computed rate; single product	Diffusion and kinetic rates included	Flux-discrete ordnance method	By mean drag only	Eulerian gas-phase Lagrangian particle phase

PCGC-2). Eight of the models were applied specifically to entrained gasifica- tion. Table 26 reviews multidimensional models with brief statements on some of the components included. Very recently, Smith and Gillis[234] have reported new work on a computationally efficient, generalized, three-dimensional com- bustion code, which is intended for application to full-scale boilers and gasifiers.

Fixed Bed Combustion Models.   In this type of process, coal is fed slowly to the top of the bed (or at one side onto a moving grate). Oxygen and other oxidants (e.g., CO_2 and H_2O) and inertants (e.g., N_2 in air) are fed into the bottom or up through the bed in this countercurrent process. Residence time of the gases is only seconds, whereas for the solids, it can be minutes or hours. The bed is

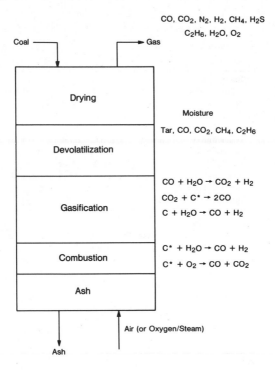

CO, CO_2, N_2, H_2, CH_4, H_2S

C_2H_6, H_2O, O_2

Coal ⟶   ⟶ Gas

Drying

Moisture

Devolatilization

Tar, CO, CO_2, CH_4, C_2H_6

Gasification

$CO + H_2O \rightarrow CO_2 + H_2$

$CO_2 + C^* \rightarrow 2CO$

$C + H_2O \rightarrow CO + H_2$

Combustion

$C^* + H_2O \rightarrow CO + H_2$

$C^* + O_2 \rightarrow CO + CO_2$

Ash

Air (or Oxygen/Steam)

Ash

FIGURE 36. Chemical reactions occurring in a fixed bed reactor.[216]

often divided into idealized zones, shown in Figure 36. The exiting hot gases dry, heat, and pyrolyze the coal. These gases then heterogeneously react with the coal char in the lower part of the bed. In direct combustion the process is dominated by oxygen–char reactions. However, in gasification, the oxygen quickly disappears in the lowest combustion region, giving rise to slower char–CO_2 or char–H_2O reactions in the gasification section. Residual ash is removed from the bottom. Operation at very high temperatures will cause this ash to melt and be removed as slag. The majority of the recent work in this area has emphasized coal gasification in fixed bed gasifiers. This type of gasifier is used commercially in various parts of the world. Operation at elevated pressure in this type of process is common.

Several fixed bed coal combustion models are summarized in Table 27.[214-219] All six are one-dimensional models, but two of the six consider transient effects. All six emphasize gasification processes, but the codes presumably apply also to direct combustion. All are for slow, vertically moving, packed beds with countercurrent gas flow. Smoot[195] provides further details on fixed bed models.

Fluidized Bed Models. The third class of coal combustion models considered is for fluidized bed combustion (FBC). Development of these models has been

TABLE 27. Summary of Fixed Bed Combustor Models

Authors	Year	Emphasis	Scope	Application	Comparison with Data	Computations
Winslow[214]	1977	Gastification packed bed countercurrent	One-dimensional, transient	In-situ gasification	Peak temperature, exit gas composition front speed	Examples for in-situ beds; subbituminous coal, 1200 K peak temperature
Amundson and Arri[215]	1978	Gasification countercurrent, char reaction model	One-dimensional, steady	Lurgi type	Limited comparison of steam/O_2 ratio on H_2/CO ratio	Parametric studies for several operational variables
Desai and Wen[216]	1978	Gasification moving bed, stirred, counter current	One-dimensional, steady	MERC air-blown	For three different coals; gas composition, % conversion, temperature results inconclusive	For several MERC gasifier cases
Yoon et al.[217]	1978	Gasification, moving bed counter current, high-pressure, slagging	One-dimensional, steady and transient with boundary layer	Lurgi, GEGAS, slagging	Exit gas comparisons, peak temperature for pilot slagging gasifier	Effects of heat loss, ash layer, oxygen rate, transient effects
Barriga and Essenhigh[218]	1979	Combustion and gasification fixed bed	Steady-state one-dimensional	Char gasification, small-scale gasifier	15 × 40 cm laboratory gasifier with O_2, CO_2, CO, temperature profiles along length coke fuel	Sensitivity analyses on heterogeneous reaction rate and char area with adjusted rate parameters
Wen et al.[219]	1982	Gasification moving bed, counter current	One-dimensional, steady with user's manual	General high-pressure gasifier (dry)	Exit temperature, carbon conversion gas composition, heating value for four coals	Several design maps for for bituminous coal
Hobbs et al.[233]	1989	Gasification, combustion counter current	One-dimensional, steady	coal, general	Laboratory and large-scale data	Parametric study; general application

pursued for the past 10–15 years, but the foundations were established by earlier research in fluidized beds. The fluidized bed models reviewed can be conveniently divided into the two classifications: (1) one-dimensional models and (2) two-dimensional models.

Sarofim and Beer[220] and, more recently and more extensively, Olofsson[221] provide reviews of one-dimensional FBC model developments that have occurred principally over the past decade. No general reviews of two-dimensional FBC models were identified. Work on two-dimensional models is more recent and less complete. The FB combustor is distinguished by low operating temperatures ($\sim$ 1100 K), high excess air levels ($\sim$ 30%), intermediate particle sizes (1–3 mm), long residence times (several minutes), and vigorous particulate motion that dominates heat transfer and reaction processes. Comprehensive modeling of this process is greatly complicated by the very complex fluid motion and interaction of the gas and particles.

TABLE 28. Summary of Selected One-Dimensional Fluidized Bed Combustor Models

Reference	Year	Current
Avedesian and Davidson[222]	1973	Early FBC model for spherical carbon-batch system
Gibbs[223]	1975	Extended model to coal and continuous feed; diffusion-controlled char reaction
Horio et al.[224]	1977	Treated sulfur capture and NO formation, and added kinetic term to char consumption, polydispersed sizes, and heat transfer tubes
Gordon et al.[225]	1978	Treated gas-phase reaction in bubbles, cooling tubes, carbon fuel
Sarofim and Beér[220]	1978	Considered NO_x formation, CO formation and reaction
Saxena et al.[226]	1978	Three-phase bed with bubbles, wake, and dispersion; ash restriction to char burnout
Wells and Kirshnan[227]	1980	Considered particle elutriation and attrition, CO_2 — C gasification, volatiles release and combustion for different coals, heat transfer surface in free-board region (user's manual)
Lewis and Tung[228]	1982	Added transient and high-pressure treatment (user's manual)

At least eight different one-dimensional FBC models were documented between 1973 and 1982, as summarized in Table 28.[222-228] A continuing, consistent approach has been used in the progressive development of one-dimensional FBC models, with new components being added by various investigators. The complex motion of particles has been neglected, but gas behavior has been divided into two or three aspects. One-dimensional, FBC models now include the following aspects: (1) gas fluid dynamics, (2) particle fluid dynamics, (3) coal devolatilization, (4) char oxidation, (5) volatiles combustion, (6) particle attrition (size reduction), (7) particle elutriation (loss from bed), (8) elevated pressure, (9) gasification reactions, (10) polydispersed particle sizes, (11) vertical and horizontal bed tubes, (12) convective and radiative heat transfer, (13) formation of oxides of nitrogen, (14) sorption of sulfur by acceptor particles (e.g., limestone or dolamite), (15) transient systems. One- dimensional FBC models with these aspects have developed significantly over the past decade to the point of application.

Development of two-dimensional FBC models was not initiated until the mid-1970s, with the principal focus on fluidized bed gasification. Parallel developments with somewhat different focus have been conducted by at least three groups of investigators.[229-232] Work has reached the stage where user's manuals are available and where independent groups are evaluating the behavior and characteristics of these codes. Smoot[195] discusses fixed and fluidized bed models in more detail.

6.6. Status of Coal Process Model Development and Application

Attempts to calculate the detailed performance of coal reaction processes have only been undertaken during the last 15 years. Prior to 1970 the best computations available were based on overall global calculations and often considered only a radiation submodel. The details of the mixing processes were not quantitative. However, the growth in capacity and speed of digital computers, and the recent awareness of the limits of energy resources, have made comprehensive calculation techniques both possible and of current interest.

Modeling of turbulent reaction process is still very much in a state of development. At least 6 fixed bed, 10 fluidized bed, and 14 pulverized coal conversion models have been developed since 1970, as summarized in this chapter. In general, the fundamental foundations of the most advanced coal combustion models are sound and often adequate for some areas of application. Even so, several unresolved fundamental issues remain; paramount among these are interactions of turbulence and chemical reactions and particle dispersion. Moreover comprehensive evaluation of these codes has been very limited, particularly for fixed and fluidized beds, and application of these codes among boiler manufacturers and utilities or engineering companies dealing with coal combustion is limited.

NOMENCLATURE*

Symbol	Units	Definition
A	m^2	area (Eq. (23))
A	sec^{-1}, other	frequency factor in Arrhenius or other rate constant (Eq. (2))
B	—	coal burnout percentage (Figure 20)
B	—	rate coefficient (Table 23)
c	—	virgin coal fraction (Eq. (5))
C	$J\ kg^{-1}\ K^{-1}$	heat capacity (Section 6)
C	$kg\ m^{-3}$	concentration of unreacted coal in solid particles (Eq. (4))
C	$kmol\ m^{-3}$	molar concentration (Eq. (13))
d	m, μm	diameter (Eq. (29))
daf	—	dry, ash free (Figure 5)
D	$m^2\ sec^{-1}$	diffusivity (Eq. (32))
E	$J\ kmol^{-1}$	activation energy (Eq. (2))
f	—	mixture fraction (Eq. (37))
H	$J\ kmol^{-1}$	molar enthalpy (Eq. (39))
h	$J\ kg^{-1}$	mass enthalpy (Eq. (50))
k	$J\ m^{-1}\ sec^{-1}\ K^{-1}$	thermal conductivity (Section 6)
k	$m\ sec^{-1}$, other	mass-transfer coefficient (Eq. (21))
k	$m\ sec^{-1}$, sec^{-1}, $g\ cm^{-2}\ sec^{-1}\ atm\ O_2^{-1}$	reaction rate coefficient (Eq. (1))
K	—	total number elements (Section 6)
K	$sec\ m^{-2}$	burning rate constant (Eq. (34))
m	kg	particle mass (Eq. (29))
m	—	true order of reaction (Figure 10)
m	$g\ cm^{-1}\ sec^{-1}$	mass addition rate (Eq. (48))
M	$kg\ kmol^{-1}$	molecular weight (Eq. (23))
n	m^{-3}	number density (Eq. (53))
n	—	apparent reaction order (Figure 10)
N	—	temperature exponent (Table 14)
P	Pa	pressure (Eq. (13))
Q	$J\ sec^{-1}\ m^3$	heat transfer rate (Eq. (50))
Q	—	constant relating quantity of high rate volatiles to proximate volatiles (Eq. (3))

*The equations noted are those where the symbol first appears. Also noted are the table, figure or chapter numbers where the symbol first appears.

r	m	radial position (Figure 35)
r	kg m^{-3} sec^{-1}, kg sec^{-1}	reaction rate (Eq. (23))
R	m	combustor radius (Figure 35)
R	kg m^{-2} sec^{-1}	specific reaction rate (Eq. (21))
R	J kmol^{-1} K^{-1}	universal gas constant (Eq. (2))
$\underline{R}$	cal gmol^{-1} K^{-1}	universal gas constant (Figure 11)
S	kg m^{-3}	concentration of unreactive char in solid particles produced by de-volatilization reaction (Eq. (4))
SR	—	stoichiometric ratio (Figure 18)
t	sec	time (Eq.(1))
T	K	temperature (Eq. (2))
v	m sec^{-1}	velocity (Eq. (53))
v	—	fraction of solid lost as volatiles (Eq. (1))
V	m^3	particle volume (Eq. (29))
V	kg m^{-3}	concentration of volatiles (Eq. (4))
w	kg sec^{-1}	mass flow rate (Eq. (48))
W	—	fraction of functional coal group re-acted during devolatilization process (Eq. (9))
x	m	coordinate direction (Eq. (48))
Y	—	devolatilization coefficient (Eq. (4))
α	kg	component mass (Eq. (53))
γ	—	particle swelling parameter (Eq. (56))
ξ	—	fraction of surface reaction heat re-leased that escapes to solid phase com-pared to gas phase (Eq. (50))
ξ	—	internal total particle surface area/ex-ternal equivalent sphere surface area (Eq. (23))
η	—	effectiveness factor (Eq. (36))
μ	kg m^{-1} sec^{-1}	viscosity (Section 6)
ν	—	surface product stoichiometric factor (Eq. (23))
ρ	kg m^{-3}	density (Figure 10)
σ	—	Gaussian coefficient for distribution of activation energies (Eq. (7))
ϕ	—	initial equivalence ratio (Table 12), char stoichiometric coefficient (Eq. (47))
χ	—	consumption rate/maximum consumption rate (Eq. (36))
ω	—	mass fraction (Eq. (48))

Subscripts

a	ash (Figure 20), apparent (Figure 10)
b	burning, (Eq. (34)), boundary (Eq. (50))
c	volatiles constant (Eq. (3)), convection (Eq. (50)), raw coal (Eq. (53))
d	diffusion (Eq. (24))
D	diffusive (Eq. (34))
f	flame (Eq. (51)), fuel (Figure 20)
g	gas (Eq. (37))
h	char (Eq. (54))
H	hydrocarbon (Eq. (10))
i	species (Section 6), intrinsic (Eq. (35)), coal functional group (Eq. (9))
j	particle (Eq. (46))
k	particle (Eq. (23)), element (Eq. (48))
l	reaction index (Eq. (47))
m	mass transfer (Eq. (21)), reaction index (Eq. (46)), mean (Eq. (25)), stoichiometric coefficient (Eq. (10)), mixture (Section 6)
n	stoichiometric coefficient (Eq. (10))
0	initial value (Eq. (56)), oxidizer (Figure 10), reference (Eq. (8))
O	oxygen (Eq. (13))
O_2	oxygen (Section 4.2)
p	particle (Figure 10), proximate (Eq. (3)), pressure (Eq. (23)), primary (Eq. (37))
r	radiation (Eq. (51)), reaction (Eq. (21))
R	reaction (Eq. (39))
s	surface (Eq. (36)), secondary (Eq. (38)), stoichiometric (Eq. (23))
t	total (Eq. (28))
w	water (Eq. (55))
ρ	recirculation (Eq. (48))
∞	maximum value (Eq. (1))
1	first reaction (Eq. (4))
2	second reaction (Eq. (4))

Superscripts

m	true reaction order (Eq. (35))
n	concentration exponent (Eq. (23))
N	temperature exponent (Table 14)
0	initial value (Eq. (9)) standard state (Eq. (39))

REFERENCES

1. H. H. Lowry, ed., *Chemistry of Coal Utilization*, Suppl. vol. Wiley, New York, 1963.
2. M. A. Elliott, ed., *Chemistry of Coal Utilization*, 2nd Suppl. Vol. Wiley, New York, 1981.

3. M. A. Field, D. W. Gill, B. B. Morgan, and P. G. W. Hawksley, *Combustion of Pulverized Coal*. British Coal Utilization Research Association, Surrey, England, 1967.

4. L. D. Smoot and D. T. Pratt, eds., *Pulverized Coal Combustion and Gasification*. Plenum, New York, 1979.

5. L. D. Smoot and P. J. Smith, *Coal Combustion and Gasification*. Plenum, New York, 1985.

6. J. Grumer, "Recent Research Concerning Extinguishment of Coal Dust Explosions." *Symp. (Int.) Combust. [Proc.]* **15**, 103 (1975).

7. D. Anson, "Fluidized Bed Combustion of Coal for Power Generation." *Prog. Energy Combust. Sci.* **2**, 61 (1976).

8. D. B. Anthony and J. B. Howard, "Coal Devolatilization and Hydrogen Gasification." *AIChE J.* **22**, 625 (1976).

9. J. M. Beér, "Fluidized combustion of Coal." *Symp. (Int.) Combust. [Proc.]* **16**, 439 (1977).

10. A. M. Godridge and A. W. Read, "Combustion and Heat Transfer in Large Boiler Furnaces." *Prog. Energy Combust. Sci.* **2**, 83 (1976).

11. A. F. Sarofim and R. C. Flagen, "NO Control for Stationary Combustion Sources." *Prog. Energy Combust. Sci.* **2**, 1 (1976).

12. B. P. Breen, "Combustion in Large Boilers: Design and Operating Effects on Efficiency and Emission." *Symp. (Int.) Combust. [Proc.]* **16**, 19 (1977).

13. R. H. Essenhigh, "Combustion and Flame Propagation in Coal Systems." *Symp. (Int.) Combust. [Proc.]* **16**, 353 (1977).

14. K. Littlewood, "Gasification: Theory and Application." *Prog. Energy Combust. Sci.* **3**, 35 (1977).

15. L. D. Smoot and M. D. Horton, "Propagation of Laminar Pulverized Coal-Air Flames." *Prog. Energy Combust. Sci.* **3**, 235 (1977).

16. R. J. Belt and L. A. Bissett, *An Assessment of Flash Pyrolysis and Hydropyrolysis*, U.S. DOE Rept., METC/RI-70/2. Morgantown Energy Technology Center, West Virginia, 1978.

17. J. Gibson, "The Constitution of Coal and Its Relevance to Coal Conversion Processes." *J. Inst. Fuel*, **51** (3), 67 (1978).

18. N. M. Laurendeau, "Heterogeneous Kinetics of Coal Char Gasification and Combustion." *Prog. Energy Combust. Sci.* **4**, 221 (1978).

19. A. F. Sarofim and H. C. Hottel, "Radiative Transfer in Combustion Chambers: Influence of Alternate Fuels." *Heat. Transfer, Int. Heat Transfer Conf., 6th, 1978*, p. 199 (1978).

20. J. C. Krazinski, R. O. Buckins, and H. Krier, "Coal Dust Flames: A Review and Development of a Model in Flame Propagation." *Prog. Energy Combust. Sci.* **5**, 31 (1979).

21. A. Macek, "Coal Combustion in Boilers: A Mature Technology Facing New Contraints." *Symp. (Int.) Combust. [Proc.]* **17**, 65 (1979).

22. T. F. Wall, A. Lowe, L. J. Wibberley, and I. McC. Stewart, "Mineral Matter in Coal and the Thermal Performance of Large Boilers." *Prog. Energy Combust. Sci.* **5**, 1 (1979).

23. L. D. Smoot, "Pulverized Coal Diffusion Flames: A Perspective Through Modeling." *Symp. (Int.) Combust.* [*Proc.*] **18**, 1185 (1981).

24. J. O. L. Wendt, "Fundamental Coal Combustion and Pollutant Formation in Furnaces." *Prog. Energy Combust. Sci.* **6**, 201 (1980).

25. J. B. Howard, "Fundamentals of Coal Pyrolysis and Hydroprolysis." In *Chemistry of Coal Utilization* (M. A. Elliott, ed.), 2nd, Suppl. Vol., p. 665. Wiley, New York, 1981.

26. T. D. Brown, "Coal Gasification Combined Cycles for Electricity Production." *Prog. Energy Combust. Sci.* **8**, 300 (1982).

27. I. W. Smith, "The Combustion Rates of Coal Chars: A Review." *Symp. (Int.) Combust.*, [*Proc.*] **19**, 1045 (1983).

28. G. A. Simons, "The Role of Pore Structure in Coal Pyrolysis and Gasification." *Prog. Energy Combust. Sci.* **9**, 269 (1983).

29. L. D. Smoot and S. C. Hill, "Critical Requirements in Combustion Research." *Prog. Energy Combust. Sci.* **9**, 77 (1983).

30. P. R. Solomon and D. G. Hamblen, "Finding Order in Coal Pyrolysis Kinetics." *Prog. Energy Combust. Sci.* **9**, 323 (1983).

31. S. S. Penner and other authors (see Table 1), "Coal Combustion and Applications," (A special issue prepared by the DOE Coal Combustion and Applications Working Group). *Prog. Energy Combust. Sci.* **10**, 87 (1984).

32. OTA, Office of Technology Assessment, *The Direct Uses of Coal.* Washington, D.C., 1979.

33. U.S. Department of Energy, *Quarterly Coal Report—October–December, 1982* DOE/EIA 0121, Office of Coal, Nuclear, Electrical and Alternative Fuels, Washington, D.C., 1983.

34. Institute of Gas Technology, *Preparation of a Coal Conversion Systems Technical Data Book*, DOE Rep., FE-2286-32. IGT, Chicago, Illinois, 1979.

35. G. J. Pitt and G. R. Millward, *Coal and Modern Coal Processing: An Introduction.* Academic Press, New York, 1979.

36. C. L. Wilson, *Coal—Bridge to the Future*, Ballinger, Cambridge, Massachusetts, 1980.

37. M. B. McNair, *Energy Data Report: Coal Distribution*, DOE/EIA-0125 (80/20). Department of Energy, Washington, D.C., 1980.

38. R. H., Essenhigh, C. Y. Wen, and E. S. Lee, *Coal Conversion Technology.* Addison Wesley, Reading, Massachusetts, 1979.

39. R. G. Schweiger, "Burning Tomorrow's Fuels." *Power* **123**, No. 2 (1979).

40. W. W. Bodle and F. C. Schora, "Coal Gasification Technology Overview." *Adv. Coal Util. Technol. Symp.*, Louisville, Kentucky, 1979.

41. P. Nowacki, *Coal Liquefaction Processes.* Noyes Data Corporation, Park Ridge, New Jersey, 1979.

42. L. D. Hansen, L. R. Phillips, N. F. Mangelson, and M. L. Lee, "Analytical Study of the Effluents from a High Temperature Entrained Flow Gasifier." *Fuel* **59**, 323 (1980).

43. D. Burgess, J. Hertzberg, J. K. Richmond, I. Liebman, K. L. Cashdollar, and C. P. Lazzara, *Combustion Extinguishment and Devolatilization in Coal Dust Explosions.* Western States Section Combustion Institute, Pittsburgh, Pennsylvania, 1979.

44. S. A. Johnson and T. M. Sommer, "Commerical Evaluation of a Low NO_x Combustion System as Applied to Coal-Fired Utility Boilers." Rep. No. WS-79-220, Electric Power Research Co., Palo Alto, California, Mu7, 1981.

45. W. Spackman, *The Characteristics of American Coals in Relation to Their Conversion into Clean Energy Fuels*, DOE Rep., FE-2030-13. Pennsylvania State University, University Park, 1980.

46. C. R. McCann, J. J. Demeter, and D. Bienstock, "Combustion of Pulverized Solvent-Refined Coal." *J. Eng. Power* **99**, 305 (1977).

47. R. D. McRainie, *Full-Scale Utility Boiler Test with Solvent Refined Coal (SRC)*, Final Rep. to U.S. DOE, Contract No. EX-76-C-01-222. Southern Company Services, Inc., Birmingham, Alabama, 1979.

48. B. Granoff and H. E. Nuttal, "Pyrolysis Kinetics and Oil-Shale Particles." *Fuel* **56**, 234 (1977).

49. *Int. Symp. Coal Slurry Combust., 4th, 1982.*

50. T. A. Hendrickson, ed., *Synthetic Fuels Data Handbook*. Cameron Engineers, *Inc., Denver, Colorado*, 1975.

51. P. H. Given, ed., *American Conference on Coal Science*, Adv. Chem. Ser. No. 55. American Chemical Society, Washington, D.C., 1964.

52. R. C. Neavel, "Coal Origin, Analysis, and Classification." 72nd *Annu. Meeting. AiChE, 1979.* Pap. No. 111b (1979).

53. D. G. Hamblen, P. R. Solomon, and R. H. Hobbs, *Physical and Chemical Characterization of Coal*, EPA Rep., 600/7-80-106. United Technologies Research Center, East Hartford, Connecticut, 1980.

54. A. F. Sarofim, J. B. Howard, and A. S. Padia, "The Physical Transformation of the Mineral Matter in Pulverized Coal Under Simulated Combustion Conditions." *Combust. Sci. Technol.* **16**, 198 (1977).

55. J. N. Cannon, G. J. Germane, S. Holst, D. Carr, D. Clark, and L. D. Smoot, *Transient Characterization of Pulverized Coal Mills with Steam Inerting*, Part VIII. Utah Power and Light Co., Brigham Young University, Provo, 1983; also see J. M. Beér, J. Chomiak, and L. D. Smoot, "Fluid Dynamics of Coal Combustion: A Review." *Prog. Energy Combust. Sci.* **10**, 177 (1984).

56. P. R. Solomon, *Characterization of Coal and Coal Thermal Decomposition*, Chapter III Rep. Advanced Fuel Research, Inc., East Hartford, Connecticut, 1980.

57. W. H. Wiser, *Preprint. Pap. Am. Chem. Soc., Div. Fuel Chem.* **20**, 122 (1975).

58. L. A. Heredy and I. Wender, *Preprint. Pap.—Am. Chem. Soc., Div. Fuel Chem.* **25**, 4 (1980).

59. P. H. Given and B. Biswas, "Dependence of Coal Liquefaction Behavior on Coal Characteristics. 2. Role of Petrographic Composition." *Fuel* **54**, 40 (1979).

60. S. P. N. Singh, J. C. Moyers, and K. R. Carr, *Coal Beneficiation—The Cinderella Technology*. Coal Combustion Applications Working Group, Oak Ridge National Laboratory, Oak Ridge, Tennessee; 1982.

61. Y. Liu, ed., *Physical Cleaning of Coal*. Dekker, New York, 1982.

62. Electric Paper Research Institute, "Coal Gasification for Electric Utilities." *EPRI J.* **4**, 6, April (1979).

63. H. H. Lowry, ed., *Chemistry of Coal Utilization*, Vol. 1. Wiley, New York, 1945.

64. L. D. Smoot and P. J. Smith, "Modeling Pulverized Coal Reaction Processes." In *Pulverized Coal Combustion and Gasification* (L. D. Smoot and D. T. Pratt, eds., p. 217, Plenum, New York, 1979.

65. D. Gray, J. G. Cogoli, and R. H. Essenhigh, "Problems in Pulverized Coal and Char Combustion." *Adv. Chem. Ser.* **131** (6), 72 (1976).

66. F. A. Williams, *Combustion Theory*, 2nd ed. Benjamin/Cummings, Menlo Park, California, 1985.

67. A. M. Kanury, *Introduction to Combustion Phenomena*, p. 90. Gordon & Breach, New York, 1975.

68. I. Glassman, *Combustion*, p. 194. Academic Press, New York, 1977.

69. M. A. Field, D. W. Gill, B. B. Morgan, and P. G. W. Hawksley, "Combustion of Pulverized Fuel. Part 6. Reaction Rate of Carbon Particles." *Br. Coal Util. Res. Assoc., Monogr. Bull.* **31**, 285 (1967).

70. J. G. Singer, *Combustion: Fossil Power Systems*, Chapter 13. Combustion Engineering, Windsor, Connecticut, 1981.

71. R. H. Essenhigh, "Fundamentals of Coal Combustion." In *Chemistry of Coal Utilization* (M. A. Elliott, ed.), 2nd Suppl. Vol., pp. 1153. Wiley, New York.

72. J. M. Kuchta, V. R. Rowe, and D. S. Burgess, "Spontaneous Combustion Susceptibility of U.S. Coals." *Rep. Invest.—U.S., Bur. Mines* **RI-8474** (1980).

73. M. Hertzberg, C. D. Litton, and R. Garloff, "Studies of Incipient Combustion and Its Detection." *Rep. Invest.—U.S., Bur. Mines* **RI-8206** (1977).

74. J. Nagy, J. G. Dorsett, and A. R. Coper, "Explosibility of Carbonaceous Dusts." *Rep. Invest.—U.S., Bur. Mines* **RI-6597**, (1965).

75. P. R. Solomon and M. B. Colket, "Coal Devolatilization." *Symp. (Int.) Combust.* [*Proc.*] **17**, 131 (1979).

76. K. Remenyi, *Combustion Stability*. Akadémia Kiadó, Budapest, 1980.

77. R. M. LaFollette, "In-situ Measurements of Coal Particle Temperatures in a Laboratory-Scale Coal Combustor." M.S. Thesis, Brigham Young University, Provo, Utah, 1984.

78. A. B. Ayling and I. W. Smith, "Measured Temperatures of Burning Pulverized-Fuel Particles and the Nature of the Primary Reaction Product." *Combust. Flame* **18**, 173 (1972).

79. K. L. Cashdollar, and M. Herzberg, "Infrared Pyrometers for Measuring Dust Explosion Temperatures." *Proc. Soc. Photo-Opt. Instrum. Eng.* **21**(1), 82 (1982).

80. D. W. Mackowski, R. A. Altenkirch, R. E. Peck, and T. W. Tong, *Infrared Pyrometer Measurement of Particle and Gas Temperatures in Pulverized-Coal Flames*. Western States Section/Combustion Institute, Spring Meeting, Salt Lake City, Utah, 1982.

81. R. E. Mitchell and W., J. McLean, *On the Temperature and Reaction Rate of Burning Pulverized Fuels*. Western States Section/Combustion Institute, Spring Meeting, Salt Lake City, Utah, 1982.

82. M. A. Nettleton, "Temperature Measurements of Burning Coal Particles in a Radiating Enclosure." *Combust. Flame* **9**, 311 (1965).

83. G. P. Starley, S. C. Manis, F. W. Bradshaw, and D. W. Pershing, "Formation and Control of NO$_x$ Emissions in Fixed-Bed Coal Combustion." Western States Section/Combustion Institute, Spring Meeting, Salt Lake City, Utah, 1982.

84. L. D. Timothy, A. F. Sarofim, and J. R. Beér "Characteristics of Single Particle Combustion." *Symp.* (*Int.*) *Combust.* [*Proc.*] **19**, 1123 (1983).

85. W. R. Seeker, G. S. Samuelsen, M. P. Heap, and J. D. Trolinger, "The Thermal Decomposition of Pulverizer Coal Particles." *Symp.* (*Int.*) *Combust.* [*Proc.*] **18**, 1213 (1981).

86. M. D. Horton, "Fast Pyrolysis." In *Pulverized Coal Combustion and Gasification*, (L. D. Smoot and D. T. Pratt, eds.), p. 133. Plenum, New York, 1979.

87. H. Kobayashi, J. B. Howard, and A. F. Sarofim, "Coal Devolatilization at High Temperatures." *Symp.* (*Int.*) *Combust.* [*Proc.*] **16**, 411 (1977).

88. E. M. Suuberg, W. A. Peters, and J. B. Howard, "Product Compositions and Formation Kinetics in Rapid Pyrolysis of Pulverized Coal—Implications for Combustion." *Symp.* (*Int.*) *Combust.* [*Proc.*] **17**, 177 (1979).

89. D. W. Blair, J. O. L. Wendt, and W. Bartok, "Evolution of Nitrogen and Other Species during Controlled Pyrolysis of Coal." *Symp.* (*Int.*) *Combust.* [*Proc.*] **16**, 475 (1977).

90. D. B. Anthony, J. B. Howard, H. C. Hottel, and H. P. Meissner, "Rapid Devolatilization of Pulverized Coal." *Symp.* (*Int.*) *Combust.* [*Proc.*] **15**, 1303 (1975).

91. S. Badzioch and P. G. W. Hawksley, "Kinetics of Thermal Decomposition of Pulverized Coal Particles." *Ind. Eng. Chem. Process Des. Dev.* **9**, 521 (1970).

92. G. M. Kimber and M. D. Gray, "Rapid Devolatilization of Small Coal Particles." *Combust. Flame* **11**, 360 (1967).

93. K. M. Sprouse and M. D. Schuman, "Predicting Lignite Devolatilization with the Multiple Parallel and Two-Competing Reaction Models." *Combust. Flame* **43**, 265 (1981).

94. P. R. Solomon and D. G. Hamblen, "Finding Order in Pyrolysis Kinetics." *Prog. Energy Combust. Sci.* **9**, 323 (1983).

95. P. R. Solomon and D. G. Hamblen, "Pyrolysis." In *Chemistry of Coal Conversion*. (R. Schlossberg, ed.). Plenum, New York, 1985.

96. P. R. Solomon, M. A. Serio, R. M. Carangelo, and J. R. Markham, "Very Rapid Coal Pyrolysis." *Fuel* **65**, 182 (1985).

97. D. McNeil, "High Temperature Coal Tar." In *Chemistry of Coal Utilization* (M. A. Elliott, ed.), 2nd Suppl. Vol., Chapter 17. Wiley, New York, 1981.

98. J. R. Thurgood and L. D. Smoot, "Volatiles Combustion." In *Pulverized Coal Combustion and Gasification* (L. D. Smoot and D. T. Pratt, eds.), p. 169. Plenum, New York, 1979.

99. D. J. Hucknall, *Chemistry of Hydrogen Combustion*. Chapman & Hall, New York, 1985.

100. U. S. Engleman, *Survey and Evaluation of Kinetic Data of Reactions in Methane/Air Combustion*, EPA-600/2-76-003. Exxon Research and Engineering Co., Linden, New Jersey, 1976.

101. R. B. Edelman and O. F. Fortune, "A Quasi-Global Chemical Kinetic Model for the Finite Rate Combustion of Hydrocarbon Fuels with Application to Turbulent Burning and Mixing in Hypersonic Engines and Nozzles." *AIAA Pap.* **69-86** (1969).

102. V. J. Siminski, F. J. Wright, R. B. Edelman, C. Economics, and O. F. Fortune, *Research on Methods of Improving the Combustion Characteristics of Liquid Hydrocarbon Fuels*, AFAPL TR 72-74, Vols. I and II. Air Force Aeropropulsion Laboratory, Wright Patterson Air Force Base, Ohio, 1972.

103. A. N. Hautman, F. L. Dryer, K. P. Schug, and I. Glassman, "A Multiple-Step Overall Kinetic Mechanism for the Oxidation of Hydrocarbons." *Combust. Sci. Technol.* **25**, 219 (1981).

104. G. G. DeSoete, "Overall Reaction Rates of NO and N_2 Formation from Fuel Nitrogen." *Symp. (Int.) Combust. [Proc.]* **15**, 1093 (1975).

105. R. H. Essenhigh, "Combustion and Flame Propagation in Coal Systems: A Review." *Symp. (Int.) Combust. [Proc.]* **16**, 372 (1977).

106. I. W. Smith, *The Combustion Rates of Pulverized Coal Char Particles*, Conf. Coal Combust. Technol. Emiss. Control. California Institute of Technology, Pasadena, 1979.

107. F. D. Skinner and L. D. Smoot, "Heterogeneous Reactions of Char and Carbon." In *Pulverized Coal Combustion and Gasification*, (L. D. Smoot and D. P. Pratt, eds.), p. 149. Plenum, New York, 1979.

108. P. L. Walker, Jr., F. Rusinko, Jr., and L. G. Austin, "Gas Reactions of Carbon." *Adv. Catal.* **11**, 135 (1959).

109. M. A. Field, "Predicting the Burning Time of the Coke Residue of Pulverized Fuel." *Br. Coal Util. Res. Assoc., Monogr. Bull.* **28**, 61 (1964).

110. W. J. Thomas, "Effect of Oxidation on the Pore Structure of Some Graphitized Carbon Blacks." *Carbon* **3**, 435 (1977).

111. G. J. Goetz, N. Y. Nsakala, K. L. Patel, and T. C. Lao, "Combustion and Gasification Kinetics of Chars from Four Commercially Significant Coals of Varying Rank." *2nd Ann. Contractor's Conf. Coal Gasification, Electr. Power Res. Inst.*, AP-3121, p. 15–1, Palo Alto, California, May, 1983.

112. I. W. Smith and R. J. Tyler, "The Reactivity of a Porous Brown Coal Char to Oxygen between 630 and 1812 K." *Combust. Sci. Technol.* **9**, 87 (1974).

113. R. H. Essenhigh, R. Froberg, and J. B. Howard, "Predicted Burning Rates of Single Carbon Particles." *Ind. Eng. Chem.* **57**, 33 (1965).

114. J. B. Howard and R. H. Essenhigh, "Mechanism of Solid-Particle Combustion with Simultaneous Gas-Phase Volatiles Combustion." *Symp. (Int.) Combust. [Proc.]* **11**, 399 (1967).

115. M. A. Nettleton, "Burning Rates of Devolatilized Coal Particles." *Ind. Eng. Chem. Fundam.* **6**, 20 (1967).

116. R. J. Hamor, I. W. Smith, and R. J. Tyler, "Kinetics of Combustion of a Pulverized Brown Coal Char between 630 and 2200 K." *Combust. Flame* **21**, 153 (1973).

117. I. W. Smith, "Kinetics of Combustion of Size-Graded Pulverized Fuels in the Temperature Range 1200–2270 K." *Combust. Flame* **17**, 303 (1971).

118. I. W. Smith and R. J. Tyler, "Internal Burning of Pulverized Semi-Anthracite: The Relation between Particle Structure and Reactivity." *Fuel* **51**, 312 (1971).

119. G. D. Sergeant and I. W. Smith, "Combustion Rates of Bituminous Coal Char in the Temperature Range 800 to 1700 K." *Fuel* **52**, 52 (1973).

120. S. Dutta and C. Y. Wen, "Reactivity of Coal and Char. 2. In Oxygen-Nitrogen Atmosphere." *Ind. Eng. Chem. Process Des. Dev.* **16**, 31 (1977).

121. G. Mandel, "Gasification of Coal Char in Oxygen and Carbon Dioxide at High Temperatures." M.S. Thesis, Massachusetts Institute of Technology, Cambridge, 1977.

122. R. W. Froberg and R. H. Essenhigh, "Reaction Order and Activation Energy on Carbon Oxidation during Internal Burning." *Symp. (Int.) Combust. [Proc.]* **17**, 179 (1979).

123. B. C. Young and I. W. Smith, "The Kinetics of Combustion of Petroleum Coke particles at 1000 to 1800 K: The Reaction Order." *Symp. (Int.) Combust. [Proc.]* **18**, 1249 (1981).

124. R. T. Yang and M. Steinburg, "A Diffusion Cell Method for Studying Heterogeneous Kinetics in the Chemical Reaction/Diffusion Controlled Region. Kinetics of $C + CO_2 - 2CO$ at 1200–1600°C." *Ind. Eng. Chem. Fundam.* **16**, 235 (1977).

125. S. Dutta, C. Y. Wen, and R. J. Belt, "Reactivity of Coal and Char. 1. In Carbon Dioxide Atmosphere." *Ind. Eng. Chem. Process Des. Dev.* **16**, 20 (1977).

126. A. Linares-Solano, O. P. Mahajan, and P. L. Walker, Jr., "Reactivity of Heat-Treated Coals in Steam." *Fuel* **58**, 327 (1979).

127. K. Otto, L. Bartosiewicz, and M. Shelef, "Catalysis of Carbon-Steam Gasification by Ash Components from Two Lignites." *Fuel* **58**, 85 (1979).

128. K. Otto, L. Bartosiewicz, and M. Shelef, "Effects of Calcium, Strontium, and Barium as Catalysts and Sulphur Scavengers in the Stream Gasification of Coal Chars." *Fuel* **58**, 565 (1979).

129. P. P. Feistel, K. H. Van Heek, and H. Juentgen, *Ger. Chem. Eng. (Engl. Transl.)* **1**, 294 (1978).

130. S. P. Chauhan, and R. R. Longenback, "Determination of the Kinetics of Hydrogasification of Char Using a Thermal Balance." *Fuel Chem.* **23**(3), 73 (1978).

131. R. W. Bryers, ed., *Ash Deposits and Corrosion due to Impurities in Combustion Gases.* Hemisphere Publ., Livingston, New Jersey, 1978.

132. H. J. Gluskoter, N. F. Shimp, and R. R. Rich, "Coal Analysis of Trace Elements and Mineral Matter." In *Chemistry of Coal Utilization* (M. A. Elliott, ed.), 2nd Suppl. Vol., p. 369. Wiley, New York, 1981.

133. W. T. Reid, "Coal Ash, Its Effect on Combustion Systems." In *Chemistry of Coal Utilization* (M. A. Elliott, ed.), 2nd Suppl. Vol., p. 1389. Wiley, New York, 1981.

134. M. A. Nettleton, "Particulate Formation in Power Station Boiler Furnaces." *Prog. Energy Combust. Sci.* **5**, 223 (1979).

135. P. L. Walker, Jr., M. Shelef, and R. T. Anderson, "Catalysis of Carbon Gasification." In *Chemistry and Physics of Carbon* (P. L. Walker, Jr., ed.), Vol. 4, p. 287. Dekker, New York, 1968.

136. W. F. Wells, S. K. Kramer, L. D. Smoot, and A. U. Blackham, "Reactivity and Combustion of Coal Chars." *Symp. (Int.) Combust. [Proc.]* **20**, 1539 (1985).

137. L. R. Radovic and P. L. Walker, Jr., "Reactivities of Chars Obtained as Residues in Selected Coal Conversion Processes." *Fuel Process. Technol.* **8**, 149 (1984).

138. D. C. Wegener, unpublished data from Phillips Petroleum Co., Bartlesville, Oklahoma, 1982. Chars provided and characterized by Combustion Laboratory, Brigham Young University, Provo, Utah, 1982.

139. I. W. Smith, "The Intrinsic Reactivity of Carbons to Oxygen." *Fuel* **57**, 409 (1978).

140. G. A. Simons and M. L. Finson, "The Structure of Coal Char. Part I. Pore Branching." *Combust. Sci. Technol.* **19**, 217 (1979).

141. G. R. Gavalas, "A Random Capillary Model with Application to Char Gasification at Chemically Controlled Rates." *AIChE J.* **26**, 577 (1980).

142. G. R. Gavalas, "Analysis of Char Combustion Including the Effect of Pore Enlargement." *Combust. Sci. Technol.* **24**, 197 (1981).

143. B. Srinivas and N. R. Amundson, "A Single-Particle Char Gasification Model." *AIChE J.* **26**, 487 (1980).

144. B. Srinivas and N. R. Amundson, "Intraparticle Effects in Char Combustion. III. Transient Studies." *Can. J. Chem. Eng.* **59**, 728 (1981).

145. K. Zygourakis, L. Arri, and N. R. Amundson, "Studies on the Gasification of a Single Char Particle." *Ind. Eng. Chem. Fundam.* **21**, 1 (1982).

146. H. E. Nuttel and G. F. Roach, *An Interdisciplinary Investigation of Coal Gasification Mechanics and Kinetics for the Optimal Development of New Mexico's Energy Resources*, Final Rep. 75–119. University of New Mexico, Albuquerque, 1978.

147. P. C. Malte and D. P. Rees, "Mechanisms and Kinetics of Pollutant Formation during Reaction of Pulverized Coal." In *Pulverized Coal Combustion and Gasification* (L. D. Smoot and D. T. Pratt. eds.), p. 182. Plenum, New York, 1979.

148. J. M. Levy, L. K. Chan, A. F. Sarofim, and Y. H. Song, *Combustion Research on the Fate of Fuel Nitrogen under Conditions of Pulverized Coal Combustion*, EPR-600/7-78-165. Massachusetts Institute of Technology, Cambridge, 1978.

149. C. A. Mims, M. Neville, R. J. Quann, and A. F. Sarofim, "Laboratory Studies of Trace Element Transformations During Coal Combustion." *87th Ann. Meet. AIChE*, Boston, Massachusetts, 1979.

150. R. D. Smith, "The Trace Element Chemistry of Coal during Combustion and the Emissions from Coal-Fired Plants." *Progr. Energy Combust. Sci.* **6**, 201 (1980).

151. M. Neville, R. J. Quam, B. S. Haynes, and A. F. Sarofim, "Vaporization and Condensation of Mineral Matter during Pulverized Coal Combustion." *Symp. (Int.) Combust. [Proc.]* **18**, 126 (1981).

152. D. D. Taylor and R. C. Flagan, "Laboratory Studies of Submicron Particles from Coal Combustion." *Symp. (Int.) Combust. [Proc.]* **18**, 1227 (1981).

153. G. D. Ulrich, "... *The Mechanism of Fly-Ash Formation in Coal-Fired Utility Boilers*, DOE Rep. FE-2205-16. University of New Hampshire, Durham, 1979.

154. M. A. Elliott and G. R. Yoke, "The Coal Industry and Coal Research and Developments in Perspective." In *Chemistry of Coal Utilization* (M. A. Elliott, ed.), 2nd Suppl. Vol., p. 1. Wiley, New York, 1981.

155. F. J. Ceely and E. L. Daman, "Combustion Process Technology." In *Chemistry of Coal Utilization* (M. A. Elliott, ed.), 2nd Suppl. Vol., p. 1313. Wiley, New York, 1981.

156. N. S. Harding, Jr., L. D. Smoot, and P. O. Hedman, "Nitrogen Pollutant Formation in a Pulverized Coal Combustor: Effect of Secondary Stream Swirl." *AIChE J.* **28**, 573 (1982).

157. B. W. Asay, L. D. Smoot, and P. O. Hedman, "Effect of Coal Moisture on Burnout and Nitrogen Oxide Formation." *Combust. Sci. Technol.* **35**, 15 (1983).

158. N. S. Harding, Jr., "Effects of Secondary Swirl and Other Burner Parameters on Nitrogen Pollution Formation in a Pulverized Coal Combustor." Ph.D. Dissertation, Brigham Young University, Provo, Utah, 1980.

159. L. D. Smoot, P. O. Hedman, and P. J. Smith, "Pulverized Coal Combustion Research at Brigham Young University." *Prog. Energy Combust. Sci.* **10**, 359 (1984).

160. L. D. Smoot and K. R. Christensen, *Data Book: For Evaluation of Pulverized Coal Reaction Models*, Final Rep., Vol. III, U.S. DOE, Contract No. DE-AC21-81MC16518. Brigham Young University, Provo, Utah, 1985.

161. B. W. Asay, "Effects of Coal Type and Moisture Content on Burnout and Nitrogeneous Pollutant Formation." Ph.D. Dissertation, Brigham Young University, Provo, Utah, 1982.

162. H. Perry, "The Gasification of Coal." *Sci. Am.* **230**, 19 (1974).

163. D. Hebden and H. J. F. Stroud, "Coal Gasification Processes." In *Chemistry of Coal Utilization* (M. A. Elliott, ed.), 2nd Suppl. Vol., p. 1599. Wiley, New York, 1981.

164. D. L. Imler and J. W. Parker, "An Update on the Great Plains Gasification Projects." *Annual AIChE Meet.*, New York, Nov. 1985.

165. Electric Power Research Institute, *Cool Water Gasification Program*, 1st Ann. Prog. Rep., EPRI-A-2487. EPRI, Palo Alto, California, 1982.

166. J. L. Johnson, "Fundamentals to Coal Gasification. In *Chemistry of Coal Utilization* (M. A. Elliott, ed.), 2nd Suppl. Vol., p. 1491. Wiley, New York, 1981.

167. L. S. Bissett, *An Engineering Assessment of Entrainment Gasification*, U.S. DOE Rep., MERC/RI-78/2. Morgantown Energy Technology Center, Morgantown, West Virginia, 1978.

168. D. R. Stull and H. Prophet, *JANAF Thermochemical Tables*, 2nd ed. National Bureau of Standards, U.S. Government Printing Office, Washington, D.C., 1971.

169. F. D. Skinner, L. D. Smoot, and P. O. Hedman, "Mixing and Gasification of Coal in an Entrained Flow Gasifier." *Am. Soc. Mech. Eng. [Pap.]* **80-WA / HT-30** (1980).

170. W. C. Edmister, H. Perry, R. C. Correy, and M. A. Elliott, "Thermodynamics of Gasification of Coal with Oxygen and Steam." *Am. Soc. Mech. Eng. J.* **74**, 621 (1952).

171. H. R. Batchelder and J. C. Sternberg, "Thermodynamic Study of Coal Gasification." *Ind. Eng. Chem.* **42**, 877 (1950).

172. P. O. Hedman, J. R. Highsmith, N. R. Soelberg, and L. D. Smoot, "Detailed Local Measurements in the BYU Entrained Gasifier." *Int. Flame Res. Found. Symp. Convers. Solid Fuels*, Newport Beach, California, 1982.

173. J. B. Howard, "Fundamentals of Coal Pyrolysis and Hydropyrolysis." In *Chemistry of Coal Utilization*. (M. A. Elliott, ed.), 2nd Suppl. Vol., p. 665. Wiley, New York, 1981.

174. B. W. Brown, L. D. Smoot, and P. O. Hedman, "Effect of Coal Type on Entrained Gasification." *Fuel* **65**, 673 (1986).

175. S. Azuhata, P. O. Hedman, and L. D. Smoot, "Carbon Conversion in an Atmospheric-Pressure Entrained Coal Gasifier." *Fuel* **65**, 212 (1986).

176. S. Azuhata, P.O. Hedman, L. S. Smoot, and W. A. Sowa, "Effects of Flame Type and Pressure on Entrained Gasification." *Fuel* **65**, 1511 (1986).

177. N. R. Soelberg, L. D. Smoot, and P. O. Hedman, "Part 1. Mixing and Reaction Processes in an Entrained Coal Gasifier from Local Measurements." *Fuel* **64**, 776 (1985).

178. J. R. Highsmith, N. R. Soelberg, P. O. Hedman, L. D. Smoot, and A. U. Blackham, "Part 2. Fate of Nitrogen and Sulfur Pollutants in an Entrained Flow Gasifier from Local Measurements." *Fuel* **64**, 782 (1985).

179. G. H. Lewis, L. D. Smoot, and P. O. Hedman, *Determination and Control of Carbon Conversion during Pulverized Coal Gasification.* Western States Section Combustion Institute, Salt Lake City, Utah, 1982.

180. G. R. Strimbeck, J. B. Cordiner, N. L. Baker, J. H. Holden, K. D. Plants, and L. D. Schmidt, *Rep. Invest.—U.S., Bur. Mines* **RI-5030** (1954).

181. K. A. Bueters, J. G. Cogoli, and W. W. Habelt, "Performance Prediction of Tangentially Fired Utility Furnaces by Computer Model." *Symp. (Int.) Combust.* [*Proc.*] **15**, 1245 (1975).

182. A. Lowe, T. F. Wall, and I. McC. Stewart, "A Zoned Heat Transfer Model of a Large Tangentially Fired Pulverized Coal Boiler." *Symp. (Int.) Combust.* [*Proc.*] **15**, 1261 (1975).

183. N. Selcuk, R. G. Siddall, and J. M. Beér, "A Comparison of Mathematical Models of the Radiative Behavior of a Large-Scale Experimental Furnace." *Symp. (Int.) Combust.* [*Proc.*] **16**, 53 (1977).

184. W. Richter and M. P. Heap, "The Impact of Heat Release Pattern and Fuel Properties on Heat Transfer in Boilers." *Am. Soc. Mech. Eng.* [*Pap.*] **81-WA / HT-27** (1981).

185. P. J. Smith and L. D. Smoot, *Mixing and Kinetic Processes in Pulverized Coal Combustors*, Vol. 2, EPRI, Final Rep. 364-1-3. Brigham Young University, Provo, Utah, 1979.

186. S. K. Ubhayakar, D. B. Stickler, C. W. von Rosenberg, and R. E. Gannon, "Rapid Devolatilization of Pulverized Coal in Hot Combustion Gases." *Symp. (Int.) Combust.* [*Proc.*] **16**, 427 (1977).

187. M. A. Field, "Rate of Combustion of Size-Graded Fractions of Char from a Low-Rank Coal Between 1200 K and 2000 K." *Combust. Flame* **13**, 237 (1969).

188. A. M. Mayers, "The Rate of Reduction of Carbon Dioxide by Graphite." *J. Am. Chem. Soc.* **56**, 70 (1934).

189. A. M. Mayers, "The Rate of Oxidation of Graphite by Steam." *J. Am. Chem. Soc.* **56**, 1879 (1934).

190. J. R. Thurgood, L. D. Smoot, and P. O. Hedman, "Rate Measurements in a Laboratory-Scale Pulverized Coal Combustor." *Combust. Sci. Technol.* **21**, 213 (1980).

191. T. Suzuki, L. D. Smoot, T. H. Fletcher, and P. J. Smith, "Prediction of High-Intensity Pulverized Coal Combustion." *Combust. Sci. Technol.* **45**, 167 (1985).

192. S. K. Kramer, L. D. Smoot, and P. J. Smith, *Combustion Characteristics of Residual Chars*. Western States Fall Section/Combustion Institute, Davis, California, 1985.

193. T. H. Fletcher, "A Two-Dimensional Model for Coal Gasification and Combustion." Ph.D. Dissertation, Department of Chemical Engineering, Brigham Young University, Provo, Utah, 1983.

194. D. P. Sloan, "Modeling of Swirl in Turbulent Systems." Ph.D. Dissertation, Department of Chemical Engineering, Brigham Young University, Provo, Utah, 1985; also see D. P. Sloan, P. J. Smith, and L. D. Smoot, "Modeling of Swirl in Turbulent Flow Systems." *Prog. Energy Combust. Sci.* **12**, 163 (1986).

195. L. D. Smoot, "Modeling of Coal Combustion Processes." *Prog. Energy Combust. Sci.* **10**, 229 (1984).

196. S. C. Hill, "Modeling of Nitrogen Pollutants in Turbulent Pulverized Coal Flames." Ph.D. Dissertation, Department of Chemical Engineering, Brigham Young University, Provo, Utah, 1983; also see P. J. Smith, S. C. Hill, and L. D. Smoot, "Theory for NO Formation in Turbulent Coal Flames." *Symp.* (*Int.*) *Combust.* [*Proc.*] **19**, 1263 (1983), P. J. Smith, S. C. Hill, and L. D. Smoot, "Effects of Swirling Flow on NO Concentration in Pulverized Coal Combustors." *AIChE J.* **32**, 1917 (1986).

197. B. Brown, "Effect of Coal Type on Entrained Gasification." Ph.D. Dissertation, Department of Chemical Engineering, Brigham Young University, Provo, Utah, 1985; also see B. S. Brown, L. D. Smoot, P. J. Smith, and P. O. Hedman, "Measurement and Prediction of Entrained Flow Gasification Processes." *AIChE J.* **34**, 435 (1988); L. D. Smoot and B. S. Brown, "Controlling Mechanisms for Gasification of Pulverized Coal." *Fuel* **66**, 1249 (1987).

198. L. L. Baxter, T. H. Fletcher, P. J. Smith, and L. D. Smoot, *Coal-Water Mixtures Combustion Model*. Western States Fall Section/Combustion Institute, Stanford, California, 1984.

199. M. M. Gibson and B. B. Morgan, "Mathematical Model of Combustion of Solid Particles in a Turbulent Stream with Recirculation." *J. Inst. Fuel* **43**, 517 (1970).

200. B. N. Mehta and R. Aris, "Computations on the Theory of Diffusion and Reaction-VII the Isothermal pth Order Reaction." *Chem. Eng. Sci.* **26**, 1699 (1971).

201. W. Richter and R. Quack, "A Mathematical Model for a Low-Volatile Pulverized Fuel Flame." In *Heat Transfer in Flames* (N. H. Afgan and J. M. Beér, eds.), p. 95. Scripta Technica, Washington, D.C., 1974.

202. T. R. Blake, D. H. Brownell, Jr., S. K. Garg, W. E. Herline, J. W. Pritchett, and G. P. Schneyer, *Computing Modeling of Coal Gasification Reactors*, Year 2, DOE/FE-1770-32. Systems, Science and Software, La Jolla, California, 1977.

203. T. R. Blake, W. E. Herline, and G. P. Schneyer, "Numerical Simulation of Coal Gasification Processes." *87th Nat. Meet. AIChE*, Boston, Massachusetts, 1979.

204. W. E. Lewellen, H. Segur, and A. K. Varma, *Modeling Two Phase Flow in a Swirl Combustor*, Final Rep., Energy Research and Development Agency, Washington, D.C., 1977. Contract No. EY-76-C-024062.

205. K. M. Sprouse, *Theory of Pulverized Coal Conversion in Entrained Flows*, Tech. Memo. for DOE/EZ-77-C-01-2518. Rockwell International, Canoga Park, California, 1977.

780   COAL AND CHAR COMBUSTION

206. S. K. Ubhayakar, D. B. Stickler, and R. G. Gannon, "Modeling of Entrained-Bed Pulverized Coal Gasifiers." *Fuel* **56**, 281 (1977).

207. M. L. Finson, G. Kothandaraman, P. F. Lewis, G. A. Simons, G. Wilemski, and K. L. Wray, *Modeling of Coal Gasification for Fuel Cell Utilization*, U.S. DOE Final Rep., SAN-1254-2. Physical Sciences Inc., Woburn, Massachusetts, 1978.

208. C. Y. Wen and T. Z. Chuang, *Entrained Coal Gasification Modeling*, DOE/FE-2274-T1. West Virginia University, Morgantown, 1978.

209. J. S. Barnhart, J. F. Thomas, and N. M. Laurendeau, *Pulverized Coal Combustion and Gasification in a Cyclone Reactor, Experimental and Models*, DOE/E-49-18. Purdue University, West Lafayette, Indiana, 1979.

210. R. K. C. Chan, C. A. Meister, M. F. Scharff, D. E. Dietrich, S. R. Goldman, H. B. Levine, and S. K. Ubahayakar, *A Computer Model for the Bigas Gasifier: Formulation of the Model*, U.S. DOE Final Rep., J-570-80-008A/2183. Jaycor Corp., Del Mar, California, 1980.

211. F. C. Lockwood, A. P. Salloja, and S. A. Syed, "A Prediction Method for Coal-Fired Furnaces." *Combust. Flame* **38**, 1 (1980).

212. P. J. Smith, T. J. Fletcher, and L. D. Smoot, "Model for Pulverized Coal-Fired Reactors." *Symp. (Int.) Combust. [Proc.]* **18**, 1285 (1981).

213. W. J. Oberjohn, D. K. Cornelious, W. A. Fiveland, R. J. Schnipke, and J. H. Wang, *Computational Tools for Pulverized Coal Combustion*, DOE/PC/40265-3. Babcock & Wilcox Co., Alliance, Ohio, 1982.

214. A. M. Winslow, "Numerical Model of Coal Gasification in a Packed Bed." *Symp. (Int.) Combust. [Proc.]* **16**, 503 (1977).

215. N. R. Amundson and L. E. Arri, "Char Gasification in a Counter Current Reactor." *AIChE J.* **24**, 87 (1978).

216. P. R. Desai and C. Y. Wen, *Computer Modeling of the MERC Fixed Bed Gasifier*, U.S. DOE Rep., MERC/CR-78/3. Morgantown Energy Technology Center, Morgantown, West Virginia, 1978.

217. H. Yoon, J. Wei, and M. M. Denn, "A Model for Moving-Bed Coal Gasification Reactors." *AIChE J.* **24**, 885 (1978).

218. A. Barriga and R. H. Essenhigh, *A Mathematical Model of a Combustion Pot.* Western States Section/Combustion Institute, Pittsburgh, Pennsylvania, 1979.

219. C. Y. Wen, H. Chen, and M. Onozaki, *User's Manual for Computer Simulation and Design of the Moving Bed Coal Gasifier*, DOE/MC/16474-1390. West Virginia University, Morgantown, 1982.

220. A. F. Sarofim and J. M. Beér, "Modeling of Fluidized Bed Combust." *Symp. (Int.) Combustion. [Proc.]* **17**, 189 (1979).

221. J. Olofsson, *Mathematical Modeling of Fluidized Bed Combustors*, NO, ICTIS/TR143. IEA Coal Research, London, 1980.

222. M. M. Avedesian and J. F. Davidson, "Combustion of Carbon Particles in a Fluidized Bed." *Trans. Inst. Chem. Eng.* **51**, 121 (1973).

223. B. M. Gibbs, *Inst. Fuel Symp. Ser. (London)* **1**, A5-1 (1975).

224. M. Horio, P. Rengarajan, R. Krishnan, and C. Y. Wen, *Fluidized Bed Combustor Modeling*, Rep. NAS3-19725. West Virginia University, Morgantown, 1977.

225. A. L. Gordon, H. S. Caram, and N. R. Amundson, "Modeling of Fluidized Bed Reactors-V." *Chem. Eng. Sci.* **33**, 713 (1978).

226. S. C. Saxena, T. P. Chen, and A. A. Jonke, *A Plug Flow Model for Coal Combustion and Desulfurization in Fluidized Beds: Theoretical Formulation*, ANL/CEN/FE-78-11. Argonne National Laboratory, Argonne, Illinois, 1978.

227. J. R. Wells and K. P. Krishnan, *Interim Annual Report for 1979*, ORNL/TM-7398. Oak Ridge National Laboratory, Oak Ridge, Tennessee, 1980.

228. J. R. Lewis and S. E. Tung (with D. Park and/or D. Lee, in selected volumes), *Modeling of Fluidized Bed Combustion of Coal*, Final Rep., Vols. I–VI, DOE/MC/16000-1294. Massachusetts Institute of Technology, Cambridge, 1982.

229. G. P. Schneyer, E. W. Peterson, P. J. Chen, J. L. Cook, D. H. Brownell, Jr., and T. R. Blake, *Computer Modeling of Coal Gasification Reactors*, U.S. DOE Final Rep., Vol. 2, DOE/ET/10242-T1. Systems, Science and Software, San Diego, California, 1981.

230. R. K.-C. Chan, M. J. Chiou, D. E. Dietrick, D. R. Dion, H. H. Klein, D. H. Laird, H. B. Levine, C. A. Meister, M. R. Scharff, and F. Srinivas, *Computer Modeling of Mixing and Agglomeration in Coal Conversion Processes*, Vols. 1 and 2, U.S. DOE Final Rep. DOE/ET/10329-1211. Jaycor Corp., Del Mar, California, 1982.

231. D. Gidaspow and B. Ettehadieh, "Fluidization in Two-Dimensional Beds with a Jet." *Ind. Eng. Chem. Fundam.* **22**, 193 (1983).

232. D. Gidaspow, C. Lin, and Y. C. Seo, "Fluidization in Two-Dimensional Beds with a Jet. 1. Experimental Porosity Distributions." *Ind. Eng. Chem. Fundam.* **22**, 198 (1983).

233. L. D. Smoot and B. S. Brewster, *Measurement and Modeling of Advanced Coal Conversion*, Annual Rep., Contract No. DE-AC21-86MC23075, for DOE, METC, Brigham Young University, Provo, Utah, Oct. 1989.

234. P. A. Gillis and P. J. Smith, *Three Dimensional Computational Fluid Dynamics Modeling in Industrial Furnaces*, Paper 89–84, Western States Section Fall Meeting, The Combustion Institute, Livermore, California, 1989.

APPENDIX Data on Fuel and Combustion Properties

PAO-CHEN WU*

and

HOYT C. HOTTEL

Department of Chemical Engineering
Massachusetts Institute of Technology
Cambridge, Massachusetts

A. Units, constants, and conversion factors
B. Bond energies for thermochemical calculations
C. Normalized enthalpy of combustion products
D. Equilibrium constants
E. Transport properties
    E.1. Thermal conductivity
    E.2. Surface tension
F. Diffusion coefficients for binary gas mixtures
G. Flame properties
    G.1. Flash point
    G.2. Flammability limits
    G.3. Autoignition temperature
    G.4. Minimum ignition energy and quenching distance
    G.5. Maximum flame velocity
    G.6. Flame temperature
H. Knock ratings of gasoline components
I. Cetane number of diesel fuel components
J. Smoke tendency of hydrocarbons
K. Rate constants of some radical reactions of combustion
L. Properties of industrial gaseous fuels
M. Liquid fuel specifications
N. Properties of representative crude oils
O. Classification of coals
P. Properties of selected U.S. coals
Q. Skeleton steam tables
    References

*Present address: Aspen Technology Inc., 251 Vassar Street, Cambridge, MA 02139.

A. UNITS, CONSTANTS, AND CONVERSION FACTORS

Although the International Bureau of Weights and Measures recommended the use of the International System of Units (SI) for scientific and technical work in 1960, different systems of units are still in use. The text of this book does not always use SI units.

Table A.1 lists the commonly used symbols of units. Table A.2 gives some physical constants. The listing of Table A.3 includes conversion factors which are encountered frequently.

Slightly different values of thermochemical properties are sometimes due to the conversion factors used. The calorie used in the appendix, the thermochemical calorie, is defined by the U.S. National Bureau of Standards and equals 4.184 joules(abs). Another commonly used unit is the International Steam Table calorie, sometimes called *cal(IST)*, which equals 4.1868 joules(abs).

For consistency, the British thermal unit (Btu) used here is the thermochemical Btu, 251.996 cal (thermochemical). In thermochemical unit, 1 cal/gm · K = 1 Btu/lb · °R.

The SI standard of energy is joule(abs). 1 joule(abs) = 0.999835 joule(Int).

References: 1, 2, 3.

TABLE A.1. Symbols of Units

Name of Unit	Symbol	Name of Unit	Symbol
ampere[a]	A	kilogram force	kgf
atmosphere	atm	kilogram mole	kg-mol
British thermal unit	Btu	kilowatt	kW
calorie	cal	meter[a]	m
candela[a]	cd	minute	min
Celsius	°C	mole[a]	mol
Fahrenheit	°F	newton	N
foot	ft	pascal	Pa
gram	gm	pound force	lbf
gram mole	gm-mol	pound force per	
horsepower	hp	square inch	psi
hour	hr	pound mass	lbm
inch	in	pound mole	lb-mol
joule	J	Rankine	°R
Kelvin[a]	K	second[a]	sec
kilocalorie	kcal	watt	W
kilogram[a]	kg		

[a]Basic International Unit (SI).

Adapted with permission from R. E. Bolz and G. L. Tuve, *Handbook of Tables for Applied Engineering Science*, 2nd. ed. (1973), © CRC Press, Inc., Boca Raton, FL.

TABLE A.2. Physical Constants

Gas law constant
  $R = 1.9872$ cal/gm-mol $\cdot$ K
    $= 82.056$ cm^{3}atm/gm-mol $\cdot$ K
    $= 8.3143 \times 10^3$ J/kg-mol $\cdot$ K
    $= 1.5453 \times 10^3$ ft $\cdot$ lbf/lb-mol $\cdot$ °R
Standard acceleration of gravity
  $g_0 = 980.665$ cm/sec^2
    $= 32.174$ ft/sec^2
    $= 6.6732 \times 10^{-11}$ N $\cdot$ m^2/kg^2
Joule's constant (mechanical equivalent of heat)
  $J_c = 4.184$ J/cal
    $= 777.65$ ft $\cdot$ lbf/Btu

Avogadro's number
  $N_0 = 6.0222 \times 10^{23}$ molecules/gm-mol
Bolzmann's constant
  $k = R/N_0$
    $= 1.3806 \times 10^{-16}$ erg/K $\cdot$ molecule
Stefan-Boltzmann constant
  $\sigma = 1.355 \times 10^{-12}$ cal/sec $\cdot$ cm^2 $\cdot$ K^4
    $= 5.6696 \times 10^{-8}$ W/m^2 $\cdot$ K^4
Planck's constant
  $h = 6.6262 \times 10^{-27}$ erg $\cdot$ sec
Speed of light in vacuum
  $c = 2.997925 \times 10^{10}$ cm/sec

Adapted with permission from R. E. Bolz and G. L. Tuve, *Handbook of Tables for Applied Engineering Science*, 2nd. ed. (1973), © CRC Press, Inc., Boca Raton, FL.

TABLE A.3. Conversion Factors

	To Convert from	To	Multiply by
Length	ft	m	0.3048
	in	cm	2.54
	m	in	39.3701
Mass	kg	lbm	2.20462
	lbm	gm	453.592
	metric ton	kg	1000
	U.S. short ton	kg	907.185
	U.S. short ton	lbm	2000
Force	dyne	N	1×10^{-5}
	kgf	N	9.80665
	lbf	N	4.44822
Temperature	°C	K	$t(K) = t(°C) + 273.15$
	°F	°R	$t(°R) = t(°F) + 459.67$
Volume	barrel	m^3	0.15899
	barrel	gal (U.S.)	42
	gal (U.S.)	in^3	231
	gal (U.S.)	ft^3	0.13368
	gal (U.S.)	liter	3.7854
	gal (U.K.)	liter	4.5461
Density	gm/cm^3	lbm/ft^3	62.428
	kg/m^3	lbm/ft^3	0.062428
	lbm/gal (U.S.)	kg/m^3	119.826
Pressure	atm	bar	1.01325
	atm	kgf/cm^2	1.03323
	atm	mmHg (torr)	760
	atm	psi	14.696
	bar	Pa	1×10^5
	lbf/ft^2	Pa	47.8803
	Pa	N/m^2	1
	psi (lbf/in^2)	Pa	6894.76

(*Continued*)

TABLE A.3. (*Continued*)

	To Convert from	To	Multiply by
Torque	ft · lbf	m · N	1.3558
	m · N	ft · lbf	0.73757
Energy	Btu	cal	251.996
	Btu	J	1054.35
	cal	J	4.184
	erg	J	1×10^{-7}
	J	N · m	1
	kcal	Btu	3.96832
	kW · hr	Btu	3414.43
	kW · hr	J	3.6×10^6
Power	Btu/hr	W	0.2929
	hp	Btu/hr	2546
	hp	cal/hr	6.416×10^5
	hp	ft · lbf/sec	550
	hp	kW	0.7457
	kcal/hr	W	1.1622
	W	cal/hr	860.42
	W	J/sec	1
Specific heat	Btu/lbm °F	J/kg °C	4184
	cal/gm · °C	Btu/lbm · °F	1
	cal/gm · °C	J/kg °C	4184
Dynamic	centipoise	gm/m · sec	1
viscosity	centipoise	lbm/ft · sec	0.6720×10^{-3}
	kg/m · sec	lbm/ft · sec	0.6720
	lbf · sec/ft²	kgf · sec/m²	4.882
	lbf · sec/ft²	N · sec/m²	47.88
	Pa · sec	N · sec/m²	1
	poise	N · sec/m²	0.1
Kinematic	centistoke	mm²/sec	1
viscosity	stoke	cm²/sec	1
	stoke	ft²/sec	1.076×10^{-3}
Mass flux	kg/sec · m²	lbm/sec · ft²	0.2048
Energy flux	Btu/hr · ft²	W/m²	3.152
	cal/hr · cm²	Btu/hr · ft²	3.687
	cal/hr · cm²	W/m²	11.62
	W/m²	J/sec · m²	1
Heat transfer	Btu/hr · ft² · °F	W/m² · °C	5.674
coefficient	cal/sec · cm² · °C	W/m² · °C	4.184×10^4
	kcal/hr · cm² · °C	Btu/hr · ft² · °F	2048.2
	W/m² · °C	J/sec · m² · °C	1
Thermal	Btu/hr · ft · °F	W/m · °C	1.730
conductivity	cal/sec · cm · °C	W/m · °C	418.4
	kcal/hr · cm · °C	Btu/hr · ft · °F	67.20
	W/m · °C	J/sec · m · °C	1
Surface tension	dyne/cm	N/m	0.001
	lbf/ft	dyne/cm	1.4594×10^4

Adapted with permission from R. E. Bolz and G. L. Tuve, *Handbook of Tables for Applied Engineering Science*, 2nd. ed. (1973), © CRC Press, Inc., Boca Raton, FL.

B. BOND ENERGIES FOR THERMOCHEMICAL CALCULATIONS

If the energy associated with a particular chemical bond in a molecule were independent of the nature of other nearby bonds, the energy of formation of a compound from dissociated elements could be determined by a summation of "standard" bond energies. Resonance among the different modes of energy storage makes this untrue, but the energy of a homopolar bond is to a first approximation independent of the nature of other groups. Though a bond table based on such an approximation is full of empiricism, it is nevertheless attractive in application to fuel use, where determination of the heat of combustion of a wide variety of fuel types is of interest.

If a suitably large number of modified values of bond energies are provided to allow for the nature of other nearby bonds in the molecule, the heats of formation can be determined to any degree of accuracy, but the table of standard bond energies thereby becomes so elaborate as to defeat its purpose of permitting calculation from a small number of constants. In the following table (Table B.1) a compromise is made between identification of so few variations of a particular bond as to make accuracy poor and so many as to make calculations unwieldy. To permit accurate prediction of the heats of combustion, carbon-carbon, carbon-oxygen, carbon-nitrogen, and a few other bonds have been partially differentiated. They represent carbon atoms to which $0, 1, \ldots, 4$ bonds have been partially differentiated. The symbols $C^0, C^i \ldots C^{iv}$, represent carbon atoms to which $0, 1 \ldots 4$, other C atoms are attached. The construction of the table is based on hydrocarbons, alcohols, aldehydes, ketones, acids, esters, ethers, and 28 compounds containing nitrogen.

Although a bond energy table (Table B.1) so constructed will permit accurate computation of the bond-break energy from molecular to completely atomic state, it will not indicate accurately the energy requirement for breaking a particular bond—a matter of great interest in considering the probability of occurrence, in combustion, of reactions involving atoms and free radicals. For example, knowledge of the energies of dissociation of H_2 and O_2 into atoms and of the heat of combustion of H_2 and O_2 to water vapor permits evaluation of the total bond energy in $H-O-H$, 221.56 kcal/gm-mol. One-half of this value is reported in the table as the average $O-H$ bond energy, but it is known that breaking the first and second $O-H$ bonds in H_2O requires 119.22 and 102.34 kcals, respectively. Another example: Knowledge of the heat of combustion of methane, heat of vaporization of carbon, and the energies of complete dissociation of H_2O and CO_2 to the atomic state permits evaluating the total energy of the bonds in CH_4 (average, 99.40, reported in the table). But breaking the first of the four bonds requires 101.9 kcal, and the second, third, and fourth require 112.5, 101.8, and 81.4, respectively.

The usual convention of signs is adopted (i.e., ΔH for the process is positive when the reaction is endothermic). Since energy is required to break bonds, all reactions of bond breaking are endothermic.

TABLE B.1. Bond Energies at 25°C[a], kcal / gm-mol

Bond	$-\Delta H$	Bond	$-\Delta H$	Bond	$-\Delta H$
H — O —	110.78[b]	C^{iii} — C^{iii}	81.96	C — I	51.5
— O — O —	34.28	C^{iii} — C^{iv}	81.11	C^0 — O	78.33
— O — N ≡	60.2	C^{iv} — C^{iv}	80.13	C^i — O	83.9
— O — N =	84.8	C_{arom} — C^i	84.6	C^{ii} — O	87.8
O = N —	107.4	C_{arom} — C^{ii}aliph	86.0	C^{iii} — O	90.5
= N — N =	41.3	C_{arom} — C^{ii}ole	87.7	C_{arom} — O	90.9
≡ N — N ≡	49.0	C — C $\Big\}$ within	91.0[e]	C^0 = O	164.0[h]
— N = N —	96.3	C = C $\Big\}$ arom. rings	150.4	C^i = O	173.6
= N = N —	93.4			C^{ii} = O	181.3
= N — H	93.4	C^i = C^i	140.9	$C_s \rightarrow C_g$	171.29[i]
= N — C^0	66.4	C^i = C^{ii}	145.0	H — H	104.21
= N — C^i	68.6	C^i = C^{iii}	148.0	O = O	119.11
= N — C^{ii}	70.3	C^{ii} = C^{ii}cis	148.1	N ≡ N	225.95
= N — C_{arom}	76.7	C^{il} = C^{ii}trans	149.1	F — F	37.77
≡ N — C^0	74.0[c]	C^{ii} = C^{iii}	150.5[f]	Cl — Cl	57.98
— N = C^0	136.2	C^{iii} = C^{iii}	150.3	Br — Br	53.46
N ≡ C^0	206.1	C^i ≡ C^i	193.8	I — I	51.03
N ≡ C^i	214.6[d]	C^i ≡ C^{ii}	200.1	H — F	135.79
C^i — C^i	79.05	C^{ii} ≡ C^{ii}	205.7	H — Cl	103.16
C^i — C^{ii}	80.17	C — H	99.4	H — Br	87.53
C^i — C^{iii}	81.45	C — S	67.7 ± 2	H — I	71.32
C^i — C^{iv}	82.15	C = S	154 ± 12		
C^{ii} — C^{ii}	81.65	C — F	104.3		
C^{ii} — C^{iii}	82.57	C — Cl	77.7		
C^{ii} — C^{iv}	82.95	C — Br	64.5		

Energy to break *all* bonds in special molecules or groups:

CO_2	384.45	N_2O	266.0	NO	151.0
CO	257.26	NO_2	224.15		

Nitrogen — oxygen bonds in nitro group, NO_2 206.5

[a]Variation of 10°C in base temperature seldom affects last significant figure in result. $-\Delta H$ of formation, gas to gas. Based on thermal data in N.B.S. "Selected Values of Chemical Thermodynamic Properties" and on the JANAF Tables (1965–1967).
[b]Use this average value for molecule-to-atom conversion; OH to O + H requires 102.34 kcal.
[c]From Sanderson.
[d]But for C_2N_2 use the N ≡ C^0 value. And in evaluating C^n — C^m in molecules containing N, count N as C in determining superscripts.
[e]Note that in forming condensed aromatic structures, three single and two double bonds are added for each added ring. Use of values given above yields $-\Delta H_{comb}$ for benzene, naphthalene and anthracene, each within less than 0.1 kcal. error.
[f]When this is *cis* or *trans*, reduce by 0.3.
[g]These values are valid for alcohols; for ethers, acids, esters; increase them by 2.5, 30, and 33.5, respectively.
[h]Increase by 29.3 when in carbonyl sulfide, by 28.22 when in carbon dioxide.
[i]Value is for graphite; 170.84 if diamond.
From Refs. 7 and 8.

TABLE B.2. Energies to Break Some Specific Bonds[a]

Bond	ΔH	Bond	ΔH
HO — H	119.22	HCO — H	74
HO_2 — H	85.13	HCO — CH_3	70
H_2O	221.56[b]	CH_3O — H	103
		CH_3CO — H	87
NO_2	224.15[b]	C_2H_5O — H	103
CH_3 — H	101.9		
		Free Radicals	
C_2H_5 — H	98	O — H	102.34
nC_3H_7 — H	98	O_2 — H	51.6[c]
iC_3H_7 — H	95	O — OH	68.37
sC_4H_9 — H	95	CH_2 — H	112.5
tC_4H_9 — H	89	CH — H	101.8
CH_2CH — H	108	C — H	81.4
CH_2CHCH_2 — H	88	N — H	75
CHC — H	125		
C_6H_5 — H	112		
$C_6H_5CH_2$ — H	85		

[a] Needed for reactions consuming or producing atoms or free radicals. ΔH of bond break (gaseous state) at 25°C and 1 atm, kcal/gm-mol.
[b] For complete atomization of the molecule.
[c] The product is H and molecular oxygen.
From R. T. Sanderson, *Chemical Bonds and Bond Energy*, 2nd ed. (1976), Academic Press, Inc.

TABLE B.3. Heats of Reaction (H_r)[a] and Activation Energies (E_0) of Some Gaseous Radical Reactions of Combustion

Reaction	H_r kcal/gm-mol	E_0 kcal/gm-mol
$H + O_2 \rightarrow OH + O$	−16.8	16.8
$H + O_2 + M \rightarrow HO_2 + M$	51.6	−1.0
$H + HO_2 \rightarrow OH + OH$	34.0	1.9
$H + H_2O \rightarrow H_2 + OH$	−15.0	20.4
$H + CH_4 \rightarrow H_2 + CH_3$	2.3	11.9
$OH + C_2H_6 \rightarrow H_2O + C_2H_5$	21.2	2.4
$OH + C_2H_4 \rightarrow H_2O + C_2H_3$	11.2	3.5
$OH + C_2H_2 \rightarrow H_2O + C_2H$	−5.8	7.0
$HO_2 + H \rightarrow OH + OH$	34.0	1.9
$HO_2 + O \rightarrow OH + O_2$	50.7	1.0
$HO_2 + HO_2 \rightarrow H_2O_2 + O_2$	33.5	1.0
$HO_2 + CO \rightarrow CO_2 + OH$	58.8	23.0

[a] $H_r \equiv -\Delta H$

An additional table is provided (Table B.2) giving the energy required to break some specific bonds in molecules or radicals. Table B.1 suffices for combustion-related molecule-to-molecule reactions. For reactions of interest in combustion kinetics—with atoms or radicals involved—one needs additional information from Table B.2.

Tables B.1 and B.2 have been used to compute the heats of reaction of some gaseous radical reactions of combustion, Table B.3. That table includes, in addition, the values of activation energy, taken from Chapter 3.

References: 4, 5, 6, 7, 8, 9.

C. NORMALIZED ENTHALPY OF COMBUSTION PRODUCTS

It has been shown[10] that the normalized molal enthalpy H^* of the combustion products of any of 20 representative gaseous, liquid, or solid fuels burned completely with air is, with an accuracy adequate for engineering use, independent of fuel type; it depends only on temperature and fractional excess air. With H representing enthalpy above the base temperature 298 K and with 1800 K taken as the normalizing temperature, H^* is defined by

$$H^* = \frac{H_T}{H_{1800}} \qquad (1)$$

The air is assumed half saturated with water vapor at 60°F (volume fraction $H_2O = 0.0088$), and dry air is assumed 21% O_2, 78% N_2, 1% Ar. H^* is given as a function of T for 15% excess air in the normalized enthalpy chart, Figure 1. The dotted line on the figure is the normalized H^* for air alone. For 0% excess air the relation

$$H^* = 0.5894\left(\frac{T-500}{1000}\right) + \sqrt{0.0383\left(\frac{T-500}{1000}\right)^2 + 0.0466} - 0.1 \qquad (2)$$

differs from individual fuel values by a maximum of 0.3% over the temperature range 300–2400 K and the fuel composition range $CH_{0.5}$ to CH_4. (Fuels containing O, N, and S have been included, with SO_2 assumed to have the same MC_p as CO_2, with negligible error). Pure H_2 shows twice the above maximum deviation at 1000 K and 2600 K; pure CO is within 0.3% from 1000 K up, but the error increases as temperature drops, to 1% at 700 K.

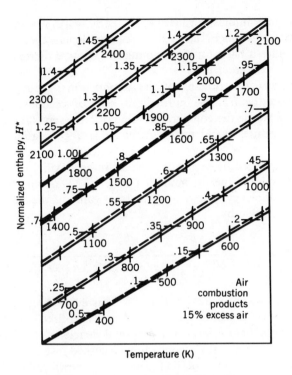

FIGURE 1. Normalized enthalpy chart.

The multiplying correction C to the value of H^* from Eq. (2) to allow for X, the percent excess air, is

$$C = 1 + \left[7.14\left(\frac{T - 2600}{1000}\right)^2 - 4.57\right]10^{-5}X. \qquad (3a)$$

When $X = 15$,

$$C = 0.9993 + 0.00107\left(\frac{T - 2600}{1000}\right)^2 \qquad (3b)$$

Equations (2) and (3a), together with fuel composition and heating value and the molal enthalpies at 1800 K of CO_2, H_2O, O_2, and N_2' (containing argon, $A : N_2 = 1 : 78$), constitute a complete description of the thermochemical properties of the combustion products of all fuels, without need for a temperature-enthalpy table for combustion products. The H_{1800} for CO_2, H_2O, O_2, N_2', and air are 18.987, 14.984, 12.354, 11.653, and 11.800 kcal/gm-mol, respectively.

A unit of energy in blast furnace gas is clearly worth less than the same unit in methane, but how much? The absence of any dependence of h^* versus T on fuel type provides a basis for determining the relative thermal value of a unit of energy in two different fuels.

Let the normalized enthalpy of the combustion products of a fuel as it enters a furnace be h_E^*, the normalized heat of combustion per mole of products. Let $h_{E,S}^*$ be the comparable value of a chosen standard of reference fuel. The subject fuel would maintain the same T-Q relation along the flow path in a furnace as the reference fuel if h_E^* were increased by $h_{E,S}^* - h_E^*$. This could be accomplished by preheating the air in an auxiliary preheater furnace fired with the subject fuel in amount $(h_{E,S}^* - h_E^*)/E$, where E is the pre-heater-furnace efficiency. The relative thermal value of the subject fuel, called its RTV, is then given by

$$ \text{RTV} = \frac{h_E^*}{h_E^* + (h_{E,S}^* - h_E^*)/E} = \frac{1}{1 + (h_{E,S}^*/h_E^* - 1)/E} \qquad (4) $$

A suggested constant value for E is 0.8. This evaluation of the subject fuel does not imply that it will in fact be used in a fuel-fired preheater; there are other ways to compensate for a fuel's thermal deficiency.

The application of the RTV concept is discussed in Ref. 10. An example is the appraisal of FLEXICOKE (Exxon trademark) tailings compared to methane; its RTV is 0.77, better than that of producer gas made from lignite.

Reference: 10.

D. EQUILIBRIUM CONSTANTS

The equilibrium constant K_p for gas reaction

$$ aA + bB = cC + dD $$

is represented by

$$ K_p = \frac{P_C^c P_D^d}{P_A^a P_B^b} $$

K_p for gas reactions is pressure-independent at low total pressures. For heterogeneous reactions, the activity of a pure solid or liquid may be taken as unity at all temperatures. Therefore, in reactions involving carbon, only gaseous species are identified in the calculation of equilibrium constants.

The equilibrium constant chart as shown in Figure 2[5] is based on *JANAF Tables.*[7] Log K_p is plotted versus $1/T$ in the range of 600–5000 K for 35 combustion-related reactions.

References: 5, 7.

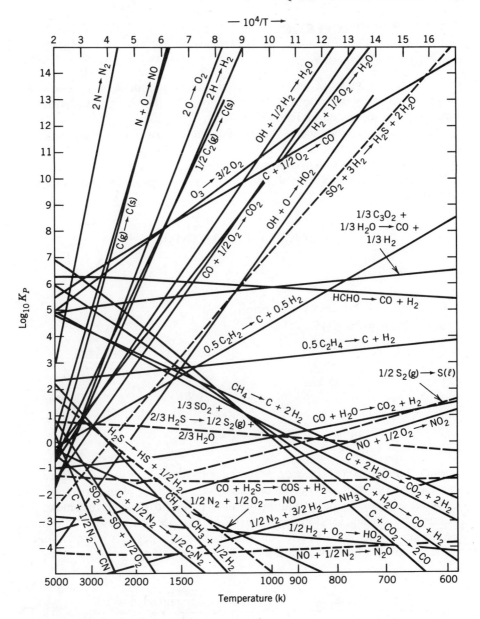

FIGURE 2. Combustion-related equilibrium constants based on *JANAF Tables*, 1960–1967: Five elements plus N, NH_3, NO, NO_2, N_2O, CN, C_2N_2, HS, H_2S, $S_2(g)$, SO, SO_2, COS, O, O_3, OH, HO_2, H_2O, CO, CO_2, C_3O_2, C(g), $C_2(g)$, CH_3, CH_4, C_2H_2, C_2H_4, HCHO.

E. TRANSPORT PROPERTIES

E.1. Thermal Conductivity

Thermal conductivities of gases are not strongly dependent on pressure. At low pressures, they increase with temperature. The thermal conductivities of common gases are listed in Table E.1.

The pressure dependence of thermal conductivity of liquids can usually be neglected. The values given in Table E.1 are for subcooled liquids under atmospheric pressure unless otherwise noted.

The thermal conductivity of petroleum oils is given by the empirical equation

$$K = 1.172 \times (1 - 0.00054t)\, d$$

where

K = thermal conductivity, mW/cm K
d = specific gravity (15.6/15.6°C)
t = temperature, °C

References: 11, 12.

E.2. Surface Tension

Surface tension is the free or potential surface energy per unit area, expressed in ergs/cm^2 or equivalently in dynes/cm. It is of major importance in emulsions and in determining the sizes of liquid droplets in smoke and clouds. The surface tension of various liquids is given in Table E.2.

The surface tension of petroleum products usually lies in a narrow range of values. Typical values at 25°C are

Material	Surface Tension (dynes/cm)
Gasoline	26
Kerosene	25–30
Gas oil	25–30
Lubricating fractions	34
Fuel oils	29–32

References: 2, 13.

TABLE E.1. Thermal Conductivity (mW / cm · K) of Pure Substances

Compound	Formula	Gas at Temperature, K, of									Liquid at Temperature, K, of			
		200	250	300	350	400	450	500	750	1000	250	300	350	400
Paraffins														
Methane	CH_4	0.218	0.277	0.343	0.412	0.484	0.578	0.671	1.137	1.69				
Ethane	C_2H_6	0.102	0.156	0.218	0.286	0.360	0.437	0.516	0.957	1.64				
Propane	C_3H_8		0.129	0.183	0.237	0.295	0.355	0.417						
n-Butane	C_4H_{10}			0.160	0.210	0.264	0.320	0.377			1.216	1.064[a]	0.913[a]	0.761[a]
Isobutane	C_4H_{10}			0.163	0.216	0.272	0.328	0.385						
n-Pentane	C_5H_{12}				0.199	0.250	0.304	0.362			1.33	1.18	1.02[b]	0.86[b]
n-Hexane	C_6H_{14}				0.176	0.232	0.292	0.355	0.768	1.36	1.35	1.230	1.109[b]	0.99[b]
n-Heptane	C_7H_{16}					0.214	0.267	0.325	0.639	0.970	1.38	1.265	1.150	1.04[b]
n-Octane	C_8H_{18}					0.226	0.278	0.334			1.46	1.32	1.17	1.03[b]
Olefin														
Ethylene	C_2H_4	0.088	0.138	0.204	0.274	0.350	0.427							
Aromatics														
Benzene	C_6H_6		0.077	0.104	0.144	0.195	0.259	0.335			1.566[c]	1.440	1.313	1.187[b]
Toluene	C_7H_8		0.116	0.146	0.189	0.240	0.295	0.349			1.467	1.340	1.209	1.09[b]

(*Continued*)

TABLE E.1. (Continued)

Compound	Formula	Gas at Temperature, K, of									Liquid at Temperature, K, of			
		200	250	300	350	400	450	500	750	1000	250	300	350	400
Miscellaneous														
Acetone	(CH$_3$)$_2$CO	0.117		0.115	0.155	0.201	0.252	0.310			1.780	1.594	1.41[b]	1.22[b]
Acetylene	C$_2$H$_2$		0.162	0.213	0.269	0.332	0.394	0.452						
Ammonia	NH$_3$		0.197	0.246	0.302	0.364	0.433	0.506	0.894		5.92[a]	4.785[a]	3.653[a]	2.52[a]
Carbon dioxide	CO$_2$	0.095	0.129	0.166	0.205	0.244	0.283	0.323	0.525	0.680	1.338[a]	C.741[a]		
Carbon monoxide	CO	0.175	0.214	0.252	0.288	0.323	0.355	0.386	0.523	0.644				
Ethyl alcohol	C$_2$H$_5$OH					0.245	0.288	0.327			1.808	1.660	1.512	1.363[b]
Hydrogen	H$_2$	1.280	1.560	1.815	2.033	2.212	2.389	2.564	3.43	4.28				
Hydrogen sulfide	H$_2$S		0.114	0.147	0.180	0.212								
Methyl alcohol	CH$_3$OH				0.196	0.249	0.300	0.351			2.164	2.022	1.879[a]	1.737[a]
Nitrogen	N$_2$	0.183	0.222	0.260	0.294	0.325	0.356	0.386	0.517	0.631				
Nitric oxide	NO	0.178	0.219	0.259	0.296	0.331	0.364	0.396	0.562	0.723				
Oxygen	O$_2$	0.182	0.225	0.267	0.306	0.342	0.377	0.412	0.574	0.717				
Water	H$_2$O			0.181	0.222	0.264	0.307	0.357	0.63		5.225[d]	6.084	6.673	6.864[b]

[a]Under saturated vapor pressures.

[b]Extrapolated for the liquid under vapor pressures, ignoring pressure dependence.

[c]Extrapolated for the supercooled liquid.

[d]Supercooled liquid.

Adapted with permission from Y. S. Touloukian, P. E. Liley, and S. C. Saxena, *Thermophysical Properties of Matter*, vol. 3 (1970), CINDAS/Purdue University.

TABLE E.2. Surface Tension of Various Liquids

Compound	Formula	In Contact with	Temperature, °C	Surface Tension, dynes/cm
Acetaldehyde	C_2H_4O	Vapor	20	21.2
Acetic acid	$C_2H_4O_2$	Vapor	10	28.8
Acetone	C_3H_6O	Air or vapor	0	26.21
	C_3H_6O	Air or vapor	20	23.70
	C_3H_6O	Air or vapor	40	21.16
Acetylene	C_2H_2	Vapor	−70.5	16.4
Ammonia	NH_3	Vapor	11.1	23.4
	NH_3	Vapor	34.1	18.1
Aniline	C_6H_7N	Vapor	20	42.9
Benzene	C_6H_6	Air	10	30.22
	C_6H_6	Air	20	28.85
n-Butyl alcohol	$C_4H_{10}O$	Air or vapor	20	24.6
Carbon bisulfide	CS_2	Vapor	20	32.33
Carbon dioxide	CO_2	Vapor	20	1.16
	CO_2	Vapor	−25	9.13
Carbon tetrachloride	CCl_4	Vapor	20	26.95
	CCl_4	Vapor	100	17.26
Carbon monoxide	CO	Vapor	−193	9.8
Chlorine	Cl_2	Vapor	20	18.4
	Cl_2	Vapor	−50	29.2
Ethyl alcohol	C_2H_6O	Air	0	24.05
	C_2H_6O	Vapor	10	23.61
	C_2H_6O	Vapor	20	22.75
Ethyl benzene	C_8H_{10}	Vapor	20	29.20
Ethylene oxide	C_2H_4O	Vapor	−20	30.8
	C_2H_4O	Vapor	0	27.6
	C_2H_4O	Vapor	20	24.3
Ethyl ether	$C_4H_{10}O$	Vapor	20	17.01
	$C_4H_{10}O$	Vapor	50	13.47
Formic acid	CH_2O_2	Air	20	37.6
Furfural	$C_5H_4O_2$	Air or vapor	20	43.5
Glycerol	$C_3H_8O_3$	Air	20	63.4
Glycol	$C_2H_6O_2$	Air or vapor	20	47.7
n-Hexane	C_6H_{14}	Air	20	18.43
Hydrogen	H_2	Vapor	−255	2.31
Hydrogen peroxide	H_2O_2	Vapor	18.2	76.1
Isopentane	C_5H_{12}	Air	20	13.72
Methyl alcohol	CH_4O	Air	0	24.49
	CH_4O	Air	20	22.61
	CH_4O	Vapor	50	20.14
Methyl ethyl ketone	C_4H_8O	Air or vapor	20	24.6
Naphthalene	$C_{10}H_8$	Air or vapor	127	28.8
Nitrobenzene	$C_6H_5NO_2$	Air or vapor	20	43.9

(*Continued*)

TABLE E.2. (*Continued*)

Compound	Formula	In Contact with	Temperature, °C	Surface Tension, dynes/cm
Nitrogen	N_2	Vapor	-183	6.6
	N_2	Vapor	-193	8.27
	N_2	Vapor	-203	10.53
Nitrous oxide	N_2O	Vapor	20	1.75
n-octane	C_8H_{18}	Vapor	20	21.8
Oxygen	O_2	Vapor	-183	13.2
Phenol	C_6H_6O	Air or vapor	20	40.9
n-Propyl alcohol	C_3H_8O	Vapor	20	23.78
Styrene	C_8H_8	Air	19	32.14
Toluene	C_7H_8	Vapor	20	28.5
Water	H_2O	Air	18	73.05
m-Xylene	C_8H_{10}	Vapor	20	28.9
o-Xylene	C_8H_{10}	Air	20	30.10
p-Xylene	C_8H_{10}	Vapor	20	28.37

F. DIFFUSION COEFFICIENTS FOR BINARY GAS MIXTURES

The diffusion flux of component A in a fluid mixture of A and B in the z direction across a plane of unit cross section perpendicular to the direction of diffusion is given by

$$\dot{N}_A = -D_{AB}\frac{dC_A}{dz} + \left(\dot{N}_A + \dot{N}_B\right)f_A \qquad (5)$$

where

$\dot{N}_A$ = molar flux of A with respect to stationary coordinates, gm-mol/cm² · sec

$\dot{N}_B$ = molar flux of B with respect to stationary coordinates, gm-mol/cm² · sec

D_{AB} = molecular diffusion coefficient of A in A, B mixture, cm²/sec

C_A = concentration of component A, gm-mol/cm³

z = distance in direction of diffusion, cm

f_A = mol fraction A in the A-B mixture

The second term on the right is the contribution of total bulk flow to the transport of A and is zero when A and B are in equimolal counterdiffusion.

If no impenetrable boundary is present, the diffusion of one component of a gas mixture will be accompanied by diffusion of one or more others in the opposite direction for molal conservation in the region under consideration. At room temperature under atmospheric pressure, the diffusion coefficients for most gases and vapors lie in the range of 0.05 to 1 cm^2/sec. In liquids of about 1 centipoise viscosity, diffusivities are of the order 10^{-4} to 10^{-5} cm^2/sec.[30]

Table F.1 gives experimental data for many pairs of gases at 1 atm. Approximate allowance for the effects of temperature T and pressure P on diffusion is given by

$$D_{T,P} = D_{T_0,P_0}\left(\frac{T}{T_0}\right)^{1.75}\left(\frac{P_0}{P}\right) \tag{6}$$

where D_{T_0,P_0} is the diffusion coefficient at a reference temperature and pressure T_0, P_0, which may be chosen arbitrarily.

Although the diffusion of component A in a multicomponent system in principle involves many diffusivity coefficients, an adequate approximation is to use a binary diffusion coefficient of A through the other components lumped together.* Values of Field et al.[16] for the calculated binary diffusion

TABLE F.1. Observed Diffusion Coefficients at a Pressure of 1 atm

Gas pair	D in cm^2/sec at Temperature		
	0°C	20°C	25°C
N_2-H_2	0.674	0.76	
N_2-O_2	0.181	0.22	
N_2-CO	0.192		
N_2-CO_2	0.144	0.16	0.165
$N_2-C_2H_6$			0.148
$N_2-C_2H_4$			0.163
$N_2-nC_4H_{10}$			0.096
N_2-iso-C_4H_{10}			0.0908
N_2-cis-2-C_4H_8			0.095
H_2-O_2	0.697		
H_2-CO	0.651		
H_2-CO_2	0.550	0.60	0.646
H_2-N_2	0.674		
H_2-CH_4	0.625		0.726
$H_2-C_2H_6$	0.459		0.537
$H_2-C_2H_4$	0.486		0.602
H_2-cis-2-C_4H_8			0.378
CO_2-O_2	0.139	0.16	
CO_2-CO	0.137		
CO_2-N_2O	0.096		
CO_2-CH_4	0.153		
$CO-O_2$	0.185		

*With f_n representing mole fraction of species B_n in an $A-B$ mixture, diffusivity of A (in low concentrations) through the mixture is given by $D_{AB} = 1/\Sigma_n(f_n/D_{AB_n})$.

TABLE F.2. Calculated Diffusion Coefficients for O_2-N_2, $CO-N_2$, CO_2-N_2 and H_2O-N_2 Gas Pairs at a Pressure of 1 atm

Temperature K	Diffusion Coefficients, cm^2/sec			
	O_2-N_2	$CO-N_2$	CO_2-N_2	H_2O-N_2
300	0.207	0.206	0.154	0.253
400	0.343	0.340	0.258	0.432
500	0.503	0.499	0.381	0.646
600	0.685	0.679	0.522	0.892
700	0.888	0.881	0.678	1.17
800	1.11	1.10	0.851	1.47
900	1.35	1.34	1.04	1.80
1000	1.61	1.60	1.24	2.15
1100	1.88	1.87	1.45	2.52
1200	2.17	2.15	1.68	2.92
1300	2.47	2.45	1.92	3.34
1400	2.79	2.77	2.17	3.78
1500	3.13	3.10	2.42	4.25
1600	3.49	3.46	2.69	4.73
1700	3.86	3.83	2.97	5.24
1800	4.26	4.22	3.26	5.76
1900	4.67	4.63	3.57	6.29
2000	5.10	5.06	3.89	6.83
2100	5.53	5.48	4.22	7.39
2200	5.97	5.91	4.56	7.97
2300	6.42	6.36	4.92	8.58
2400	6.88	6.82	5.29	9.20
2500	7.36	7.29	5.67	9.84
2600	7.86	7.79	6.07	10.5
2700	8.37	8.29	6.46	11.2
2800	8.89	8.81	6.86	11.9
2900	9.43	9.35	7.26	12.6
3000	9.99	9.90	7.67	13.3

TABLE F.3. Diffusion Coefficient and Schmidt Number for Water Vapor into Air

Temperature °C	Diffusion Coefficient, D cm²/sec	ft²/hr	Schmidt Number[a] ($\mu/\rho D$)
0	0.218	0.844	0.608
10	0.232	0.898	0.610
20	0.246	0.952	0.612
30	0.260	1.01	0.614
40	0.275	1.06	0.615
50	0.290	1.12	0.616
60	0.305	1.18	0.618
70	0.321	1.24	0.619
80	0.337	1.30	0.619

[a] The values of $\mu/\rho D$ were calculated using the viscosity and density of dry air. Thus the values apply only when the diffusing water vapor is very dilute.[3]
Adapted with permisson from R. E. Bolz and G. L. Tuve, *Handbook of Tables for Applied Engineering Science*, 2nd ed. (1973), © CRC Press, Inc., Boca Raton, FL.

coefficients of oxygen, carbon monoxide, carbon dioxide and water through nitrogen over a range of gas temperatures up to 3000°K are shown in Table F.2. The accuracy of these values is probably in the region of 10% as indicated by Field et al.[16]

The diffusion coefficients of water vapor into air and of other gases and vapors into air, together with the Schmidt numbers, are given in Tables F.3 and F.4, respectively. The Schmidt number, the ratio of kinematic viscosity to the molecular diffusion coefficient (0.2 to 3.0 for most gas pairs), is a dimensionless group useful in mass-transfer correlations.

In turbulent flow, diffusion is primarily by eddy mixing. The eddy diffusion coefficient E_{AB} is customarily defined by

$$\left(\dot{N}_A\right)_E = -E_{AB}\frac{dC_A}{dz} \tag{7}$$

where

$(\dot{N}_A)_E$ = rate of eddy diffusion of component A into mixture, gm-mol/cm² · sec

E_{AB} = Eddy diffusion coefficient, cm²/sec

E_{AB} is roughly proportional to the eddy scale and to the magnitude of the velocity fluctuations. It is much larger than the molecular diffusion coefficients, usually varying from 3 to 40 cm²/sec over a range of Reynolds numbers from 10,000 to 175,000.

References: 3, 4, 14, 15, 16, 17, 18.

TABLE F.4. Diffusion Coefficient and Schmidt Number for Gases and Vapors into Air at 1 atm

Substance	Diffusion Coefficient, D, cm²/sec		Diffusion Coefficient, D, ft²/hr		$\dfrac{\mu}{\rho D}$ [a]	
	0°C	25°C	0°C	25°C	0°C	25°C
H_2	0.611	0.712	2.37	2.76	0.217	0.216
NH_3	0.198	0.229	0.766	0.886	0.669	0.673
N_2	0.178		0.691		0.744	
O_2	0.178	0.206	0.689	0.80	0.744	0.748
CO_2	0.142	0.164	0.550	0.635	0.933	0.940
CS_2	0.094	0.107	0.36	0.414	1.41	1.44
Methyl alcohol	0.132	0.159	0.513	0.615	1.00	0.969
Formic acid	0.131	0.159	0.509	0.615	1.01	0.969
Acetic acid	0.106	0.133	0.411	0.515	1.25	1.16
Ethyl alcohol	0.102	0.119	0.394	0.461	1.30	1.29
Chloroform	0.091		0.352		1.46	
Diethylamine	0.0884	0.105	0.342	0.406	1.50	1.47
n-Propyl alcohol	0.085	0.100	0.329	0.387	1.56	1.54
Propionic acid	0.0846	0.099	0.328	0.383	1.57	1.56
Methyl acetate	0.0840	0.100	0.325	0.387	1.58	1.54
Butylamine	0.0821	0.101	0.318	0.391	1.61	1.53
Ethyl ether	0.0786	0.093	0.304	0.360	1.69	1.66
Benzene	0.0751	0.088	0.291	0.341	1.76	1.75
Ethyl acetate	0.0715	0.085	0.277	0.330	1.85	1.81
Toluene	0.0709	0.084	0.274	0.325	1.87	1.83
n-Butyl alcohol	0.0703	0.090	0.272	0.348	1.88	1.71
i-Butyric acid	0.0679	0.081	0.263	0.313	1.95	1.90
Chlorobenzene		0.073		0.283		2.11
Aniline	0.0610	0.072	0.236	0.279	2.17	2.14
Xylene	0.059	0.071	0.228	0.275	2.25	2.17
Amyl alcohol	0.0589	0.070	0.228	0.271	2.25	2.20
n-Octane	0.0505	0.060	0.195	0.232	2.62	2.57
Naphthalene	0.0513	0.052	0.199	0.20	2.58	2.96

[a] Based on $\mu/\rho = 0.1325$ cm²/sec for air at 0°C and 0.1541 cm²/sec for air at 25°C; applies only when the diffusing gas or vapor is very dilute.[3]

Adapted with permisson from R. E. Bolz and G. L. Tuve, *Handbook of Tables for Applied Engineering Science*, 2nd ed. (1973), © CRC Press, Inc., Boca Raton, FL.

G. FLAME PROPERTIES

Some of the properties of flames relate to safety. Most of these properties are apparatus dependent. Table G.1 gives a summary of some representative data for the flash point, flammability limits, autoignition temperature, minimum ignition energy, quenching distance, and maximum flame velocity of some selected fuels. Adiabatic temperatures of some typical fuel-air mixtures with and without dissociation are listed in Tables G.2 and G.3.

G.1. Flash Point

The flash point is defined as the minimum liquid temperature at which the vapor gives a visible flash when a small flame is applied in a prescribed type of apparatus. Tabulated in Table G.1 are values determined by closed-cup methods.

References: 19, 20.

G.2. Flammability Limits

Table G.1 gives the lean and rich limits of flame propagation in fuel-air mixtures of some selected fuels. Most mixtures are inflammable when the fuel-air volume ratio lies between 50% and 300% of the stoichiometric value. Hydrogen and carbon disulfide are inflammable in much richer mixtures.

References: 19, 20, 21, 22.

G.3. Autoignition Temperature

The autoignition temperature (AIT), or the minimum spontaneous ignition temperature, is the lowest temperature at which ignition occurs. Because of the fact that such a temperature is extremely dependent on material of construction, apparatus configuration, and test procedure, the reported test values from literature vary. Table G.1 gives autoignition temperatures and identifies the methods used for most hydrocarbons.

References: 19, 20, 23.

G.4. Minimum Ignition Energy and Quenching Distance

The minimum ignition energy is the smallest quantity of energy which will ignite a mixture. The quenching distance is the minimum-sized channel, or minimum electrode spacing, which will allow a flame to propagate. The values listed in Table G.1 are for stoichiometric fuel-air mixtures at atmospheric pressure.

References: 20, 21, 24, 25.

G.5. Maximum Flame Velocity

The flame velocity is the velocity with which a plane flame front propagates through the mixture relative to the combustible mixture at rest. Flame velocity depends primarily on the initial composition, temperature and pressure of the system, and is secondarily affected by such variables as flow rate and geome-

TABLE G.1. Flame Properties of Selected Fuel Components[a]

Compound	Formula	Flash Point °C	Flammability Limits Vol. % in Air — Lean	Flammability Limits Vol. % in Air — Rich	Autoignition Temperature, °C	Stoichiometric Mixture — Vol % Fuel	Stoichiometric Mixture — Minimum Ignition Energy 10^{-5} J	Stoichiometric Mixture — Quenching Distance, mm	Flame Velocity — Vol % Fuel	Flame Velocity — $S_{U,max}$ cm/sec
Paraffins										
Methane	CH_4	−188 −180	5.0 5.35	14.2 15.0	595(2), 537(4)	9.47	33	1.9	9.96	33.8
Ethane	C_2H_6	−135 −130	2.9 3.3	10.6 13.0	515(2), 472(4)	5.64	42	2.0	6.28	40.1
Propane	C_3H_8	−104	2.0 2.3	7.3 9.5	493(1), 470(2)	4.02	40	2.1	4.54	39.0
n-Butane	C_4H_{10}	−76 −60	1.5 2.0	6.5 9.0	430(1), 365(1)	3.12	76	3.0	3.52	37.9
2-Methylpropane (Isobutane)	C_4H_{10}	−82 −81	1.8 1.9	8.4 8.5	462(1), 460(4)	3.12			3.48	34.9
n-Pentane	C_5H_{12}	−49 < −20	1.4 1.6	7.8 8.3	284(1), 285(2)	2.55	82	3.3	2.92	38.5
2-Methylbutane (Isopentane)	C_5H_{12}	−56 < −20	1.3 1.6	7.6 8.0	420(1)	2.55	96	3.6	2.89	36.6
2,2-Dimethylpropane	C_5H_{12}	−65 < −7	1.3 1.4	7.5	456(1), 450(4)	2.55	157	4.3	2.85	33.3
n-Hexane	C_6H_{14}	−30 < −20	1.1 1.5	6.9 7.7	261(1), 233(2)	2.16	95	3.6	2.51	38.5
2,2-Dimethylbutane (Neohexane)	C_6H_{14}	−48 < −20	1.2	7.0	440(1), 405(3)	2.16	164	4.6	2.43	35.7
n-Heptane	C_7H_{16}	−4 −1	1.0 1.3	6.7 7.0	247(1), 215(2)	1.87	110	3.8	2.26	38.6
2,2,3-Trimethylbutane (Triptane)	C_7H_{16}	< 0	1.0	7.0	454(1), 412(3)	1.87	100	3.5	2.15	35.9
n-Octane	C_8H_{18}	10 15	0.8 1.1	3.2 6.5	240(1), 206(3)	1.65				
2,2,4-Trimethylpentane (Isooctane)	C_8H_{18}	−12 < 22	0.95 1.15	6.0	447(1), 418(2)	1.65	29	2.0	1.90	34.6

Olefins

Ethylene	C_2H_4	-136	-120	2.7	3.1	28.5	36.0	425(2), 490(4)	6.52	9.6	1.3	7.40	68.3
Propylene	C_3H_6		-108	2.0	2.4	10.3	11.7	455(4)	4.44	28.2	2.0	5.04	43.8
1-Butene	C_4H_8		-80	1.6	1.7	9.3	10.0	440(2), 385(4)	3.37			3.87	43.2
cis-2-Butene	C_4H_8	-73	< -7	1.6	2.0	9.0	10.0	325(4)	3.37				
trans-2-Butene	C_4H_8	< -73	< -7	1.6	2.0	9.7	10.0	325(4)	3.37				
2-Methylpropene (Isobutene)	C_4H_8	-76	< -7	1.6	1.8	8.8	10.0	465(2)	3.37			3.83	37.5
1-Pentene	C_5H_{10}	-51	-18	1.3	1.6	7.7	10.0	298(1), 275(4)	2.71			3.07	42.6
1-Hexene	C_6H_{12}	-26	< -7	1.2	1.3	8.4	9.0	272(1), 253(3)	2.27			2.67	42.1
Naphthanes													
Cyclopentane	C_5H_{10}	-42	6.5		1.4		8.4	385(1)	2.71	83	3.3	3.16	37.3
Cyclohexane	C_6H_{12}	-20	-17		1.3	7.8	8.4	270(1)	2.27		4.1	2.65	38.7
Methylcyclopentane	C_6H_{12}	-27	-25		1.2		8.4	323(1)	2.27			2.75	36.0
Methylcyclohexane	C_7H_{14}	-4	-1	1.2	1.5	7.1		265(1)	1.95			2.43	37.5
Aromatics													
Benzene	C_6H_6		-11	1.3	1.5	5.5	9.2	592(1)	2.71	79	2.8	2.94	44.6
Toluene	C_7H_8	4.5	7	0.9	1.4	4.6	7.4	568(1)	2.27			2.39	38.8
Ethylbenzene	C_8H_{10}		15		1.0	6.7		460(1)	1.95				
o-Xylene	C_8H_{10}	17	32	1.0	1.1	6.0	7.0	501(1)	1.95			2.12	34.4
m-Xylene	C_8H_{10}	23.2	25	0.9	1.1	6.1	7.0	563(1)	1.95				
p-Xylene	C_8H_{10}	23	27		1.1	6.6	7.0	564(1)	1.95				
Styrene	C_8H_8	31	32		1.1	6.1		490	2.05				
Naphthalene	$C_{10}H_8$	79	80		0.9	5.9		526	1.72				

TABLE G.1. (*Continued*)

Compound	Formula	Flash Point °C	Flammability Limits Vol. % in Air — Lean	Rich	Autoignition Temperature, °C	Stoichiometric Mixture — Vol % Fuel	Minimum Ignition Energy 10^{-5} J	Quenching Distance, mm	Flame Velocity — Vol % Fuel	$S_{U,max}$, cm/sec
					Miscellaneous					
Acetaldehyde	CH_3CHO		4	60	175	7.72	37.6	2.3		
Acetic acid	CH_3COOH		5.4		465	9.47				
Acetone	$(CH_3)_2CO$	−18	2.6	13	465	4.97	115	3.8		
Acetylene	C_2H_2	−18	2.5	80	305	7.72	3	0.8	10.1	141
Ammonia	NH_3		15	28	651	21.81				
Aniline	$C_6H_5NH_2$		1.2	8.3	615	2.63				
Carbon disulfide	CS_2	−6	1.3	50	90	6.52	1.5	0.5	~6.7	49.5
Carbon monoxide	CO		12.5	74	609	29.50			~50	39
Ethyl alcohol	C_2H_5OH	12	3.3	19	365	6.52				
Ethylene oxide	CH_2CH_2O		3.6	100	429	7.72	10.5	1.3		
Formaldehyde	$HCHO$		7	73		17.30				
Hydrogen	H_2		4	75	400	29.50	2	0.6	~50	264.8
Hydrogen sulfide	H_2S		4	44	290	12.24	7.7	1.1		
Methyl alcohol	CH_3OH	11	6.7	36	385	12.24	21.5	1.3	~12.4	47.6
n-Propyl alcohol	C_3H_7OH	15	2.2	14	440	4.44				
Propylene oxide	C_3H_6O		2.8	37		4.97	19	1.3		
Tetralin	$C_{10}H_{12}$	71 77	0.8	5	385	1.58			1.6	36.2

[a] For flash points (closed cup) and flammability limitsY 53F, the minimum and maximum values are shown in cases where more than one value was determined. For autoignition temperatures of hydrocarbons, listed in the table are the lowest reported values and the determination method identified in parenthesis as follows: (1) glass or quartz flask, volume less than the standard 200 mL. (usually 125 mL, conical), (2) standard conical glass or quartz flask, volume 200 mL, (3) glass or quartz flask, volume greater than the standard 200 mL (usually 1000 mL, spherical), (4) nonglass or unspecified flask.

TABLE G.2. Adiabatic Flame Temperature of Some Typical Fuels (No Dissociation)[a]

Fuel	Adiabatic Flame Temperature, K	Fuel	Adiabatic Flame Temperature, K
H_2	2525	$n-C_4H_{10}$	2400
CO	2660	$n-C_8H_{18}$	2410
CH_4	2325	C_6H_6	2510
C_2H_6	2380	C_2H_5OH	2320
C_2H_4	2565	NO	2850
C_2H_2	2910		
C_3H_8	2390		
C_3H_6	2505		

[a]Pressure: 1 atm. Base temperature: 25°C. Oxidizer: air. State: gas. Equivalence ratio: 1.0.

TABLE G.3. Calculated Flame Temperature and Combustion Products of Some Fuel-Oxygen / Air Mixtures[a]

Fuel	Adiabatic Flame Temperature, K	Equilibrium Composition, mol %									
		H_2O	CO_2	CO	O_2	H_2	OH	H	O	NO	N_2
Oxidizer: Oxygen											
H_2	3083	57			5	16	10	8	4		
CH_4	3010	37	12	15	7	7	14	5	3		
C_2H_4	3170	24	14	24	10	6	9	6	7		
C_2H_2	3325			61		21		18			
C_3H_8	3091	31.3	13.4	20.0	9.8	6.4	8.7	5.3	5.1		
C_3H_6	3146	24.7	14.0	23.8	10.8	5.7	8.8	5.8	6.4		
$n-C_4H_{10}$	3096	30.1	13.9	20.8	10.0	6.2	8.6	5.2	5.2		
C_4H_8	3138	25.0	14.2	23.7	10.8	5.7	8.8	5.6	6.2		
$n-C_8H_{18}$	3082	26	14	22	9	5	14	5	5		
CO	2973		46	35	15				4		
C_2N_2[b]	4850			66					0.8	0.03	32
Oxidizer: Air											
H_2	2373	32			2	1					65
CH_4	2222	18	8.5	0.9	0.4	0.4	0.3	0.04	0.02	0.2	70.9
C_2H_2	2523	7	12	4	2		1			1	73

[a]Pressure: 1 atm. Base temperature: 25°C. State: gas. Equivalence ratio: 1.0 except for acetylene-oxygen and cyanogen-oxygen. Acetylene-oxygen has a molal ratio of 1 (40% stoichiometric air), the maximum-temperature mixture contains a bit more oxygen. Cyanogen-oxygen is an equimolal mixture.

[b]Remaining products; are mostly N.

Adapted with permission from the Oxford University Press from I. I. Berenblut and A. B. Downes, *Tables for Petroleum Gas/Oxygen Flames Combustion Products and Thermodynamic Properties* (1960) and the authors, A. G. Gaydon and H. G. Wolfhard, *Flames: Their Structures, Radiation and Temperature*, Chapman and Hall Ltd., 1960.

try. It will reach a maximum at a gas composition about 10 percent on the rich side of the stoichiometric value (see table).

The most used methods for measuring flame velocity are the Bunsen burner method and the tube method. The data given in Table G.1 are for fuel-air mixtures, and are based on the tube method and the use of the actual area of the propagating flame cusp.

References: 20, 21, 26, 27.

G.6. Flame Temperature

Flame temperatures are usually calculated from thermochemical data for a fuel burnt adiabatically with air as a function of the equivalence ratio. The calculation of the adiabatic flame temperature of various fuels without dissociation is straightforward. For lean to stoichiometric mixtures,* complete combustion is assumed; the products contain only carbon dioxide, water, nitrogen, oxygen, and sulfur dioxide; some results are given in Table G.2. Heat loss by radiation, thermal conduction, or diffusion to the walls is not considered.

At high temperatures, evaluation of true adiabatic flame temperatures necessitates allowance for dissociation. The values of the calculated temperature and composition of burnt gases with allowance for dissociation of some typical flames taken from Gaydon and Wolfhard[28] and from Berenblut and Downes[29] (data of C_3H_8, C_3H_6, nC_4H_{10}, C_4H_8), are listed in Table G.3. These values are sometimes based on older, slightly inaccurate, thermochemical data.

References: 28, 29, 30.

H. KNOCK RATINGS OF GASOLINE COMPONENTS

The octane number is a measure of the resistance of a fuel to knock. It is the percent of isooctane (2,2,4 trimethyl pentane) in a mixture with *n*-heptane which produces the same intensity of knock as the fuel being tested. Secondary reference fuel blends are commonly used. There are two methods generally used, the Research Method (ASTM D 2699) and the Motor Method (ASTM D 2700). The Research and Motor octane numbers, as well as blending octane numbers, of some hydrocarbons are given in Table H.1.

The Research Method is representative of city driving at low speed; the Motor Method represents high-speed performance on the highway. Road vehicles generally rate gasolines between the two methods.

Empirical correlations have been developed which permit calculation of vehicle antiknock performance based on the following general equation:

$$AI = k_1(RON) + k_2(MON) + k_3$$

---

*For rich mixtures assume all oxygen deficiency is in CO and H_2, obtainable from water–gas equilibrium, $(CO_2)(H_2)/(CO)(H_2O) = 0.2$.

TABLE H.1. Knock Ratings of Hydrocarbons[a]

Compound	Research Octane Number (CRC-F1)				Motor Octane Number (CRC-F2)			
	mL TEL/U.S. gal			Blending Octane Number	mL TEL/U.S. gal			Blending Octane Number
	0.0	1.0	3.0		0.0	1.0	3.0	
Paraffins								
n-Pentane	61.70	74.9	88.7	62	61.9	77.1	83.6	67
2-Methylbutane (Isopentane)	92.3	+0.4	+1.0	100	90.3	> 100		104
2,2-Dimethylpropane (Neopentane)	85.5	97.4	+0.1	100	80.2	93.0	99.9	90
n-Hexane	24.8	43.4	65.3	19	26.0	51.1	65.2	22
2,2-Dimethylbutane (Neohexane)	91.8	+0.0	+0.6	89	93.4	+0.6	+2.1	97
n-Heptane	0.0	10.0	43.5	0	0.0	25.4	46.9	0
2,2,3-Trimethylbutane (Triptane)	+1.8			113	+0.1		+3.07	113
n-Octane		0.0	24.8	−19		0.7	28.1	−16
2,2,4-Trimethylpentane (Isooctane)	100.0	+1.0	+3.0	100	100.0	+1.0	+3.0	100
n-Nonane				−18				−20
Olefins								
1-Butene	97.5			144	79.9			126
2-Butene (cis)	100			154	83.5			130
1-Pentene	90.9		98.6	118	77.1	81.3	82.9	109
2-Methyl-1-butene	+0.2		+0.3	146	81.9		84.2	133
1-Hexene	76.4		91.7	97	63.4		76.3	94
2,3,3-Trimethyl-1-butene	+0.5	+0.8	+1.2	145	90.5	92.3	93.7	130
1-Octene	28.7	43.8	63.5		34.7	46.6	57.7	
2-Octene	56.3	69.9	78.7	75	56.5	67.9	73.0	68
1-Nonene				35				22
Naphthenes								
Cyclopentane	+0.1		+0.9	141	85.0	91.4	95.2	141
Cyclohexane	83.0	92.9	97.4	110	77.2	85.4	87.3	97
Methylcyclopentane	91.3	99.5	+0.5	107	80.0	89.4	93.0	99
Methylcyclohexane	74.8	83.5	88.2	104	71.1	82.0	86.2	84
Ethylcyclohexane	45.6	54.0	65.1	43	40.8	52.3	65.4	40
Aromatics								
Benzene				99	+2.8			91
Toluene	+5.8			124	+0.3	+1.0	+1.7	112
Ethylbenzene	+0.8	+0.8	+0.8	124	97.9	100.0	+0.2	107
o-Xylene				120	100.0		100.0	103
m-Xylene	+4.0		+6.0	145	+2.8		+6.0	124
p-Xylene	+3.4		> +6	146	+1.2		+5.1	127
Isopropylbenzene (Cumene)	+2.1	+3.4	+4.3	132	99.3	+0.2	+0.5	124

[a]Figures prefixed by a + sign represents mL TEL/U.S. gal of isooctane to give an equivalent knock rating.

where

$$AI = \text{antiknock index}$$
$$RON = \text{research octane number}$$
$$MON = \text{motor octane number}$$
$$k_1, k_2, k_3 = \text{constants}$$

The values of k_1, k_2, and k_3 vary from vehicle to vehicle, depending on the operating conditions as well as engine and transmission characteristics. The current version of the equation in ASTM Specification D 439 is

$$AI = 0.5(RON + MON)$$

The blending octane number in the table is obtained from ASTM Motor and ASTM Research ratings of blends of pure hydrocarbons. The ratings were made on blends of 20% hydrocarbon plus 80% of 60 : 40 mixture of isooctane and n-heptane (octane number 60). The blending octane number is then a hypothetical value obtained by extrapolation from the rating at 20% concentration to a rating at 100% concentration of the hydrocarbon. The value indicates the blending value of the hydrocarbon in this particular base stock only and cannot be used to predict blending values in other mixtures because of the widely variable blending value of each hydrocarbon and its dependency on the fuel type of the mixture in which it is blended.

References: 31, 32.

I. CETANE NUMBER OF DIESEL FUEL COMPONENTS

The ignition quality of diesel fuels is numerically indicated by the cetane number. The cetane number and octane number are opposites with respect to the effect of molecular configuration. Table I.1 lists the cetane number and boiling point of some hydrocarbons, mostly in the diesel fuel range. Normal paraffins have the highest cetane numbers. As the molecular weight of the normal paraffin increases, so does the cetane number, which levels off at 110. Branched chain paraffins have lower cetane numbers than normal paraffins. The more the branching, the lower the cetane number is. Olefins have lower cetane numbers than paraffins of corresponding structure, and follow the rules similar to paraffins with respect to branching. Naphthenes follow olefins in cetane number. Aromatics have especially low cetane numbers. Ring condensation and side-chain branching on rings cause even lower cetane values.

TABLE I.1. Cetane Number and Boiling Point of Pure Hydrocarbons

Hydrocarbon	Formula	Boiling Point, °C	Cetane Number
n-Paraffins			
n-Heptane	C_7H_{16}	98.4	56
n-Decane	$C_{10}H_{22}$	174.1	76
n-Dodecane	$C_{12}H_{26}$	216.3	80
n-Tetradecane	$C_{14}H_{30}$	253.5	93
n-Pentadecane	$C_{15}H_{32}$	270.6	95
n-Hexadecane (cetane)	$C_{16}H_{34}$	286.8	100
n-Heptadecane	$C_{17}H_{36}$	302.2	105
n-Octadecane	$C_{18}H_{38}$	316.7	110
n-Nonadecane	$C_{19}H_{40}$	330.6	110
n-Eicosane	$C_{20}H_{42}$	343.8	110
Branched Chain Paraffins			
2-Methylpentane	C_6H_{14}	60.3	33
2,2,4-Trimethylpentane	C_8H_{18}	99.3	12
3-Ethyldecane	$C_{12}H_{26}$	202	48
4,5-Diethyloctane	$C_{12}H_{26}$	193	20
2,2,4,6,6-Pentamethylheptane	$C_{12}H_{26}$	180[a]	9
8-Propylpentadecane	$C_{18}H_{38}$	206	48
5,6-Dibutyldecane	$C_{18}H_{38}$	156.5	30
Olefins			
1-Tetradecene	$C_{14}H_{28}$	251.1	79
1-Hexadecene	$C_{16}H_{32}$	284.9	88
Tetraisobutylene	$C_{16}H_{32}$	245[a]	4.5
8-Propyl-8-pentadecene	$C_{18}H_{36}$	320[a]	45
Di-isobutylene	C_8H_{16}	102.6	10
Naphthenes			
Cyclohexane	C_6H_{12}	80.7	13
Methylcyclohexane	C_7H_{14}	100.9	20
Bicyclohexyl	$C_{12}H_{22}$	238.5	53
3-Cyclohexylhexane	$C_{12}H_{24}$	220[a]	36
2-Methyl-3-cyclohexyl-nonane	$C_{16}H_{32}$	345[a]	70
2-Cyclohexyltetradecane	$C_{20}H_{40}$	255[a]	57
Decalin	$C_{10}H_{18}$	187.3	48
n-Propyldecalin	$C_{13}H_{24}$	250[a]	35
n-Butyldecalin	$C_{14}H_{26}$	264	31
n-Octyldecalin	$C_{18}H_{34}$	327[a]	31
n-Propyltetralin	$C_{13}H_{18}$	257	8
n-Butyltetralin	$C_{14}H_{20}$	269.5	18
n-Octyltetralin	$C_{18}H_{28}$	326[a]	18

TABLE I.1. (*Continued*)

Hydrocarbon	Formula	Boiling Point, °C	Cetane Number
Aromatics			
n-Amylbenzene	$C_{11}H_{16}$	205.4	8
n-Hexylbenzene	$C_{12}H_{18}$	226.1	26
n-Heptylbenzene	$C_{13}H_{20}$	246.0	35
n-Octylbenzene	$C_{14}H_{22}$	264.4	31
n-Nonylbenzene	$C_{15}H_{24}$	282	50
n-Octylxylene	$C_{16}H_{26}$		20
n-Tetradecylbenzene	$C_{20}H_{34}$	353.9	72
Diphenyl	$C_{12}H_{10}$	256	21
α-Methylnaphthalene	$C_{11}H_{10}$	244.7	0
β-tert-Butylnaphthalene	$C_{14}H_{16}$	273.5	3

[a]Calculated from equilibrium temperature at vapor pressure lower than 1 atm.
From J. W. Rose and J. R. Cooper, "Cetane Number of Pure Hydrocarbons," *Technical Data on Fuel*, The British National Committee of the World Energy Conference, 1977.

The petroleum refining industry is more concerned about cetane number of diesel fuels rather than that of pure compounds. The cetane number requirement for commercial diesel fuels is usually above 40, whereas low-atmospheric temperatures as well as engine operation at high altitude may require use of fuels with higher cetane ratings.

Over the years, a number of cetane number prediction methods have been advocated. ASTM issued a Calculated Cetane Index formula (ASTM D 976) as a means for directly estimating the ASTM cetane number of distillate fuels from API gravity and mid-boiling point. It is particularly applicable to straight-run fuels, catalytically cracked stocks, and blends of the two. The Calculated Cetane Index is determined from the following equation:

$$\text{Calculated Cetane Index} = -420.34 + 0.016G^2 + 0.192\,G\log M + 65.01(\log M)^2 - 0.0001809M^2.$$

where

G = API gravity, determined by Method D 287 or D 1298
M = mid-boiling temperature, °F, determined by Method D 86 and corrected to standard barometric pressure.

The cetane number can then be estimated from the following equation:

$$\text{Cetane number} = 21.843 - 0.33924(\text{CI}) + 0.018669(\text{CI})^2.$$

where CI is the Calculated Cetane Index.

References: 13, 32, 33, 34.

J. SMOKE TENDENCY OF HYDROCARBONS

Soot formation due to combustion in aviation turbines depends both on the design of atomizer and the chemical structure of the fuel. Aromatics tend to form soot, multiring automatics more readily than single-ring aromatics, and they exhibit both greater smoking tendency and greater flame luminosity. Normal paraffins form minimum soot. Other hydrocarbon structures exhibit smoke and radiation characteristics between extremes of multiring aromatics and normal paraffins. Table J.1 lists certain properties related to the smoking tendency of hydrocarbons in combustion.

The aniline point is a sensitive indicator of the aromatic content of liquid fuels. The hydrogen content is specified in military jet fuels for the control of fuel-burning quality. The heat of combustion correlates reasonably well with hydrogen content. Fuels with higher net heat of combustion also have higher hydrogen content. The luminometer number and smoke point are used in aviation turbine fuel specifications for burning-quality control (see Table M.4). The luminometer number is a measure of flame temperature related to the color temperature in the green-yellow region of the spectrum. High luminometer number fuel forms little soot in combustion. The smoke point is also a measure of smoking tendency of fuels. It is the height in millimeters of the flame that can be produced in a standard lamp without causing smoking. The luminometer number and smoke point are correlated by the following relation (ASTM D 1740):

$$SP = 4.16 + 0.331(LN) + 0.000648(LN)^2$$

where

SP = smoke point, mm
LN = luminometer number

References: 21, 32, 35, 36.

K. RATE CONSTANTS OF SOME RADICAL REACTIONS OF COMBUSTION

The kinetic mechanisms of free atom and radical reactions of combustion have been discussed in Chapter 3. This section gives a graphical representation (Fig. 3) of the rate constants as a function of temperature of some representative radical reactions of combustion, listed in Table K.1. The rate constant and activation energy values are taken from Chapter 3.

For the reaction

$$A + B + M \longrightarrow \text{Products}$$

TABLE J.1. Properties of Hydrocarbons in Connection with Smoking Tendency in Combustion

Compound	Formula	Boiling Point, °C	Aniline Point, °C	Net Heat of Combustion,[a] kcal/gm-mol	Hydrogen Content, wt %	Luminometer Number	ASTM Smoke Point, mm
Normal Paraffins							
n-Hexane	C$_6$H$_{14}$	68.74	68.6	921.37	16.38	226	over 50
n-Heptane	C$_7$H$_{16}$	98.43	69.7	1067.11	16.10	209	over 50
n-Octane	C$_8$H$_{18}$	125.67	70.6	1212.86	15.88	197	over 50
n-Decane	C$_{10}$H$_{22}$	174.12	77.0	1504.34	15.59	170	over 50
n-Dodecane	C$_{12}$H$_{26}$	216.28	83.8	1795.83	15.39	166	over 50
n-Tetradecane	C$_{14}$H$_{30}$	253.52	89.3	2087.30	15.24	152	over 50
n-Hexadecane (Cetane)	C$_{16}$H$_{34}$	286.79	94.8	2378.78	15.13	149	over 50
Isoparaffins							
3-Methylpentane	C$_6$H$_{14}$	63.28	66.3	920.61	16.38	177	over 50
2,2-Dimethylbutane (Neohexane)	C$_6$H$_{14}$	49.74	50.7	917.88	16.38		48.8
2,2,4-Trimethylpentane (Isooctane)	C$_8$H$_{18}$	99.2	79.5	1210.61	15.88	100	41.6
2,3,4-Trimethylpentane	C$_8$H$_{18}$	113	67.9	1211.60	15.88	120	50
2,2,4,6,8,8,-Heptamethylnonane	C$_{16}$H$_{34}$	236.6	93.4	2405.52	15.13	74	33.2

Olefins							
1-Hexene	C_6H_{12}	63.49	22.8	893.81	14.37	105	40.1
1-Octene	C_8H_{16}	121.28	32.5	1185.30	14.37	110	47.8
2,2,4-Trimethyl-2-pentene	C_8H_{16}	101.7	42.6	1179.21	14.37	63	25.9
Cyclohexene	C_6H_{10}	83	−20.0	844.02	12.27	71	31.1
4-Vinyl-cyclohexene-1	C_8H_{12}	127.78	−13.9[b]	1076.46	11.18	37	18.8
Naphthenes							
Cyclohexane	C_6H_{12}	80.74	31.0	873.75	14.37	130	over 50
Methylcyclopentane	C_6H_{12}	71.81	33.0	878.03	14.37	76	33.9
Tetracyclododecane	$C_{12}H_{18}$	230.0	30.9	1371.11	11.18	16	10.9
Isopropyl bicyclohexyl	$C_{15}H_{28}$	277.2	61.2	2125.65	13.54	54	23.0
Aromatics							
Benzene	C_6H_6	80.10	−18.9[b]	749.42	7.74	11	8.1
Toluene	C_7H_8	110.63	−40.0[b]	892.42	8.75	3	7.7
Ethylbenzene	C_8H_{10}	136.19	−30.0	1038.43	9.49	−2	7.4
p-Xylene	C_8H_{10}	138.35	−28.9[b]	1035.56	9.49	2	6.8
Styrene	C_8H_8	145.10	−23.3[b]	1008.44	7.74	2.7	6.2
Sec-Butylbenzene	$C_{10}H_{14}$	173.34	−18.9[b]	1331.88	10.51	15	8.2
Tetralin	$C_{10}H_{12}$	207.6	−31.1[b]	1278.73	9.15	0	6.6
1-Methylnaphthalene	$C_{11}H_{10}$	244.44	−30.0[b]	1335.88	7.09	−14	5.1

[a] – ΔH_c at 25°C and 1 atm, compound in liquid state.
[b] Crystals at this temperature.
From J. W. Rose and J. R. Cooper, "Cetane Number of Pure Hydrocarbons," *Technical Data on Fuel*, The British National Committee of the World Energy Conference, 1977.

TABLE K.1. Some Representative Free Atom and Radical Reactions[a]

Reaction	log A	E_a, kcal
$H + O + M \rightarrow OH + M$	16.00	0.00
$H + O_2 \rightarrow OH + O$	14.34	16.79
$H + O_2 + M \rightarrow HO_2 + M$	15.2	−1.0
$H + HO_2 \rightarrow OH + OH$	14.40	1.9
$H + HO_2 \rightarrow H_2 + O_2$	13.40	0.7
$H + H_2O \rightarrow OH + H_2$	13.97	20.37
$H + H_2O_2 \rightarrow HO_2 + H_2$	12.23	3.78
$OH + H_2O_2 \rightarrow H_2O + HO_2$	13.00	1.80
$O + CO + M \rightarrow CO_2 + M$	15.77	4.10
$O + H_2O \rightarrow OH + OH$	13.83	18.36
$HO_2 + CO \rightarrow CO_2 + OH$	14.00	23.00
$HO_2 + HO_2 \rightarrow H_2O_2 + O_2$	13.00	1.00
$H_2O_2 + M \rightarrow OH + OH + M$	17.08	45.50
$H_2O_2 + O_2 \rightarrow HO_2 + HO_2$	13.6	42.6
$CO + O_2 \rightarrow CO_2 + O$	11.50	37.60
$CH_4 + H \rightarrow CH_3 + H_2$	14.1	11.9
$CH_4 + O \rightarrow CH_3 + OH$	13.2	9.2
$CH_4 + M \rightarrow CH_3 + H + M$	17.1	88.4

[a] $k = AT^n \exp(-E_a/RT)$, in cm³-gm·mol-sec-kcal units. $n = 0$ for all the reactions.

From J. W. Rose and J. R. Cooper, "Cetane Number of Pure Hydrocarbons," *Technical Data on Fuel*, The British National Committee of the World Energy Conference, 1977.

the rate expression is

$$\frac{-dN_A}{V \, dt} = kC_A C_B C_M$$

where

N_A = molal mass of reactant A, gm-mol

V = volume, cm³

t = time, sec

k = rate constant, cm³/gm-mol · sec (second order), cm⁶/gm-mol² sec (third order)

C_A = concentration of reactant A, gm-mol/cm³

C_B = concentration of reactant B, gm-mol/cm³

C_M = concentration of the third body, gm-mol/cm³ (delete C_M if M is absent)

In Figure 3, log k is plotted versus the reciprocal of temperature. The dashed line (second order) and dotted line (third order) give the k neces-

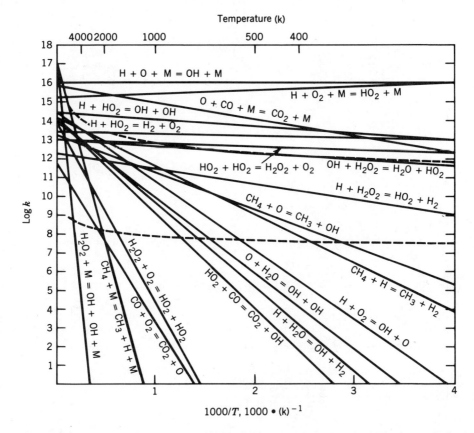

FIGURE 3. Rate constants of some radical reactions of combustion. The lower dashed line (second order) and upper dashed line (third order) give k necessary to cause 1% fractional decrease in A or B per millisecond, when partial pressure of other reactant is 0.01 atm and, if a three-body process, P_M is 1 atm.

sary to cause a 1% fractional decrease in A or B per millisecond, when the partial pressure of other reactant is 0.01 atm and, if a three-body process, P_M is 1 atm.

References: 5, 37.

L. PROPERTIES OF INDUSTRIAL GASEOUS FUELS

Table L.1 lists typical analyses of some representative industrial gaseous fuels, together with pure hydrogen, carbon monoxide, and methane for comparison.

TABLE L.1. Properties of Industrial Gaseous Fuels

Type of Gas	Specific Gravity	Heating Value (Dry), Btu/ft^3, 30"Hg, 60°F		Composition, vol %								
		Gross	Net	CO_2	CO	H_2	CH_4	C_2H_6	C_3H_8	other C_mH_n	N_2	O_2
Blast furnace gas	1.04	81	80	15.6	23.4	1.6	0.1				59.3	
Water gas												
blue, from coke	0.54	300	273	5.1	40.2	50.0	0.7				4.0	
carbureted, normal operation	0.64	540	462	3.4	30.0	31.7	12.2			8.4	13.1	1.2
carbureted, high Btu	0.69	850	791	1.6	21.3	28.0	20.7	4.3		18.9	5.0	0.2
Air-blown producer gas												
from coke		129.4	122.7	5.8	26.0	12.1	0.5				55.3	0.3
Lurgi, from brown coal	0.80		165.6	14.0	16.0	25.0	5.0				40.0	
Wellman-Galusha, from coke	0.84		161.6	3.0	29.0	15.0	3.0				50.0	
Winkler, from lignite	0.91		112.7	10.0	22.0	12.0	1.0				55.0	
Oxygen-blown producer gas												
GKT, from brown coal	0.69		275.8	7.0	56.0	35.0					2.0	
Lurgi, from brown coal	0.75		289.5	33.0	13.0	37.0	16.0				1.0	
Wellman-Galusha, from bit. coal	0.73		266.5	12.0	52.0	33.0	1.0				2.0	
Winkler, from lignite	0.71		249.0	20.0	35.0	40.0	3.0				2.0	
Coal gas												
continuous vertical retort	0.42	532	477	3.0	10.9	54.5	24.2			2.8	4.4	0.2
horizontal retort	0.47	542	486	2.4	7.4	48.0	27.1			3.0	11.3	0.8
Coke oven gas, by-product	0.40	580	523	2.0	6.2	53.2	26.7			4.0	7.0	0.9
Synthetic gas, Lurgi crude, from coke	0.72	247		29.3	21.9	44.0	3.3				1.5	
Oil gas												
Portland, from residuum	0.45	586	544	3.0	9.5	46.1	29.9	0.6		5.3	5.2	0.4
high Btu, from gas oil	0.78	963	831	6.0	1.5	14.6	33.4			24.2	19.4	0.9
Refinery dry gas, Philadelphia	0.89	1388		1.1	5.4	12.7	28.1	17.1	14.1	21.5[a]		
Natural gas												
Oklahoma, Hugoton	0.71	1043		0.1			75.3	6.4	3.7	2.0	12.5	
Texas, Panhandle	0.68	1090					81.8	5.6	3.4	2.2	6.9	
Louisiana, Monroe	0.61	997		0.3			91.3	1.5	0.7	0.8	5.4	
Reformed	0.44	464	424	1.2	22.3	49.9	21.9			0.8	3.8	
Catalytically cracked	0.50	300	270	4.5	18.0	49.0	8.5				20.0	0.1
Hydrogen	0.07		274.1			100.0						
Carbon monoxide	0.97		320.8		100.0							
Methane	0.55		909.5				100.0					

[a]Including 2.6% H_2S. Adapted with permission from C. G. Segeler (ed.), *Gas Engineers' Handbook* (1965), Industrial Press

Blast furnace gas, a by-product of pig iron production, is a low heating value fuel used within the steel plant.

Water gas, made by passing steam over hot coke or other carbonaceous material, is mostly CO and H_2. It is usually carbureted to a high Btu gas for sale.

Producer gas is generated when air or oxygen is passed through a hot coal or coke bed. Steam is also added to reduce clinker formation. Table L.1 shows that gas analyses vary with different equipment and fuels used.

Coal gas is made by the distillation of the volatile matter from coal with some steaming of the coke to produce water gas. It contains large amount of H_2 and CH_4, with lesser amounts of CO and illuminants.

Coke oven gas is made in by-product coke ovens by the distillation of the volatile matter from the coal.

Synthesis gas is made by various processes from coal, coke, or hydrocarbons. It consists of H_2 and CO in an approximate ratio of 2. The crude gas also contains high percentages of CO_2.

Oil gases are made by the thermal cracking of oils from naphtha, gas oil, or heavy residuum. They contain large amounts of H_2 and CO, and have heating values of 550 to 1000 Btu/ft^3. The lower-Btu gases are distributed by utilities; the high-Btu gases are used as a peak-loud supplement by natural gas companies.

Refinery gas is produced in petroleum refineries. Its composition depends on the refinery schemes. A typical gas composition is given in Table L.1. Depending on the amount of hydrocarbons not removed from the recovery system, the heating value may vary from 1400 to 2000 Btu/ft^3.

Three natural gas compositions are listed in Table L.1, with different CH_4 contents. Crude natural gas usually contains small amounts of helium, from zero to less than 0.3%, and some hydrogen sulfide, both of which are not shown in the table.

Reformed natural gas is obtained by the thermal cracking of natural gas or other petroleum gases in water gas generator or similar special equipment. It contains H_2, CO, and hydrocarbons. Catalytically cracked gas is made by cracking natural gas (or light hydrocarbon liquid) over a nickel oxide catalyst with steam. The gas contains CO and H_2 with appreciable amounts of N_2 and CO_2.

References: 10, 13, 38.

M. LIQUID FUEL SPECIFICATIONS

The most common liquid fuels in use today are gasoline, kerosene, diesel fuel, gas turbine fuel, and fuel oil. Liquefied petroleum gas (LPG), which is a mixture of propane and butane, is used extensively as a domestic fuel.

TABLE M.1. Requirements for Liquefied Petroleum Gases (D1835-87)

	Product Designation				
	Commercial Propane	Commercial Butane	Commercial PB Mixtures	Special-Duty Propane[a]	ASTM Test Methods
Vapor pressure at 100°F (37.8°C), max, psig	208	70	[b]	208	D1267 or
kPa	1430	485		1430	D 2598[c]
Volatile residue					
evaporated temperature, 95%, max, °F	−37	36	36	−37	D 1837
°C	−38.3	2.2	2.2	−38.3	D 2163
butane and heavier, max, vol %	2.5	...	...	2.5	D 2163
pentane and heavier, max, vol %	...	2.0	2.0	...	D 2163
Propylene content, max, vol %	...	...	...	5.0	D 2163
Residual matter					
residue on evaporation 100 mL, max, mL	0.05	0.05	0.05	0.05	D 2158
oil stain observation	pass[d] e	pass[d] e	pass[d] e	pass[d]	D 2158
Relative density (specific gravity) at 60/60°F (15.6/15.6°C)	...	...	...	...	D 1657 or D 2598
Corrosion, copper, strip, max	No. 1	No. 1	No. 1	No. 1	D 1838
Sulfur, ppmw	185	140	140	123	D 2784
Moisture content	pass	...	...	pass	D 2713
Free water content	...	none[f]	none[f]	...	

[a]Equivalent to Propane HD-5 of GPA Standard 2140.
[b]The permissible vapor pressures of products classified as PB mixtures must not exceed 208 psig (1430 kPa) and additionally must not exceed that calculated from the following relationship between the observed vapor pressure and the observed specific gravity: Vapor pressure, max = 1167 − 1880 (sp gr 60/60°F) or 1167 − 1880 (density at 15°C). A specific mixture shall be designated by the vapor pressure at 100°F in pounds per square inch gage. To comply with the designation, the vapor pressure of the mixture shall be within +0 to −10 psi of the vapor pressure specified.
[c]In case of dispute about the vapor pressure of a product, the value actually determined by Test Method D 1267 shall prevail over the value calculated by Method D 2598.
[d]An acceptable product shall not yield a persistent oil ring when 0.3 mL of solvent residue mixture is added to a filter paper, in 0.1-mL increments and examined in daylight after 2 min as described in Test Method D 2158.
[e]Although not a specific requirement, the specific gravity must be determined for other purposes and should be reported. Additionally, the specific gravity of PB mixture is needed to establish the permissible maximum vapor pressure (see Footnote b).
[f]The presence or absence of water shall be determined by visual inspection of the samples on which the gravity is determined.
Adapted with permission from American Society for Testing and Materials, *Annual Book of ASTM Standards*, Sect. 5 (1989). © ASTM.

TABLE M.2. Requirements for Automotive Gasoline (D439-88b)

Volatility Class	Distillation Temperatures, °C (°F), at Percent Evaporated[a,b]					Distillation Residue, Vol %, max	Vapor/Liquid Ratio Test Temperature °C (°F)
	10 Vol %, max	50 Vol % min	50 Vol % max	90 Vol %, max	End Point, max		
A	70 (158)	77 (170)	121 (250)	190 (374)	225 (437)	2	60 (140)
B	65 (149)	77 (170)	118 (245)	190 (374)	225 (437)	2	56 (133)
C	60 (140)	77 (170)	116 (240)	185 (365)	225 (437)	2	51 (124)
D	55 (131)	77 (170)	113 (235)	185 (365)	225 (437)	2	47 (116)
E	50 (122)	77 (170)	110 (230)	185 (365)	225 (437)	2	41 (105)

Volatility Class	Reid Vapor Pressure, max,[b] kPa (psi)	Lead Content, max, g/L (g/gal) Unleaded[c]	Lead Content, max, g/L (g/gal) Leaded[d]	Copper Strip Corrosion, max	Existent Gum, max, mg/100 mL	Sulfur, max, Mass % Unleaded	Sulfur, max, Mass % Leaded	Oxidation Stability, Minimum, Minutes
A	62 (9.0)	0.013 (0.05)	1.1 (4.2)	No. 1	5	0.10	0.15	240
B	69 (10.0)	0.013 (0.05)	1.1 (4.2)	No. 1	5	0.10	0.15	240
C	79 (11.5)	0.013 (0.05)	1.1 (4.2)	No. 1	5	0.10	0.15	240
D	93 (13.5)	0.013 (0.05)	1.1 (4.2)	No. 1	5	0.10	0.15	240
E	103 (15.0)	0.013 (0.05)	1.1 (4.2)	No. 1	5	0.10	0.15	240

[a]At 101.3 kPa pressure (760 mm Hg).
[b]If Federal legislation or regulatory action restricts Reid Vapor Pressure to a level lower than the volatility classes specified in Table 2 of ASTM D439 for a given time and place, the distillation temperature limits shall be consistent with the corresponding Reid Vapor Pressure in this table. If the Reid Vapor Pressure limit is between the two classes, the distillation temperature limits of either class are acceptable.
[c]The international addition of lead or phosphorus compounds is not permitted. U.S. Environmental Protection Agency (EPA) regulations limit their maximum concentrations to 0.05 g of lead per gallon and 0.005 g of phosphorus per gallon (by Test Method D 3231), respectively.
[d]EPA regulations limit the lead concentration in leaded gasoline to no more than 0.1 g/gal (0.026 g/L) averaged for quarterly production by refinery. Adapted with permission from American Society for Testing and Materials, *Annual Book of ASTM Standards*, Sect. 5 (1989), © ASTM.

TABLE M.3. Gasoline Antiknock Indexes (D439-88b)

Antiknock Index (RON + MON)/2, min$^{a,\,b}$	Application
Leaded Gasoline (for vehicles that can or must use leaded gasoline)	
87	Meets antiknock requirements of most 1971 and later model vehicles that can use leaded gasoline and of vehicles with low antiknock requirements.
88	Meets antiknock requirements of most 1970 and prior model vehicles that were designed to operate on leaded and of 1971 and later model vehicles that can use leaded gasoline and have high antiknock requirements.
89	Meets antiknock requirements of medium and heavy duty trucks that require higher octane leaded gasoline.
92	Suitable for most vehicles with very high antiknock requirements that can use leaded gasoline.
Unleaded Gasoline (for vehicles that can or must use unleaded gasoline)	
85	For vehicles with low antiknock requirements.
87^c	Meets antiknock requirements of most 1971 and later model vehicles.
90	For most 1971 and later model vehicles with high antiknock requirements.

^{a}Reductions for seasonal variations are allowed in accordance with Fig. 1 of ASTM D439.
^{b}Reductions for altitude are allowed in accordance with Fig. 2 of ASTM D439.
^{c}In addition, Motor octane number must not be less than 82.0.
Adapted with permission from American Society for Testing and Materials, *Annual Book of ASTM Standards*, Sect. 5 (1989), © ASTM.

The limiting requirements for various liquid fuels included in this section are shown below:

Table	Liquid fuel	Standard
M.1	Liquefied petroleum gases	ASTM D 1835-87
M.2	Automotive gasoline	ASTM D 439-88b
M.3	Gasoline antiknock indexes	ASTM D 439-88b
M.4	Aviation turbine fuels—civil	ASTM D 1655-88a
M.5	Aviation turbine fuels—military	
	Military JP-4, JP-5	Mil-T-5624-K
	Military JP-7	Mil-T-38219
	Military JP-8	Mil-T-83133
M.6	Diesel fuel oils	ASTM D 975-88
M.7	Gas turbine fuel oils	ASTM D 2880-88
M.8	Fuel oils	ASTM D 396-86

References: 32, 36.

TABLE M.4. Requirements for Civil Aviation Turbine Fuels (D1655-88a)[a]

Property	Jet A or Jet A-1	Jet B	ASTM Test Method[b]
Acidity, total max, mg KOH/g	0.1	. . .	D 974 or D 3242
Aromatics, max, vol %	20 [c]	20 [c]	D 1319
Sulfur, mercaptan,[d] max, weight %	0.003	0.003	D 3227
Sulfur, total max, weight %	0.3	0.3	D 1266 or D 1552 or D 2622
Distillation temperature, °C:			
10% recovered, max, temp	205	. . .	D 86
20% recovered, max, temp	. . .	145	
50% recovered, max, temp	report	190	
90% recovered, max, temp	report	245	
Final boiling point, max, temp	300	. . .	
Distillation residue, max, %	1.5	1.5	
Distillation loss, max, %	1.5	1.5	
Flash point, min, °C	38	. . .	D 56 or D 3828[e]
Density at 15°C, kg/m³	775 to 840	751 to 802	D 1298 or D 4052
Vapor pressure, 38°C, max, kPa	. . .	21	D 323
Freezing point, max, °C	−40 Jet A[f] −47 Jet A-1[f]	−50 [f]	D 2386
Viscosity −20°C, max, mm²/sec[g]	8.0		D 445
Net heat of combustion, min, MJ/kg	42.8[h]	42.8[h]	D 4529, D 2382, or D 3338
Combustion properties: one of the following requirements shall be met:			
(1) Luminometer number, min, or	45	45	D 1740
(2) Smoke point, min, mm, or	25	25	D 1322
(3) Smoke point, min, mm, and	20 [i]	20 [i]	D 1322
Naphthalenes, max, vol, %	3	3	D 1840
Corrosion, copper strip, 2 h at 100°C, max	No. 1	No. 1	D 130
Thermal stability: one of the following requirements shall be met:			
(1) Filter pressure drop, max, mm Hg	76 [j]	76 [j]	D 1660[k]
Preheater deposit less than	Code 3	Code 3	
(2) Filter pressure drop, max, mm Hg	25 [l]	25 [l]	D 3241[m]
Tube deposit less than	Code 3	Code 3	
Existent gum, max, mg/100 mL	7	7	D 381
Water reaction			
Separation rating, max	(2)	(2)	D 1094
Interface rating, max	1b	1b	D 1094
Additives	See 5.2	See 5.2	
Electrical conductivity, pS/m	[n]	[n]	D 2624 or D 4308

[a] The requirements herein are absolute and are not subject to correction for tolerance of the test methods. If multiple determinations are made, average results shall be used.
[b] The test methods indicated in this table are referred to in Section 9 of Annual Book of ASTM Standards.
[c] Fuels with an aromatic content over 20 volume % but not exceeding 25 volume % are permitted provided the supplier (seller) notifies the purchaser of the volume, distribution and aromatic content within 90 days of date of shipment unless other reporting conditions are agreed to by both parties.
[d] The mercaptan sulfur determination may be waived if the fuel is considered sweet by the doctor test described in 4.2 of Specification D 235.

(continued)

TABLE M.4. (*Continued*)

[e]Results obtained by Test Method D 3828 may be up to 2°C lower than those obtained by Test Method D 56 which is the preferred method. In case of dispute, Test Method D 56 will apply.

[f]Other freezing points may be agreed upon between supplier and purchaser.

[g]1 mm^2/sec = 1 centistoke.

[h]For all grades use either the Eq 1 or the Table 1 in Test Method D 4529 or Eq 2 in Test Method D 3338. Test Method D 2382 may be used as an alternative. In case of dispute Test Method D 2382 must be used.

[i]Fuels having a smoke point less than 20 but not less than 18 and a maximum of 3 vol % of naphthalenes are permitted provided the supplier (seller) notifies the purchaser of the volume, distribution and smoke point and naphthalenes content within 90 days of date of shipment unless other reporting conditions are agreed to by both parties.

[j]Preferred SI units are 10 kPa, max.

[k]Thermal Stability test shall be conducted for 5 h at 149°C preheater temperature, 204.5°C filter temperature and at a flow rate of 2.7 kg/h.

[l]Preferred SI units are 3.3 kPa, max.

[m]Thermal stability test (JFTOT) shall be conducted for 2.5 h at a control temperature of 260°C, but if the requirements of Table 1 are not met, the test may be conducted at 245°C. Results at both temperatures shall be reported in this case. Tube deposits shall always be reported by the Visual Method: a rating by the Tube Deposit Rating (TDR) optical density method is desirable but not mandatory.

[n]A limit of 50 to 450 conductivity units (pS/m) applies only when an electrical conductivity additive is used and under the condition at point of use. 1 pS/m = 1×10^{-12} Ω^{-1} m^{-1}.

Adapted with permission from American Society for Testing and Materials, *Annual Book of ASTM Standards*, Sect. 5 (1989), © ASTM.

TABLE M.5. Selected Specification of Military Aviation Gas Turbine Fuels[a]

| Characteristic | Mil-T-5624-K | | Mil-T-38219 | Mil-T-83133 |
	JP-4 Wide-Cut U.S.A.F.	JP-5 Kerosene U.S.N.	JP-7 Kerosene U.S.A.F.	JP-8 Kerosene U.S.A.F.
Composition				
aromatics, vol % max	25	25	5	25
sulfur, wt % max	0.4	0.4	0.1	0.4
Volatility				
dist. ⎰10% rec'd		205	196	205
temp ⎱50% rec'd	190			
max °C				
⎩end pt	270	290	288	300
flash pt, °C min		60	60	38
vapor pressure at 38°C, kPa max (psi)	14–21 (2–3)			
density at 15°C, kg/m^3	751–802	788–845	779–806	775–840
Fluidity				
freezing pt, °C max	−58	−46	−43	−50
viscosity at −20°C, mm^2/sec max (= centistoke)		8.5	8.0	8.0
Combustion				
heat content, MJ/kg, min	42.8	42.6	43.5	42.8
smoke pt, mm, min	20	19	35	20
H$_2$ content, wt % min	13.6	13.5	14.2	13.6
Stability				
test temp, °C min	260	260	350	260

[a]Full specification requires other tests. From Ref. 36.

N. PROPERTIES OF REPRESENTATIVE CRUDE OILS

Crude oils vary considerably in property as well as in composition. They can be paraffinic, naphthenic, or aromatic in nature; they can contain large amounts of lighter fractions, or almost none. Crude oils also vary in their contents of sulfur, nitrogen, as well as metals like Ni, V, and Fe, important as far as processing is concerned.

Analyses of typical crude oils found in representative areas of the United States are given in Table N.1. The crudes found in eastern and mideastern sections of the United States are predominantly sweet and paraffinic; those found along the Gulf Coast usually are naphthenic; those in the inland southwest are sour and naphthenic; and those along the west coast are asphaltic.

Analyses of some crude oils found outside the United States are given in Table N.2, which illustrates the variety of existent crudes but is not intended to give a full representation of worldwide petroleum compositions.

References: 39, 40, 41.

O. CLASSIFICATION OF COALS

The coal classification system given in Table O.1 was developed from the joint project of the American Standards Association and the ASTM. The higher-rank coals having 69% or more fixed carbon on a dry mineral-matter-free basis are classified according to fixed carbon, regardless of the heating value. The lower-rank coals are classified by heating value on the moist, mineral-matter free-basis. Some overlap between bituminous coal and sub-bituminous coal occurs, and is resolved on the basis of the agglomerating properties. High-volatile C bituminous coal with heating values of 24.4–26.7 MJ/kg (5832–6373 kcal/kg) is always agglomerating. All other bituminous coals are commonly agglomerating. Anthracite and subbituminous coal and lignite are nonagglomerating.

This classification does not include a few coals, principally nonbanded varieties, which have unusual physical and chemical properties and which come within the limits of fixed carbon or heating value of the high-volatile bituminous and subbituminous ranks. All these coals either contain less than 48% dry, mineral-matter-free fixed carbon or have more than 36.0 MJ/kg (8610 kcal/kg), calculated on the moist, mineral-free basis.

References: 32, 42.

TABLE M.6. Requirements for Diesel Fuel Oils (D975-88)[a,b]

Grade of Diesel Fuel Oil	Flash Point, °C (°F) Min	Cloud Point, °C Max	Water and Sediment, Vol % Max	Carbon Residue on 10% Residuum, % Max	Ash, Weight % Max	Distillation Temperatures, °C (°F) 90% Point Min	Max	Viscosity Kinematic, cSt (= mm²/sec) at 40°C Min	Max	Viscosity Saybolt, SUS at 100°F Min	Max	Sulfur,[c] Weight % Max	Copper Strip Corrosion Max	Cetane Number[d] Min
No. 1-D A volatile distillate fuel oil for engines in service requiring frequent speed and load changes	38 (100)	[e]	0.05	0.15	0.01	...	288 (550)	1.3	2.4	...	34.4	0.50	No. 3	40[f]
No. 2-D A distillate fuel oil of lower volatility for engines in industrial and heavy mobile service	52 (125)	[e]	0.05	0.35	0.01	282[g] (540)	338 (640)	1.9	4.1	32.6	40.1	0.50	No. 3	40[f]

| No. 4-D A fuel oil for low and medium speed engines. | 55 | (130) | [e] | 0.50 | ... | 0.10 | ... | 5.5 | 24.0 | 45.0 | 125.0 | 2.0 | ... | 30[f] |

[a]To meet special operating conditions, modifications of individual limiting requirements may be agreed upon between purchaser, seller, and manufacturer.

[b]The values stated in SI units are to be regarded as the standard. The values in inch-pound units are for information only.

[c]In countries outside the United States, other sulfur limits may apply.

[d]Where cetane number by Method D 613 is not available, Method D 976 may be used as an approximation. Where there is disagreement, Method D 613 should be the referee method.

[e]It is unrealistic to specify low-temperature properties that will ensure satisfactory operation on a broad basis. Satisfactory operation should be achieved in most cases if the cloud point (or wax appearance point) is specified at 6°C above the tenth percentile minimum ambient temperature for the area in which the fuel will be used.

[f]Low-atmospheric temperatures as well as engine operation at high altitudes may require use of fuels with higher cetane ratings.

[g]When cloud point less than −12°C (10°F) is specified, the minimum viscosity shall be 1.7 cSt (or mm^2/sec) and the 90% point shall be waived.

Adapted with permission from American Society for Testing and Materials, *Annual Book of ASTM Standards*, Sect. 5 (1989), © ASTM.

TABLE M.7. Requirements for Ground Gas Turbine Fuel Oils (D2880-88)^a

Designation^b	Grade of Gas Turbine Fuel Oil	Flash Point, °C (°F)^c min	Pour Point, °C (°F)^c max	Water and Sediment, vol % max	Carbon Residue on 10% Residuum, wt % max	Ash, wt % max	Distillation Temperature, 90% Point^c °C(°F) min	max	Kinematic Viscosity, cSt (= mm²/sec) at 40°C (104°F) min	max	at 50°C (122°F) max	Saybolt Viscosity, S^c Universal at 38°C (100°F) min	max	Furol at 50°C (122°F) max
No. 0-GT	A naphtha or other low-flash hydrocarbon liquid.	^{d}		0.05	0.15	0.01	^{d}							
No. 1-GT	A distillate for gas turbines requiring a fuel that burns cleaner than No. 2-GT.	38 (100)	−18^e (0)	0.05	0.15	0.01		288 (550)	1.3	2.4			(34.4)	
No. 2-GT	A distillate fuel of low ash suitable for gas turbines not requiring No. 1-GT.	38 (100)	−6^e (20)	0.05	0.35	0.01	282 (540)	338 (640)	1.9	4.1		(32.6)	(40.2)	

No. 3-GT	A low-ash fuel that may contain residual components.	55 (130)	1.0	0.03	5.5	638 (45)	(300)
No. 4-GT	A fuel containing residual components and having higher vanadium content than No. 3-GT.	66 (150)	1.0		5.5	638 (45)	(300)

[a] Gas turbines with waste heat recovery equipment may require sulfur limits in the fuel to prevent cold-end corrosion.

[b] No. 0-GT includes naphtha, Jet B fuel, and other volatile hydrocarbon liquids. No. 1-GT corresponds in general to Specification D 396 Grade No. 1 fuel and Classification D 975 Grade No. 1-D diesel fuel in physical properties. No. 2-GT corresponds in general to Specification D 396 Grade No. 2 fuel and Classification D 975 Grade No. 2-D diesel fuel in physical properties. No. 3-GT and No. 4-GT viscosity range brackets Specification D 396 Grade No. 4, No. 5 (light), No. 5 (heavy), and No. 6 and Classification D 975 Grade No. 4-D diesel fuel in physical properties.

[c] Values in parentheses are for information only and may be approximate.

[d] When flash point is below 38°C, or when kinematic viscosity is below 1.3 cSt at 40°C, or when both conditions exist, the turbine manufacturer should be consulted with respect to safe handling and fuel system design.

[e] For cold weather operation, the pour point should be specified 6°C below the ambient temperature at which the turbine is to be operated except where fuel heating facilities are provided. When a pour point less than −18°C is specified for Grade No. 2-GT, the minimum viscosity shall be 1.7 cSt, and the minimum 90% point shall be waived.

Adapted with permission from American Society for Testing and Materials, *Annual Book of ASTM Standards*, Sect. 5 (1989), © ASTM.

TABLE M.8. Requirements for Fuel Oils (D396-86)

	No. 1	No. 2	No. 4 (Light)	No. 4	No. 5 (Light)	No. 5 (Heavy)	No. 6
Grade Description	A distillate oil intended for vaporizing pot-type burners and other burners requiring this grade of fuel	A distillate oil for general purpose heating for use in burners not requiring No. 1 fuel oil	Preheating not usually required for handling or burning	Preheating not usually required for handling or burning	Preheating may be required depending on climate and equipment	Preheating may be required for burning and, in cold climates may be required for handling	Preheating required for handling and burning
Specific gravity, 60/60°F (deg API), max	0.8499 (35 min)	0.8762 (30 min)	0.8762[a] (30 max)				
min							
Flash point, °C (°F) min	38 (100)	38 (100)	38 (100)	55 (130)	55 (130)	55 (130)	60 (140)
Pour point, °C (°F) max	−18[b] (0)	−6[b] (20)	−6[b] (20)	−6[b] (20)			c
Kinematic viscosity, mm^2/sec (cSt)[d]							
At 38°C (100°F), min	1.4	2.0[b]	2.0	5.8	> 26.4	> 65	
max	2.2	3.6	5.8	26.4[e]	65[e]	194[e]	
At 40°C (104°F), min	1.3	1.9[b]		5.5	> 24.0	> 58	
max	2.1	3.4		24.0[e]	(58)[e]	(168)[e]	
At 100°C (212°F), min					5.0	9.0	15.0
max					8.9[e]	14.9[e]	50.0
Saybolt Viscosity:[d]							
Universal at 38°C (100°F), min		(32.6)	(32.6)	(45)	(> 125)	(> 300)	(> 900)
max		(37.9)	(45)	(125)	(300)	(900)	(9000)
Furol at 50°C (122°F), min						(23)	(> 45)
max						(40)	(300)

Property							
Distillation Temperature, °C (°F)							
10% Point max	215 (420)						
90% Point min		282[b] (540)					
max	288 (550)	338 (640)					
Sulfur content, % mass, max	0.5	0.5[f]					
Corrosion copper strip, max	3	3					
Ash, % mass, max			0.05	0.10	0.15	0.15	
Carbon residue, 10% b; % m, max	0.15	0.35					
Water and sediment, % vol, max	0.05	0.05	(0.50)[g]	(0.50)[g]	(1.00)[g]	(1.00)[g]	(2.00)[g]

[a] This limit guarantees a minimum heating value and also prevents misrepresentation and misapplication of this product as Grade No. 2.

[b] Lower or higher pour points may be specified whenever required by conditions of storage or use. When pour point less than −18°C (0°F) is specified, the minimum viscosity for grade No. 2 shall be 1.7 cSt (31.5 SUS) and the minimum 90% point shall be waived.

[c] Where low sulfur fuel oil is required, Grade 6 fuel oil will be classified as low pour +15°C (60°F) max or high pour (no max). Low pour fuel oil should be used unless all tanks and lines are heated.

[d] Viscosity values in parentheses are for information only and not necessarily limiting.

[e] Where low sulfur fuel oil is required, fuel oil falling in the viscosity range of a lower numbered grade down to and including No. 4 may be supplied by agreement between purchaser and supplier. The viscosity range of the initial shipment shall be identified and advance notice shall be required when changing from one viscosity range to another. This notice shall be in sufficient time to permit the user to make the necessary adjustments.

[f] In countries outside the United States other sulfur limits may apply.

[g] The amount of water by distillation plus the sediment by extraction shall not exceed the value shown in the table. For Grade No. 6 fuel oil, the amount of sediment by extraction shall not exceed 0.50 weight %, and a deduction in quantity shall be made for all water and sediment in excess of 1.0 weight %.

Adapted with permission from American Society for Testing and Materials, *Annual Book of ASTM Standards*, Sect. 5 (1989), © ASTM.

TABLE N.1. Analyses of Representative U.S. Crude Oils

Property	Swanson River, Alaska	San Ardo, California	Opelousas, Louisiana	McComb, Mississippi	Grayburg-Jackson, New Mexico	Sho-Vel-Tum, Oklahoma	East Texas	Garza, Texas	Little Buffalo Basin, Wyoming
Specific gravity	0.878	0.992	0.799	0.827	0.836	0.864	0.828	0.843	0.930
Pour point, °C	< -15	27	< -15	4	< -15	-4	< -15	-9	< -15
Sulfur, wt %	0.16	2.25	0.06	0.09	0.89	1.00	0.32	0.51	3.14
Nitrogen, wt %	0.203	0.913	0.020	0.050	0.077	0.203	—	—	0.38
Viscosity, Saybolt Universal at 100°F, sec[a]	61	> 6000	40	40	43	74	40	42	320
Carbon residue, Conradson, wt %	8.1	3.4	0.3	1.7	1.8	2.0	1.7	2.5	9.5
Volume %									
Gasoline and naphtha	27.4	—	41.9	33.2	35.6	23.7	35.6	34.5	12.8
Kerosine distillate	9.1	—	18.2	11.4	5.1	9.3	9.7	5.0	6.8
Gas oil	15.4	11.7	8.6	17.3	18.6	13.5	16.1	18.8	10.8
Nonviscous lubricating distillate	9.3	8.5	9.8	12.4	10.0	11.2	10.0	8.9	10.3
Medium lubricating distillate	7.4	4.6	5.5	6.7	6.6	3.7	4.0	5.5	5.7
Viscous lubricating distillate	—	1.6	1.0	—	0.8	—	—	2.4	4.2
Residuum	31.4	72.0	12.8	18.6	20.8	37.4	22.9	23.9	49.2
Distillation loss	0	1.6	2.2	0.4	2.5	1.2	1.7	1.0	0.2
Carbon residue, Conradson, residuum, wt %	22.3	4.6	2.3	7.9	7.6	4.8	6.6	8.2	17.6

[a] Saybolt Universal viscosity (in sec) can be converted to kinematic viscosity (in centistoke) by the following approximate equation:

$$v = 0.219t - \frac{149.7}{t}$$

where

v = kinematic viscosity, centistoke (mm^2/sec), and

t = Saybolt Universal viscosity, sec.

Both measurements are at the same temperature.
From Ref. 39.

TABLE N.2. Analyses of Representative World Crude Oils

Property	Hassi Messaoud, Algeria	Sarir, Libya	Seria, Borneo	Minas, Sumatra	Agha Jari, Iran	Ghawar, Saudi Arabia	Golden Spike, Canada	La Ceibita, Venezuela	Lama, Venezuela
Specific gravity	0.812	0.840	0.841	0.861	0.852	0.861	0.839	0.829	0.915
Pour point, °C	< −15	21	−12	24	−7	< −15	< −15	7	< −15
Sulfur, wt %	0.15	0.15	0.10	0.10	1.43	2.14	0.24	0.41	1.47
Nitrogen, wt %	0.018	0.076	—	0.130	0.015	0.114	—	0.055	0.203
Viscosity, Saybolt Universal at 100°F, sec[a]	33	62	35	92	46	52	42	35	142
Carbon residue, Conradson, wt %	0.9	2.9	0.3	1.5	3.0	2.5	2.2	1.4	4.0
Volume %									
Gasoline and naphtha	43.9	24.9	37.1	18.6	28.8	28.1	31.0	43.5	13.2
Kerosine distillate	11.9	13.3	—	14.7	10.2	9.5	4.5	5.1	3.2
Gas oil	13.7	11.2	39.9	10.8	14.6	13.2	18.3	18.3	14.9
Nonviscous lubricating distillate	9.6	12.2	11.0	18.1	9.6	9.5	9.6	9.4	10.9
Medium lubricating distillate	5.6	6.9	4.5	—	5.6	6.5	7.0	5.2	7.4
Viscous lubricating distillate	0.8	—	7.4	—	—	—	—	—	6.4
Residuum	13.6	30.2	0.1	37.3	28.9	30.9	25.9	17.6	42.6
Distillation loss	0.9	1.3	0.1	0.5	2.3	2.3	3.7	0.9	1.4
Carbon residue, Conradson, residuum, wt %	5.7	8.4	3.8	3.7	9.1	7.1	7.4	6.7	8.7

[a] Saybolt Universal viscosity (in sec) can be converted to kinematic viscosity (in centistoke) by the following approximate equation:

$$v = 0.219t - \frac{149.7}{t}$$

where

v = kinematic viscosity, centistoke (mm²/sec), and

t = Saybolt Universal viscosity, sec.

Both measurements are at the same temperature.
From Ref. 40.

TABLE O.1. ASTM Classification for Coals[a]

Class	Group	Fixed Carbon Dry %	Fixed Carbon Moist %	Volatile Matter Dry %	Volatile Matter Moist %	Natural Moisture %	Heating Values Dry Basis MJ/kg	Dry Basis kcal/kg	Moist Basis MJ/kg	Moist Basis kcal/kg	Agglomerating Character
I. Anthracite	1. Meta-anthracite	> 98	> 92	< 2	< 2	6	32.4	7740	31.4	7500	Nonagglomerating
	2. Anthracite	92–98	89–95	2–8	2–8	3	35.5	8000	35.5	8000	
	3. Semianthracite	86–92	81–89	8–14	8–15	3	34.7	8300	34.6	8275	
II. Bituminous	1. Low-volatile	78–86	73–81	14–22	13–21	5	36.6	8740	35.8	8550	
	2. Medium-volatile	69–78	65–73	22–31	21–29	7	36.2	8640	34.6	8275	Commonly agglomerating
	3. High-volatile A	< 69	58–65	> 31	> 30	5	34.2	8160	> 32.5	> 7775	
	4. High-volatile B	57	53	43	40	7	28.3–34.2	6750–8160	30.2–32.5	7220–7775	
	5. High-volatile C	54	45	46	40	16	31.0–35.1	7410–8375	26.7–30.2	6390–7220	
							28.3–31.0	6765–7410	24.4–26.7	5830–6390	Agglomerating
III. Subbituminous	1. Subbituminous A	55	45	45	38	18	28.8–31.6	6880–7540	24.4–26.7	5830–6390	Nonagglomerating
	2. Subbituminous B	56	43	44	35	24	27.4–30.3	6540–7230	22.1–24.4	5280–5830	
	3. Subbituminous C	53	37	47	36	30	25.1–28.7	5990–6860	19.3–22.1	4610–5280	
IV. Lignite	1. Lignite A	52	32	48	35	38	20.2–26.6	4830–6360	14.7–19.3	3500–4610	Nonagglomerating
	2. Lignite B	52	26	48	32	50	< 22	< 5250	< 14.7	< 3500	

[a] All analyses are on mineral-matter-free basis. Moist refers to coal containing its natural inherent moisture but not including visible water on the surface of the coal.

TABLE P.1. Analyses of Typical U.S. Solid Fuels[a]

Classification	State, County	Proximate analysis, %				Ultimate analysis, %					Heating value (gross)	
		Moisture	Volatile matter	Fixed carbon	Ash	Sulfur	Hydrogen	Carbon	Nitrogen	Oxygen	MJ/kg	kcal/kg
Meta-anthracite	Rhode Island, Newport	13.2	2.6	65.3	18.9	0.3	1.9	64.2	0.2	14.5	21.64	5170
			3.8	96.2		0.4	0.6	94.7	0.3	4.0	31.89	7620
Anthracite	Pennsylvania, Lackawanna	4.3	5.1	81.0	9.6	0.8	2.9	79.7	0.9	6.1	29.93	7155
			5.9	94.1		0.9	2.8	92.5	1.0	2.8	34.81	8320
Semianthracite	Arkansas, Johnson	2.6	10.6	79.3	7.5	1.7	3.8	81.4	1.6	4.0	32.26	7710
			11.7	88.3		1.9	3.9	90.6	1.8	1.8	35.86	8570
Low-volatile bituminous	West Virginia, Wyoming	2.9	17.7	74.0	5.4	0.8	4.6	83.2	1.3	4.7	33.47	8000
			19.3	80.7		0.8	4.6	90.7	1.4	2.5	36.46	8715
Medium-volatile bituminous	Pennsylvania, Clearfield	2.1	24.4	67.4	6.1	1.0	5.0	81.6	1.4	4.9	33.26	7950
			26.5	73.5		1.1	5.2	88.9	1.6	3.2	36.23	8660
High-volatile A bituminous	West Virginia, Marion	2.3	36.5	56.0	5.2	0.8	5.5	78.4	1.6	8.5	32.63	7800
			39.5	60.5		0.8	5.7	84.8	1.7	7.0	35.28	8430
High-volatile B bituminous	Kentucky, Muhlenburg	8.5	36.4	44.3	10.8	2.8	5.4	65.1	1.3	14.6	27.14	6490
			45.0	55.0		3.4	5.5	80.6	1.7	8.8	33.61	8030
High-volatile C bituminous	Illinois, Sangamon	14.4	35.4	40.6	9.6	3.8	5.8	59.7	1.0	20.1	25.12	6005
			46.6	53.4		5.0	5.6	78.6	1.3	9.5	33.07	7905
Subbituminous A	Wyoming, Sweetwater	16.9	34.8	44.7	3.6	1.4	6.0	60.4	1.2	27.4	24.75	5915
			43.7	56.3		1.8	5.2	76.0	1.5	15.5	31.12	7440
Subbituminous B	Wyoming, Sheridan	22.2	33.2	40.3	4.3	0.5	6.9	53.9	1.0	33.4	22.33	5340
			45.2	54.8		0.6	6.0	73.4	1.3	18.7	30.40	7265
Subbituminous C	Colorado, El Paso	25.1	30.4	37.7	6.8	0.3	6.2	50.5	0.7	35.5	19.89	4755
			44.6	55.4		0.5	5.0	74.1	1.1	19.3	29.19	6975
Lignite	North Dakota, McLean	36.8	27.8	29.5	5.9	0.9	6.9	40.6	0.6	45.1	16.27	3890
			48.4	51.6		1.6	5.0	70.9	1.1	21.4	28.42	6795

Classification	Proximate analysis, %				Ultimate analysis, %					Heating value (gross)	
	Moisture	Volatile matter	Fixed carbon	Ash	Sulfur	Hydrogen	Carbon	Nitrogen	Oxygen	MJ/kg	kcal/kg
High-temperature coke	5.0	1.3	83.7	10.0	0.8	0.5	82.0	1.0	0.7	28.35	6775
Low-temperature coke	2.8	15.1	72.1	10.0	1.8	3.2	74.5	1.6	6.1	29.28	7000
Beehive coke	0.5	1.8	86.0	11.7	1.0	0.7	84.4	1.2	0.5	29.11	6960
By-product coke	0.8	1.4	87.1	10.7	1.0	0.7	85.0	1.3	0.5	29.49	7050
High-temperature coke breeze	12.0	4.2	65.8	18.0	0.6	1.2	66.8	0.9	0.5	23.70	5665
Petroleum coke	1.1	7.0	90.7	1.2	0.8	3.2	90.8	0.8	2.1	35.00	8365
Pitch coke	0.3	1.1	97.6	1.0	0.5	0.6	96.6	0.7	0.3	32.76	7830

[a]Analyses are "as received," except where moisture and ash percentages are omitted.

Adapted with permission from R. E. Bolz and G. L. Tuve, *Handbook of Tables for Applied Engineering Science*, 2nd ed. (1973), © CRC Press, Inc., Boca Raton, FL.

TABLE P.2. Detailed Analyses of Three U.S. Coals

Analysis item	Ohio #6, Washed	Indiana #3, Cleaned	Illinois #6, Unground
Bulk density, kg/L, mf		0.798	0.759
Compacted density, kg/L, mf		0.837	0.772
Density, kg/L, mf	1.357	1.432	1.447
Ultimate analysis, maf, %			
C	82.0	80.1	79.8
H	5.5	5.6	5.2
S	4.4	5.5	3.3
N	1.5	1.5	1.6
O by difference	6.6	7.3	10.1
Ultimate analysis, mf, %			
C	76.7	68.2	70.2
H	5.1	4.8	4.6
S	4.1	4.7	2.9
N	1.4	1.3	1.4
ash	6.5	14.8	12.0
O by difference	6.2	6.2	8.9
Raw			
Total water, %	2.15	13.7	16.1
Proximate analysis, maf, %			
Volatile matter	42.9	48.1	41.9
Fixed carbon	57.1	51.9	58.1
Proximate analysis, mf, %			
Volatile matter	40.1	41.0	36.9
Fixed carbon	53.4	44.2	51.1
Ash	6.5	14.8	12.0
Grindability			
Degree Hardgrove	46.5	51.0	48.0
at % H_2O	2.2	7.0	9.2
Coke reactivity	medium	good	good
$CO/100\ CO_2$	84.2	101.3	125.7
Gross heating value			
MJ/kg, mf	32.615	28.939	29.161
MJ/kg, maf	34.884	33.963	33.139
Net heating value			
MJ/kg, mf	31.489	27.880	28.148
MJ/kg, maf	33.679	32.724	31.987
Ash feasibility (A.S. No.)	1330	1330a	1330b
Deformation point, °C	1100	1115	1105
Hemisphere point, °C	1130	1180	1235
Flow point, °C	1200	1275	1275
Ash viscosity (VISCOS No.)	239	240	241
T at 100 poise, °C	1350	1410	1420
250 poise, °C	1320	1350	1330
500 poise, °C	1290	1325	1270

TABLE P.2. (*Continued*)

Analysis item	Ohio #6, Washed	Indiana #3, Cleaned	Illinois #6, Unground
Sulfur detail (wt %, mf)			
Combustible	4.07	4.71	2.85
Incombustible	0.01	0.17	0.41
Total	4.08	4.88	3.26
CO_2 content (wt %, mf)	0.09	0.34	0.59
Cl content (wt %, mf)	0.18	< 0.01	0.06
Ash composition			
Fe_2O_3	58.1	20.4	13.4
SiO_2	20.6	44.5	44.8
Al_2O_3	18.0	24.4	19.1
CaO	1.2	3.3	8.5
MgO	0.3	0.7	0.9
Na_2O	0.3	0.3	0.7
K_2O	0.4	1.7	1.6
TiO_2	0.9	0.9	0.9
P_2O_5	0.1	0.4	0.1
SO_3	0.4	3.0	8.6

P. PROPERTIES OF SELECTED U.S. COALS

Table P.1 gives analyses of samples of common banded coal in the United States, classified according to the rank and group of coal as defined in Table O.1. Also included are analyses of various kinds of commercial cokes.

Table P.2 shows detailed analyses of three U.S. coals, all in the high-volatile bituminous coal group.

References: 3, 5.

Q. SKELETON STEAM TABLES

Skeleton steam tables are abridged from the 1969 Edition of Steam Tables (Metric Units) by Keenan, Keyes, Hill and Moore.[43] Three tables are included. Table Q.1 gives saturated steam properties. Tables Q.2 and Q.3 list specific enthalpy and specific volume of superheated steam, respectively.

Reference: 43.

TABLE Q.1. Saturated Steam Properties

Temperature, T °C	Pressure, P bars	Specific volume, cm³/g		Enthalpy, J/g		Entropy, J/gm K	
		Saturated Liquid,[a] V_f	Saturated Vapor,[b] V_g	Saturated Liquid,[a] H_f	Saturated Vapor,[b] H_g	Saturated Liquid,[a] S_f	Saturated Vapor,[b] S_g
0	0.006109	1.0002	206278	−0.2	2501.3	−0.0001	9.1565
0.01	0.006113	1.0002	206136	0.01	2501.4	0.0000	9.1562
1	0.006567	1.0002	192577	4.16	2503.2	0.0152	9.1299
4	0.008131	1.0001	157232	16.78	2508.7	0.0610	9.0514
5	0.008721	1.0001	147120	20.98	2510.6	0.0761	9.0257
10	0.012276	1.0004	106379	42.01	2519.8	0.1510	8.9008
15	0.017051	1.0009	77926	62.99	2528.9	0.2245	8.7814
18	0.020640	1.0014	65038	75.58	2534.4	0.2679	8.7123
20	0.02339	1.0018	57791	83.96	2538.1	0.2966	8.6672
25	0.03169	1.0029	43360	104.89	2547.2	0.3674	8.5580
30	0.04246	1.0043	32894	125.79	2556.3	0.4369	8.4533
40	0.07384	1.0078	19523	167.57	2574.3	0.5725	8.2570
50	0.12349	1.0121	12032	209.33	2592.1	0.7038	8.0763
60	0.19940	1.0172	7671	251.13	2609.6	0.8312	7.9096
70	0.3119	1.0228	5042	292.98	2626.8	0.9549	7.7553
80	0.4739	1.0291	3407	334.91	2643.7	1.0753	7.6122
90	0.7014	1.0360	2361	376.92	2660.1	1.1925	7.4791
100	1.0135	1.0435	1672.9	419.04	2676.1	1.3069	7.3549
110	1.4327	1.0516	1210.2	461.30	2691.5	1.4185	7.2387
120	1.9853	1.0603	891.9	503.71	2706.3	1.5276	7.1296
130	2.701	1.0697	668.5	546.31	2720.5	1.6344	7.0269
140	3.613	1.0797	508.9	589.13	2733.9	1.7391	6.9299
150	4.758	1.0905	392.8	632.20	2746.5	1.8418	6.8379
160	6.178	1.1020	307.1	675.55	2758.1	1.9427	6.7502
180	10.021	1.1274	194.05	763.22	2778.2	2.1396	6.5857
200	15.538	1.1565	127.36	852.45	2793.2	2.3309	6.4323
220	23.18	1.1900	86.19	943.62	2802.1	2.5178	6.2861
240	33.44	1.2291	59.76	1037.32	2803.8	2.7015	6.1437
260	46.88	1.2755	42.21	1134.37	2796.9	2.8838	6.0019
280	64.12	1.3321	30.17	1135.99	2779.6	3.0668	5.8571
300	85.81	1.4036	21.67	1344.0	2749.0	3.2534	5.7045
320	112.74	1.4988	15.488	1461.5	2700.1	3.4480	5.5362
340	145.86	1.6379	10.797	1594.2	2622.0	3.6594	5.3357
360	186.51	1.8925	6.945	1760.5	2481.0	3.9147	5.0526
374.136	220.9	3.155	3.155	2099.3	2099.3	4.4298	4.4298

[a] The subscript f refers to a property of liquid in equilibrium with vapor.
[b] The subscript g refers to a property of vapor in equilibrium with liquid.

TABLE Q.2. Superheated Steam: Specific Volume V_g in cm^3/g

Temperature, °C	Pressure, bars (Saturated Temperature, °C)													
	0.1 (45.81)	0.2 (60.06)	0.4 (75.87)	0.6 (85.94)	0.8 (93.50)	1.0 (99.63)	1.5 (111.37)	2.0 (120.23)	5.0 (151.86)	10 (179.91)	20 (212.42)	30 (233.90)	40 (250.40)	50 (263.99)
50	14869													
60	15336													
70	15801	7883												
80	16267	8117	4043											
90	16731	8351	4161	2764										
100	17196	8585	4279	2844	2127	1695.8								
120	18123	9051	4515	3003	2247	1792.9	1187.8							
140	19050	9516	4749	3160	2365	1888.7	1253.0	935.0						
160	19975	9980	4982	3317	2484	1983.8	1317.3	984.1	383.6					
180	20900	10444	5215	3473	2601	2078.2	1381.1	1032.4	404.5	194.5				
200	21825	10907	5448	3628	2718	2172	1444.3	1080.3	424.9	206.0				
220	22749	11370	5680	3783	2835	2266	1507.3	1127.9	444.9	216.9	102.15			
240	23674	11832	5912	3938	2951	2359	1570.0	1175.2	464.6	227.5	108.45	68.20		
260	24598	12295	6143	4093	3068	2453	1632.5	1222.3	484.1	237.8	114.36	72.85	51.74	
280	25521	12757	6375	4248	3184	2546	1694.8	1269.3	503.4	248.0	120.01	77.12	55.46	42.24
300	26445	13219	6606	4402	3300	2639	1757.0	1316.2	522.6	257.9	125.47	81.14	58.84	45.32
400	31063	15529	7763	5174	3879	3103	2067	1549.3	617.3	306.6	151.20	99.36	73.41	57.81
500	35679	17838	8918	5944	4458	3565	2376	1781.4	710.9	354.1	175.68	116.19	86.43	68.57
600	40295	20147	10072	6714	5035	4028	2685	2013	804.1	401.1	199.60	132.43	98.85	78.69
700	44911	22455	11227	7484	5613	4490	2993	2244	896.9	447.8	223.2	148.38	110.95	88.49
800	49526	24763	12381	8254	6190	4952	3301	2475	989.6	494.3	246.7	164.14	122.87	98.11
900	54141	27070	13535	9023	6767	5414	3609	2706	1082.2	540.7	270.0	179.80	134.69	107.62
1000	58757	29378	14689	9792	7344	5875	3917	2937	1174.7	587.1	293.3	195.41	146.45	117.07
1100	63372	31686	15843	10562	7921	6337	4225	3168	1267.2	633.5	316.6	210.98	158.17	126.48
1200	67987	33994	16997	11331	8498	6799	4532	3399	1359.6	679.8	339.8	226.52	169.87	135.87

TABLE Q.3. Superheated Steam: Specific Enthalpy H_g in kJ / kg

Temperature, °C	— Pressure, bars (Saturated Temperature, °C)													
	0.1 (45.81)	0.2 (60.06)	0.4 (75.87)	0.6 (85.94)	0.8 (93.50)	1.0 (99.63)	1.5 (111.37)	2.0 (120.23)	5.0 (151.86)	10 (179.91)	20 (212.42)	30 (233.90)	40 (250.40)	50 (263.99)
50	2592.6													
60	2611.5													
70	2630.5	2628.8												
80	2649.5	2647.9	2644.8											
90	2668.5	2667.1	2664.3	2661.5										
100	2687.5	2686.2	2683.8	2681.3	2678.8	2676.2								
120	2725.6	2724.6	2722.6	2720.6	2718.6	2716.6	2711.4							
140	2763.8	2763.0	2761.4	2759.8	2758.2	2756.5	2752.4	2748.1						
160	2802.2	2801.6	2800.2	2798.9	2797.6	2796.2	2792.8	2789.3	2767.4					
180	2840.8	2840.2	2839.1	2838.0	2836.9	2835.8	2832.9	2830.1	2812.0	2778.3				
200	2879.5	2879.1	2878.1	2877.2	2876.2	2875.3	2872.9	2870.5	2855.4	2827.9				
220	2918.5	2918.1	2917.3	2916.3	2915.7	2914.8	2912.8	2910.7	2897.9	2874.9	2821.7			
240	2957.7	2957.3	2956.6	2955.9	2955.2	2954.5	2952.7	2950.9	2939.9	2920.4	2876.5	2824.3		
260	2997.0	2996.7	2996.1	2995.5	2994.9	2994.3	2992.7	2991.1	2981.5	2964.6	2927.7	2885.5	2836.3	
280	3036.7	3036.4	3035.8	3035.3	3034.7	3034.2	3032.8	3031.4	3022.9	3008.2	2976.4	2941.3	2901.8	2856.8
300	3076.5	3076.3	3075.8	3075.3	3074.8	3074.3	3073.1	3071.8	3064.2	3051.2	3023.5	2993.5	2960.7	2924.5
400	3279.6	3279.4	3279.1	3278.8	3278.5	3278.2	3277.4	3276.6	3271.9	3263.9	3247.6	3230.9	3213.6	3195.7
500	3489.1	3489.0	3488.8	3488.6	3488.3	3488.1	3487.6	3487.1	3483.9	3478.5	3467.6	3456.5	3445.3	3433.8
600	3705.4	3705.3	3705.2	3705.0	3704.9	3704.7	3704.3	3704.0	3701.7	3697.9	3690.1	3682.3	3674.4	3666.5
700	3928.7	3928.6	3928.5	3928.4	3928.3	3928.2	3927.9	3927.6	3925.9	3923.1	3917.4	3911.7	3905.9	3900.1
800	4159.0	4159.1	4158.9	4158.8	4158.7	4158.6	4158.4	4158.2	4156.9	4154.7	4150.3	4145.9	4141.5	4137.1
900	4396.4	4396.4	4396.3	4396.3	4396.2	4396.1	4396.0	4395.8	4394.7	4392.9	4389.4	4385.9	4382.3	4378.8
1000	4640.6	4640.6	4640.5	4640.4	4640.4	4640.3	4640.2	4640.0	4639.1	4637.6	4634.6	4631.6	4628.7	4625.7
1100	4891.2	4891.2	4891.1	4891.1	4891.0	4891.0	4890.8	4890.7	4889.9	4888.6	4885.9	4883.3	4880.6	4878.0
1200	5147.8	5147.8	5147.7	5147.7	5147.6	5147.6	5147.5	5147.3	5146.6	5145.4	5142.9	5140.5	5138.1	5135.7

REFERENCES

1. E. A. Mechtly, *The International System of Units, Physical Constants and Conversion Factors.* NASA [Spec. Publ.] SP-7012 (1969).
2. R. C. Weast, ed., *Handbook of Chemistry and Physics*, 68th ed. CRC Press, Boca Raton, Florida, 1987–1988.
3. R. E. Bolz and G. L. Tuve, *Handbook of Tables for Applied Engineering Science*, 2nd ed. Chemical Rubber Company, Cleveland, Ohio, 1973
4. R. C. Reid, J. M. Prausnitz, and B. E. Poling, *The Properties of Gases and Liquids*, 4th ed. McGraw-Hill, New York, 1987.
5. H. C. Hottel, M.I.T. Notes in 10.70, 1971.
6. R. T. Sanderson, *Chemical Bonds and Bond Energy*, 2nd ed. Academic Press, New York, 1976.
7. *JANAF Thermochemical Tables*, PB 168 370, U.S. Government Printing Office, Washington, D.C., 1965.
8. *JANAF Thermochemical Tables*, 3rd ed., NSRDS-NBS37, U.S. Government Printing Office, Washington, D.C., 1986.
9. N. N. Semenov, *Some Problems in Chemical Kinetics and Reactivity* (M. Boudart, transl.) Princeton University Press, Princeton, New Jersey, 1958.
10. H. C. Hottel, "The Relative Thermal Value of Tomorrow's Fuels." *Pap., Eng. Found. Conf.*, Santa Barbara, California, Nov. 7–12, 1982; *Ind. Eng. Chem. Funadam.* **22**, 271 (1983).*
11. J. G. Speight, *The Chemistry and Technology of Petroleum.* Dekker, New York, 1980.
12. Y. S. Touloukian, P. E. Liley, and S. C. Saxena, *Thermophysical Properties of Matter*, Vol. 5. IFI/Plenum, New York, 1970.
13. J. W. Rose and J. R. Cooper, *Technical Data on Fuel,* 7th ed. The British National Committee of the World Energy Conference, London, 1977.
14. J. O. Hirschfelder, C. F. Curtiss, and R. B. Bird, *Molecular Theory of Gases and Liquids.* Wiley, New York, 1954.
15. R. B. Bird, W. E. Stewart, and E. N. Lightfoot, *Transport Phenomena*, Wiley, New York, 1960.
16. M. A. Field, D. W. Gill, B. B. Morgan, and P. G. W. Hawksley, *Combustion of Pulverized Coal*, British Coal Utilization Research Association, Surrey, England, 1967.
17. *International Critical Tables of Numerical Data, Physics, Chemistry and Technology*, Vol. 5. McGraw-Hill, New York, 1929.
18. *McGraw-Hill Encyclopedia of Science and Technology*, 5th ed., Vol. 4. McGraw-Hill, New York, 1982.
19. *Fire Hazard Properties: Flash Points, Flammability Limits and Autoignition Temperatures.* Eng. Sci. Data. Item 82030 (1982).
20. D. J. McCracken, *Hydrocarbon Combustion and Physical Properties*, Rep. No. 1496. Ballistic Research Laboratories (September 1970).

---

*Sadly, this ref. contains a confusing typographical error. In its Table II, Col. (9), all numbers except 1.039 and 1.000 should have 0 before the decimal instead of 1.

21. H. C. Barnett and R. R. Hibbard, *Basic Considerations in the Combustion of Hydrocarbon Fuels with Air*. NASA Tech. Rep. 1300 (1959).

22. B. Lewis and G. von Elbe, *Combustion Flames and Explosions of Gases*. Academic Press, New York, 1961.

23. M. G. Zabetakis, *Flammability Characteristics of Combustible Gases and Vapors*. Bull.—Bur Mines 627 (1965).

24. M. V. Blanc, P. G. Guest, G. von Elbe, and B. Lewis, *Third Symposium, Combustion and Flame and Explosion Phenomena*, Williams & Wilkins (1949).

25. H. F. Colcote, C. A. Gregory, C. M. Barnett, and R. B. Gilmer, "Spark Ignition—Effect of Molecular Structure." *Ind. Eng. Chem.* **44**, 2656 (1952).

26. D. M. Simon, "Flame Propagation: Active Particle Diffusion Theory." *Ind. Eng. Chem.* **43**, 2718 (1951).

27. M. Gerstein, O. Levine and E. L. Wong, *J. Am. Chem. Soc.* **73**, 1 (1951).

28. A. G. Gaydon and H. G. Wolfhard, *Flames—Their Structure, Radiation and Temperature*. Chapman & Hall, London, 1970.

29. I. I. Berenblut and A. B. Downes, *Tables for Petroleum Gas/Oxygen Flames Combustion Products and Thermodynamic Properties*. Oxford University Press, London and New York, 1960.

30. R. M. Frisrtrom and A. A. Westenberg, *Flame Structure*. McGraw-Hill, New York, 1965, p. 55.

31. American Petroleum Institute, *Knocking Characteristics of Pure Hydrocarbons*. Res. Proj. No. 45, ASTM Special Tech. Publ. 225 (1958).

32. American Society for Testing and Materials, *Annual Book of ASTM Standards*, Sect. 5, ASTM, Philadelphia, Pennsylvania, 1989.

33. J. M. Collins and G. H. Unzelman, "Diesel Trends Put New Emphasis on Economics and Fuel Quality." *Oil Gas J.* **80**(22), 87 (1982).

34. J. M. Collins and G. H. Unzelman, "Better Cetane Prediction on Equations Developed." *Oil Gas J.* **80**(23), 148 (1982).

35. R. M. Schirmer and E. W. Aldrich, *Microburner Studies of Flame Radiation as Related to Hydrocarbon Structure*. Res. Div. Rep. 3752-62R, Phillips Petroleum Company, Prog. Rep. No. 4 (1964).

36. M. Grayson and D. Eckroth, eds., *Kirk-Othmer Encyclopedia of Chemical Technology*, 3rd ed., Vol. 11. Wiley, New York, 1980.

37. D. L. Baulch, D. D. Drysdale, and A. C. Lloyd, *Critical Evaluation of Phase Rate Data for Homogeneous, Gas-Phase Reactions of Interest in High-Temperature Systems*. School of Chemistry, The University of Leeds, 1968.

38. C. G. Segeler, ed., *Gas Engineers Handbook*, 1st. ed. Industrial Press, New York, 1966.

39. C. M. McKinney, E. P. Ferrero, and W. J. Wenger, *Analyses of Crude Oils from 546 Important Oil Fields in the United States*, Bur. Mines Rep. Invest. 6819 (1966).

40. E. P. Ferrero and D. T. Nichols, *Analyses of 169 Crude Oils from 122 Foreign Oil Fields*, Bur. Mines Inf. Circ. 8542 (1972).

41. P. E. Considine, ed., *van Nostrand's Scientific Encyclopedia*, 6th ed. van Nostrand Reinhold, New York, 1983.

42. M. Grayson and D. Eckroth, eds., *Kirk-Othmer Encyclopedia of Chemical Technology*, 3rd ed., Vol. 6. Wiley, New York, 1980.

43. J. H. Keenan, F. G. Keyes, P. G. Hill, and J. G. Moore, *Steam Tables: Thermodynamic Properties of Water Including Vapor, Liquid, and Solid Phases*, International Edition—Metric Units. Wiley, New York, 1969.

INDEX

Abstraction, 143
  constant ratio:
    n-butane, 148–149
    n-butyl, 145–146
Acetylene flame:
  ionization and temperature profiles, 299
  PAH kinetics, 273–274
  peak ion concentrations, 298
  sooting, PAH ions, 278, 280–282
  surface growth, 304–306
  temperature, 16
Acid rain problem, 13–14
Acoustic instabilities, 510–511
Activation energy, 789
  asymptotics, 484
  estimation, 98
  overall, 181
  premixed-flame structure, 485
  radical reactions, 816
Activation entropy, C_2H_5Cl pyrolysis, 94–96
Activation parameters, $O + CH_4$, 115
Active centers, 125–127
Addition reactions, 91, 161
Additives:
  diesel fuel, 91–92
  domestic heating oil, 44
Additivity rules, 53–71
  atom additivity, 54
  bond additivity, 52, 57, 59
  group additivity, 59, 63, 65–67, 69–71
  group values:
    halogen-containing compounds, 62–63
    hydrocarbons, 53–54
    nitrogen-containing compounds, 60–61
    organoboron groups, 67
    organometallic compounds, 64–65
    organophosphorus groups, 66
    oxygen-containing compounds, 56–59
    sulfur-containing compounds, 68–70
  notation for groups, 63
  ring-compound estimates corrections, 55
  thermochemistry and equilibrium constants
    from, 85
Aerodynamics, laminar flame, 361–364

A-factor:
  estimation, 92–93
  HOCO decomposition, 103–104
Agglomerates, 309
Agglomeration, 301–303
Air-assist atomizers, 555–568
Airblast atomizers, 567–582
  advantages, 569
  analysis of drop-size relationships, 579–
    581
  cone angle, 540
  Cranfield atomizers, 574
  design, 571
  drop-size equations, 568
  drop size measurement, 571
  gas turbines, 583
  light-scattering technique, 571–572
  mean drop size, 576–579
    air properties, 579
    influence of air/liquid ratio, 573
    liquid properties, 576, 579
  Nukiyama-Tanasawa, 569–570
  plain-jet, 574–576
  prefilming:
    drop-size equations, 577
    SMD, 580
    type, 571–572
  SMD, 569–570, 572–576
Air/liquid mass ratio, 579
Air pressure, effect on SMD, 556, 579
Air quality, policy of delegating
  responsibility to states, 13
Air temperature, effect on SMD, 556–557,
  579
Aldehydes, formation, 141
Aliphatics:
  combustion, 262–264
  ring formation by, 274–275
Alkanes:
  low-temperature hydrocarbon oxidation,
    135–136
  oxidation, 191–193
Alkylated aromatics, properties, 34–35
Alkyls, 143–144, 148

Ammonia:
  oxidation, 244–246
  pyrolysis, 247
Aniline point, 813–815
Anthracite, 661
Antiknock index, gasoline, 822
Antiknock performance, *see also* Knock
    ratings
  equation, 808, 810
Arenes, group values, 63
Aromatics:
  alkylated, properties, 34–35
  antiknock level, 28
  in domestic heating oil, 44
  intermediate temperature chemistry, 151–152
  jet fuel content, 8
Array theory, 621
Arrhenius rate factor, 470–472
Ash, *see also* Airblast atomizers; Simplex
    atomizer; Spray
  coal combustion effects, 702–703
  composition, 668–669
Association reactions, 91
ASTM coal classification, 661, 665
ASTM D 1, 821–823
ASTM D 396–86, 830–831
ASTM D 439, 810, 821–822
ASTM D 975–88, 826–827
ASTM D 976, 812
ASTM D 1655, 823
ASTM D 2699, 808
ASTM D 2700, 808
ASTM D 2880–88, 828–829
Atom:
  additivity, 54
  conservation, 490–492
Atomization, 530–596
  air-assist atomizers, 555–568
  diesel engines, 530, 586–589
  electrostatic atomizers, 592–594
  fan-spray atomizers, 564–565
  gas turbines, 582–586
    combustion chambers, 531
    gas injection, 584
    premix-prevaporize combustor, 584
    slinger system, 585–586
    vaporizers, 583–584
  jet breakup, 531–532
  oil-fired furnaces, 531
  plain-orifice atomizer, 530, 546–548
  power systems, 589–592
    twin-fluid atomizer, 592
    Y-jet atomizers, 590–591

  Rayleigh mechanism, 531
  rotary atomizers, 565–566
  sheet breakup, 532–533
  ultrasonic atomizer, 594–596
  Weber mechanism, 531
  whistle atomizers, 596
  wide-range atomizers, 560–564
  Z number, 532
Autoignition, 139
  n-butane, 141
  temperature, 386–389, 803–806
    definition, 387
    determination, 388–389
    induction time, 386–388
    relation to flammability limits, 389
  transition, 410–411
Automobiles:
  carburetor icing, 21–23, 26
  cold starting, 18–20
  gasoline requirements, 835
  knock, 24–30
  vapor lock, 20
  warm-up performance, 19–20
Aviation turbine fuels, specifications, 823–824
  military, specification, 824

Barr and Mullins' equation, 430
Benzaldehyde formation, 154–155
Benzene:
  addition to nonaromatic, 276
  combustion, 264–265
  destruction, 283–285
  higher aromatics from, 275–283
    ionic mechanisms, 278, 280–283
    radical mechanisms, 275–279
  intermediate temperature oxidation, 155, 159
  oxidation, species evolution, 155, 158–159
  phenyl radical additions, 277
Benzene flame:
  ionization and temperature profiles, 299
  PAH kinetics, 273
  ring rupture, 283
  soot behavior, 293–294
Benzo[*a*]pyrene, emissions, 270–271
Beta function, conserved scalar, 499–500
Beta scission, 144, 146–148
Bimolecular atom-transfer reaction,
    potential-energy function, 86–87
Binary gas mixtures, diffusion coefficients,
    798–802
Biomass, 5

Bituminous coal, 661
  combustion, element release, 721–723
Blast furnace gas, 819
Blowing effect, 618
Blow off, 361–363
  coherent structures in turbulent
    combustion, 522
  diffusion flames, 453–454
  turbulent diffusion flames, 505–506
Blowout, spray combustion, 632
Boiling point, pure hydrocarbons, 811–812
Bond additivity, 52, 57, 59
Bond bending, frequencies assigned to, 75–
  78
Bond energies, 82–83, 787–789
Bonds:
  energy required to break, 789
  stretching, frequencies assigned to, 75–78
Branched chain reactions, 126–127
Bubble concept, 13
Burke and Schumann theory, 427–431
  flame height, 437–438
Burning velocity:
  auto ignition transition, 410
  flat-flame burner technique, 343
  initial temperature of mixture, 356
  laminar flame, 351, 361–362
  mixture composition effect, 344
  premixed flames, 341–342
  pressure exponents and, 355
  relation to preheat zone thickness, 346
  slot-burner technique, 342–343
  soap-bubble technique, 343
Burnout:
  carbon monoxide, 221–222
  char particles, 705–706
  coal, 751
n-Butane:
  autoignition, 141
  oxidation, species evolution, 149–151, 155
Butane-air flame, height, 439–440
n-Butyl, cleavage, 145

Carbon, concentration profile vs. axial
    distance, 443
Carbonaceous solids, intrinsic reactivity,
    708–709
Carbon conversion:
  coal particle diameter effect, 736
  during entrained coal gasification, 736
  entrained gasifier data, 738–739
  O_2/coal ratio effects, 738
  oxygen/coal ratio and flame type effect,
    737–738

Carbon dioxide:
  emissions, 10
  production, low-temperature hydrocarbon
    oxidation, 141–142
Carbon monoxide, 216–228
  burnout, 221–222
  chemical kinetic modeling, 165
  concentration predictions in exhaust gas,
    226
  emissions, 222
    characteristic time, gas turbine
      combustor, 222–223
    estimates, 216
  flame temperature, 16
  formation, 218
    modeling, 218–221
    premixed methane combustion, 220
    premixed methanol combustion, 220
  fractions in lean quenched combustion
    products, 226–227
  oxidation, 162, 163
    detailed kinetic model, 222–223, 225–
      226
    modeling, 221
    overall rate expressions, 224
    rate, 226
    rate coefficients, 221–222
    Westenberg model, 224–225
  partial equilibrium model, 226–227
  premixed methane-air combustion, 216–
    217
  production, 142–143
  quenching, in combustion gases, 225
  reaction with OH, 102–104
Carburetor icing, 21–23, 26
  ambient air temperature and humidity, 21
  cryoscopes, 21–23
  surfactants, 23, 26
  underhood configuration, 21
C—C bond, cleavage, 144
Cellular flames, 357–360
Cetane improvers, 41
Cetane index, calculated, 812
Cetane number, 39–40
  diesel fuel components, 810–811
  equation, 812
  pure hydrocarbons, 811–812
Chain reactions:
  branching, 126–127, 140–141, 169, 356
    degenerate, 127–128
  propagation, 125
  unbranched, 125–126
Chain theory, 125
Chapman-Jouguet detonation, 338–339, 378

Chapman-Jouguet Mach number, 370–371
Chapman-Jouguet point, 335, 338, 369
Chapman-Jouguet pressure, 370–371
Char:
  oxidation, 696–707
    diffusion limited combustion, 706–707
    global modeling, 703–706
    mineral matter effect, 701–703
    oxidizer diffusion rate, 704
    particle burning times, 707
    particle burnout, 705
    particle reaction rate, 704
    rate-controlling regimes, 695–696
    rate equation parameters, 700–701
    reaction rate coefficients, 697, 699
    reaction rate parameters, 697
    time required, 705
  proximate and ultimate analysis, 667, 670,
    710
  reactions, 695–716
    aspects, 715–716
    coal flames, 735
    global reaction rates, *see* Char, oxidation
    reaction variables, 695–696
  reaction rates, 708–714
    background, 708
    coefficients in CO_2, 701–702
    intrinsic data, 708–713
    intrinsic reaction rate coefficient, 708,
      713
    macroscopic modeling, 713
    microscopic pore char models, 714
    reactivity, 708–709, 711–712
    reactivity modeling, 713–714
  residual, oxidation, 675–676
Characteristic time, 162, 465–466
  gas turbine combustor, 222–223
  model, pollutant emissions, 222–224
  NO formation, 233
  turbulent reacting flows, 465–466
C_2H_5Cl pyrolysis, activation entropy, 94–96
Chemical heat release, average rates, 503–
  505
Chemical inhibitors, 365
Chemical kinetic modeling, 163–165
  deriving and validating detailed
    mechanisms, 165–181
  brute force fitting techniques, 170
  carbon monoxide oxidation, 165
  comparison of predictions, 170–172
  comprehensive model, 169–170
  concentration profiles, 173, 176
  feature sensitivity analysis, 179–180
  gradient sensitivity analysis, 173–175

local sensitivity analysis, 178
methanol, 165–168
normalized sensitivity coefficients, 175–
  177
rate constants, 173
second-order sensitivity coefficients,
  178–179
sensitivity coefficients, 171–173
falloff, 164
hierarchical process, 163
Lindeman mechanism, 164
models as sources of reduced mechanisms,
  194–202
modified four-step, reversible mechanisms,
  195–198
number of species required, 164
pressure-dependent, 164
theoretical methods, 163–164
three-body reaction efficiencies, 201–202
Chemical production term, modeling
  approximations, 489
Chemical species, conservation equation,
  491, 503
CH_2OH, rapid decomposition, 169
Circumferential fuel distribution, *see*
  Patternation
Clausius-Clapeyron equation, steady-state
  evaporation, 603
Cleavage, H atoms, 155
Cloud point, diesel fuel, 41
Coagulation, 301–303
Coal:
  analyses, 698
  availability and uses, 658–659
  burnout, 751
  classification, 661, 665, 825, 834
  combustion:
    direct, compared to gasification, 733
    summary of surveys, 656–657
  elemental and proximate analyses, 710
  formation and variation, 660–661
  future use of, 6
  increased use issues, 660, 664
  mineral matter removal, 673
  modeling of processes, 740–765
    background and scope, 740–742
    classifications, 744
    elements and premises, 742–744
    fixed bed combustion models, 761–763
    fluidized bed models, 763, 764–765
    multidimensional flame model, *see*
      PCGC-2
    plug-flow model, *see* Plug-flow model
    properties, 743

pulverized coal models, 758–761
status of model development and
    application, 765
particle ignition and devolatilization, 673–
    694
  combustion of volatiles, 689–695
  devolatilization, 675, 682–691
  global hydrocarbon reaction parameters,
    693
  global reaction rates, 692, 693–694
  ignition, 676–678
  local equilibrium, 692
  oxidation of residual char, 675–676
  particle heat-up, 678–682
  reaction rates, 674, 692
  temperatures of reacting particles, 679
physical and chemical properties, 665–672,
    835–838
  free-swelling index values, 666
  proximate and ultimate analyses, 667,
    670
  size distributions, pulverized coal, 670–
    672
  specific gravity, 666
  specific heat, 666
  thermal conductivity, 666
processes, 658, 660, 662–663
pulverized, nitrogenous emissions, 248
reaction processes, 654–655
structural characteristics, 671, 673
sulfur levels, 252
types, 661
Coal combustor, characteristics, 719
Coal flames, 716–740
  background and scope, 716
  bed type and coal size, 718–719
  classification, 717–718
  diffusion flame, 718
  gasification process, 729
    applications, 729–730
    basic reactions, 730–732
    carbon conversion, 736–739
    comparison with combustion, 733
    entrained gasifier data, 735–740
    feed ratio effect on coal gas efficiency,
      730–731
    gaseous reactions, 731–732
    physical and chemical mechanisms, 734–
      735
    species profile, 739–740
    types, 732–733
  premixed flame, 718
  processes, 720
  pulverized, 741–742

pulverized coal diffusion flames,
    720–729
  burnout and particle mixture fraction,
    725–726, 728
  characteristics, 720
  CO and CO_2 concentrations,
    727–729
  combustion data, 720–725
  combustor profile data, 725–729
  element release, 721–723
  NO and HCN + NH_3 concentrations,
    728–729
  O_2 concentration and gas mixture
    fraction, 726–728
  test variable effects, 724
  undried and dried, combustion
    characteristics, 722–724
  well-stirred reactor, 717
Coal furnaces, modeling, 741
Coal gas, 819
Coalification, 661
Coaxial spray diffusion flame, 625–626
CO/CO_2 equilibrium, pressure dependence,
    188
Coherent structures, methods identifying,
    473–474
Coke oven gas, 819
Collection efficiency, 542
Combustible, definition, 385
Combustible pollutants, reduction, 9–10,
Combustion:
  efficiency, spray combustion, 629–632
  instabilities, 404, 406
  products, 10
    fuel-oxygen/air mixtures, 807–808
    normalized enthalpy, 790, 791
  single-drop, 434–437
Combustors:
  coal, 719
    profile data, 725–729
  design, impact of coherent structures, 523
  gas turbine, characteristic time, 222–223
  modeling, 123–124
  numerical models, 123–124
  premix-prevaporize, gas turbines, 584
Conditioned probability density functions,
    480–481
Connective heat transfer, plug-flow model,
    749
Conradson carbon, in fuel oils, 46
Conservation equations, moments of
    conserved scalar, 494–495
Conservation of energy, 434
Conservation of momentum, 434, 487

Conserved scalars, probability density functions, 498–500
Conserved scalar, *see* Turbulent diffusion flames
Continuity of species, 433–434
Conversion factors, 785–786
Cool flames, 131–134
    equivalence ratio effect, 131, 134
Correlation coefficient, 481
Countergradient diffusion, 509
Coupling functions, 492
Cross correlation, 481
Cross flows, diffusion flames, 456
Crude oils, properties, 825, 832–833
Cryoscopes, 21–23
Curvature, turbulent flame speed effects, 516–518
Cyclic paraffins, antiknock level, 29
Cylindrical shock wave initiation, detonation, 380–383

Damköhler number, 466–467, 484
    laminar flame, 506
Deflagrative branch, 334
Degenerate branching, 127–128, 139
Delta functions, 478, 504
    Gaussian functions with, 499
Dense sprays, combustion, 623–624
Density ratio, 337
Detonation:
    branch, 334–335
    failure, 384–385
    propagation, gas phase, 369–385
        Chapman-Jouguet Mach number, 370–371
        Chapman-Jouguet pressure, 370–371
        critical power density for initiation by spark, 380–381
        free radical initiation, 383–384
        frontal structure, 373–378
        gross behavior, 369–371
        Hugoniot curve fit data, 370
        induction delay time, 384–385
        inherent instability, 372–373
        mechanism, 378
        minimum transverse wave spacings, 384
        orifice diameter, 384
        planar initiation, 378–380
        power density, 381–383
        spherical or cylindrical shock wave initiation, 380–383
        Zel'dovich, von Neumann, and Döring theory, 371–372
    transition, deflagration to, 408, 410–412

Devolatilization, coal, 749, 751
    particles, 682–691
        cracking of molecule, 688
        experimental studies, 680–685
        models, 685–689
        pyrolysis product distributions, 682–684
        pyrolysis product yield, 689–691
        techniques and conditions, 680–681
        weight losses, 682–683, 686–687
    raw, 675
1-DICOG, *see* Plug-flow model
Diesel engine, 36–37
Diesel fuel, 36–42
    additives, 91–92
    cetane number, 39–40, 810–812
    desired characteristics, 39
    gravity, 40
    pour and cloud points, 41
    properties, 37
    quality and uses, 37–39
    storage stability, 40–41
    sulfur content, 40
    types, 37–39
    viscosity, 40
    volatility, 40
Diesel fuel oils, requirements for, 826–827
Diesel injection, atomization, 586–589
    droplet size distribution, 588–589
    fuel spray penetration, 586–587
Diffusion coefficient, 439, 447
    binary gas mixtures, 798–802
    gases and vapors into air, 802
    through nitrogen, 800
    water vapor into air, 801
Diffusion equation, 434
Diffusion flames, 423–456. *See also* Laminar diffusion flames; Turbulent diffusion flames
    coal flames, 718
    condensed phase, 425
    configurations, 423–445
    cross flows, 456
    equilibrium, 469–470
    flame lift and resettle hysteresis, 453–454
    flame strength, 455–456
    gaseous, flame zones, 424–426
    gas jet, 424
    laminar, *see* Laminar diffusion flames
    opposed jet, 455–456
    pulverized coal, 720–729
    Reynolds numbers, 454–455
    shapes, 424
    spray, 625–626
    stability, 453–456

three-zone structure, 486
traverse, 425
turbulent, *see* Turbulent diffusion flames
Diffusion limited combustion, char oxidation, 706–707
Diffusion velocity, 347
Diffusive-thermal instability, premixed turbulent flames, 511–512
Discharge coefficient, 546
  relationship with atomizer geometry, 552–553
  swirl atomizer, 550–552
Dispersion, spray, 540
Displacement distance theory, 430, 432–433
Distillate fuels, 31–44
  diesel fuel, 36–42
  jet fuel, *see* Jet fuel
  kerosene, 31
D^2 law, 598, 614
Domestic heating oil, 42–44
  additives, 44
  aromatic content, 44
  desired characteristics, 42
  hydrocarbon analysis, 43
  quality and uses, 37–39
  sediment presence, 43
Domestic oil burners, 37
Double-flash photography, 545
Drag coefficient:
  acceleration effect, 617–618
  evaporating and burning drops, 618–620
  solid sphere, 615–617
  under intense mass transfer, 619
Droplet burning, 612–613
Droplet evaporation, 542, 596–620
  convective effects, 607–609
  D^2 law, 614
  droplet burning, 612–613
  drop lifetime, 604–605, 610, 612
  drop transport in sprays, 615–620
  effective evaporation constant, 609–612
  factors affecting lifetime, 609
  forced convection, 608–609
  heat transfer number, 601–602, 613
  heat transfer rate, 607–608
  heat-up period, 606–607, 610
  high temperature, 605–606
  mass rate, 600, 602, 604
  mass transfer number, 600–601
  multicomponent fuel drops, 614–615
  Nusselt number, 607–608
  steady-state, 597–604
  wet bulb temperature, 606

Droplet group theory, 621–622
Drop lifetime, 604–605, 610, 612
  burning droplet, 613
  factors affecting, 609
Drop size:
  analysis:
    of relationships, 579–581
    simplex atomizer, 558–559
  effect on weak extinction limits, 634
  equations:
    airblast atomizers, 568
    plain-jet airblast atomizers, 578
    prefilming airblast atomizers, 577
  mean, 533–534
    atomizer scale effect, 574
    influence of air/liquid ratio, airblast atomizers, 573
    mathematical definitions, 534
    simplex atomizer, 555–557
  measurement:
    airblast atomizers, 571
    direct method, 541
    Dobbins method, 543–544
    laser-scattering and intensity-rationing techniques, 545
    Malvern particle analyzer, 544
    nitrogen-freezing technique, 543
    optical methods, 543
    photographic techniques, 545
    Rosen-Rammler distribution, 544
    spray, 541–545
    wax droplet technique, 542–543
  relationships, simplex atomizer, 558
  ultrasonic atomizer, 595
Drop-size distribution, 534–539
  cumulative distribution curve, 535–537
  frequency-distribution curve, 535–536
  Nukiyama-Tanasawa equation, 537–538
  powders, 537
  Sauter mean diameter, 533, 538
  upper limit function, 538–539
  volume distribution equation, 538
Drop transport, sprays, 615–620
  acceleration effect on drop drag coefficient, 617–618
  drag coefficient of solid sphere, 615–617
  evaporating and burning drops, drag coefficient, 618–620
  motion equations, 615
  Reynolds number, 616–617
Drop velocities, spray combustion, 627–628
Dual-orifice atomizer, 562–563, 582
Duplex nozzle, 560–561
Duplex spray atomizer, 562–563, 582

Eddy diffusion coefficient, 801
Eddy diffusivity, 447
Efficiency factor, 100–101
Einstein's diffusion equation, 432
Electrical pressure, electrostatic atomizer,
    592
Electric utility steam generators, new,
    emission standards, 12–13
Electrostatic atomizers, 592–594
Empirical modeling concepts, 181–202
  auto ignition transition, 410–411
  in enclosures:
    with low L/D, 412–413
    with obstacles or large L/D, 413–414
    venting, 414
  multistep global and quasi-global
      mechanisms, 191–194
    alkanes, 191–193
    NO_x emission predictions, 193
    paraffin, 192
  one- and two-step global mechanisms,
      182–190
    adiabatic flame temperature, 188
    dissociation, 188
    flame properties, 185–186
    flame speed pressure dependence, 186,
        188
    flammability limit, 186–187, 189
    local equilibrium model, 189–190
    methane oxidation, 183–185
    single-step irreversible reaction, 185
    single-step reaction rate parameters,
        186–187
    two-step reaction parameters, 189
  overall specific rate constant, 181
  shock and detonation, front trajectories
      downstream of orifice plate, 412–413
  unconfined explosions, 414–415
Energy conservation, 8
Energy spectrum, turbulent reacting flows,
    463–464
Engulfment, stretching, coherence,
    interdiffusion, and moving observer,
    521
Ensemble averages, probability density
    functions, 481–482
Enthalpy, 72
  bond dissociation energies, 82–83
  conservation equation, 494
  laminar flame, 348–349
  normalized, 790
  pi-bond energies, 83
  pressure fluctuation neglect, 492–494
  strain energy, 83

Enthalpy equation, 433
Entropy:
  additivity contributions, 65
  free rotors, 75, 78
  as functions of barrier height, temperature
      and partition function, 79
  harmonic oscillator, 74–75
  molecular model compounds, 80–82
  structural estimates, 85
  vibrational contribution, 71
Epoxidation, 138
Equation of state, ideal gas, 330
Equilibrium constants, 85, 792
Equivalence ratio, 340
Ergodic theorems, 482
ESCIMO approach, 521
Ethane, 143–144
Ethene, addition reactions, 161
Ethylene, 143
Ethylene flame:
  diffusion, surface growth region, 316
  flat diffusion, absorption and emission
      spectra, 440–441
  laminar diffusion, soot distribution, 313,
      315
  peak ion concentrations, 298
  soot behavior, 293–294
Evaporation, single-drop, 434–437
Evaporation constant:
  drop lifetime, 604–605
  effective, 609–612
  stagnant fuel-air mixtures, 600
Explosion dynamics, 401–415
  central ignition, spherical vessel, 401–
      404
  combustion instabilities, 404, 406
  deflagration to detonation transition, 408,
      410–412
  gas velocity ahead of flame ball, 403
  Kelvin-Helmholtz, 404, 406, 408
  maximum pressure rise variation, 403,
      405
  overpressures, 406, 408–409
  Taylor instability, 404, 406–408
  venting, 402–403
  vent ratio, 403
  vessel processes complications, 404, 406–
      408
Explosion limit, lower, 364
Extinction:
  laminar flame, 364–365, 367–368
  local, flame stretch, 505–506
Extinction limits, weak, spray combustion,
    632–634

Fan-spray atomizers, 564–565
Favre averages, moments based on, 489–490
Fick's law approximation, 503
Fick's law of diffusion, 428
Fick's second law of diffusion, 432
First law of thermodynamics, differential
    form, 336
Fission, 91
Fixed bed combustion models, 761–763
Flame:
    chemistry, 440–442
    front, 429, 436
        curved, 396–398
    laminar pre-mixed properties, modeling,
        185–186
    maximum velocity, 803–806, 808
    propagation:
        closed end of long rectangular duct,
            410–411
        ignition energy effects, 394–395
    properties, 339–340
    speed, pressure dependence, 186
Flame height:
    fuel and air velocity effects, 437–438
    fuel concentration effect, 443
    laminar diffusion flames, 437–440
    turbulent diffusion flames, 447–448
    variation with pressure, 438–439
    variation with volumeric flow rate, 439–
        440
Flame holding, 361–364
    coherent structures in turbulent
        combustion, 522
    mechanism, 361–362
Flame profiles, 50–51
Flame strength, 455–456
Flame stretch, local extinctions, 505–506
Flame-stretch factor, 467–468
Flame temperature, 16, 808
    adiabatic, 188, 807–808
    fuel concentration effect, 443
Flame zones, gaseous diffusion flames, 424–
    426
Flammability, gas, 16–17
Flammability limit, 186–187, 803–806
    determination, 365
    initial temperature and pressure, 365
    laminar flame, 364–365, 367–368
    lower, 364
    single-step reaction mechanisms, 187
    temperature effect, 389
    two-step reaction mechanism, 189
Flammable, definition, 385
Flashback, 361–363

Flash point, 385–386, 804–806
    definition, 803
    measurement techniques, 386
Flat-flame burner technique, 343
Flow-number, simplex atomizer, 552–
    553
Fluidized bet models, 762, 764–765
Fossil fuels:
    nitrogen and sulfur contents, 230
    reserves and resources, 4–5
Free atom reactions, rate constants, 816
Free radicals:
    group additives, 83–84
    initiation, 383–384
Free rotation, 75, 78–79
Frequency factor, overall, 181
Frontal structure, detonation, 373–378
    Mach-stem configurations, 373–374
    mode number, 376
    single spin, 373
    smoke-foil records, 374–375
    transverse cell spacing, 376–377
    transverse wave spacing, 375
Froude number, 465
Fuel:
    adiabatic flame temperature, 807–808
    density, 553
    interchangeability, laminar premixed flame
        burners, 366–369
    mass fraction, 496
    nitrogen, conversion to NO_x, 236, 238
    PAH, 291
    properties, 368, 639
    thermophysical properties, 603
    use:
        future, 5–8
        history, 3–14
        importation, 6
        sources of fuel, 4
Fuel-air mixtures:
    combustion products, 807–808
    flame temperature, 807–808
    stagnant, evaporation constant, 600
Fuel injection, 530, 557
Fuel oils, 44–46
    analysis, 44, 46
    Conradson carbon and metal contents, 46
    requirements for, 830–831
Fuel-oxidizer mixtures, critical energy for
    direct initiation, 383

Gas:
    classifications, 401
    composition and properties, 14–17

Gas (*Continued*)
  diffusion coefficient into air, 802
  equation of state, 330
  flammability, 16–17
  postflame, adiabatic equilibrium condition,
    188
  Schmidt number, 802
  volumetric heating value, 16
Gas-air mixtures, properties, 403–404
Gaseous diffusion flame theory.
  Burke and Schumann theory, 427–431
  comprehensive theories, 433–434
  concentration profiles, 428
  displacement distance theory, 430, 432–
    433
Gaseous radical reactions, 789
Gas injection, atomization, 584
Gas jet diffusion flame, 424
Gasoline, 17–31
  antiknock index, 822
  carburetor icing, 21–23, 26
  characteristics that depend on fuel
    composition, 17–18
  cold starting, 18–20
  composition, 22–23, 26
  components, knock ratings, 808–810
  final boiling point, 26
  future of use, 6–7, 31
  gas chromatograph analysis, 22
  knock, 24–30
  octane number, 26–30
  refinery blending components, octane
    number and volatility characteristics,
    30
  tests for determining antiknock quality,
    24–25, 28
  vapor lock, 20
  volatility characteristics, 18
  warm-up performance, 19–20
Gas phase:
  energy balance, 747
  oxidation, hydrocarbons, 129
  PCGC–2, 753
Gas turbine:
  combustion chambers, atomization, 531
  NO and NO₂ exhaust fractions, 231–232
  combustors, CO and NO emission
    characteristic times, 222–223
  fuel oils, ground, specifications, 828–829
Gaussian functions, with delta functions, 499
Gibbs free energy, 72
Gradient sensitivity analysis, CO—H₂—O
    mechanism, 173–175
Gravity, diesel fuel, 40
Group additives, free radicals, 83–84

Group additivity, 59, 63, 65–67, 69–71,
    limitations, 69
Group combustion number, 621–622

Halogen-containing compounds, group
    contribution, 62–63
Harmonic oscillator, entropy, 74–75
H atoms, cleavage, 155
HCN:
  decomposition, 243–244
  formation, 236, 243, 247
  global oxidation rates, 694
Heat, 80
Heat capacity:
  as function of barrier temperature,
    partition function, 80
  molecular model compounds, 80–82
  structural estimates, 85
Heating value:
  gas, 16
  industrial gaseous fuels, 818
Heat loss theory, 364
Heat transfer equations, 437
Heat transfer number, 601–602, 613
Heat-up period:
  drop diameter, 610
  droplet evaporation, 606–607
Helmholtz free energy, 72
n-Hexane, diffusion flame:
  benzene and acetylene concentrations,
    287–289
  soot distribution, 313–314
High-temperature reaction phenomena, 160–
    162
  characteristic time, 162
  chemical initiation, 161
  constant ratio, 160
HOCO, decomposition, *A*-factors, 103–104
Homogeneity, probability density functions,
    482
Hot ignition, 143
Hugoniot equations, differentiated, 336
Hydrocarbon-air flame, nitric oxide
    formation rates, 231
Hydrocarbon flames:
  gaseous, global reaction rates, 693–694
  pressure exponent, 355
Hydrocarbons:
  boiling point, 811–812
  cetane number, 811–812
  combustion:
    aliphatics, 262–264
    benzene, 264–265
  emissions, from engines, 264–265, 269
  gas phase oxidation, 129

group values, 53–54
ions, soot formation, 299–300
knock ratings, 809
low-temperature oxidation mechanism,
    134–142
    chain branching, 140–141
    CO_2 production, 141–142
    epoxidation, 138
    hydroperoxide formation, 137–139
    hydroxyl radical reaction, 137
    initiation reaction, 135
    isomerization, 140–141
    propagation reactions, 136
    turnover temperature, 139–140
oxidation:
    chain-branching reaction, 356
    recombination reaction, 356
oxygenated, octane number, 30
smokey tendency, 813–815
Hydrogen:
    abstraction, 143
    flame temperature, 16
    as ideal fuel, 10
    jet fuel content, 34–35
Hydrogen-oxygen flame, composition profile,
    450, 452
Hydrogen peroxide, decomposition, 142–
    143
Hydroperoxide:
    formation, 137–139, 153–154
    low-temperature phenomena, 129
    turnover temperature, 154
Hydroperoxy:
    formation, isomerization, 139–140
    in H_2O_2 decomposition, 142–143

Ignition:
    aliphatic combustion, 262
    coal particles, 676–678
    kernel, decay or growth behavior, 395–
        396
    methane, 186
    spray combustion, 634–635
    temperatures, coal, 676–677
Ignition delay, 162, 168, 183
Ignition energy:
    determination, 390–391
    minimum, 390–398, 803–806
        definition, 390
        effect on flame propagation, 394–395
        electrode flanging effect, 391
        ignition kernel decay or growth
            behavior, 395–396
        relation to quenching distance, 392–393,
            395

space velocity, 397–398
spray combustion, 635–638
vs. stoichiometry, 396–397
Ignition theory, spray combustion, 635–639
Induction period, 127
    temperature and equivalence ratio effect,
        130, 134
Induction time, 179–181
    auto ignition temperature, 386–388
Inductive spark ignition, 401
Industrial gaseous fuels, properties, 817–
    819
Industrial steam generators, new, emission
    standards, 12–13
Inert gases, 365
Inorganic particulates, emission, 10
Instability, premixed turbulent flames, 510–
    512
Integral scale, 462–463
Intermediate temperature reaction
        phenomena, 142–160
    abstraction constant ratio, 145–146, 148–
        149
    alkyl radical production, 148
    aromatics, 152
    equivalence ratio effects, 150–151
    H_2O_2 decomposition, 142–143
    hot ignition, 143
    hydrogen atom yield, 149
    isoenergetic region, 149
    isomeric structure effects, 150
    negative temperature coefficient, 143
    species evolution:
        benzene oxidation, 155, 158–159
        n-butane oxidation, 149–151, 155
        iso-butane oxidation, 149, 152
        iso-octane, 149, 154
        n-octane oxidation, 149–151, 153, 156
        propane oxidation, 149–150
        toluene, 154–155, 157
    thermal decomposition, 144
Internal combustion engine:
    hydrocarbon emissions, 264–265, 269
    wall quenching, 264–265, 269
Inviscid fluid, one-dimensional steady flow,
    330–339
    density ratio, 337
    equations, 330–331
    physically viable solutions, 339
    Rayleigh and Hugoniot equations, 331–
        332
    solution with full equilibrium chemistry,
        336–339
    working fluid-heat addition model, 332–
        336

Ions:
  polycyclic aromatic hydrocarbons, 278
    280–282
  soot formation, 297–301
Iso-butane, oxidation, species evolution, 149,
    152
Isomerization, 91, 140–141, 148
Iso-octane, oxidation, species evolution, 149,
    154
Iso-octyl, beta scission fragmentation
    patterns, 147–148

Jacobian matrix, 190
Jet breakup, 531–532, 547
Jet fuel:
  aromatics:
    content, 8
    properties, 34–35
  determination of composition, 34
  ERBS, 32–33
  final boiling point, 35–36
  freezing point limitations, 32
  hydrogen content:
    effect on linear temperature and smoke
      number, 33–34
    variation with aromatic content, 36
  quality and uses, 37–39
  specifications, 32–33
  tank temperatures, 35–36
Joint probability density functions, 479–480
JP4, burning rate curves, 599–600

Kelvin-Helmholtz, 404, 406, 408
k-ϵ modeling, 487–488, 623
Kerosene, 31
  burning rate curves, 599–600
  quality and uses, 37–39
Ketones, formation, 141
Kinetics:
  bimolecular reactions, 90–91
  bound intermediates, 100–104
  efficiency factor, 100–101
  mechanisms, 122–123
  potential-energy profile, 86, 88
  reaction coordinate, 86–88
  reaction OH + CO → products, 102–104
  transition state theory, 86–90
  unimolecular reactions, 91
Knock, *see* Antiknock performance
Knock ratings:
  determination methods, 808
  gasoline components, 808–810
  hydrocarbons, 809
Kolmogorov length, 464

Kolmogorov scale, 464–465
  eddy, 467–468
Kolmogorov time, 466
Kurtosis, 476–477

Lambent flames, 424
Laminar burning velocity, 16
Laminar diffusion flames, 427–445
  Burke and Schumann theory, 427–431
  comprehensive theories, 433–434
  displacement distance theory, 430, 432–
    433
  flame chemistry, 440–442
  flame height, 437–440
  nozzle velocity effect on flame appearance,
    445–446
  polycyclic aromatic hydrocarbons, 287–289
  shapes, 430–431
  single-drop evaporation and combustion,
    434–437
  smoke, 442–445
  smoke-free fuel, 443–444
  soot, 313–318
Laminar flame:
  additive effects, 365
  aerodynamics, 361–364
  attachment point, 361
  blow off, 361–363
  critical velocity gradient, 363–364
  Damköhler number, 506
  flammability limits and extinction, 364–
    365, 367–368
  flashback, 361–363
  fuel interchangeability, 366–369
  lean limit, 367
  Le Chatelier's rule, 367
  propagation, 339–369
    burning velocity, 341–342
    cellular instability, 357–359
    chain-branching reaction, 356
    CO production, 367
    equivalence ratio, 340
    initial temperature of mixture, 356
    laminar flame theory, *see* Laminar flame
      theory
    oblique one-dimensional flame, 342
    physical observations, 339–344
    pressure exponent, 355
    Rayleigh equation, 339
    recombination reaction, 356
    relationship between burning velocity
      and preheat zone thickness, 346
    theoretical models and real flame
      behavior, 355–361

quenching, 362–363
  diameter, 365
  temperature, fuel-air and fuel-oxygen
    mixtures, 340–341
  velocity, temperature and pressure
    sensitivity, 340–341
Laminar flame theory, 344–354
  assumptions, 345
  burning velocity, 351
  cellular structure, 357, 359–360
  diffusion velocity, 347
  flame structure, 345, 352–353
  high-temperature region, 357
  Lewis number, 350–352
  mass-balance equation, 347–348
  preheat zone thickness, 345–346
  rate equation, 350, 354
  solutions to equations, 350–352
  total enthalpy flux, 348–349
Landau instabilities, premixed turbulent
    flames, 511
Le Chatelier's rule, 367
Lewis number, 350–352, 492–494, 602
Lift, diffusion flames, 453–454
Lift-off, turbulent diffusion flames, 505–
    506
Light-oil combustion, swirled, PAH levels,
    289–290
Lignite, 661
Lindeman mechanism, 97–98, 164
Liquefied petroleum gases, requirements for,
    820
Liquid fuels, *see also* specific fuels
  limiting requirements, 822
  specifications, 819–824
  surface tension, 797–798
  use, 7
Liquid jet, disintegration, 532
Local equilibrium model, 189–190
Low-temperature phenomena, 128–134
  cool flames, 131–134
  diffuse luminosity, 129, 132
  propane-oxygen system, 129, 134
Luminometer number, 813

Mach-stem configurations, 373–374
Malvern particle size analyzer, 544
Marginal probability density functions, 480
Marine diesel fuel, quality and uses, 37–39
Mass-balance equation, 347–348
Mass fraction of species equations, 487
Mass transfer number:
  droplet evaporation, 600–601
  relevant fuel properties, 639

Maximum experimental safe gap, 398–401
  chamber volume effect, 400
  measurement, 398–400
  National Electric Code, 401
  pressure piling, 400
Metals, in fuel oils, 46
Metathesis reactions, 90
Methane:
  chemical kinetic modeling, 163
  CO formation, 220
  disappearance, rate constants, 183–184
  global mechanisms, 184–185
  ignition, 186
  ignition delay time, 183
  nitrogen oxides, NO fractions, 228–229
  oxidation:
    multistep mechanisms, 198–201
    postinduction, 183–184
    reaction pathways, 262–263
  reaction with oxygen:
    activation parameters, 115
    Arrhenius parameters, 116
Methane flame:
  CO levels, 217–218
  composition profiles and reaction rate,
    451–452
  diffusion flame, carbon monoxide levels,
    219
  postflame composition, 202
  stable fixed nitrogen species product
    distribution, 240–241
Methanol:
  chemical kinetic modeling, 165
  CO formation, 220
  fuel-lean conditions, 168
  nitrogen oxide control, 11
  oxidation:
    comparison of predictions, 170–172
    Westbrook and Dryer comprehensive
      mechanism, 165–168
Methyl, decomposition, 143
Methyl substitution pathways, PAH
    formation, 278–279
Microscopic pore char models, 714
Mineral matter, removal from coal, 673
Mixing time, 222
Mixture fraction, 490–492
Mixture ratio, 507
Modeling, 122–125
  amount of detail required, 51
  combustors, 123–124
  goals, 50
  kinetic mechanisms, 123
  numerical, 122

Molecular model compounds, 80–82
Moment of inertia, free rotors, 75, 78
Moments:
  calculations, 475–479
  conservation equations, conserved scalar, 494–495
Multicomponent fuel drops, 614–615
Mutagenicity, PAHs, 269–270

Naglo and Strickland-Constable formula, 307–308
National Ambient Air Quality Standards, 11–12
National Electric Code, classifications for gases or vapors, 401
Natural gas, 819
Navier-Stokes equations, 521
Negative temperature coefficient, 139, 143
Nitric oxide, formation rates near flame zone, 231
Nitrogen, fossil fuel content, 230
Nitrogen-containing compounds, group values, 60–61
Nitrogen-freezing technique, 543
Nitrogen oxides, 228–252
  control with methanol, 11
  emissions:
    characteristic time, gas turbine combustor, 223
    diesel engine, 37
    estimates, 216
    gas turbine, prediction, 193
    importance of prompt NO, 231
    new steam generators, 12–13
    reduction, 248–249, 256
    sources, 228–229
  exhaust fractions, 229, 231
  fractions, premixed methane-air combustion, 228–229
  fuel-NO formation, 236, 238–249
    conversion in practical devices, 239
    correlation with flame conditions, 238
    detailed kinetic model, 241–243
    HCN decomposition, 243–244
    HCN formation, 243, 247
    inert pyrolysis, 242
    NH_3 oxidation, 244–246
    NH_3 pyrolysis, 247
    reaction paths, 241–242
    reburning, 248–249
    selective reduction, 249
    stable fixed nitrogen species product distribution, 240–241
    staged combustion, 248

hydrocarbon-N_2 and hydrocarbon-NO reactions, 236–237
  partial equilibrium model, 239
  production, Zel'dovich mechanism, 502–503
  prompt, 231
    formation, 235–236
    importance in NO_x emissions, 231
  simplified kinetic model, 238–239
  thermal formation, 233–235
    characteristic times, 233
    maximum rate, 233–234
    rate coefficients, 233
    temperature reductions, 234
Nitrogen peroxide:
  formation, 249–250
  impact as pollutant, 231
Nitrous oxide:
  in fossil-fuel-fired power plant stack gases, 231–232
  formation, 250–252
No. 2 heating oil, *see* Domestic heating oil
Nonreacting turbulence, 497
Nozzle, velocity effect on flame appearance, 445–446
Nukiyama-Tanasawa atomizer, 569–570
Nukiyama-Tanasawa equation, 537
Nusselt number, 607–608

n-Octane:
  combustion, 182, 187
  oxidation, species evolution, 149–151, 153, 156
Octane number:
  definition, 808
  oxygenated hydrocarbons, 30
n-Octyl, beta scission fragmentation patterns, 146, 148
OH, reaction with CO, 102–104
Oil, recoverable resources, 6
Oil-fired furnaces, atomization, 531
Oil gases, 819
Olefin, production, 138–139
Opposed jet diffusion flames, 455–456
Organoboron groups, group values, 67
Organometallic compounds, group values, 64–65
Organophosphorus groups, group values, 66
O—H radical, pool in flame zone and sulfur presence, 256
  recombination, presence of sulfur, 255
Overdriven detonations, 335, 339
Oxidation:
  alkanes, quasi-global modeling, 192–193

carbon monoxide, 162, 163
    modeling, 221
char, *see* Char, oxidation
HCN, global rates, 694
hydrocarbons, 356
intermediate temperature, 143
methane:
    postinduction, 183–184
    reaction pathways, 262–263
multistep global modeling, paraffin, 192
n-paraffin fuels, 188
PAHs, 284–285
phenyl, 153
residual char, 675–676
soot, 307–309, 316
sulfur compounds, 254
Oxidizer, mass fraction, 496
Oxygen:
    reaction with CH_4, 115–116
    superequilibrium concentrations, 502
Oxygenated hydrocarbons, octane number, 30
Oxygen-containing compounds, group values, 56–57
Oxygen index, effect on PAH emissions, 288–289

Paraffins:
    antiknock level, 27–28
    cetane number, 39
    oxidation, 188
        multistep global modeling, 192
Partial bond contributions, 52, 57
Partial equilibrium model, carbon monoxide, 226–227
Particle flow, PCGC-2, 753
Particle heat-up, coal, 678–682
Particle phase energy, 747
Patternation, 539–540, 546
PCGC-2, 752–757
    application and observations, 754–760
    basis, 752–753
    gas phase, 753
    particle flow, 753
    particle reaction, 754
    predictions:
        CO_2, CO, and H_2 concentrations, 754–755, 757
        data comparisons, 754–755
    radial profiles, 754, 756
    radiation, 753
Penetration:
    simplex atomizer, 559–560
    spray, 541

Perforated sheet disintegration, 532–533
Peroxy, low-temperature hydrocarbon oxidation, 136–137
Perturbation methods, 473, 501–503
Petroleum:
    oils, thermal conductivity, 794
    products, surface tension, 794
Phenyls:
    intermediate temperature oxidation, 155, 159
    oxidation, 153
Physical constants, 785
Pi-bond energies, 83
Plain-jet air-blast atomizer, 574–576
Plain-orifice atomizer, 530, 546–548
    dispersion, 540
    flow rate, 546
    penetration, 547–548
    Sauter mean diameter, 547
Planar initiation, detonation, 378–380
Plug-flow model, 744–752
    assumptions, 745–746
    auxiliary equations, 748–749
    coal particle burnout, 751
    differential equations, 746–748
    gas mole fractions, 751–752
    gas-phase energy balance, 747
    gas-phase mass flow rate, 748
    kinetic parameters, 749–750
    limitations, 746
    mass rate of change, 746–747
    particle phase energy, 747
    predictions and comparisons, 749, 751–752
Pollutants, *see also* Carbon monoxide; Nitrogen oxides; Sulfur oxides
    ambient air quality standards, 11–12
    categories, 9–10
    classes, 215
    combustible, *see* Combustible pollutants
    emissions:
        characteristic time model, 222–224
        estimates, 216
    transport between states, 13
Pollution:
    history of legislation, 9
    nonattainment area, 13
Polyacetylenes, combustion, 263
Polycyclic aromatic hydrocarbons, 269–291
    combustion kinetics, 272–287
        acetylene flame, 273–274, 285
        aliphatic ring formation, 274–275
        benzene destruction, 283–285
        benzene flame, 273
        beyond primary reaction zone, 285–287

Polycyclic aromatic hydrocarbons
(*Continued*)
condensed-ring compounds, 276–277
decomposition in postflame gases, 284
destruction, 283–285
fluorescence, 285–286
higher aromatics from benzene, 275–283
methyl substitution pathways, 278–279
oxygen index effect, 288–289
postflame formation, 286–287
reactions with $C_3H_3^+$, 283
emissions, 270–271
coal and liquid fuels, 291
fate, 270–271
smoking ethylene diffusion flames, 288–289
ions, 278, 280–282
laminar diffusion flames, 287–289
mutagenicity, 269–270
oxidation, 284–285
sampling and analysis, 271–272
structures, 266–268
turbulent diffusion flames, 289–291
Potential-energy function, bimolecular atom-transfer reaction, 86–87
Potential-energy profile, 86, 88
Pour point, diesel fuel, 40
Power density, detonation initiation, 381–383
Preferential diffusion, 396–398
Preheat zone thickness, 345–346
Premixed flames, *see also* Turbulent reacting flows; Soot
burning velocity, 341–342
coal, 718
two-zone structure, 484–486
Premix-prevaporize combustion, atomization, 584
Pressure:
histories, equivalence ratio effect, 131, 134
rate of rise, temperature and equivalence ratio effect, 130, 134
Pressure atomizer, design, 589–590
Pressure exponent, hydrocarbon flames, 355
Pressure piling, 400
Pressure-swirl atomizer, *see* Simplex atomizer
Prevention of significant deterioration, 12
Probability density functions, 469–470, 474–482
approximations using moments, 472
calculation:
evolution, 472–473
random variable, 475–479
conserved scalar, *see* Turbulent diffusion flames

Delta functions, 478
ensemble averages, stationarity, and homogeneity, 481–482
joint, marginal, and conditioned, 479–481
kurtosis, 476–477
moment methods disregarding, 470–472
premixed turbulent flames, 507–509
countergradient diffusion, 509
modeling assumptions, 508–509
reaction progress variable, 507–508
random variable, 474–475
skewness, 476–477
statistical independence, 481
stochastic processes, 475
strange attractor, 475
temperature and concentrations, 500–501
turbulent diffusion flames, 452
variance, 476–477
Producer gas, 819
Propane:
laminar flame:
energy release rate, 195–196, 198
flame speed and quench distances, 197–198
modified four-step reversible mechanism, 195–198
oxidation:
low-temperature hydrocarbon, 136
species evolution, 149–150
Propane-oxygen system, cool flame regions, kinetics, 129, 134
Propene, addition reactions, 161
Pulverized coal models, 758–761
Pyrene, decomposition, in postflame gases, 284
Pyrolysis:
C_2H_5Cl, 94–96
coal:
product distributions, 682–684
product yield, 689–691
inert, fuel-nitrogen compounds, 242
NH_3, 247

QRRK theory, 102
Quasi-global mechanisms, 191–194
Quenching:
carbon monoxide, in combustion gases, 225
combustion gases, CO concentration predictions, 226–227
diameter, 365
distance, 390–398, 803–806
definition, 390
determination, 390–391

relation to minimum ignition energy, 392–393, 395
relation to quenching tube geometry, 392
spray combustion, 635–636
laminar flame, 362–363

Radial fuel distribution, measurement, 545–546
Radiation, PCGC-2, 753
Radiative heat transfer, plug-flow model, 748–749
Radical reactions:
  activation energy, 830
  rate constants, 813, 816–817
Railroad diesel fuel, quality and uses, 37–39
Random variable, 474–475
  probability density functions calculation, 475–479
Random vortex method, 521–522
Rate coefficient:
  CO oxidation, 221–222
  HCN removal reactions, 244
  hydrocarbon-N_2 and -NO reactions, 236–237
  NO_2 reactions, 250
  N_2O reactions, 251
  SO_3 reactions, 255
  thermal NO formation reactions, 233
Rate constant:
  effect of perturbing, 173
  overall specific, 181
  methane disappearance, 183–184
  radical reactions, 813, 816–817
  surface growth, 304–306
  thermal, 89
Rayleigh equation, 339
Rayleigh line, 331–332, 334, 336–337
Rayleigh mechanism, atomization, 531
Reaction coordinate, 86–88
Reaction order, overall, 181
Reburning, NO, 248–249
Recombination reaction, 356
Recoverable resources, 4–5
Refinery gas, 819
Reflected shock, initiation, 378–380
Reformed natural gas, 819
Reynolds decomposition, 449
Reynolds number, 462
  drag coefficient, 616–617
  stability and, 454–455
  turbulence, 462
  turbulent diffusion flames, 495

Rosin-Rammler distribution, 537–538, 544
Rotary atomizers, 565–566
RRKM theory, 98

Safety, 385–401
  auto ignition temperature, 386–389
  flash point, 385–386
  inductive spark ignition, 401
  maximum experimental safe gap, 398–401
  minimum ignition energy and quenching distance, 390–398
Saturated steam, properties, 839
Sauter mean diameter, 533
  airblast atomizers, 569–570, 572–576
  air pressure and temperature effects, 555–556, 579
  atomizer dimension effect, 557, 579
  diesel sprays, 586, 589
  drop size measurement, 543
  factors governing, 579–580
  fan-spray atomizers, 564
  fuel-injection pressure, 557
  kerosene sprays with closed spill line, 563
  plain-orifice atomizers, 547
  rotary atomizers, 566
  surface tension effect, 555–556, 579
  swirl atomizers, 630–631
  ultrasonic atomizers, 596
  viscosity effect, 556, 576
Saybolt Universal viscosity, 832–833
Scale of turbulence, 445
Schmidt number:
  gases and vapors into air, 802
  turbulent, 488
  water vapor into air, 801
Selective reduction, NO, 249
Sensitivity coefficients, 171–173
  normalized, 175–177
  normalized second-order elementary, 178–179
Sheet breakup, 532–533
Shock waves:
  amplification by coherent energy release, 383, 410, 414
  detonation, 372–373
Simplex atomizer, 584–560. See also Swirl atomizer
  cone-angle, 554–555
  designs, 550
  drop-size analysis, 558–559
  flow-number, 552–553
  gas turbine, 582
  geometry, relation to cone angle, 554

Simplex atomizer (*Continued*)
  mean drop size, 555–557
    air properties, 556–557
    atomizer dimensions, 557
    fuel-injection pressure, 557
    liquid properties, 555–556
  relationship of geometry and discharge
    coefficient, 552–553
  spray:
    development, 548–549
    penetration, 559–560
Single spin detonation, 373
Skeleton steam tables, 838–839
Skewness, 476–477
Slinger system, 585–586
Slot-burner flame, nonsteady cellular, 357,
  359–360
Slot-burner technique, 342–343
Smoke:
  diffusion flames, characteristics, 442
  laminar diffusion flames, 442–445
Smoke-free fuel, flow *vs.* reciprocal pressure,
  443–444
Smoke point, 317–318, 813
Smokey tendency, hydrocarbons, 813–815
Smoluchowski equation, 302
Soap-bubble technique, 343
Solid fuels, *see also* Coal
  conversion to liquids or gases, 6
  properties, 835–836
Solid sphere, drag coefficient, 615–617
Soot, 291–320
  additive effects, 319–320
  aerosol surface, 306
  agglomerates, 292, 309
  agglomeration, 301–303
  burnout region, 317, 319
  diffusion flames, inert additives, 319
  laminar diffusion flames, 313–318
    burnout region, 317
    distribution, 313–316
    loading, 315
    overventilated flame, 316
    smoke point, 317–318
    surface growth, 315
    temperature and pressure effects, 316–
      317
    tendency to soot, 317
  oxidation, 307–309, 316
    Nagle and Strickland-Constable formula,
      307–308
  premixed flames, 309–313
    C/O ratio, 311–312

    pressure and temperature effects, 311–
      312
    sooting limit, 309–311
    tendency to soot, 311
    threshold soot index, 309, 311
    yield, 311–312
  structure, 292–293
  tempering, 304
  turbulent diffusion flames, 318–319
Soot formation, 813
  aerosol, 293–295
  coagulation, 293–294, 301–303
    particle number, 303
    rate, 302
  mechanism in sooting region, 300
  particle inception, 293, 295–301
    benzene and acetylene flames, 295–
      297
    ionic mechanisms, 297–301
    negative ions, 299–300
    particle size, 299
    peak ion concentrations, 298
    radical mechanisms, 296–297
    signal intensities, 296
  phenomenology, 295
  propensity for, 306
  reactivity of surface, 304
  second ion peak, 300
  Smoluchowski equation, 302
  surface growth, 293–294, 303–307
    acetylene flame, 304–306
    laminar diffusion flames, 315
    rate constant, 304–306
    reactivity, 306
  transition from coalescent to chain-forming
    collisions, 301–302
  transition from particle inception to
    growth, 294–295
  volume fraction profiles, 301
Sooting limit, 309–311
Space velocity, *vs.* radius, 397–398
Species:
  concentration as function of reaction time,
    132–134
  conservation equation, 433
Specific gravity:
  coal, 666
  industrial gaseous fuels, 818
Specific heats, coal, 666
Specific volume, superheated steam, 840
Spherical shock wave initiation, detonation,
  380–383
Spill atomizer, 563–564

Spray:
  characteristics, diesel fuel, 586
  cone angle, *see* Spray cone angle
  dispersion, 540
  drop-size distribution, 534–539
  drop transport, 615–620
  mean drop size, 533–534
  measurement:
    circumferential fuel distribution, 546
    cone angle, 545
    drop size, 541–545
    radial fuel distribution, 545–546
  patternation, 539–540
  penetration, 541
Spray combustion, 620–639
  array theory, 621
  blowout, 632
  combustion efficiency, 629–632
  dense sprays, 623–624
  droplet arrays, groups, and sprays, 621–622
  droplet group theory, 621–622
  drop-size distribution, 636
  experiments, 624–628
  fuel/air ratio, 629
  fuel vaporization rate, 629
  group combustion number, 621–622
  ignition, 634–635
  ignition theory, 635–639
  mathematical models, 622–623
  minimum ignition energy, 635–638
  quenching distance, 635–636
  weak extinction limits, 632
Spray cone angle:
  effects of fuel-injection pressure and
    ambient pressure, 554–555
  measurement, 545
  relation to atomizer geometry, 554
  simplex atomizer, 554–555
  spray, 540, 545
  turbulent flame propagation, 547
Spray flame:
  diffusion flame, 625–626
  drop velocities, 627–628
  heavy oil, 625
  idealized model, 626–627
  vaporization rates, 627–628
  zones, 624–625
Spray Triode atomizer, 593–594
Stability, diffusion flames, 453–456
Staged combustion, NO, 248
Stationarity, probability density functions, 482

Statistical thermodynamics, 71–80
  frequencies assigned to normal and partial
    bond bending and stretching motions,
    75–78
  molar thermodynamic quantities, 72
  rotational degrees of freedom, 73–74
  translational contribution, 72
  vibrational degree of freedom, 73
Steady-state evaporation, 597–604
  burning rate curves for kerosene and JP4,
    599–600
  Clausius-Clapeyron equation, 603
  D^2 law, 598
  heat transfer number, 601–602
  mass transfer number, 600–601
  rate calculation, 602–604
  rate measurement, 598–599
  theoretical background, 599–602
Steam:
  saturated properties, 839
  superheated, 840–841
Stem-cavity-atomizer, 596
Stochastic processes, probability density
    functions, 475
Storage stability, diesel fuel, 40–41
Strain energy, 83
Strange attractor, 475
Streamwise Eulerian displacement, 512–513
Stretch, turbulent flame speed effects, 516–518
Strong driven detonations, 335
Subbituminous coal, 661
  coal burnout and particle mixture fraction,
    725–726, 728
  CO and CO_2 concentration, 727–729
  NO and HCN + NH_3 concentrations,
    728–729
  O_2 concentration and gas mixture fraction,
    726–728
Sulfur:
  diesel fuel content, 40
  fossil fuel content, 230
  group values, 68–70
  in hydrocarbon flames, NO_x emission
    reduction, 256
  oxidation, 254
Sulfur oxides, 252–256
  emission estimates, 252
  emissions for new steam generators, 12–13
  formation, 254–255
  kinetic models, 255
  product distributions, 252–253
  profiles, H_2S—O_2—N_2 flame, 253

Superheated steam, 840–841
Surface growth, soot, 303–307
Surface tension, 794
   effect on SMD, 555–556, 579
   liquids, 797–798
Surfactants, carburetor icing, 23, 26
SWACER effect, 383, 410, 414
Swirl atomizer:
   cone angle, 540, 545
   discharge coefficient, 550–552
   relation between cone angle and viscosity,
     554
   Sauter mean diameter, 630–631
   spray combustion, 630
Synthesis gas, 819

Taylor instability, 404, 406–408
   premixed turbulent flames, 511
Taylor scale, 465
Temperature, profiles, as function of reaction
   time, 132–134
Thermal conductivity, 666, 795–796
Thermal decomposition, 142, 144, 688–689
Thermal rate constant, 89
Thermochemical properties, 52–84
   additivity rules, *see* Additivity rules
   partial bond contributions, 52, 57
Thermochemistry, from additivity rules, 85
Thermophysical properties, 603
Third body efficiencies, 160
Three-zone structure, diffusion flames, 486
Threshold soot index, 309, 311
Toluene, oxidation, species evolution, 154–
   155, 157
Torsion barriers, free rotation, 79
Trace species, perturbation approaches, 501–
   503
Transfer number, 436–437
Transition state theory, 86–90
   A-factor estimation, 92–93, 95
   application, 91–100
     metathesis reaction, 91–93
     surrogate models, 93
     unimolecular reactions, 93–95
   correction factor, 98
   partition function, 98–99
   thermal rate constant, 89
   unimolecular reactions, pressure
     dependence, 97–100
Transverse wave spacing, detonation, 376
Turbulence:
   nondimensional parameters, 461–465

   modeling approximations, 470
   scale, 518–519
Turbulent diffusion flames, 445–453
   blow off, 505–506
   chemical heat release average rates, 503–
     505
   coherent structures, impact on combustor
     design, 523
   conserved scalar, 490–500
     approximations of equilibrium and
       irreversible chemistry, 495–497
     atom conservation and mixture fraction,
       490–492
     conservation equations for moments,
       494–495
     fuel and oxidizer mass fractions, 496
     Lewis number of unity, 492–494
     neglect of pressure fluctuations in
       enthalpy, 492–494
     nonreacting turbulence, 497
   eddy diffusivity, 447
   flame height, 447–448
   global reaction rate, 449, 451
   lift-off, 505–506
   local extinctions by flame stretch, 505–506
   nozzle velocity effect on flame appearance,
     445–446
   PAHs, 289–291
   perturbation approaches, 501–503
   probability density functions, 452
     conserved scalars, 498–500
     temperature and concentrations, 500–
       501
   Reynolds decomposition, 449
   scale of turbulence, 445
   simplified model, 447
   soot, 318–319
   species concentration, 452
   velocity, 445
Turbulent flame:
   propagation, 518–520
     cone angle, 540, 547
     discharge coefficient, 546
     flame speeds in different regimes, 519–
       520
     intensity and scale parameters, 518–519
     prediction by perturbation methods, 520
   speed:
     prediction, 512–516
     regimes, 519–520
     stretch and curvature effects, 516–518
     wrinkled laminar flame, 513
Turbulent motion, 461–465

Turbulent reacting flows, *see also* Probability
density functions
approximations of probability density
functions using moments, 472
Arrhenius rate factor, 470–472
asymptotic expansions, 483–486
expansion parameters, 484
matched, 483–484
three-zone structure of diffusion flames,
486
two-zone structure of premixed flames,
484–486
utility of limit processes, 483
calculation of evolution of probability
density functions, 472–473
coherent structures, 520–523
ESCIMO approach, 521
nonpremixed combustion, 520
predictive methods, 521
premixed combustion, 520
random vortex method, 521–522
relevance to flame holding and blow
out, 522
relevance to other flame properties,
522–523
methods identifying coherent structures,
473–474
moments, 487–490
based on Favre averages, 489–490
based on ordinary averages, 487–489
chemical production term, 489
methods disregarding probability density
functions, 470–472
modeling approximations, 490
motivation and terminology, 459–470
characteristic time, 465–466
complexities, 469
countercascade phenomena, 464–465
Damköhler number, 466–467
energy spectrum, 463–464
equilibrium diffusion flames, 469–470
flame-strength factor, 467–468
Froude number, 465
integral scale, 462–463
intractable problems, 461
Kolmogorov length, 464
Kolmogorov-scale eddy, 467–468
limitations of coverage, 470
nondimensional parameters, 461–465
practical problems, 459–461
probability density function, 469–470
Reynolds number, 462
Taylor scale, 464

tractable problems, 460–461
perturbation methods beginning with
known structures, 473
premixed flames:
flame speed prediction, 512–516
instabilities, 510–512
mean reactant flux prediction, 518
small-gradient turbulence, 509–518
streamwise Eulerian displacement, 512–
513
stretch and curvature effects on flame
speed, 516–518
wrinkled-laminar-flame structure, 509–
510
subsequent coverage, 474
velocity component, 474–475
weakly turbulent pre-mixed flames, 469–
470
Turnover temperature, 139–140, 143,
154
Two-zone structure, premixed flames, 484–
486

Ultrasonic atomizer, 594–596
Unbranched chain reactions, 125–126
Unconfined explosions, 414–415
Unimolecular reactions, 91, 97–100
Units, symbols, 784

Vaporization-condensation process, 10
Vaporization rates:
droplet, 604
spray combustion, 627–628
Vaporizers, atomization, 583–584
Vapor lock, 20
Vapors:
classifications, 401
diffusion coefficient into air, 802
Schmidt number, 802
Velocity correlation tensor, 463
Velocity gradient, critical, 363–364
Venting, explosions, 414
Vent ratio, 403
Vessel explosion processes, complications,
404, 406–408
Viscosity:
diesel fuel, 40
effect on SMD, 556, 576
Saybolt Universal, 832–833
Volatility, diesel fuel, 40
von Neumann spike, 373

Vortex flames, 424

Wall quenching, 264–265, 269
Water gas, 819
Water vapor, diffusion coefficient, 801
Wax droplet technique, 542–543
Weber number, 532, 557–558
Well-stirred reactor, coal flames, 717
Whistle atomizers, 596
Wide range atomizers, 560–561
    dual-orifice, 562–563
    duplex, 560–561
    spill, 563–564
Working fluid-heat addition model, 332–336
    enthalpy-temperature relationships, 332–
        333

Hugoniot relationship, 333–334
    pressure-volume plot, 334–335
    Rayleigh line, 334–335
Wrinkled-flame limit, 467–468
Wrinkled laminar flame:
    structure, premixed turbulent flames, 509–
        510
    turbulent flame speed, 513

Y-jet atomizers, 590–591

Zel'dovich mechanism, 502
Zel'dovich number, 484
Zel'dovich, von Neumann, and Döring
        theory, 371–372
Z number, 532